Université Joseph Fourier

Les Houches

Session LXXXVII

2007

String Theory and the Real World:

From Particle Physics to Astrophysics

Lecturers who contributed to this volume

Ignatios Antoniadis
Jose L.F. Barbón
Marcus K. Benna
Thibault Damour
Frederik Denef
Fabiola Gianotti
Gian-Franco Giudice
Kenneth Intriligator
Elias Kiritsis
Igor R. Klebanov
Marc Lilley
Juan M. Maldacena
Eliezer Rabinovici
Nathan Seiberg
Angel M. Uranga
Pierre Vanhove

ÉCOLE D'ÉTÉ DE PHYSIQUE DES HOUCHES

SESSION LXXXVII, 2 JULY–27 JULY 2007

ÉCOLE THÉMATIQUE DU CNRS

STRING THEORY AND THE REAL WORLD:
FROM PARTICLE PHYSICS TO ASTROPHYSICS

Edited by

C. Bachas, L. Baulieu, M. Douglas, E. Kiritsis, E. Rabinovici,
P. Vanhove, P. Windey and L.F. Cugliandolo

ELSEVIER

Amsterdam – Boston – Heidelberg – London – New York – Oxford
Paris – San Diego – San Francisco – Singapore – Sydney – Tokyo

Elsevier
Radarweg 29, PO Box 211, 1000 AE Amsterdam, The Netherlands
Linacre House, Jordan Hill, Oxford OX2 8DP, UK

First edition 2008

Library of Congress Cataloging-in-Publication Data
A catalog record for this book is available from the Library of Congress

British Library Cataloguing in Publication Data
A catalogue record for this book is available from the British Library

ISBN: 978-0-0805-4813-5
ISSN: 0924-8099

For information on all Elsevier publications
visit our website at books.elsevier.com

Transferred to Digital Printing 2009

Working together to grow
libraries in developing countries

www.elsevier.com | www.bookaid.org | www.sabre.org

ELSEVIER   BOOK AID International   Sabre Foundation

ÉCOLE DE PHYSIQUE DES HOUCHES

Service inter-universitaire commun
à l'Université Joseph Fourier de Grenoble
et à l'Institut National Polytechnique de Grenoble

Subventionné par le Ministère de l'Éducation Nationale,
de l'Enseignement Supérieur et de la Recherche,
le Centre National de la Recherche Scientifique,
le Commissariat à l'Énergie Atomique

Previous sessions

Publishers:
- Session VIII: Dunod, Wiley, Methuen
- Sessions IX and X: Herman, Wiley
- Session XI: Gordon and Breach, Presses Universitaires
- Sessions XII–XXV: Gordon and Breach
- Sessions XXVI–LXVIII: North Holland
- Session LXIX–LXXVIII: EDP Sciences, Springer
- Session LXXIX–LXXXVII: Elsevier

Organizers

BACHAS Costas, LPT-ENS, Paris, France
BAULIEU Laurent, LPTHE, Paris, France
DOUGLAS Michael, Rutgers University, USA, and IHES, France
KIRITSIS Elias, CPHT, Palaiseau, France
RABINOVICI Eliezer, Racah Institute of Physics, Jerusalem, Israel
VANHOVE Pierre, IPhT, Saclay, France
WINDEY Paul, LPTHE, Paris, France
CUGLIANDOLO Leticia, Université Pierre et Marie Curie, Paris VI, France

Lecturers

ANTONIADIS Ignatios, CERN, Switerland
ARKANI-HAMED Nima, IAS, Princeton, USA
BÁRBON, J., UAM/CSIC, Madrid, Spain
BAULIEU Laurent, LPTHE, Paris, France
DAMOUR Thibault, IHES, France
DENEF Frederik, Université de Louvain, Belgium
ELLIS John, CERN, Switzerland
GIANOTTI Fabiola, CERN, Switzerland
GIUDICE Gian-France, CERN, Switzerland
GLOVER Nigel, IPP, Durham, England
INTRILIGATOR Kenneth, University of San Diego, USA
KLEBANOV Igor, Princeton University, USA
KIRITSIS Elias, CPHT, Palaiseau, France
MALDACENA Juan, IAS, Princeton, USA
RABINOVICI Eliezer, Racah Institute of Physics, Jerusalem, Israel
SHENKER Steve, Stanford University, USA
URANGA Angel, CERN, Switerland
VANHOVE Pierre, IPhT, Saclay, France
WIEDEMANN Urs, CERN, Switzerland

Participants

ANGUELOVA Lilia, Queen Mary Univ. of London, UK

ANTONIADIS Ignatios, CERN, Switerland

AREAN Daniel, Universidade De Santiago, Spain

ARKANI-HAMED Nima, IAS, Princeton, USA

BACHAS Costas, LPT-ENS, France

BAO Ling, Chalmers Univ of Tech, Sweden

BAULIEU, Laurent, Université Pierre et Marie Curie, France

BEDOYA DELGADO Oscar, IFT-Unesp, Brazil

BENNA Marcus K., Princeton University, USA

BJORNSSON Jonas, Physics Department, Sweden

BOURJAILY Jacob, Princeton University, USA

CARDELLA Matteo, Hebrew University, Israel

CICOLI Michele, DAMTP, Cambridge, UK

CLOSSET Cyril, ULB, Belgium

CREMONESI Stefano, SISSA, Trieste, Italy

DAMOUR Thibault, IHES, France

DENEF Frederik, Université De Louvain, Belgium

DOMOKOS Sophia, University of Chicago, USA

DOUGLAS Michael, Rutgers University, USA and IHES, France

ELLIS John, CERN, Switerland

FAULKNER Thomas, MIT-CTP, USA

FERRO Livia, Torino University, Italy

FRANCIA Dario, Chalmers University, Sweden

GARCIA-ETXEBARRIA Inaki, IFT, Madrid, Spain

GIANOTTI Fabiola, CERN, Switerland

GLOVER Nigel, IPP, Durham, England

GORBONOS Dan, The Hebrew University, Israel

GUIDICE Gian-France, CERN, Switzerland

GWYN Rhiannon, Mcgill University, Canada

HAQUE Sheikh Shajidul, UW-Madison, USA

HAUPT Alexander, Imperial College London, UK
HOOVER Doug, Mcgill University, Canada
INTRILIGATOR Kenneth, University of San Diego, USA
JOHNSON Matthew, UC Santa Cruz, USA
KANITSCHEIDER Ingmar, ITF Amsterdam, Netherlands
KELLER Christoph, ETH Zurich, Switzerland
KIRITSIS Elias, CPHT, France
KLEBANOV Igor, Princeton University, USA
KLEVTSOV Sam, Rutgers University, USA
KNAPP Johanna, CERN, Switzerland
KREFL Daniel, MPI and Lmu Munich, Germany
LARFORS Magdalena, Uppsala University, Sweden
LILLEY Marc, IAP, France
MALDACENA Juan, IAS, Princeton, USA
MANN Nelia, University of Chicago, USA
MARSANO Joseph, Caltech, USA
MARTIN Alexis, LPTHE, Paris Vi-Vii, France
MASON John, Uc Santa Cruz, USA
METHER Lotta, University of Helsinki, Finland
MONTEIRO Ricardo, DAMTP, Cambridge, UK
MOURA Cesar, LPTHE, France
MUKHOPADHYAY Ayan, HRI, India
NACIRI Mohamed, Mohamed V Univ. Morocco, Morocco
NIARCHOS Vasilis, CPHT, Ecole Polytechnique, France
PASSERINI Filippo, Perimeter Institute, Canada
PETERSSON Christoffer, Chalmers University, Sweden
PLAUSCHINN Erik, MPI For Physics, Munich, Germany
QUIGLEY Callum, University of Chicago, USA
RABINOVICI Eliezer, Racah Institute of Physics, Hebrew University, Israel
RICCO Giovanni, Universita Di Pisa, Italy
SAHOO Bindusar, HRI, India
SANTOS Jorge, DAMTP, Cambridge, UK
SEKINO Yasuhiro, Okayama Institute, Japan

SHENKER Steve, Stanford University, USA

SONNER Julian, DAMTP, UK

URANGA Angel, CERN, Switerland

VAN DEN BLEEKEN Dieter, ITF Kuleuven, Belgium

VANHOVE Pierre, Institut de Physique Théorique, CEA, France

VICEDO Benoit, DAMTP, Cambridge, UK

WINDEY Paul, Université Pierre et Marie Curie, France

YAMADA Daisuke, Hebrew University, Israel

Foreword

The summer of 2007 is a propitious moment in the history of physics. While the Standard Model, developed in the early 70's and largely confirmed by the end of the decade, remains the cornerstone of our understanding of particle physics, many believe its days are numbered.

Experiments to begin in 2008 at CERN will probe energies almost an order of magnitude beyond what we can reach today. Not only can we expect to discover the Higgs boson of the original Standard Model, there are compelling theoretical reasons to believe that far more novel physics is there, connected with the large ratio between the energy scales of electroweak symmetry breaking, and those of more fundamental physics such as gravity. This connection gives us hope that new discoveries will bear directly on levels of structure associated with far higher energies than we can probe directly.

Many, many speculations have been made about this new physics. Two of the best known are supersymmetry and large extra dimensions, and there are many others. These ideas now exist in a plethora of variations, giving rise to a rich subject of "Beyond the Standard Model" physics. But the moment of truth is approaching, when many of these speculations will be cut down to size.

Fortunately, the lack in recent years of striking new discoveries from particle physics, has been more than made up for by a wealth of new data from observational cosmology. Deep sky surveys, maps of the cosmic microwave background, and other developments have led to what many consider to be a "Standard Model of cosmology," the Λ CDM inflationary model. Inflation (and its competitors) gives us another window on physics at higher energies, and it is reasonable to hope that given new discoveries at LHC, some sort of grand synthesis between our two main (avenues of investigation) of fundamental physics will emerge.

Many physicists feel that the best hope for such a synthesis lies within string theory, arguably the grandest speculation of all. For decades, the problem of reconciling quantum mechanics and general relativity was considered unsolvable, a rock on which even Einstein foundered. The solution of this problem by string theory, and the general failure of other approaches, has led to a widespread belief that string theory must contain essential clues about fundamental physics. We can even imagine that string theory is the complete theory from which all the rest of physics will someday be derived.

For a variety of reasons, it has been very difficult to propose decisive tests which would settle this claim. Most importantly, almost all specific predictions of string theory depend crucially on which particular solution we consider. Since string theory is ten dimensional, modelling four dimensional physics requires choosing a six dimensional compactification manifold, as well as specific background fields and other structures on that manifold, and the number of possibilities is vast. The development of M theory and duality in the 1990's, far from solving this problem, has only enlarged the set of possibilities, and increased our confidence that this analysis of the situation is valid.

An apparent death blow to the hope of a unique preferred solution has been dealt by the compelling evidence of recent years that our universe contains dark energy, well modeled by a non-zero positive cosmological constant. At present there is only one generally accepted theoretical explanation for this, namely the anthropic argument of Weinberg, as realized in string theory by Bousso and Polchinski. This explanation depends crucially on having a vast number of vacuum configurations, certainly more than 10^{60}. More precise estimates suggest that there are at least 10^{500} distinct solutions, and forming any picture of the range of their possible predictions to guide us in testing the theory is a formidable challenge.

Nevertheless we must try. Might it happen that the upcoming experiments at LHC will cut the grand speculation of string theory down to size as well? Or might we discover evidence for strings, or other predictions of the theory? Or could indirect arguments, combining the evidence from particle physics, cosmology and elsewhere, somehow point to a particular class of solutions, which would then make predictions we could test in future experiments?

Or might we be better off making contact between string theory and more accessible physics, such as that of QCD and nuclear physics? Of course, this was the original inspiration for string theory, and recently great strides have been made in this direction, in part fueled by theoretical developments such as string-inspired techniques for perturbative computation as well as the celebrated AdS/CFT correspondence, and in part by developments in heavy ion collider physics, which has revealed surprising new collective behaviors of matter with simple models in terms of the new theoretical ideas. While the relevance of this for fundamental physics remains unclear, the historical progress of string theory came from considering a wide variety of possible applications, and this will surely continue.

This then, the contact between "String theory and the real world," was the theme of our school. Our general approach was "bottom-up", in that most of the lecturers tried to explain the known facts and prospects for discovery in some established area of particle or nuclear physics, or of astrophysics and cosmology, and then move towards the question of how string theory might make contact with these developments.

Let us begin with the lectures and seminars more closely related to particle and nuclear physics. The general topic of physics beyond the standard model was surveyed by John Ellis, who described both the capabilities of LHC and some of the more popular models for BSM physics. An in-depth description of LHC and its experiments was given by Fabiola Gianotti. Finally, Nigel Glover described the status of the perturbative QCD computations which will be necessary to interpret LHC data. This area has seen an influx of new ideas and techniques from string theory, and Glover's even-handed appraisal of the old and new techniques combined with his clear description of what was needed by experimentalists was of great value.

Several lecturers covered ideas for beyond the standard model (BSM) physics in detail. Perhaps the most popular class of BSM models postulate low energy supersymmetry. Supersymmetric phenomenology was covered in depth by Gian Giudice, while the question of how supersymmetry breaking arises from the dynamics of a theory was discussed by Ken Intriligator. Another popular class of models postulates observable consequences of the extra dimensions of string/M theory. This was capably surveyed by Ignatios Antoniadis.

Any contact between these more phenomenological ideas, and string/M theory, will necessarily involve detailed string model building. This is a large subject and rather than survey it all, we chose to provide in-depth lectures on a few of its better understood parts. Angel Uranga covered the subject of brane constructions of the Standard Model in type II superstring theory. Elias Kiritsis discussed similar issues using the very different techniques of CFT and Gepner models.

Frederik Denef described flux compactification and moduli stabilization, and how its study has led to the current picture of the string landscape. Michael Douglas raised various general questions about the landscape whose better understanding might lead to significant progress.

Finally, the study of gauge/gravity duality, particularly, the AdS/CFT correspondence, has led to important technical developments in nonperturbative gauge theory, which may have important applications to QCD. Igor Klebanov introduced and reviewed this general subject, surveying most of the developments of recent years and covering his work on models of confinement in detail. Juan Maldacena added to this with a lecture on recent developments in finding an integrable sector within $N = 4$ super Yang–Mills theory.

A particularly interesting recent development is the suggestion that the collective phenomena seen in heavy ion collisions at RHIC, and to be studied at LHC as well, can be modeled using AdS/CFT techniques. This area was described by Urs Wiedemann, starting with the original experimental discoveries, and explaining the arguments according to which these indicate novel collective phenomena, before going into the recent theoretical developments.

We now turn to lectures primarily focusing on astrophysics and cosmology. Juan Maldacena began his lectures with an introduction to inflationary cosmology, and a detailed explanation of how to compute the spectrum of fluctuations in the cosmic microwave background. He went on to explain nongaussianity, which led into a discussion of what a speculative "de Sitter/CFT" duality would look like. Finally he discussed how inflation might arise in string models.

Steve Shenker's lectures covered inflation at a more conceptual level. The primary goal was to explain the many confusing issues surrounding the notion of "eternal inflation," which is a central part of arguments for the string theory landscape. He concluded with a survey of proposals for the measure factor in cosmology.

Nima Arkani-Hamed described anthropic considerations at length. He then explained recent work on the limits of effective field theory in cosmology. Rabinovici described spontaneous breaking of space time symmetries and the possible role of broken scale invariance in solving the cosmological constant problem. Jose Barbón discussed how topology change in space-time is related the Hagedorn regime in critical string theories using the string/black hole correspondence principle and the holography of radiation corrections in various systems including Little String Theory. Pierre Vanhove described the recent progresses in the analysis of the ultra-violet behaviour of maximally extended supergravities and reviewed the dualities arguments indicating that such quantum field theories of gravity could be perturbatively finite in four dimensions. Laurent Baulieu described new techniques for handling the off-shell properties of super-Yang–Mills theories.

Finally, Thibault Damour began with a detailed explanation of the membrane description of black hole horizons, which has found recent application in AdS/CFT computations of collective phenomena. He then surveyed the present status of experimental tests of general relativity. He concluded by explaining how primordial cosmic strings might be detected at gravitational wave observatories, perhaps leading to direct evidence for fundamental string theory.

Only time will tell which if any of these ideas will find experimental support. Perhaps, in a few years, many of the specific proposals we discussed at the school will seem irrelevant, or naive. If even one hits the mark squarely, we will have reason to celebrate. But whatever their fate, we suspect that the fundamental ideas behind them will retain their interest, and we hope these lectures will guide our students and readers to meet the coming challenges and contribute in their turn.

We would like to thank Giora Mikenberg as well as other members of the ATLAS collaboration for having allowed the students of this school to visit the

ATLAS detector, which is an essential part of the experimental apparatus of the LHC campaign. We wish to address our warmest thanks to our main Sponsor, the European Science Foundation, which has agreed to fund a series of summer schools in theoretical physics linked to elementary particles, astroparticles and cosmology. We also thank the Bureau de la formation permanente du CNRS and the Les Houches summer school for their financial support. We are grateful to Brigitte Rousset and Isabelle Lelièvre for their invaluable help with the local organisation of the school.

Paris, 2008

Costas Bachas
Laurent Baulieu
Michael Douglas
Elias Kiritsis
Eliezer Rabinovici
Pierre Vanhove
Paul Windey
Leticia Cugliandolo

CONTENTS

Course 8. *Non-renormalisation theorems in superstring and supergravity theories, by Pierre Vanhove*   *301*

Course 9. *Preparing the LHC experiments for first data, by F. Gianotti* — *353*

Course 10. *String theory, gravity and experiment, by Thibault Damour and Marc Lilley* — *371*

*Course 13. Gauge–string dualities and some applications,
by Marcus K. Benna and Igor R. Klebanov*   *611*

Course 1

TOPICS IN STRING PHENOMENOLOGY

I. Antoniadis

Department of Physics, CERN—Theory Division, 1211 Geneva 23, Switzerland

On leave from CPHT (UMR CNRS 7644) Ecole Polytechnique, F-91128 Palaiseau

C. Bachas, L. Baulieu, M. Douglas, E. Kiritsis, E. Rabinovici, P. Vanhove, P. Windey and L.F. Cugliandolo, eds.
Les Houches, Session LXXXVII, 2007
String Theory and the Real World: From Particle Physics to Astrophysics
© *2008 Published by Elsevier B.V.*

1

Contents

3

1. Introduction

During the last few decades, physics beyond the Standard Model (SM) was guided from the problem of mass hierarchy. This can be formulated as the question of why gravity appears to us so weak compared to the other three known fundamental interactions corresponding to the electromagnetic, weak and strong nuclear forces. Indeed, gravitational interactions are suppressed by a very high energy scale, the Planck mass $M_P \sim 10^{19}$ GeV, associated to a length $l_P \sim 10^{-35}$ m, where they are expected to become important. In a quantum theory, the hierarchy implies a severe fine tuning of the fundamental parameters in more than 30 decimal places in order to keep the masses of elementary particles at their observed values. The reason is that quantum radiative corrections to all masses generated by the Higgs vacuum expectation value (VEV) are proportional to the ultraviolet cutoff which in the presence of gravity is fixed by the Planck mass. As a result, all masses are "attracted" to become about 10^{16} times heavier than their observed values.

Besides compositeness, there are three main ideas that have been proposed and studied extensively during the last years, corresponding to different approaches of dealing with the mass hierarchy problem. (1) Low energy supersymmetry with all superparticle masses in the TeV region. Indeed, in the limit of exact supersymmetry, quadratically divergent corrections to the Higgs self-energy are exactly cancelled, while in the softly broken case, they are cutoff by the supersymmetry breaking mass splittings. (2) TeV scale strings, in which quadratic divergences are cutoff by the string scale and low energy supersymmetry is not needed. (3) Split supersymmetry, where scalar masses are heavy while fermions (gauginos and higgsinos) are light. Thus, gauge coupling unification and dark matter candidate are preserved but the mass hierarchy should be stabilized by a different way and the low energy world appears to be fine-tuned. All these ideas are experimentally testable at high-energy particle colliders and in particular at LHC. Below, I discuss their implementation in string theory.

The appropriate and most convenient framework for low energy supersymmetry and grand unification is the perturbative heterotic string. Indeed, in this theory, gravity and gauge interactions have the same origin, as massless modes of the closed heterotic string, and they are unified at the string scale M_s. As a result, the Planck mass M_P is predicted to be proportional to M_s:

$$M_P = M_s/g, \tag{1.1}$$

where g is the gauge coupling. In the simplest constructions all gauge couplings are the same at the string scale, given by the four-dimensional (4d) string coupling, and thus no grand unified group is needed for unification. In our conventions $\alpha_{\text{GUT}} = g^2 \simeq 0.04$, leading to a discrepancy between the string and grand unification scale M_{GUT} by almost two orders of magnitude. Explaining this gap introduces in general new parameters or a new scale, and the predictive power is essentially lost. This is the main defect of this framework, which remains though an open and interesting possibility [1].

The other two ideas have both as natural framework of realization type I string theory with D-branes. Unlike in the heterotic string, gauge and gravitational interactions have now different origin. The latter are described again by closed strings, while the former emerge as excitations of open strings with endpoints confined on D-branes [2]. This leads to a braneworld description of our universe, which should be localized on a hypersurface, i.e. a membrane extended in p spatial dimensions, called p-brane (see Fig. 1). Closed strings propagate in all nine dimensions of string theory: in those extended along the p-brane, called parallel, as well as in the transverse ones. On the contrary, open strings are attached on the p-brane. Obviously, our p-brane world must have at least the three known

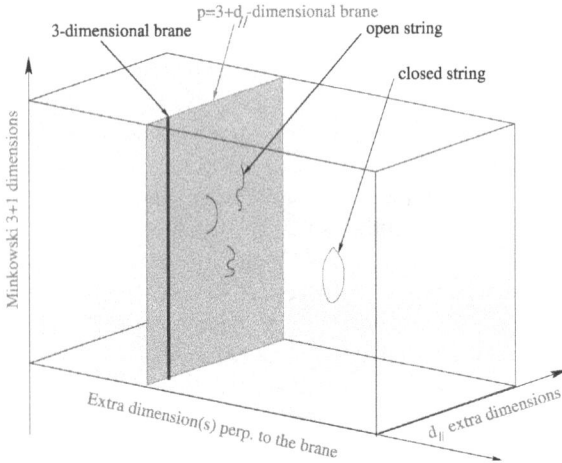

Fig. 1. In the type I string framework, our Universe contains, besides the three known spatial dimensions (denoted by a single blue line), some extra dimensions ($d_\parallel = p - 3$) parallel to our world p-brane (green plane) where endpoints of open strings are confined, as well as some transverse dimensions (yellow space) where only gravity described by closed strings can propagate.

dimensions of space. But it may contain more: the extra $d_\parallel = p - 3$ parallel dimensions must have a finite size, in order to be unobservable at present energies, and can be as large as $\text{TeV}^{-1} \sim 10^{-18}$ m [3]. On the other hand, transverse dimensions interact with us only gravitationally and experimental bounds are much weaker: their size should be less than about 0.1 mm [4]. In the following, I review the main properties and experimental signatures of low string scale models [5,6].

These lectures have two parts. In the first part, contained in Sections 2 to 6, I describe the implementation, the properties and the main physical properties of low scale string theories. In the second part, contained in the following sections, starting from Section 7, I discuss a simple framework of toroidal type I string compactifications with in general high string scale, in the presence of magnetized branes, that can be used for moduli stabilization, model building and supersymmetry breaking.

2. Framework of low scale strings

In type I theory, the different origin of gauge and gravitational interactions implies that the relation between the Planck and string scales is not linear as (1.1) of the heterotic string. The requirement that string theory should be weakly coupled, constrain the size of all parallel dimensions to be of order of the string length, while transverse dimensions remain unrestricted. Assuming an isotropic transverse space of $n = 9 - p$ compact dimensions of common radius $R_\perp$, one finds:

$$M_P^2 = \frac{1}{g^4} M_s^{2+n} R_\perp^n, \quad g_s \simeq g^2, \tag{2.1}$$

where g_s is the string coupling. It follows that the type I string scale can be chosen hierarchically smaller than the Planck mass [5,7] at the expense of introducing extra large transverse dimensions felt only by gravity, while keeping the string coupling small [5]. The weakness of 4d gravity compared to gauge interactions (ratio M_W/M_P) is then attributed to the largeness of the transverse space $R_\perp$ compared to the string length $l_s = M_s^{-1}$.

An important property of these models is that gravity becomes effectively $(4+n)$-dimensional with a strength comparable to those of gauge interactions at the string scale. The first relation of Eq. (2.1) can be understood as a consequence of the $(4+n)$-dimensional Gauss law for gravity, with

$$M_*^{(4+n)} = M_s^{2+n}/g^4 \tag{2.2}$$

the effective scale of gravity in $4 + n$ dimensions. Taking $M_s \simeq 1$ TeV, one finds a size for the extra dimensions $R_\perp$ varying from 10^8 km, 0.1 mm, down

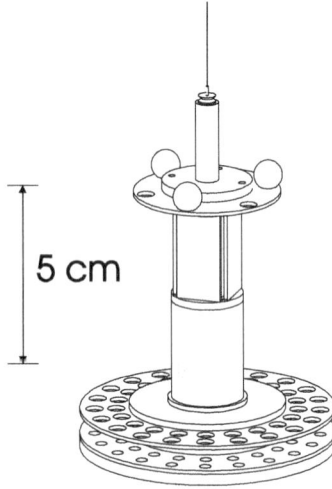

Fig. 2. Torsion pendulum that tested Newton's law at 55 μm.

to a Fermi for $n = 1, 2,$ or 6 large dimensions, respectively. This shows that while $n = 1$ is excluded, $n \geq 2$ is allowed by present experimental bounds on gravitational forces [4, 8]. Thus, in these models, gravity appears to us very weak at macroscopic scales because its intensity is spread in the "hidden" extra dimensions. At distances shorter than $R_\perp$, it should deviate from Newton's law, which may be possible to explore in laboratory experiments (see Fig. 2).

The main experimental implications of TeV scale strings in particle accelerators are of three types, in correspondence with the three different sectors that are generally present: (i) new compactified parallel dimensions, (ii) new extra large transverse dimensions and low scale quantum gravity, and (iii) genuine string and quantum gravity effects. On the other hand, there exist interesting implications in non accelerator table-top experiments due to the exchange of gravitons or other possible states living in the bulk.

3. Experimental implications in accelerators

3.1. World-brane extra dimensions

In this case $RM_s \gtrsim 1$, and the associated compactification scale $R_\parallel^{-1}$ would be the first scale of new physics that should be found increasing the beam energy [3, 9, 10]. There are several reasons for the existence of such dimensions. It is a logical possibility, since out of the six extra dimensions of string theory only two are needed for lowering the string scale, and thus the effective p-brane of our world has in general $d_\parallel \equiv p - 3 \leq 4$. Moreover, they can be used to address several physical problems in braneworld models, such as obtaining different SM gauge couplings, explaining fermion mass hierarchies due to different localization points of quarks and leptons in the extra dimensions, providing calculable mechanisms of supersymmetry breaking, etc.

The main consequence is the existence of Kaluza–Klein (KK) excitations for all SM particles that propagate along the extra parallel dimensions. Their masses are given by:

$$M_m^2 = M_0^2 + \frac{m^2}{R_\parallel^2}; \quad m = 0, \pm 1, \pm 2, \ldots, \tag{3.1}$$

where we used $d_\parallel = 1$, and M_0 is the higher dimensional mass. The zero-mode $m = 0$ is identified with the 4d state, while the higher modes have the same quantum numbers with the lowest one, except for their mass given in (3.1). There are two types of experimental signatures of such dimensions [9,11,12]: (i) virtual exchange of KK excitations, leading to deviations in cross-sections compared to the SM prediction, that can be used to extract bounds on the compactification scale; (ii) direct production of KK modes.

On general grounds, there can be two different kinds of models with qualitatively different signatures depending on the localization properties of matter fermion fields. If the latter are localized in 3d brane intersections, they do not have excitations and KK momentum is not conserved because of the breaking of translation invariance in the extra dimension(s). KK modes of gauge bosons are then singly produced giving rise to generally strong bounds on the compactification scale and new resonances that can be observed in experiments. Otherwise, they can be produced only in pairs due to the KK momentum conservation, making the bounds weaker but the resonances difficult to observe.

When the internal momentum is conserved, the interaction vertex involving KK modes has the same 4d tree-level gauge coupling. On the other hand, their couplings to localized matter have an exponential form factor suppressing the interactions of heavy modes. This form factor can be viewed as the fact that the branes intersection has a finite thickness. For instance, the coupling of the KK excitations of gauge fields $A^\mu(x, y) = \sum_m A_m^\mu \exp i\frac{my}{R_\parallel}$ to the charge density $j_\mu(x)$ of massless localized fermions is described by the effective action [13]:

$$\int d^4x \sum_m e^{-\ln 16 \frac{m^2 l_s^2}{2R_\parallel^2}} j_\mu(x) A_m^\mu(x). \tag{3.2}$$

After Fourier transform in position space, it becomes:

$$\int d^4x \, dy \, \frac{1}{(2\pi \ln 16)^2} e^{-\frac{y^2 M_s^2}{2\ln 16}} j_\mu(x) \, A^\mu(x, y), \tag{3.3}$$

from which we see that localized fermions form a Gaussian distribution of charge with a width $\sigma = \sqrt{\ln 16} \, l_s \sim 1.66 \, l_s$.

To simplify the analysis, let us consider first the case $d_\parallel = 1$ where some of the gauge fields arise from an effective 4-brane, while fermions are localized states on brane intersections. Since the corresponding gauge couplings are reduced by the size of the large dimension $R_\parallel M_s$ compared to the others, one can account for the ratio of the weak to strong interactions strengths if the $SU(2)$ brane extends along the extra dimension, while $SU(3)$ does not. As a result, there are 3 distinct cases to study [12], denoted by (t, l, l), (t, l, t) and (t, t, l), where the three positions in the brackets correspond to the three SM gauge group factors $SU(3) \times SU(2) \times U(1)$ and those with l (longitudinal) feel the extra dimension, while those with t (transverse) do not.

In the (t, l, l) case, there are KK excitations of $SU(2) \times U(1)$ gauge bosons: $W_\pm^{(m)}$, $\gamma^{(m)}$ and $Z^{(m)}$. Performing a χ^2 fit of the electroweak observables, one finds that if the Higgs is a bulk state (l), $R_\parallel^{-1} \gtrsim 3.5$ TeV [14]. This implies that LHC can produce at most the first KK mode. Different choices for localization of matter and Higgs fields lead to bounds, lying in the range 1–5 TeV [14].

In addition to virtual effects, KK excitations can be produced on-shell at LHC as new resonances [11] (see Fig. 3). There are two different channels, neutral Drell–Yan processes $pp \to l^+l^- X$ and the charged channel $l^\pm \nu$, corresponding to the production of the KK modes $\gamma^{(1)}$, $Z^{(1)}$ and $W_\pm^{(1)}$, respectively. The discovery limits are about 6 TeV, while the exclusion bounds 15 TeV. An interesting observation in the case of $\gamma^{(1)} + Z^{(1)}$ is that interferences can lead to a "dip" just before the resonance. There are some ways to distinguish the corresponding signals from other possible origin of new physics, such as models with new gauge bosons. In fact, in the (t, l, l) and (t, l, t) cases, one expects two resonances located practically at the same mass value. This property is not shared by most of other new gauge boson models. Moreover, the heights and widths of the resonances are directly related to those of SM gauge bosons in the corresponding channels.

In the (t, l, t) case, only the $SU(2)$ factor feels the extra dimension and the limits set by the KK states of $W^\pm$ remain the same. On the other hand, in the (t, t, l) case where only $U(1)_Y$ feels the extra dimension, the limits are weaker and the exclusion bound is around 8 TeV. In addition to these simple possibilities, brane constructions lead often to cases where part of $U(1)_Y$ is t and part is l. If $SU(2)$ is l the limits come again from $W^\pm$, while if it is t then it will be difficult

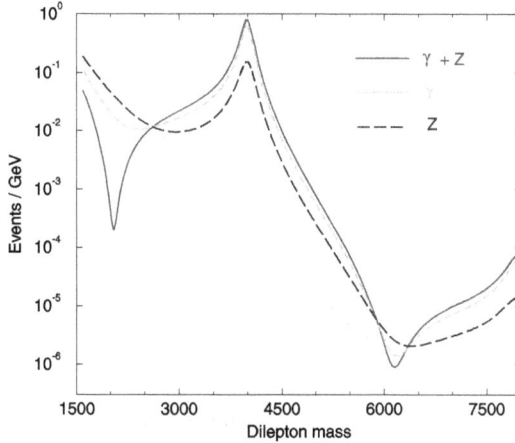

Fig. 3. Production of the first KK modes of the photon and of the Z boson at LHC, decaying to electron-positron pairs. The number of expected events is plotted as a function of the energy of the pair in GeV. From highest to lowest: excitation of $\gamma + Z$, γ and Z.

to distinguish this case from a generic extra $U(1)'$. A good statistics would be needed to see the deviation in the tail of the resonance as being due to effects additional to those of a generic $U(1)'$ resonance. Finally, in the case of two or more parallel dimensions, the sum in the exchange of the KK modes diverges in the limit $R_\| M_s \gg 1$ and needs to be regularized using the form factor (3.2). Cross-sections become bigger yielding stronger bounds, while resonances are closer implying that more of them could be reached by LHC.

On the other hand, if all SM particles propagate in the extra dimension (called universal),[1] KK modes can only be produced in pairs and the lower bound on the compactification scale becomes weaker, of order of 300–500 GeV. Moreover, no resonances can be observed at LHC, so that this scenario appears very similar to low energy supersymmetry. In fact, KK parity can even play the role of R-parity, implying that the lightest KK mode is stable and can be a dark matter candidate in analogy to the LSP [15].

3.2. Extra large transverse dimensions

The main experimental signal is gravitational radiation in the bulk from any physical process on the world-brane. In fact, the very existence of branes breaks translation invariance in the transverse dimensions and gravitons can be emitted

---

[1] Although interesting, this scenario seems difficult to be realized, since 4d chirality requires non-trivial action of orbifold twists with localized chiral states at the fixed points.

Table 1

Limits on $R_\perp$ in mm

Experiment	$n = 2$	$n = 4$	$n = 6$
	Collider bounds		
LEP 2	5×10^{-1}	2×10^{-8}	7×10^{-11}
Tevatron	5×10^{-1}	10^{-8}	4×10^{-11}
LHC	4×10^{-3}	6×10^{-10}	3×10^{-12}
NLC	10^{-2}	10^{-9}	6×10^{-12}
	Present non-collider bounds		
SN1987A	3×10^{-4}	10^{-8}	6×10^{-10}
COMPTEL	5×10^{-5}	–	–

from the brane into the bulk. During a collision of center of mass energy $\sqrt{s}$, there are $\sim (\sqrt{s} R_\perp)^n$ KK excitations of gravitons with tiny masses, that can be emitted. Each of these states looks from the 4d point of view as a massive, quasi-stable, extremely weakly coupled (s/M_P^2 suppressed) particle that escapes from the detector. The total effect is a missing-energy cross-section roughly of order:

$$\frac{(\sqrt{s} R_\perp)^n}{M_P^2} \sim \frac{1}{s} \left(\frac{\sqrt{s}}{M_s} \right)^{n+2}. \tag{3.4}$$

Explicit computation of these effects leads to the bounds given in Table 1. However, larger radii are allowed if one relaxes the assumption of isotropy, by taking for instance two large dimensions with different radii.

Figure 4 shows the cross-section for graviton emission in the bulk, corresponding to the process $pp \rightarrow jet + graviton$ at LHC, together with the SM background [16]. For a given value of M_s, the cross-section for graviton emission decreases with the number of large transverse dimensions, in contrast to the case of parallel dimensions. The reason is that gravity becomes weaker if there are more dimensions because there is more space for the gravitational field to escape. There is a particular energy and angular distribution of the produced gravitons that arise from the distribution in mass of KK states of spin-2. This can be contrasted to other sources of missing energy and might be a smoking gun for the extra dimensional nature of such a signal.

In Table 1, there are also included astrophysical and cosmological bounds. Astrophysical bounds [17, 18] arise from the requirement that the radiation of gravitons should not carry on too much of the gravitational binding energy released during core collapse of supernovae. In fact, the measurements of Kamiokande and IMB for SN1987A suggest that the main channel is neutrino fluxes. The best cosmological bound [19] is obtained from requiring that decay of bulk gravitons

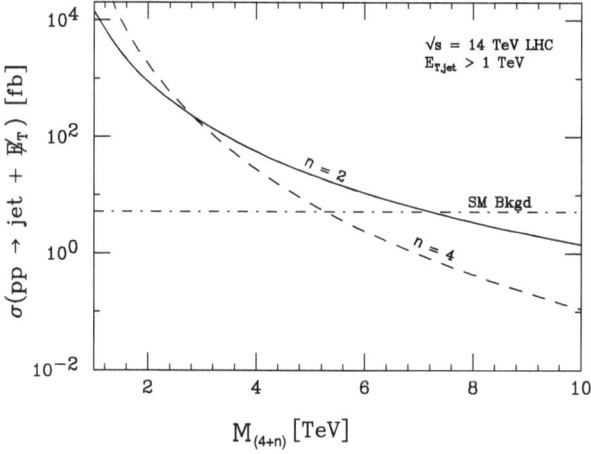

Fig. 4. Missing energy due to graviton emission at LHC, as a function of the higher-dimensional gravity scale M_*, produced together with a hadronic jet. The expected cross-section is shown for $n = 2$ and $n = 4$ extra dimensions, together with the SM background.

to photons do not generate a spike in the energy spectrum of the photon background measured by the COMPTEL instrument. Bulk gravitons are expected to be produced just before nucleosynthesis due to thermal radiation from the brane. The limits assume that the temperature was at most 1 MeV as nucleosynthesis begins, and become stronger if temperature is increased.

3.3. String effects

At low energies, the interaction of light (string) states is described by an effective field theory. Their exchange generates in particular four-fermion operators that can be used to extract independent bounds on the string scale. In analogy with the bounds on longitudinal extra dimensions, there are two cases depending on the localization properties of matter fermions. If they come from open strings with both ends on the same stack of branes, exchange of massive open string modes gives rise to dimension eight effective operators, involving four fermions and two space-time derivatives [13,20]. The corresponding bounds on the string scale are then around 500 GeV. On the other hand, if matter fermions are localized on non-trivial brane intersections, one obtains dimension six four-fermion operators and the bounds become stronger: $M_s \gtrsim$ 2–3 TeV [6, 13]. At energies higher than the string scale, new spectacular phenomena are expected to occur, related to string physics and quantum gravity effects, such as possible micro-black hole production [21–23]. Particle accelerators would then become the best tools for studying quantum gravity and string theory.

4. Supersymmetry in the bulk and short range forces

4.1. Sub-millimeter forces

Besides the spectacular predictions in accelerators, there are also modifications of gravitation in the sub-millimeter range, which can be tested in "table-top" experiments that measure gravity at short distances. There are three categories of such predictions:

(i) Deviations from the Newton's law $1/r^2$ behavior to $1/r^{2+n}$, which can be observable for $n = 2$ large transverse dimensions of sub-millimeter size. This case is particularly attractive on theoretical grounds because of the logarithmic sensitivity of SM couplings on the size of transverse space [24], that allows to determine the hierarchy [25].

(ii) New scalar forces in the sub-millimeter range, related to the mechanism of supersymmetry breaking, and mediated by light scalar fields φ with masses [5, 26]:

$$m_\varphi \simeq \frac{m_{susy}^2}{M_P} \simeq 10^{-4} \text{--} 10^{-6} \text{ eV}, \qquad (4.1)$$

for a supersymmetry breaking scale $m_{susy} \simeq 1\text{--}10$ TeV. They correspond to Compton wavelengths of 1 mm to 10 μm. m_{susy} can be either $1/R_\parallel$ if supersymmetry is broken by compactification [26], or the string scale if it is broken "maximally" on our world-brane [5]. A universal attractive scalar force is mediated by the radion modulus $\varphi \equiv M_P \ln R$, with R the radius of the longitudinal or transverse dimension(s). In the former case, the result (4.1) follows from the behavior of the vacuum energy density $\Lambda \sim 1/R_\parallel^4$ for large $R_\parallel$ (up to logarithmic corrections). In the latter, supersymmetry is broken primarily on the brane, and thus its transmission to the bulk is gravitationally suppressed, leading to (4.1). For $n = 2$, there may be an enhancement factor of the radion mass by $\ln R_\perp M_s \simeq 30$ decreasing its wavelength by an order of magnitude [25].

The coupling of the radius modulus to matter relative to gravity can be easily computed and is given by:

$$\sqrt{\alpha_\varphi} = \frac{1}{M}\frac{\partial M}{\partial \varphi}; \quad \alpha_\varphi = \begin{cases} \dfrac{\partial \ln \Lambda_{QCD}}{\partial \ln R} \simeq \dfrac{1}{3} & \text{for } R_\parallel \\[2mm] \dfrac{2n}{n+2} = 1\text{--}1.5 & \text{for } R_\perp, \end{cases} \qquad (4.2)$$

where M denotes a generic physical mass. In the longitudinal case, the coupling arises dominantly through the radius dependence of the QCD gauge coupling [26], while in the case of transverse dimension, it can be deduced from the rescaling of the metric which changes the string to the Einstein frame and depends slightly on the bulk dimensionality ($\alpha = 1\text{--}1.5$ for $n = 2\text{--}6$) [25]. Such a force can be tested in microgravity experiments and should be contrasted with the change of Newton's law due the presence of extra dimensions that is observable only for $n = 2$ [4,8]. The resulting bounds from an analysis of the radion effects are [27]:

$$M_* \gtrsim 6\,\text{TeV}. \tag{4.3}$$

In principle there can be other light moduli which couple with even larger strengths. For example the dilaton, whose VEV determines the string coupling, if it does not acquire large mass from some dynamical supersymmetric mechanism, can lead to a force of strength 2000 times bigger than gravity [28].

(iii) Non universal repulsive forces much stronger than gravity, mediated by possible abelian gauge fields in the bulk [17,29]. Such fields acquire tiny masses of the order of M_s^2/M_P, as in (4.1), due to brane localized anomalies [29]. Although their gauge coupling is infinitesimally small, $g_A \sim M_s/M_P \simeq 10^{-16}$, it is still bigger that the gravitational coupling E/M_P for typical energies $E \sim 1$ GeV, and the strength of the new force would be 10^6–10^8 stronger than gravity. This is an interesting region which will be soon explored in micro-gravity experiments (see Fig. 5). Note that in this case supernova constraints impose that there should be at least four large extra dimensions in the bulk [17].

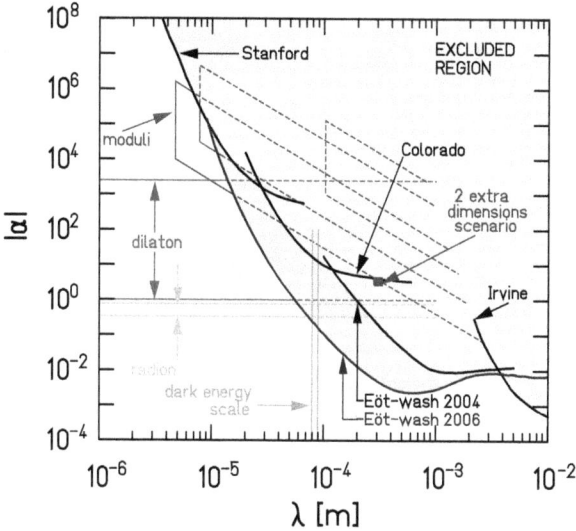

Fig. 5. Present limits on new short-range forces (yellow regions), as a function of their range λ and their strength relative to gravity α. The limits are compared to new forces mediated by the graviton in the case of two large extra dimensions, and by the radion.

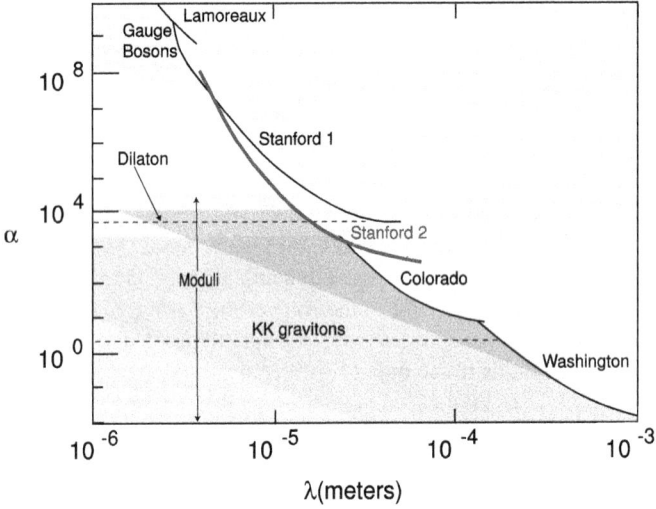

Fig. 6. Bounds on non-Newtonian forces in the range 6–20 μm (see S.J. Smullin et al. [8]).

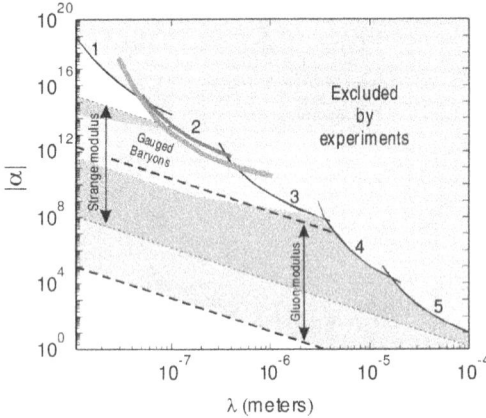

Fig. 7. Bounds on non-Newtonian forces in the range of 10–200 nm (see R.S. Decca et al. in Ref. [8]). Curves 4 and 5 correspond to Stanford and Colorado experiments, respectively, of Fig. 6 (see also J.C. Long and J.C. Price of Ref. [8]).

In Fig. 5 we depict the actual information from previous, present and up-coming experiments [8, 25]. The solid lines indicate the present limits from the experiments indicated. The excluded regions lie above these solid lines. Measuring gravitational strength forces at short distances is challenging. The horizontal lines correspond to theoretical predictions, in particular for the graviton in the case $n = 2$ and for the radion in the transverse case. These limits are compared to those obtained from particle accelerator experiments in Table 1. Finally, in Figs. 6 and 7, we display recent improved bounds for new forces at very short distances by focusing on the left hand side of Fig. 5, near the origin [8].

4.2. Brane non-linear supersymmetry

When the closed string sector is supersymmetric, supersymmetry on a generic brane configuration is non-linearly realized even if the spectrum is not supersymmetric and brane fields have no superpartners. The reason is that the gravitino must couple to a conserved current locally, implying the existence of a goldstino on the brane world-volume [30]. The goldstino is exactly massless in the infinite (transverse) volume limit and is expected to acquire a small mass suppressed by the volume, of order (4.1). In the standard realization, its coupling to matter is given via the energy momentum tensor [31], while in general there are more terms invariant under non-linear supersymmetry that have been classified, up to dimension eight [32, 33].

An explicit computation was performed for a generic intersection of two brane stacks, leading to three irreducible couplings, besides the standard one [33]: two of dimension six involving the goldstino, a matter fermion and a scalar or gauge field, and one four-fermion operator of dimension eight. Their strength is set by the goldstino decay constant κ, up to model-independent numerical coefficients which are independent of the brane angles. Obviously, at low energies the dominant operators are those of dimension six. In the minimal case of (non-supersymmetric) SM, only one of these two operators may exist, that couples the goldstino χ with the Higgs H and a lepton doublet L:

$$\mathcal{L}_\chi^{int} = 2\kappa (D_\mu H)(L D^\mu \chi) + h.c., \tag{4.4}$$

where the goldstino decay constant is given by the total brane tension

$$\frac{1}{2\kappa^2} = N_1 T_1 + N_2 T_2; \quad T_i = \frac{M_s^4}{4\pi^2 g_i^2}, \tag{4.5}$$

with N_i the number of branes in each stack. It is important to notice that the effective interaction (4.4) conserves the total lepton number L, as long as we assign to the goldstino a total lepton number $L(\chi) = -1$ [34]. To simplify the analysis, we will consider the simplest case where (4.4) exists only for the first generation and L is the electron doublet [34].

The effective interaction (4.4) gives rise mainly to the decays $W^\pm \to e^\pm \chi$ and $Z, H \to \nu\chi$. It turns out that the invisible Z width gives the strongest limit

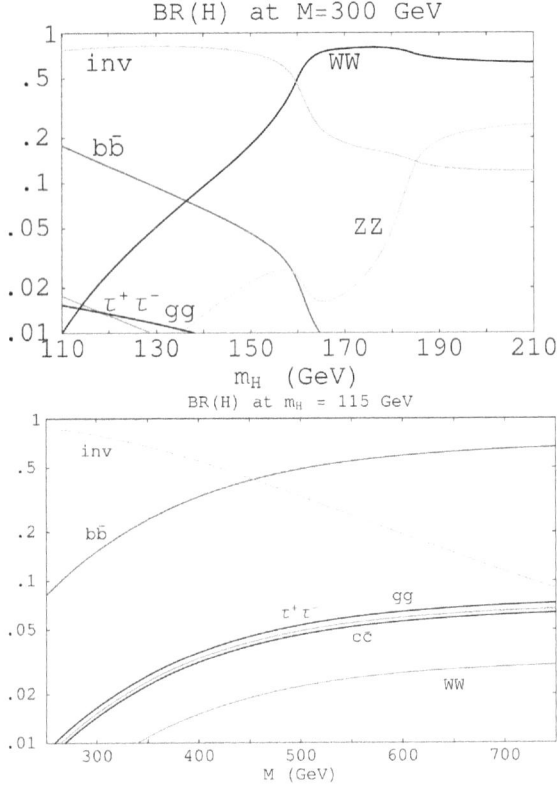

Fig. 8. Higgs branching rations, as functions either of the Higgs mass m_H for a fixed value of the string scale $M_s \simeq 2M = 600$ GeV, or of $M \simeq M_s/2$ for $m_H = 115$ GeV.

on κ which can be translated to a bound on the string scale $M_s \gtrsim 500$ GeV, comparable to other collider bounds. This allows for the striking possibility of a Higgs boson decaying dominantly, or at least with a sizable branching ratio, via such an invisible mode, for a wide range of the parameter space (M_s, m_H), as seen in Fig. 8.

5. Electroweak symmetry breaking

Non-supersymmetric TeV strings offer also a framework to realize gauge symmetry breaking radiatively. Indeed, from the effective field theory point of view, one expects quadratically divergent one-loop contributions to the masses of scalar fields. The divergences are cut off by M_s and if the corrections are negative, they can induce electroweak symmetry breaking and explain the mild hierarchy between the weak and a string scale at a few TeV, in terms of a loop factor [35]. More precisely, in the minimal case of one Higgs doublet H, the scalar potential is:

$$V = \lambda (H^\dagger H)^2 + \mu^2 (H^\dagger H), \tag{5.1}$$

where λ arises at tree-level. Moreover, in any model where the Higgs field comes from an open string with both ends fixed on the same brane stack, it is given by an appropriate truncation of a supersymmetric theory. Within the minimal spectrum of the SM, $\lambda = (g_2^2 + g'^2)/8$, with g_2 and g' the $SU(2)$ and $U(1)_Y$ gauge couplings. On the other hand, μ^2 is generated at one loop:

$$\mu^2 = -\varepsilon^2 g^2 M_s^2, \tag{5.2}$$

where ε is a loop factor that can be estimated from a toy model computation and varies in the region $\epsilon \sim 10^{-1}$–10^{-3}.

Indeed, consider for illustration a simple case where the whole one-loop effective potential of a scalar field can be computed. We assume for instance one extra dimension compactified on a circle of radius $R > 1$ (in string units). An interesting situation is provided by a class of models where a non-vanishing VEV for a scalar (Higgs) field ϕ results in shifting the mass of each KK excitation by a constant $a(\phi)$:

$$M_m^2 = \left(\frac{m + a(\phi)}{R} \right)^2, \tag{5.3}$$

with m the KK integer momentum number. Such mass shifts arise for instance in the presence of a Wilson line, $a = q \oint \frac{dy}{2\pi} g A$, where A is the internal component of a gauge field with gauge coupling g, and q is the charge of a given state under the corresponding generator. A straightforward computation shows that the ϕ-dependent part of the one-loop effective potential is given by [36]:

$$V_{eff} = -Tr(-)^F \frac{R}{32 \pi^{3/2}} \sum_n e^{2\pi i n a} \int_0^\infty dl \, l^{3/2} f_s(l) \, e^{-\pi^2 n^2 R^2 l}, \tag{5.4}$$

where $F = 0, 1$ for bosons and fermions, respectively. We have included a regulating function $f_s(l)$ which contains for example the effects of string oscillators. To understand its role we will consider the two limits $R \gg 1$ and $R \ll 1$. In the first case only the $l \to 0$ region contributes to the integral. This means that the effective potential receives sizable contributions only from the infrared (field theory) degrees of freedom. In this limit we would have $f_s(l) \to 1$. For example, in the string model considered in [35]:

$$f_s(l) = \left[\frac{1}{4l} \frac{\theta_2}{\eta^3} \left(il + \frac{1}{2} \right) \right]^4 \to 1 \quad \text{for } l \to 0, \tag{5.5}$$

and the field theory result is finite and can be explicitly computed. As a result of the Taylor expansion around $a = 0$, we are able to extract the one-loop contribution to the coefficient of the term of the potential quadratic in the Higgs field. It is given by a loop factor times the compactification scale [36]. One thus obtains $\mu^2 \sim g^2/R^2$ up to a proportionality constant which is calculable in the effective field theory. On the other hand, if we consider $R \to 0$, which by T-duality corresponds to taking the extra dimension as transverse and very large, the one-loop effective potential receives contributions from the whole tower of string oscillators as appearing in $f_s(l)$, leading to squared masses given by a loop factor times M_s^2, according to Eq. (5.2).

More precisely, from the expression (5.4), one finds:

$$\varepsilon^2(R) = \frac{1}{2\pi^2} \int_0^\infty \frac{dl}{(2l)^{5/2}} \frac{\theta_2^4}{4\eta^{12}} \left(il + \frac{1}{2} \right) R^3 \sum_n n^2 e^{-2\pi n^2 R^2 l}, \tag{5.6}$$

which is plotted in Fig. 9. For the asymptotic value $R \to 0$ (corresponding upon T-duality to a large transverse dimension of radius $1/R$), $\varepsilon(0) \simeq 0.14$, and the effective cut-off for the mass term is M_s, as can be seen from Eq. (5.2). At large R,

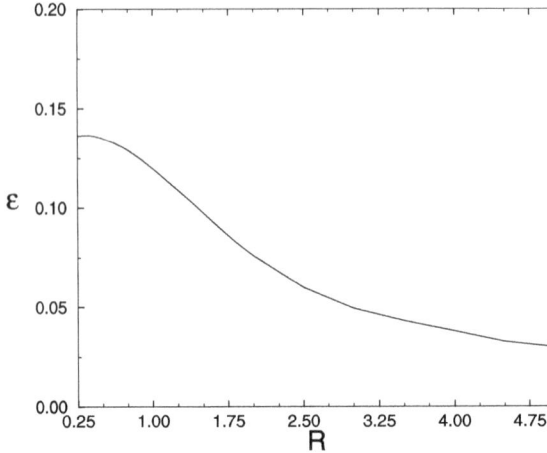

Fig. 9. The coefficient ε of the one loop Higgs mass (5.2).

$\mu^2(R)$ falls off as $1/R^2$, which is the effective cut-off in the limit $R \to \infty$, as we argued above, in agreement with field theory results in the presence of a compactified extra dimension [26, 37]. In fact, in the limit $R \to \infty$, an analytic approximation to $\varepsilon(R)$ gives:

$$\varepsilon(R) \simeq \frac{\varepsilon_\infty}{M_s R}, \quad \varepsilon_\infty^2 = \frac{3\,\zeta(5)}{4\,\pi^4} \simeq 0.008. \tag{5.7}$$

The potential (5.1) has the usual minimum, given by the VEV of the neutral component of the Higgs doublet $v = \sqrt{-\mu^2/\lambda}$. Using the relation of v with the Z gauge boson mass, $M_Z^2 = (g_2^2 + g'^2)v^2/4$, and the expression of the quartic coupling λ, one obtains for the Higgs mass a prediction which is the Minimal Supersymmetric Standard Model (MSSM) value for $\tan\beta \to \infty$ and $m_A \to \infty$: $m_H = M_Z$. The tree level Higgs mass is known to receive important radiative corrections from the top-quark sector and rises to values around 120 GeV. Furthermore, from (5.2), one can compute M_s in terms of the Higgs mass $m_H^2 = -2\mu^2$:

$$M_s = \frac{m_H}{\sqrt{2}\,g\varepsilon}, \tag{5.8}$$

yielding naturally values in the TeV range.

6. Standard Model on D-branes

The gauge group closest to the Standard Model one can easily obtain with D-branes is $U(3) \times U(2) \times U(1)$. The first factor arises from three coincident "color" D-branes. An open string with one end on them is a triplet under $SU(3)$ and carries the same $U(1)$ charge for all three components. Thus, the $U(1)$ factor of $U(3)$ has to be identified with *gauged* baryon number. Similarly, $U(2)$ arises from two coincident "weak" D-branes and the corresponding abelian factor is identified with *gauged* weak-doublet number. Finally, an extra $U(1)$ D-brane is necessary in order to accommodate the Standard Model without breaking the baryon number [38]. In principle this $U(1)$ brane can be chosen to be independent of the other two collections with its own gauge coupling. To improve the predictability of the model, we choose to put it on top of either the color or the weak D-branes [39]. In either case, the model has two independent gauge couplings g_3 and g_2 corresponding, respectively, to the gauge groups $U(3)$ and $U(2)$. The $U(1)$ gauge coupling g_1 is equal to either g_3 or g_2.

Let us denote by Q_3, Q_2 and Q_1 the three $U(1)$ charges of $U(3) \times U(2) \times U(1)$, in a self explanatory notation. Under $SU(3) \times SU(2) \times U(1)_3 \times U(1)_2 \times$

$$U(2)_L$$

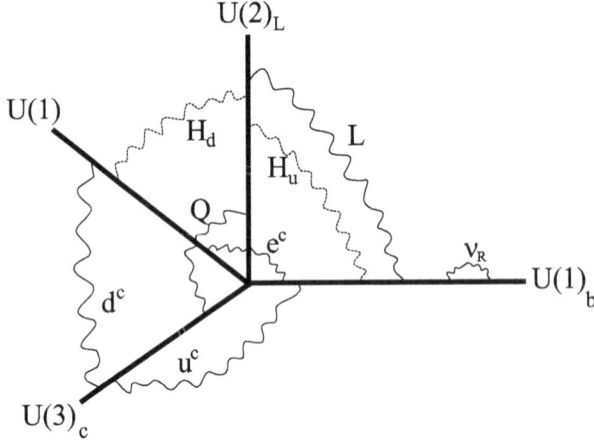

Fig. 10. A minimal Standard Model embedding on D-branes.

$U(1)_1$, the members of a family of quarks and leptons have the following quantum numbers:

$$
\begin{aligned}
Q & \quad (\mathbf{3}, \mathbf{2}; 1, w, 0)_{1/6} \\
u^c & \quad (\bar{\mathbf{3}}, \mathbf{1}; -1, 0, x)_{-2/3} \\
d^c & \quad (\bar{\mathbf{3}}, \mathbf{1}; -1, 0, y)_{1/3} \\
L & \quad (\mathbf{1}, \mathbf{2}; 0, 1, z)_{-1/2} \\
l^c & \quad (\mathbf{1}, \mathbf{1}; 0, 0, 1)_1
\end{aligned}
\tag{6.1}
$$

The values of the $U(1)$ charges x, y, z, w will be fixed below so that they lead to the right hypercharges, shown for completeness as subscripts.

It turns out that there are two possible ways of embedding the Standard Model particle spectrum on these stacks of branes [38], which are shown pictorially in Fig. 10. The quark doublet Q corresponds necessarily to a massless excitation of an open string with its two ends on the two different collections of branes (color and weak). As seen from the figure, a fourth brane stack is needed for a complete embedding, which is chosen to be a $U(1)_b$ extended in the bulk. This is welcome since one can accommodate right handed neutrinos as open string states on the bulk with sufficiently small Yukawa couplings suppressed by the large volume of the bulk [40]. The two models are obtained by an exchange of the up and down antiquarks, u^c and d^c, which correspond to open strings with one end on the color branes and the other either on the $U(1)$ brane, or on the $U(1)_b$ in the bulk. The lepton doublet L arises from an open string stretched between the weak branes and $U(1)_b$, while the antilepton l^c corresponds to a string with one end on the $U(1)$ brane and the other in the bulk. For completeness, we also show the two possible Higgs states H_u and H_d that are both necessary in order to give tree-level masses to all quarks and leptons of the heaviest generation.

6.1. Hypercharge embedding and the weak angle

The weak hypercharge Y is a linear combination of the three $U(1)$'s:

$$Y = Q_1 + \frac{1}{2}Q_2 + c_3 Q_3; \quad c_3 = -1/3 \text{ or } 2/3, \tag{6.2}$$

where Q_N denotes the $U(1)$ generator of $U(N)$ normalized so that the fundamental representation of $SU(N)$ has unit charge. The corresponding $U(1)$ charges appearing in Eq. (6.1) are $x = -1$ or 0, $y = 0$ or 1, $z = -1$, and $w = 1$ or -1, for $c_3 = -1/3$ or $2/3$, respectively. The hypercharge coupling g_Y is given by:[2]

$$\frac{1}{g_Y^2} = \frac{2}{g_1^2} + \frac{4c_2^2}{g_2^2} + \frac{6c_3^2}{g_3^2}. \tag{6.3}$$

It follows that the weak angle $\sin^2 \theta_W$, is given by:

$$\sin^2 \theta_W \equiv \frac{g_Y^2}{g_2^2 + g_Y^2} = \frac{1}{2 + 2g_2^2/g_1^2 + 6c_3^2 g_2^2/g_3^2}, \tag{6.4}$$

where g_N is the gauge coupling of $SU(N)$ and $g_1 = g_2$ or $g_1 = g_3$ at the string scale. In order to compare the theoretical predictions with the experimental value of $\sin^2 \theta_W$ at M_s, we plot in Fig. 11 the corresponding curves as functions of M_s. The solid line is the experimental curve. The dashed line is the plot of the function (6.4) for $g_1 = g_2$ with $c_3 = -1/3$ while the dotted-dashed line corresponds to $g_1 = g_3$ with $c_3 = 2/3$. The other two possibilities are not shown because they lead to a value of M_s which is too high to protect the hierarchy. Thus, the second case, where the $U(1)$ brane is on top of the color branes, is compatible with low energy data for $M_s \sim 6$–8 TeV and $g_s \simeq 0.9$.

From Eq. (6.4) and Fig. 11, we find the ratio of the $SU(2)$ and $SU(3)$ gauge couplings at the string scale to be $\alpha_2/\alpha_3 \sim 0.4$. This ratio can be arranged by an appropriate choice of the relevant moduli. For instance, one may choose the color and $U(1)$ branes to be D3 branes while the weak branes to be D7 branes. Then, the ratio of couplings above can be explained by choosing the volume of the four compact dimensions of the seven branes to be $V_4 = 2.5$ in string units. This being larger than one is consistent with the picture above. Moreover it predicts

---

[2] The gauge couplings $g_{2,3}$ are determined at the tree-level by the string coupling and other moduli, like radii of longitudinal dimensions. In higher orders, they also receive string threshold corrections.

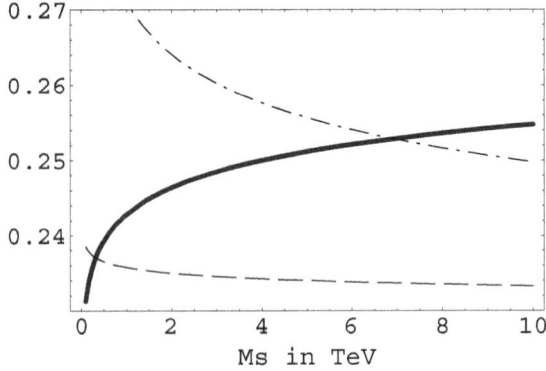

Fig. 11. The experimental value of $\sin^2 \theta_W$ (thick curve), and the theoretical predictions (6.4).

an interesting spectrum of KK states for the Standard model, different from the naive choices that have appeared hitherto: the only Standard Model particles that have KK descendants are the W bosons as well as the hypercharge gauge boson. However, since the hypercharge is a linear combination of the three $U(1)$'s, the massive $U(1)$ KK gauge bosons do not couple to the hypercharge but to the weak doublet number.

6.2. The fate of $U(1)$'s, proton stability and neutrino masses

It is easy to see that the remaining three $U(1)$ combinations orthogonal to Y are anomalous. In particular there are mixed anomalies with the $SU(2)$ and $SU(3)$ gauge groups of the Standard Model. These anomalies are cancelled by three axions coming from the closed string RR (Ramond) sector, via the standard Green–Schwarz mechanism [41]. The mixed anomalies with the non-anomalous hypercharge are also cancelled by dimension five Chern–Simons type of interactions [38]. An important property of the above Green–Schwarz anomaly cancellation mechanism is that the anomalous $U(1)$ gauge bosons acquire masses leaving behind the corresponding global symmetries. This is in contrast to what would had happened in the case of an ordinary Higgs mechanism. These global symmetries remain exact to all orders in type I string perturbation theory around the orientifold vacuum. This follows from the topological nature of Chan–Paton charges in all string amplitudes. On the other hand, one expects non-perturbative violation of global symmetries and consequently exponentially small in the string coupling, as long as the vacuum stays at the orientifold point. Thus, all $U(1)$ charges are conserved and since Q_3 is the baryon number, proton stability is guaranteed.

Another linear combination of the $U(1)$'s is the lepton number. Lepton number conservation is important for the extra dimensional neutrino mass suppression mechanism described above, that can be destabilized by the presence of a large Majorana neutrino mass term. Such a term can be generated by the lepton-number violating dimension five effective operator $LLHH$ that leads, in the case of TeV string scale models, to a Majorana mass of the order of a few GeV. Even if we manage to eliminate this operator in some particular model, higher order operators would also give unacceptably large contributions, as we focus on models in which the ratio between the Higgs vacuum expectation value and the string scale is just of order $\mathcal{O}(1/10)$. The best way to protect tiny neutrino masses from such contributions is to impose lepton number conservation.

A bulk neutrino propagating in $4 + n$ dimensions can be decomposed in a series of 4d KK excitations denoted collectively by $\{m\}$:

$$S_{kin} = R_\perp^n \int d^4x \sum_{\{m\}} \left\{ \bar{v}_{Rm} \not{\partial} v_{Rm} + \bar{v}_{Rm}^c \not{\partial} v_{Rm}^c + \frac{m}{R_\perp} v_{Rm} v_{Rm}^c + c.c. \right\},$$

(6.5)

where v_R and v_R^c are the two Weyl components of the Dirac spinor and for simplicity we considered a common compactification radius $R_\perp$. On the other hand, there is a localized interaction of v_R with the Higgs field and the lepton doublet, which leads to mass terms between the left-handed neutrino and the KK states v_{Rm}, upon the Higgs VEV v:

$$S_{int} = g_s \int d^4x\, H(x) L(x) v_R(x, y = 0) \quad \rightarrow \quad \frac{g_s v}{R_\perp^{n/2}} \sum_m v_L v_{Rm}$$

(6.6)

in strings units. Since the mass mixing $g_s v / R_\perp^{n/2}$ is much smaller than the KK mass $1/R_\perp$, it can be neglected for all the excitations except for the zero-mode v_{R0}, which gets a Dirac mass with the left-handed neutrino

$$m_v \simeq \frac{g_s v}{R_\perp^{n/2}} \simeq v \frac{M_s}{M_p} \simeq 10^{-3}\text{--}10^{-2} \text{ eV},$$

(6.7)

for $M_s \simeq 1\text{--}10$ TeV, where the relation (2.1) was used. In principle, with one bulk neutrino, one could try to explain both solar and atmospheric neutrino oscillations using also its first KK excitation. However, the later behaves like a sterile neutrino which is now excluded experimentally. Therefore, one has to introduce three bulk species (at least two) v_R^i in order to explain neutrino oscillations in a 'traditional way', using their zero-modes v_{R0}^i [42]. The main difference with the usual seesaw mechanism is the Dirac nature of neutrino masses, which remains an open possibility to be tested experimentally.

7. Internal magnetic fields

We now consider type I string theory, or equivalently type IIB with orientifold 9-planes and D9-branes [2]. Upon compactification in four dimensions on a Calabi–Yau manifold, one gets $\mathcal{N} = 2$ supersymmetry in the bulk and $\mathcal{N} = 1$ on the branes. We then turn on internal magnetic fields [43,44], which, in the T-dual picture, amounts to intersecting branes [45,46]. For generic angles, or equivalently for arbitrary magnetic fields, supersymmetry is spontaneously broken and described by effective D-terms in the four-dimensional (4d) theory [43]. In the weak field limit, $|H|\alpha' < 1$ with α' the string Regge slope, the resulting mass shifts are given by:

$$\delta M^2 = (2k + 1)|q H| + 2q H \Sigma; \quad k = 0, 1, 2, \ldots, \tag{7.1}$$

where H is the magnetic field of an abelian gauge symmetry, corresponding to a Cartan generator of the higher dimensional gauge group, on a non-contractible 2-cycle of the internal manifold. Σ is the corresponding projection of the spin operator, k is the Landau level and $q = q_L + q_R$ is the charge of the state, given by the sum of the left and right charges of the endpoints of the associated open string. We recall that the exact string mass formula has the same form as (7.1) with $q H$ replaced by:

$$q H \longrightarrow \theta_L + \theta_R; \quad \theta_{L,R} = \arctan(q_{L,R} H \alpha'). \tag{7.2}$$

Obviously, the field theory expression (7.1) is reproduced in the weak field limit.

The Gauss law for the magnetic flux implies that the field H is quantized in terms of the area of the corresponding 2-cycle A:

$$H = \frac{m}{nA}, \tag{7.3}$$

where the integers m, n correspond to the respective magnetic and electric charges; m is the quantized flux and n is the wrapping number of the higher dimensional brane around the corresponding internal 2-cycle. In the T-dual representation, associated to the inversion of the compactification radius along one of the two directions of the 2-cycle, m and n become the wrapping numbers around these two directions.

For simplicity, we consider first the case where the internal manifold is a product of three factorized tori $\prod_{i=1}^{3} T^2_{(i)}$. Then, the mass formula (7.1) becomes:

$$\delta M^2 = \sum_i (2k_i + 1)|q H_i| + 2q H_i \Sigma_i, \tag{7.4}$$

where Σ_i is the projection of the internal helicity along the i-th plane. For a ten-dimensional (10d) spinor, its eigenvalues are $\Sigma_i = \pm 1/2$, while for a 10d vector $\Sigma_i = \pm 1$ in one of the planes $i = i_0$ and zero in the other two ($i \neq i_0$). Thus, charged higher dimensional scalars become massive, fermions lead to chiral 4d zero modes if all $H_i \neq 0$, while the lightest scalars coming from 10d vectors have masses

$$M_0^2 = \begin{cases} |q\,H_1| + |q\,H_2| - |q\,H_3| \\ |q\,H_1| - |q\,H_2| + |q\,H_3| \\ -|q\,H_1| + |q\,H_2| + |q\,H_3| \end{cases} . \tag{7.5}$$

Note that all of them can be made positive definite, avoiding the Nielsen–Olesen instability, if all $H_i \neq 0$. Moreover, one can easily show that if a scalar mass vanishes, some supersymmetry remains unbroken [44, 45].

8. Minimal Standard Model embedding

We turn on now several abelian magnetic fields H_I^a of different Cartan generators $U(1)_a$, so that the gauge group is a product of unitary factors $\prod_a U(N_a)$ with $U(N_a) = SU(N_a) \times U(1)_a$. In an appropriate T-dual representation, it amounts to consider several stacks of D6-branes intersecting in the three internal tori at angles. An open string with one end on the a-th stack has charge ± 1 under the $U(1)_a$, depending on its orientation, and is neutral with respect to all others.

In this section, we perform a general study of SM embedding in three brane stacks with gauge group $U(3) \times U(2) \times U(1)$ [47], and present an explicit example having realistic particle content and satisfying gauge coupling unification [48]. We consider in general non oriented strings because of the presence of the orientifold plane that gives rise mirror branes with opposite magnetic fluxes $m \to -m$ in Eq. (7.3). An open string stretched between a brane stack $U(N)$ and its mirror transforms in the symmetric or antisymmetric representation, while the multiplicity of chiral fermions is given by their intersection number.

The quark and lepton doublets (Q and L) correspond to open strings stretched between the weak and the color or $U(1)$ branes, respectively. On the other hand, the u^c and d^c antiquarks can come from strings that are either stretched between the color and $U(1)$ branes, or that have both ends on the color branes (stretched between the brane stack and its orientifold image) and transform in the anti-symmetric representation of $U(3)$ (which is an anti-triplet). There are therefore three possible models, depending on whether it is the u^c (model A), or the d^c (model B), or none of them (model C), the state coming from the antisymmetric representation of color branes. It follows that the antilepton l^c comes in a similar

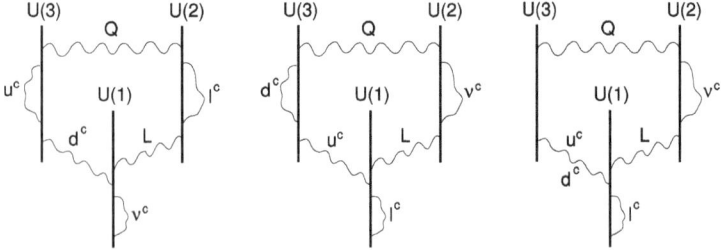

Fig. 12. Pictorial representation of models A, B and C.

way from open strings with both ends either on the weak brane stack and transforming in the antisymmetric representation of $U(2)$ which is an $SU(2)$ singlet
(in model A), or on the abelian brane and transforming in the "symmetric" representation of $U(1)$ (in models B and C). The three models are presented pictorially
in Fig. 12.

Thus, the members of a family of quarks and leptons have the following quantum numbers:

	Model A	Model B	Model C	
Q	$(\mathbf{3}, \mathbf{2}; 1, 1, 0)_{1/6}$	$(\mathbf{3}, \mathbf{2}; 1, \varepsilon_Q, 0)_{1/6}$	$(\mathbf{3}, \mathbf{2}; 1, \varepsilon_Q, 0)_{1/6}$	
u^c	$(\bar{\mathbf{3}}, \mathbf{1}; 2, 0, 0)_{-2/3}$	$(\bar{\mathbf{3}}, \mathbf{1}; -1, 0, 1)_{-2/3}$	$(\bar{\mathbf{3}}, \mathbf{1}; -1, 0, 1)_{-2/3}$	
d^c	$(\bar{\mathbf{3}}, \mathbf{1}; -1, 0, \varepsilon_d)_{1/3}$	$(\bar{\mathbf{3}}, \mathbf{1}; 2, 0, 0)_{1/3}$	$(\bar{\mathbf{3}}, \mathbf{1}; -1, 0, -1)_{1/3}$	(8.1)
L	$(\mathbf{1}, \mathbf{2}; 0, -1, \varepsilon_L)_{-1/2}$	$(\mathbf{1}, \mathbf{2}; 0, \varepsilon_L, 1)_{-1/2}$	$(\mathbf{1}, \mathbf{2}; 0, \varepsilon_L, 1)_{-1/2}$	
l^c	$(\mathbf{1}, \mathbf{1}; 0, 2, 0)_1$	$(\mathbf{1}, \mathbf{1}; 0, 0, -2)_1$	$(\mathbf{1}, \mathbf{1}; 0, 0, -2)_1$	
ν^c	$(\mathbf{1}, \mathbf{1}; 0, 0, 2\varepsilon_\nu)_0$	$(\mathbf{1}, \mathbf{1}; 0, 2\varepsilon_\nu, 0)_0$	$(\mathbf{1}, \mathbf{1}; 0, 2\varepsilon_\nu, 0)_0,$	

where the last three digits after the semi-column in the brackets are the charges
under the three abelian factors $U(1)_3 \times U(1)_2 \times U(1)$, that we will call Q_3,
Q_2 and Q_1 in the following, while the subscripts denote the corresponding hypercharges. The various sign ambiguities $\varepsilon_i = \pm 1$ are due to the fact that the
corresponding abelian factor does not participate in the hypercharge combination
(see below). In the last lines, we also give the quantum numbers of a possible
right-handed neutrino in each of the three models. These are in fact all possible
ways of embedding the SM spectrum in three sets of branes.

The hypercharge combination is:

$$\text{Model A:} \quad Y = -\frac{1}{3}Q_3 + \frac{1}{2}Q_2$$

$$\text{Model B, C:} \quad Y = \frac{1}{6}Q_3 - \frac{1}{2}Q_1 \tag{8.2}$$

leading to the following expressions for the weak angle:

$$\text{Model A:} \quad \sin^2 \theta_W = \frac{1}{2 + 2\alpha_2/3\alpha_3} = \frac{3}{8} \Big|_{\alpha_2 = \alpha_3}$$

$$\text{Model B, C:} \quad \sin^2 \theta_W = \frac{1}{1 + \alpha_2/2\alpha_1 + \alpha_2/6\alpha_3}$$

$$= \frac{6}{7 + 3\alpha_2/\alpha_1} \Big|_{\alpha_2 = \alpha_3} \tag{8.3}$$

In the second part of the above equalities, we used the unification relation $\alpha_2 = \alpha_3$, that can be imposed if for instance $U(3)$ and $U(2)$ branes are coincident, leading to a $U(5)$ unified group. Alternatively, this condition can be generally imposed under mild assumptions [48]. It follows that model A admits natural gauge coupling unification of strong and weak interactions, and predicts the correct value for $\sin^2 \theta_W = 3/8$ at the unification scale M_{GUT}. On the other hand, model B corresponds to the flipped $SU(5)$ where the role of u^c and d^c is interchanged together with l^c and ν^c between the **10** and **5̄** representations [49].

Besides the hypercharge combination, there are two additional $U(1)$'s. It is easy to check that one of the two can be identified with $B - L$. For instance, in model A choosing the signs $\varepsilon_d = \varepsilon_L = -\varepsilon_\nu = -\varepsilon_H = \varepsilon_{H'}$, it is given by:

$$B - L = -\frac{1}{6}Q_3 + \frac{1}{2}Q_2 - \frac{\varepsilon_d}{2}Q_1. \tag{8.4}$$

Finally, the above spectrum can be easily implemented with a Higgs sector, since the Higgs field H has the same quantum numbers as the lepton doublet or its complex conjugate.

9. Moduli stabilization

Internal magnetic fluxes provide a new calculable method of moduli stabilization in four-dimensional (4d) type I string compactifications [50–52]. In fact, moduli stabilization in the presence of 3-form closed string fluxes led to significance progress over the last years [53,54] but presents some drawbacks: (i) it has no exact string description and thus relies mainly on the low energy supergravity approximation; (ii) in the generic case, it can fix only the complex structure and the dilaton [55], while for the Kähler class non-perturbative effects have to be used [54]. On the other hand, constant internal magnetic fields can stabilize mainly Kähler moduli [50,56] and are thus complementary to 3-form closed string fluxes. Moreover, they can also be used in simple toroidal compactifications, stabilizing all geometric moduli in a supersymmetric vacuum using only magnetized $D9$-branes that have an exact perturbative string description [43,57]. They have also a natural implementation in intersecting D-brane models.

Here, we make use of the conventions given in Appendix A of Ref. [51], for the parametrization of the torus T^6, as well as for the general definitions of the Kähler and complex structure moduli. In particular, the coordinates of three factorized tori: $(T^2)^3 \in T^6$ are given by x_i, y_i $i = 1, 2, 3$ with periodicities: $x^i = x^i + 1$, $y^i \equiv y^i + 1$, and a volume normalization:

$$\int dx_1 \wedge dy_1 \wedge dx_2 \wedge dy_2 \wedge dx_3 \wedge dy_3 = 1. \tag{9.1}$$

The 36 moduli of T^6 correspond to 21 independent deformations of the internal metric and 15 deformations of the two-index antisymmetric tensor C_2 from the RR closed string sector. They form nine complex parameters of Kähler class and nine of complex structure. Indeed, the geometric moduli decompose in a complex structure variation which is parametrized by the matrix τ_{ij} entering in the definition of the complex coordinates

$$z_i = x_i + \tau_{ij} y^j, \tag{9.2}$$

and in the Kähler variation of the mixed part of the metric described by the real $(1, 1)$-form $J = i \delta g_{i\bar{j}} dz^i \wedge d\bar{z}^j$. The later is complexified with the corresponding RR two-form deformation.

The stacks of $D9$-branes are characterized by three independent sets of data:

(a) Their multiplicities N_a, that describe the rank of the unitary gauge group $U(N_a)$ on each $D9$ stack.

(b) The winding matrices $W_\alpha^{\hat{\alpha}, a}$ describing the covering of the world-volume of each stack-a of $D9$-branes on the compactified ambient space. They are defined as $W_\alpha^{\hat{\alpha}} = \partial \xi^{\hat{\alpha}} / \partial X^\alpha$ for $\alpha, \hat{\alpha} = 1, \ldots, 6$, where $\xi^{\hat{\alpha}}$ and X^α are the six internal coordinates on the world-volume and space-time, respectively. For simplicity, in the examples we consider here, the winding matrix $W_\alpha^{\hat{\alpha}}$ is chosen to be diagonal, implying that the world-volume and target space T^6 coordinates are identified, up to a winding multiplicity factor n_α^a for each brane stack-a:

$$n_\alpha^a \equiv W_\alpha^{\hat{\alpha}, a}. \tag{9.3}$$

(c) The first Chern numbers $m_{\hat{\alpha}\hat{\beta}}^a$ of the $U(1)$ background on the branes world-volume. In other words, for each stack $U(N_a) = U(1)_a \times SU(N_a)$, the $U(1)_a$ has a constant field strength on the covering of the internal space, which is a 6×6 antisymmetric matrix. These are subject to the Dirac quantization condition which implies that all internal magnetic fluxes $F^a_{\hat{\alpha}\hat{\beta}}$, on the world-volume of each stack of $D9$-branes, are integrally quantized. Explicitly, the world-volume fluxes $F^a_{\hat{\alpha}\hat{\beta}}$ and the corresponding target space induced fluxes $p^a_{\alpha\beta}$ are quantized as (see (7.3))

$$
\begin{aligned}
F^a_{\hat{\alpha}\hat{\beta}} &= m^a_{\hat{\alpha}\hat{\beta}} \in \mathbb{Z} \\
p^a_{\alpha\beta} &= (W^{-1})^{\hat{\alpha}, \, a}_\alpha (W^{-1})^{\hat{\beta}, \, a}_\beta \, m^a_{\hat{\alpha}\hat{\beta}} \in \mathbb{Q}.
\end{aligned}
\tag{9.4}
$$

The complexified fluxes in the basis (9.2) can be written as

$$
F^a_{(2,0)} = (\tau - \bar{\tau})^{-1}{}^T \left[\tau^T p^a_{xx} \tau - \tau^T p^a_{xy} - p^a_{yx} \tau + p^a_{yy}\right](\tau - \bar{\tau})^{-1}, \tag{9.5}
$$

$$
F^a_{(1,1)} = (\tau - \bar{\tau})^{-1}{}^T \left[-\tau^T p^a_{xx} \bar{\tau} + \tau^T p^a_{xy} + p^a_{yx} \bar{\tau} - p^a_{yy}\right](\tau - \bar{\tau})^{-1}, \tag{9.6}
$$

where the matrices $(p^a_{x^i x^j})$, $(p^a_{x^i y^j})$ and $(p^a_{y^i y^j})$ are the quantized field strengths in target space, given in Eq. (9.4). The field strengths $F^a_{(2,0)}$ and $F^a_{(1,1)}$ are 3×3 matrices that correspond to the upper half of the matrix $\mathcal{F}^a$:

$$
\mathcal{F}^a \equiv -(2\pi)^2 i\alpha' \begin{pmatrix} F^a_{(2,0)} & F^a_{(1,1)} \\ -F^a{}^\dagger_{(1,1)} & F^{a*}{}_{(2,0)} \end{pmatrix}, \tag{9.7}
$$

which is the total field strength in the cohomology basis $e_{i\bar{j}} = i dz^i \wedge d\bar{z}^j$.

9.1. Supersymmetry conditions

The supersymmetry conditions then read [50, 51]:

1. $$F^a_{(2,0)} = 0 \quad \forall a = 1, \ldots, K, \tag{9.8}$$

    for K brane stacks, stating that the purely holomorphic flux vanishes. For given flux quanta and winding numbers, this matrix equation restricts the complex structure τ. Using Eq. (9.5), it imposes a restriction on the parameters of the complex structure matrix elements τ:

    $$F^a_{(2,0)} = 0 \quad \to \quad \tau^T p^a_{xx} \tau - \tau^T p^a_{xy} - p^a_{yx} \tau + p^a_{yy} = 0, \tag{9.9}$$

    giving rise to at most six complex equations for each brane stack a.

2. $$\mathcal{F}_a \wedge \mathcal{F}_a \wedge \mathcal{F}_a = \mathcal{F}_a \wedge J \wedge J, \tag{9.10}$$

    that gives rise to one real equation restricting the Kähler moduli. This can be understood as a D-flatness condition. In the 4d effective action, the magnetic fluxes give rise to topological couplings for the different axions of the compactified field theory. These arise from the dimensional reduction of the

Wess Zumino action. In addition to the topological coupling, the $\mathcal{N} = 1$ supersymmetric action yields a Fayet–Iliopoulos (FI) term of the form:

$$\frac{\xi_a}{g_a^2} = \frac{1}{(4\pi^2\alpha')^3} \int_{T^6} \left(\mathcal{F}_a \wedge \mathcal{F}_a \wedge \mathcal{F}_a - \mathcal{F}_a \wedge J \wedge J\right). \tag{9.11}$$

The D-flatness condition in the absence of charged scalars requires then that $\langle D_a \rangle = \xi_a = 0$, which is equivalent to Eq. (9.10). In the case where T^6 is a product of three orthogonal 2-tori, this condition becomes

$$H_1 + H_2 + H_3 = H_1 H_2 H_3 \quad \Leftrightarrow \quad \theta_1 + \theta_2 + \theta_3 = 0, \tag{9.12}$$

in terms of the magnetic fields H_i along the internal planes defined in Section 7, or equivalently in terms of the angles of D6-branes with respect to the orientifold axis.

3. $\det W_a \left(J \wedge J \wedge J - \mathcal{F}_a \wedge \mathcal{F}_a \wedge J\right) > 0,$ (9.13)

which can also be understood from a 4d viewpoint as the positivity of the $U(1)_a$ gauge coupling g_a^2. Indeed, its expression in terms of the fluxes and moduli reads

$$\frac{1}{g_a^2} = \frac{1}{(4\pi^2\alpha')^3} \int_{T^6} \left(J \wedge J \wedge J - \mathcal{F}_a \wedge \mathcal{F}_a \wedge J\right). \tag{9.14}$$

In toroidal models with NS-NS vanishing B-field backround, the net generation number of chiral fermions is in general even [58]. Thus, it is necessary to turn on a constant B-field in order to obtain a Standard Model like spectrum with three generations. Due to the world-sheet parity projection, the NS-NS two-index field $B_{\alpha\beta}$ is projected out from the physical spectrum and constrained to take the discrete values 0 or $1/2$ (in string units) along a 2-cycle $(\alpha\beta)$ of T^6 [59]. Its effect is simply accounted for by shifting the target space flux matrices p^a by $p^a + B$ in all formulae.

The main ingredients for the moduli stabilization are [50,51]:

• A set of nine magnetized $D9$-branes is needed to stabilize all 36 moduli of the torus T^6 by the supersymmetry conditions [44,60]. This follows from the second condition (9.10) above, in order to fix all nine Kähler class moduli. At the same time, all nine corresponding $U(1)$ brane factors become massive by absorbing the RR partners of the Kähler moduli [44,50]. This is due to a kinetic mixing between the $U(1)$ gauge fields A^a and the RR axions, arising from the 10d Chern-Simons coupling involving the RR two-form C_2 along its internal components: $dC_2 \wedge \star(A^a \wedge \langle \mathcal{F}^a \rangle)$.

• At least six of the magnetized brane stacks must have oblique fluxes given by mutually non-commuting matrices, in order to fix all off-diagonal components of the metric. The fluxes however can be chosen so that the metric is fixed in a diagonal form, as we will see below. At the same time, the complex structure RR moduli get stabilized by a potential generated through mixing with the metric moduli from the NS-NS (Neuveu–Schwarz) closed string sector [52].

• The non-linear part of Dirac–Born–Infeld (DBI) action which is needed to fix the overall volume. This is only valid in 4d compactifications (and not in higher dimensions). Indeed, in six dimensions, the condition (9.10) becomes $\mathcal{F}^a \wedge J = 0$ which is homogeneous in J and thus cannot fix the internal volume.

Below, we give an explicit example of nine magnetized D-brane stacks stabilizing all T^6 moduli in a way that the metric is fixed in a diagonal form [51]. The winding matrix W^a is chosen to the identity, for simplicity. The first six $U(1)$ branes with oblique fluxes are presented in Table 2. They fix all moduli except the areas of the three factorized 2-torii. These are fixed by adding three diagonal brane stacks displayed in the upper part of Table 3 (stacks ♯7, ♯8 and ♯9). These give the following restrictions on the diagonal Kähler moduli:

$$
\begin{pmatrix} \mathcal{F}_1^7 & \mathcal{F}_2^7 & \mathcal{F}_3^7 \\ \mathcal{F}_1^8 & \mathcal{F}_2^8 & \mathcal{F}_3^8 \\ \mathcal{F}_1^9 & \mathcal{F}_2^9 & \mathcal{F}_3^9 \end{pmatrix} \begin{pmatrix} J_2 J_3 \\ J_1 J_3 \\ J_1 J_2 \end{pmatrix} = \begin{pmatrix} \mathcal{F}_1^7 \mathcal{F}_2^7 \mathcal{F}_3^7 \\ \mathcal{F}_1^8 \mathcal{F}_2^8 \mathcal{F}_3^8 \\ \mathcal{F}_1^9 \mathcal{F}_2^9 \mathcal{F}_3^9 \end{pmatrix},
\tag{9.15}
$$

where we the subscript $i = 1, 2, 3$ denotes the diagonal element $i\bar{i}$. It follows that the moduli are fixed to the values:

$$
\tau_{ij} = i\delta_{ij}; \ J_{i\bar{j}} = 0; \ (J_{x_1 y_1}, J_{x_2 y_2}, J_{x_3 y_3}) = 4\pi^2 \alpha' \sqrt{\frac{3}{22}} (44, 66, 19). \tag{9.16}
$$

Table 2

Six $U(1)$ branes with oblique magnetic fluxes

Stack ♯	Fluxes	Fixed moduli	5-brane localization
♯1 $N_1 = 1$	$(F_{x_1 y_2}^1, F_{x_2 y_1}^1) = (1, 1)$	$\tau_{31} = \tau_{32} = 0$ $\tau_{11} = \tau_{22}$ $\mathrm{Re} J_{1\bar{2}} = 0$	$[x_3, y_3]$
♯2 $N_2 = 1$	$(F_{x_1 y_3}^2, F_{x_3 y_1}^2) = (1, 1)$	$\tau_{21} = \tau_{23} = 0$ $\tau_{11} = \tau_{33}$ $\mathrm{Re} J_{1\bar{3}} = 0$	$[x_2, y_2]$
♯3 $N_3 = 1$	$(F_{x_1 x_2}^3, F_{y_1 y_2}^3) = (1, 1)$	$\tau_{13} = 0, \ \tau_{11}\tau_{22} = -1$ $\mathrm{Im} J_{1\bar{2}} = 0$	$[x_3, y_3]$
♯4 $N_4 = 1$	$(F_{x_2 x_3}^4, F_{y_2 y_3}^4) = (1, 1)$	$\tau_{12} = 0$ $\mathrm{Im} J_{2\bar{3}} = 0$	$[x_1, y_1]$
♯5 $N_5 = 1$	$(F_{x_1 x_3}^5, F_{y_1 y_3}^5) = (1, 1)$	$\mathrm{Im} J_{1\bar{3}} = 0$	$[x_2, y_2]$
♯6 $N_6 = 1$	$(F_{x_2 y_3}^6, F_{x_3 y_2}^6) = (1, 1)$	$\mathrm{Re} J_{2\bar{3}} = 0$	$[x_1, y_1]$

Table 3

Brane stacks with diagonal magnetic fluxes

Stack ♯	Multiplicity	Fluxes
♯7	$N_7 = 1$	$(F^7_{x_1y_1}, F^7_{x_2y_2}, F^7_{x_3y_3}) = (-4, -4, 3)$
♯8	$N_8 = 2$	$(F^8_{x_1y_1}, F^8_{x_2y_2}, F^8_{x_3y_3}) = (-3, 1, 1)$
♯9	$N_9 = 3$	$(F^9_{x_1y_1}, F^9_{x_2y_2}, F^9_{x_3y_3}) = (-2, 3, 0)$
♯10	$N_{10} = 2$	$(F^{10}_{x_1y_1}, F^{10}_{x_2y_2}, F^{10}_{x_3y_3}) = (5, 1, 2)$
♯11	$N_{11} = 2$	$(F^{11}_{x_1y_1}, F^{11}_{x_2y_2}, F^{11}_{x_3y_3}) = (0, 4, 1)$

Note that for every solution, an infinite discreet family of vacua can be found in general by appropriate rescaling of fluxes and volumes. For instance, a uniform rescaling of all fluxes by the same (integer) factor Λ leads to new solutions where all areas J_i are rescaled by the same factor, $J_i \to \Lambda J_i$. These are large volume solutions that remain compatible with tadpole cancellation, as we will see below.

9.2. Tadpole cancellation conditions

In toroidal compactifications of type I string theory, the magnetized $D9$-branes induce 5-brane charges as well, while the 3-brane and 7-brane charges automatically vanish due to the presence of mirror branes with opposite flux. For general magnetic fluxes, RR tadpole conditions can be written in terms of the Chern numbers and winding matrix [51,52] as:

$$16 = \sum_{a=1}^{K} N_a \det W_a \equiv \sum_{a=1}^{K} Q^{9,a}, \tag{9.17}$$

$$0 = \sum_{a=1}^{K} N_a \det W_a \, \epsilon^{\alpha\beta\delta\gamma\sigma\tau} p^a_{\delta\gamma} p^a_{\sigma\tau} \equiv \sum_{a=1}^{K} Q^{5,a}_{\alpha\beta}, \quad \forall \alpha, \beta = 1, \ldots, 6. \tag{9.18}$$

The l.h.s. of Eq. (9.17) arises from the contribution of the $O9$-plane. On the other hand, in toroidal compactifications there are no $O5$-planes and thus the l.h.s. of Eq. (9.18) vanishes.

In the example presented above, all induced 5-brane tadpoles are diagonal despite the presence of oblique fluxes. Their localization is shown in the last column of Table 2. It turns out however that the conditions of supersymmetry and tadpole cancellation cannot be satisfied simultaneously in toroidal compactifications, as can also be seen in our example. Our strategy is therefore to add extra branes in order to satisfy the RR tadpole conditions. These branes are not supersymmetric and generate a potential for the dilaton, which is the only remaining closed string modulus not fixed by the supersymmetry conditions of the first nine stacks, from the FI D-terms (9.11). One is then has two possibilities to obtain a consistent vacuum with stabilized moduli:

1. Keep supersymmetry by turning on VEVs for charged scalars on the extra brane stacks. In their presence, the D-flatness supersymmetry condition (9.10) gets modified and in the low energy approximation, it reads:

$$D_a = -\left(\sum_\phi q_a^\phi |\phi|^2 + M_s^2 \xi_a\right) = 0, \tag{9.19}$$

where ξ_a is given by Eqs. (9.11) and (9.14). The sum is extended over all scalars ϕ charged under the a-th $U(1)_a$ with charge q_a^ϕ. When one of these scalars acquire a non-vanishing VEV $\langle|\phi|\rangle^2 = v_\phi^2$, the condition (9.10) is modified to:

$$q_a v_a^2 \int_{T^6} (J \wedge J \wedge J - \mathcal{F}_a \wedge \mathcal{F}_a \wedge J) = M_s^2 \int_{T^6} (\mathcal{F}_a \wedge J \wedge J - \mathcal{F}_a \wedge \mathcal{F}_a \wedge \mathcal{F}_a). \tag{9.20}$$

Note that this is valid for small values of v_a (in string units), since the inclusion of charged scalars in the D-term is in principle valid only perturbatively.

Indeed, the model presented above can be implemented by two extra stacks ♯10 and ♯11 with diagonal fluxes, presented in the lower part of Table 3, so that all 9- and 5-brane RR tadpoles are cancelled [51]. These stacks can be made supersymmetric only in the presence of non-trivial VEV's for open string states charged under the corresponding $U(1)$ gauge bosons. Let us then switch on VEV's for the fields ϕ_{10} and ϕ_{11}, v_{10} and v_{11} respectively, transforming in the antisymmetric representations of the corresponding $SU(2)$ gauge groups and charged under the $U(1)$'s of the last two stacks. From the quanta given in Table 3 and the values for the Kähler moduli (9.16), the positivity conditions (9.13) for these branes are satisfied. Moreover, since the Kähler form is already fixed, the supersymmetry conditions (9.20) determine the values of v_{10} and v_{11} as:

$$v_{10}^2 l_s^2 \simeq \frac{0.71}{q} \simeq 0.35; \quad v_{11}^2 l_s^2 \simeq \frac{0.31}{q} \simeq 0.15, \tag{9.21}$$

where we used that the $U(1)$ charge of the fields in the antisymmetric representation is $q = 2$. These VEV's break the two $U(1)$ factors and the final gauge group of the model becomes $SU(3) \times SU(2)^3$. Finally, the above values of the VEV's are reasonably small in string units, consistently with our perturbative approach of including the charged scalar fields in the D-terms.

We have thus presented a model where the open string moduli corresponding to charged scalar VEV's are also fixed by the magnetic fluxes. In principle, the same method can be applied for stabilizing other open string moduli, as well. Note also that the discrete family of large volume solutions is still valid for fixed v_a. All of them have the same gauge symmetry but different couplings (9.14) and matter spectra.

2. Break supersymmetry by D-terms in a anti-de Sitter vacuum, by going "slightly" off-criticality and thus generating a tree-level bulk dilaton potential that can also fix the dilaton at weak string coupling [61]. If this breaking of supersymmetry arises on brane stacks independent from the Standard Model, its mediation involves gauge interactions and is of particular D-type. In particular, gauginos can acquire Dirac masses at one loop without breaking the R-symmetry, due to the extended supersymmetric nature of the gauge sector [62]. A more detail discussion is done in the next section.

9.3. Spectrum

For completeness, here we present the spectrum of magnetized branes in a toroidal background. The gauge sector of the spectrum follows from the open strings starting and ending on the same brane stack. The gauge symmetry group is given by a product of unitary groups $\otimes_a U(N_a)$, upon identification of the associated open strings attached on a given stack with the ones attached on its orientifold mirror. In addition to these vector bosons, the massless spectrum contains adjoint scalars and fermions forming $\mathcal{N} = 4, d = 4$ supermultiplets.

In the matter sector, the massless spectrum is obtained from the following open string states [44, 46]:

1. Open strings stretched between the a-th and b-th stack give rise to chiral spinors in the bifundamental representation $(N_a, \bar{N}_b)$ of $U(N_a) \times U(N_b)$. Their multiplicity I_{ab} is given by [52]:

$$I_{ab} = \frac{\det W_a \det W_b}{(2\pi)^3} \int_{T^6} \left(q_a F^a_{(1,1)} + q_b F^b_{(1,1)} \right)^3, \tag{9.22}$$

where $F^a_{(1,1)}$ (given in Eqs. (9.6) and (9.7)) is the pullback of the integrally quantized world-volume flux $m^a_{\hat{\alpha}\hat{\beta}}$ on the target torus in the complex basis (9.2), and q_a is the corresponding $U(1)_a$ charge; in our case $q_a = +1 \, (-1)$ for the fundamental (anti-fundamental representation).

For factorized toroidal compactifications $T^6 = (T^2)^3$ with only diagonal fluxes $p_{x^i y^i}$ $(i = 1, 2, 3)$, the multiplicities of chiral fermions, arising from strings starting from stack a and ending at b or vice versa, take the simple form

$$(N_a, \overline{N}_b) : I_{ab} = \prod_i (\hat{m}_i^a \hat{n}_i^b - \hat{n}_i^a \hat{m}_i^b),$$

$$(N_a, N_b) : I_{ab^*} = \prod_i (\hat{m}_i^a \hat{n}_i^b + \hat{n}_i^a \hat{m}_i^b). \tag{9.23}$$

where the integers $\hat{m}_i^a$, $\hat{n}_i^a$ are defined by:

$$\hat{m}_i^a \equiv m_{x^i y^i}^a, \quad \hat{n}_1^a \equiv n_1^a n_2^a; \quad \hat{n}_2^a \equiv n_3^a n_4^a, \quad \hat{n}_3^a \equiv n_5^a n_6^a, \tag{9.24}$$

in terms of the magnetic fluxes m^a and winding numbers n^a of Eqs. (9.4) and (9.3), respectively.

2. Open strings stretched between the a-th brane and its mirror a^* give rise to massless modes associated to I_{aa^*} chiral fermions. These transform either in the antisymmetric or symmetric representation of $U(N_a)$. For factorized toroidal compactifications $(T^2)^3$, the multiplicities of chiral fermions are given by;

$$\text{Antisymmetric:} \quad \frac{1}{2}\left(\prod_i 2\hat{m}_i^a\right)\left(\prod_j \hat{n}_j^a + 1\right),$$

$$\text{Symmetric:} \quad \frac{1}{2}\left(\prod_i 2\hat{m}_i^a\right)\left(\prod_j \hat{n}_j^a - 1\right). \tag{9.25}$$

In generic configurations, where supersymmetry is broken by the magnetic fluxes, the scalar partners of the massless chiral spinors in twisted open string sectors (i.e. from non-trivial brane intersections) are massive (or tachyonic). Moreover, when a chiral index I_{ab} vanishes, the corresponding intersection of stacks a and b is non-chiral. The multiplicity of the non-chiral spectrum is then determined by extracting the vanishing factor and calculating the corresponding chiral index in higher dimensions.

9.4. A supersymmetric SU(5) GUT with stabilized moduli

A more realistic model of moduli stabilization with three generations of quarks an leptons can be obtained by realizing in the above framework the model A of Section 8 with $U(3)$ and $U(2)$ coincident, giving rise to an $SU(5)$ GUT [63]. To elaborate further, the model is described by twelve stacks of branes, namely $U_5, U_1, O_1, \ldots, O_8, A$, and B, whose role is described below:

• The $SU(5)$ gauge group arises from the open string states of stack-U_5 containing five magnetized branes. The remaining eleven stacks contain only a single magnetized brane. Also, the stack-U_5 containing the GUT gauge sector, contributes to the GUT particle spectrum through open string states which either start and end on itself (or on its orientifold image) or on the stack-U_1, having only a single brane and therefore contributing an extra $U(1)$. More precisely, open strings stretched in the intersection of $U(5)$ with its orientifold image give rise to 3 chiral generations in the antisymmetric representation **10** of $SU(5)$, while the intersection of $U(5)$ with the orientifold image of $U(1)$ gives 3 chiral states transforming as $\bar{\bf 5}$. Finally, the intersection of $U(5)$ with the $U(1)$ is non chiral, giving rise to Higgs pairs $\bf{5} + \bar{\bf 5}$. The magnetic fluxes along the various branes are constrained by the fact that the chiral fermion spectrum, mentioned above, of the $SU(5)$ GUT should arise from these two sectors only.

• The eight single brane stacks $O_1, \ldots, O_8$, contain oblique fluxes and generalize the set of the six stacks ♯1–♯6 of the previous toy model, in the presence of a B-field background needed to obtain odd number (three) of chiral fermions. A crucial property of these 'oblique' branes is that the combined induced 5-brane charge lies only along the three diagonal directions $[x_i, y_i]$.

• The eight 'oblique' branes together with U_5 fix all geometric moduli by the supersymmetry conditions. The holomorphicity condition (9.8) stabilizes the complex structure moduli to the identity matrix, as in (9.16), while the D-flatness condition (9.10) for the nine stacks $U_5, O_1, \ldots, O_8$, imposing the vanishing of the FI terms ξ_a (9.11), fix the nine Kähler moduli in a diagonal form. The residual diagonal 5-brane tadpoles of the branes in the stacks $U_5, U_1, O_1, \ldots, O_8$ are then cancelled by introducing the last two single brane stacks A and B, satisfying also the required 9-brane charge.

• The D-flatness conditions for the brane stacks U_1, A and B can also be satisfied, provided some VEVs of charged scalars living on these branes are turned on to cancel the corresponding FI parameters, according to Eqs. (9.19) and (9.20). They all take values smaller than the string scale, consistently with their perturbative treatment, and break the three $U(1)$ symmetries. On the other hand, the remaining nine $U(1)$ brane factors become massive by absorbing the RR partners of the Kähler class moduli. As a result, all extra $U(1)$'s are broken and the only leftover gauge symmetry is an $SU(5)$ GUT. Furthermore, the intersections of the $U(5)$ stack with any additional brane used for moduli stabilization are non-chiral, yielding the three families of quarks and leptons in the $\bf{10} + \bar{\bf 5}$ representations as the only chiral spectrum of the model (gauge non-singlet).

10. Gaugino masses and D-term gauge mediation

Here, we study the possibility of breaking supersymmetry by magnetic fluxes in a part of the theory, instead of turning on charged scalar VEVs, such as in the brane stacks ♯10 and ♯11 of the toy model of Section 9.1, or in the stacks U_1, A and B of the $SU(5)$ model discussed above. Since this breaking of supersymmetry is induced by D-terms, gaugino masses are vanishing at the tree-level, because they are protected by a (chiral) R-symmetry. This symmetry is broken in general in the presence of gravity by the gravitino mass, as well as by higher order in α' string corrections (on the branes). Both effects generate gaugino masses radiatively from a diagram involving at least one boundary, where the gauginos are localized, and having effective 'genus' 3/2 [64].

For oriented strings, there are two possibilities: (1) one boundary and one handle, corresponding to one gravitational loop in the effective supergravity; (2) three boundaries, corresponding to two loops in the effective gauge theory. In the limit of small supersymmetry breaking compared to the string scale, both diagrams are reduced to topological amplitudes receiving contributions only from massless states:

(1) The one loop gravitational contribution of the first diagram leads to gaugino masses $m_{1/2}$ scaling as the third power of the gravitino mass $m_{3/2}$:

$$m_{1/2} \propto g_s^2 \frac{m_{3/2}^3}{M_s^2}. \qquad (10.1)$$

On the other hand, scalars on the brane acquire generically one-loop mass corrections m_0 from the annulus diagram [65]: $m_0 \gtrsim g_s m_{3/2}^2 / M_s$, implying that gaugino masses are suppressed relative to scalar masses:

$$m_{1/2}^2 \lesssim g_s \frac{m_0^3}{M_s}. \qquad (10.2)$$

Fixing $m_{1/2}$ in the TeV range, one then finds that scalars are much heavier $m_0 \gtrsim 10^8$ GeV. Thus, this mechanism leads to a hierarchy between scalar and gaugino masses of the type required by split supersymmetry [66,67].

(2) Similarly, the gauge contribution of the second diagram leads to even larger hierarchy:

$$m_{1/2} \propto g_s^2 \frac{m_0^4}{M_s^3}, \qquad (10.3)$$

with the proportionality constant given by the open string topological partition function $F_{(0,3)}$ [68]. This result can be understood from the supersymmetric dimension seven ciral operator in the effective field theory: $\int d^2\theta \, \mathcal{W}^2 \text{Tr} \, W^2$, when the magnetized $U(1)$ gauge superfield $\mathcal{W}$ acquires an expectation value along its D-auxiliary component: $\langle \mathcal{W} \rangle = \theta \langle D \rangle$ with $\langle D \rangle \sim m_0^2$. Thus, the gauginos appearing in the lowest component of the (non-abelian) gauge superfield W acquire the Majorana mass (10.3), which is in the TeV region when the scalar masses are of order 10^{13}–10^{14} GeV.

An alternative way to generate gaugino masses is by giving Dirac type masses. Indeed, in the toroidal models we studied above, we mentioned already that the gauge sector on the branes comes into multiplets of $\mathcal{N} = 4$ extended supersymmetry and thus gauginos can be paired into Dirac massive fermions without breaking the R-symmetry [69]. This leads to the possibility of a new gauge mediation mechanism [70]. A prototype model can be studied with the following setup, based on two sets of magnetized brane stacks: the observable set $\mathcal{O}$ and the hidden set $\mathcal{H}$ [62,69].

• The Standard Model gauge sector corresponds to open strings that propagate with both ends on the same stack of branes that belong to $\mathcal{O}$: it has therefore an extended $\mathcal{N} = 4$ or $\mathcal{N} = 2$ supersymmetry. Similarly, the 'secluded' gauge sector corresponds to strings with both ends on the hidden stack of branes $\mathcal{H}$.

• The Standard Model quarks and leptons come from open strings stretched between different stacks of branes in $\mathcal{O}$ that intersect at fixed points of the internal six-torus T^6 and have therefore $\mathcal{N} = 1$ supersymmetry.

• The Higgs sector on the other hand corresponds to strings stretched between different stacks of branes in $\mathcal{O}$ that intersect at fixed points of a T^4 and are parallel along a T^2: it has therefore $\mathcal{N} = 2$ supersymmetry and the two Higgs doublets form a hypermultiplet. Finally, the messenger sector contains strings stretched between stacks of branes in $\mathcal{O}$ and the hidden branes $\mathcal{H}$, that form also $\mathcal{N} = 2$ hypermultiplets. Moreover, the two stacks of branes along the T^2 are separated by a distance $1/M$, which introduces a supersymmetric mass M to the hypermultiplet messengers. The latter are also charged under the magnetized $U(1)$(s) that break supersymmetry in the 'secluded' sector $\mathcal{H}$ via D-terms.

The main properties of this mechanism are:

1. The gauginos obtain Dirac masses at one loop given by:

$$m_{1/2}^D \sim \frac{\alpha}{4\pi} \frac{\mathrm{D}}{M}, \tag{10.4}$$

where α is the corresponding gauge coupling constant.

2. Scalar quarks and leptons acquire masses by one-loop diagrams involving Dirac gauginos in the effective theory where messengers have been integrated out (three-loop diagrams in the underlying theory). Their contributions are finite and one-loop suppressed with respect to gaugino masses [71,72].

3. The tree-level Higgs potential gets modified because of its $\mathcal{N} = 2$ structure.

$$V = V_{\text{soft}} + \frac{1}{8}(g^2 + g'^2)(|H_1|^2 - |H_2|^2)^2 + \frac{1}{2}(g^2 + g'^2)|H_1 H_2|^2, \tag{10.5}$$

where $H_{1,2}$ are the two Higgs doublets, g and g' are the $SU(2)$ and $U(1)$ couplings, and the last term is a genuine $\mathcal{N} = 2$ contribution which is absent in the MSSM. It follows that the lightest Higgs behaves as in the (non supersymmetric) Standard Model with no $\tan \beta$ dependence on its couplings to fermions. On the other hand, the heaviest Higgs plays no role in electroweak symmetry breaking and does not couple to the Z-boson. In fact, the model behaves as the MSSM at large $\tan \beta$ and the 'little' fine-tuning problem is significantly reduced [62].

4. The supersymmetric flavor problem is solved as in usual gauge mediation. Moreover, there is a common supersymmetry breaking scale in the observable sector, the masses of all supersymmetric particles being proportional to powers of gauge couplings. Finally, there are distinct collider signals different from that of the MSSM.

In conclusion, the framework of toroidal string compactifications with magnetized branes described above, starting from Section 7, offers an interesting self-consistent setup for string phenomenology, in which one can build simple calculable models of particle physics with stabilized moduli and implement low energy supersymmetry breaking that can be studied directly at the string level.

Acknowledgements

This work was supported in part by the European Commission under the RTN contract MRTN-CT-2004-503369, and in part by the INTAS contract 03-51-6346.

References

[1] Fore a review, see e.g. K.R. Dienes, *Phys. Rept.* **287** (1997) 447 [arXiv:hep-th/9602045]; and references therein.

[2] C. Angelantonj and A. Sagnotti, *Phys. Rept.* **371** (2002) 1 [Erratum-ibid. **376** (2003) 339] [arXiv:hep-th/0204089].

[3] I. Antoniadis, *Phys. Lett.* B **246** (1990) 377.

[4] D.J. Kapner, T.S. Cook, E.G. Adelberger, J.H. Gundlach, B.R. Heckel, C.D. Hoyle and H.E. Swanson, *Phys. Rev. Lett.* **98** (2007) 021101.

[5] N. Arkani-Hamed, S. Dimopoulos and G.R. Dvali, *Phys. Lett.* B **429** (1998) 263 [arXiv:hep-ph/9803315]; I. Antoniadis, N. Arkani-Hamed, S. Dimopoulos and G.R. Dvali, *Phys. Lett.* B **436** (1998) 257 [arXiv:hep-ph/9804398].

[6] For a review see e.g. I. Antoniadis, *Prepared for NATO Advanced Study Institute and EC Summer School on Progress in String, Field and Particle Theory, Cargese, Corsica, France (2002)*; and references therein.

[7] J.D. Lykken, *Phys. Rev.* D **54** (1996) 3693 [arXiv:hep-th/9603133].

[8] J.C. Long and J.C. Price, *Comptes Rendus Physique* **4** (2003) 337; R.S. Decca, D. Lopez, H.B. Chan, E. Fischbach, D.E. Krause and C.R. Jamell, *Phys. Rev. Lett.* **94** (2005) 240401; R.S. Decca et al., arXiv:0706.3283 [hep-ph]; S.J. Smullin, A.A. Geraci, D.M. Weld, J. Chiaverini, S. Holmes and A. Kapitulnik, arXiv:hep-ph/0508204; H. Abele, S. Haeßler and A. Westphal, in 271th WE-Heraeus-Seminar, Bad Honnef (2002).

[9] I. Antoniadis and K. Benakli, *Phys. Lett.* B **326** (1994) 69.

[10] K.R. Dienes, E. Dudas and T. Gherghetta, *Phys. Lett.* B **436** (1998) 55 [arXiv:hep-ph/9803466]; *Nucl. Phys.* B **537** (1999) 47 [arXiv:hep-ph/9806292].

[11] I. Antoniadis, K. Benakli and M. Quirós, *Phys. Lett.* B **331** (1994) 313 and *Phys. Lett.* B **460** (1999) 176; P. Nath, Y. Yamada and M. Yamaguchi, *Phys. Lett.* B **466** (1999) 100 T.G. Rizzo and J.D. Wells, *Phys. Rev.* D **61** (2000) 016007; T.G. Rizzo, *Phys. Rev.* D **61** (2000) 055005; A. De Rujula, A. Donini, M.B. Gavela and S. Rigolin, *Phys. Lett.* B **482** (2000) 195.

[12] E. Accomando, I. Antoniadis and K. Benakli, *Nucl. Phys.* B **579** (2000) 3.

[13] I. Antoniadis, K. Benakli and A. Laugier, *JHEP* **0105** (2001) 044.

[14] P. Nath and M. Yamaguchi, *Phys. Rev.* D **60** (1999) 116004; *Phys. Rev.* D **60** (1999) 116006; M. Masip and A. Pomarol, *Phys. Rev.* D **60** (1999) 096005; W.J. Marciano, *Phys. Rev.* D **60** (1999) 093006; A. Strumia, *Phys. Lett.* B **466** (1999) 107; R. Casalbuoni, S. De Curtis, D. Dominici and R. Gatto, *Phys. Lett.* B **462** (1999) 48; C.D. Carone, *Phys. Rev.* D **61** (2000) 015008; A. Delgado, A. Pomarol and M. Quirós, *JHEP* **1** (2000) 30.

[15] G. Servant and T.M.P. Tait, *Nucl. Phys.* B **650** (2003) 391.

[16] G.F. Giudice, R. Rattazzi and J.D. Wells, *Nucl. Phys.* B **544** (1999) 3; E.A. Mirabelli, M. Perelstein and M.E. Peskin, *Phys. Rev. Lett.* **82** (1999) 2236; T. Han, J.D. Lykken and R. Zhang, *Phys. Rev.* D **59** (1999) 105006; K. Cheung and W.-Y. Keung, *Phys. Rev.* D **60** (1999) 112003; C. Balázs et al., *Phys. Rev. Lett.* **83** (1999) 2112; L3 Collaboration (M. Acciarri et al.), *Phys. Lett.* B **464** (1999) 135 and **470** (1999) 281: J.L. Hewett, *Phys. Rev. Lett.* **82** (1999) 4765.

[17] N. Arkani-Hamed, S. Dimopoulos and G. Dvali, *Phys. Rev.* D **59** (1999) 086004.

[18] S. Cullen and M. Perelstein, *Phys. Rev. Lett.* **83** (1999) 268; V. Barger, T. Han, C. Kao and R.J. Zhang, *Phys. Lett.* B **461** (1999) 34.

[19] K. Benakli and S. Davidson, *Phys. Rev.* D **60** (1999) 025004; L.J. Hall and D. Smith, *Phys. Rev.* D **60** (1999) 085008.

[20] E. Dudas and J. Mourad, *Nucl. Phys.* B **575** (2000) 3 [arXiv:hep-th/9911019]; S. Cullen, M. Perelstein and M.E. Peskin, *Phys. Rev.* D **62** (2000) 055012; D. Bourilkov, *Phys. Rev.* D **62** (2000) 076005; L3 Collaboration (M. Acciarri et al.), *Phys. Lett.* B **489** (2000) 81.

[21] P.C. Argyres, S. Dimopoulos and J. March-Russell, *Phys. Lett.* B **441** (1998) 96 [arXiv:hep-th/9808138]; T. Banks and W. Fischler, arXiv:hep-th/9906038.

[22] S.B. Giddings and S. Thomas, *Phys. Rev.* D **65** (2002) 056010 [arXiv:hep-ph/0106219]; S. Dimopoulos and G. Landsberg, *Phys. Rev. Lett.* **87** (2001) 161602 [arXiv:hep-ph/0106295].

[23] P. Meade and L. Randall, arXiv:0708.3017 [hep-ph].

[24] I. Antoniadis, C. Bachas, *Phys. Lett.* B **450** (1999) 83.

[25] I. Antoniadis, K. Benakli, A. Laugier and T. Maillard, *Nucl. Phys.* B **662** (2003) 40 [arXiv:hep-ph/0211409].

[26] I. Antoniadis, S. Dimopoulos and G. Dvali, *Nucl. Phys.* B **516** (1998) 70; S. Ferrara, C. Kounnas and F. Zwirner, *Nucl. Phys.* B **429** (1994) 589.

[27] E.G. Adelberger, B.R. Heckel, S. Hoedl, C.D. Hoyle, D.J. Kapner and A. Upadhye, *Phys. Rev. Lett.* **98** (2007) 131104.

[28] T.R. Taylor and G. Veneziano, *Phys. Lett.* B **213** (1988) 450.

[29] I. Antoniadis, E. Kiritsis and J. Rizos, *Nucl. Phys.* B **637** (2002) 92.

[30] E. Dudas and J. Mourad, *Phys. Lett.* B **514** (2001) 173 [arXiv:hep-th/0012071].

[31] D.V. Volkov and V.P. Akulov, *JETP Lett.* **16** (1972) 438 and *Phys. Lett.* B **46** (1973) 109.

[32] A. Brignole, F. Feruglio and F. Zwirner, *JHEP* **9711** (1997) 001; T.E. Clark, T. Lee, S.T. Love and G. Wu, *Phys. Rev.* D **57** (1998) 5912; M.A. Luty and E. Ponton, *Phys. Rev.* D **57** (1998) 4167; I. Antoniadis, K. Benakli and A. Laugier, *Nucl. Phys.* B **631** (2002) 3.

[33] I. Antoniadis and M. Tuckmantel, *Nucl. Phys.* B **697** (2004) 3.

[34] I. Antoniadis, M. Tuckmantel and F. Zwirner, *Nucl. Phys.* B **707** (2005) 215 [arXiv:hep-ph/0410165].

[35] I. Antoniadis, K. Benakli and M. Quirós, *Nucl. Phys.* B **583** (2000) 35.

[36] I. Antoniadis, K. Benakli and M. Quiros, *New Jour. Phys.* **3** (2001) 20.

[37] I. Antoniadis, C. Muñoz and M. Quirós, *Nucl. Phys.* B **397** (1993) 515; I. Antoniadis and M. Quirós, *Phys. Lett.* B **392** (1997) 61; A. Pomarol and M. Quirós, *Phys. Lett.* B **438** (1998) 225; I. Antoniadis, S. Dimopoulos, A. Pomarol and M. Quirós, *Nucl. Phys.* B **544** (1999) 503; A. Delgado, A. Pomarol and M. Quirós, *Phys. Rev.* D **60** (1999) 095008; R. Barbieri, L.J. Hall and Y. Nomura, *Phys. Rev.* D **63** (2001) 105007.

[38] I. Antoniadis, E. Kiritsis and T.N. Tomaras, *Phys. Lett.* B **486** (2000) 186; I. Antoniadis, E. Kiritsis, J. Rizos and T.N. Tomaras, *Nucl. Phys.* B **660** (2003) 81.

[39] G. Shiu and S.-H.H. Tye, *Phys. Rev.* D **58** (1998) 106007; Z. Kakushadze and S.-H.H. Tye, *Nucl. Phys.* B **548** (1999) 180; L.E. Ibáñez, C. Muñoz and S. Rigolin, *Nucl. Phys.* B **553** (1999) 43.

[40] K.R. Dienes, E. Dudas and T. Gherghetta, *Nucl. Phys.* B **557** (1999) 25 [arXiv:hep-ph/9811428]; N. Arkani-Hamed, S. Dimopoulos, G.R. Dvali and J. March-Russell, *Phys. Rev.* D **65** (2002) 024032 [arXiv:hep-ph/9811448]; G.R. Dvali and A.Y. Smirnov, *Nucl. Phys.* B **563** (1999) 63.

[41] A. Sagnotti, *Phys. Lett.* B **294** (1992) 196; L.E. Ibáñez, R. Rabadán and A.M. Uranga, *Nucl. Phys.* B **542** (1999) 112; E. Poppitz, *Nucl. Phys.* B **542** (1999) 31.

[42] H. Davoudiasl, P. Langacker and M. Perelstein, *Phys. Rev.* D **65** (2002) 105015 [arXiv:hep-ph/0201128].

[43] C. Bachas, arXiv:hep-th/9503030.

[44] C. Angelantonj, I. Antoniadis, E. Dudas and A. Sagnotti, *Phys. Lett.* B **489** (2000) 223 [arXiv:hep-th/0007090].

[45] M. Berkooz, M.R. Douglas and R.G. Leigh, *Nucl. Phys.* B **480** (1996) 265 [arXiv:hep-th/9606139].

[46] R. Blumenhagen, L. Goerlich, B. Kors and D. Lust, *JHEP* **0010** (2000) 006 [arXiv:hep-th/0007024]; G. Aldazabal, S. Franco, L.E. Ibanez, R. Rabadan and A.M. Uranga, *J. Math. Phys.* **42** (2001) 3103 [arXiv:hep-th/0011073]; N. Ohta and P.K. Townsend, *Phys. Lett.* B **418** (1998) 77 [arXiv:hep-th/9710129].

[47] I. Antoniadis, E. Kiritsis and T.N. Tomaras, *Phys. Lett.* B **486** (2000) 186 [arXiv:hep-ph/0004214]; I. Antoniadis, E. Kiritsis, J. Rizos and T.N. Tomaras, *Nucl. Phys.* B **660** (2003) 81 [arXiv:hep-th/0210263]; R. Blumenhagen, B. Kors, D. Lust and T. Ott, *Nucl. Phys.* B **616** (2001) 3 [arXiv:hep-th/0107138]; M. Cvetic, G. Shiu and A.M. Uranga, *Nucl. Phys.* B **615**, 3 (2001) [arXiv:hep-th/0107166]; I. Antoniadis and J. Rizos, 2003 unpublished work.

[48] I. Antoniadis and S. Dimopoulos, *Nucl. Phys.* B **715**, 120 (2005) [arXiv:hep-th/0411032].

[49] S.M. Barr, *Phys. Lett.* B **112** (1982) 219; J.P. Derendinger, J.E. Kim and D.V. Nanopoulos, *Phys. Lett.* B **139** (1984) 170; I. Antoniadis, J.R. Ellis, J.S. Hagelin and D.V. Nanopoulos, *Phys. Lett.* B **194** (1987) 231.

[50] I. Antoniadis and T. Maillard, *Nucl. Phys.* B **716** (2005) 3, [arXiv:hep-th/0412008].

[51] I. Antoniadis, A. Kumar, T. Maillard, arXiv: hep-th/0505260; *Nucl. Phys.* B **767** (2007) 139, [arXiv:hep-th/0610246].

[52] M. Bianchi and E. Trevigne, *JHEP* **0508** (2005) 034, [arXiv:hep-th/0502147] and *JHEP* **0601** (2006) 092, [arXiv:hep-th/0506080].

[53] S.B. Giddings, S. Kachru and J. Polchinski, *Phys. Rev.* D **66** (1997) 106006, [arXiv:hep-th/0105097].

[54] R. Kallosh, S. Kachru, A. Linde and S. Trivedi, *Phys. Rev.* D **68** (2003) 046005, [arXiv:hep-th/0301240].

[55] S. Kachru, M.B. Schulz and S. Trivedi, *JHEP* **0310** (2003) 007, [arXiv: hep-th/0201028]; A. Frey and J. Polchinski, *Phys. Rev.* D **65** (2002) 126009, [arXiv:hep-th/0201029].

[56] R. Blumenhagen, D. Lust and T.R. Taylor, *Nucl. Phys.* B **663** (2003) 319 [arXiv:hep-th/0303016]; J.F.G. Cascales and A.M. Uranga, *JHEP* **0305** (2003) 011 [arXiv:hep-th/0303024].

[57] E.S. Fradkin and A.A. Tseytlin, *Phys. Lett.* B **163** (1985) 123; A. Abouelsaood, C.G. . Callan, C.R. Nappi and S.A. Yost, *Nucl. Phys.* B **280** (1987) 599.

[58] R. Blumenhagen, B. Kors and D. Lust, *JHEP* **0102** (2001) 030 [arXiv:hep-th/0012156].

[59] M. Bianchi, G. Pradisi and A. Sagnotti, *Nucl. Phys.* B **376** (1992) 365; C. Angelantonj, *Nucl. Phys.* B **566** (2000) 126 [arXiv:hep-th/9908064]; C. Angelantonj and A. Sagnotti, arXiv:hep-th/0010279;

[60] M. Marino, R. Minasian, G. W. Moore and A. Strominger, *JHEP* **0001** (2000) 005, [arXiv:hep-th/9911206].

[61] I. Antoniadis, J.-P. Derendinger and T. Maillard, to appear.

[62] I. Antoniadis, K. Benakli, A. Delgado, M. Quiros and M. Tuckmantel, *Nucl. Phys.* B **744** (2006) 156 [arXiv:hep-th/0601003]; I. Antoniadis, K. Benakli, A. Delgado and M. Quiros, arXiv:hep-ph/0610265.

[63] I. Antoniadis, A. Kumar and B. Panda, arXiv:0709.2799 [hep-th].

[64] I. Antoniadis and T.R. Taylor, *Nucl. Phys.* B **695** (2004) 103 [arXiv:hep-th/0403293] and *Nucl. Phys.* B **731** (2005) 164 [arXiv:hep-th/0509048].

[65] I. Antoniadis, E. Dudas and A. Sagnotti, *Nucl. Phys.* B **544** (1999) 469 [arXiv:hep-th/9807011].

[66] N. Arkani-Hamed and S. Dimopoulos, *JHEP* **0506** (2005) 073 [arXiv:hep-th/0405159]; G.F. Giudice and A. Romanino, *Nucl. Phys.* B **699** (2004) 65 [Erratum-ibid. B **706** (2005) 65] [arXiv:hep-ph/0406088].

[67] I. Antoniadis and S. Dimopoulos, *Nucl. Phys.* B **715** (2005) 120 [arXiv:hep-th/0411032].

[68] I. Antoniadis, K.S. Narain and T.R. Taylor, *Nucl. Phys.* B **729** (2005) 235 [arXiv:hep-th/0507244].

[69] I. Antoniadis, A. Delgado, K. Benakli, M. Quiros and M. Tuckmantel, *Phys. Lett.* B **634** (2006) 302 [arXiv:hep-ph/0507192] and *Nucl. Phys.* B **744** (2006) 156 [arXiv:hep-th/0601003].

[70] I. Antoniadis, K. Benakli, A. Delgado and M. Quiros, arXiv:hep-ph/0610265.

[71] I. Antoniadis and K. Benakli, *Phys. Lett.* B **295** (1992) 219 [Erratum-ibid. B **407** (1997) 449] [arXiv:hep-th/9209020].

[72] P.J. Fox, A.E. Nelson and N. Weiner, *JHEP* **0208** (2002) 035 [arXiv:hep-ph/0206096].

Course 2

ORIENTIFOLDS, AND THE SEARCH FOR THE STANDARD MODEL IN STRING THEORY

Elias Kiritsis

CPHT, Ecole Polytechnique, 91128, Palaiseau, France
(UMR du CNRS 7644)
and
Department of Physics, University of Crete
71003 Heraklion, Greece

C. Bachas, L. Baulieu, M. Douglas, E. Kiritsis, E. Rabinovici, P. Vanhove, P. Windey
and L.F. Cugliandolo, eds.
Les Houches, Session LXXXVII, 2007
String Theory and the Real World: From Particle Physics to Astrophysics

Contents

1. Introduction

Ever since the Standard Model (SM) was accepted as the correct description of particle physics phenomena in the accessible energy range, its fundamental (theoretical) limitations were obvious and a new quest was launched for a more fundamental theory. This theory would share the low-energy successes of the SM while it would extend its range of validity to the ultimate energy frontier. As GUTs suggested, this ultimate frontier must at least reach the Planck scale and therefore the fundamental theory should include a quantum theory of gravity as well. Subsequent inclusion of supersymmetry in order to manage the hierarchy problem only made the quest for quantum (super) gravity even more inevitable.

Such observations opened Pandora's box as it was already widely appreciated that even defining perturbative gravity theories was so far intractable. Interestingly, perturbatively well-defined quantum theories of gravity were known (as superstring theories) as was first pointed out by Scherk and Schwarz [1] and independently Yoneya [2] in the mid 70s. Almost nobody paid however attention to this observation until 1984, when superstring theories came to the forth. The reason was advocated to be anomaly cancelation [3] and uniqueness. After almost forty years of research in string theory (since the original Veneziano paper [4]), we know today that uniqueness is, to put it mildly, the most elusive property of the theory. On the other hand, the fact that it contains a perturbatively well-defined quantum theory of gravity, (at least when supersymmetry is unbroken) is non-trivial. Never the less, the theory seems incapable of addressing phenomena close or beyond the Planck scale, unless it is defined (around very specific backgrounds) via large-N gauge theories/matrix models. There is even a dichotomy of point of view: is string theory a single theory with many different vacua, as probably most people hope/advocate? Or is it a collection of different theories, defined holographically via large-N field theories as a minority would argue?

Whatever the answer to the previous question, the original challenge remains: can we construct a theory that contains a controllable perturbative theory of quantum gravity and reduces to the standard model at low energies? This question was addressed in recurring waves in the context of the heterotic theories in the 80s and early 90s without a clean answer (see for example [6]). Model building in the heterotic string turned out to be very difficult, at least compared with field theory model building. The reasons are clear: In field theory, at least in perturbation theory, the input needed to construct a fundamental theory is a gauge group, fermions and bosons in specific representations, and a finite number of couplings (Yukawa and scalar), that reflect very closely the phenomenology we want to reproduce: the spectrum of particles and their low energy interactions.

In the heterotic string however the input needed for model building is very remotely connected to the spectrum and the phenomenology we want to achieve. In particular the input is the geometry and other background fields of the internal space, and in its most abstract form, the CFT describing the compact part of space and time. Therefore, producing the spectrum and gauge groups we like is very difficult, and matching the couplings, most of the time intractable. Although heterotic vacua were found that come close to the SM, none does it very well. The reasons for the difficulties advocated above can be used to argue that the SM is probably there (maybe in more than one incarnation), we just have a hard time to find it.

The return of open string vacua and the advent of orientifolds in the early-to-mid 90s [7] has created a new string arena for "model building/searching". Moreover, it was realized [8–10] that the natural distinction of open and closed string sectors, was helpful in turning the search for SM into a modular enterprise, where one could put together his favorite open string sector, generating the SM and its extensions and postpone the stringy compatibility conditions till the end. This so-called "bottom-up" approach has yielded today a far richer set of SM-like orientifold vacua (see [12] for reviews) although a clone of the (non-supersymmetric) SM still remains to be found.

Building the SM spectrum using D-branes involves several peculiarities that were realized along the way. Generically speaking we obtain product gauge groups. Unified models are typically difficult to obtain (both E_6 and $SO(10)$ are not possible perturbatively). $SU(5)$ as we will see is possible in various forms, however its detailed phenomenology, especially the quark masses, needs "fine-tuning". Once we move to product groups, very few of the possible ways of realizing the standard model have been so far explored. The reason is that product groups are mostly "unmotivated" from the theory point of view.[1] It has been already observed [8] that there several different ways of embedding the hypercharge inside a product gauge group containing the SM model. Moreover, different hypercharge embeddings lead to different phenomenology at or beyond the SM regime [13].

It is therefore an important task to examine the different ways of embedding successfully the SM spectrum into the Chan–Paton (CP) gauge groups of orientifolds and this is what we will review in this lecture.

---

[1]There are however exceptions: the Pati–Salam $SU(4) \times SU(2)^2$ model or the trinification $SU(3)^3$ are in this class.

The strategy for building orientifold vacua can be described as follows. One starts from a type II vacuum, typically based on a solvable (bulk) CFT. This vacuum describes a "compactification" of the type II theory to four dimensions with (or without) space-time supersymmetry. One then builds the BCFT associated to this bulk CFT. In particular this involves the construction of boundary states, that can be intuitively thought of as possible branes to be placed in the closed string background (if they carry non-zero CP multiplicities). One then chooses an orientifold projection, to define the open string sector (invariant boundary states). At this stage, we can choose CP multiplicities of appropriate branes (boundary states) in view of reproducing the SM gauge group. The spectrum is then checked, and once we are happy with it we try to solve the tadpole conditions. Rarely this works without adding other branes. We must therefore add other branes that must generate "hidden sectors" of the theory (if done appropriately). We continue this until the tadpole conditions are solved.

This is the algorithm that we will use, combined with a few extra ingredients. The first is a list of rational CFTs that can be used as building blocks of the bulk type II partition function. The second is an algorithm using simple currents to generate many more modular invariant partition functions starting from a left-right diagonal one [14]. The third involves general formulae for boundary and crosscap coefficients for such BCFTs developed in a series of papers by Schellekens and collaborators [15]. The final ingredient is that all of the above can be algorithmized so that the search can be done numerically. This is an enormous advantage as the vacua that can be searched are so numerous that millions of spectra can be produced that match chirally the SM spectrum. In the first application of this numerical algorithm to an extended search for a concrete brane configuration realizing the SM model 200000 different successful spectra were found in millions of copies, [17] using Gepner models [18] as building blocks.[2] The kind of bottom up models considered in [17] were variations on the "Madrid" model first proposed in [16]. They are characterized by four stacks of branes with a Chan–Paton group $U(3)_\mathbf{a} \times U(2)_\mathbf{b} \times U(1)_\mathbf{c} \times U(1)_\mathbf{d}$, with the standard model generator Y embedded as $Y = \frac{1}{6}Q_\mathbf{a} - \frac{1}{2}Q_\mathbf{b} - \frac{1}{2}Q_\mathbf{c}$. The variations include the possibility of choosing the second and third Chan–Paton factor real, and allowing the $B - L$ Abelian vector boson to be either massive or massless in the exact string theory. These models have a perturbatively unbroken baryon and lepton number.

In the search that we review here the only feature assumed is the most robust part of what we presently know about the Standard Model: that there are three chiral families of quarks and leptons in the familiar representations of $SU(3) \times SU(2) \times U(1)$. In practice, we still have to make a few concessions. In particular,

---

[2]Orientifolds using Gepner models were developed in [19–24], although getting chiral spectra proved to be a difficult exercise. This was first achieved in [22].

we will have to limit the number of participating branes and forbid non-chiral mirror pairs of arbitrary charge. This will be discussed in more detail in the next section.

The features that are allowed include
- Anti-quarks realized as anti-symmetric tensors of $U(3)$
- Charged leptons and neutrinos realized as anti-symmetric tensors
- Non-standard embeddings of the Y-charge
- Embeddings of Y in non-Abelian groups
- Strong-Weak unification (e.g. $SU(5)$)
- Baryon-lepton unification (e.g. Pati–Salam models)
- Trinification
- Baryon and/or lepton number violation
- Family symmetries

Not all of these features are desirable, but the strategy is to allow as many possibilities in an early stage, and leave the final selection to the last stage, so that it will not be necessary to restart the entire search procedure if new insights emerge.

Some of these options may address unsolved problems that occur for the Madrid realization [16] of the standard model. For example, the perturbatively unbroken lepton number of these models makes it hard to implement a see-saw like mechanism to give small masses to neutrinos.[3] Coupling constant unification, if it is indeed a fundamental feature of nature and not a semi-coincidence, is not automatic in the standard realization, but it would be in $SU(5)$ models. This does not mean that the Madrid realization cannot accommodate the current experimental values of the couplings constants, but only that the fact that they presently appear to converge (with gaugino contributions taken into account) would be a mere coincidence. We will find some really simple and elegant realizations of $SU(5)$ models, but instantons need to be advocated to generate up-quark masses [11, 28, 32, 35]. We will comment on this in Section 7.

One of our goals is to analyse which model can be built from a bottom-up point of view, and how many of them can be realized as top-down models. By "bottom-up" we mean here a brane realization that produces the correct chiral standard model spectrum if the gauge group is reduced to $SU(3) \times SU(2) \times U(1)$ (without assuming a particular mechanism for that reduction). On the "top-down" side two types of concepts should be distinguished: standard model brane configurations and solutions to the tadpole conditions. The focus in this lecture is on the former, i.e. choices of boundary labels[4] **a**, **b**, **c** and **d** such that with an appropriate choice

---

[3] Instantons may bypass this difficulty, however, [31, 33].

[4] We label the complete set of boundaries of a given modular invariant partition function of a CFT as $a, b, c, d, \ldots$. The specific boundaries that participate in a Standard Model configuration are denoted as **a**, **b**, **c** and **d**. We allow a maximum of four (plus a hidden sector), with the first two corresponding to $SU(3)_{color}$ and $SU(2)_{weak}$.

of the Chan–Paton gauge group and the appropriate embedding of $SU(3) \times SU(2) \times U(1)$ one obtains the standard model. Here we also require that the standard model $U(1)$ generator does not acquire mass due to bilinear axion couplings.

Given such a standard model configuration, there may still be uncanceled tadpoles in RR closed string one-point functions on the disk and the crosscap. Within this context, the only way to cancel them is to add additional hidden matter, except in a few cases where they already cancel among the standard model branes. To see if this can happen is an extremely time-consuming, and ultimately unsolvable problem. Furthermore for any given brane configuration there may be many ways of canceling the tadpoles. In the continuum theory, background fluxes, not considered here, contribute to the tadpoles. But perhaps more importantly, the set of boundary states we consider here is limited by the choice of rational CFT. We consider the complete set of boundaries allowed by the RCFT, i.e. all boundaries that respect its chiral algebra. But that chiral algebra is larger than the $N = 2$ world-sheet algebra required to describe a geometric Calabi–Yau compactification. Since we get the $c = 9$ chiral algebra as a tensor product of minimal $N = 2$ algebras, the chiral algebra also contains all differences of the $N = 2$ algebras of the factors. If we would reduce the chiral algebra, additional boundary states are allowed, and could contribute to tadpole cancelation.

It is essentially impossible to conclude, with RCFT techniques alone, that the tadpoles of a certain standard model configuration cannot be canceled. Positive results, on the other hand, imply that one has a valid supersymmetric string vacuum. We see tadpole cancelation therefore mainly as an existence proof of a given string vacuum. Once that proof has been given, we do not continue searching for additional tadpole solutions for the same chiral configuration. This gives an enormous cut-off in computer time. One should keep in mind that for the most frequent chiral model considered in [17], we found a total of 16 million tadpole solutions (about 110000 of them distinct). We now keep only one of those solutions. This also implies that we cannot provide meaningful statistical results regarding tadpole solutions, but only regarding brane configurations.

We summarize briefly the results:
• We develop a detailed classification of allowed embeddings of the SM hypercharge inside the orientifold gauge group. To do this, we classify brane stacks according to how they contribute to the hypercharge. The hypercharge embedding is then characterized by a real variable x which is quantized in half-integral units in genuine non-orientable vacua.
• We produce 19345 chirally distinct top-down SM spectra (before tadpole cancelation) and 1900 chirally distinct models solving the tadpole conditions and realizing the different embeddings.
• We find that the $x = \frac{1}{2}$ hypercharge embedding dominates by far all other choices. The Madrid embedding [16] belongs to this class.

• The presence of chiral symmetric and antisymmetric tensors is highly suppressed. For some hypercharge embeddings, such tensors are crucial for anomaly cancelation and they may produce anti-quarks and other weak singlets. This implies the associated suppression of such embeddings.

• We produce the first examples of supersymmetric $SU(5)$ and flipped $SU(5)$ orientifold vacua, with the correct chiral spectrum (no extra gauge groups and no exotic G_{CP} chiral states). However, as we argue, all such orientifold models, as well as models with quarks in the antisymmetric representation have a serious phenomenological problem associated with masses.

• We find some minimal supersymmetric Pati–Salam and trinification vacua.

• We have examples of spectra (but no tadpole solutions yet) with extended ($\mathcal{N} = 4$ or $\mathcal{N} = 8$) supersymmetry in the bulk and $\mathcal{N} = 1$ supersymmetry on the branes.

• We have found SM spectra solving the tadpole conditions on a relative of the quintic CY.

Most of the work that is reviewed here, has appeared in [11].

2. What we are looking for

Our goal is to search for the most general embedding of the standard model in the Chan–Paton gauge group of Gepner Orientifolds.

We first introduce some notation. We denote the full Chan–Paton group as G_{CP}. This is the group obtained directly from the multiplicities of the branes, without taking into account masses generated by two-point axion-gauge boson couplings. We require that the standard model gauge group, $G_{SM} = SU(3) \times SU(2) \times U(1)_Y$ is a subgroup of G_{CP}. Furthermore we require that the generator of $U(1)_Y$ does not get a mass from axion-gauge boson couplings.

The main condition we impose on the spectrum is the presence of three families of quarks and leptons, and the absence of chiral exotics. Since chirality can be defined with respect to various groups, and the term "exotics" is used in different senses in the literature, we will define this more precisely. Group-theoretically, the standard-model-like spectra we allow are described as follows. Denote the full set of massless representations of G_{CP} as R_{CP}. The subset of these representations that is chiral with respect to G_{CP} is denoted R_{CP}^{chir}. The reduction of these representations to the group G_{SM} are denoted as R_{SM} and R_{SM}^{chir} respectively. By "reduction" we mean here only that we decompose representations in terms of representations of a subgroup. No assumptions are made at this point regarding dynamical mechanisms (like the Brout–Englert–Higgs mechanism) to achieve such a reduction. Consider now the subset of either R_{SM} or R_{SM}^{chir} that is chiral with respect to G_{SM}. The result is required to be precisely the following set of left-handed fermions (all fermions will be in left-handed form in this paper)

$$3 \times \left[\left(3, 2, \frac{1}{6}\right) + \left(3^*, 1, -\frac{2}{3}\right) + \left(3^*, 1, \frac{1}{3}\right) + \left(1, 2, -\frac{1}{2}\right) + (1, 1, 1) \right]. \quad (2.1)$$

Any other particles must be non-chiral with respect to G_{SM}. This may include left-handed anti-neutrinos in the representation $(1, 0, 0)$ and MSSM Higgs pairs, $(1, 2, \frac{1}{2}) + (1, 2, -\frac{1}{2})$. Anything else will be called exotic.

The foregoing describes the most general configuration one could reasonably call an embedding of the standard model without chiral exotics, but we will have to impose a few additional constraints to make a search feasible. First of all we require that the standard model groups $SU(3)$ and $SU(2)$ come each from a single stack of branes, denoted **a** and **b** respectively. This forbids diagonal embeddings of these groups in more than one CP factor. In general by a stack we mean a single label for a real (orthogonal or symplectic) boundary, or a pair of conjugate labels for complex, unitary branes. The CP factor yielding $SU(3)$ must be $U(3)$, whereas the weak gauge symmetry $SU(2)$ can come from either $U(2)$ or $Sp(2)$. The group $O(3)$ is not allowed, because one cannot get spinor representations of orthogonal groups in perturbative open string constructions.

The hypercharge generator Y is a linear combination of the unitary phase factors of $U(3)$, $U(2)$ (if available) and any other generator of one of the other factors in G_{CP}. All representations $(3, 2)$ must necessarily come from bi-fundamentals of the **a** and **b** stacks, but not all anti-quarks can come from those stacks. Although there can be anti-quarks due to chiral anti-symmetric tensors of $SU(3)$, they all have the same hypercharge. Hence there must be at least one other stack of branes, labeled **c**.

In principle there could be any number of additional stacks of branes, but for purely practical reasons we allow at most one more stack (labeled **d**) to contribute to the standard model representation (2.1).[5] Additional branes may be present, and may be required for tadpole cancellation. They will be referred to as the "hidden sector". If stack **d** does not contribute to (2.1) at all we regard it as part of the hidden sector. The standard model branes **a**, **b**, **c** (and **d**, if present) will be called the "observable sector". Note that left-handed anti-neutrinos[6] are not listed in (2.1). We do not impose an *a priori* constraint[7] on the number of left-handed anti-neutrinos, although in some cases a certain number of such states

---

[5] In general we also expect that the number of exotics to rise fast with the number of additional stacks participating in the SM group.

[6] Since our convention is to represent all matter in terms of left-handed fermions, right-handed neutrinos are referred to as left-handed anti-neutrinos.

[7] The minimum number is two in order to accommodate the experimental data. We will comment further on neutrino masses in Section 7.4.

is required by anomaly cancellation in G_{CP}. They may in fact come from the hidden sector or the observable sector, or even from strings stretching between the two sectors.

Our next condition concerns the precise definition of the standard model generator Y. We allow it to be embedded in the most general way possible in the Chan–Paton factors of brane **c** and **d** (in addition to the unitary phases of **a** and **b**). In principle it could also have components in the hidden sector without affecting any of the foregoing, as long as all particles charged under those components of Y are massive or at least non-chiral. One could even try to use this as a mechanism to cancel bilinear axion coupling of Y, which would give the Y-boson a mass.[8] We will not consider that possibility here. This is equivalent to a restriction to standard model realizations with at most four participating branes, except for one intriguing possibility: a three brane realization with a fourth brane added purely to fix the axion couplings of Y, without contributing to quarks or leptons. This possibility was not included in our search. It should be mentioned however, that a qualitatively similar situation does indeed arise. There are orientifold vacua where there is a $U(1)$ arising from the SM stack of branes, under which all SM particles are neutral. In this case there is a continuous family of possible hypercharge embeddings. In some cases, the masslessness condition breaks the degeneracy. This provides a string realization of the field theory models in [40]. In other cases, even the masslessness condition does not lift the degeneracy.

The general form of Y is

$$Y = \sum_{\alpha} t_\alpha Q_\alpha + W_{\mathbf{c}} + W_{\mathbf{d}}, \tag{2.2}$$

where α runs over the values **a**, **b**, **c**, **d**, Q_α is the brane charge of brane α ($+1$ for a complex brane, -1 for its conjugate, and 0 for a real brane), and $W_{\mathbf{c}}$ and $W_{\mathbf{d}}$ are generators from the non-Abelian part of the Chan–Paton group. Therefore $W_{\mathbf{c}}$ and $W_{\mathbf{d}}$ are traceless. Such contributions to Y occur for example in Pati–Salam and trinification models, and therefore we want to allow this possibility.

There is one more condition we impose for practical reasons, namely that $R_{\text{CP}}^{\text{chir}}$ may only yield representations of standard model particles or their mirrors. The main purpose of this condition (as we will see in more detail below) is to prevent an unlimited proliferation of G_{CP}-chiral, but G_{SM} non-chiral representations such as $(1, 1, q) + (1, 1, -q)$, with q arbitrary. In addition, this condition also forbids triplets of $SU(2)_{\text{weak}}$, which can be chiral with respect to $U(2)_{\mathbf{b}}$.

One may distinguish three types of matter in these models: OO, OH and HH, where the two letters indicate if the endpoints of the open string are in the observable or hidden sector. All conditions on OO matter were already formulated

---

[8]Anomalous $U(1)$ masses have been calculated for general orientifolds in [39].

above. The "no chiral exotics" constraint formulated above allows HH matter to be chiral with respect to G_{CP}. For OH matter we impose a somewhat stronger constraint, namely that there cannot be any bi-fundamentals between the standard model and the hidden sector that are chiral with respect to G_{CP}. This is a stronger condition because the "no chiral exotics" constraint allows SM-Hidden sector bi-fundamentals as long as they are non-chiral with respect to G_{SM}. For example a mirror quark pair $(3, V) + (3^*, V)$, where V is a vector in a hidden sector $U(N)$ group, could be allowed under the more general rules. The resulting $U(N)$ anomalies can be cancelled in various ways.

We will allow the brane stacks **a**, **b**, **c**, **d** to have identical labels, with the exception of **c** and **d** (if they are identical, we might as well regard them as a single brane stack with a larger CP multiplicity). By allowing identical labels we are able to obtain examples of unified models, such as (flipped) $SU(5)$ or Pati–Salam like models. In the case of identical labels, we count them as follows: the QCD and weak group count as one stack each, and the branes that remain after removing the QCD and weak groups count as additional stacks, such that the total does not exceed four. For example, we can get $U(5)$ models with at most two additional CP-factors (plus any number of hidden sector branes).

We conclude this section with a summary of the kind of "exotics" (plus singlets and Higgs candidates) that may occur in generic models, indicating which kind we do and do not allow. We split G_{CP} into an observable and a hidden part as $G_O \times G_H$. In all cases we combine representations into non-chiral sets (usually, but not always pairs) if possible. We can distinguish the following possibilities

1. Matter of type OO

    (a) Non-chiral with respect to G_{CP}. This may include symmetric and anti-symmetric tensors or adjoints of $SU(3)$ or of $SU(2)$, mirror pairs of quarks and leptons, as well as bi-fundamentals with unusual and in a few cases even irrational charges. All particles in this class are allowed by our conditions.

    (b) Chiral with respect to G_{CP}, non-chiral with respect to G_{SM}. Examples are symmetric tensors of $U(2)_{weak}$, mirror pairs of quark and lepton doublets that are chiral with respect to $U(2)_{weak}$, mirror pairs where one member of the pair is a rank-2 tensor and the other member a bi-fundamental. We do allow such particles, except the symmetric $U(2)_{weak}$ tensors, and non-chiral pairs of quarks and leptons with non-standard charges.

    (c) Chiral with respect to G_{CP}, chiral with respect to G_{SM}, non-chiral with respect to QED $\times$ QCD. An example of such exotics would be a fourth family. Exotics of this type are not allowed by our conditions.

(d) Chiral with respect to G_{CP}, chiral with respect to G_{SM}, and chiral with respect to QED $\times$ QCD. Clearly this is not acceptable.

A mass term for exotics of type 1a is allowed by the full gauge symmetry, and hence it is possible that such a term is generated by shifting the moduli of the model. It is an interesting question whether the appearance of such exotics is a special feature of RCFT, or if they persist outside the rational points. It should be possible to get some insight in this question by analysing the coupling of these particles to the moduli, but this is beyond the scope of this paper. Exotics of type 1b may get a mass without invoking the standard model Higgs mechanism, and hence may become more massive than standard quarks and leptons. However, this will always require some additional dynamical mechanism beyond perturbative string theory. Exotics of type 1c require the standard model Higgs mechanism to get a mass. This may not be sufficient, since the Higgs couplings themselves may be forbidden by string symmetries, in which case additional mechanisms must be invoked. In any case it would be hard to argue that such particles would be considerably more massive than known quarks and leptons.

2. Matter of type HH. These are standard model singlets. No constraints are imposed on this kind of matter. One may distinguish two kinds.

   (a) Non-chiral with respect to G_{CP}. These particles may get a mass from continuous deformations of the model, as above.

   (b) Chiral with respect to G_{CP}, non-chiral with respect to G_H. These particles may get a mass from hidden sector dynamics.

3. Matter of type OH. In many cases particles in this class have half-integer charge. This occurs if the electromagnetic charge gets a contribution $\frac{1}{2}$ from each observable brane, which turns out to be the most frequently occurring kind of model. There are many possibilities for the chiralities, which we list here for completeness. We use a notation $(\chi_{G_{CP}}, \chi_{G_H}, \chi_{G_O}, \chi_{G_{SM}}, \chi_{QED \times QCD})$, where each χ indicates chirality, and can be Y (yes) or N (no).

   (a) (N,N,N,N,N),

   (b) (Y,N,N,N,N),

   (c) (Y,Y,N,N,N),

   (d) (Y,N,Y,N,N),

   (e) (Y,N,Y,Y,N),

   (f) (Y,N,Y,Y,Y),

(g) (Y,Y,Y,N,N),

(h) (Y,Y,Y,Y,N),

(i) (Y,Y,Y,Y,Y).

An example of type 3b, chiral with respect to the full Chan–Paton group, but not with respect to any of its subgroups, is $(3, 0, V) + (3^*, 0, V) + 3 \times (1, 1, V^*) + 3 \times (1, -1, V^*)$ in $U(3) \times U(1) \times U(N)$, with the first two factors from G_O and the last from G_H. Of all these possibilities, only 3a is allowed by our criteria. Types 3b, 3c and 3g might be tolerated on more general grounds, and types 3f and 3i are clearly unacceptable.

3. Classification of bottom-up embeddings

Here we will discuss the possible values of the coefficients t_α that occur in the brane decomposition of Y. We will use the following expression for Y:

$$Y = \sum_\alpha x_\alpha Q_\alpha, \tag{3.1}$$

where Q_α is the $U(1)$ charge of brane α. In contrast to (2.2) the sum is here not *a priori* restricted to a definite number of branes. In our search we will allow also the possibility that diagonal Lie algebra generators W of $SO(N)$, $Sp(2N)$ or $SU(N)$ groups contribute to Y, but this can always be taken into account by splitting those groups into $U(m)$ factors according to the W eigenvalues e_i. For example, if there are two distinct eigenvalues[9] we get for symplectic groups $Sp(2N)$ a contribution $W_\alpha = \text{diag}(N \times (e), N \times (-e))$, which can be accommodated by splitting $Sp(2N)$ into conjugate brane stacks with a CP group $U(N)$ and a contribution $e Q_\alpha$. Geometrically, this means that the $2N$ symplectic branes are moved off the orientifold plane. The same reasoning applies to $O(2N)$ branes. If there are $O(2N+1)$ stacks, the assumption of at most two distinct eigenvalues only allows the traceless generator $W = 0$ in (2.2), and hence such branes cannot contribute to Y at all. Finally, $U(N)$ branes can contribute $t_\alpha Q_\alpha + \text{diag}(n_1 \times e_1, n_2 \times e_2)$, with $n_1 + n_2 = N$, $n_1 e_1 + n_2 e_2 = 0$. This can be regarded as two stacks $U(n_1) \times U(n_2)$ contributing $(t_\alpha + e_1) Q_{\alpha_1} + (t_\alpha + e_2) Q_{\alpha_2}$, so that $x_{\alpha_1} = t_\alpha + e_1$ and $x_{\alpha_2} = t_\alpha + e_2$ Therefore formula (3.1) covers all cases.

The brane configurations we consider here are subject to two constraints: the spectrum must match that of the standard model in the chiral sense, with chirality defined with respect to $SU(3) \times SU(2) \times U(1)$. Furthermore all cubic

---

[9] Two is the maximum we allow. If there are more, this necessarily yields unconventional quark or lepton charges. For more details, see Appendix A.

anomalies in each factor of the full Chan–Paton group must cancel. This must be true because we want to be able to cancel tadpoles, and tadpole cancellation imposes cubic anomaly cancellation (mixed anomalies are cancelled by the generalized Green–Schwarz mechanism). The tadpoles are usually cancelled by adding hidden sectors, which adds new massless states to the spectrum. We do not allow these to be chiral with respect to $SU(3) \times SU(2) \times U(1)$, and hence they cannot alter the cubic anomalies. The cubic anomaly cancellation conditions that are derived from tadpole cancellation are the usual ones for the non-Abelian subgroups of $U(N)$, $N > 2$. Vectors contribute 1, symmetric tensors $N + 4$ and anti-symmetric tensors $N - 4$, and conjugates contribute with opposite signs. But the same condition emerges even if $N = 1$ and $N = 2$. This means that for example a combination of three vectors and an anti-symmetric tensor is allowed in a $U(1)$ factor. This is counter-intuitive, because the anti-symmetric tensor does not even contribute massless states, so that one is left with just three chiral massless particles, all with charge 1. The origin of the paradox is that it is incorrect to call this condition "anomaly cancellation" if $N = 1$ and $N = 2$ and if chiral tensors are present. It is simply a consequence of tadpole cancellation; the anomaly introduced by the three charge 1 particles is factorizable, and cancelled by the Green–Schwarz mechanism.

One might entertain the thought that this peculiar $U(1)$ cancellation might have something to do with the fact that we have three families of standard model particles. For example, one could assign the same $U(1)$ charge to all quarks or leptons of a certain type, and then cancel this anomaly with anti-symmetric tensors. This would require this particle type to appear with a multiplicity divisible by three. Because the $U(1)$ is anomalous, it would acquire a mass via the Green–Schwarz term. However, although configurations of this kind can indeed be constructed, they are complicated and unlikely to occur. We did indeed find examples of $U(1)$ anomaly cancellations due to anti-symmetric tensors, but usually with a more complicated family structure that does not admit such an interpretation.

3.1. Orientable configurations

Let us now return to our goal of determining the possibilities for Y. We begin by demonstrating that in principle all real values of the leading coefficient $x_{\mathbf{a}}$ are allowed. Using the quark doublet charges we may write Y as follows

$$Y = \left(x - \frac{1}{3} \right) Q_{\mathbf{a}} + \left(x - \frac{1}{2} \right) Q_{\mathbf{b}} + \text{rest.} \tag{3.2}$$

Here we assume (without loss of generality) that the quark doublets all come from bi-fundamentals (V, V^*) stretching between the QCD and the weak brane. The second entry could also be a V, but then we can conjugate $U(2)$ to obtain

(V, V^*). A mixture of V and V^* is however not allowed if we want x to take all real values; neither is a chiral anti-symmetric tensor in either $U(3)$ or $U(2)$, or the option of using $Sp(2)$ instead of $U(2)$. Here and in the following all representations are in terms of left-handed spinors.

Now we need lepton doublets. They can only be bi-fundamentals ending on the $U(2)$. The other end must be on a brane that contributes to Y in such a way that the total charge is either $-\frac{1}{2}$ or $\frac{1}{2}$. The latter value is considered because in addition to lepton doublets, we also allow mirrors, or MSSM Higgs pairs. Again we will write these bi-fundamentals exclusively as (V, V^*) (the first entry corresponds to $U(2)$). Mixtures of (V, V) and (V, V^*) between the same branes would fix x, and if there are no mixtures we can convert all bi-fundamentals to the form (V, V^*). The multiplicities of these bi-fundamentals may be negative, in which case we interpret them as (V^*, V).

Since we only allow $SU(2)$ doublets with charges $\pm\frac{1}{2}$, the possibilities for the charge coefficients of the new branes are x or $x - 1$. We refer to branes with these charges as "type C" and "type D" respectively (the QCD and weak branes are defined to be of type A and B respectively. We use small letters $\mathbf{a}, \mathbf{b}, \mathbf{c}, \mathbf{d}, \mathbf{e}, \ldots$ to label different stacks, and capitals A, B, C, $\ldots$ to label their types, with respect to the hypercharge embedding. Branes $\mathbf{a}$ and $\mathbf{b}$ are always of type A and B, but there is no one-to-one correspondence for the other branes). Note that these types C and D become equivalent (up to conjugation) if and only if $x = \frac{1}{2}$. We are not requiring that the type C or D branes are identical for all leptons or Higgs, or each other's conjugate, even if their charges would allow that.

Let n_1 be the net number of chiral states between brane $\mathbf{b}$ and all of the C-type branes, and n_2 the same for type D. To be precise:

$$n_1 = \sum_i \left[(N(V, V^*)_{\mathbf{b}C_i} - N(V^*, V)_{\mathbf{b}C_i} \right], \tag{3.3}$$

where N is the absolute number of massless states with given properties. We now impose anomaly cancellation in $U(2)$ (for three families)

$$-9 + n_1 + n_2 = 0, \tag{3.4}$$

because no chiral tensors are allowed for generic x. We also impose the requirement of having three chiral lepton doublets

$$n_1 - n_2 = 3, \tag{3.5}$$

which can be solved to yield $n_1 = 6$ and $n_2 = 3$. Note that the anomaly conditions for the Chan–Paton factors at the other end can aways be satisfied for some of the solutions. This is because the solution allows all multiplicities of $N(V, V^*)$ as well as $N(V^*, V)$ to be multiples of three. If we make three open strings end on the same $U(1)$ brane, the corresponding $U(1)$ anomalies can always be cancelled by anti-symmetric tensors.

Next we need anti-quarks. Since for general x anti-symmetric $U(3)$ tensors are not allowed, they must be bi-fundamentals between the $U(3)$ stack and other branes. If we introduce new branes for the anti-quark strings to end on, we can always arrange the configuration so that the anti-quarks are of the form (V^*, V). Then we need a brane of type C for down anti-quarks and a brane of type D for up anti-quarks. One may also use one of the already present branes of type C and D for this purpose, provided that only combinations (V, V^*) or (V^*, V) are used. Anything else implies a condition on x. Even if one uses distinct branes for all particle types, there are many ways to cancel the $U(1)$ anomalies using anti-symmetric tensors.

Finally we need charged lepton singlets and their mirrors. They can occur in four different ways for generic x:

1. With both ends on an existing brane of types C and D.

2. With one end on a previous C or D brane and one end on a new one. This would require new branes with various possible charges. In particular, it allows the following new charges: $x + 1$, $x - 2$ and their conjugates. We refer to these as types E and F. For $x = \frac{1}{2}$ these are each other's conjugates, and for $x = \frac{3}{2}, 1, 0$ and $-\frac{1}{2}$ some of the types C,D,E and F are equivalent.

3. With both ends on the same, new brane. This requires a new brane with $t_\alpha = \pm\frac{1}{2}$. We call this type G, unless it coincides with a previous type.

4. With both ends on two distinct new branes. This would in principle allow two new branes with contributions y and $1 - y$ to Y. Such branes (if they do not coincide with any previous type) will be called type H.

There are even more possibilities if one allows arbitrary numbers of additional branes for charged leptons. For example, one can connect new branes to types E and F with charge contributions $x - 2$ or $x + 3$, connect new branes to types G and H or add more branes of type H. By allowing mirror leptons one can build arbitrarily long chains of branes in this manner. However, this is too baroque[10] to consider seriously, and can in any case not be realized with at most four branes, a restriction we will ultimately impose. Already the fourth option is then impossible.

---

[10]It should be kept in mind that as the number of branes participating in the SM configuration increases, the number of chiral exotics, fractionally charged particles and other unwanted states increases exponentially fast. It is possible that the lower success rate may be compensated by the potentially larger number of such configurations. It is still true however, that the effective field theory of such vacua, will be very complicated or maybe intractable.

Options three and four split the standard model into two chirally disconnected sectors (i.e. there are no chiral strings connecting the two). This implies that the Y anomaly does not cancel in each sector separately, and hence the two components of the would-be Y-boson must have Green–Schwarz couplings to axions that give it a mass. In principle these contributions could cancel for Y, but that seems improbable, and hence reduces the statistical likelihood of this sort of configuration in a search. Furthermore lepton Yukawa couplings are perturbatively forbidden in such models.

The same four options exist for left-handed anti-neutrinos, but we do not impose any requirements on our construction with regard to their multiplicity. If they come from strings not attached to any of the previous branes, we regard them as part of the hidden sector.[11] Furthermore, we do not allow Y to have contributions from branes that do not couple to charged quarks and leptons. Otherwise one could extend Y by arbitrarily large linear combinations that only contribute non-chiral states. This implies that we regard a brane configuration as complete (prior to tadpole cancellation) if all charged quark and leptons exist chirally, and if all cubic $U(N)$ anomalies cancel. This configuration may already contain a few candidate right-handed neutrinos, and additional ones may appear, after tadpole cancellation, from hidden sector states, or strings between the standard model and the hidden sector.

Clearly this still leaves a huge number of possibilities to realize this kind of configuration, but there is an obvious maximally economical choice, namely identifying all branes of equal charge with each other, and the brane with opposite charge with its conjugate. This then results in a $U(3) \times U(2) \times U(1) \times U(1)$ model with the following chiral spectrum

$$
\begin{array}{rcl}
3 & \times & (V, V^*, 0, 0), \\
3 & \times & (V^*, 0, V, 0), \\
3 & \times & (V^*, 0, 0, V), \\
6 & \times & (0, V, V^*, 0), \\
3 & \times & (0, V, 0, V^*), \\
3 & \times & (0, 0, V, V^*).
\end{array}
$$

Although we anticipated the possible need for anti-symmetric tensors, it turns out that they are not needed at all in this particular configuration. All anomalies are already cancelled. This is a consequence of standard model anomaly cancellation. The formula for Y is

$$
Y = \left(x - \frac{1}{3}\right) Q_{\mathbf{a}} + \left(x - \frac{1}{2}\right) Q_{\mathbf{b}} + x Q_{\mathbf{c}} + (x - 1) Q_{\mathbf{d}}. \tag{3.6}
$$

---

[11] In the actual search we have relaxed this condition slightly, and allowed a brane **d** that just yields anti-neutrinos.

This model has the feature that it can be realized entirely in terms of oriented strings, which of course implies that x is not fixed. The converse is not true because one can allow $U(1)$ anti-symmetric tensors; they do not yield massless particles and hence give no restriction on x. By construction, this is the minimal realization of the standard model in terms of oriented strings. Oriented configurations (although more complicated than the one shown above) were considered earlier in [9,41] in the context of type-II theories.

One can generalize these orientable models further by allowing stack **c** and/or **d** to consist of several type C and D branes. The most general configuration can be denoted as $U(3) \times U(2) \times U(p_1 + q_1) \times U(p_2 + q_2)$, where p_1 is the number of type C branes on stack c, etc. To achieve this split we allow non-trivial generators $W_\mathbf{c}$ and $W_\mathbf{d}$ in the definition of Y. This gives an infinite set of solutions, all with at least three Higgs pairs (this follows from $U(2)$ anomaly cancellation). All these models have in fact precisely the same structure as the basic four-stack model above, except for an additional possibility that occurs if type C or D branes are in different positions (i.e. have different boundary labels). If in total three open strings are needed ending on brane C to get three anti-quarks, then if there are several type C branes the total number of such strings must be three. However, each multiplicity can be positive and negative, and hence cancellations are possible, that show up in the spectrum as additional mirrors on top of the basic configuration.

One of these cases corresponds to the "trinification" model [42,43]. One starts with a gauge group $SU(3)_{\text{color}} \times SU(3)_L \times SU(3)_R$ and matter in three copies of the representation $(V, V^*, 0) + (V^*, 0, V) + (0, V, V^*)$. This configuration fits into our construction by starting with four stacks (**a**, **b**, **c**, **d**) with a CP group $U(3) \times U(2) \times U(1) \times U(3)$, and $Y = -\frac{1}{6}Q_\mathbf{b} + \frac{1}{3}Q_\mathbf{c} + W_\mathbf{d}$, where $W_\mathbf{d}$ is the $SU(3)_\mathbf{d}$ generator $\text{diag}(\frac{1}{3}, \frac{1}{3}, -\frac{2}{3})$. The spectrum is three times $(V, V^*, 0, 0) + (V, 0, V^*, 0) + (V^*, 0, 0, V) + (0, V, 0, V^*) + (0, 0, V, V^*)$. The trinification model is obtained by putting stacks **b** and **c** on top of each other. In terms of the foregoing discussion, this model has $x = \frac{1}{3}$, and three branes of type C (one from stack **c** and two from stack **d**) plus one brane of type D (from stack **d**). The value $x = \frac{1}{3}$ can easily be understood as follows: in a standard trinification model Y is embedded entirely in $SU(3)$ factors, and cannot have components in the brane charges. Therefore in particular it cannot have any component in $U(3)_\mathbf{a}$.

The foregoing orientable standard model configurations can be realized in principle in non-orientable string theories. In these realizations the value of x is often fixed by the requirement that Y does not get a mass due to bilinear couplings with axions. Sometimes this yields rather bizarre looking solutions. For example, in our set of solutions there is one with $t_a = \frac{1}{33}$. There are also cases where Y remains massless for any value of x.

3.2. Charge quantization

There are further constraints on x if one considers unoriented models. First of all, for generic values of x the non-chiral part of the string spectrum contains states of fractional or even irrational charge, from (V, V) bi-fundamentals or from rank-2 tensors. Since such states are always non-chiral, they may be massive, or become massive under perturbations of the model. They would however be stable and are not confined by additional gauge interactions, because they live entirely within the standard model sector. Therefore, although this possibility cannot be completely ruled out, it certainly seems preferable to avoid it.

The foregoing discussion is quite general, and can be used to analyse charge quantization for non-standard-model states in any brane realization of the standard model. The dependence on Q_a and Q_b in (3.6) is the most general one possible, up to an irrelevant sign choice. The complete string spectrum contains states with charges of all sums and differences of the components of Y, as well as all values multiplied by 2. It is easy to see that just from branes **a** and **b**, we get the charge quantization condition

$$x = 0 \bmod \frac{1}{2}, \qquad (3.7)$$

if we require that all massive open string states from bi-fundamentals and rank two tensors between standard model branes **a** and **b** to have integer charges (taking into account QCD confinement). Clearly this condition also implies charge integrality if branes of types C,D,E and F are present. Only if charged leptons come from a chirally decoupled sector (the third or fourth case listed earlier) further conditions may be needed.

A second type of fractional charges that may occur are those coming from strings with a single end on a standard model brane, and the other end on a hidden sector brane. Even if these states are non-chiral, they certainly exist as massive excitations. In principle, such charges could be confined by hidden sector gauge groups, but to avoid them altogether, the following condition must hold

$$x = 0 \bmod 1. \qquad (3.8)$$

Also this condition can be derived from just the **a** and **b** branes. If it is satisfied, branes of types C, D, E and F satisfy the hidden sector charge quantization condition, but types G and H do not, in general.

Note that the first charge quantization condition (absence of fractional charge within the standard model sector) is automatically satisfied in oriented strings for any x, because the strings that might violate it simply do not exist in oriented string theories. However, quantization conditions do arise if one wishes to include hidden branes. These should not contribute to Y. This imposes the second charge quantization condition, $x = 0 \bmod 1$, for oriented strings.

3.3. Non-orientable configurations

The foregoing restrictions were necessary if one wishes to avoid non-chiral fractionally charged matter. More severe restrictions apply if some of the quarks and leptons themselves come from states that break the orientability of the open string theory.

Note first of all that in most cases both type C and type D branes are needed, in order to get up and down anti-quarks. The only way out is to get either all down anti-quarks or all up anti-quarks from anti-symmetric $U(3)$ tensors. The former possibility requires $x = \frac{1}{2}$, and then types C and D are the same. This possibility is realized in flipped $SU(5)$ models, of which we will give examples later. The second option leads to $x = 0$. Then no type D brane is needed for the quarks, and type C branes do not contribute to Y. This possibility finds a natural realization in $SU(5)$ GUT models. For all other values of x at least one type C and one type D brane is needed in addition to branes $\mathbf{a}$ and $\mathbf{b}$.

Consider now the possibility that a chiral state (a quark or lepton, or a mirror) breaks the orientability of the configuration. Obviously this sort of analysis applies to each chirally decoupled subsector separately (connected components of quiver diagrams), and we will only consider the component connected to the $\mathbf{a}$ and $\mathbf{b}$ branes.

The possibilities for such a chiral state, and the resulting restrictions on x are as follows

- Chiral anti-symmetric tensors on brane $\mathbf{a}$; $x = 0$ or $\frac{1}{2}$.
- Chiral anti-symmetric tensors on brane $\mathbf{b}$; $x = 0$, $\frac{1}{2}$ or 1.
- (V, V) between on branes $\mathbf{a}$ and $\mathbf{b}$; $x = \frac{1}{2}$.
- Chiral tensors on a brane of type C; $x = 0$, $\frac{1}{2}$ or $-\frac{1}{2}$.
- Chiral tensors on a brane of type D; $x = \frac{3}{2}$, 1 or $\frac{1}{2}$.
- (V, V) between brane $\mathbf{a}$ or $\mathbf{b}$ and a type C brane; $x = 0$ or $\frac{1}{2}$.
- (V, V) between brane $\mathbf{a}$ or $\mathbf{b}$ and a type D brane; $x = \frac{1}{2}$ or 1.
- (V, V) between type C and a type D brane; $x = 0$, $\frac{1}{2}$ or 1.

Note that the occurrence of (V, V) is automatic if one of the endpoint branes is real, and that (V, V) between two distinct type C or type D branes is equivalent to chiral tensors on a single such brane. We can extend this list further by including branes of types E and F, but this will just give similar numbers modulo half-integers. Note that in all cases the quantization condition (3.7) is satisfied.

One important general observation can be made now. For values of x other than 0, $\frac{1}{2}$ and 1 all quarks and lepton doublets must be realized exactly as in the orientable four-stack model discussed above, because anti-quark weak singlets can only come from bi-fundamentals, and $U(2)$ anomaly cancellation cannot be fixed with anti-symmetric tensors. This only leaves some freedom for the leptonic weak singlets. On the other hand, for $x = 0$, $\frac{1}{2}$ and 1 the $U(2)$ anomaly condition can always be satisfied by adding anti-symmetric tensors. They contribute ± 2 to the anomaly, but since the total number of doublets is even, so is the chiral number of doublets (the number of V's minus the number of V^*). (Note that is true for any $U(2)$ because of cancellation of global anomalies).

If we limit ourselves to four stacks, the number of possibilities is even smaller. For values of x other than 0 and $\frac{1}{2}$ branes of both types C and D are needed. This means that there is no room for E or F branes and the more exotic values for x they might allow. This is true even if branes C and D are "unified" into a single Chan–Paton group. In order to get a value of x outside the range $-\frac{1}{2}, \ldots, \frac{3}{2}$ in a non-orientable configuration, it must be the chiral strings between the unified C/D brane and E or F type branes that break the orientability, i.e. both (V, V) and (V, V^*) must occur. But it is easy to see that in that case such states necessarily give rise to leptons with charges ± 2, because they must couple to both the type C and the type D brane.

This reduces the allowed range for x to $-\frac{1}{2} \ldots \frac{3}{2}$, and one can read off from the list which orientation breaking chiral states are allowed in each case. In the following sections we will show how to construct four-stack non-orientable realizations of any of these, at least as "bottom up" brane configurations.

3.4. The cases $x = -\frac{1}{2}$ or $x = \frac{3}{2}$

To get the largest and smallest numbers in this range, the only orientation breaking chiral states must be chiral tensors on a type C or type D brane, respectively. This implies that the first five representations (3.6) (those yielding quarks and lepton doublets) must be identical to those of the four-stack orientable model (up to mirror pairs due to distributing type C and D branes over various positions, as discussed above for the orientable configuration). In particular it means that we can only vary the open string origin of the charged leptons. The values $-\frac{1}{2}$ and $\frac{3}{2}$ are essentially "dual" to each other under interchange and conjugation of the type C and D branes.

To construct a non-orientable $x = -\frac{1}{2}$ configuration we start with four stacks $(\mathbf{a}, \mathbf{b}, \mathbf{c}, \mathbf{d})$ generating a CP group $U(3) \times U(2) \times U(1) \times U(1)$, with the latter two are type C and D branes respectively. The only allowed deviation in comparison to the orientable configuration are $S_{\mathbf{c}}$ symmetric tensors on brane $\mathbf{c}$, m bi-fundamentals (V, V^*) between branes $\mathbf{c}$ and $\mathbf{d}$, $A_{\mathbf{c}}$ anti-symmetric tensors on brane **c** and $A_\mathbf{d}$ on brane **d**. Although the anti-symmetric tensor can occur only in non-orientable strings, they do not break the orientability in the sense of fixing x, because they do not yield massless particles imposing constraints on x. Their only rôle is to cancel chiral anomalies.

We get the following conditions from cubic anomaly cancellation and the requirement that the net number of positively charged leptons must be three:

$$5S_\mathbf{c} + m - 3A_\mathbf{c} = 3$$
$$-m - 3A_\mathbf{d} = -3$$
$$m - S_\mathbf{c} = 3.$$

The solution is $S_\mathbf{c} = -3A_\mathbf{d}$, $m = 3 - 3A_\mathbf{d}$, $A_\mathbf{c} = -6A_\mathbf{d}$. Hence m and $S_\mathbf{c}$ must be multiples of 3, and since $S_\mathbf{c} = 0$ brings us back to an orientable configuration, the simplest non-trivial solution is $S_\mathbf{c} = -3$, $m = 0$, $A_\mathbf{c} = -6$ and $A_\mathbf{d} = 1$. The analysis for $x = \frac{3}{2}$ is analogous, interchanging the rôles of branes C and D.

Another set of possibilities (for $x = -\frac{1}{2}$) is obtained by putting three type-C branes in stack c, with the CP multiplicity providing the multiplicities of the anti-quarks and the lepton doublets. Now anti-symmetric tensors on brane **c** produce chiral particles, and fix x. A simple sequence of solutions is obtained for $S_\mathbf{c} = 0$, $m = 1 - A_\mathbf{d}$, $A_\mathbf{c} = -A_\mathbf{d}$. This is a $U(3) \times U(2) \times U(3) \times U(1)$ solution with one anti-symmetric conjugate tensor on brane **c** (which provides the charged leptons) and an anti-symmetric tensor on brane **d**, just to cancel anomalies.

One can generalize this further by allowing (p_1, q_1) type (C,D) branes on stack **c**, and (p_2, q_2) type (C,D) branes. This is accomplished by having CP gauge groups $U(p_1 + q_1)_\mathbf{c}$ and $U(p_2 + q_2)_\mathbf{d}$, and splitting up their contribution to Y by means of non-trivial generator $W_\mathbf{c}$ and $W_\mathbf{d}$ in (2.2). Since there must be both type C and type D branes, and they cannot come all from the same stack, we may require $p_1 > 0$ and $q_2 > 0$. Solving the constraints then yields solutions only in the following cases: $p_1 = 1$ or 3, $q_2 = 0$, $q_2 = 1$ and arbitrary p_2, each with a sequence of allowed values for the representation multiplicities. The spectra with $p_2 \neq 0$ are rather unappealing: they either have G_{CP}-chiral pairs of mirror anti-quarks, or large numbers of rank-2 tensors. The ones with $p_2 = 0$ were already discussed above.

3.5. The case $x = 1$

A simple way to obtain a configuration with $x = 1$ is to replace the fourth CP group in the orientable configuration by $O(1)$ in order to break the orientability. In addition, there is a possibility of allowing k anti-symmetric tensors of $U(2)$, yielding k charged leptons. If brane **c** has a Chan–Paton group $U(1)$, the most general structure is, with CP-group $U(3) \times U(2) \times U(1) \times O(1)$ is

$$
\begin{array}{rcl}
3 & \times & (V, V^*, 0, 0), \\
3 & \times & (V^*, 0, V, 0), \\
3 & \times & (V^*, 0, 0, V), \\
m & \times & (0, V, V^*, 0), \\
n & \times & (0, V, 0, V), \\
l & \times & (0, 0, V, V), \\
k & \times & (0, A, 0, 0), \\
t & \times & (0, 0, A, 0).
\end{array}
$$

with the conditions

$$
\begin{array}{rcl}
m - n & = & 3 \\
-9 + m + n - 2k & = & 0 \\
k + l & = & 3 \\
9 - 2m + l - 3t & = & 0.
\end{array}
$$

These are respectively the requirements of three lepton doublets, $U(2)$ anomaly cancellation, three charged leptons and brane **c** anomaly cancellation. This yields a one-parameter set of solutions, $m = 6 + k, n = 3 + k, l = 3 - k, t = -k$. There are many more possibilities if we allow larger CP-factors for **c** and **d**. It is also possible to use a $U(1)$ CP-factor for **d**. This leads to an additional anomaly constraint, but there are many ways to satisfy it by replacing some of the vectors by their conjugates, and adding anti-symmetric and/or symmetric tensors. The latter yield singlet neutrinos. The complete solution is too complicated to present here.

3.6. Realizations with three brane stacks for $x = 0$

The cases $x = 0$ and $x = \frac{1}{2}$ allow far more possibilities. We will solve them here in general, in the special case that they are realized with just three branes, yielding a group $U(3) \times U(2) \times U(p, q)$, where p and q are the number of eigenvalues x and $x - 1$.

Consider first $x = 0$. We assume that there are t chiral rank-2 tensors on brane **a**. Then the most general choice of bi-fundamentals for anti-quarks and lepton doublets is as follows

$$
\begin{array}{rcl}
n & \times & (V^*, 0, V), \\
m & \times & (V^*, 0, V^*), \\
k & \times & (0, V, V^*), \\
l & \times & (0, V, V).
\end{array}
$$

Furthermore we allow r chiral anti-symmetric $U(2)$ tensor, and a and s chiral anti-symmetric and symmetric $U(p, q)$ tensors. The latter are allowed only if $q = 0$ (since otherwise one gets charge 2 leptons), and if $q > 1$ no $U(p, q)$ tensors are allowed at all. Furthermore we must require $mq = lq = 0$ to prevent particles with unacceptable charges. To get three lepton doublets we need $k(p - q) = 3$, i.e. $p - q = \pm 3$ or ± 1. The total number of charged leptons is $-r - apq$.

Let us assume first that $q > 1$. Then $a = s = 0$, and $r = -3$, and $m = l = 0$. $U(2)$ anomaly cancellation then implies $(p + q)k - 2r - 9 = 0$, and hence $(p + q)k = 3$. But we have already seen that $k(p - q) = 3$, and hence this is not consistent with the assumption. Now assume $q = 1$. Also in this case m and l must vanish. Then the condition for getting three anti-down-quarks is $np = 3$. This allows $p = 1$ or $p = 3$, but neither is consistent with $p - q = \pm 3$ or ± 1.

Hence the only possibility is $q = 0$. Then $r = -3$. The third brane does not contribute to Y, and the distinction between V and V^* on that brane is irrelevant for all hypercharges. The conditions for getting the right number of anti-down quarks is $(n + m)p = 3$, and for lepton doublets it is $(k + l)p = 3$. Hence p is either 1 or 3. Anti-up quarks can only come from the t anti-symmetric $U(3)$ tensors. Hence $t = 3$. In the $U(3) \times U(2)$ subgroup we find the representation $3 \times (A, 0) + 3 \times (V, V^*) + 3 \times (0, A^*)$, which of course fits precisely in $3 \times (10)$ of $U(5)$. The $U(1)$ generators Y becomes an $SU(5)$ generator. Hence the only possibility for $x = 0$ and at most three participating branes is broken $U(5)$. This can be reduced to two participating branes by putting the **a** and **b** branes on top of each other, to get unbroken $U(5)$. The CP group on the third brane can be $U(1)$ or $U(3)$, but since this brane does not contribute to Y one can also allow $O(1)$ or $O(3)$. In that case there are no anomaly constraints to worry about. If the c-brane group is unitary, the total anomaly is $3(n - m) + 2(l - k)$. This leaves many possible values, and this anomaly can be cancelled in many ways using symmetric or anti-symmetric tensors. In the spectrum, these appear as standard model singlets, i.e. candidate anti-neutrinos.

3.7. Realizations with three brane stacks for $x = \frac{1}{2}$

Consider now $x = \frac{1}{2}$. Then if $p = q$ the third brane could be orthogonal or symplectic, in which case there is no anomaly cancellation condition for it. Furthermore the weak group can then be $Sp(2)$. This makes little difference, because $U(2)$ anomalies can be cancelled by means of anti-symmetric tensors, which in this case are standard model singlets (right-handed neutrinos) which we do not constrain *a priori*.

We assume that there are t chiral rank-2 tensors on brane **a**. Then the most general structure is as follows

$$n \quad \times \quad (V^*, 0, V),$$
$$m \quad \times \quad (V^*, 0, V^*),$$
$$k \quad \times \quad (0, V, V^*),$$
$$l \quad \times \quad (0, V, V).$$

We have to require

$$t + np + mq \;=\; 3$$
$$nq + mp \;=\; 3$$
$$kp + lq - kq - lp \;=\; 3$$

for getting the right anti-up, anti-down and lepton doublet count. The first two equations imply $(n - m)(p - q) = -t$, and the last one $(k - l)(p - q) = 3$. Hence $p \neq q$, and brane **c** cannot be real. The only allowed values for $p - q$ are $-3, -1, 1, 3$, and t must be a multiple of $p - q$. Given these four values, we can compute $n - m$ and $k - l$. To cancel the anomalies on brane **c** and to provide charged leptons we introduce a anti-symmetric and b symmetric tensors. The conditions for anomaly cancellation on brane **c**, and a net number of 3 charged leptons can be combined to yield

$$3(n - m)(p - q) - 2(k - l)(p - q) - 3(a - b)(p - q) = -6, \qquad (3.9)$$

which together with the previous conditions implies $a - b = n - m$. The remaining equations are

$$(n + m)(p + q) \;=\; 6 - t, \qquad\qquad\qquad\qquad (3.10)$$

$$(a + b)(p + q) \;=\; (n - m) + 2(k - l) = \frac{6 - t}{p - q}. \qquad (3.11)$$

From their ratio we see that $(n + m) = (p - q)(a + b)$. Furthermore we see that $p + q$ and $p - q$ must both be divisors of $6 - t$. This allows a limited number of values for $p + q$, and then $(a + b)$ and $(n + m)$ are determined. Hence all solutions are specified in terms of t plus a limited number of values for $p + q$ and $p - q$. There are three more parameters that are not yet specified: $k + l$, the number of anti-symmetric tensors on brane **b**, and the difference between the number of (V, V^*) and (V, V) quark doublets. One linear relation between them is imposed by $U(2)$ anomaly cancellation; in the $Sp(2)$ case there is no constraint.

3.8. Solutions with type E and F branes

Type E and F branes contribute to Y with coefficients $x + 1$ and $x - 2$ respectively. They cannot contribute to quarks or lepton doublets. We assume here that their contribution includes at least one (V, V^*) bi-fundamental; if they produce valid quarks or lepton doublets (or mirrors) only as (V, V) bi-fundamentals we conjugate the E/F brane, and redefine its coefficients. Depending on the actual value of x an E or F brane then becomes a brane of type C or D, and is already included in our foregoing discussion.

Furthermore an E/F brane must be connected, by definition, via $(0, 0, V, V^*)$ bi-fundamentals to the **c**-brane. As discussed above, in a four-stack configuration E or F branes can only be allowed in principle for $x = 0$ or $x = \frac{1}{2}$. As in the rest of the paper, we allow the **c** and **d** stacks to consist of two brane types, with eigenvalues differing by one unit. The options are then **c** = (C,D), **d** = (E,C) or **c** = (C,D), **d** = (D,F), where each type can occur with an arbitrary multiplicity, and E and F have to occur at least once. However, in all cases one of the two branes on stack **c** would give rise to a charge-2 lepton. This reduces the possibilities to **c** = (C), **d** = (E,C) for $x = \frac{1}{2}$ (and its conjugate, **c** = (D), **d** = (D,F)) or **c** = (D), **d** = (D,F) for $x = 0$. However, the latter possibility is ruled out, since at least one C-type brane is needed to produce d^c anti-quarks. The next constraint is anomaly cancellation for stack **d**. Since it only shares bi-fundamentals $(0, 0, V, V^*)$ with brane **c** and nothing with any other brane, the anomalies of the V^*'s must be cancelled by rank-2 tensors. This forbids two distinct Y-eigenvalues on stack **d**, since the sums of these eigenvalues would appears as invalid charges in the spectrum. It also limits the multiplicity of the E or F branes to 1, and only allows anti-symmetric tensors to cancel the anomaly. The multiplicity of $(0, 0, V, V^*)$ must then be a multiple of three.

Configurations of this type can indeed be constructed. The **c**-group can either be $U(1)$ or $U(3)$. In the former case, there is a two-parameter series of solutions labelled by the number of $SU(3)_\mathbf{a}$ anti-symmetric tensors, and the number of $(0, 0, V, V^*)$. The $U(1)_\mathbf{c}$ anomalies are cancelled by anti-symmetric and/or symmetric tensors, and the latter also contribute charged leptons. If **c**-group is $U(3)$, there must be three anti-symmetric *conjugate* tensors of $SU(3)_\mathbf{a}$ (yielding three left-handed down quarks, which must be combined with six left-handed down anti-quarks from $(V^*, 0, V, 0)$), and there can be charged leptons from $(0, 0, V, V^*)$ as well as anti-symmetric $U(3)$ tensors.

Furthermore, one may use both $U(2)$ and $Sp(2)$ as the Chan–Paton group of brane **b**.

None of these models have appeared in our top-down search.

3.9. Solutions with type G branes

Type-G branes are defined as branes that contribute non-trivially to Y but that contribute to the chiral spectrum only through rank-2 tensors. This implies that their Y-coefficient must be $\pm\frac{1}{2}$. If $x = \frac{1}{2}$, this can be viewed as just a standard type C or D brane. These cases are taken into account in our bottom-up construction as standard $x = \frac{1}{2}$ models. They do indeed occur as brane configurations, although rarely. For example, we have generated all brane configurations with four unitary CP factors, at most three Higgs pairs, at most three G_{CP} exotics and at most six G_{CP} chiral singlets. Of the 10820995 unitary models with $x = \frac{1}{2}$, only 338 have type-G branes, i.e. a brane with only chiral tensors and no bi-fundamentals.

A more interesting situation occurs when $x = 0$ (the only other value of x where type-G branes might occur). In that case the type-G brane has a non-canonical contribution $\pm \frac{1}{2}$ to Y (the canonical value is 0 or ± 1).

However, the foregoing analysis of three brane realizations with $x = 0$ shows that this possibility does not exist. The only three-brane models are (broken) $SU(5)$ with a set of neutral C-type branes. This result was obtained without requiring any particular value for the number of charged leptons. The latter came out uniquely as three. Since the **c** stack is neutral, it cannot provide charged leptons or mirrors either. Hence all three-stack models already have precisely three charged leptons, and all the G-brane could still do is add mirror pairs. This could happen even with a chiral **d** stack, for example with three anti-symmetric tensors and a symmetric tensor of $U(2)$, with $W_{\mathbf{d}} = \mathrm{diag}(\frac{1}{2}, -\frac{1}{2})$. However, this is not of much interest, and furthermore these models are equivalent to those where brane **d** does not contribute to Y at all, and brane **d** just yields G_{CP}-chiral neutrinos.

4. Statistics of bottom-up configurations

In this section we will provide an enumeration of bottom-up configurations, providing some numbers to the theoretical analysis of the previous section. We will consider for simplicity the **c** and **d** groups to be Abelian. We will also impose (generalized) anomaly cancellation.

The associated statistics is shown and compared in Table 5 of the next section, where detailed definitions are also given.

4.1. Three stacks: the $U(3) \times U(2) \times U(1)$ models

We first consider three-stack models. We will consider the possible realizations of MSSM-like Higgs pairs, and the presence of baryon and lepton number symmetries. We also indicate the total number of configurations of a given type. In our search, we can also include the right-handed neutrino ν^c which may appear as an open string with both ends on the weak or other branes.

Requiring that the particles have the proper hypercharge there are two possible ways to embed the Standard Model in this D-brane system of three stacks, [25]:

$$Y = -\frac{1}{3}Q_a - \frac{1}{2}Q_b, \quad Y = \frac{1}{6}Q_a + \frac{1}{2}Q_c. \tag{4.1}$$

For the first embedding, $Y = -\frac{1}{3}Q_a - \frac{1}{2}Q_b$, we obtain the following allowed spectra, (by $\tilde{R}$ we indicate that both the representation R or the conjugate representation R^* can be a valid choice)

Q:	$(V, V, 0)$
u^c:	$(A, 0, 0)$
d^c:	$(V^*, 0, \tilde{V})$
L:	$(0, V^*, \tilde{V}^*)$
l^c:	$(0, A, 0)$
H:	$(0, V, \tilde{V})$
H':	$(0, V^*, \tilde{V})$.

From the above charge assignments we can construct families and search for triplets of these families which form anomaly-free models. For that embedding $(Y = -\frac{1}{3}Q_a - \frac{1}{2}Q_b)$ there are 10 different anomaly-free spectra that describe the SM. If the anti-neutrino ν^c also arises from strings stretching inside this stack, it will be of the form $(0, 0, \tilde{S})$.

For the second embedding $Y = \frac{1}{6}Q_a + \frac{1}{2}Q_c$ we have the following allowed spectra

Q:	$(V, \tilde{V}, 0)$		
u^c:	$(V^*, 0, V^*)$		
d^c:	$(A, 0, 0)$	or	$(V^*, 0, V)$
L:	$(0, \tilde{V}, V^*)$		
l^c:	$(0, 0, A)$		
H:	$(0, \tilde{V}, V)$		
H':	$(0, \tilde{V}, V^*)$.		

There are 24 different anomaly-free models. If the anti-neutrino ν^c also arises from strings stretching inside this stack, it will be of the form $(0, \tilde{A}, 0)$. Notice the ambiguity of the representations (with tilde) when a brane does not contribute to the hypercharge and also the two different possibilities for the charges of d^c: $(V^*, 0, V)$ or $(A, 0, 0)$ in the second case.

The baryon number $B = Q_a/3$ is a gauge symmetry only in models where d^c arises from a string with the two ends onto different branes. In none of the models above, lepton number is a symmetry.

4.2. Four stacks: $U(3) \times U(2) \times U(1) \times U(1)'$ models

In this section, we study four-stack realizations of the Standard Model. We continue with the statistics of fours-stack models.

- Hypercharge $Y = (x - \frac{1}{3})Q_a + (x - \frac{1}{2})Q_b + x Q_c + (x - 1)Q_d$

Notice that in order for x to remain arbitrary, the right-handed neutrino must necessarily arise in the hidden sector. The corresponding charge assignments are:

Q: $(V, V^*, 0, 0)$

u^c: $(V^*, 0, 0, V)$

d^c: $(V^*, 0, V, 0)$

L: $(0, V, V^*, 0)$ or $(0, V^*, 0, V)$

l^c: $(0, 0, V, V^*)$

H: $(0, V, 0, V^*)$ or $(0, V^*, V, 0)$

H': $(0, V, V^*, 0)$ or $(0, V^*, 0, V)$.

Following the same spirit as in the tree-stack models, we can form families from the above charge assignments and require that triplets of them are free of irreducible anomalies. For the present hypercharge embedding there is only one anomaly-free model which can describe the SM and given by three copies of the previous assignments; it is (3.6) shown in the previous section

- Hypercharge $Y = -\frac{1}{3}Q_a - \frac{1}{2}Q_b + Q_d$

The corresponding charge assignments are:

Q: $(V, V^*, 0, 0)$

u^c: $(A, 0, 0, 0)$ or $(V^*, 0, 0, V)$

d^c: $(V^*, 0, \tilde{V}, 0)$

L: $(0, V, \tilde{V}, 0)$ or $(0, V^*, 0, V^*)$

l^c: $(0, A^*, 0, 0)$ or $(0, 0, \tilde{V}, V^*)$

H: $(0, V^*, \tilde{V}, 0)$ or $(0, V, 0, V^*)$

H': $(0, V^*, 0, V)$ or $(0, V, \tilde{V}, 0)$.

If ν^c is coming from the hidden sector, there are 302 anomaly-free models which can describe the SM particles. Among them, there are 62, 72, 96 and 72 models with three, two, one and none chiral Higgs pairs.

On the other hand, if v^c is attached onto branes of the above stacks, it can only be charged under the $U(1)_c$ which does not contribute to the hypercharge. Therefore, it will transform as $(0, 0, \tilde{S}, 0)$. In that case, there are 1208 different anomaly-free models which can describe the SM particles (including v^c). Among them, there are 240, 384, 288 and 248 models with tree, two, one and none chiral Higgs pairs.

When u^c is not described by an antisymmetric representation, the baryon number $B = Q_a/3$ is conserved.

- Hypercharge $Y = \frac{2}{3}Q_a + \frac{1}{2}Q_b + Q_c$

The corresponding charge assignments are:

Q: $(V, V^*, 0, 0)$
u^c: $(V^*, 0, 0, \tilde{V})$
d^c: $(V^*, 0, V, 0)$
L: $(0, V^*, 0, \tilde{V})$ or $(0, V, V^*, 0)$
l^c: $(0, A, 0, 0)$ or $(0, 0, V, \tilde{V})$
H: $(0, V^*, V, 0)$ or $(0, V, 0, \tilde{V})$
H': $(0, V^*, 0, \tilde{V})$ or $(0, V, V^*, 0)$.

In total, there are 6 different anomaly-free models which can describe the SM particles with chiral Higgs-pairs.

A v^c which is a string attached onto these stacks of branes would be of the form $(0, 0, 0, \tilde{S})$. In that case, there are 24 different anomaly-free models with chiral Higgs-pairs (including v^c) and they all have baryon number $B = Q_a/3$.

- Hypercharge $Y = \frac{1}{6}Q_a + \frac{1}{2}Q_c - \frac{1}{2}Q_d$

The corresponding charge assignments are:

Q: $(V, \tilde{V}, 0, 0)$
u^c: $(V^*, 0, V^*, 0)$ or $(V^*, 0, 0, V)$
d^c: $(A, 0, 0, 0)$ or $(V^*, 0, V, 0)$ or $(V^*, 0, 0, V^*)$
L: $(0, \tilde{V}, V^*, 0)$ or $(0, \tilde{V}, 0, V)$
l^c: $(0, 0, S, 0)$ or $(0, 0, V, V^*)$ or $(0, 0, 0, S^*)$
H: $(0, \tilde{V}, 0, V^*)$ or $(0, \tilde{V}, V, 0)$
H': $(0, \tilde{V}, 0, V)$ or $(0, \tilde{V}, V^*, 0)$.

In that case, there are 8552 different anomaly-free models with chiral Higgs pairs which can describe the SM particles.

Some models have lepton number. There are four independent combinations:

- $Q_L = \frac{1}{2}Q_a + \frac{1}{2}Q_b - \frac{1}{2}Q_c - \frac{1}{2}Q_d$:

  $3 \times (V, V^*, 0, 0)$,
  $3 \times (V^*, 0, V^*, 0)$,
  $3 \times (V^*, 0, 0, V^*)$,
  $3 \times (0, V, V^*, 0)$,
  $3 \times (0, V, V, 0)$,
  $3 \times (0, V, 0, V)$,
  $3 \times (0, 0, S, 0)$.

- $Q_L = Q_d$:

  $3 \times (V, V^*, 0, 0)$,
  $3 \times (V^*, 0, V^*, 0)$,
  $\{m \times (V^*, 0, V, 0), \ n \times (A, 0, 0, 0)\}$,
  $3 \times (0, V, 0, V)$,
  $3 \times (0, V, V, 0)$,
  $3 \times (0, V, V^*, 0)$,
  $3 \times (0, 0, V, V^*)$,

where $m, n \in [0, 1, 2, 3]$ and $m + n = 3$. Therefore, d^c in each family can be either a string which is attached onto the **a** and **c** stacks or a string with both ends on the **a** stack.

- $Q_L = -Q_c$:

  $3 \times (V, V^*, 0, 0)$,
  $3 \times (V^*, 0, 0, V)$,
  $\{m \times (V^*, 0, 0, V^*), \ n \times (A, 0, 0, 0)\}$,
  $3 \times (0, V, V^*, 0)$,
  $3 \times (0, V, 0, V^*)$,
  $3 \times (0, V, 0, V)$,
  $3 \times (0, 0, V, V^*)$,

where again $m, n \in [0, 1, 2, 3]$ and $m + n = 3$.

- $Q_L = \frac{1}{2}Q_a + \frac{1}{2}Q_b + \frac{1}{2}Q_c + \frac{1}{2}Q_d$:

  $3 \times (V, V^*, 0, 0)$,
  $3 \times (V^*, 0, 0, V)$,
  $3 \times (V^*, 0, V, 0)$,
  $3 \times (0, V, 0, V)$,
  $3 \times (0, V, 0, V^*)$,
  $3 \times (0, V, V^*)$,
  $3 \times (0, 0, 0, S^*)$.

If the right-handed neutrino ν^c is attached onto the SM branes, it can be described by $(0, \tilde{A}, 0, 0)$ or $(0, 0, \tilde{V}, \tilde{V})$. Including ν^c, there are 150672 different anomaly-free models. Among them, there are 29360, 61344, 48800 and 11168 models with tree, two, one and none chiral Higgs pairs.

If d^c is not described by an antisymmetric representation, there is baryon number $B = Q_a/3$.

- Hypercharge $Y = \frac{1}{6}Q_a + \frac{1}{2}Q_c - \frac{3}{2}Q_d$

  The corresponding charge assignments are:

  Q: $(V, \tilde{V}, 0, 0)$
  u^c: $(V^*, 0, V^*, 0)$
  d^c: $(V^*, 0, V, 0)$ or $(A, 0, 0, 0)$
  L: $(0, \tilde{V}, V, 0)$
  l^c: $(0, 0, V^*, V)$ or $(0, 0, S, 0)$
  H: $(0, \tilde{V}, V, 0)$
  H': $(0, \tilde{V}, V^*, 0)$.

In that case, there are 4 different anomaly-free models with chiral Higgs pairs which can describe the SM.

A ν^c which is stretched between the four stacks can be of the form $(0, \tilde{A}, 0, 0)$. Including ν^c, the number of different charge assignments is 24 (8 of them have two chiral Higgs pairs and the other 16 have non chiral Higgs pairs). Half of these states have baryon number $Q_B = Q_a/3$ and in none lepton number is a symmetry. All models have one non-anomalous $U(1)$.

- Hypercharge $Y = -\frac{1}{3}Q_a - \frac{1}{2}Q_b$

  The corresponding charge assignments are:

  Q: $(V, V^*, 0, 0)$
  u^c: $(A, 0, 0, 0)$

d^c: $(V^*, 0, \tilde{V}, 0)$ or $(V^*, 0, 0, \tilde{V})$

L: $(0, V^*, \tilde{V}, 0)$ or $(0, V^*, 0, \tilde{V})$

l^c: $(0, A^*, 0, 0)$

H: $(0, V, \tilde{V}, 0)$

H': $(0, V^*, \tilde{V}, 0)$

with 936 anomaly-free models. Among them, there are 256, 120, 120 and 440 models with tree, two, one and none chiral Higgs pairs.

A ν^c which will be stretched between the four branes will be of the form $(0, 0, 0, \tilde{S})$ or $(0, 0, \tilde{S}, 0)$ or $(0, 0, \tilde{V}, \tilde{V})$. Including ν^c, there are 106792 different anomaly-free models. Among them, there are 15072, 32332, 36228 and 23160 models with tree, two, one and none chiral Higgs pairs.

- Hypercharge $Y = -\frac{5}{6}Q_a - Q_b - \frac{1}{2}Q_c + \frac{3}{2}Q_d$

The above hypercharge embedding is allowed only in cases where the right-handed neutrino is coming from the hidden sector. The corresponding charge assignments are:

Q: $(V, V^*, 0, 0)$

u^c: $(V^*, 0, 0, V)$

d^c: $(V^*, 0, V, 0)$

L: $(0, V^*, 0, V)$ or $(0, V, V^*, 0)$

l^c: $(0, 0, S^*, 0)$ or $(0, 0, V, V^*)$

H: $(0, V^*, V, 0)$ or $(0, V, 0, V^*)$

H': $(0, V^*, 0, V)$ or $(0, V, V^*, 0)$.

In that case, there are 2 different anomaly-free models which can describe the SM:

$3 \times (V, V^*, 0, 0)$,

$3 \times (V^*, 0, 0, V)$,

$3 \times (V^*, 0, V, 0)$,

$6 \times (0, V, V^*, 0)$,

$3 \times (0, V, 0, V^*)$,

$\{3 \times (0, 0, S^*, 0)$ or $3 \times (0, 0, V, V^*)\}$,

and they have baryon number $Q_B = Q_a/3$. Lepton number is not a symmetry.

- Hypercharge $Y = \frac{7}{6}Q_a + Q_b + \frac{3}{2}Q_c + \frac{1}{2}Q_d$

The above hypercharge embedding is allowed only in cases where the right-handed neutrino is coming from the hidden sector. The corresponding charge assignments are:

$$
\begin{array}{ll}
Q: & (V, V^*, 0, 0) \\
u^c: & (V^*, 0, 0, V) \\
d^c: & (V^*, 0, V, 0) \\
L: & (0, V, V^*, 0) \quad \text{or} \quad (0, V^*, 0, V) \\
l^c: & (0, 0, 0, S) \quad \text{or} \quad (0, 0, V, V^*) \\
H: & (0, V^*, V, 0) \quad \text{or} \quad (0, V, 0, V^*) \\
H': & (0, V, V^*, 0) \quad \text{or} \quad (0, V^*, 0, V).
\end{array}
$$

In that case, there are 2 different anomaly-free models which can describe the SM particles:

$$
\begin{array}{l}
3 \times (V, V^*, 0, 0), \\
3 \times (V^*, 0, 0, V), \\
3 \times (V^*, 0, V, 0), \\
6 \times (0, V, V^*, 0), \\
3 \times (0, V, 0, V^*), \\
\{3 \times (0, 0, 0, S) \quad \text{or} \quad 3 \times (0, 0, V, V^*)\},
\end{array}
$$

and they have baryon number $Q_B = Q_a/3$. Lepton number is not a symmetry.

5. Top-down configurations and SM spectra

5.1. Scope of the top-down search

The set of models we are able to search in principle consists of all three and four-stack combinations of all boundaries of all simple current orientifolds [15] of all simple current MIPFs [14] of the 168 $c = 9$ tensor products of $N = 2$ minimal models. We denote these as $(k_1, \ldots, k_m)$, where k_i is the $SU(2)$ level, which ranges from 1 to ∞. The total number of MIPFs is 5403, and the total number of orientifolds 49304. Some of these have zero-tension O-planes, which means that there is no possibility of cancelling tadpoles between D-branes and O-planes. This leaves 33012 orientifold models. Of the 168 Gepner models, 5 are non-chiral $K_3 \times T_2$ compactifications, which need not be considered because they can never yield a chiral spectrum.[12] These non-chiral theories contribute in total 88 MIPFs and 228 orientifolds.

---

[12]Note that all boundaries we consider respect the full chiral algebra of the tensor product, and all partition functions are expressed in terms of the characters of that algebra, which are space-time non-chiral. One may also consider orbifold projections of these theories, which reduce the chiral algebra,

Table 1

Total number of three and four stack configurations of various types

Type	Total	This paper
UUU	1252013821335020	1443610298034
UUO, UOU	99914026743414	230651325566
UUS, USU	14370872887312	184105326662
USO	2646726101668	74616753980
USS	1583374270144	73745220170
UUUU	21386252936452225944	366388370537778
UUUO	2579862977891650682	105712361839642
UUUS	187691285670685684	82606457831286
UUOO	148371795794926076	19344849644848
UUOS	17800050631824928	26798355134612
UUSS	4487059769514536	13117152729806
USUU	93838457398899186	41211176252312
USUO	17800050631824928	26798355134612
USUS	8988490411916384	26418410786274

The number of boundary states in a complete set can range from a few hundred to 108612 for tensor product $(1, 5, 82, 82)$. In that case the number of unitary brane pairs is 53046 and 52920 for the two orientifold choices. The number of combinations one needs to consider for a four-stack configuration grows with the fourth power of the number of pairs. In [17] almost all these cases were searched. This was possible because the standard model configuration searched for was more limited. For example, no chiral rank-2 tensors were allowed, reducing the number of choices for the **a,b,c** and **d** branes dramatically. Furthermore the configuration of [16] is such that branes **a** and **d** have a different multiplicity (3 and 1) but identical intersection numbers with the other branes. This can be used to reduce the power behavior of the search algorithm essentially from four to three.

Neither of these shortcuts help us here, and therefore a full search is practically impossible at present. Here we limit ourselves to MIPFs with at most 1750 boundaries. This limits us to 4557 of the 5403 MIPFS and 29257 of the 33012 non-zero tension orientifolds. We can now work out how many brane configurations exist in total. To do this really correctly, unitary, orthogonal and symplectic branes must be distinguished.

Table 1 lists the total number of configurations for all combinations of unitary, orthogonal and symplectic branes, without taking into account the additional freedom of assigning Chan–Paton multiplicities. The second column gives the

and may introduce chiral characters, but our methods do not apply to that case. We do allow the inverse of this: a chiral theory with a non-chiral extension. Indeed, we found some standard model configurations for such theories.

grand total for all 163 chiral Gepner models and non-zero tension orientifolds. It is the maximal number of three and four-stack configurations of given type that we have at our disposal for Standard Model searches. The third column gives the size of the subset actually searched in this paper.

The precise counting is as follows. Denote the number of unitary brane pairs as N_U. Then the total number of UUUU configurations with distinct **c** and **d** branes is $(2N_U)(N_U) \times \frac{1}{2} N_U(N_U - 1)$, etc. The choices for **a**, **b** and **c** are independent, since we allow all these stacks to coincide, but if **c** and **d** coincide we regard it as a three-stack configuration. Furthermore both conjugates of the **a** brane are counted, because they give rise to conjugate $SU(3)$ representations, and hence yield distinct spectra. Conjugations of the **b**, **c** and **d** branes can always be compensated by changing the sign of the coefficients of Y, and hence do not yield new possibilities.

Obviously, although we cover a substantial fraction of MIPFs and orientifolds, only a small fraction of possible brane configurations has been searched, because the missing MIPFs are the ones with the largest number of branes. Nevertheless, in our previous search [17], which was more extensive, the MIPFs we are not considering in the present paper produced relatively few SM-configurations and tadpole solutions. Part of the reason for the latter is that probably there are many more candidate branes in the hidden sector, making the tadpole equations harder to solve.

5.2. *Standard model brane configurations found*

Of the 4557 MIPFs, 1639 contained at least one standard model spectrum, without taking into account tadpole cancellation. In Table 2 we list the total number of brane configurations with a chiral standard model spectrum sorted according to x. In [17] only a subset of the possible $x = \frac{1}{2}$ models was considered, but for a much larger set of MIPFs. This produced a total of about 45 million such configurations, whereas now we find about 124 million, in both cases before attempting to solve the tadpole conditions. In column 1, a $*$ indicates that the value of x is

Table 2

Number of standard model configurations sorted by the value of x

x	Total occurrences	Without $SU(3)$ tensors
$-1/2$	0	0
0	21303612	202108
1/2	124006839	115350426
1	12912	12912
3/2	0	0
$*$	1250080	1250080

not fixed by the quark and lepton charges, as is the case in orientable models. In these models, the value of x may or may not be fixed by the zero-mass condition for Y. If it is fixed, it can in principle have any real value. In Table 2 this distinction is not taken into account, but we do treat these models as distinct in the complete list, Table 6, to be discussed below.

Apart from the $x = *$ cases, all other models are categorized with the value of x that follows from the quark and lepton charges as well as the zero mass condition for Y. In some cases, the quark and lepton charges alone might allow more than one value of x even for unorientable models. For example, in $SU(5)$ GUT models one can get the correct spectrum for $x = 0$ (standard $SU(5)$) and $x = \frac{1}{2}$ (flipped $SU(5)$). The zero-mass condition for Y always allows the former option (since Y is a generator of the non-Abelian group $SU(5)$) and may or may not allow the latter. If both are allowed, both are taken into account in Table 2. Finally, if a model with $x = *$ gets x fixed to a half-integer value by the Y-mass condition, it is counted once as an $x = *$ model, and once for the actual value of x.

In the third column we list how many of the configurations have no anti-quarks realized as anti-symmetric $SU(3)$ tensors. As we will discuss later, it is nearly impossible to get mass terms or Yukawa couplings for such tensors, and therefore they should be regarded as implausible. Note that anti-symmetric $SU(3)$ tensors are only allowed for $x = 0$ and $x = 1/2$. In the former case, it turns out that about 99% of the configurations have such tensors, whereas for $x = 1/2$ only a few per cent have them.

Table 3 summarizes all 19345 top-down distinct spectra we have observed after considering all three and four stacks counted in the last column of Table 1. The spectra are distinguished on the basis of the chiral numbers of rank-2 tensors and bi-fundamentals, the decomposition of Y, the presence and embedding of additional massless (i.e. not acquiring mass from axion couplings) $U(1)$-gauge bosons from the **a**, **b**, **c**, **d** stacks and brane unification among the **a**, **b**, **c**, **d** branes. The columns contain the following data:

- 1. The value of x. An asterisk indicates that any value is allowed. In all other cases the value of x is the one determined from the "zero Y-mass" condition.
- 2. Number of participating branes and their property:
- U: Unitary (complex)
- S: Symplectic
- R: Real (Symplectic or Orthogonal)
- N: Neutral (see below for a definition).
- 3. Composition of stack **c** in terms of branes of types C and D.
- 4. Composition of stack **d** in terms of branes of types C and D.
- 5. Total number of distinct (in the sense defined above) spectra of the type specified in the first four columns.

Table 3

Number of standard model configurations and tadpole solutions according to type. Columns **c** and **d** indicate the type of branes comprising the **c** and **d** stacks. Column "Top" contains the total number of MIPFs for which spectra of given type were found

x	Config.	**c**	**d**	Cases	Total occ.	Top	Solved
1/2	UUUU	C,D	C,D	1732	1661111	8011	110(1,0)*
1/2	UUUU	C	C,D	2153	2087667	10394	145(43,5)*
1/2	UUUU	C	C	358	586940	1957	64(42,5)*
1/2	UUU	C,D	–	2	28	2	0
1/2	UUU	C	–	7	13310	74	3(3,2)*
1/2	UUUN	C,D	–	2	60	2	0
1/2	UUUN	C	–	11	845	28	0
1/2	UUUR	C,D	C,D	1361	3242251	12107	128(1,0)*
1/2	UUUR	C	C,D	914	3697145	12294	105(72,6)*
1/2	USUU	C,D	C,D	1760	4138505	14829	70(2,0)*
1/2	USUU	C	C,D	1763	8232083	17928	163(47,5)*
1/2	USUU	C	C	201	4491695	3155	48(39,7)*
1/2	USU	C,D	–	5	13515	384	5(2,0)
1/2	USU	C	–	2	222	4	0
1/2	USUN	C,D	–	29	46011	338	2(2,0)
1/2	USUN	C	–	1	32	1	0
1/2	USUR	C,D	C,D	944	45877435	34233	130(4,0)*
1/2	USUR	C	C,D	207	49917984	11722	70(54,10)*
0	UUUU	C,D	C,D	20	7950	110	2(2,0)
0	UUUU	C	C,D	164	50043	557	8(0,0)
0	UUUU	D	C,D	5	4512	40	0
0	UUUU	C	C	1459	999122	5621	119(40,3)*
0	UUUU	C	D	26	6830	54	0
0	UUU	C	–	11	17795	225	3(3,3)*
0	UUUN	C	–	31	5989	133	0
0	UUUR	C,D	C	90	195638	702	4(4,0)
0	UUUR	C	C	4411	7394459	24715	392(112,2)*
0	UUUR	D	C	24	50752	148	0
0	UUR	C	–	8	233071	1222	6(6,0)
0	UURN	C	–	37	260450	654	4(4,0)
0	UURR	C	C	1440	12077001	15029	218(44,0)
1	UUUU	C,D	C,D	5	212	8	0
1	UUUU	C	C,D	6	7708	21	0
1	UUUU	D	C,D	4	7708	11	0
1	UUUR	C,D	D	1	1024	2	0
1	UUUR	C	D	1	640	4	0
*	UUUU	C,D	C,D	109	571472	1842	19(1,0)*
*	UUUU	C	C,D	32	521372	1199	7(7,0)
*	UUUU	D	C,D	8	157232	464	0
*	UUUU	C	D	1	4	1	0

- 6. Total number of spectra of given type. This is the grand total of all such spectra found after scanning all the three and four brane configurations in the last column of Table 1, and assigning Chan–Paton multiplicities in order to get the Standard Model gauge group and spectrum.
- 7. Total number of MIPFs for which spectra of given type were found.
- 8. Number of distinct spectra for which tadpole solutions were found. Between parenthesis we specify how may of these solutions have at most three mirror pairs, three MSSM Higgs pairs and six singlet neutrinos, and how many have no mirror pairs, at most one Higgs pairs, and precisely three singlet neutrinos. An asterisk indicates that at least one solution was found without additional hidden branes.

In column 2, "Neutral" means that this brane does not participate to Y, and that there are no chiral bi-fundamentals ending on it. The latter fact implies that there must be chiral rank-2 tensors in this brane (which in particular implies that it must be unitary), or otherwise it would violate condition 5b of the search algorithm. Such a brane can only give singlet neutrinos. We found a total of 111 such cases. They are anomaly free by having (a multiple of) $-(N-4)$ symmetric tensors and $(N+4)$ antisymmetric ones (for $N = 4$ the anti-symmetric tensors are actually real, and should strictly speaking have been omitted.) An N-brane can always be removed to get a valid three-stack model, which of course satisfies all our search criteria by itself. Note that branes of this kind are in any case allowed to exist in the hidden sector, and therefore from the point of view of classification it is most natural to view these models as three-stack models with one additional hidden sector brane. The reason we explicitly allowed them is that singlet neutrinos from separate branes might be of interest for understanding the neutrino mass problem (see also Section 7.4). In the following analysis we will omit these 111 cases.

5.3. Bottom-up versus top-down

In Tables 4 and 5 we compare the bottom-up and top-down results. This can only be done by imposing some restrictions on the spectra. In addition to three families of quarks and leptons and fully non-chiral matter (which we ignore) there can be G_{CP}-chiral matter that is G_{SM} non-chiral. The possibilities are mirror pairs of fermions, singlet neutrino's and MSSM Higgs pairs. Denote these three quantities as M, N and H. If we leave them unrestricted, there is an infinite number of bottom up solutions. Given the current experimental knowledge, the optimal values for getting the standard model would appear to be $M = 0$, $N = 3$ and $H = 1$. However, if there is a surplus of these particles, one can assume that they get a standard-model-allowed mass above the weak scale. On the other hand, if there is a shortage ($H = 0$ or $N < 3$), there still remains a possibility that the missing particles can come from G_{CP} non-chiral matter, or (in the case

Table 4

Bottom-up versus top-down results for spectra with at most three mirror pairs, at most three MSSM
Higgs pairs, and at most six singlet neutrinos. The column B-U contains the bottom-up constructions
while the column T-D contains the top-down constructions

x	Config.	**c**	**d**	B-U	T-D	Occurrences	Solved
1/2	UUUU	C,D	C,D	27	9	5194	1
1/2	UUUU	C	C,D	103441	434	1056708	31
1/2	UUUU	C	C	10717308	156	428799	24
1/2	UUUU	C	F	351	0	0	0
1/2	UUU	C,D	–	4	1	24	0
1/2	UUU	C	–	215	5	13310	2
1/2	UUUR	C,D	C,D	34	5	3888	1
1/2	UUUR	C	C,D	185520	221	2560681	31
1/2	USUU	C,D	C,D	72	7	6473	2
1/2	USUU	C	C,D	153436	283	3420508	33
1/2	USUU	C	C	10441784	125	4464095	27
1/2	USUU	C	F	184	0	0	0
1/2	USU	C	–	104	2	222	0
1/2	USU	C,D	–	8	1	4881	1
1/2	USUR	C	C,D	54274	31	49859327	19
1/2	USUR	C,D	C,D	36	2	858330	2
0	UUUU	C,D	C,D	5	5	4530	2
0	UUUU	C	C,D	8355	44	54102	2
0	UUUU	D	C,D	14	2	4368	0
0	UUUU	C	C	2890537	127	666631	9
0	UUUU	C	D	36304	16	6687	0
0	UUU	C	–	222	2	15440	1
0	UUUR	C,D	C	3702	39	171485	4
0	UUUR	C	C	5161452	289	4467147	32
0	UUUR	D	C	8564	22	50748	0
0	UUR	C	–	58	2	233071	2
0	UURR	C	C	24091	17	8452983	17
1	UUUU	C,D	C,D	4	1	1144	1
1	UUUU	C	C,D	16	5	10714	0
1	UUUU	D	C,D	42	3	3328	0
1	UUUU	C	D	870	0	0	0
1	UUUR	C,D	D	34	1	1024	0
1	UUUR	C	D	609	1	640	0
3/2	UUUU	C	D	9	0	0	0
3/2	UUUU	C,D	D	1	0	0	0
3/2	UUUU	C, D	C	10	0	0	0
3/2	UUUU	C,D	C,D	2	0	0	0
*	UUUU	C,D	C,D	2	2	5146	1
*	UUUU	C	C,D	10	7	521372	3
*	UUUU	D	C,D	1	1	116	0
*	UUUU	C	D	3	1	4	0

Table 5

Bottom-up versus top-down results for spectra without mirror pairs, at most one MSSM Higgs pair, and precisely three singlet neutrinos. Only cases that have been found in the top-down search are shown

x	Config.	c	d	B-U	T-D	Occurrences	Solved
1/2	UUU	C	–	8	2	13242	1
1/2	UUUU	C	C	10670	16	81985	4
1/2	UUUU	C	C,D	148	8	378418	3
1/2	UUUR	C	C,D	495	13	641485	3
1/2	USUU	C	C,D	314	6	2757164	3
1/2	USUU	C	C	10816	6	4037872	4
1/2	USUR	C	C,D	434	3	47689675	3
0	UUUU	C	C,D	23	1	6	0
0	UUUU	C	C	1996	5	17301	2
0	UUUU	C	D	91	4	4227	0
0	UUU	C	–	9	1	15282	1
0	UUUR	C	C	5136	15	63051	1

of neutrinos) from additional branes (other than **a**, **b**, **c** or **d**). Note for example that most of the models of [17] have no G_{CP}-chiral Higgses, but usually a large number of fully non-chiral Higgs candidates. Since we have to impose cuts on M, N and H to make the comparison, we present the comparison for two cases: a loose cut (with $M \leq 3$, $N \leq 6$, $H \leq 3$) and a tight cut ($M = 0$, $H \leq 1$ and $N = 3$). The former comparison is in Table 4 and the latter in Table 5. In both tables, the number of bottom-up configurations satisfying the criteria is listed in column 5. In column 6, we list the number of those bottom-up configurations that was encountered in our search, and in column 7 the total number of occurrences of the given class[13] of configurations, summed over all three or four brane combination considered in the search. This is the same information as in column 6 of Table 3, but with the limit on the numbers M, N and H imposed. In column 8 we list the number of distinct configurations for which the tadpole conditions were solved. In these tables the top-down spectra are only distinguished on the basis of criteria that can be directly compared to the bottom-up approach. Brane unification is ignored and the masses of $U(1)$ vector bosons are not taken into account. This means that some models that were distinct in the previous table are considered identical here, because they merely differ by branes that are not on top of each other, or by different embeddings of an additional massless $U(1)$ factor. This affects column 6 and column 8, but not column 7, which is simply the sum of all occurrences within the class. Note for example the in the class ($x = *$, UUUU, **c** = C, **d** = (C,D)) there is a total number of occurrences of

---

[13]By "class" we mean here all brane configurations that match the criteria in the first four columns.

521372 in both tables. This implies that all models satisfy the constraints on the number of Higgs, mirrors and neutrinos. In Table 1 these models correspond to 32 distinct cases with 7 distinct solutions, whereas in Table 4 they form only 7 distinct models with 3 distinct solutions.

Some bottom-up solutions can exist for more than one value of Y. The most obvious example is the class $x = *$, which can exist for all values of Y. In making the comparison we have used the actual massless linear combination of Y allowed by the axion-gauge boson couplings in the top-down Gepner model. Only for the $x = *$ case we have ignored the precise form of Y, because this would split this class into an indefinite number of subclasses. However, in those cases where Y was of the form corresponding to $x = 0, \frac{1}{2}$ or 1, we have compared those top-down models twice: once in the $x = *$ class, and once in the class given by Y. This explains the tadpole solution indicated in the last column of Table 4 for an $x = 1$ model. Actually, this model has $x = *$, but x is fixed to 1 by the Y-mass condition.

The bottom-up numbers in these tables cannot be directly compared with those in Section 4 because here we allow several branes of types C and D on the same stack, whereas in Section 4 we assumed that stack **c** consists only of a single type-C brane, and stack **d** of a single type-D brane. Furthermore in Section 4 both G_{CP} chiral and G_{CP} non-chiral Higgses are counted. We do not do that here because the top-down search G_{CP} non-chiral Higgses were ignored.

Table 6 contains all 19345 distinct models we found. Unfortunately the full table would be more than 500 pages, and is too long to include, so we have only displayed the top and some entries of interest.[14] The table is ordered according to the total number of occurrences (listed in column 2) of a given spectrum. Column 3 gives the number of MIPFs for which it occurs. This gives some more indication how rare a certain spectrum is. In column 4 we give the Chan–Paton group, with factors combined if some of the branes are on the same position. In column 5 we give a rough indication of the spectrum. Here "V" means that a CP-factor only contributes bi-fundamentals, "S"("A") that there is at least one (anti-)symmetric tensor and "T" that both occur. Column 6 gives the value of x, and the last column indicates if a solution to the tadpole conditions was found ("Y"), and if a solution was found without additional branes ("Y!").

The first 25 models are all relatives of the $U(3) \times Sp(2) \times U(1) \times U(1)$ models that dominated the search results of [17]. The variations include replacing the third factor by $O(2)$ or $Sp(2)$, absorbing the family multiplicity of some of the quarks or leptons in the Chan–Paton multiplicities of the **c** and **d** branes, unifying the baryon and lepton brane to get a Pati–Salam-like structure, and other brane unifications. Models 17 and 18 occur with the same frequency because they are

---

[14]However, the full list is available on request.

Table 6

The list of 19345 models sorted according to frequency. The column "occ." tabulates the total number of occurrences, column "mipf" tabulates the number of MIPFs at which they were found, column "spec." tabulates the type of spectrum they carry, and column "S" indicates whether a tadpole solution was found

Nr.	occ.	mipf	Chan–Paton group	spec.	x	S
1	9801844	648	$U(3) \times Sp(2) \times Sp(6) \times U(1)$	VVVV	1/2	Y!
2	8479808	675	$U(3) \times Sp(2) \times Sp(2) \times U(1)$	VVVV	1/2	Y!
3	5775296	821	$U(4) \times Sp(2) \times Sp(6)$	VVV	1/2	Y!
4	4810698	868	$U(4) \times Sp(2) \times Sp(2)$	VVV	1/2	Y!
5	4751603	554	$U(3) \times Sp(2) \times O(6) \times U(1)$	VVVV	1/2	Y!
6	4584392	751	$U(4) \times Sp(2) \times O(6)$	VVV	1/2	Y
7	4509752	513	$U(3) \times Sp(2) \times O(2) \times U(1)$	VVVV	1/2	Y!
8	3744864	690	$U(4) \times Sp(2) \times O(2)$	VVV	1/2	Y!
9	3606292	467	$U(3) \times Sp(2) \times Sp(6) \times U(3)$	VVVV	1/2	Y
10	3093933	623	$U(6) \times Sp(2) \times Sp(6)$	VVV	1/2	Y
11	2717632	461	$U(3) \times Sp(2) \times Sp(2) \times U(3)$	VVVV	1/2	Y!
12	2384626	560	$U(6) \times Sp(2) \times O(6)$	VVV	1/2	Y
13	2253928	669	$U(6) \times Sp(2) \times Sp(2)$	VVV	1/2	Y!
14	1803909	519	$U(6) \times Sp(2) \times O(2)$	VVV	1/2	Y!
15	1676493	517	$U(8) \times Sp(2) \times Sp(6)$	VVV	1/2	Y
16	1674416	384	$U(3) \times Sp(2) \times O(6) \times U(3)$	VVVV	1/2	Y
17	1654086	340	$U(3) \times Sp(2) \times U(3) \times U(1)$	VVVV	1/2	Y
18	1654086	340	$U(3) \times Sp(2) \times U(3) \times U(1)$	VVVV	1/2	Y
19	1642669	360	$U(3) \times Sp(2) \times Sp(6) \times U(5)$	VVVV	1/2	Y
20	1486664	346	$U(3) \times Sp(2) \times O(2) \times U(3)$	VVVV	1/2	Y!
21	1323363	476	$U(8) \times Sp(2) \times O(6)$	VVV	1/2	Y
22	1135702	350	$U(3) \times Sp(2) \times Sp(2) \times U(5)$	VVVV	1/2	Y!
23	1050764	532	$U(8) \times Sp(2) \times Sp(2)$	VVV	1/2	Y
24	956980	421	$U(8) \times Sp(2) \times O(2)$	VVV	1/2	Y
25	950003	449	$U(10) \times Sp(2) \times Sp(6)$	VVV	1/2	Y
26	910132	51	$U(3) \times U(2) \times Sp(2) \times O(1)$	AAVV	0	Y
...						
34	869428	246	$U(3) \times Sp(2) \times U(1) \times U(1)$	VVVV	1/2	Y!
153	115466	335	$U(4) \times U(2) \times U(2)$	VVV	1/2	Y
225	71328	167	$U(3) \times U(3) \times U(3)$	VVV	1/3	
303	47664	18	$U(3) \times U(2) \times U(1) \times U(1)$	AAVA	1/2	Y
304	47664	18	$U(3) \times U(2) \times U(1) \times U(1)$	AAVA	0	Y
343	40922	63	$U(3) \times Sp(2) \times U(1) \times U(1)$	VVVV	1/2	Y!
411	31000	17	$U(3) \times U(2) \times U(1) \times U(1)$	AAVA	0	Y
417	30396	26	$U(3) \times U(2) \times U(1) \times U(1)$	AAVS	0	Y
495	23544	14	$U(3) \times U(2) \times U(1) \times U(1)$	AAVS	0	
509	22156	17	$U(3) \times U(2) \times U(1) \times U(1)$	AAVS	0	Y
519	21468	13	$U(3) \times U(2) \times U(1) \times U(1)$	AAVA	0	Y
543	20176(*)	38	$U(3) \times U(2) \times U(1) \times U(1)$	VVVV	1/2	Y
617	16845	296	$U(5) \times O(1)$	AV	0	Y
671	14744(*)	29	$U(3) \times U(2) \times U(1) \times U(1)$	VVVV	1/2	
761	12067	26	$U(3) \times U(2) \times U(1)$	AAS	1/2	Y!
762	12067	26	$U(3) \times U(2) \times U(1)$	AAS	0	Y!

(continued on next page)

Table 6

(continued)

Nr.	occ.	mipf	Chan–Paton group	spec.	x	S
1024	7466	7	$U(3) \times U(2) \times U(2) \times U(1)$	VAAV	1	
1125	6432	87	$U(3) \times U(3) \times U(3)$	VVV	*	Y
1201	5764(*)	20	$U(3) \times U(2) \times U(1) \times U(1)$	VVVV	1/2	
1356	5856(*)	10	$U(3) \times U(2) \times U(1) \times U(1)$	VVVV	1/2	Y
1725	2864	14	$U(3) \times U(2) \times U(1) \times U(1)$	VVVV	1/2	Y
1886	2381	115	$U(6) \times Sp(2)$	AV	1/2	Y!
1887	2381	115	$U(6) \times Sp(2)$	AV	0	Y!
1888	2381	115	$U(6) \times Sp(2)$	AV	1/2	Y!
2624	1248	3	$U(3) \times U(2) \times U(2) \times U(3)$	VAAV	1	
2880	1049	34	$U(5) \times U(1)$	AS	1/2	Y!
2881	1049	34	$U(5) \times U(1)$	AS	0	Y!
2807	1096(*)	8	$U(3) \times U(2) \times U(1) \times U(1)$	VVVV	1/2	
2919	1024	2	$U(3) \times U(2) \times U(2) \times O(3)$	VAAV	1	
4485	400(*)	2	$U(3) \times U(2) \times U(1) \times U(1)$	VVVV	1/2	
4727	352	3	$U(3) \times U(2) \times U(1) \times U(1)$	VVVV	1/2	
4825	332	20	$U(4) \times U(2) \times U(2)$	VAS	1/2	Y!
4902	320(*)	1	$U(3) \times U(2) \times U(1) \times U(1)$	VVVV	1/2	Y
4996	304	30	$U(3) \times Sp(2) \times U(1) \times U(1)$	VVVV	1/2	Y
6993	128(**)	1	$U(3) \times U(2) \times U(2) \times U(1)$	VVVV	1/2	
7053	124	4	$U(3) \times U(2) \times U(2) \times U(1)$	VASV	1/2	Y!
7241	116(**)	4	$U(3) \times U(2) \times U(2) \times U(1)$	VVVV	1/2	
7280	114	3	$U(3) \times Sp(2) \times U(1)$	AVS	1/2	
7464	108	1	$U(3) \times Sp(2) \times U(1)$	VVT	1/2	
7905	96(*)	1	$U(3) \times U(2) \times U(1) \times U(1)$	VVVV	1/2	
8747	68(**)	3	$U(3) \times U(2) \times U(1) \times U(1)$	VVVV	1/2	
8773	68	4	$U(3) \times U(2) \times U(1) \times U(1)$	VVVV	1/2	
11347	32(**)	1	$U(3) \times U(2) \times U(1) \times U(1)$	VVVV	1/2	
11462	32(*)	1	$U(3) \times U(2) \times U(1) \times U(1)$	VVVV	1/2	
12327	24	1	$U(3) \times U(3) \times U(3)$	VVV	1/2	
15824	8	1	$U(3) \times U(2) \times U(1) \times U(1)$	VVVV	0	
15846	8	1	$U(3) \times U(2) \times U(1) \times U(1)$	VVVV	1/2	
16674	6	1	$U(3) \times U(2) \times U(1)$	AVT	1/2	Y!
17055	4	1	$U(3) \times U(2) \times U(1) \times U(1)$	VVVV	*	
19345	1	1	$U(5) \times U(2) \times O(3)$	ATV	0	

closely related. They only differ by a traceless generator diag$(\frac{1}{3}, \frac{1}{3}, -\frac{2}{3})$ from the $U(3)$ factor contributing to Y, changing the distribution of some quarks and leptons. There are several other cases of closely related models with identical frequencies, and one such set, nrs. 1886 . . . 1888 will be discussed in more detail in Section 6.5. In the bottom part of the table we display several lines of special interest, which will be discussed in more detail below.

Entry nr. 26 in the table is the first one that cannot be viewed as a relative of the "Madrid model". It has $x = 0$ and three anti-symmetric tensors on the QCD and the weak brane. It can be viewed as a broken $SU(5)$ model.

There exist several infinite series of models. In the top of the list one can observe the beginning of the series $U(2n) \times Sp(2) \times G, n > 2$, where G can be $O(2)$, $O(6)$, $Sp(2)$ or $Sp(6)$, with a chiral spectrum consisting of $\frac{6}{N_c}(V, 0, V) + 3(V, V, 0)$.

In column 2 we indicate between parentheses if a certain type of model was searched for in [17], and how often it was found. It is interesting to compare this with Table 1. Observe that the number of four-stack configurations we consider in the present paper is considerably smaller than in [17], but nevertheless we recover a large fraction of the standard model configurations of that paper. For example, in [17], 2.8×10^{15} configurations of type USUS were examined, in the present paper only 26×10^{14}, ten times less. Nevertheless, we have already found about half of the standard model configurations. This is because the number of brane configurations is dominated by cases with a large number of branes, but very few standard model spectra. This in particular true for the charge conjugation invariant (the simplest case, for which the boundary coefficients were derived by Cardy [44]) which in essentially all cases has by far the largest number of boundaries. The explanation may be that a non-trivial MIPF tends to fold over a Calabi–Yau manifold several times, thus increasing the typical intersection numbers, and causing the number three to occur more frequently.

There are in total three cases with an $SU(3) \times Sp(2) \times U(1) \times U(1)$ Chan–Paton group and only bi-fundamentals, namely nr. 30, nr. 343 and nr. 4996. The first two were also searched for in [17], and we find most of them back. They are distinguished by having a massless (nr. 30) or massive (nr. 343) $B - L$ gauge boson. The third one differs in the way quarks and leptons end on branes **c** and **d**. It does not have a lepton number symmetry, and was not considered in [17]. We show this case in more detail in the next section, as a curiosity.

The remaining models considered in [17] have a $U(2)_\mathbf{b}$ group instead of $Sp(2)_\mathbf{b}$. Here a direct comparison is harder, because this splits into many subclasses, which differ in the way the doublets are divided into (2) and (2*) representations of $U(2)$. The cases indicated by a single (*) are models considered in [17] that have a massless $B - L$ boson. In total 131704 such configurations were found in that paper. For three of them we found tadpole solutions; they correspond to the three "type-1" models in table 4 of [17]. The ones indicated by (**) have a massive $B - L$ boson. Only 1306 of these were found in [17], and in no case the tadpole conditions could be solved.

Perhaps the most standard Chan–Paton group for standard model realizations is $U(3) \times U(2) \times U(1) \times U(1)$. The total number of spectra with that CP-group on the complete list is 281. Of these, 19 have a purely bi-fundamental spectrum, and among these 19 there are 17 with $x = \frac{1}{2}$, one with $x = 0$ and one with $x = *$. Of the 17 $x = \frac{1}{2}$ models, 13 are variations on the "Madrid" model, discussed above. The fourth $x = \frac{1}{2}$ model with a tadpole solution is discussed below in Section 6.5. All these 19 purely bi-fundamental models are shown in Table 6. In addition we show all $U(3) \times U(2) \times U(1) \times U(1)$ configurations that occur more frequently than the first purely bi-fundamental model, nr. 543. These are models with anti-symmetric $U(3)$ tensors. Note that they occur more frequently despite the fact that models with rank-2 tensors are suppressed, as will be discussed below. All of them are broken $SU(5)$ models, except nr. 303, which is a broken flipped $SU(5)$ variation of nr. 304.

5.4. Standard model brane configurations not found

Note that only a very small fraction of the allowed bottom-up models is actually realized as top-down configurations.[15] This can be explained in part by the fact that the bottom up models can have several chiral tensors instead of chiral bi-fundamentals. In Fig. 1 we plot the distribution of the number of standard model top-down configurations we have found versus the total number of chiral tensors in the spectrum. This distribution is sharply peaked at zero. This implies that models in which some quarks and leptons are realized as rank-2 tensors are considerably harder to find in the part of the landscape we are exploring here. In itself, this does not mean much for the actual realization of the standard model in our universe. After all, the suppression of models with tensors is by several

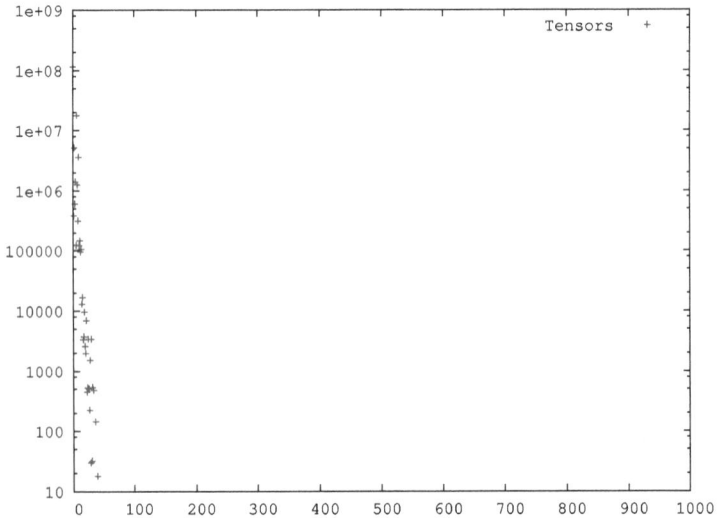

Fig. 1. Chiral tensor distribution for all standard model configurations.

[15] All results in this section concern brane configurations prior to tadpole cancellation.

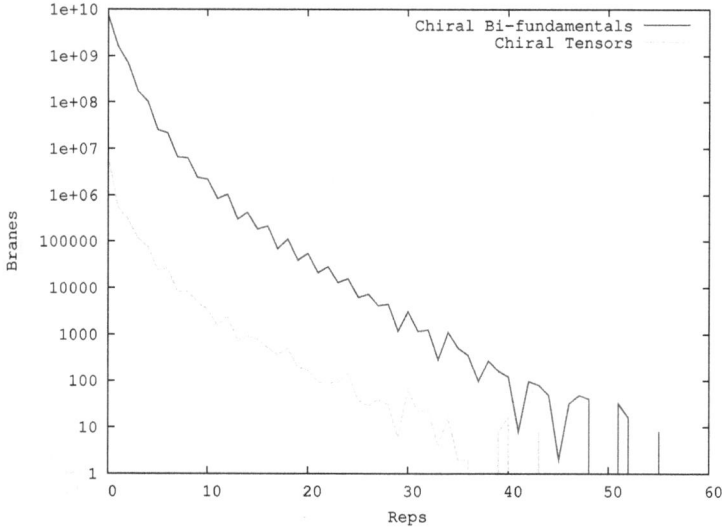

Fig. 2. Number of chiral tensors and bi-fundamentals for a selection of branes.

factors of ten only, and this does not seem very significant in comparison to the total number of models in the landscape.

A partial understanding of this strong chiral tensor suppression can be gained as follows. In Fig. 2 we plot for all branes of a sample of 18001 orientifolds the distribution of chiral bi-fundamentals and chiral tensors. On the horizontal axis is the absolute value of the chirality, and on the vertical axis the total number of occurrences. Clearly—and not unexpectedly—the number bi-fundamentals is much greater than the number of chiral tensors. This can be intuitively understood by realizing that a brane has a much bigger chance intersecting with *any* brane yielding a bi-fundamental than intersecting with one specific brane (namely itself), yielding a chiral tensor.

One can also make an interesting observation regarding the occurrence of chiral tensors in comparison to non-chiral ones. In Fig. 3 we list for all branes in all 33012 non-zero tension orientifolds the distribution of chiral and non-chiral tensors (separately for adjoints and the other rank-2 tensors). Note that this includes *all* branes in all Gepner orientifolds with non-zero-tension O-planes, not just those considered in the present paper. Clearly the chiral distribution falls off much faster than the non-chiral ones.

Although some other qualitative observations can be made, we do not have a good understanding of the absence of certain models. Hypercharge embeddings with $x = -1/2, 3/2$ were not found at all. The full list of 19345 configurations

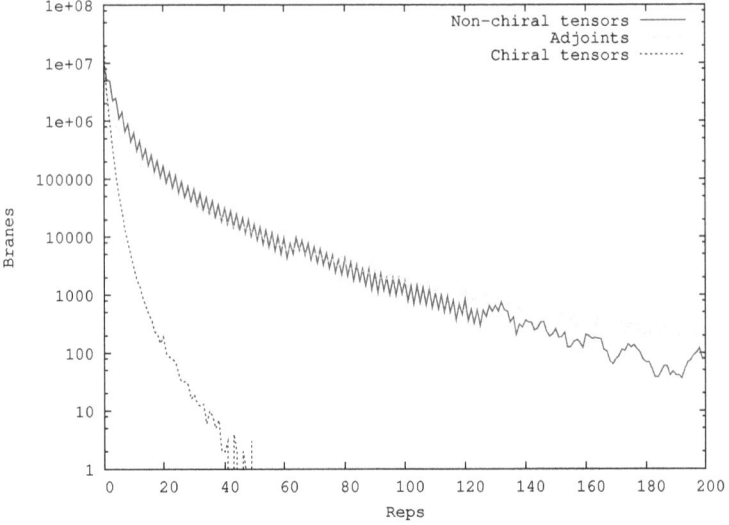

Fig. 3. Number of chiral and non-chiral tensors for all single branes.

does contain some genuine $x = 1$ models, with x fixed to that value by the quark and lepton charges. There is a total of 17 distinct ones (for none of these we found a solution to the tadpole conditions). Only one of these, nr. 2919, has an orthogonal group on the **d**-stack, but it is not identical to one of the simple models written down in Section 3.5. It has a Chan–Paton group $U(3) \times U(2) \times U(2) \times O(3)$, with both a C and a D brane on stack **c**. This model was found a total of 1024 times for just two MIPFs. The purely unitary $x = 1$ models 1024 and 2624 occur more frequently. Another noteworthy absence in this class is the type B,B' model introduced in [8, 10]. These models have a Chan–Paton group $U(3) \times U(2) \times U(1) \times U(1)$, and the type-B model only has bifundamentals, whereas type-B' has anti-symmetric tensor on $U(2)_{\mathbf{b}}$. However, all $x = 1$ models we found have a $U(2)$ group on brane **c**, and all have anti-symmetric tensors both on branes **b** and **c**. Some of these are similar to the models of [8, 10], but not identical. Note that the type B,B' models of [8, 10], in to order to be free of cubic anomalies in the two $U(1)$ factors and the $U(2)$, need $U(2)_{\mathbf{b}}$-chiral Higgs pairs and anti-symmetric $U(1)$ tensors, as discussed in Section 3.5. This suppresses their statistical likelihood.

Another model proposed in the literature that did not emerge in our search is model C of [25]. This is a $U(3) \times U(2) \times U(1)$ model with three G_{CP}-chiral neu-trinos appearing as anti-symmetric tensors of $U(2)$. However, model nr. 7464 in Table 6 is similar to it. It has exactly the same structure as model C of [25], after replacing $U(2)$ by $Sp(2)$. Then such neutrinos necessarily become non-chiral, and the anomaly cancellation condition for the $U(2)$ factor becomes irrelevant, increasing the chances of finding an example. Model nr. 7464 occurred only 108 times (and without tadpole solutions). Its presence suggests that there is no fundamental obstacle to finding model C, but that it is simply statistically disfavored. In other situations, replacing $U(2)$ by $Sp(2)$ increases the number of occurrences by factors of about 40 to 80, and hence we would expect at most a few examples of model C. This is consistent with finding none.

On the full list of 19345 models there are 150 of the class $x = *$. All of them are truly orientable, i.e. the possibility of having anti-symmetric $U(1)$ tensors that do not contribute massless states does not occur. Only one has Chan–Paton group $U(3) \times U(2) \times U(1) \times U(1)$. It is indeed precisely the model (3.6) shown in Section 3. Amazingly this simple model occurs only four times (nr. 17055), and just for one MIPF (and without any tadpole solution to tadpole cancellation). This is especially surprising since there are many other $U(3) \times U(2) \times U(1) \times U(1)$ configurations with only bi-fundamentals that do occur much more frequently, as discussed above. For example nr. 543 in Table 6 occurs 20176 times. This is a standard "Madrid"-type configuration.

5.5. *Higgs, neutrino and mirror distributions*

Figures 4, 5 and 6 and show the distribution in terms of the number of Higgs, right-handed neutrinos and mirror pairs. On the vertical axis we show the total number of three and four-brane configurations that have a chiral standard model spectrum, plus the number of Higgses/neutrinos/mirrors indicated on the horizontal axis. Just as all data in this section, these numbers refer to brane configurations prior to tadpole cancellation. The Higgses/neutrinos/mirrors are G_{CP} chiral but of course G_{SM} non-chiral. In addition to these particles, the massless spectrum may contain G_{CP}-non-chiral particles with the same standard model transformation properties. Since we classify models modulo full non-chiral matter, we have no general information about such particles. The mirror count is the total of all mirror pairs of quark and charged lepton weak singlets, as well as quark doublets (in this case mirrors can occur only for $x = \frac{1}{2}$). The Higgs count refers to $(1, 2, \frac{1}{2}) + (1, 2, -\frac{1}{2})$ pairs; for example the MSSM has one such pair. Note that these pairs could also be viewed as lepton doublet mirror pairs. The distinction can be made in models with a well-defined lepton number, but since we are not insisting on that we simply count all such pairs as candidate Higgs. Once one (or more) of these candidates acquires a v.e.v, one may discuss if lepton number violation is absent or acceptably small.

Finally Fig. 5 shows the distribution of the total number of standard model singlets in the G_{CP}-chiral spectrum.

E. Kiritsis

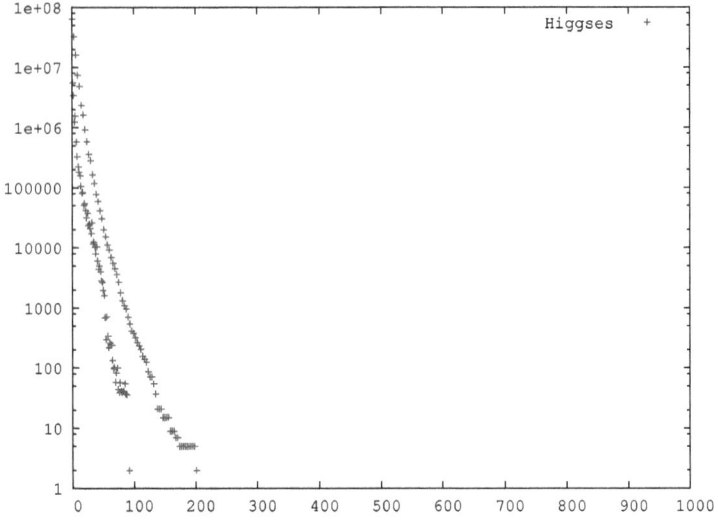

Fig. 4. Higgs pair distribution for all standard model configurations.

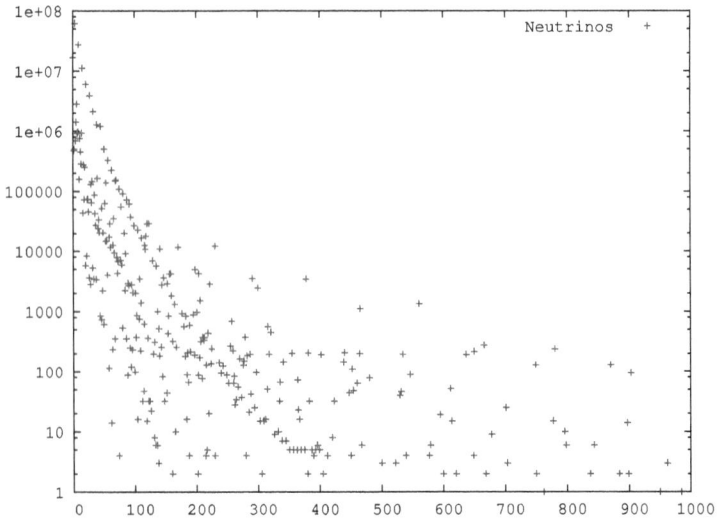

Fig. 5. Right-handed neutrino distribution for all standard model configurations.

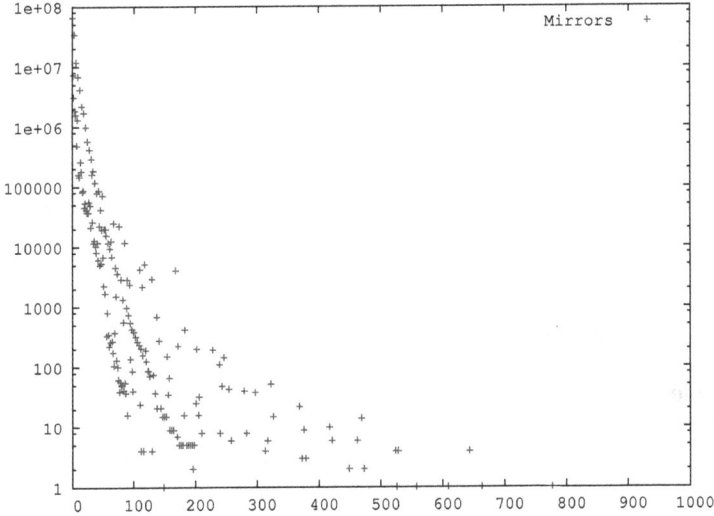

Fig. 6. Mirror distribution for all standard model configurations.

In all three plots two lines are visible. The top line corresponds to multiplicities that are 0 mod 3, and the lower to multiplicities that are not 0 mod 3. The former occur more frequently due to anomaly cancellation and the fact that we require the presence of three chiral families. In some classes of models this imposes a mod 3 constraint on the multiplicities of Higgses, mirror or neutrinos. This feature is clearest in the Higgs plot, because the Higgs is in a definite, and non-trivial standard model representation with few G_{CP} realizations. It is less clear in the neutrino plot, because there are often many ways of making neutrinos. The models with huge numbers of (right-handed) neutrino candidates usually contain a large factor G_c or G_d, with neutrinos coming from rank-2 tensors.

6. Solutions to the tadpole conditions

In this section we present some examples of solutions to complete set of tadpole solutions that we have found. All solutions that we present also satisfy the probe brane constraints for the absence of global anomalies [45], as discussed in [46] for this class of models. We emphasize that we have collected at most two tadpole solutions for each chiral model, one with additional branes, and one without additional branes. This means, for example, that as soon as one solution was found for one of the 9785532 $SU(3) \times SU(2) \times Sp(6) \times U(1)$ models that appears as nr. 1 in Table 6, no further attempt was made for any of the others with the same chiral spectrum. This is a very different strategy than the one of [17], where all tadpole solutions were collected for models with distinct *non*-chiral spectra. In the examples below we present the full massless spectrum of the actual tadpole solution, including non-chiral states. The non-chiral states are however specific to the example we present, and solutions with different non-chiral multiplicities for a given chiral multiplicity certainly exist. Indeed, for spectrum nr. 2 in Table 6, which was included in the search presented in [17], more than 100000 non-chirally distinct samples with tadpole solutions were found.

We only present a small selection of the 1900 tadpole solution we have collected. They should be viewed merely as existence proofs of a certain type of model, and not as a statement that one of these is likely to survive further phenomenological constraints. Whenever possible, we present examples without hidden branes, not because we believe these are more viable (indeed, hidden sector branes may be required for a variety of phenomenological reasons), but simply because they can be written down more easily.

6.1. Hypercharge embeddings of the tadpole solutions

Let us first make a few more comments on the models that do or do not occur in the list of 1900 tadpole solutions. We have seen in the previous section that most bottom-up models of Section 3 and 4 do not occur on the list of brane configurations, and it is therefore clear that most are also absent from the list of tadpole solutions (see Section 5.4). Furthermore, in many top-down tadpole solutions, the hypercharge appears to be a combination of more than one of the hypercharge embeddings of the bottom-up models in Section 4. First consider the "pure" models

- 762 top-down configurations have hypercharge of the form $Y = -\frac{1}{3}Q_{\mathbf{a}} - \frac{1}{2}Q_{\mathbf{b}}$. This is related to a small subclass of bottom-up models in Section 4.2. In Table 4 these models have $x = 0$ and both **c, d** branes are of the C type (or are real, or absent).

- 1095 top-down configurations have hypercharge of the form $Y = -\frac{1}{6}Q_{\mathbf{a}} + \frac{1}{2}Q_{\mathbf{c}} - \frac{1}{2}Q_{\mathbf{d}}$ which is related to a subclass of the bottom-up models in Section 4.2.

The rest of the configurations appear with the hypercharge to be described by two different embeddings. This is due to the contribution of traceless generators in the hypercharge. These "mixed" models are distributed as follows:

- 17 top-down configurations have a combined hypercharge of the type: $Y = -\frac{1}{3}Q_{\mathbf{a}} - \frac{1}{2}Q_{\mathbf{b}} + Q_{\mathbf{d}}$ (Section 4.2 and corresponding to models A,A' in [8,10,47]) and $Y = -\frac{1}{3}Q_{\mathbf{a}} - \frac{1}{2}Q_{\mathbf{b}}$ (Section 4.2). These are hypercharges with $x = 0$ but **c** and **d** branes are of the type C and D, C respectively.

- 2 top-down configurations appear with a combined hypercharge with $x = 1$ (Section 4.2 and corresponding to models B,B' in [8, 10, 47]), but **c** and **d** branes are of the type C, D and D respectively.

Here we used the hypercharge values as determined from the quark and lepton charges as well as the Y-mass condition. The two mixed $x = 1$ models mentioned above actually have $x = *$, with x fixed to 1 by the Y-mass condition. One of those appears in Table 4; the other has too many neutrinos and lies outside the limits used for that table. A total of 20 out of the 1900 tadpole solutions have $x = *$, but x fixed to a non-canonical value by the Y-mass condition. Finally there are 4 with x completely unfixed by any condition.

6.2. Notation

The notation of the examples is as follows. Minimal model tensor products are denoted as $(k_1, \ldots, k_m)$, where k_i is the $SU(2)$ level. Their modular invariant partition functions are labelled by an integer, which is assigned sequentially as they are computed. This labelling can be resolved in terms of more precise data: the simple current subgroup and the rational matrix X defining the MIPF (as defined in [14]). We omit these data here, but they are available on request. To help identify the MIPF we will provide the Hodge numbers of the corresponding Calabi–Yau manifold, and the number of singlets that occur in the spectrum of heterotic strings compactified on such a manifold. Orientifolds are also labelled by a sequential integer assigned by the computer program.

Representations are denoted as $(r_a, \ldots r_d, \ldots)$, where each entry refers to one of the branes (**a**, **b**, **c**, **d** and hidden), and r can be V for vector, A for anti-symmetric tensor, S for symmetric tensor and Adj for Adjoint. An asterisk indicates complex conjugation. All representations refer to left-handed fermions. Multiplicities of complex representations are denoted as

$$N \times (r_a, \ldots)_M,$$

where N is the total number of times a representation plus its conjugate appears, and M is the chirality, the difference of the multiplicity of the representation that is listed, and its conjugate. The subscript is omitted for non-chiral representations.

6.3. $U(3) \times U(2) \times U(1)$ models

Here we list all tadpole solutions we found with a Chan–Paton group which is exactly $U(3) \times U(2) \times U(1)$ (or less, if some combinations of the unitary phase factors—other than Y—get a mass from axion couplings).

The first two examples are nr. 761 and 762 from the list. They are respectively broken versions of $SU(5)$ and flipped $SU(5) \times U(1)$ unifications, with $SU(5)$

broken by splitting the stack of five branes into three plus two. These models occurred for MIPF 31 of $(1, 1, 1, 1, 7, 16)$ (there is just one orientifold choice). The $U(3) \times U(2) \times U(1)$ spectrum is

$$
\begin{array}{rcl}
3 & \times & (A, 0, 0)_3, \\
3 & \times & (0, A, 0)_3, \\
5 & \times & (V, V, 0)_3, \\
25 & \times & (0, 0, S)_3, \\
9 & \times & (V, 0, V)_{-3}, \\
3 & \times & (0, V, V)_{-3}, \\
4 & \times & (Ad, 0, 0), \\
1 & \times & (0, Ad, 0), \\
16 & \times & (0, 0, Ad), \\
6 & \times & (0, 0, A), \\
8 & \times & (S, 0, 0), \\
14 & \times & (V, 0, V^*), \\
4 & \times & (0, V, V^*).
\end{array}
$$

The possible choices for Y are the $SU(5)$ embedding $Y = -\frac{1}{3}Q_a + \frac{1}{2}Q_b$ and the flipped embedding $Y = \frac{1}{6}Q_a + \frac{1}{2}Q_c$ for nr. 562 and 561 respectively. In both cases an additional $U(1)$, the independent linear combination of these two, also remains massless.

There is a second, far less standard example of a $U(3) \times U(2) \times U(1)$ model, which occurred for invariant 28 of 441010, orientifold 0. This is nr. 16674 on the list, which occurred only six times in total (and only for this MIPF), but against all odds a tadpole solution was found for at least one of the six occurrences. The embedding of Y is as for the flipped $SU(5)$ model above, but only two of the three down quarks are due to anti-symmetric tensors, and there are no anti-symmetric tensors in $U(2)$. Furthermore there are three candidates for Higgs bosons, but unfortunately no symmetry like lepton number to distinguish them from the lepton doublets. This implies that there are no singlet neutrino candidates from the standard model branes, and that with a suitable Higgs boson chosen from the three candidates mentioned above, all up quarks and one of the down quarks can acquire a mass. The exact spectrum is as follows

$$
\begin{array}{rcl}
9 & \times & (0, V, V^*)_{-3}, \\
3 & \times & (0, 0, S)_3, \\
6 & \times & (A, 0, 0)_2,
\end{array}
$$

$3 \quad \times \quad (0, 0, A)_1,$

$6 \quad \times \quad (0, V, V)_{-6},$

$7 \quad \times \quad (V, V, 0)_3,$

$7 \quad \times \quad (V, 0, V^*)_{-1},$

$3 \quad \times \quad (V, 0, V)_{-3},$

$3 \quad \times \quad (Ad, 0, 0),$

$6 \quad \times \quad (0, A, 0),$

$7 \quad \times \quad (0, Ad, 0),$

$8 \quad \times \quad (0, S, 0),$

$8 \quad \times \quad (V, V^*, 0),$

$4 \quad \times \quad (0, 0, Ad).$

The gauge group is exactly $SU(3) \times SU(2) \times U(1)$, because all Abelian gauge bosons other than Y acquire a mass.

Somewhat surprisingly, there were no tadpole solutions for $U(3) \times Sp(2) \times U(1)$ models, even though usually replacing $U(2)$ by $Sp(2)$ greatly increases the frequency of a model.

6.4. Unification

In general we can speak of (partial) unification if some of the stacks **a**, **b**, **c** and **d** coincide. One can distinguish the following possibilities

1. **a = b**. In this case the bi-fundamentals that yield quark doublets must necessarily come from anti-symmetric tensors on the combined stack. There must therefore be three anti-symmetric tensors, and the combined gauge group is $U(5)$. Hence this leads to $SU(5)$ GUT models. The $SU(3)$ anti-symmetric tensors can be u^c or d^c quarks. The first case corresponds to standard $SU(5)$, the second to flipped $SU(5)$. There must be at least one more brane stack to accommodate the anti-quarks of the other charge. Hence these models can be realized with just two stacks.

2. **a = c**. In this case the weak brane remains separate, but the QCD brane is extended. The best-known example is the Pati–Salam model, where $U(3)_\mathbf{a}$ is extended with a lepton-number $U(1)$. The Pati–Salam model requires three stacks, but it is possible to realize unifications of this type with just two stacks. An example is (one of the variations of) the $U(6) \times Sp(2)$ discussed below.

3. **b = d**. In this case the weak brane is part of a larger group. An example is trinification: here $U(2)_\mathbf{b}$ is embedded in a $U(3)$. Without loss of generality, we may choose stack **d** as the one that merges with the weak brane. The

trinification model then needs one additional brane stack, $U(3)_c$. All models in this class must in fact have a third brane stack, in order to get anti-quarks as bi-fundamentals; at least one of the two anti-quarks charges must be realized as a bi-fundamental.

4. $\mathbf{a} = \mathbf{b} = \mathbf{d}$. An example will be given below.

Here it is assumed that no more branes coincide than those indicated. If $\mathbf{c}$ and $\mathbf{d}$ coincide this would be regarded as a single stack denoted $\mathbf{c}$. If $\mathbf{c}$ coincides with $\mathbf{a}$ or $\mathbf{b}$ we switch the rôles of $\mathbf{c}$ and $\mathbf{d}$. This limits the possibilities to those listed here.

$SU(5)$ models

The following is an example of an $SU(5)$ model. It is item 617 in Table 6 and despite having a hidden sector, this model has as its gauge group precisely $SU(5)$ and nothing more! The standard model part consists of an $U(5)$ complex stack and a single real $O(1)$ brane. This is needed for the endpoints of the strings yielding the representation (5^*). In addition this example has one extra $O(1)$ brane that serves as a hidden sector. The example occurs for tensor product $(1,4,4,4,4)$ and MIPF nr. 63 in our classification, which is characterized by Hodge numbers $(h_{21}, h_{11}) = (7, 31)$, and yields 237 singlets if one uses this MIPF to construct a heterotic string. The total number of boundaries is 246. The orientifold is the one with maximal O-plane tension. The precise spectrum is as follows

$$
\begin{aligned}
3 &\times (A, 0, 0)_3, \\
11 &\times (V, V, 0)_{-3}, \\
8 &\times (S, 0, 0), \\
3 &\times (Ad, 0, 0), \\
1 &\times (0, A, 0), \\
3 &\times (0, V, V), \\
8 &\times (V, 0, V), \\
2 &\times (0, S, 0), \\
4 &\times (0, 0, S), \\
4 &\times (0, 0, A).
\end{aligned}
$$

We emphasize that this just one sample of many such models. There are 16845 configurations of this kind (i.e. with the same first two CP-factors $U(5) \times O(1)$ and the same chiral spectrum). The other 16844 configurations may differ from the one shown here by having, for example, different numbers of $U(5)$ adjoints or (V, V) mirror pairs. Some of these 16845 configurations are identical to the one shown here, because of surviving discrete symmetries of the $(1, 4, 4, 4, 4)$ tensor product. But the fact that this chiral spectrum was found for 296 different MIPFs essentially guarantees that many different versions exist.

This model has one hidden sector brane. According to our strategy, outlined in the beginning of this section, none of the remaining models of this type was checked for tadpole cancellation *with* hidden branes after this tadpole solution was found. All 16845 configurations were checked for tadpole cancellation *without* hidden branes, and no solutions were found. It is straightforward to re-examine all these 16845 model and check for further possibilities of tadpole cancellation, in order to obtain different non-chiral spectra or different hidden sectors. But there are many other models of potential interest, including many more $SU(5)$ models.

Flipped $SU(5)$ models

The simplest flipped $SU(5)$ we found occurs for invariant 52 of (1,4,4,4,4), orientifold 0, with characteristics (3,51,253). It solves all tadpole equations with just two brane stacks, the minimal number needed to realize flipped $SU(5)$. The full Chan–Paton group is $U(5) \times U(1)$, and the spectrum is

$$
\begin{array}{rcl}
11 & \times & (0, S)_3, \\
3 & \times & (A, 0)_3, \\
5 & \times & (V, V)_{-3}, \\
8 & \times & (S, 0), \\
9 & \times & (Ad, 0), \\
5 & \times & (0, Ad), \\
4 & \times & (0, A), \\
12 & \times & (V, V^*).
\end{array}
$$

In terms of $(\mathbf{a}, \mathbf{b}, \mathbf{c}, \mathbf{d})$ branes this model is of the form $U(3)_a \times U(2)_b \times U(1)_c$ with $\mathbf{a} = \mathbf{b}$ and no $\mathbf{d}$ brane, and $Y = \frac{1}{6}(1, 0, 3)$. The way the $U(1)$ anomalies cancel is noteworthy. Per family, there are five $U(1)$ anti-vector representations, contribution -5 to the cubic anomaly. This anomaly is cancelled by a symmetric tensor, which contributes $+5$ in a $U(1)$ theory. The chiral part of the spectrum yields exactly the standard model spectrum, with 3 right-handed neutrinos from the three chiral symmetric tensors. There are no G_{CP}-chiral Higgs candidates.

This is model nr. 2880 in Table 6. As explained earlier, such a flipped $SU(5)$ model always has a standard $SU(5)$ counterpart, because the masslessness of the extra $U(1)$ of flipped $SU(5)$ is an additional constraint not needed for standard $SU(5)$. This is model nr. 2881 in Table 6.

To the best of our knowledge, these are the first exact chiral, supersymmetric $SU(5)$ and flipped $SU(5)$ models in the literature. Their chiral spectrum, directly obtained in string theory, without postulating further Higgs effects or non-perturbative physics, is exactly $3 \times (10) + 3 \times (5^*)$. By contrast, the models found in [48] contain additional (15)'s of $SU(5)$. The models found recently in [49] have G_{CP} mirror pairs of (5) and (5^*), which must be made massive by postulating an additional Higgs mechanism breaking part of the additional gauge symmetry. We emphasize that the mirror pairs shown above in the explicit spectrum are non-chiral with respect to the full Chan–Paton group, and hence require no gauge symmetry breaking to acquire a mass.

In addition, the model shown above is obviously the simplest one possible, apart from the $U(5) \times O(1)$ of the previous subsection, if one could find a realization without hidden sector.

However, both the standard and the flipped $SU(5)$ model have a serious problem with either the (u, c, t) or (d, s, b) Yukawa coupling. We will discuss this in detail in Section 7.2.

For other work discussing aspects of (flipped) $SU(5)$ model building along similar lines, see [48–53]. For other issues in $SU(5)$ model building with branes and the associated problems see [28, 54].

Pati–Salam models

The simplest Pati–Salam model is nr. 4 on the list, and is therefore one of the most frequent ones. A tadpole solution was found for invariant 57 of (2, 10, 10, 10), orientifold 3. The gauge group is $U(4) \times Sp(2) \times Sp(2)$, and the spectrum is as follows

$$
\begin{aligned}
5 &\times (V, 0, V)_{-3}, \\
3 &\times (V, V, 0)_3, \\
2 &\times (Ad, 0, 0), \\
2 &\times (0, A, 0), \\
7 &\times (0, 0, A), \\
4 &\times (A, 0, 0), \\
2 &\times (0, S, 0), \\
5 &\times (0, 0, S), \\
7 &\times (0, V, V).
\end{aligned}
$$

The embedding of Y is as $Y = \frac{1}{6} Q_{\mathbf{a}} - \frac{1}{2} Q_{\mathbf{d}} + W_{\mathbf{c}}$, where $W_{\mathbf{c}} = \frac{1}{2} \sigma_3$. Brane **a** and **d** are unified to $U(4)$.

The following model is of interest because it is a $U(4) \times U(2) \times U(2)$ Pati–Salam model that satisfies all tadpole conditions without hidden branes, because it has some chiral rank-2 tensors in its spectrum, and because it occurs for a MIPF related to the "quintic" Calabi–Yau, namely MIPF 6 of (3, 3, 3, 3, 3), the trivial orientifold (the only one possible). It is nr. 4825 on the list Table 6. It has precisely one G_{CP} chiral MSSM Higgs pair, plus a G_{CP}-chiral charged lepton mirror pair, and four right-handed neutrinos. There is one massless $U(1)$ in addition to Y, namely the diagonal combination of the phase factors of the $U(2)$'s. The Chan–Paton group is $U(4) \times U(2) \times U(2)$, and the representations are

$$
\begin{array}{rcl}
3 & \times & (V, 0, V^*)_{-1}, \\
2 & \times & (V, 0, V)_{-2}, \\
1 & \times & (0, 0, S)_1, \\
5 & \times & (0, A, 0)_1, \\
5 & \times & (V, V^*, 0)_1, \\
6 & \times & (V, V, 0)_2, \\
3 & \times & (0, V, V)_{-1}, \\
4 & \times & (0, S, 0), \\
4 & \times & (S, 0, 0), \\
3 & \times & (Ad, 0, 0), \\
5 & \times & (0, Ad, 0), \\
1 & \times & (0, 0, Ad), \\
2 & \times & (0, V, V^*).
\end{array}
$$

There also exist a broken version of this model, with $U(4)$ split into $U(3) \times U(1)$ already in the exact string theory. This is nr. 7053 in Table 6.

There is also a $U(4) \times U(2) \times U(2)$ Pati–Salam model (nr. 153) which has a standard, purely bi-fundamental spectrum. For this model we only found a tadpole solution with hidden branes, which is a bit too complicate to display here. It has a hidden sector group $U(6) \times U(2)^3 \times O(2)^2 \times Sp(2)$.

Orientifolds exhibiting a Pati–Salam realization of the SM have been considered before, [55–57]. Bottom-up configurations, investigating also gauge couplings and the issue of masses, have been also considered, [58,59].

Trinification models

Trinification models are built out of three factors $SU(3)$ with purely bi-fundamental matter. At first sight this would seem to be an ideal configuration for intersecting brane models, but in fact it is surprisingly rare.

In a genuine trinification model the generator Y is embedded in $SU(3)_\mathbf{a} \times SU(3)_\mathbf{b} \times SU(3)_\mathbf{d}$ as $Y = \frac{1}{6}W_\mathbf{b} - \frac{1}{3}W_\mathbf{d}$, where $W_\mathbf{b} = W_\mathbf{d} = \text{diag}(1, 1, -2)$. However, a trinification model is in our classification a model with $x = *$, which allows arbitrary shifts in the choices of Y. For any other choice of Y this implies that a combination of the unitary phases contributes to Y. The canonical choice of Y has no contribution from $U(3)_\mathbf{a}$ and hence would correspond to $x = \frac{1}{3}$, a non-standard choice. Although the quark and lepton charges do not fix x, this may be done by the zero Y-mass condition.

In Table 6 three distinct models with this characteristic appear. The most frequent one, nr. 225, has a fixed value of Y of the canonical trinification type, with $x = \frac{1}{3}$. However, we did not find solutions to the tadpole conditions for any of these 71328 models. The second one, nr. 1125, has a completely free Y; even the zero mass condition for Y does not fix it. This type of model occurred 6432 times and for at least one of these we found a solution to all tadpole conditions. The third one, nr. 12327, occurred only 24 times, and for none of them the tadpoles were solved. It has Y fixed to a value which does not correspond to standard trinification ($x = \frac{1}{2}$).

The aforementioned tadpole solution occurred for invariant 11 of tensor product $(1, 16, 16, 16)$, orientifold 0 (with $(h_{21}, h_{11}, S) = (9, 111, 481)$). It has a rather large hidden sector gauge group $U(3) \times U(3) \times U(3) \times O(4) \times O(2) \times U(6) \times U(12) \times O(12) \times U(12) \times O(4)$, with respect to which the spectrum is as follows:

$$3 \times (V, V, 0, 0, 0, 0, 0, 0, 0, 0)_3,$$
$$3 \times (V, 0, V, 0, 0, 0, 0, 0, 0, 0)_{-3},$$
$$3 \times (0, V, V^*, 0, 0, 0, 0, 0, 0, 0)_{-3},$$
$$1 \times (0, 0, 0, V, 0, V, 0, 0, 0, 0)_{-1},$$
$$1 \times (0, 0, 0, 0, 0, S, 0, 0, 0, 0)_1,$$
$$5 \times (0, 0, 0, 0, 0, 0, 0, V, V, 0)_1,$$
$$3 \times (0, 0, 0, 0, 0, 0, 0, 0, S, 0)_1,$$
$$1 \times (0, 0, 0, 0, 0, A, 0, 0, 0, 0)_{-1},$$
$$2 \times (0, 0, 0, 0, 0, 0, 0, 0, A, 0)_{-2},$$
$$1 \times (0, 0, 0, V, 0, 0, 0, 0, V, 0)_1,$$
$$1 \times (0, 0, 0, 0, V, 0, 0, 0, V, 0)_1,$$
$$1 \times (0, 0, 0, 0, 0, V, 0, V, 0, 0)_1,$$
$$1 \times (0, 0, 0, 0, 0, V, 0, 0, V, 0)_{-1},$$
$$1 \times (0, 0, 0, 0, 0, 0, V, V, 0, 0)_1,$$
$$1 \times (0, 0, 0, 0, 0, 0, V, 0, V, 0)_{-1},$$

1 × $(0, 0, 0, 0, 0, V, 0, 0, 0, V)_{-1}$,
1 × $(0, 0, 0, V, V, 0, 0, 0, 0, 0)$,
1 × $(0, 0, 0, 0, S, 0, 0, 0, 0, 0)$,
1 × $(0, 0, 0, 0, 0, Ad, 0, 0, 0, 0)$,
1 × $(0, 0, 0, 0, 0, 0, Ad, 0, 0, 0)$,
3 × $(0, 0, 0, 0, 0, 0, 0, S, 0, 0)$,
3 × $(0, 0, 0, 0, 0, 0, 0, 0, Ad, 0)$,
1 × $(0, 0, 0, 0, 0, 0, 0, 0, 0, S)$,
2 × $(0, 0, 0, 0, V, V, 0, 0, 0, 0)$,
1 × $(0, 0, 0, 0, V, 0, 0, V, 0, 0)$,
2 × $(0, 0, 0, 0, 0, V, 0, 0, V^*, 0)$,
2 × $(0, 0, 0, 0, 0, 0, V, 0, V^*, 0)$,
1 × $(0, 0, 0, 0, V, 0, 0, 0, 0, V)$,
1 × $(0, 0, 0, 0, 0, 0, 0, V, 0, V)$.

Bottom-up trinification models and their phenomenology has been discussed in [43].

6.5. Curiosities

A non-standard $U(3) \times Sp(2) \times U(1) \times U(1)$ model

The following spectrum was found for 17, orientifold 2 of the tensor product $(2, 2, 2, 6, 6)$. It has a hidden sector group $U(2)$ which is completely decoupled from all massless matter: both OH as HH matter is absent. The main reason for listing it here is however that it is an alternative to the standard lepton-number conserving configurations. This is nr. 4996 in Table 6.

The full Chan–Paton group is $U(3) \times Sp(2) \times U(1) \times U(1) \times U(2)$, with the following spectrum

3 × $(V, V, 0, 0, 0)_3$,
3 × $(0, 0, V, V, 0)_{-3}$,
1 × $(V, 0, 0, V^*, 0)_{-1}$,
2 × $(V, 0, V, 0, 0)_{-2}$,
2 × $(0, V, 0, V, 0)_2$,
3 × $(V, 0, 0, V, 0)_{-1}$,
3 × $(0, V, V, 0, 0)_1$,

2 × $(V, 0, V^*, 0, 0)_{-2}$,
1 × $(0, 0, V, V^*, 0)_1$,
4 × $(A, 0, 0, 0, 0)$,
2 × $(0, 0, 0, S, 0)$.

The Y-embedding is $Y = \frac{1}{6}Q_{\mathbf{a}} - \frac{1}{2}Q_{\mathbf{c}} - \frac{1}{2}Q_{\mathbf{d}}$. There is no additional massless $U(1)$ factor from the standard model branes (we did not compute the mass of the Abelian factor of $U(2)$). Note that the endpoints of the quarks and lepton doublet bi-fundamentals are distributed over the $\mathbf{c}$ and $\mathbf{d}$ branes, making it impossible to assign a lepton number. Indeed, there are perturbatively allowed lepton-number violating couplings of the type (Q, L, d^c) or (L, L, l^c), but further CFT computations would be needed to verify if these couplings do indeed occur. The G_{CP}-chiral spectrum has no Higgs candidates and just one right-handed neutrino candidate.

We have also found a similar model with $U(2)_{\mathbf{b}}$ instead of $Sp(2)_{\mathbf{b}}$, and a slightly more complicated hidden sector. It combines two features not encountered together in [17]: a group $U(3) \times U(2) \times U(1) \times U(1)$ of which only $U(1)_Y$ survives as an Abelian vector boson. Unfortunately this is achieved at a price that is presumably to high: the reason is that lepton number cannot be written in terms of the brane charges. As a result, no linear combination of B and L is anomaly free. Model nr. 1725 is of the same kind, but with $Sp(2)$ replaced by $U(2)$. A tadpole solution exists for that model with an $O(2) \times O(2)$ hidden sector.

A $U(6)$ **model**

The following examples were found for invariant 79 of $(1, 4, 4, 4, 4)$, orientifold 0, corresponding to an orientifold with Calabi–Yau characteristics (6, 60, 288). These are exact standard model realizations with just two branes stacks, a complex and a real one. In fact, this single model can accommodate the standard model spectrum in three distinct ways. The unified gauge group is $U(6) \times Sp(2)$. The spectrum is as follows

9 × $(A, 0)_3$,
9 × $(V, V)_{-3}$,
8 × $(Ad, 0)$,
1 × $(0, A)$,
7 × $(0, S)$.

The first standard model realization is obtained by splitting $U(6)$ so that the full $(\mathbf{a}, \mathbf{b}, \mathbf{c}, \mathbf{d})$ configuration becomes $U(3)_{\mathbf{a}} \times U(2)_{\mathbf{b}} \times Sp(2)_{\mathbf{c}} \times U(1)_{\mathbf{d}}$, with

a, b and **d** belonging to the same stack. The choice of Y is $\frac{1}{6}(1, 0, 0, -3) + W_{\mathbf{c}}$, where $W_{\mathbf{c}}$ is the diagonal Pauli matrix $\frac{1}{2}\sigma_3$ in $Sp(2)_{\mathbf{c}}$. The first term of Y is part of the non-Abelian group $SU(4)$ formed by the **a** and **d** branes, and hence automatically massless. If the breaking pattern is interpreted as $U(6) \to U(5) \times U(1) \to U(3)_{\mathbf{a}} \times U(2)_{\mathbf{b}} \times U(1)_{\mathbf{c}}$ the second step is a flipped $SU(5)$ model; if the breaking is interpreted as $U(6) \to U(4) \times U(2) \to U(3)_{\mathbf{a}} \times U(2)_{\mathbf{b}} \times U(1)_{\mathbf{c}}$ the intermediate stage is Pati–Salam-like.

The second realization appears if we split $U(6)$ in the same way as $U(3)_{\mathbf{a}} \times U(2)_{\mathbf{b}} \times Sp(2)_{\mathbf{c}} \times U(1)_{\mathbf{d}}$, but now with Y is $(-\frac{1}{3}, \frac{1}{2}, 0, 0)$. This amounts to a standard $SU(5)$ embedding of the standard model. The $Sp(2)$ group does not contribute to Y in this case.

Finally there is the possibility of using $Sp(2)$ as a **b**-type stack for the weak interactions. To achieve this we split $U(6)$ as $U(3)_{\mathbf{a}} \times U(3)_{\mathbf{c}}$, and write Y as $\frac{1}{6}(1, 0, -1) + W_{\mathbf{c}}$, where $W_{\mathbf{c}}$ is the $SU(3)$-generator $(\frac{2}{3}, -\frac{1}{3}, -\frac{1}{3})$. There is no **d**-stack.

All three models have three candidate Higgs pairs an three down quarks mirror pairs, as well as six right-handed neutrino candidates, which are chiral with respect to $U(6)$. The first two are nrs. 1886 and 1887 in Table 6, and the third one is nr. 1888.

7. Phenomenological implications and the problem of masses

In this section we will address, in a rather general fashion, some phenomenological aspects of SM brane configurations. In particular, we are going to discuss the problem of masses in theories with anti-quarks in the antisymmetric representation of $SU(3)$, as well as the nature of potential family symmetries and neutrino masses.

7.1. Antisymmetric anti-quarks and the problem of quark masses

There is a generic potential phenomenological problem, when one of the anti-quarks originate from anti-symmetrized strings starting and ending on the color branes. Although for $SU(3)$, $\boxbar = \overline{\square}$, the antisymmetric representation has charge 2, under the $U(1)_{\mathbf{a}}$ instead of the -1 for $\square$.

We are using the language of left-handed fermions where

$$\bar{\psi}_R^c \equiv \psi_L^T C, \quad C^{-1}\gamma^\mu C = -(\gamma^\mu)^T, \tag{7.1}$$

where C is the charge conjugation matrix. $\bar{\psi}_R^c$ is a right-handed Weyl fermion transforming in the same representation of the gauge group as ψ_L. The mass terms can be therefore be written in terms of fermion bilinears

$$\bar{\psi}_R^c \, \chi_L + h.c. \tag{7.2}$$

Consider the (color singlet) quark mass operator[16] $(\bar{Q}^c)_a^I q^J$, where Q denotes the quarks (3,2) and q stands for the anti-quarks in the ▢ of $SU(3)$. I, J indices from now on will collectively indicate any other index except color and weak indices. a is a weak doublet index. $(\bar{Q}^c)^I q^J$ transforms as a weak doublet, and has charge 3 under $U(1)_{\mathbf{a}}$. Therefore it must be coupled to a weak Higgs doublet that should also carry charge -3. However a single field in orientifolds cannot carry charge -3. Therefore, a product of scalar fields must be involved. The minimal case involves scalars H_a^I transforming as $(\bar{3}, 2, -1)$ under $SU(3) \times SU(2) \times U(1)_{\mathbf{a}}$ and A^K transforming as $(\overline{▢}, 1, -2)$. The putative mass term would then be

$$\delta\mathcal{L}_1 = h_{I,J,K,L} \left((\bar{Q}^c)_a^I Q^J \right)(H^{Ka} A^L), \tag{7.3}$$

where the parentheses indicate the color contractions. Non-minimal couplings would include

$$\delta\mathcal{L}_2 = \tilde{h}_{I,J,K,L} \left((\bar{Q}^c)_a^I Q^J \right) G^a (F^K A^L), \tag{7.4}$$

$$\delta\mathcal{L}_2 = \hat{h}_{I,J,K,L,M} \left((\bar{Q}^c)_a^I Q^J \right) G^a (F^K F^L F^M), \tag{7.5}$$

where G^a is a standard Higgs (weak doublet), F^I transforms as $(\bar{3}, 1, -1)$ and in the last case an antisymmetric color coupling of three triplets is implied. There might also be additional constraints, due to the fact that the F_I scalars come from strings that have one end point in the c, d branes.

The crucial point is that in order to generate the quark mass terms, the scalar combinations in (7.3) and (7.5) must acquire expectation values. This necessarily implies that the scalars H^{Ia} or F^I or A^I must have vevs, and this necessarily breaks the color symmetry to $SU(2)_{\text{color}}$ (along with $U(1)_{\mathbf{a}}$ of course). This seems incompatible with current data. Moreover, this conclusion is robust, and is valid independent of the presence or not of supersymmetry.[17]

There are two a priori possibilities in order to avoid the previous impasse. The first is that non-perturbative effects break the associated global symmetry. It is well known that anomalous $U(1)$'s have always mixed anomalies with non-Abelian groups. Therefore, there are always gauge instantons and their string theory generalization, that violate the global symmetry non-perturbatively

---

[16] We work with left-handed spinors only. $\bar{Q}^c$ is the proper conjugate of a left-handed spinor.

[17] A related fact is that a $U(N)$ D-brane on a CY manifold is generically expected to have it gauge symmetry broken to $U(N-1)$ because of the D-terms. The gauge symmetry may be enhanced back to $U(N)$ at orbifold points.

(see [60]). There are two distinct possibilities, but only one is relevant here: the case when the non-Abelian gauge group is unbroken at low energy.[18] This is indeed the case with the color group. In this case only terms involving a minimum number of fermions can be generated. This minimum number is required in order to soak up the zero modes of instantons. It is always larger than two in realistic situations. Therefore, it is not relevant for generating mass terms for the fermions.

The other option is to start from a higher gauge-group, that is eventually broken to the color $SU(3)$, giving masses to the standard quarks. Let us entertain first the case of $SU(4)$. We should use the following facts, [61]: A scalar in the adjoint of $SU(4)$ obtaining a vev may break the gauge symmetry $SU(4) \rightarrow SU(2) \times U(2)$ or $SU(3)$ depending on the type of vev. A scalar in the $\Box\Box$, breaks the gauge symmetry as $SU(4) \rightarrow O(4)$ or $SU(3)$ depending on the type of vev. A scalar in the $\Box$ breaks the gauge symmetry as $U(4) \rightarrow Sp(4)$ or $SU(4) \rightarrow SU(2)$ depending on the type of vev. Finally a scalar transforming in the $(4, 2)$ of $SU(4) \times SU(2)$ breaks the symmetry as $SU(4) \times SU(2) \rightarrow SU(3)$, or $SU(2) \times SU(2)$ depending on the type of vev. Although this may be acceptable from the color point of view, the breaking of the weak $SU(2)$ group is acceptable only if the bi-fundamental scalar carries the correct SM hypercharge.

Therefore the scalar vevs that preserve an $SU(3)$ color subgroup $SU(4)$ transformations are[19]

$$\text{adjoint} \sim \Phi_\alpha{}^\beta \sim \begin{pmatrix} 1 & 0 & 0 & 0 \\ 0 & 0 & 0 & 0 \\ 0 & 0 & 0 & 0 \\ 0 & 0 & 0 & 0 \end{pmatrix} \sim \phi_{\alpha\beta} \sim \Box\Box, \tag{7.6}$$

$$\Box \sim F_\alpha \sim \begin{pmatrix} 1 \\ 0 \\ 0 \\ 0 \end{pmatrix}, \quad (4, 2) \sim H_\alpha^a \sim \begin{pmatrix} 1 & 0 & 0 & 0 \\ 0 & 1 & 0 & 0 \end{pmatrix}. \tag{7.7}$$

The last operator breaks also to the $SU(2)$. and they must be aligned. This poses strong constraints on the appropriate scalar potentials. In particular, no antisymmetric vev is allowed.

We may now go through the potential mass terms and show that none is acceptable. We suppress all other indices but $SU(4)$ color and write

$$\mathcal{O}_1 = (\bar{Q}^c)_\alpha q_{\beta\gamma} F^\alpha F_I^\beta F_J^\gamma, \quad \mathcal{O}_2 = (\bar{Q}^c)_\alpha q_{\beta\gamma} \phi^{\alpha\beta} F^\gamma, \tag{7.8}$$

$$\mathcal{O}_3 = \epsilon^{\alpha\beta\gamma\delta} (\bar{Q}^c)_\alpha q_{\beta\gamma} F_\delta \, \epsilon_{\alpha'\beta'\gamma'\delta'} F_I^{\alpha'} F_J^{\beta'} F_K^{\gamma'} F_L^{\delta'}, \tag{7.9}$$

---

[18] The other case concerns a spontaneously broken group. This is qualitatively distinct since more terms in the effective action can be generated.

[19] We use Greek indices from the beginning of the alphabet for color.

where a lower $SU(4)$ index transforms as $\square$ and an upper one as $\overline{\square}$. The operators $\mathcal{O}_i$ moreover transform as weak doublets and have $U(1)_{\mathbf{a}}$ charge zero. There are also operators which involve adjoint scalars but they have no new features. It is straightforward to check that operators $\mathcal{O}_{1,2}$ fail to provide mass operators for any of the fermions after the breaking $SU(4) \to SU(3)$ by the vevs in (7.6) and (7.7). Operator $\mathcal{O}_3$ gives masses to the standard $SU(3)$ quarks, but leaves the rest massless. One of the fundamentals in $\mathcal{O}_i$ can be substituted with the H_α^a scalar. This will provide a weak singlet. Moreover as we have seen this vev breaks $SU(4) \times SU(2) \to SU(3)$, and if the hypercharge of the scalar is 1/2, then it will provide at the same time the proper, electroweak symmetry breaking. However, the same considerations as above indicate than no reasonable mass terms are generated.

The final case to be considered is the possibility to include a scalar vev in the antisymmetric representation, $R^{\alpha\beta}$. In this case we must start from $SU(5)$, which the vev will break to $SU(3)$. Upon choosing a convenient basis this vev is

$$\square \sim R^{\alpha\beta} \sim \begin{pmatrix} 0 & 1 & 0 & 0 & 0 \\ -1 & 0 & 0 & 0 & 0 \\ 0 & 0 & 0 & 0 & 0 \\ 0 & 0 & 0 & 0 & 0 \\ 0 & 0 & 0 & 0 & 0 \end{pmatrix}. \tag{7.10}$$

We also assume that there are fundamentals F^α with a vev in the 4 and 5 directions, so that it does not break $SU(3)$ further. Then we may write the following operators

$$\mathcal{O}_4 = (\bar{Q}^c)_\alpha q_{\beta\gamma} F^\alpha R^{\beta\gamma}, \tag{7.11}$$

$$\mathcal{O}_5 = \epsilon^{\alpha\beta\gamma\delta\epsilon} (\bar{Q}^c)_\alpha q_{\beta\gamma} \rho_{\delta\epsilon} \, \epsilon_{\alpha'\beta'\gamma'\delta'\epsilon'} F^{\alpha'} R^{\beta'\gamma'} R^{\delta'\epsilon'}. \tag{7.12}$$

The operator $\mathcal{O}_4$ provides masses for the various singlets after the breaking. Operator $\mathcal{O}_5$ provides masses for the standard quarks. However the two extra triplets emerging from the $\square$ of $SU(5)$ will remain massless.

It therefore seems that orientifold models with anti-quarks in antisymmetric representations are phenomenologically untenable.

7.2. Masses in $SU(5)$ and flipped $SU(5)$ vacua and instanton effects

The case of standard $U(5)$ group deserves special attention.[20] The SM particles are in the antisymmetric representation $\psi^{\alpha\beta}$ as well as the anti-fundamental, ψ_α. The minimal set of scalar needed for symmetry breaking is an adjoint $\Phi^\alpha{}_\beta$

---

[20]Several of the remarks below were independently put forward recently in [28].

whose expectation value diag$(2V, 2V, 2V, -3V, -3V, -3V)$ breaks $SU(5) \rightarrow$ $SU(3) \times SU(2) \times U(1)_Y$ and a fundamental, H^α whose expectation value $(0, 0, 0, 0, v)$ breaks $SU(2) \times U(1)_Y \rightarrow U(1)_{em}$ The standard mass terms

$$\mathcal{O}_1 \sim (\bar{\psi}^c)_\alpha \psi^{\alpha\beta} H_\beta, \quad \mathcal{O}_2 \sim \epsilon_{\alpha\beta\gamma\delta\epsilon} (\bar{\psi}^c)^{\alpha\beta} \psi^{\gamma\delta} H^\epsilon \tag{7.13}$$

give masses to all SM fermions. However here, $\mathcal{O}_2$ which gives masses to up-type quarks is not allowed, since it carries charge $+5$ under the overall $U(1)$ of the $U(5)$. This charge can be cancelled by multiplication by $\epsilon^{\alpha\beta\gamma\delta\epsilon} H_\alpha^I H_\beta^J H_\gamma^K$ $H_\delta^K H_\epsilon^L$, which however requires the presence of 5 fundamental Higgs scalars with vevs that are aligned, and of the order of the electroweak scale. However, such a mass is suppressed by a factor $\prod_{I=1}^{5} \frac{v_I}{M_s}$. Since all $v_I \lesssim M_Z$, we obtain an unacceptable suppression factor of 10^{-50}. The other possibility is the presence of symmetric or antisymmetric scalars that acquire vevs. An antisymmetric vev cannot preserve the $SU(3) \times U(1)_{em}$ group of low energy physics. A symmetric one, $R^{\alpha\beta}$ is fine provided it is aligned as in (7.6). Its vev $\tilde{V}$ must be smaller than the EW vev as it contributes to the W,Z masses. Again, although $\mathcal{O}_2$ can be neutralized, it gives too small a contribution to up quark masses. There are new operators we may write now like

$$\mathcal{O}_3 \sim (\bar{\psi}^c)_{\alpha\beta} \psi_{\gamma\delta} R^{\alpha\gamma} R^{\beta\delta}. \tag{7.14}$$

However, such an operator does not contribute to fermion masses.

We can imagine of two non-perturbative loopholes to the previous arguments. A first non-perturbative possibility is based on breaking the offending $U(1)$ symmetry by a vacuum condensate. An example will be a Chan–Paton group that contains $U(5) \times SO(5) \times SO(5)$, and we have extra scalars (denoted Q_α^A) in the representation (V,V,0)+(V*,0,V), so that the $U(5)$ anomalies cancel. If the dynamics is favorable, we may imagine that one of the $SO(5)$'s creates a composite out of five scalars, of the form

$$\epsilon^{\alpha\beta\gamma\delta\epsilon} \epsilon_{ABCDE} Q_\alpha^A Q_\beta^B \cdots Q_\epsilon^E, \tag{7.15}$$

where $\alpha, \beta, \dots$ are $SU(5)$ indices and $A, B, \dots SO(5)$ indices. If the condensate gets a vev at the GUT scale, but not the individual fields Q_α^A, it breaks the $U(1)$ of $U(5)$, and upon coupling to the $U(5)$ quarks and leptons can generate the appropriate masses. It should be however mentioned that such a dynamical setup seems unlikely.

The final possibility, is a non-perturbative breaking of the overall (anomalous) $U(1)$ symmetry because of spacetime instantons. This seems a realistic possibility as stringy instantons are reasonably well understood by now. They first appeared in string theory along with the duality conjectures, and a lot has been learned about them from their correspondence with perturbative effects (namely world-sheet instantons) in dual string theories (see [29] for a review). The instanton calculus associated with D-branes was developed eventually from first principles [30, 31]. In supersymmetric orientifold vacua, the relevant instanton effects can be two-fold. One is a stringy version of the standard gauge instantons of field theory. The other is a stringy instanton that has no counterpart in field theory, and resembles octonionic instantons [32]. In one of its incarnations it is a Euclidean D_1-instanton, [37]. A stringy instanton for a given gauge group, may be a gauge instanton for another gauge group.

Such instantons can give corrections to the superpotential that can include Yukawa couplings necessary for masses. Generic instantons do not contribute to the superpotential as they have 4 zero modes, [33]. Orientifold projections can reduce the zero modes to two, and then the instanton (called an $O(1)$ instanton) can contribute to the superpotential. The first nontrivial example of this, with full tadpole cancellation was described in [32]. Moreover in this example, in one region of the moduli space of vacua, the relevant gauge group is $SU(5)$ with three 10' and 3 $\bar{5}$, plus other non-chiral particles. The stringy instantons in that vacuum generate a non-perturbative mass matrix for the 10s. Moreover at the minimum of the open string superpotential, the vacuum energy is non-zero and if closed moduli are already stabilized that would break supersymmetry [34, 36]. This is an incarnation of gaugino condensation, driven by stringy instantons. Further examples were described in [35, 36].

It is therefore possible to generate masses in $SU(5)$ orientifolds by stringy instantons.

7.3. Family symmetries

We have allowed extra non-Abelian groups to participate in the local SM collection of branes. In particular standard model particles are charged under such groups. This setup is very reminiscent of the idea of family symmetries. The purpose of the introduction of family symmetry in the past was to explain/organize the existence of three generations and the hierarchy of masses of the SM particles. There are two relevant questions in this context:

(a) Can such symmetries play the role of family symmetries? Can they help achieve realistic mass matrices for the SM particles?

(b) Are there cases where the presence of such symmetries forbids realistic mass matrices?

In following we will make some comments on these two questions. Although our setup is reminiscent of family symmetries, it incorporates a radical departure from that idea as well. The reason is that the quark (3,2) states, cannot be charged under any other gauge symmetry. This is unlike any other family symmetry introduced in the literature. Since the quarks are necessarily not-charged under such symmetries, there are non-trivial considerations concerning the potential mass matrices and the existence of realistic patterns.

At this stage, we are not fully prepared to calculate three and higher point couplings in the superpotential. We can however derive some selection rules on couplings, especially renormalizable ones (three-point couplings) that are allowed by the gauge symmetries. Such selection rules can have non-trivial consequences because

(i) Extra non-Abelian symmetries, although broken, may be more or less constraining, due to the possible symmetry breaking vevs

(ii) The presence of several (anomalous) $U(1)$s provides further constraints, especially if the corresponding global symmetries remain intact in perturbation theory.

From now on we will call for concreteness the non-Abelian group G distinct from $SU(2)$ and $S(3)$, the family symmetry group. Let us consider the case where the anti-quarks q_i transform in a non-trivial representation R of the group G. Then the potential mass term $(\bar{Q}^c)^{I,a}q_i$ transform as a doublet of $SU(2)$ and as R of G (I is a extra index labeling the three quark generations, while a is a weak doublet index). At the cubic level the existence of a scalar Φ_a^i transforming in the $(2, R)$ of $SU(2) \times G$, gives rise to the Yukawa coupling

$$(\bar{Q}^c)^{I,a}q_i \; \Phi_a^i. \tag{7.16}$$

Up to base change there are two types of vevs for Φ, [61]. The first type breaks the symmetry $SU(2) \times G \rightarrow G'$, with $G' = O(N-1)$ if $G = O(N)$ and $G' = SU(N-1)$ if $G = SU(N)$. Therefore, the electroweak symmetry is broken while the family symmetry is not fully broken. For this to be realistic, further vev's should break both the $U(1)_Y$ symmetry and the leftover family symmetry. The pattern then becomes complicated and deserves a detailed study. The second type breaks $SU(2) \times G \rightarrow SU(2) \times G'$ with $G' = O(N-2)$ if $G = O(N)$ and $G' = SU(N-2)$ if $G = SU(N)$. Here the family symmetry is completely broken if $N = 2$. This is the case for example of a Pati–Salam group. If there is a leftover family symmetry, further symmetry breaking is necessary. The Φ vev identifies the weak and the G index and provides a mass matrix for quarks that is degenerate. The existence of several copies of Φ does not improve the situation.

We can contemplate higher dimension terms involving a weak double H^a and a scalar Φ_i in the fundamental of G

$$(\bar{Q}^c)^{I,a}q_i \; H^a \Phi^i. \tag{7.17}$$

In such a case a vev of Φ of the order of M_s will give a mass matrix of order of the electroweak scale but it will be degenerate. Moreover the G symmetry is partly broken. Several scalars H_i^I with couplings

$$\frac{g_{IJ}}{M_s} \, (\bar{Q}^c)^{I,a} q_i \, H^a \Phi_J^i \tag{7.18}$$

could fare better. First non-aligned expectation values can break a larger portion or all of the G group. Second, for generic couplings g_{IJ} the mass matrix after electroweak breaking will be non-degenerate. Therefore, in this case, a reasonable non-zero mass matrix is viable.

There are more complicated possibilities of the occurrence of quasi-family symmetries and the charge assignments of SM particles under them. We have studied in some indicative examples, the relevant issues present. A full study of all possibilities is a major task and it will not be undertaken here.

7.4. Neutrino masses

In our search we have not explicitly constrained the presence of anti-neutrinos. A priori, any SM singlet fermion can play that role. Of course, for a realistic pattern of masses to emerge, important constraints on the interactions are appropriate.

There are two mechanisms that so far have been successful in producing neutrino masses of acceptable magnitude. The first, relies on the see-saw mechanism and is appropriate for vacua with high values of the string scale. An important ingredient for its operations is that lepton number is not conserved. Moreover at least two (and typically three) antineutrinos are necessary for accommodating present data. As we have discussed earlier, the presence of lepton number cannot be directly tracked until a formal separation of doublets into leptons and Higgses is possible. Therefore in this context, the question of neutrino masses remains a question to be addressed in concrete string ground states.

The second mechanism involves a brane wrapping one (or several) large dimensions and is necessarily operative in string vacua with a low string scale. In this context the neutrinos mix with antineutrinos emerging from the "bulk" brane, and the masses are suppressed by the volume of large dimensions. For this mechanism to succeed large Majorana masses should be forbidden. Therefore it is important that lepton number is a good symmetry. Moreover, the minimal implementation involves a single antineutrino and its KK tower of states and leads to predictions marginally compatible with current data [10]. More comfortable constructions involve at least two antineutrinos.

8. Dependence of the results on the Calabi–Yau topology

Table 7 lists the MIPFs for which the standard model spectrum was found, and how often it occurred. The table is ordered according to standard model frequency, that is the total number of standard model configurations divided by the total number of three and four brane configurations. Note that this does not take into account tadpole cancellation, since we have not systematically solved the tadpole conditions for all standard model configurations. Column 2 gives the MIPF id-number using the same sequential labeling used in [17]. We can provide further details on these MIPFs on request. To help identifying them, we list in columns 3, 4 and 5 the resulting heterotic Calabi–Yau spectrum (Hodge numbers and the number of E_6 singlets). In columns 6, 7 and 8 we list the total number of configurations for each value of x. The last column gives the frequency.

The complete table has 1639 cases with non-zero frequency. Therefore we only present the top of the table here, which starts with a frequency as high as 0.2%. The last three entries are modular invariants of the tensor $(3, 3, 3, 3, 3)$, corresponding to the quintic. They occur much further down the list, but are shown here because the quintic is a well-studied Calabi–Yau manifold. The lowest non-zero frequency we encountered is 3.5×10^{-12} (for a total of 4 configurations found).

In column 2 an asterisk indicates that at least one tadpole solution was found for that MIPF in [17]. Note that we did not perform an exhaustive search for tadpole solutions in the present work. Indeed, if all brane configurations occurring for a given MIPF are of a type for which the tadpoles have already been solved before (for a different MIPF), no further attempts are made to solve them. Therefore we cannot make definitive conclusions about the non-existence of tadpole solutions for a given MIPF from our present results.

Note the presence of models with Hodge numbers $(20, 20)$. The corresponding Calabi–Yau manifolds are in fact of the form $K_3 \times T_2$. There is also a case with $h_{11} = h_{12} = 0$, which is in fact a torus compactification. The fact that these are (partly) torus compactifications is not in contradiction with the fact that the spectrum is chiral. Each MIPF can be thought of as a an extension of the chiral algebra of the original tensor product, modified by an automorphism. This extension may lead to a non-chiral torus compactification. However the boundary states that are admitted are a complete set with respect to the original unextended chiral algebra, which always corresponds to a chiral compactification (except for five non-chiral tensor products that we do not consider). Hence a non-chiral bulk extension may have chiral boundary states. It is possible that the $K_3 \times T_2$ models are related to models discussed in [62]; this will require further investigation. In any case we did not find tadpole solutions for any of these torus or $K_3 \times T_2$ models (but again with the caveat that we did not search for them exhaustively).

Table 7
Standard model success rate for various MIPFs

Tensor product	mipf	h_{11}	h_{12}	Scal	$x = 0$	$x = \frac{1}{2}$	$x = *$	rate
(1, 1, 1, 1, 7, 16)	30	11	35	207	1698	388	0	2.1×10^{-3}
(1, 1, 1, 1, 7, 16)	31	5	29	207	890	451	0	1.35×10^{-3}
(1, 4, 4, 4, 4)	53	20	20	150	2386746	250776	0	4.27×10^{-4}
(1, 4, 4, 4, 4)	54	3	51	213	5400	5328	4248	3.92×10^{-4}
(6, 6, 6, 6)	37	3	59	223	0	946432	0	2.79×10^{-4}
(1, 1, 1, 1, 10, 10)	50	12	24	183	1504	508	36	2.63×10^{-4}
(1, 1, 1, 1, 10, 10)	56	4	40	219	244	82	0	2.01×10^{-4}
(1, 1, 1, 1, 8, 13)	5	20	20	140	328	27	0	1.93×10^{-4}
(1, 1, 1, 1, 7, 16)	26	20	20	140	157	14	0	1.72×10^{-4}
(1, 1, 7, 7, 7)	9	7	55	276	7163	860	0	1.59×10^{-4}
(1, 1, 1, 1, 7, 16)	32	23	23	217	135	20	0	1.56×10^{-4}
(1, 4, 4, 4, 4)	52	3	51	253	110493	8303	0	1.02×10^{-4}
(1, 4, 4, 4, 4)	13	3	51	250	238464	168156	0	1.01×10^{-4}
(1, 1, 1, 2, 4, 10)	44	12	24	225	704	248	0	1.01×10^{-4}
(1, 1, 1, 1, 1, 2, 10)	21	20	20	142	2	1	0	1.00×10^{-4}
(1, 1, 1, 1, 1, 4, 4)	124	0	0	78	729	0	0	9.8×10^{-5}
(4, 4, 10, 10)	79	7	43	215	0	57924	0	9.39×10^{-5}
(4, 4, 10, 10)	77	5	53	232	0	1068926	0	8.29×10^{-5}
(1, 4, 4, 4, 4)	77	3	63	248	0	1024	0	8.12×10^{-5}
(4, 4, 10, 10)	74	9	57	249	0	1480812	0	8.06×10^{-5}
(1, 1, 1, 1, 1, 2, 10)	24	20	20	142	0	0	6	7.87×10^{-5}
(1, 2, 4, 4, 10)	67	11	35	213	0	14088	1008	7×10^{-5}
(1, 1, 1, 1, 5, 40)	5	20	20	140	303	36	0	6.73×10^{-5}
(2, 8, 8, 18)	8	13	49	249	0	1506776	0	6.03×10^{-5}
(1, 1, 7, 7, 7)	7	22	34	256	2700	68	0	5.5×10^{-5}
(1, 4, 4, 4, 4)	78	15	15	186	20270	6792	0	5.39×10^{-5}
(2, 8, 8, 18)	28	13	49	249	0	670276	0	5.25×10^{-5}
(1, 2, 4, 4, 10)	75	5	41	212	304	580	244	4.87×10^{-5}
(1, 1, 7, 7, 7)	17	10	46	220	1662	624	108	4.76×10^{-5}
(2, 2, 2, 6, 6)	106	3	51	235	0	201728	0	4.74×10^{-5}
(1, 1, 1, 16, 22)	7	20	20	140	244	19	0	4.67×10^{-5}
(1, 2, 4, 4, 10)	65	6	30	196	0	1386	0	4.41×10^{-5}
(4, 4, 10, 10)	66	6	48	223	0	61568	0	4.33×10^{-5}
(1, 4, 4, 4, 4)	57	4	40	252	0	266328	58320	4.19×10^{-5}
(1, 4, 4, 4, 4)	80	7	37	200	0	1968	1408	4.15×10^{-5}
(6, 6, 6, 6)	58	3	43	207	0	190464	0	3.93×10^{-5}
(1, 1, 1, 1, 10, 10)	36	20	20	140	266	26	6	3.82×10^{-5}
(1, 1, 1, 4, 4, 4)	125	12	24	214	351	0	0	3.62×10^{-5}
(4, 4, 10, 10)	14	4	46	219	0	114702	0	3.3×10^{-5}
(1, 1, 1, 1, 10, 10)	33	20	20	140	47	5	0	3.21×10^{-5}
...								...
(3, 3, 3, 3, 3)	6	21	17	234	0	192	0	6.54×10^{-6}
...								
(3, 3, 3, 3, 3)	4	5	49	258	0	24	0	8.17×10^{-7}
...								
(3, 3, 3, 3, 3)	2	49	5	258	6	27	6	1.65×10^{-9}
...								...

We did find tadpole solutions for one of the MIPFs of the quintic, namely MIPF nr. 6. These solutions are the broken and unbroken Pati–Salam $U(4) \times U(2) \times U(2)$ models discussed above.

Acknowledgements

The work reviewed here was done in collaboration with Pascal Anastasopoulos, Tim Dijkstra and Bert Schellekens. I would like to thank them for a very enjoyable and fruitful collaboration. I would like to thank I. Antoniadis, M. Bianchi, R. Blumenhagen, M. Cvetic, M. Douglas, F. Gmeiner, G. Honecker, D. Lüst, F. Quevedo and H. Verlinde for discussions. Thanks to Laurent Baulieu, Mike Douglas and Eliezer Rabinovici for organizing a superb school in a great place on a hot topic. It was a pleasure to be back to Les Houches after 19 years. Thanks also to Pierre Vanhove for the lots of work he put in and his patience.

This work was partially supported by ANR grant NT05-1-41861, INTAS grant, 03-51-6346, RTN contracts MRTN-CT-2004-005104 and MRTN-CT-2004-503369, CNRS PICS 2530 and 3059, the program FP 52 and FP 57 of the Foundation for Fundamental Research of Matter (FOM), Project BFM2002-03610 of the Spanish "Ministerio de Ciencia y Tecnología" and by a European Excellence Grant, MEXT-CT-2003-509661.

References

[1] J. Scherk and J.H. Schwarz, Dual models for nonhadrons, *Nucl. Phys. B* **81** (1974) 118.

[2] T. Yoneya, Connection of dual models to electrodynamics and gravidynamics, *Prog. Theor. Phys.* **51** (1974) 1907.

[3] M.B. Green and J.H. Schwarz, Anomaly cancellation in supersymmetric $D = 10$ gauge theory and superstring theory, *Phys. Lett. B* **149** (1984) 117.

[4] G. Veneziano, Construction of a crossing—symmetric, Regge behaved amplitude for linearly rising trajectories, *Nuovo Cim. A* **57** (1968) 190.

[5] A.N. Schellekens, The landscape 'avant la lettre', [ArXiv:physics/0604134];
L. Susskind, The anthropic landscape of string theory, [ArXiv:hep-th/0302219];
F. Denef and M.R. Douglas, Computational complexity of the landscape. I, [ArXiv:hep-th/0602072].

[6] W. Lerche, A.N. Schellekens and N.P. Warner, Lattices and strings, *Phys. Rept.* **177** (1989) 1;
F. Quevedo, Lectures on superstring phenomenology, [ArXiv:hep-th/9603074];
K.R. Dienes, String theory and the path to unification: A review of recent developments, *Phys. Rept.* **287** (1997) 447 [ArXiv:hep-th/9602045].

[7] G. Pradisi and A. Sagnotti, Open string orbifolds, *Phys. Lett. B* **216** (1989) 59;
P. Horava, Strings on world sheet orbifolds, *Nucl. Phys. B* **327** (1989) 461;
E.G. Gimon and J. Polchinski, Consistency conditions for orientifolds and D-manifolds, *Phys. Rev. D* **54** (1996) 1667 [ArXiv:hep-th/9601038];

C. Angelantonj, M. Bianchi, G. Pradisi, A. Sagnotti and Y.S. Stanev, Chiral asymmetry in four-dimensional open-string vacua, *Phys. Lett. B* **385** (1996) 96 [ArXiv:hep-th/9606169].

[8] I. Antoniadis, E. Kiritsis and T.N. Tomaras, A D-brane alternative to unification, *Phys. Lett. B* **486** (2000) 186 [ArXiv:hep-ph/0004214];
D-brane standard model, *Fortsch. Phys.* **49** (2001) 573 [ArXiv:hep-th/0111269].

[9] G. Aldazabal, L.E. Ibanez, F. Quevedo and A.M. Uranga, D-branes at singularities: A bottom-up approach to the string embedding of the standard model, *JHEP* **0008** (2000) 002 [ArXiv:hep-th/0005067].

[10] I. Antoniadis, E. Kiritsis, J. Rizos and T.N. Tomaras, D-branes and the standard model, *Nucl. Phys. B* **660** (2003) 81 [ArXiv:hep-th/0210263].

[11] P. Anastasopoulos, T.P.T. Dijkstra, E. Kiritsis and A.N. Schellekens, Orientifolds, hypercharge embeddings and the standard model, *Nucl. Phys. B* **759** (2006) 83 [arXiv:hep-th/0605226].

[12] E. Kiritsis, D-branes in standard model building, gravity and cosmology, *Fortsch. Phys.* **52** (2004) 200 [*Phys. Rept.* **421** (2005) (ERRAT, 429, 121-122.2006) 105] [ArXiv:hep-th/0310001];
R. Blumenhagen, M. Cvetic, P. Langacker and G. Shiu, Toward realistic intersecting D-brane models, *Ann. Rev. Nucl. Part. Sci.* **55** (2005) 71 [ArXiv:hep-th/0502005];
R. Blumenhagen, B. Kors, D. Lust and S. Stieberger, Four-dimensional string compactifications with D-branes, orientifolds and fluxes, *Phys. Rept.* **445** (2007) 1 [ArXiv:hep-th/0610327];
A.M. Uranga, Intersecting brane worlds, *Class. Quant. Grav.* **22** (2005) S41; The standard model in string theory from D-branes, *Nucl. Phys. Proc. Suppl.* **171** (2007) 119.

[13] E. Kiritsis and P. Anastasopoulos, The anomalous magnetic moment of the muon in the D-brane realization of the standard model, *JHEP* **0205** (2002) 054 [ArXiv:hep-ph/0201295];
C. Coriano', N. Irges and E. Kiritsis, On the effective theory of low scale orientifold string vacua, *Nucl. Phys. B* **746** (2006) 77 [ArXiv:hep-ph/0510332];
P. Anastasopoulos, M. Bianchi, E. Dudas and E. Kiritsis, Anomalies, anomalous $U(1)$'s and generalized Chern–Simons terms, *JHEP* **0611** (2006) 057 [ArXiv:hep-th/0605225];
D. Berenstein and S. Pinansky, The minimal quiver standard model, *Phys. Rev. D* **75** (2007) 095009 [ArXiv:hep-th/0610104];
C. Coriano, N. Irges and S. Morelli, Stueckelberg axions and the effective action of anomalous Abelian models. II: A $SU(3)C \times SU(2)W \times U(1)Y \times U(1)B$ model and its signature at the LHC, *Nucl. Phys. B* **789** (2008) 133 [ArXiv:hep-ph/0703127];
J. Kumar, A. Rajaraman and J.D. Wells, Probing the Green–Schwarz mechanism at the large hadron collider, [ArXiv:0707.3488][hep-ph];
R. Armillis, C. Coriano and M. Guzzi, Trilinear anomalous gauge interactions from intersecting branes and the neutral currents sector, [ArXiv:0711.3424][hep-ph].

[14] B. Gato-Rivera and A.N. Schellekens, Complete classification of simple current modular invariants for $(Z(p))^{**}k$, *Commun. Math. Phys.* **145** (1992) 85;
M. Kreuzer and A.N. Schellekens, Simple currents versus orbifolds with discrete torsion: A complete classification, *Nucl. Phys. B* **411** (1994) 97 [arXiv:hep-th/9306145].

[15] J. Fuchs, L.R. Huiszoon, A.N. Schellekens, C. Schweigert and J. Walcher, Boundaries, cross-caps and simple currents, *Phys. Lett. B* **495** (2000) 427 [ArXiv:hep-th/0007174].

[16] L.E. Ibanez, F. Marchesano and R. Rabadan, Getting just the standard model at intersecting branes, *JHEP* **0111** (2001) 002 [ArXiv:hep-th/0105155].

[17] T.P.T. Dijkstra, L.R. Huiszoon and A.N. Schellekens, Chiral supersymmetric standard model spectra from orientifolds of Gepner models, *Phys. Lett. B* **609** (2005) 408 [ArXiv:hep-th/0403196];
Supersymmetric standard model spectra from RCFT orientifolds, *Nucl. Phys. B* **710** (2005) 3 [ArXiv:hep-th/0411129].

[18] D. Gepner, Space-time supersymmetry in compactified string theory and superconformal models, *Nucl. Phys. B* **296** (1988) 757.

[19] C. Angelantonj, M. Bianchi, G. Pradisi, A. Sagnotti and Y.S. Stanev, Comments on Gepner models and type I vacua in string theory, *Phys. Lett. B* **387** (1996) 743 [ArXiv:hep-th/9607229].

[20] R. Blumenhagen and A. Wisskirchen, Spectra of 4D, $N = 1$ type I string vacua on non-toroidal CY threefolds, *Phys. Lett. B* **438** (1998) 52 [ArXiv:hep-th/9806131].

[21] G. Aldazabal, E.C. Andres, M. Leston and C. Nunez, Type IIB orientifolds on Gepner points, *JHEP* **0309** (2003) 067 [ArXiv:hep-th/0307183].

[22] I. Brunner, K. Hori, K. Hosomichi and J. Walcher, Orientifolds of Gepner models, [ArXiv:hep-th/0401137].

[23] R. Blumenhagen and T. Weigand, Chiral supersymmetric Gepner model orientifolds, *JHEP* **0402** (2004) 041 [ArXiv:hep-th/0401148].

[24] G. Aldazabal, E.C. Andres and J.E. Juknevich, Particle models from orientifolds at Gepner-orbifold points, *JHEP* **0405** (2004) 054 [ArXiv:hep-th/0403262].

[25] I. Antoniadis and S. Dimopoulos, Splitting supersymmetry in string theory, *Nucl. Phys. B* **715** (2005) 120 [ArXiv:hep-th/0411032].

[26] R. Blumenhagen, M. Cvetic, P. Langacker and G. Shiu, Toward realistic intersecting D-brane models, [ArXiv:hep-th/0502005].

[27] G. Honecker and T. Ott, Getting just the supersymmetric standard model at intersecting branes on the Z_6-orientifold, *Phys. Rev. D* **70** (2004) 126010 [Erratum-ibid. *D* **71** (2005) 069902] [ArXiv:hep-th/0404055].

[28] D. Berenstein, Branes vs. GUTS: Challenges for string inspired phenomenology, [ArXiv:hep-th/0603103].

[29] E. Kiritsis, Duality and instantons in string theory, [ArXiv:hep-th/9906018].

[30] M. Billo, M. Frau, I. Pesando, F. Fucito, A. Lerda and A. Liccardo, Classical gauge instantons from open strings, *JHEP* **0302** (2003) 045 [ArXiv:hep-th/0211250].

[31] R. Blumenhagen, M. Cvetic and T. Weigand, Spacetime instanton corrections in 4D string vacua—the seesaw mechanism for D-brane models, *Nucl. Phys. B* **771** (2007) 113 [ArXiv:hep-th/0609191];
L.E. Ibanez and A.M. Uranga, Neutrino Majorana masses from string theory instanton effects, *JHEP* **0703** (2007) 052 [ArXiv:hep-th/0609213].

[32] M. Bianchi and E. Kiritsis, Non-perturbative and flux superpotentials for type I strings on the Z_3 orbifold, *Nucl. Phys. B* **782** (2007) 26 [ArXiv:hep-th/0702015].

[33] L.E. Ibanez, A.N. Schellekens and A.M. Uranga, Instanton induced neutrino Majorana masses in CFT orientifolds with MSSM-like spectra, *JHEP* **0706** (2007) 011 [ArXiv:0704.1079][hep-th].

[34] E. Kiritsis, talk at the String Phenomenology Conference, Roma 4–7 June 2007.

[35] R. Blumenhagen, M. Cvetic, D. Lust, R. Richter and T. Weigand, Non-perturbative Yukawa couplings from string instantons, [ArXiv:0707.1871][hep-th].

[36] L.E. Ibanez and A.M. Uranga, Instanton induced open string superpotentials and branes at singularities, [ArXiv:0711.1316][hep-th].

[37] C. Bachas, C. Fabre, E. Kiritsis, N.A. Obers and P. Vanhove, Heterotic/type-I duality and D-brane instantons, *Nucl. Phys. B* **509** (1998) 33 [ArXiv:hep-th/9707126];
E. Kiritsis and N.A. Obers, Heterotic/type-I duality in $D < 10$ dimensions, threshold corrections and D-instantons, *JHEP* **9710** (1997) 004 [ArXiv:hep-th/9709058];
C. Bachas, Heterotic versus type I, *Nucl. Phys. Proc. Suppl.* **68** (1998) 348 [ArXiv:hep-th/9710102].

[38] F. Gmeiner, R. Blumenhagen, G. Honecker, D. Lust and T. Weigand, One in a billion: MSSM-like D-brane statistics, *JHEP* **0601** (2006) 004 [ArXiv:hep-th/0510170].

[39] I. Antoniadis, E. Kiritsis and J. Rizos, Anomalous U(1)s in type I superstring vacua, *Nucl. Phys. B* **637** (2002) 92 [ArXiv:hep-th/0204153].

[40] B. Kors and P. Nath, "A Stückelberg extension of the standard model, *Phys. Lett. B* **586** (2004) 366, [ArXiv:hep-ph/0402047];
A supersymmetric Stueckelberg $U(1)$ extension of the MSSM, *JHEP* **0412** (2004) 005 [ArXiv:hep-ph/0406167];
Aspects of the Stückelberg extension, [ArXiv:hep-ph/0503208].

[41] D. Berenstein, V. Jejjala and R.G. Leigh, The standard model on a D-brane, *Phys. Rev. Lett.* **88** (2002) 071602 [ArXiv:hep-ph/0105042].

[42] A. de Rújula, H. Georgi, and S.L. Glashow, in *Fifth Workshop on Grand Unification*, edited by K. Kang, H. Fried, and P. Frampton (World Scientific, Singapore, 1984);
Y. Achiman and B. Stech, in *New Phenomena in Lepton-Hadron Physics*, edited by D.E.C. Fries and J. Wess (Plenum, New York, 1979).

[43] G.K. Leontaris and J. Rizos, A D-brane inspired $U(3)_C \times U(3)_L \times U(3)_R$ model, *Phys. Lett. B* **632** (2006) 710 [ArXiv:hep-ph/0510230];
A D-brane inspired trinification model, [ArXiv:hep-ph/0603203].

[44] J.L. Cardy, Boundary conditions, fusion rules and the Verlinde formula, *Nucl. Phys. B* **324** (1989) 581.

[45] A.M. Uranga, D-brane probes, RR tadpole cancellation and K-theory charge, *Nucl. Phys. B* **598** (2001) 225 [ArXiv:hep-th/0011048].

[46] B. Gato-Rivera and A.N. Schellekens, Remarks on global anomalies in RCFT orientifolds, *Phys. Lett. B* **632** (2006) 728 [ArXiv:hep-th/0510074].

[47] C. Corianó, N. Irges and E. Kiritsis, On the effective theory of low scale orientifold string vacua, [ArXiv:hep-th/0510332].

[48] M. Cvetic, I. Papadimitriou and G. Shiu, Supersymmetric three family $SU(5)$ grand unified models from type IIA orientifolds with intersecting D6-branes, *Nucl. Phys. B* **659** (2003) 193 [Erratum-ibid. *B* **696** (2004) 298] [ArXiv:hep-th/0212177].

[49] C.M. Chen, T. Li and D.V. Nanopoulos, Flipped and unflipped $SU(5)$ as type IIA flux vacua, [ArXiv:hep-th/0604107].

[50] R. Blumenhagen, B. Kors, D. Lust and T. Ott, The standard model from stable intersecting brane world orbifolds, *Nucl. Phys. B* **616** (2001) 3 [ArXiv:hep-th/0107138].

[51] J.R. Ellis, P. Kanti and D.V. Nanopoulos, Intersecting branes flip $SU(5)$, *Nucl. Phys. B* **647** (2002) 235 [ArXiv:hep-th/0206087].

[52] M. Axenides, E. Floratos and C. Kokorelis, $SU(5)$ unified theories from intersecting branes, *JHEP* **0310** (2003) 006 [ArXiv:hep-th/0307255].

[53] C.M. Chen, G.V. Kraniotis, V.E. Mayes, D.V. Nanopoulos and J.W. Walker, A K-theory anomaly free supersymmetric flipped $SU(5)$ model from intersecting branes, *Phys. Lett. B* **625** (2005) 96 [ArXiv:hep-th/0507232].

[54] Y.E. Antebi, Y. Nir and T. Volansky, Solving flavor puzzles with quiver gauge theories, *Phys. Rev. D* **73** (2006) 075009 [ArXiv:hep-ph/0512211].

[55] R. Blumenhagen, L. Gorlich and T. Ott, Supersymmetric intersecting branes on the type IIA T^6/Z_4 orientifold, *JHEP* **0301** (2003) 021 [ArXiv:hep-th/0211059].

[56] G. Honecker, Chiral supersymmetric models on an orientifold of $Z_4 \times Z_2$ with intersecting D6-branes, *Nucl. Phys. B* **666** (2003) 175 [ArXiv:hep-th/0303015].

[57] M. Cvetic, T. Li and T. Liu, Supersymmetric Pati–Salam models from intersecting D6-branes: A road to the standard model, *Nucl. Phys. B* **698** (2004) 163 [ArXiv:hep-th/0403061].

[58] G.K. Leontaris and J. Rizos, A Pati–Salam model from branes, *Phys. Lett. B* **510** (2001) 295 [ArXiv:hep-ph/0012255].

[59] T. Dent, G. Leontaris and J. Rizos, Fermion masses and proton decay in string-inspired $SU(4) \times SU(2)^{\times} U(1)_X$, *Phys. Lett. B* **605** (2005) 399 [ArXiv:hep-ph/0407151].

[60] E. Kiritsis, D-branes in standard model building, gravity and cosmology, *Fortsch. Phys.* **52** (2004) 200 [*Phys. Rept.* **421** (2005) 105] [ArXiv:hep-th/0310001].

[61] L.F. Li, Group theory of the spontaneously broken gauge symmetries, *Phys. Rev. D* **9** (1974) 1723.

[62] G. Aldazabal, E. Andres and J.E. Juknevich, On SUSY standard-like models from orbifolds of $D = 6$ Gepner orientifolds, [ArXiv:hep-th/0603217].

[63] B. Gato-Rivera and A.N. Schellekens, Non-supersymmetric Tachyon-free Type-II and Type-I closed strings from RCFT, *Phys. Lett. B* **656** (2007) 127 [ArXiv:0709.1426][hep-th].

Course 3

LECTURES ON SUPERSYMMETRY BREAKING

Kenneth Intriligator[1] and Nathan Seiberg[2]

[1] *Department of Physics, University of California, San Diego, La Jolla, CA 92093 USA*
[2] *School of Natural Sciences, Institute for Advanced Study, Princeton, NJ 08540 USA*

*C. Bachas, L. Baulieu, M. Douglas, E. Kiritsis, E. Rabinovici, P. Vanhove, P. Windey
and L.F. Cugliandolo, eds.*
Les Houches, Session LXXXVII, 2007
String Theory and the Real World: From Particle Physics to Astrophysics
© *2008 Published by Elsevier B.V.*

Contents

1. Introduction

With the advent of the LHC it is time to review old model building issues leading to phenomena which could be discovered, or disproved, by the LHC. Supersymmetry (SUSY) is widely considered as the most compelling new physics that the LHC could discover. It gives a solution to the hierarchy problem, leads to coupling constant unification and has dark matter candidates.

Clearly, the standard model particles are not degenerate with their superpartners, and therefore supersymmetry should be broken. To preserve the appealing features of supersymmetry, this breaking must be spontaneous, rather than explicit breaking. This means that the Lagrangian is supersymmetric, but the vacuum state is not invariant under supersymmetry.

Furthermore, as was first suggested by Witten [1], we would like the mechanism which spontaneously breaks supersymmetry to be dynamical. This means that it arises from an exponentially small effect, and therefore it naturally leads to a scale of supersymmetry breaking, M_s, which is much smaller than the high energy scales in the problem M_{cutoff} (which can be the Planck scale or the grand unified scale):

$$M_s = M_{cutoff} e^{-c/g(M_{cutoff})^2} \ll M_{cutoff}. \tag{1.1}$$

This can naturally lead to hierarchies. For example, the weak scale m_W can be dynamically generated, explaining why $m_W/m_{Pl} \sim 10^{-17}$.

In these lectures, we will focus on the key conceptual issues and mechanisms for supersymmetry breaking, illustrating them with the simplest examples. We will not discuss more detailed model building questions, such as the question of how the supersymmetry breaking is mediated to the MSSM, and what the experimental signatures of the various mediation schemes are. These are very important topics, which deserve separate sets of lectures. Also, we will not discuss supersymmetry breaking by Fayet–Iliopoulos terms [2].

We will assume that the readers (and audience in the lectures) have some basic familiarity with supersymmetry. Good textbooks are [3–7].

As seen from the supersymmetry algebra,

$$\{Q_\alpha, \overline{Q}_{\dot\alpha}\} = 2P_{\alpha\dot\alpha}, \tag{1.2}$$

the vacuum energy

$$\langle \psi | \mathcal{H} | \psi \rangle \propto \sum_{\alpha} \left| Q_\alpha | \psi \rangle \right|^2 + \sum_{\dot\alpha} \left| \overline{Q}_{\dot\alpha} | \psi \rangle \right|^2 \geq 0 \tag{1.3}$$

is an order parameter for supersymmetry breaking. Supersymmetry is spontaneously broken if and only if the vacuum has non-zero energy,[1]

$$V_{vac} = M_s^4. \tag{1.4}$$

In the case of dynamical supersymmetry breaking (DSB), the scale M_s is generated by dimensional transmutation, as in (1.1).

As with the spontaneous breaking of an ordinary global symmetry, the broken supersymmetry charge Q does not exist in an infinite volume system. Instead, the supersymmetry current S exists, and its action on the vacuum creates a massless particle—the Goldstino. (The supercharge tries to create a zero momentum Goldstino, which is not normalizable.) In the case of supergravity, where the symmetry (1.2) is gauged, we have the standard Higgs mechanism and the massless Goldstino is "eaten" by the gravitino.

There are many challenges in trying to implement realistic realizations of dynamical supersymmetry breaking. A first challenge, which follows from the Witten index [8], is that dynamical supersymmetry breaking, where the true vacuum is static and has broken supersymmetry, seems non-generic, requiring complicated looking theories. On the other hand, accepting the possibility that we live in a metastable vacuum improves the situation. As even very simple theories can exhibit metastable dynamical supersymmetry breaking, it could be generic [9]. (Particular models of metastable supersymmetry breaking have been considered long ago, e.g. a model [10], which we review below.)

Another challenge is the relation [11] between R-symmetry and broken supersymmetry. Generically, there is broken supersymmetry if and only if there is an R-symmetry. As we will also discuss, there is broken supersymmetry in a metastable state if and only if there is an approximate R-symmetry. For building realistic models, an unbroken R-symmetry is problematic. It forbids Majorana gaugino masses. Having an exact, but spontaneously broken R-symmetry is also problematic, it leads to a light R-axion (though including gravity can help).[2] We are thus led to explicitly break the R-symmetry. Ignoring gravity, this then means that we should live in a metastable state!

---

[1] In these lectures we focus on global SUSY, $M_{pl} \to \infty$. In supergravity we can add an arbitrary negative constant to the vacuum energy, via $\Delta W = $ const., so the cosmological constant can still be tuned to the observed value.

[2] Including gravity, the R-symmetry needs to be explicitly broken, in any case, by the $\Delta W = $ const., needed to get a realistic cosmological constant. It is possible that this makes the R-axion sufficiently massive [12].

The outline of these lectures is as follows. In the next section, we consider theories in which the supersymmetry breaking can be seen semiclassically. Such theories can arise as the low energy theory of another microscopic theory. Various general points about supersymmetry breaking (or restoration) are illustrated, via several simple examples.

In Section 3, we give a lightning review of $\mathcal{N} = 1$ supersymmetric QCD (SQCD), with various numbers of colors and flavors. Here we will be particularly brief. The reader can consult various books and reviews, e.g. [6, 7, 13–16], for more details.

In Section 4, we discuss dynamical supersymmetry breaking (DSB), where the supersymmetry breaking is related to a dynamical scale Λ, and thus it is non-perturbative in the coupling. Using the understood dynamics of SQCD, it is possible to find an effective Lagrangian in which supersymmetry breaking can be seen semiclassically. We will discuss only four characteristic examples, demonstrating four different mechanisms of DSB.

2. Semiclassical spontaneous supersymmetry breaking

In this section we consider theories with chiral superfields Φ^a, a smooth Kähler potential $K(\Phi, \overline{\Phi})$ and a superpotential $W(\Phi)$. For simplicity we will ignore the possibility of adding gauge fields. A detailed analysis of their effect will be presented in [17]. The Kähler potential leads to the metric on field space

$$g_{a\bar{a}} = \partial_a \partial_{\bar{a}} K, \tag{2.1}$$

which determines the Lagrangian of the scalars

$$
\begin{aligned}
\mathcal{L}_{scalars} &= g_{a\bar{a}} \partial_\mu \Phi^a \partial^\mu \overline{\Phi}^{\bar{a}} - V(\Phi, \overline{\Phi}) \\
V &= g^{a\bar{a}} \partial_a W \partial_{\bar{a}} \overline{W}.
\end{aligned}
\tag{2.2}
$$

It is clear from the scalar potential V that supersymmetric ground states, which must have zero energy, are related to the critical points of W; i.e. points where we can solve

$$\partial_a W(\Phi^a) = 0 \quad \forall a. \tag{2.3}$$

If no such point exists, it means that the system does not have supersymmetric ground states.

However, before we conclude in this case that supersymmetry is spontaneously broken we should also exclude the possibility that the potential slopes to zero at infinity. Roughly, in this case the system has "a supersymmetric state at infinity." More precisely, it does not have a ground state at all!

2.1. The simplest example

Consider a theory of a single chiral superfield X, with linear superpotential with coefficient f (with units of mass square),

$$W = fX, \tag{2.4}$$

and canonical Kähler potential

$$K = K_{can} = \overline{X}X. \tag{2.5}$$

Supersymmetry is spontaneously broken by the expectation value of the F-component of X, $\overline{F}_X = -f$. Using (2.2) the potential is $V = |f|^2$. It is independent of X, so there are classical vacua for any $\langle X \rangle$.

Supersymmetric theories often have a continuous manifold of supersymmetric vacua which are usually referred to as "moduli space of vacua." However, in the case where supersymmetry is broken, such a space is not robust: this nonsupersymmetric degeneracy of vacua is often lifted once radiative corrections are taken into account. Therefore, we prefer to refer to this space as *pseudomoduli space of vacua*. The example we study here is free, and therefore the space of vacua remains present even in the quantum theory. We will see below examples of the more typical situation, in which the classical theory has a pseudomoduli space of nonsupersymmetric vacua, but the quantum corrections lift the degeneracy.

The exactly massless Goldstino is ψ_X, and its complex scalar partner X is the classically massless pseudomodulus. Note that there is a $U(1)_R$ symmetry, with $R(X) = 2$. For $\langle X \rangle \neq 0$ it is spontaneously broken, and the corresponding massless Goldstone boson is the phase of the field X.

Deforming (2.4) by any superpotential interactions, say a degree n polynomial in X, leads to $n - 1$ supersymmetric vacua. For example, if we add $\Delta W = \frac{1}{2}\epsilon X^2$, there is a vacuum with unbroken supersymmetry at $\langle X \rangle = -f/\epsilon$. This deformation lifts the pseudomoduli space by creating a potential $|f + \epsilon X|^2$ over it. We can also see that supersymmetry is not broken from the fact that ψ_X now has mass ϵ, so there is no massless Goldstino. Note also that any such ΔW deformations of (2.4) explicitly break the $U(1)_R$ symmetry; the fact that they lead to supersymmetric vacua illustrates a general connection between R-symmetry and supersymmetry breaking, which will be developed further below.

2.2. The simplest example but with more general Kähler potential

Consider again the theory of Section 2.1 with superpotential (2.4), but with a general Kähler potential $K(X, \overline{X})$. Of course, this theory is not renormalizable. It should be viewed either as a classical field theory or as a quantum field theory with a cutoff Λ. More physically, such a theory can be the low energy approximation of another, microscopic theory, which is valid at energies larger than Λ.

The potential,

$$V = K_{X\overline{X}}^{-1}|f|^2 \tag{2.6}$$

lifts the degeneracy along the pseudomoduli space of the previous example. Let us suppose that the Kähler potential K is smooth. (Non-smooth K signals the need to include additional degrees of freedom, in the low-energy effective field theory at the singularity. An example of this case is discussed in the next subsection.) For smooth K, the potential (2.6) is non-vanishing, and thus there is no supersymmetric vacuum.

Before concluding that supersymmetry is spontaneously broken, we should consider the behavior at $|X| \to \infty$. If there is any direction along which $\lim_{|X| \to \infty} K_{X\overline{X}}$ diverges, then V slopes to zero at infinity and the system does not have a ground state. If $\lim_{|X| \to \infty} K_{X\overline{X}}$ vanishes in all directions, the potential rises at infinity and it has a supersymmetry breaking global minimum for some finite X. Finally, if there are directions along which $\lim_{|X| \to \infty} K_{X\overline{X}}$ is finite, the potential approaches a constant along these directions and the global minimum of the potential needs a more detailed analysis.

Consider the behavior of the system near a particular point, say $X \approx 0$. Let

$$K = X\overline{X} - \frac{c}{|\Lambda|^2}(X\overline{X})^2 + \cdots, \tag{2.7}$$

with positive c.[3] Then there is a locally stable nonsupersymmetric vacuum at $X = 0$. In this vacuum, the scalar component of X gets mass $m_X^2 = 4c|f|^2/|\Lambda|^2$. The fermion ψ_X is the exactly massless Goldstino. Note also that if $K(X, \overline{X})$ depends only on $X\overline{X}$, then there is a $U(1)_R$ symmetry, which is unbroken if the vacuum is at $X = 0$. This ground state can be the global minimum of the potential. Alternatively, it can be only a local minimum, with either another minimum of lower energy or no minimum at all if the system runs away to infinity.

If $X = 0$ is not the global minimum of the potential, the state at $X = 0$ is metastable. If the theory is sufficiently weakly coupled, the tunneling out of this vacuum can be highly suppressed and this vacuum can be very long lived. We see that it is easy to find examples where supersymmetry is broken in a long lived metastable state. (Though we have not yet demonstrated what physical dynamics leads to such features in the Kähler potential.)

---

[3]The parameter Λ in (2.7) determines the scale of the features in the potential. When this theory arises as the low energy approximation of another theory, this parameter Λ is typically the scale above which the more microscopic theory is valid.

Let us consider again the theory with Kähler potential (2.7), but deform the superpotential (2.4) to

$$W = fX + \frac{1}{2}\epsilon X^2, \tag{2.8}$$

taking ϵ as a small parameter. There is now a supersymmetric vacuum at

$$\langle X \rangle_{susy} = -f/\epsilon, \tag{2.9}$$

which is very far from the origin. On the other hand, for X near the origin, we find for the potential

$$V(X, \overline{X}) = (K_{X\overline{X}})^{-1} |f + \epsilon X|^2 \tag{2.10}$$

$$= |f|^2 + \overline{f}\epsilon X + f\overline{\epsilon}\overline{X} + \frac{4c|f|^2}{|\Lambda|^2}|X|^2 + \cdots \quad (X \approx 0, \; \epsilon \ll 1).$$

There is a local minimum, with broken supersymmetry, at

$$\langle X \rangle_{meta} = -\frac{\overline{\epsilon}|\Lambda|^2}{4c\overline{f}}. \tag{2.11}$$

For $|\epsilon| \ll \sqrt{c}|f/\Lambda|$, this supersymmetry breaking vacuum is very far from the supersymmetric vacuum (2.9). The metastable state (2.11) can thus be very long lived.

At first glance, there is a small puzzle with the broken supersymmetry vacuum (2.11). The superpotential (2.8) gives a mass ϵ to the fermion ψ_X, whereas any vacuum with broken supersymmetry must have an *exactly massless* Goldstino. The Goldstino must be exactly massless, regardless of whether the supersymmetry breaking state is a local or global minimum of the potential. The resolution of the apparent puzzle is that

$$\int d^4\theta K \supset K_{XX\overline{X}} \overline{F}_X \psi_X \psi_X \tag{2.12}$$

and evaluating this term in the vacuum (2.11), with $\overline{F}_X \approx -f$, exactly cancels the $\epsilon\psi\psi$ term coming from the superpotential. So there is indeed an exactly massless Goldstino, ψ_X, consistent with the supersymmetry breaking in the metastable state.

2.3. Additional degrees of freedom can restore supersymmetry

Let us consider a renormalizable theory of two chiral superfields, X and q, with canonical Kähler potential, $K = X\overline{X} + q\overline{q}$. We modify the example of Section 2.1

by coupling the field X to the additional field q via

$$W = \frac{1}{2}hXq^2 + fX, \tag{2.13}$$

where h is the coupling constant. The field q gets a mass from an X expectation value (an added mass term $\Delta W = \frac{1}{2}Mq^2$ can be eliminated by a shift of X). There is a $U(1)_R$ symmetry, with $R(X) = 2$, and $R(q) = 0$, and also a $\mathbf{Z}_2$ symmetry $q \to -q$.

The potential

$$V = |hXq|^2 + \left| \frac{1}{2}hq^2 + f \right|^2 \tag{2.14}$$

does not break supersymmetry. There are two supersymmetric vacua, at

$$\langle X \rangle_{susy} = 0, \qquad \langle q \rangle_{susy} = \pm\sqrt{-2f/h}. \tag{2.15}$$

The additional degrees of freedom, q, as compared with the example of Section 2.1, have restored supersymmetry.

Note that the potential (2.14) also has a supersymmetry breaking pseudoflat direction with $\langle q \rangle = 0$, and arbitrary $\langle X \rangle$, with $V = |f|^2$. It reflects the fact that for large X the q fields are massive, can be integrated out, and the low energy theory is then the same as that of Section 2.1. The spectrum of the massive q fields depends on X, and is given by

$$m_0^2 = |hX|^2 \pm |hf|; \qquad m_{1/2} = hX. \tag{2.16}$$

We see, however, that this pseudomoduli space has a tachyon for

$$|X|^2 < \left| \frac{f}{h} \right|. \tag{2.17}$$

In the region (2.17), the potential can decrease along the $\langle q \rangle$ direction, down to the supersymmetric vacua (2.15).

2.4. *An example with a runaway [18]*

Consider a renormalizable theory of two chiral superfields, X and Y, with canonical Kähler potential, and superpotential

$$W = \frac{1}{2}hX^2Y + fX. \tag{2.18}$$

There is a $U(1)_R$ symmetry, with $R(X) = 2$, and $R(Y) = -2$. The potential is

$$V = \left| \frac{1}{2}hX^2 \right|^2 + |hXY + f|^2 . \tag{2.19}$$

It is impossible for both terms to vanish, so the theory does not have super-symmetric ground states. As usual, before concluding that supersymmetry is spontaneously broken, we must examine for runaway directions. Indeed, taking $X = -f/hY$ the potential has a runaway direction as $Y \to \infty$:

$$V \to \left| \frac{f^2}{2hY^2} \right|^2 \to 0. \tag{2.20}$$

There is no static vacuum, but supersymmetry is asymptotically restored as $Y \to \infty$.

For large $|Y|$ the supersymmetry breaking is small, and the mass of X is large, so we can describe the theory by a supersymmetric low-energy effective La-grangian with X integrated out. Integrating out X in (2.18) we find the effective superpotential

$$W_{eff} = -\frac{f^2}{2hY} \tag{2.21}$$

which is consistent with the R-symmetry, and leads to the potential (2.20).

2.5. O'Raifeartaigh-type models

Here we discuss models of supersymmetry breaking which arise in renormal-izable field theories; i.e. unlike the example of Section 2.2, we will examine classical theories with a canonical Kähler potential (for a recent analysis of such models see e.g. [19]).

The simplest version of this class of models has three chiral superfields, X_1, X_2, and ϕ, with canonical Kähler potential

$$K_{cl} = \overline{X}_1 X_1 + \overline{X}_2 X_2 + \overline{\phi}\phi \tag{2.22}$$

and superpotential

$$W = X_1 g_1(\phi) + X_2 g_2(\phi) \tag{2.23}$$

with quadratic polynomials $g_{1,2}(\phi)$. This theory has a $U(1)_R$ symmetry, with $R(X_1) = R(X_2) = 2$, and $R(\phi) = 0$. The tree-level potential for the scalars is

$$V_{tree} = |F_{X_1}|^2 + |F_{X_2}|^2 + |F_\phi|^2 \tag{2.24}$$

with

$$-\overline{F}_{X_1} = \partial_{X_1} W \tag{2.25}$$
$$= g_1(\phi), \quad -\overline{F}_{X_2} = g_2(\phi), \quad -\overline{F}_\phi = X_1 g_1'(\phi) + X_2 g_2'(\phi).$$

We are interested in the minima of this potential.

We can always choose X_1 and X_2 to set $F_\phi = 0$. But, for generic functions $g_1(\phi)$ and $g_2(\phi)$, we cannot simultaneously solve $g_1(\phi) = 0$ and $g_2(\phi) = 0$, so F_{X_1} or F_{X_2} is non-zero, and hence supersymmetry is generically broken. There is a one-complex dimensional classical pseudomoduli space of non-supersymmetric vacua, since only one linear combination of X_1 and X_2 is constrained by the condition that $F_\phi = 0$. Setting $F_\phi = 0$ ensures that the vacuum satisfies the X_1 and X_2 equations of motion, $\partial_{X_i} V_{tree} = 0$. We still need to impose $\partial_\phi V_{tree} = 0$, which requires that $\langle \phi \rangle$ solve

$$\overline{g_1(\phi)} g_1'(\phi) + \overline{g_2(\phi)} g_2'(\phi) = 0. \tag{2.26}$$

Expanding to quadratic order in δX_1, δX_2, and $\delta \phi$ yields the mass matrix m_0^2 of the massive scalars; the eigenvalues of this matrix must all be non-negative, of course, if we are expanding around a (local) minimum of the potential. The fermion mass terms are given by

$$\mathcal{L} \supset (X_1 g_1''(\phi) + X_2 g_2''(\phi)) \psi_\phi \psi_\phi + (g_1'(\phi)\psi_{X_1} + g_2'(\phi)\psi_{X_2})\psi_\phi. \tag{2.27}$$

It is easy to see that there is a massless eigenvector, corresponding to the massless Goldstino.

Example 1—the basic O'Raifeartaigh model [20]

As a special case of the above class of models, consider[4] $g_1(\phi) = \frac{1}{2}h\phi^2 + f$, $g_2(\phi) = m\phi$. It is characterized by the discrete $\mathbf{Z}_2$ symmetry under which ϕ and X_2 are odd.

For convenience, let us also write it as

$$W = \frac{1}{2}hX\phi_1^2 + m\phi_1\phi_2 + fX, \tag{2.28}$$

where we denote $X = X_1$, $\phi_2 = X_2$, and $\phi_1 = \phi$. Note that, for $m \to 0$, the field ϕ_2 decouples, and what remains in (2.28) is the theory of Section 2.3, which we have seen does not break supersymmetry. For $m \neq 0$, it does break supersymmetry, as in the general case discussed above, as there is no simultaneous solution

---

[4]If, instead, $g_{1,2}$ are even quadratic polynomials: $g_i(\phi) = \frac{1}{2}h_i\phi^2 + f_i$, a simple change of variables shows that the theory decouples to a free field which breaks supersymmetry as in Section 2.1 and the example of Section 2.3.

of $g_1(\phi_1) = \frac{1}{2}h\phi_1^2 + f = 0$ and $g_2(\phi_1) = m\phi_1 = 0$. The potential rises for large ϕ_1 and ϕ_2, so these fields do not have runaway directions. The minima of the potential form a one-complex dimensional pseudomoduli space of degenerate, non-supersymmetric vacua, with $\langle X \rangle$ arbitrary.

The equation (2.26) is a cubic equation for ϕ_1. The solution with minimum energy depends on the parameter

$$y \equiv \left| \frac{hf}{m^2} \right| . \tag{2.29}$$

Consider the case $y < 1$. Then the potential is minimized[5] by $F_{\phi_2} = 0$, with value

$$V_{min} = |F_X|^2 = |f|^2, \tag{2.30}$$

at $\phi_1 = \phi_2 = 0$ and arbitrary X.

The fermion ψ_X is the exactly massless Goldstino. The scalar component of X is a classical pseudomodulus. The classical mass spectrum of the ϕ_1 and ϕ_2 fields can be easily computed. For the two, two-component fermions, the eigenvalues are

$$m_{1/2}^2 = \frac{1}{4}\left(|hX| \pm \sqrt{|hX|^2 + 4|m|^2} \right)^2, \tag{2.31}$$

and for the four real scalars the mass eigenvalues are

$$m_0^2 = \left(|m|^2 + \frac{1}{2}\eta|hf| + \frac{1}{2}|hX|^2 \right.$$
$$\left. \pm \frac{1}{2}\sqrt{|hf|^2 + 2\eta|hf||hX|^2 + 4|m|^2|hX|^2 + |hX|^4} \right), \tag{2.32}$$

where $\eta = \pm 1$. We see that, as in (2.16), the spectrum changes along the pseudomoduli space parameterized by X; these vacua are physically distinct.

The parameter y sets the relative size of the mass splittings, corresponding to supersymmetry being broken, between (2.31) and (2.32). For $y \ll 1$, the spectrum (2.31) and (2.32) is approximately supersymmetric, whereas for $y \sim 1$ supersymmetry is badly broken. (In particular, for $y = 1$, there is a massless real scalar in (2.32) for all X, whereas the fermions (2.31) are all massive.)

We can write (2.28) as $W = \frac{1}{2}M_{ij}\phi^i\phi^j + fX$, where $M = \begin{pmatrix} hX & m \\ m & 0 \end{pmatrix}$, and the supersymmetry breaking can be seen from the fact that $\det M = -m^2$ is non-zero

---

[5]There is a second order phase transition at $y = 1$, where this minimum splits to two minima and a saddle point. Here we will not analyze the phase $y > 1$. See e.g. [9] for a detailed analysis.

and X independent. This can be generalized to similar models, with more fields ϕ^i, and M_{ij} such that det M is non-zero and independent of X [9].

Example 2—supersymmetry breaking in a metastable state [10]

We noted above that the theory (2.23) breaks supersymmetry for generic functions $g_1(\phi)$ and $g_2(\phi)$, because we generically cannot solve $g_1(\phi) = g_2(\phi) = 0$. Let us consider the case of a *non-generic* superpotential, where there is a solution $\langle\phi\rangle_{susy}$ of $g_1(\phi) = g_2(\phi) = 0$. In this case, there are supersymmetric vacua. There can still, however, be metastable vacua with broken supersymmetry.

As a particular example, consider

$$g_1(\phi) = h\phi(\phi - m_1), \qquad g_2(\phi) = m_2(\phi - m_1). \tag{2.33}$$

(This theory was first analyzed in [10] and was recently reexamined in [19].) There is a moduli space of supersymmetric vacua at

$$\langle\phi\rangle_{susy} = m_1; \qquad \langle X_2\rangle_{susy} = -\frac{hm_1}{m_2}\langle X_1\rangle_{susy}, \tag{2.34}$$

with arbitrary $\langle X_1\rangle_{susy}$. The equation (2.26) is a cubic equation for ϕ, and this moduli space of supersymmetric vacua corresponds to one root of this cubic equation. For $|hm_1/m_2|^2 > 8$, there is also a pseudomoduli space of supersymmetry violating minima of the potential at

$$\langle\phi_1\rangle_{meta} \approx \left|\frac{m_2}{hm_1}\right|^2 m_1, \qquad \langle X_2\rangle_{meta} \approx \frac{hm_1}{m_2}\langle X_1\rangle_{meta} \quad \text{for} \quad \left|\frac{hm_1}{m_2}\right| \gg 1 \tag{2.35}$$

with arbitrary $\langle X_1\rangle_{meta}$. These metastable false vacua, in which supersymmetry is broken, become parametrically long lived as $|hm_1/m_2|$ is increased [10]. (The third root of the cubic equation (2.26) is a saddle point.)

2.6. *Metastable SUSY breaking in a modified O'Raifeartaigh model [17]*

Let us modify the original, basic O'Raifeartaigh model by adding to the superpotential (2.28) a small correction

$$W = \frac{1}{2}hX\phi_1^2 + m\phi_1\phi_2 + fX + \frac{1}{2}\epsilon m\phi_2^2 \tag{2.36}$$

with $|\epsilon| \ll 1$. This added term breaks the $U(1)_R$ symmetry. It has an interesting effect: it leads to metastable supersymmetry breaking. A similar model, but with the ϵ term in (2.36) replaced with $\frac{1}{2}\epsilon mX^2$ was considered in [21], with similar conclusions to ours here. (Note that adding $\Delta W = \frac{1}{2}b\phi_1^2$ has no physical effect; it can simply be eliminated by shifting X by an appropriate constant.)

The potential is now

$$V_{tree} = |F_X|^2 + |F_{\phi_1}|^2 + |F_{\phi_2}|^2 \tag{2.37}$$

with

$$-\overline{F}_X = \frac{1}{2}h\phi_1^2 + f, \quad -\overline{F}_{\phi_1} = hX\phi_1 + m\phi_2,$$
$$-\overline{F}_{\phi_2} = m\phi_1 + \epsilon m\phi_2. \tag{2.38}$$

Because of the modification of the superpotential by the last term in (2.36) two new supersymmetric minima appear at

$$\langle\phi_1\rangle_{susy} = \pm\sqrt{-2f/h}, \quad \langle\phi_2\rangle_{susy} = \mp\frac{1}{\epsilon}\sqrt{-2f/h}, \quad \langle X\rangle_{susy} = \frac{m}{h\epsilon}. \tag{2.39}$$

However, for small ϵ and $y = \left|\frac{hf}{m^2}\right| < 1$, the potential near the previous supersymmetry breaking minimum $\phi_1 = \phi_2 = 0$ is not modified a lot.

Strictly, this theory does not break supersymmetry—it has supersymmetric ground states at (2.39). However, the generalization of the eigenvalues (2.32), to include ϵ, remains non-tachyonic for

$$\left|X - \frac{m}{h\epsilon}\right|^2 > \left(\frac{1}{|\epsilon|^2} + 1\right)\left|\frac{f}{h}\right|. \tag{2.40}$$

Therefore, most of the pseudomoduli space of vacua of the $\epsilon = 0$ theory remains locally stable, and the tachyon exists only in a neighborhood of the supersymmetric value (2.39). In particular, for small ϵ and $y < 1$, the region near $X = 0$ is locally stable.

As $\epsilon \to 0$ the supersymmetry preserving vacua (2.39) are pushed to infinity until finally, for $\epsilon = 0$ they are not present, and we are left with only the pseudomoduli space of nonsupersymmetric vacua. A more detailed analysis will be presented in [17].

2.7. Supersymmetry breaking by rank condition [9]

Our final example in this section is more complicated. In involves several fields transforming under a large symmetry group. The fields X_i in (2.23) are replaced by a matrix of fields. Apart from the intrinsic interest in this example, it will also be useful in our discussion in Section 4.

Consider a theory with fields φ, $\widetilde{\varphi}$, Φ, and parameters f, with global[6] symmetries

	$SU(n)$	$SU(N_f)_L$	$SU(N_f)_R$	$U(1)_V$	$U(1)_R$	$U(1)_A$	
φ	**n**	$\overline{\mathbf{N_f}}$	**1**	1	0	1	
$\widetilde{\varphi}$	$\overline{\mathbf{n}}$	**1**	$\mathbf{N_f}$	-1	0	1	(2.41)
Φ	**1**	$\mathbf{N_f}$	$\overline{\mathbf{N_f}}$	0	2	-2	
f	**1**	$\overline{\mathbf{N_f}}$	$\mathbf{N_f}$	0	0	2	

We will take

$$n < N_f. \tag{2.42}$$

We take the Kähler potential K to be canonical, and the superpotential is

$$W = h\,\text{Tr}\,\Phi\varphi\widetilde{\varphi}^T + \text{Tr}\,f\Phi, \tag{2.43}$$

where h is a coupling constant and the trace is over the global symmetry indices. The last term in (2.43) respects the symmetries in (2.41) because of the transformation laws of the parameter f. Alternatively, the parameter f breaks $SU(N_f) \times SU(N_f)$ to a subgroup, and breaks $U(1)_A$, but it does not break the $SU(n)$ symmetry or the R-symmetry.

Supersymmetry is broken when (2.42) is satisfied. Consider the F-component of Φ

$$-F_\Phi^\dagger = h\varphi\widetilde{\varphi}^T + f \tag{2.44}$$

(here we use $\dagger$ even in the classical theory because of the flavor indices of Φ). This is an $N_f \times N_f$ matrix relation. Because of (2.42), the first term is a matrix of rank n. On the other hand, we can take f to have rank larger than n, up to rank N_f. Therefore, if the rank of f is larger than n, and in particular if f is proportional to the unit matrix $\mathbb{1}_{N_f}$, then (2.44) cannot vanish, $F_\Phi \neq 0$, and supersymmetry is broken.

When (2.42) is not satisfied, there are supersymmetric vacua, as in the example (2.13), which is similar to the case $n = N_f = 1$. The difference is that, when (2.42) is satisfied, there are not enough additional degrees of freedom, φ and $\widetilde{\varphi}$, at $\Phi = 0$ to restore supersymmetry.

---

[6]For our discussion in Section 4, we will take the $SU(n)$ symmetry to be gauged, but IR free. In that case, the $U(1)_R$ symmetry below is anomalous (a linear combination of $U(1)_R$ and $U(1)_A$ is anomaly free, but broken by the parameter f), but is restored as an approximate, accidental symmetry in the IR. Also, the $SU(n)$ D-terms will vanish in the vacua. The results discussed here will be completely unaffected by the weak gauging of $SU(n)$ in Section 4.

For simplicity, we take $f \equiv -h\mu^2 \mathbb{1}_{N_f}$, proportional to the unit matrix. The minimum of the potential is then at

$$V = (N_f - n)|h\mu^2|^2 \tag{2.45}$$

and it occurs along the pseudomoduli space

$$\Phi = \begin{pmatrix} 0 & 0 \\ 0 & \Phi_0 \end{pmatrix}, \quad \varphi = \begin{pmatrix} \varphi_0 \\ 0 \end{pmatrix}, \quad \tilde{\varphi} = \begin{pmatrix} \tilde{\varphi}_0 \\ 0 \end{pmatrix}, \quad \text{with } \varphi_0 \tilde{\varphi}_0^T = \mu^2 \mathbb{1}_n, \tag{2.46}$$

and arbitrary Φ_0, φ_0 and $\tilde{\varphi}_0$ (subject to the constraint in (2.46)). The first entries in (2.46) are the first n components, and the second are the remaining $N_f - n$ components, so e.g. Φ_0 is a $(N_f - n) \times (N_f - n)$ square matrix. The non-zero F terms are $F_{\Phi_0} = \overline{h}\overline{\mu}^2 \mathbb{1}_{N_f - n}$. The massless Goldstino comes from the fermionic components of Φ_0.

2.8. One-loop lifting of pseudomoduli

As we have seen in the examples above, models of tree-level spontaneous supersymmetry breaking generally have classical moduli spaces of degenerate, non-supersymmetric, vacua. Indeed, the massless Goldstino is in a chiral superfield (for F-term breaking), whose scalar component is a classical pseudomodulus. The example of Section 2.3 shows that this is the case even if this space of classical vacua becomes unstable in a region in field space. The example of Section 2.7 (2.46) shows that there can be additional pseudomoduli. We said above that we should use the term "pseudomoduli" space for the space of classical non-supersymmetric vacua, because the degeneracy between these vacua is usually lifted once quantum corrections are taken into account. In this section, we review how this comes about.

We will be interested in the one-loop effective potential (the Coleman–Weinberg potential) for the pseudomoduli (such as X), which comes from computing the one-loop correction to the vacuum energy

$$
\begin{aligned}
V_{eff}^{(1)} &= \frac{1}{64\pi^2} \text{STr} \left(\mathcal{M}^4 \log \frac{\mathcal{M}^2}{M_{cutoff}^2} \right) \\
&\equiv \frac{1}{64\pi^2} \left[\text{Tr} \left(m_B^4 \log \frac{m_B^2}{M_{cutoff}^2} \right) - \text{Tr} \left(m_F^4 \log \frac{m_F^2}{M_{cutoff}^2} \right) \right],
\end{aligned}
\tag{2.47}
$$

where m_B^2 and m_F^2 are the tree-level boson and fermion masses, as a function of the expectation values of the pseudomoduli, and M_{cutoff} is a UV cutoff. In (2.47), $\mathcal{M}^2$ stands for the classical mass-square matrix of the various fields of the theory.

We would like to make two comments about the divergences in this expression:

1. In non-supersymmetric theories the effective potential includes also a quartic divergent term proportional to $M_{cutoff}^4 \, \mathrm{STr}\, \mathbb{1}$ and a quadratic divergent term proportional to $M_{cutoff}^2 \, \mathrm{STr}\, \mathcal{M}^2$. They vanish in supersymmetric theories.

2. The logarithmic divergent term $(\log M_{cutoff}) \, \mathrm{STr}\, \mathcal{M}^4$ in (2.47) can be absorbed into the renormalization of the coupling constants appearing in the tree-level vacuum energy V_0 (see below). In particular, $\mathrm{STr}\, \mathcal{M}^4$ is independent of the pseudomoduli.

For completeness, we recall the standard expressions for these masses. For a general theory with k chiral superfields, Φ^a, with canonical classical Kähler potential, $K = \Phi^a \overline{\Phi}^a$, and superpotential $W(\Phi^a)$:

$$
m_0^2 = \begin{pmatrix} \overline{W}^{ac} W_{cb} & \overline{W}^{abc} W_c \\ W_{abc} \overline{W}^c & W_{ac} \overline{W}^{cb} \end{pmatrix}, \qquad
m_{1/2}^2 = \begin{pmatrix} \overline{W}^{ac} W_{cb} & 0 \\ 0 & W_{ac} \overline{W}^{cb} \end{pmatrix}, \quad (2.48)
$$

with $W_c \equiv \partial W / \partial Q^c$, etc., and m_0^2 and $m_{1/2}^2$ are $2k \times 2k$ matrices. Note that

$$
\mathrm{STr}\, \mathcal{M}^2 = 0. \tag{2.49}
$$

We will be interested in situations where we integrate out some massive fields Φ^a whose superpotential is locally of the form

$$
W = \frac{1}{2}\Phi^a M_{ab}\Phi^b + \cdots, \tag{2.50}
$$

where M_{ab} can depend on various massless fields X. Integrating out Φ^a leads to the one loop effective Kähler potential

$$
K_{eff}^{(1)} = -\frac{1}{32\pi^2} \mathrm{Tr}\big[M M^\dagger \log\big(M M^\dagger / M_{cutoff}^2 \big)\big]. \tag{2.51}
$$

If the supersymmetry breaking is small, we can use the effective Kähler potential to find the effective potential. For example, if M_{ab} depends on one pseudomodulus X, the effective potential is

$$
V_{trunc} = \big(K_{eff\ X,\overline{X}} \big)^{-1} |\partial_X W|^2. \tag{2.52}
$$

However, as we will discuss below, (2.52) gives the correct expression for the effective potential (2.47) only to leading order in $F_X = -\frac{\overline{\partial_X W}}{K_{eff\ X,\overline{X}}}$. (It is verified in [9] that (2.52) and (2.47) agree to order $\mathcal{O}(F_X \overline{F}_X)$.) Higher powers of F_X arise from terms in the low energy effective Lagrangian with more superspace covariant derivatives, e.g. terms of the form

$$
\int d^4\theta \, H(X, \overline{X})(DX)^2 + c.c. \tag{2.53}
$$

for some function $H(X, \overline{X})$. They cannot be ignored when the supersymmetry breaking is large. The full effective potential (2.47) includes all these higher order corrections.

Example 1—the theory of Section 2.3

As a first application, we compute the one-loop potential on the supersymmetry breaking pseudomoduli space mentioned in Section 2.3. Recall that this space exists for X outside of the range (2.17) where there is a tachyon, so we limit ourselves to $|X|^2 > |f/h|$. We treat the pseudomodulus X as a background, and use the masses (2.16) in (2.47). This yields

$$
\begin{aligned}
V^{(1)}(|X|) &= \frac{1}{64\pi^2}\big[-2|hf|^2 \log M_{cutoff}^2 - 2|hX|^4 \log|hX|^2 \\
&\quad + i(|hX|^2 - |hf|)^2 \log(|hX|^2 - |hf|) \\
&\quad + (|hX|^2 + |hf|)^2 \log(|hX|^2 + |hf|)\big] \\
&= \frac{|hf|^2}{32\pi^2}\left[\log\left|\frac{hX}{M_{cutoff}}\right|^2 + \frac{3}{2} + v(z)\right], \\
z &\equiv \left|\frac{f}{hX^2}\right|, \\
v(z) &\equiv \frac{1}{2}\big(z^{-2}(1+z)^2 \log(1+z) + z^{-2}(1-z)^2 \log(1-z) - 3\big) \\
&= -\frac{z^2}{12} + \mathcal{O}(z^4),
\end{aligned}
\tag{2.54}
$$

where the shift by $\frac{3}{2}$ is for later convenience.

The potential (2.54) lifts the degeneracy along the pseudomoduli space. It is an increasing function of $|X|$. It pushes X into the region (2.17); i.e. toward the region with a tachyon (where the expression (2.54) no longer makes sense). From there, the theory falls into its supersymmetric vacua (2.15).

We will now use this simple example, and result (2.54), to clarify and illustrate a number of technical points. Similar statements will apply to other examples.

Let us clarify the nature of the semiclassical limit. We take $h \to 0$ (the coupling h is IR free) with f, X, $q \sim h^{-1}$ (and therefore $z \sim h^0$). In this limit the classical Lagrangian, based on canonical Kähler potential and the superpotential (2.13), scales like h^{-2}. The one loop corrections, in particular (2.54), are of order h^0. We can neglect higher loop terms, which are order h^2 and higher.

Next, we want to understand the dependence on the UV cutoff M_{cutoff}. We define the running coupling

$$
f(\mu) = f_{bare}\left(1 + \frac{|h^2|}{64\pi^2}\left(\frac{3}{2} + \log\frac{\mu^2}{M_{cutoff}^2}\right) + \mathcal{O}(h^4)\right),
\tag{2.55}
$$

where we have set an additive constant to a convenient value. In terms of this running f the potential (2.54) is independent of the UV cutoff M_{cutoff}

$$V(X) = |f(|hX|)|^2 \left(1 + \frac{|h^2|}{32\pi^2} v(z) + \mathcal{O}(h^4)\right). \tag{2.56}$$

Here $f(\mu = |hX|)$ is the running coupling (2.55) at the scale of the massive fields q.

Equivalently, we can remember that in supersymmetric theories there is only wavefunction renormalization. The potential arises from F_X, and therefore at the leading order only Z_X can affect the potential. The renormalization of f in (2.55) can be understood as coming from Z_X, as

$$V = Z_X^{-1}|\partial_X W|^2 + \text{finite} = Z_X^{-1}|f|^2 + \text{finite}. \tag{2.57}$$

We thus have

$$-\frac{\partial V}{\partial \ln M_{cutoff}^2} = \gamma_X |f|^2 = \frac{1}{64\pi^2} \text{Str} \mathcal{M}^4 + \mathcal{O}(h^2), \tag{2.58}$$

where we recognize γ_X as the anomalous dimension of X.

A special situation arises when the supersymmetry breaking mass splittings are effectively small. This happens when $z \equiv |f/hX^2| \ll 1$; i.e. either for small $|f|$, or for large $|X|$. Expanding (2.54) we find

$$V \approx |f|^2 + \frac{|hf|^2}{32\pi^2} \left[\log \left|\frac{hX}{M_{cutoff}}\right|^2 + \frac{3}{2}\right] + \mathcal{O}(h^4) = |f(hX)|^2. \tag{2.59}$$

This can be interpreted as arising from renormalization of the Kähler potential

$$K_{ren} = |X|^2 - \frac{|hX|^2}{32\pi^2} \left(\log \left|\frac{hX}{M_{cutoff}}\right|^2 - \frac{1}{2}\right) + \mathcal{O}(|h|^4). \tag{2.60}$$

Note that this expression for the renormalized K is valid also for $f = 0$, where supersymmetry is not broken along the moduli space parameterized by $\langle X \rangle$.

We should also comment that since as $X \to 0$ the coupling constant h is renormalized to zero, the expression (2.60) becomes accurate for small X (though still outside of the tachyonic range (2.17)).

We have just seen that for small z we can study a supersymmetric low energy theory with superpotential $W = fX$ and an effective Kähler potential given by (2.60). This is a special case of the discussion above about the Kähler potential (2.51). Using $M = hX$ in (2.51) and $W = fX$, the approximate effective potential (2.52) agrees with (2.54)e.

As discussed around (2.52), the supersymmetric effective potential (2.52) is valid only when the supersymmetry breaking is small. The correct one-loop effective potential is given by (2.47) (which in our simple example is given by (2.54)), whether or not the supersymmetry breaking is small. In general, additional contributions which are not included in (2.52) are higher orders in $|f|$ in (2.54) (i.e. the function $v(z)$ in (2.54)).

Example 2—the basic O'Raifeartaigh model (Section 2.5)

We now compute the one loop correction to the pseudomodulus potential in the O'Raifeartaigh model, example 1 of Section 2.5. The classical flat direction of the classical pseudomodulus X is lifted by a quantum effective potential, $V_{eff}(X)$ [22].

We again treat the pseudomodulus X as a background. The one-loop effective potential $V_{eff}(X)$ is given by the expression (2.47), using the classical masses (2.31) and (2.32). As follows from the R-symmetry, $V_{eff}(X)$ depends only on $|X|$. We find that the potential $V_{eff}(X)$ is a monotonically increasing function of $|X|$, with the following asymptotic behavior at small and large $|X|$:

$$
V_{eff}(X) = \begin{cases} V_0 + m_X^2 |X|^2 + \mathcal{O}(|X|^4) & X \approx 0, \\[2mm] |f|^2 \left(1 + \gamma_X \left(\log \left| \dfrac{hX}{M_{cutoff}} \right|^2 + \dfrac{3}{2} \right) + \mathcal{O}\left(h^4, \dfrac{\log |X|}{|X|^4} \right) \right) & X \to \infty, \end{cases}
$$

$$(2.61)$$

where the constants are

$$
\begin{aligned}
V_0 &= |f|^2 \left[1 + \frac{|h^2|}{32\pi^2} \left(\log \frac{|m|^2}{M_{cutoff}^2} + \frac{3}{2} + v(y) \right) + \mathcal{O}(h^4) \right], \\[3mm]
y &= \left| \frac{hf}{m^2} \right|, \\[3mm]
v(y) &= \frac{1}{2} \left(y^{-2}(1+y)^2 \log(1+y) + y^{-2}(1-y)^2 \log(1-y) - 3 \right) \\[2mm]
&= -\frac{y^2}{12} + \mathcal{O}(y^4), \\[3mm]
m_X^2 &= \frac{1}{32\pi^2} \left| \frac{h^4 f^2}{m^2} \right| v(y) + \mathcal{O}(h^4), \\[3mm]
v(y) &= y^{-3}\left((1+y)^2 \log(1+y) - (1-y)^2 \log(1-y) - 2y \right) \\[2mm]
&= \frac{2}{3} + \mathcal{O}(y^2), \\[3mm]
\gamma_X &= \frac{|h|^2}{32\pi^2} + \mathcal{O}(h^4).
\end{aligned}
$$

$$(2.62)$$

The function $v(y)$ is as in (2.54) but its argument here, y, depends only on the coupling constants, and is independent of the pseudomodulus X. Recall that we take the parameter y, defined in (2.29), to be in the range $0 \leq y \leq 1$.

As in the previous example, the semiclassical limit is $h \to 0$ (the coupling h is IR free) with f, X, $\phi_{1,2} \sim h^{-1}$ and $m \sim h^0$ (and therefore $y \sim h^0$).

Also, as in that example, the running coupling constant

$$f(\mu) = f_{bare} \left(1 + \frac{|h^2|}{64\pi^2} \left(\frac{3}{2} + \log \frac{\mu^2}{M_{cutoff}^2} \right) + \mathcal{O}(h^4) \right), \qquad (2.63)$$

removes the dependence on the UV cutoff M_{cutoff}

$$V(x) \quad = \quad \begin{cases} V_0 + m_X^2 |X|^2 + \mathcal{O}(|X|^4) & X \approx 0, \\ |f(hX)|^2 + \cdots & X \to \infty, \end{cases} \qquad (2.64)$$

$$V_0 \quad = \quad |f(m)|^2 \left(1 + \frac{|h^2|}{32\pi^2} v(y) + \mathcal{O}(h^4) \right). \qquad (2.65)$$

Let us discuss the effective potential in the two limits $X \approx 0$ and $|X| \to \infty$. The sign of the mass square in (2.62) is positive, signaling that the potential has a minimum at $X = 0$. The behavior for large X is dominated by the renormalization group running of the effective coupling constant at the scale $|hX|$, which is the scale of the masses in the problem. Finally, it is easy to show using the full expression from (2.47) that the one loop potential is monotonic between these two limits, and therefore $X = 0$ is the global minimum of the potential.

Again, as in the previous example, for $y \equiv |hf/m^2| \ll 1$, the supersymmetry breaking is small. Then, the effective potential can alternatively be computed in the supersymmetric low-energy effective theory, with K given by (2.51) and $W = fX$, leading to the effective potential (2.52). The potential (2.47) applies more generally.

For example, expanding around the minimum at $X = 0$, (2.52) only reproduces the leading order term in the expansion in $y \ll 1$ for m_X^2 in (2.62). It fails to reproduce the answer for larger values of y, e.g.

$$m_X^2 = \frac{|h^3 f|}{16\pi^2} (\log 4 - 1) \quad \text{for } |hf| = |m|^2; \qquad y = 1. \qquad (2.66)$$

On the other hand, even if y is not small, the higher order F terms are insignificant far from the origin of the pseudomoduli space, and indeed there the truncated potential (2.52) agrees with the full effective potential (2.61):

$$V^{(1)} \to \gamma_X^{(1)} \log \left(\frac{|hX|^2}{M_{cutoff}^2} \right) |f|^2 \quad \text{for } hX \text{ large.} \tag{2.67}$$

Let us now consider the modified model of Section 2.6, where we add $\frac{1}{2}h\epsilon\phi_2^2$ to the superpotential (2.36). As we saw, there are then two supersymmetric states at (2.39), and there can also be a metastable state near $X = 0$. Including the ϵ correction to the mass eigenvalues, the one-loop potential (2.47) now has a linear term in X (a tadpole) at $X = 0$, with coefficient $\mathcal{O}(\epsilon)$. The quadratic term in X is not much changed by the $O(\epsilon)$ correction, so the upshot is a local minimum of the one-loop potential at $X \sim \epsilon$.

To summarize this example, we found in Section 2.6 that the theory with nonzero f and ϵ has a classical pseudomoduli space of nonsupersymmetric vacua, which is sensible in the range (2.40) (which includes the region around $X = 0$), where there are no tachyonic modes. Now we have shown that the one-loop effective potential lifts this pseudomoduli space, and stabilizes X near the origin. For $\epsilon \ll 1$, the tachyonic direction down to the supersymmetric vacua (2.39) only appears at large X, so the metastable vacuum near the origin, with broken supersymmetry, can be parametrically long lived.

It is straightforward to repeat the computation of the one-loop effective potential for the model where supersymmetry is broken by the rank condition (Section 2.7). Again, we set $f = -h\mu^2\mathbb{1}$, and then we find that most of the degeneracy along the classical pseudomoduli space (2.46) is removed by the one-loop effective potential (2.47). The masses of the fluctuations of Φ, φ and $\widetilde{\varphi}$, as a function of the pseudomoduli in (2.46), are found to be similar to those of the O'Raifeartaigh model given in (2.31) and (2.32), with $m^2 = hf \equiv -h\mu^2$ (so $y = 1$ in (2.29)). The $SU(n)$ gauge fields do not contribute to (2.47), since their spectrum is supersymmetric to this order. Up to symmetry transformations, the vacua are found to be at

$$\Phi = \begin{pmatrix} 0 & 0 \\ 0 & 0 \end{pmatrix}, \qquad \varphi = \widetilde{\varphi} = \begin{pmatrix} \mu\mathbb{1}_n \\ 0 \end{pmatrix}. \tag{2.68}$$

The vacua (2.68) spontaneously break the global symmetry, $G \to H$. Associated with that, the vacua (2.68) actually form a compact moduli space of vacua, $\mathcal{M}_{vac} = G/H$, parameterized by the massless Goldstone bosons. Since this space of vacua is associated with an exact global symmetry breaking it is robust, and the degeneracy is not lifted by higher order corrections. In particular, these vacua cannot become tachyonic. The one-loop potential computed from (2.47) gives non-tachyonic masses to *all* other pseudomoduli, so the vacua (2.68) are true local minima of the effective potential [9].

2.9. Relation to R-symmetry [11]

Consider a generic theory and ask for a condition for broken supersymmetry. This means that we cannot solve all the equations

$$\partial_a W(\Phi) = 0 \quad \text{for all } a = 1 \ldots k. \tag{2.69}$$

But if W is a generic superpotential, then (2.69) involves k equations for the k quantities Φ^a, so generally they can all be solved. Non-R flavor symmetries do not help. Consider for example a global non-R $U(1)$ symmetry. Then, the equations (2.69) can be written as $k - 1$ independent equations for $k - 1$ independent unknowns, as seen by writing

$$W = W\left(t^a = \Phi^a \Phi_1^{-q_a/q_1}\right) \quad a = 2 \ldots k \tag{2.70}$$

(q_a is the $U(1)$ charge of Φ^a). But if there is an R-symmetry, then we can write

$$W = T f\left(t^a = \Phi^a \Phi_1^{-r_a/r_1}\right) \quad T = \Phi_1^{2/r_1}, \tag{2.71}$$

(r_a is the R-charge of Φ^a), and then in terms of T and t^a for generic f the equations (2.69) set $T = 0$ which is a singular point. Away from $T = 0$ the equations are over-constrained: they are k equations for $k - 1$ independent unknowns, so generically they cannot be solved. Exceptions occur either for a non-generic f, or when a solution with $T = 0$ and therefore $\Phi_1 = 0$ is allowed. This is the case when $r_1 = 2$ and all other $r_a = 0$. Then there is a $k - 2$ dimensional space of supersymmetric vacua, at $\Phi_1 = 0$, $f(\Phi_a) = 0$. (More generally, there are exceptional cases with supersymmetry unbroken for fields at the origin, when all fields, for which the Kähler potential is smooth, have non-negative R-charges less than 2.)

These observations about the relation between R-symmetry and supersymmetry breaking fit with the examples above.

The simplest theory (Section 2.1) with $W = fX$ has an R-symmetry and broken supersymmetry. Adding e.g. $\Delta W = \frac{1}{2} \epsilon X^2$ breaks the R-symmetry, and restores supersymmetry.

This is also true for its generalization with more complicated K of Section 2.2, which depends only on $X\overline{X}$. If K depends separately on X and $\overline{X}$ (not only through the combination $X\overline{X}$), the theory does not have an R-symmetry but supersymmetry is still broken. This shows that we can have broken supersymmetry without R-symmetry. Here it happens because the superpotential is not a generic function of X.

The addition of light fields as in Section 2.3 preserves the R-symmetry, but restores supersymmetry. This demonstrates that having an R-symmetry does not guarantee that supersymmetry is broken. This example realizes the exceptional case, $r_1 = 2, r_{a\neq1} = 0$, mentioned above.

The example of Section 2.4 has a $U(1)_R$ symmetry, and indeed there is no static supersymmetric vacuum. But there is a runaway direction, along which supersymmetry is asymptotically restored. This illustrates the need to still check for runaway directions.

The O'Raifeartaigh type models of Section 2.5 have an R-symmetry, and broken supersymmetry for generic $g_1(\phi)$ and $g_2(\phi)$. The example 2 there, with non-generic $g_1(\phi)$ and $g_2(\phi)$, illustrates that having an R-symmetry does not guarantee broken symmetry, if the superpotential is not generic.

The deformation (2.36) of the O'Raifeartaigh model in Section 2.6 breaks the R-symmetry, and indeed restores supersymmetry. However, for small ϵ there is an approximate R-symmetry which is related to supersymmetry breaking in the metastable state.

Finally, the models based on the rank condition of Section 2.7 have an R-symmetry and correspondingly they have broken supersymmetry, for $n < N_f$. (For $n \geq N_f$, supersymmetry is not broken, by a generalization of the comment following (2.71) about the case $r_1 = 2$, with all other $r_a = 0$.) As mentioned in footnote 6, we will later discuss this model with the $SU(n)$ symmetry gauged, but IR free. The $U(1)_R$ symmetry is then only an approximate symmetry. Correspondingly, the supersymmetry breaking (with $n < N_f$) will be in metastable vacua [9].

To summarize, generically there is broken supersymmetry if and only if there is an R-symmetry. There is broken supersymmetry in a metastable state if and only if there is an approximate R-symmetry. For realistic models of supersymmetry breaking, we need to break the R-symmetry, to get gaugino masses. To avoid having a massless R-axion if the symmetry is spontaneously broken it should also be explicitly broken. Gravity effects can help [12], but ignoring gravity, we conclude that realistic and generic models of supersymmetry breaking require that we live in a metastable state.

3. Supersymmetric QCD

In this section we will discuss the dynamics of supersymmetric QCD (SQCD) for various numbers of colors and flavors. This section will be brief. We refer the reader to the books and reviews of the subject, e.g. [6,7,13–16], for more details.

3.1. Super Yang–Mills theory—$N_f = 0$

A pure gauge theory is characterized by a scale Λ. At energy of order Λ, it confines and leads to nonzero gluino condensation, breaking a discrete R-symmetry.

For $SU(N_c)$ gauge theory we define the gauge invariant chiral operator

$$
\begin{aligned}
S &\equiv -\frac{1}{32\pi^2}\text{Tr}\, W^\alpha W_\alpha \\
&= \frac{1}{32\pi^2}\text{Tr}\left(\lambda\lambda + \cdots + \theta\theta\left(\frac{1}{2}F^{\mu\nu}F_{\mu\nu} + \cdots\right)\right),
\end{aligned}
\tag{3.1}
$$

which can be interpreted as a "glueball" superfield. Here we follow the Wess and Bagger notation [3] where $\lambda\lambda \equiv \lambda^\alpha \lambda_\alpha$. The dynamics leads to gaugino condensation:

$$
\langle S \rangle = \frac{1}{32\pi^2}\langle \text{Tr}\, \lambda\lambda \rangle = (\Lambda^{3N_c})^{\frac{1}{N_c}},
\tag{3.2}
$$

where branches of the fractional power in (3.2) represent the values in the N_c different supersymmetric vacua. The theory has an anomaly free $\mathbf{Z}_{2N_c}$ discrete symmetry (left unbroken by instantons), and (3.2) implies that it is spontaneously broken to $\mathbf{Z}_2$.

The N_c supersymmetric vacua with (3.2) are those counted by the Witten index, $\text{Tr}(-1)^F = N_c$ [8]. Since $\lambda\lambda$ is the first component of the chiral superfield S, the expectation values (3.2) do not break supersymmetry.

The relation (3.2) is exact. This can be seen by promoting Λ to an expectation value of a background chiral superfield [23, 24], which is assigned charge $R(\Lambda) = 2/3$ to account for the anomaly. There is no correction to (3.2) compatible with this R charge assignment and holomorphy.[7]

The gaugino condensation can be represented as a nontrivial superpotential

$$
W_{eff} = N_c(\Lambda^{3N_c})^{\frac{1}{N_c}}.
\tag{3.3}
$$

Comments:

1. The superpotential (3.3) is independent of fields. It is meaningful when coupling to supergravity, or if Λ is a background field source.

2. Equation (3.3) can be used to find the tension of domain walls interpolating between these vacua labelled by k_1 and k_2 [26]

$$
T_{k_1,k_2} = \left| N_c(\Lambda^{3N_c})^{\frac{1}{N_c}}\left(e^{\frac{2\pi i k_1}{N_c}} - e^{\frac{2\pi i k_2}{N_c}}\right)\right|.
\tag{3.4}
$$

3. Thinking of $3N_c \log \Lambda$ as a source for the operator $S \sim \text{Tr}\,W_\alpha^2$ we can find

$$
\langle S \rangle = \frac{1}{3N_c}\partial_{\log \Lambda} W_{eff} = (\Lambda^{3N_c})^{\frac{1}{N_c}}.
\tag{3.5}
$$

---

[7]The non-zero value of the coefficient in (3.2) can be set to one in a particular renormalization scheme. See [25] for discussion, and comparison with various instanton calculations.

4. Using this observation we can perform a Legendre transform to derive the Veneziano–Yankielowicz superpotential [27]

$$W_{eff}(S) = N_c S(1 - \log S/\Lambda^3). \tag{3.6}$$

It should be stressed that S is not a light fields and therefore this expression is not a term in the Wilsonian effective action. It is a term in the 1PI action and therefore it can be used only to find $\langle S \rangle$ and tensions of domain walls. However, there is no particle-like excitation (e.g. a glueball) which is described by the field S.

3.2. Semiclassical SQCD

We consider $SU(N_c)$ gauge theory with N_f quarks Q and N_f anti-quarks $\widetilde{Q}$. The gauge and global symmetries are

	$SU(N_c)$	$[SU(N_f)_L$	$SU(N_f)_R$	$U(1)_B$	$U(1)_R$	$U(1)_A]$
Q	$\mathbf{N_c}$	$\mathbf{N_f}$	$\mathbf{1}$	1	$1 - \frac{N_c}{N_f}$	1
$\widetilde{Q}$	$\mathbf{\overline{N}_c}$	$\mathbf{1}$	$\mathbf{\overline{N}_f}$	-1	$1 - \frac{N_c}{N_f}$	1

$$\tag{3.7}$$

Here the global symmetries are denoted by $[\ldots]$. The $U(1)_A$ symmetry is anomalous and the other symmetries are anomaly free. We also assign charges to the coupling constants: regarding them as background chiral superfields leads to useful selection rules [23],

	$SU(N_c)$	$[SU(N_f)_L$	$SU(N_f)_R$	$U(1)_B$	$U(1)_R$	$U(1)_A]$
m	1	$\mathbf{\overline{N}_f}$	$\mathbf{N_f}$	0	$2\frac{N_c}{N_f}$	-2
$\Lambda^{3N_c - N_c}$	1	1	1	0	0	$2N_f$

$$\tag{3.8}$$

Here m is a possible mass term that we can add, $W_{tree} = \mathrm{Tr}\, m\widetilde{Q}Q$, and Λ is the dynamical scale, related to the running gauge coupling as

$$\Lambda^{3N_c - N_f} = e^{-\frac{8\pi^2}{g^2(\mu)} + i\theta} \mu^{3N_c - N_f}. \tag{3.9}$$

Instanton amplitudes come with the factor of $\Lambda^{3N_c - N_f}$, and their violation of the $U(1)_A$ symmetry is accounted for by the charge assignment in (3.8).

As seen from (3.9), the theory is UV free for $N_f < 3N_c$, i.e. $g^2(\mu) \to 0$ for $\mu \gg |\Lambda|$. On the other hand, for $N_f \geq 3N_c$, the theory is IR free, i.e. $g^2(\mu) \to 0$ for $\mu \ll |\Lambda|$ (for $N_f = 3N_c$ the beta function vanishes at one loop, but at two loops it is IR free).

In the rest of this subsection, we take $W_{tree} = 0$. The classical potential is then

$$V \sim \sum_a (D^a)^2 = \sum_a \left(\text{Tr} \left(Q T^a Q^\dagger - \tilde{Q}^* T^a \tilde{Q}^T \right) \right)^2 \tag{3.10}$$

(T^a are the $SU(N_c)$ generators). It leads to flat directions which we refer to as the classical moduli space of vacua $\mathcal{M}_{cl}$. As is always the case, $\mathcal{M}_{cl}$ can be understood in terms of gauge invariant monomials of the chiral superfields, and the light moduli in $\mathcal{M}_{cl}$ can be understood as the chiral superfields that are left uneaten by the Higgs mechanism.

For $N_f < N_c$ up to gauge and flavor rotations, $\mathcal{M}_{cl}$ is given by [28]

$$Q = \tilde{Q} = \begin{pmatrix} a_1 & & & \\ & a_2 & & \\ & & \cdot & \\ & & & a_{N_f} \end{pmatrix}. \tag{3.11}$$

Its complex dimension is $\dim_{\mathbf{C}} \mathcal{M}_{cl} = N_f^2$. The gauge invariant description is $\mathcal{M}_{cl} = \{ M_{\tilde{g}}^f = (\tilde{Q} Q^T)_{\tilde{g}}^f \}$, $f, \tilde{g} = 1 \dots N_f$. The gauge group is broken on $\mathcal{M}_{cl}$ as $SU(N_c) \to SU(N_c - N_f)$. The classical Kähler potential on $\mathcal{M}_{cl}$ is

$$K_{cl} = 2 \text{Tr} \sqrt{M^\dagger M}. \tag{3.12}$$

(To see that, write the D-term equations as $Q^\dagger Q = \tilde{Q}^T \tilde{Q}^*$, and use it find $M^\dagger M = Q^* \tilde{Q}^\dagger \tilde{Q} Q^T = (Q^* Q^T)^2$. Then the Kähler potential is $\text{tr} \, Q^\dagger Q + \text{tr} \, \tilde{Q}^\dagger \tilde{Q} = 2 \text{tr} \sqrt{M^\dagger M}$.) This is singular near the origin. As always, singularities in the low-energy effective theory signal new light fields, which should be included for a smooth description of the physics. Here the singularities of K_{cl} occur at subspaces where some of the $SU(N_c)/SU(N_c - N_f)$ gauge bosons become massless, and they need to be included in the description.

For $N_f \geq N_c$ we have $\dim_{\mathbf{C}} \mathcal{M}_{cl} = 2 N_c N_f - (N_c^2 - 1)$. Up to gauge and flavor rotations [28],

$$Q = \begin{pmatrix} a_1 & & & \\ & a_2 & & \\ & & \cdot & \\ & & & a_{N_c} \end{pmatrix}, \quad \tilde{Q} = \begin{pmatrix} \tilde{a}_1 & & & \\ & \tilde{a}_2 & & \\ & & \cdot & \\ & & & \tilde{a}_{N_c} \end{pmatrix}, \tag{3.13}$$

$$|a_i|^2 - |\tilde{a}_i|^2 = \text{independent of } i.$$

The gauge invariant description is given by the fields $M = \tilde{Q}Q^T$, $B = Q^{N_c}$ (contracted with the epsilon-symbol), $\tilde{B} = \tilde{Q}^{N_c}$, subject to various classical relations,

$$\mathcal{M}_{cl} = \{M, B, \tilde{B}|\, C_i(M, B, \tilde{B}) = 0\}. \tag{3.14}$$

The functions C_i, giving the classical relations, are of course compatible with the symmetries (3.7), including $U(1)_A$. For example for $N_f = N_c$, we have [29]

$$\mathcal{M}_{cl} = \{M_{\tilde{g}}^f, B, \tilde{B}|\, \det M - B\tilde{B} = 0\}, \tag{3.15}$$

where the constraint follows from $\det M = \det Q \det \tilde{Q} = B\tilde{B}$. The spaces (3.14), for all $N_f \geq N_c$, are singular at the origin, $M = B = \tilde{B} = 0$, because it is possible to set all $C_i = 0$, and also all variations $\delta C_i = 0$ there. The classical interpretation is that the $SU(N_c)$ gauge fields, which are massless at the origin, need to be included for the low-energy effective theory to be non-singular.

For $N_f > N_c$, among other constraints, the $N_f \times N_f$ matrix $M = \tilde{Q}Q^T$ satisfies

$$\text{rank}(M) \leq N_c \qquad \text{classically.} \tag{3.16}$$

3.3. Adding large quark mass terms

Consider adding quark masses, via the tree-level superpotential

$$W_{tree} = \text{Tr}\, m\tilde{Q}Q^T \equiv \text{Tr}\, mM. \tag{3.17}$$

For large m (more precisely, the eigenvalues of m are much larger than $|\Lambda|$) we can integrate out the quarks and the low energy theory is a pure gauge theory. Its scale Λ_L is determined at one loop as

$$\Lambda_L^{3N_c} = \det m\, \Lambda^{3N_c - N_f}. \tag{3.18}$$

Gluino condensation in this theory leads, as in (3.3), to

$$W_{eff} = N_c \left(\det m\, \Lambda^{3N_c - N_f}\right)^{\frac{1}{N_c}}; \tag{3.19}$$

it follows from holomorphy and symmetries that (3.19) is the exact effective superpotential. The superpotential (3.19) can be interpreted as part of the generating functional for correlation functions, with the mass m in (3.17) acting as the source for the operator M, and $\log \Lambda^{3N_c - N_f}$ as the source for the operator $S \sim \text{Tr} W_\alpha W^\alpha$ [24, 30]. We can thus use (3.19) to find

$$\langle M \rangle_{susy} = \partial_m W_{eff} = \left(\det m \; \Lambda^{3N_c-N_f}\right)^{\frac{1}{N_c}} \frac{1}{m},$$

$$\langle S \rangle_{susy} = \partial_{\log \Lambda^{3N_c-N_f}} W_{eff} = \left(\det m \; \Lambda^{3N_c-N_f}\right)^{\frac{1}{N_c}}.$$

The subscript emphasizes that these are the expectation values in the supersymmetric vacua. Note that there are N_c solutions in (3.20), differing by a N_c-th root of unity phase, which correspond to the $\mathrm{Tr}(-1)^F = N_c$ supersymmetric vacua of the low-energy super-Yang–Mills theory. The result (3.20) is valid for all N_f. It is interesting to note that, for $N_f > N_c$, the matrix $\langle M \rangle$ in (3.20) does not satisfy the classical constraint (3.16) of the theory with massless flavors; however, taking $m \to 0$ in (3.20) does bring $\langle M \rangle$ back to $\mathcal{M}_{cl}$.

Performing a Legendre transform between m and M, we can use (3.19) to derive the 1PI effective action

$$W_{eff}(M) = (N_c - N_f)\left(\frac{\Lambda^{3N_c-N_f}}{\det M}\right)^{1/(N_c-N_f)} + \mathrm{Tr}\, m M. \tag{3.20}$$

One might be tempted to interpret (3.20) also as a Wilsonian effective action for the light field M. However, as we will discuss below, this is not always correct.

Finally we can introduce the field S into (3.20) by performing a Legendre transform with respect to its source $\log \Lambda^{3N_c-N_f}$ to find [31]

$$W_{eff}(M, S) = S\left((N_c - N_f) - \log \frac{S^{N_c-N_f} \det M}{\Lambda^{3N_c-N_f}}\right) + \mathrm{Tr}\, m M. \tag{3.21}$$

Again, this expression can be used to find the expectation values (3.20) and to study domain wall tensions, but it should not be viewed as a term in a Wilsonian effective action.

3.4. $N_f < N_c$ massless flavors [28]

We have seen that the classical theory has a moduli space of supersymmetric vacua $\mathcal{M}_{cl}$. We now explore the low energy effective Lagrangian along $\mathcal{M}_{cl}$ and examine whether a superpotential can be generated there. The symmetries (3.7) constrain the superpotential to be of the form [32]

$$W_{dyn} \propto \left(\frac{\Lambda^{3N_c-N_f}}{\det M}\right)^{1/(N_c-N_f)}. \tag{3.22}$$

Therefore, we face a dynamical question of determining the coefficient in (3.22). Note that (3.22) is non-perturbative, because of the positive power of $\Lambda \sim \exp(-8\pi^2/(3N_c - N_f)g^2)$.

Recall that the gauge group is Higgsed to $SU(N_c - N_f)$ on the classical moduli space. For $N_f = N_c - 1$, the gauge group is completely Higgsed, and then there are finite action (constrained) instantons which generate (3.22). For $N_f < N_c - 1$, (3.22) is instead associated with gaugino condensation in the unbroken $SU(N_c - N_f)$—that is the reason for the fractional power in (3.22). Finally, comparing with (3.20) we see that the coefficient in (3.22) must be $N_c - N_f$

$$W_{dyn} = (N_c - N_f) \left(\frac{\Lambda^{3N_c - N_f}}{\det M} \right)^{1/(N_c - N_f)}. \tag{3.23}$$

For $N_f \geq N_c$, (3.22) does not make sense. For $N_f = N_c$, the exponent diverges. For $N_f > N_c$, the constraint (3.16) implies $\det M = 0$. Therefore, for $N_f \geq N_c$ massless flavors, the quantum theory has a moduli space of inequivalent vacua.

3.5. $N_f = N_c$ massless flavors [29]

Here the vacuum degeneracy cannot be lifted by W_{dyn}, so the moduli space is still parameterized by the gauge invariant fields M, B and $\tilde{B}$. But the classical constraint (3.15) they satisfy is modified (consistent with the symmetries (3.7) and (3.8))

$$\mathcal{M}_{qu} = \left\{ M_{\tilde{g}}^f, \, B, \, \tilde{B} \middle| \det M - B\tilde{B} = \Lambda^{2N_c} \right\}. \tag{3.24}$$

Note that this is a nonperturbative effect, proportional to a positive power of Λ. So, as is appropriate, the deformation is important only near the origin, and is negligible at large fields, relative to Λ, where the theory is weakly coupled. Indeed, the power in (3.24) is precisely that associated with a one instanton correction to the constraint in (3.15). The constraint (3.24) can be seen from (3.20), which for $N_f = N_c$ has $\det M = \Lambda^{2N_c}$, independent of m. (One can introduce sources for the operators B and $\tilde{B}$, to get the full constraint (3.24).) The space $\mathcal{M}_{cl}$ in (3.15) was singular at $M = B = \tilde{B} = 0$, but the space (3.24) is everywhere smooth. The only light degrees of freedom of the low-energy effective theory are the moduli of (3.24).

The theory with the modified constraint can be described using a Lagrange multiplier X and a superpotential

$$W = X \left(\det M - B\tilde{B} - \Lambda^{2N_c} \right), \tag{3.25}$$

but it should be stressed that this is not a term in a Wilsonian action. There is no light field X and similarly, the mode of M, B and $\tilde{B}$ which is proportional to $\det M - B\tilde{B}$ are not light. However, (3.25) is still a useful way to implement the constraint.

3.6. $N_f > N_c$ [33]

The vacuum degeneracy of the theory with massless flavors again cannot be lifted by W_{dyn}. Moreover, for all $N_f > N_c$, the classical moduli space constraints (3.14) cannot be deformed because no deformation would be compatible with holomorphy and the symmetries in (3.7) and (3.8). So there is a quantum moduli space of vacua, coinciding with the classical moduli space (3.14), $\mathcal{M}_q = \mathcal{M}_{cl}$. The singularity of these spaces at the origin indicates additional, massless degrees of freedom there. Their nature is clarified by a duality.

The original $SU(N_c)$ theory, with N_f flavors, is dual to another gauge theory based on the gauge group $SU(n = N_f - N_c)$ with spectrum of fields and couplings

	$SU(n)$	$[SU(N_f)_L$	$SU(N_f)_R$	$U(1)_B$	$U(1)_R$	$U(1)_A]$
φ	$\mathbf{n}$	$\overline{\mathbf{N}}_{\mathbf{f}}$	$\mathbf{1}$	$\frac{N_c}{n}$	$1 - \frac{n}{N_f}$	1
$\tilde{\varphi}$	$\overline{\mathbf{n}}$	$\mathbf{1}$	$\mathbf{N}_{\mathbf{f}}$	$-\frac{N_c}{n}$	$1 - \frac{n}{N_f}$	1
Φ	$\mathbf{1}$	$\mathbf{N}_{\mathbf{f}}$	$\overline{\mathbf{N}}_{\mathbf{f}}$	0	$2\frac{n}{N_f}$	-2
f	$\mathbf{1}$	$\overline{\mathbf{N}}_{\mathbf{f}}$	$\mathbf{N}_{\mathbf{f}}$	0	$2 - 2\frac{n}{N_f}$	2
$\Lambda^{3n - N_f}$	$\mathbf{1}$	$\mathbf{1}$	$\mathbf{1}$	0	0	$2N_f$

$$(3.26)$$

(again, the group in [. . .] is a global symmetry) with canonical K for the fields φ, $\tilde{\varphi}$, and Φ, and superpotential

$$W = h \operatorname{Tr} \Phi \varphi \tilde{\varphi}^T + \operatorname{Tr} f \Phi. \qquad (3.27)$$

As we will discuss, the coupling f is proportional to the mass of the electric quarks. In particular, if $m = 0$ in the electric theory, then $f = 0$ in the magnetic theory. $U(1)_A$ in (3.26) is anomalous but the other symmetries are not. The scale $\tilde{\Lambda}$ of the magnetic theory can be taken to be the same as the Λ of the electric theory, as we indicate in (3.26).

We refer to the original theory (3.7) as electric and to (3.26) as magnetic. This duality between the electric and the magnetic theories states that these two different theories have the same IR behavior. Better agreement between the two theories is obtained if we modify the Kähler potential by higher order terms.

Comments:

1. The anomaly free symmetries of the electric and the magnetic theories are the same. All 'tHooft anomaly matching conditions of these symmetries are satisfied.

2. The relations between the variables of the electric and magnetic descriptions are

$$M = \tilde{Q}Q^T = \alpha\Lambda\Phi, \qquad B = Q^{N_c} = \beta^n \Lambda^{2N_c - N_f} \varphi^n \qquad (3.28)$$

with some dimensionless constants α and β. (Below we will determine α.) It is easy to check that the identification of operators (3.28) is consistent with the anomaly free symmetries. (An alternative description was given in [13], where the scales of the electric and magnetic theories were taken to be different; the descriptions are equivalent, as reviewed, e.g. in [9].)

3. For $\frac{3}{2}N_c < N_f < 3N_c$, the electric and magnetic theories are both UV free, and they differ in the UV. The two different UV free starting points flow under the renormalization group (RG) to the same interacting RG fixed point in the IR. A detailed discussion of this RG flow can be found, e.g. in [16].

4. For $N_c + 2 \le N_f \le \frac{3}{2}N_c$ the magnetic theory is IR free, with irrelevant interactions. The UV free electric theory flows at long distance to the IR free magnetic theory.

5. For $N_f = N_c + 1$ we can still use the variables in (3.26) but without the magnetic gauge fields and with the addition of a term proportional to $\det \Phi$ to the superpotential [29].

6. Turning on mass terms $\operatorname{Tr} m Q\tilde{Q} = \operatorname{Tr} mM$ in the electric theory is described by adding to the magnetic superpotential $\Lambda\alpha\operatorname{Tr} m\Phi$. We will analyze it in detail in the next subsection.

3.7. Adding small mass terms

We again add (3.17)

$$W_{tree} = \operatorname{Tr} m\tilde{Q}Q^T = \operatorname{Tr} mM \qquad (3.29)$$

but this time we take the masses (eigenvalues of m) small compared with $|\Lambda|$. Now, we should be able to reproduce the expectation values (3.20) from our low energy effective theory.

For $N_f < N_c$, the low energy theory has $W_{exact} = W_{dyn} + W_{tree}$, which gives precisely the superpotential (3.20). The Legendre transform in (3.20) ensures that setting $F_M^\dagger = -\partial_M W_{exact} = 0$ yields the N_c supersymmetric vacua at $\langle M \rangle$ given in (3.20).

As we mentioned above, for $N_f \ge N_c$, (3.20) is not meaningful as a superpotential on the moduli space. Rather, it should be viewed as a superpotential on a larger field space, where M is arbitrary rather than subject to (3.16), and which is meaningful only for nonzero m. As we are going to discuss, the dual theory provides an interpretation of this.

For $N_f = N_c$ (3.20) does not make sense. Instead, we can find $\langle M \rangle$ using the superpotential (3.25).

For $N_f = N_c + 1$ we have to add (3.20) to the superpotential (as commented after (3.28)).

For $N_f > N_c + 1$ the meaning of (3.20) is slightly more subtle. Consider moving the field $\Phi \sim M$ away from its expectation value. The superpotential (3.27) gives masses to the dual quarks φ. Using an expression like (3.3) for gluino condensation in the magnetic gauge group leads to

$$W = n\left(h^{N_f} \det \Phi \Lambda^{3n-N_f}\right)^{\frac{1}{n}}, \tag{3.30}$$

where we set the scales of the magnetic and electric theories to be the same Λ. This agrees with (3.20) provided

$$h^{N_f} \det \Phi \Lambda^{3n-N_f} = (-1)^{N_f - N_c} \frac{\det M}{\Lambda^{3N_c - N_f}} \tag{3.31}$$

which fixes the coefficient α in (3.28)

$$M = (-1)^{1 - \frac{N_c}{N_f}} h \Lambda \Phi. \tag{3.32}$$

Correspondingly, the coefficient f in (3.27) is related to the electric mass by

$$f = \alpha \Lambda m = (-1)^{1 + \frac{N_c}{N_f}} m h \Lambda. \tag{3.33}$$

4. Dynamical supersymmetry breaking

We will now consider four typical examples of DSB. The common feature of these examples is that at low energies they can be given a semiclassical supersymmetric description as in the examples in Section 2. The first three examples which are based on the dynamics of $N_f < N_c$, $N_f = N_c$ and $N_f > N_c$ were found in the 80s, 90s and 00s respectively. The fourth example, which is based on the dynamics of $N_f = 0$, allows us to easily convert any example in Section 2 to a model of DSB.

Many other examples of DSB are known. Some of them are strongly coupled and do not admit a semiclassical supersymmetric description involving an effective Kähler potential and an effective superpotential (examples are $SU(5)$ or $SO(10)$ gauge theories with a single generation of quarks and leptons [34, 35]). In other situations the question of supersymmetry breaking is inconclusive (e.g. an $SU(2)$ gauge theory with matter in the four dimensional representation [36]). In addition, many variants of the examples below are known and they exhibit various interesting features (see, e.g. [37–47]). Additional review and references can be found in e.g. [6,7,48,49].

4.1. The (3,2) model [38]

The gauge group is

$$SU(3) \times SU(2) \tag{4.1}$$

and we have chiral superfields: Q in $(\mathbf{3}, \mathbf{2})$, $\tilde{u}$ in $(\bar{\mathbf{3}}, \mathbf{1})$, $\tilde{d}$ in $(\bar{\mathbf{3}}, \mathbf{1})$, L in $(\mathbf{1}, \mathbf{2})$. For $W_{tree} = 0$, the classical moduli space is given by arbitrary expectation values of the gauge invariants

$$X_1 = Q\tilde{d}L, \qquad X_2 = Q\tilde{u}L, \qquad Z = QQ\tilde{u}\tilde{d}. \tag{4.2}$$

Both gauge groups are Higgsed on this classical moduli space. We add to the model a tree level superpotential

$$W_{tree} = \lambda Q\tilde{d}L = \lambda X_1. \tag{4.3}$$

This theory has a $U(1)_R$ symmetry, with $R(Q) = -1$, $R(\tilde{u}) = R(\tilde{d}) = 0$, $R(L) = 3$. A crucial aspect of (4.3) is that it lifts all of the classical D-flat directions. Therefore, the theory does not have any runaway directions.

Using the global symmetries (including those under which the couplings, treated as background chiral superfields, are charged), the exact superpotential for the fields (4.2) is

$$W_{exact} = \frac{\Lambda_3^7}{Z} + \lambda X_1. \tag{4.4}$$

The first term in (4.4) is W_{dyn}, which is generated by an $SU(3)$ instanton. This theory dynamically breaks supersymmetry.[8]

For $\lambda \ll 1$, the vacuum is at large expectation value for the fields. Since the gauge groups are Higgsed at a high energy scale, their running coupling is weak. Because the theory is weakly coupled for the fields in this limit, we have $K \approx K_{classical}$, so the Kähler potential is under control. It is then easy to find that the field expectation values and the vacuum energy density at the minimum are

$$v \sim \Lambda_3/\lambda^{1/7}; \qquad V = M_S^4 \sim |\lambda^{10/7}\Lambda_3^4| \tag{4.5}$$

---

[8] A quick way to see that is to note that W_{dyn} pushes Z away from the origin, which spontaneously breaks the $U(1)_R$ symmetry. There is thus a compact moduli space of vacua, whose modulus is the massless Goldstone boson. If supersymmetry were unbroken, the Goldstone boson would have a scalar superpartner, which would lead to a non-compact moduli space—but that cannot be the case, because W_{tree} lifts all of the classical flat directions [34].

(the precise coefficient can be computed, using $K = K_{cl}$). Note that, to justify $K \approx K_{cl}$, we need $v \gg \Lambda_3$ and also $v \gg \Lambda_2$, and the latter condition requires $\Lambda_3 \gg \lambda^{1/7}\Lambda_2$. In addition to the massless Goldstino, there is a massless Goldstone boson, because the vacuum spontaneously breaks the $U(1)_R$ symmetry.

The above analysis is valid when $\Lambda_3 \gg \Lambda_2$. As seen from the expressions above, in this limit the $SU(2)$ gauge dynamics scale Λ_2 does not appear directly in the approximate answers (4.5). The $SU(2)$ gauge group is weakly coupled at the scale Λ_3, and the role of the $SU(2)$ gauge symmetry is simply to restrict the possible superpotential couplings, and its classical gauge potential lifts certain directions in field space thus avoiding runaway. The fact that Λ_2 does not enter into (4.4) fits with the fact that the $SU(2)$ gauge group has $N_f = N_c$. So, as reviewed in Section 3.5, it does not contribute to W_{dyn}, but instead leads to the quantum modified moduli space constraint [29] of (3.24). The quantum modified moduli space is neglected in the analysis above, and that is justified when $\Lambda_3 \gg \Lambda_2$.

On the other hand, in the limit $\Lambda_2 \gg \Lambda_3$, the $SU(2)$ group becomes strong first in the RG flow to the IR, and it is then essential to include the quantum modified moduli space constraint. Below the scale Λ_2, the light fields are $q = QL/\Lambda_2$, in the $\mathbf{3}$ of $SU(3)$, and $\tilde{q} = Q^2/\Lambda_2$, and $\tilde{u}$ and $\tilde{d}$, all in the $\bar{\mathbf{3}}$, subject to the quantum constraint $q\tilde{q} = \Lambda_2^2$. The constraint breaks $SU(3)$ to $SU(2)' \subset SU(3)$, at the scale Λ_2, and q and $\tilde{q}$ are Higgsed. The fields $\tilde{u}$ and $\tilde{d}$ each decompose as $\mathbf{3}, \bar{\mathbf{3}} \to \mathbf{2} + \mathbf{1}$ under $SU(3) \to SU(2)'$, so we have $SU(2)'$ with $N_f = 1$ flavor, plus two singlets. In the limit, we obtain a superpotential which is similar to (4.4), but with a different interpretation of the terms. In particular, the λX_1 term is interpreted as $\lambda\Lambda_2^2 S_d$, where S_d is the $SU(2)'$ singlet from $\tilde{d}$. In the $\lambda^{1/7}\Lambda_2 \gg \Lambda_3$ limit, the $SU(2)' \subset SU(3)$ dynamics is insignificant, and we have $M_S^4 = \alpha|\lambda^2\Lambda_2^4|$, where α is a positive $\mathcal{O}(1)$ Kähler potential coefficient, $K \supset \frac{1}{\alpha}S_d\bar{S}_d$ that cannot be directly calculated [50].

4.2. Modified moduli space example [50, 51]

Consider the $SU(N_c)$ theory with $N_f = N_c$ and add fields $S_a^{\tilde{a}}$, b and $\tilde{b}$ and a superpotential (up to coupling constants)

$$W_{tree} = \operatorname{tr} S\tilde{Q}Q^T + b \det \tilde{Q} + \tilde{b} \det Q. \tag{4.6}$$

Classically $Q = \tilde{Q} = 0$. In the quantum theory we get the effective superpotential (see (3.25))

$$W_{effective} = \operatorname{tr} SM + b\tilde{B} + \tilde{b}B + X(\det M - B\tilde{B} - \Lambda^{2N_c}) \tag{4.7}$$

which breaks SUSY. This breaking is dynamical. It depends on the IR confinement of the $N_f = N_c$ theory, from quarks and gluons in the UV, into the composite fields M and B and $\tilde{B}$ in the IR and on the quantum deformation of the moduli space by Λ^{2N_c} in (3.24).

Let us specialize to $N_f = N_c = 2$, where the fundamentals and anti-fundamentals can be written as $2N_f = 4$ fundamentals Q^{fc}, $f = 1 \ldots 4, c = 1, 2$. The gauge invariants are $U^{fg} = Q^{fc} Q^{gd} \epsilon_{cd}$, in the **6** of the global $SU(4) \cong SO(6)$ flavor symmetry. To emphasize that it is an $SO(6)$ vector we will also express it as

$$\vec{V} = \left(V^1 = \frac{1}{2}(U^{12} + U^{34}), V^2 = \frac{i}{2}(U^{12} - U^{34}), \ldots \right). \tag{4.8}$$

The quantum moduli space constraint (3.25) for this case is [29]

$$\text{Pf} \, U = U^{12}U^{34} - U^{13}U^{24} + U^{14}U^{23} = \vec{V} \cdot \vec{V} = \Lambda^4. \tag{4.9}$$

We add singlets $\vec{S}$, also in the **6** of the global flavor $SO(6)$, with superpotential

$$W_{tree} = \frac{1}{2}hS_{fg}Q^{fc}Q^{gd}\epsilon_{cd} = 2h\vec{S} \cdot \vec{V}, \tag{4.10}$$

where S_{fg} is related to $\vec{S}$ as in (4.8) and the factor of 2 arises from this change of notation. Unlike (4.6), (4.7), here we have explicitly exhibited the coupling constant h. There is a conserved $U(1)_R$ symmetry, with $R(Q) = 0$, and hence $R(\vec{V}) = 0$, and $R(\vec{S}) = 2$. Because $\vec{F}_{\vec{S}} = -2h\vec{V}$, the constraint (4.9) implies that $F_{\vec{S}} \neq 0$, so SUSY is broken.

Let us analyze it in more detail. We start with the classical theory. The superpotential coupling $\frac{1}{2}hS_{fg}Q^{fc}Q^{gd}\epsilon_{cd}$ lifts all the flat directions with nonzero Q. So the classical moduli space is the space of $\vec{S}$. Moving far out along these flat directions the fundamental quarks are massive and can be integrated out. The low energy $SU(2)$ gauge theory has scale $\Lambda_L^6 = \Lambda^4 h^2 \vec{S} \cdot \vec{S}$, and its gluino condensation generates

$$W_{low} = 2\left(\Lambda_L^6\right)^{1/2} = 2\left(h^2\Lambda^4\vec{S} \cdot \vec{S}\right)^{\frac{1}{2}}. \tag{4.11}$$

Using the symmetries and holomorphy it is easy to see that (4.11) is exact. Now it is clear that for any nonzero $\vec{S}$ the superpotential is not stationary, and the point $\vec{S} = 0$ is singular and needs to be examined in detail.

Before we conclude that supersymmetry is broken away from the origin we have to examine the potential at infinity to make sure that there is no runaway.

Using the classical Kähler potential for $\vec{S}$ which is canonical, the superpotential (4.11) leads to

$$V_{cl} = 4|h\Lambda^2|^2 \frac{\vec{S} \cdot \vec{\bar{S}}}{|\vec{S} \cdot \vec{S}|}. \tag{4.12}$$

Depending on the direction in the space this expression either diverges at infinity or asymptotes to a constant $4|h\Lambda^2|^2$. It is straightforward to include the one loop correction to this expression. This situation is very similar to the discussion around (2.54). The fundamental quarks Q are massive and their loop leads to logarithmic corrections to the potential which makes it grow at infinity. We conclude that the pseudoflat directions with broken supersymmetry in (4.12) is lifted and pushes the system to smaller values of $\vec{S}$.

When $|h\vec{S}| \ll |\Lambda|$ the superpotential (4.10) gives the quarks small masses and they cannot be integrated out so easily. But then we can use our understanding of the macroscopic theory, where the $SU(2)$ gauge fields and matter of the microscopic theory are replaced in the IR with the fields $\vec{V}$, subject to the constraint (4.9). We solve this constraint as

$$\vec{V} = \Lambda\left(\sqrt{\Lambda^2 - \vec{v}^2}, \vec{v}\right), \tag{4.13}$$

where $\vec{v}$ is an $SO(5)$ vector. We will assume that $|\vec{v}| \ll |\Lambda|$. This assumption is valid up to symmetry transformations near the origin of the classical theory, where we expect to find our ground state. Similarly, we write $\vec{S} \equiv (S_1, \vec{s})$, where $\vec{s}$ is an $SO(5)$ vector. Then (4.10) is

$$W = 2h\Lambda S_1\sqrt{\Lambda^2 - \vec{v}^2} + 2h\Lambda\vec{v} \cdot \vec{s} \approx 2h\Lambda^2 S_1 - hS_1\vec{v}^2 + 2h\Lambda\vec{v} \cdot \vec{s}. \tag{4.14}$$

The Kähler potential for the fields S_1, $\vec{s}$, and $\vec{v}$ is smooth, and can be taken to be

$$K = S_1\bar{S}_1 + \vec{s} \cdot \vec{\bar{s}} + \frac{1}{\alpha}\vec{v} \cdot \vec{\bar{v}} + \mathcal{O}\left(\frac{1}{|\Lambda|^2}\right), \tag{4.15}$$

where α is an $\mathcal{O}(1)$ coefficient that we cannot determine.

Up to symmetry transformations, the vacua have arbitrary $\langle S_1 \rangle$, and $\vec{v} = \vec{s} = 0$. This leads to a seven real dimensional pseudomoduli space. Its dimensions include the two non-compact directions given by $\langle S_1 \rangle$, and five real Goldstone bosons living on $SO(6)/SO(5) \cong S^5$, coming from components of $\vec{v}$ and $\vec{s}$.

We can integrate out the massive modes of $\vec{v}$ to find an effective superpotential. For $\vec{s} = 0$ it is $W_{eff} = 2h\Lambda^2 S_1$, and more generally, it is given by $W_{eff} = 2(h^2\Lambda^4\vec{S} \cdot \vec{S})^{\frac{1}{2}}$ which agrees with (4.11).

Supersymmetry is broken by $-\overline{F}_{S_1} = 2h\Lambda^2 \neq 0$. Since F_{S_1} is generated by dimensional transmutation, the supersymmetry breaking is dynamical. The massless Goldstino comes from S_1.

We should now examine how this pseudomoduli space is lifted in the quantum theory. This is easily done using the low energy theory based on the superpotential (4.14) and the Kähler potential (4.15) by noticing that it is a multi-field analog of the $y = 1$ O'Raifeartaigh model. The one-loop potential (2.47) lifts the degeneracy and leads to a supersymmetry breaking minimum at $\vec{S} = 0$ [52]. At this vacuum the global $SO(6)$ symmetry is spontaneously broken to $SO(5)$ by the constraint (4.9), but the $U(1)_R$ symmetry is unbroken. So there is a five real dimensional, compact space of supersymmetry breaking vacua, given by the Goldstone boson manifold $SO(6)/SO(5) \cong S^5$.

For $h \ll 1$, we can have large S_1 and still use the low energy effective theory provided

$$|hS_1| \ll |\Lambda| \ll |S_1|. \tag{4.16}$$

In this limit, the behavior of the one-loop potential (2.47), computed in the low-energy effective field theory, asymptotes as in (2.67) to

$$V^{(1)} \to \gamma^{(1)}_{macro} \log\left(\frac{|2hS_1|^2}{M^2_{cutoff}}\right) |2h\Lambda^2|^2. \tag{4.17}$$

As we have reviewed, the dependence on M_{cutoff} can be absorbed into the renormalization of h. The coefficient in (4.17) is the anomalous dimension of the pseudomodulus, computed in the macroscopic theory. It depends on the $\mathcal{O}(1)$ unknown constant α in (4.15). Since $\gamma^{(1)}_{macro} > 0$, the potential (4.17) is an increasing function of $|S_1|$.

On the other hand, as we remarked above, if $|\Lambda| \ll |hS_1|$, then, we should instead use the microscopic theory. The result for the potential is similar to (4.17), though with a different, but again positive, numerical coefficient $\gamma^{(1)}_{micro}$ for the one-loop anomalous dimension of S_1, computed from the microscopic Q fields running in the loop [53]. We cannot compute the potential in the intermediate range, $|hS_1| \sim |\Lambda|$, but in all calculable regions the potential slopes toward the origin, $S_1 = 0$.

Deforming the model

Consider adding a $U(1)_R$ breaking, but $SO(6)$ invariant, term

$$\Delta W = \frac{1}{2}\epsilon \vec{S}^2 \tag{4.18}$$

to (4.10). Adding this to (4.11) or (4.14), the theory has a five complex dimensional, non-compact, moduli space of supersymmetric vacua

$$\vec{S} = -\frac{2h}{\epsilon} \vec{V}; \qquad \vec{V}^2 = \Lambda^4. \qquad (4.19)$$

For $|\epsilon| \gg |\Lambda|$, the fields $\vec{S}$ are heavy and can be integrated out. The low energy theory is simply the $SU(2)$ theory with four massless doublets and no superpotential (the cubic couplings of (4.10) do not lead to a quartic superpotential when $\vec{S}$ is integrated out). This has a moduli space which is reproduced by (4.19).

For $|\epsilon| \ll |\Lambda|$, the $\vec{S}$ fields are light, and need to be included in the low energy theory; i.e. we add (4.18) to (4.14). As we take $\epsilon \to 0$, the SUSY vacua (4.19) run off to infinity. In addition to these supersymmetric ground states at large $|\vec{S}|$, we still have the compact moduli space of supersymmetry breaking vacua discussed following (4.14), with $\vec{S}$ near the origin. For $|\epsilon| \ll |\Lambda|$ these metastable, supersymmetry breaking states are very long lived. Finally, as $\epsilon \to 0$ the supersymmetric states disappear from the Hilbert space and we are left with only the metastable states.

Note that these theories provide examples of nonchiral theories that dynamically break supersymmetry. How is that compatible with the Witten index [8]? The argument based on the Witten index relies on adding mass terms to the theory and tracking the supersymmetric states as the mass is removed. In this problem we can add two possible mass terms. First, we can add mass terms for the fundamental quarks. This is done in the effective theory by adding $\vec{m} \cdot \vec{V}$ to the superpotential. But this has no effect because $\vec{m}$ can be absorbed in a shift of $\vec{S}$. Second, if we add (4.18), $\vec{S}$ is massive. For large mass it leads to the noncompact moduli space of supersymmetric states (4.19). For small mass we also find the compact moduli space of supersymmetry breaking metastable states, and as $\epsilon \to 0$ the supersymmetric states disappear from the Hilbert space and supersymmetry is broken.

4.3. Metastable states in SQCD [9]

Consider SQCD with $N_c + 1 \le N_f < \frac{3}{2}N_c$, with small quark masses

$$|\text{Eigenvalues}(m)| \ll |\Lambda|. \qquad (4.20)$$

The range of N_f is such that the magnetic dual [33] of Section 3.6 is the IR free, low-energy effective field theory. We thus analyze the groundstates in the magnetic dual, with superpotential

$$h \text{Tr} \, \Phi \varphi \widetilde{\varphi} + \alpha \Lambda \text{Tr} \, m \Phi. \qquad (4.21)$$

This is the same as the theory we studied in (2.41), (2.43) with the identification[9]

$$\alpha \Lambda m = f. \tag{4.22}$$

For simplicity, we will take m (and therefore also f) to be proportional to the unit matrix, thus preserving the global $SU(N_f)$.

As discussed following (2.41), this low energy theory has a supersymmetry breaking minimum (2.68). All non-Goldstone modes have non-tachyonic masses there, from the one-loop potential, which is computed via (2.47) in the low-energy dual theory. The fact that the magnetic theory is IR free ensures that higher loops are suppressed, and in particular cannot invalidate the results from the one-loop potential.

We thus conclude that SQCD has metastable dynamical supersymmetry breaking vacua. In terms of the microscopic electric SQCD theory, the DSB vacua (2.68) have zero expectation value for the meson fields, $\langle M \rangle = 0$, and non-zero expectation value of some baryon fields, $\langle B \rangle \neq 0$ and $\langle \widetilde{B} \rangle \neq 0$, which follow from the non-zero $\langle \varphi \rangle$ and $\langle \widetilde{\varphi} \rangle$ in (2.68). In terms of the IR dual magnetic theory, these vacua are semi-classical, but in terms of the microscopic, electric SQCD they are not, they are strongly quantum-mechanical.

As noted after (2.68), the supersymmetry breaking vacua (2.68) spontaneously break the global symmetries, from $G = SU(N_f) \times U(1)_B$ to $H = SU(N_f - N_c) \times SU(N_c) \times U(1)$. Associated with that, there is a compact moduli space of vacua, the manifold of massless Goldstone bosons,[10] $\mathcal{M}_{vac} = G/H$. Note that the DSB vacua have an assortment of massless fields: the G/H Goldstone bosons and a number of massless fermions including the Goldstino, which come from the fermionic components of the fields Φ_0 in (2.46). This is to be contrasted with the naive expectation that there should be no massless fields (and, in particular, no candidate Goldstino for DSB to occur), since the quarks Q all have a mass m, and the low-energy SYM gets a mass gap. The dual magnetic theory shows that this naive expectation is incorrect.

SQCD also has N_c supersymmetric vacua, with mass gap and $\langle M \rangle \sim \langle \Phi \rangle \neq 0$, and $\langle B \rangle = \langle \widetilde{B} \rangle = 0$. These supersymmetric vacua arise from the effective interaction (3.30) which, as explained earlier, are obtained from gluino condensation in the magnetic theory. Thus, in terms of the magnetic dual theory, supersymmetry is non-perturbatively restored, in a theory that breaks supersymmetry at tree-level. Indeed, from the point of view of the theory (2.41), (2.43), the R-symmetry is anomalous and is explicitly broken (this is manifest with the inter-

---

[9]The global vector $U(1)$ symmetry in (2.41) is normalized differently than the baryon number symmetry in (3.26). Also, the $U(1)_R$ symmetry in (3.26) is anomaly free but it is broken by the mass term, while in (2.41) we took $U(1)_R$ to preserve the term linear in Φ but it is anomalous.

[10]In various generalizations of this example, these compact moduli spaces of DSB vacua can support topological solitons, which can be (meta) stable, see [54] for a fuller discussion.

action (3.30)), and therefore supersymmetry is restored. As long as N_f is in the free magnetic range, $N_f < \frac{3}{2}N_c$, the supersymmetry restoring interaction (3.30) is irrelevant at the DSB vacua near $\Phi = 0$. Then the DSB and the SUSY vacua are sufficiently separated for the DSB vacua to be meaningful.

The small mass condition (4.20) has the following useful consequences:

1. It ensures that the analysis within the low-energy effective field theory (the magnetic dual) is valid: the superpotential coupling $f \sim m\Lambda$ is then safely below the UV cutoff, Λ, of the magnetic dual theory.

2. It ensures that effects from the microscopic (electric) theory do not invalidate the macroscopic analysis of supersymmetry breaking and the one loop stabilization of the vacua (2.68). A way to see this is to note that the one-loop potential gives all (non-Goldstone) pseudomoduli mass squares of order $|f| \sim |m\Lambda|$ (much as in (2.66)) which is non-analytic in the superpotential coupling $f \sim m\Lambda$. This reflects the fact that it comes from integrating out modes which become massless in this limit. On the other hand, any effects from the microscopic theory must be analytic in m, and then (4.20) ensures that such effects are subleading to (2.66).

3. The condition (4.20) also ensures that the supersymmetric vacua (3.20) can be seen in the magnetic effective theory, as then (3.20) is safely below its cutoff, $|\langle M \rangle| \ll |\Lambda|$.

4. It ensures that the metastable state is parametrically long lived. The tunneling probability is $\sim \exp(-S_{bounce})$, where $S_{bounce} \sim \Delta\Phi^4/V_{meta}$, with $\Delta\Phi$ the separation in field space between the metastable and the supersymmetric vacua, and $V_{meta} = M_s^4$. For small masses (4.20), S_{bounce} is parametrically large, and thus the metastable DSB vacua can be made parametrically arbitrarily long lived.

This kind of DSB appears generic. It exists also in similar $SO(N_c)$ and $SP(N_c)$ gauge theories [9], and many generalizations of it were found recently (see e.g. [21, 55–64]). Also, the early universe favors populating the DSB vacua over the SUSY vacua. One reason for that is the large degeneracy of the Goldstone boson moduli space of DSB vacua, versus the discrete N_c mass gapped supersymmetric vacua. Another reason is that the DSB vacua are closer to the origin of the moduli space than the supersymmetric vacua, and that is favored by the thermal effective potential [65–68].

4.4. Naturalizing (retrofitting) models [21, 60]

As we stressed in the introduction (around equation (1.1)), in order for a model of supersymmetry breaking to be fully natural, all scales which are much smaller than the UV cutoff M_{cutoff} should arise via dimensional transmutation. To be fully natural, the Lagrangian cannot have any super-renormalizable (relevant) operators, since they are naturally of order a positive power of M_{cutoff}. The Lagrangian should have only renormalizable (marginal) operators and non-renormalizable (irrelevant) operators, which are suppressed by inverse powers of M_{cutoff}. Any needed relevant operators should then arise dynamically, with exponentially suppressed coefficients, as in (1.1).

A simple way to achieve that is the following. Consider an "unnatural model" of supersymmetry breaking like one of the models in Section 2, with superpotential terms like $W_{tree} \supset f\mathcal{O}_1 + m\mathcal{O}_2$, where $\mathcal{O}_1$ is some dimension one operator, $\mathcal{O}_2$ is a dimension two operator, and $f \equiv \mu^2$. We want the mass scales m and μ to be much less than M_{cutoff}. Such a model can easily be naturalized (or retrofitted) by removing these couplings from the theory and replacing them with interactions with the operator $S \equiv -\text{Tr}\, W_\alpha^2/32\pi^2$ of some added, but otherwise decoupled, pure Yang–Mills theory (with no charged matter):

$$\int d^2\theta \left[-\frac{8\pi^2}{g^2(M_{cutoff})} + \frac{a_1}{M_{cutoff}}\mathcal{O}_1 + \frac{a_2}{M^2_{cutoff}}\mathcal{O}_2 \right] S, \qquad (4.23)$$

where $a_{1,2}$ are dimensionless coefficients of order one, so the couplings in (4.23) are natural.

The pure Yang–Mills theory entering in (4.23) has a dynamically generated scale Λ, which satisfies $\Lambda \ll M_{cutoff}$, as in (1.1). For energies below the scale Λ, the added Yang–Mills theory becomes strong and leads to gaugino condensation $\langle S \rangle = \Lambda^3$. Substituting this in (4.23) we find

$$\int d^2\theta \left[\frac{a_1\Lambda^3}{M_{cutoff}}\mathcal{O}_1 + \frac{a_2\Lambda^3}{M^2_{cutoff}}\mathcal{O}_2 \right]. \qquad (4.24)$$

Thus we generate super-renormalizable couplings in the superpotential with $\mu^2 \sim \Lambda^3/M_{cutoff} \ll M^2_{cutoff}$ and $m \sim \Lambda^3/M^2_{cutoff} \ll M_{cutoff}$. For example, the O'Raifeartaigh model of Section 2.5 can be naturalized by replacing (2.28) with

$$\int d^2\theta \left[\frac{1}{2}hX\phi_1^2 + \left(-\frac{8\pi^2}{g^2(M_{cutoff})} + \frac{a_1}{M_{cutoff}}X + \frac{a_2}{M^2_{cutoff}}\phi_1\phi_2 \right) S \right]. \quad (4.25)$$

More generally, we can use couplings like (4.23) with different gauge groups or with couplings with higher powers of W_α. This way, every unnatural model can be easily naturalized.

This naturalization procedure is not unique. A given macroscopic theory can be naturalized in more than one way. Consider, for example, the macroscopic models based on the rank condition of Section 2.6. One way to naturalize them is to replace the last term in (2.43) with $\frac{1}{M_{cutoff}} \mathrm{Tr}\, \Phi \, \mathrm{Tr}\, W_\alpha'^2$, where W_α' is the field strength of some other pure Yang–Mills theory, with scale Λ'; this leads to $f \sim \Lambda'^3 / M_{cutoff}$. Alternatively, we can first view this theory as the low energy approximation of a SQCD theory, as in Section 4.3. This theory is not yet fully natural because of the existence of the quark mass term $m\,\mathrm{Tr}\,\tilde{Q}Q^T$ in the Lagrangian. As in (4.22), this leads to $f \sim m\Lambda$, which is dynamical, but not yet fully natural because we need (4.20), $|m| \ll |\Lambda| \ll M_{cutoff}$. It can be made fully natural by replacing the mass term of the UV Lagrangian with $\frac{1}{M_{cutoff}^2} \mathrm{Tr}\, \tilde{Q}Q^T \, \mathrm{Tr}\, W_\alpha'^2$ [63]. This leads to $m \sim \Lambda'^3 / M_{cutoff}^2$, so $|m| \ll |\Lambda|$ is natural, and $f \sim \Lambda\Lambda'^3 / M_{cutoff}^2$.

Throughout this analysis, we have viewed the theory in an expansion in powers of M_{cutoff}^{-1}. For example, in (4.25) we did not consider higher dimension operators like $\frac{X^2}{M_{cutoff}^2} W_\alpha^2$. As another example, gluino condensation in (4.25) does not simply replace $\left(-\frac{8\pi^2}{g^2} + \frac{X}{M_{cutoff}}\right) S$ with $\frac{X}{M_{cutoff}} \Lambda^3$. More precisely, following the analysis in Section 3.1, for an $SU(N_c)$ gauge theory it replaces it with

$$N_c \Lambda^3 \exp\left(\frac{X}{N_c M_{cutoff}}\right) \approx N_c \Lambda^3 + \frac{X}{M_{cutoff}} \Lambda^3, \qquad (4.26)$$

where we neglected higher order terms in M_{cutoff}^{-1} in the latter expression.

This expansion in powers of M_{cutoff}^{-1} is significant. It is well known that one can trigger supersymmetry breaking by coupling a chiral superfield to a Yang–Mills theory via higher dimension operators and using gluino condensation [69–71]. This usually leads to runaway behavior, as is clear from the first expression in (4.26). However, since we content ourselves with finding supersymmetry breaking only in a metastable state, we can focus on a particular region in field space and ignore possible vacua elsewhere in field space. This focusing on a region in field space is achieved by the expansion in M_{cutoff}^{-1} we mentioned above. Therefore, this naturalization procedure leads to acceptable, metastable, dynamical supersymmetry breaking.

Acknowledgements

We would like to thank the organizers of the various schools, and also the participants for their questions and comments. We thank our many colleagues and friends for useful discussions about these topics. In particular, we would like to thank our collaborators on these and related subjects: I. Affleck, M. Dine, R. Leigh, A. Nelson, P. Pouliot, S. Shenker, D. Shih, M. Strassler, S. Thomas and E. Witten. The research of NS is supported in part by DOE grant DE-FG02-90ER40542. The research of KI is supported in part by UCSD grant DOE-FG03-97ER40546.

References

[1] E. Witten, Dynamical breaking of supersymmetry, *Nucl. Phys. B* **188** (1981) 513.

[2] P. Fayet and J. Iliopoulos, Spontaneously broken supergauge symmetries and Goldstone spinors, *Phys. Lett. B* **51** (1974) 461.

[3] J. Wess and J. Bagger, Supersymmetry and supergravity.

[4] S.J. Gates, M.T. Grisaru, M. Rocek and W. Siegel, Superspace, or one thousand and one lessons in supersymmetry, *Front. Phys.* **58** (1983) 1 [arXiv:hep-th/0108200].

[5] S. Weinberg, The quantum theory of fields. Vol. 3: Supersymmetry.

[6] J. Terning, Modern supersymmetry: Dynamics and duality.

[7] M. Dine, Supersymmetry and String Theory: Beyond the Standard Model.

[8] E. Witten, Constraints on supersymmetry breaking, *Nucl. Phys. B* **202** (1982) 253.

[9] K. Intriligator, N. Seiberg and D. Shih, Dynamical SUSY breaking in meta-stable vacua, *JHEP* **0604** (2006) 021 [arXiv:hep-th/0602239].

[10] J.R. Ellis, C.H. Llewellyn Smith and G.G. Ross, Will the Universe become supersymmetric? *Phys. Lett. B* **114** (1982) 227.

[11] A.E. Nelson and N. Seiberg, R symmetry breaking versus supersymmetry breaking, *Nucl. Phys. B* **416** (1994) 46 [arXiv:hep-ph/9309299].

[12] J. Bagger, E. Poppitz and L. Randall, The R axion from dynamical supersymmetry breaking, *Nucl. Phys. B* **426** (1994) 3 [arXiv:hep-ph/9405345].

[13] K.A. Intriligator and N. Seiberg, Lectures on supersymmetric gauge theories and electric-magnetic duality, *Nucl. Phys. Proc. Suppl.* **45BC** (1996) 1 [arXiv:hep-th/9509066].

[14] M.E. Peskin, Duality in supersymmetric Yang–Mills theory, arXiv:hep-th/9702094.

[15] M.A. Shifman, Nonperturbative dynamics in supersymmetric gauge theories, *Prog. Part. Nucl. Phys.* **39** (1997) 1 [arXiv:hep-th/9704114].

[16] M.J. Strassler, An unorthodox introduction to supersymmetric gauge theory, arXiv:hep-th/0309149.

[17] K. Intriligator, N. Seiberg and D. Shih, Supersymmetry breaking, R-symmetry breaking and metastable vacua, arXiv:hep-th/0703281.

[18] E. Witten, Mass hierarchies in supersymmetric theories, *Phys. Lett. B* **105** (1981) 267.

[19] S. Ray, Some properties of meta-stable supersymmetry-breaking vacua in Wess-Zumino models, *Phys. Lett. B* **642** (2006) 137 [arXiv:hep-th/0607172].

[20] L. O'Raifeartaigh, Spontaneous symmetry breaking for chiral scalar superfields, *Nucl. Phys. B* **96** (1975) 331.

[21] M. Dine, J.L. Feng and E. Silverstein, Retrofitting O'Raifeartaigh models with dynamical scales, *Phys. Rev. D* **74** (2006) 095012 [arXiv:hep-th/0608159].

[22] M. Huq, On spontaneous breakdown of fermion number conservation and supersymmetry, *Phys. Rev. D* **14** (1976) 3548.

[23] N. Seiberg, Naturalness versus supersymmetric nonrenormalization theorems, *Phys. Lett. B* **318** (1993) 469 [arXiv:hep-ph/9309335].

[24] K.A. Intriligator, R.G. Leigh and N. Seiberg, Exact superpotentials in four-dimensions, *Phys. Rev. D* **50** (1994) 1092 [arXiv:hep-th/9403198].

[25] D. Finnell and P. Pouliot, Instanton calculations versus exact results in four-dimensional SUSY gauge theories, *Nucl. Phys. B* **453** (1995) 225 [arXiv:hep-th/9503115].

[26] G.R. Dvali and M.A. Shifman, Domain walls in strongly coupled theories, *Phys. Lett. B* **396** (1997) 64 [Erratum-ibid. *B* **407** (1997) 452] [arXiv:hep-th/9612128].

[27] G. Veneziano and S. Yankielowicz, An effective Lagrangian for the pure $N = 1$ supersymmetric Yang–Mills theory, *Phys. Lett. B* **113** (1982) 231.

[28] I. Affleck, M. Dine and N. Seiberg, Dynamical supersymmetry breaking in supersymmetric QCD, *Nucl. Phys. B* **241** (1984) 493.

[29] N. Seiberg, Exact results on the space of vacua of four-dimensional SUSY gauge theories, *Phys. Rev. D* **49** (1994) 6857 [arXiv:hep-th/9402044].

[30] K.A. Intriligator, 'Integrating in' and exact superpotentials in 4-d, *Phys. Lett. B* **336** (1994) 409 [arXiv:hep-th/9407106].

[31] T.R. Taylor, G. Veneziano and S. Yankielowicz, Supersymmetric QCD and its massless limit: An effective Lagrangian analysis, *Nucl. Phys. B* **218** (1983) 493.

[32] A.C. Davis, M. Dine and N. Seiberg, The massless limit of supersymmetric QCD, *Phys. Lett. B* **125** (1983) 487.

[33] N. Seiberg, Electric—magnetic duality in supersymmetric non-Abelian gauge theories, *Nucl. Phys. B* **435** (1995) 129 [arXiv:hep-th/9411149].

[34] I. Affleck, M. Dine and N. Seiberg, Dynamical supersymmetry breaking in chiral theories, *Phys. Lett. B* **137** (1984) 187.

[35] Y. Meurice and G. Veneziano, SUSY vacua versus chiral fermions, *Phys. Lett. B* **141** (1984) 69.

[36] K.A. Intriligator, N. Seiberg and S.H. Shenker, Proposal for a simple model of dynamical SUSY breaking, *Phys. Lett. B* **342** (1995) 152 [arXiv:hep-ph/9410203].

[37] I. Affleck, M. Dine and N. Seiberg, Calculable nonperturbative supersymmetry breaking, *Phys. Rev. Lett.* **52** (1984) 1677.

[38] I. Affleck, M. Dine and N. Seiberg, Dynamical supersymmetry breaking in four-dimensions and its phenomenological implications, *Nucl. Phys. B* **256** (1985) 557.

[39] H. Murayama, Studying noncalculable models of dynamical supersymmetry breaking, *Phys. Lett. B* **355** (1995) 187 [arXiv:hep-th/9505082].

[40] E. Poppitz, Y. Shadmi and S.P. Trivedi, Supersymmetry breaking and duality in $SU(N) \times SU(N - M)$ theories, *Phys. Lett. B* **388** (1996) 561 [arXiv:hep-th/9606184].

[41] C. Csaki, L. Randall, W. Skiba and R.G. Leigh, Supersymmetry breaking through confining and dual theory gauge dynamics, *Phys. Lett. B* **387** (1996) 791 [arXiv:hep-th/9607021].

[42] K.A. Intriligator and S.D. Thomas, Dual descriptions of supersymmetry breaking, arXiv:hep-th/9608046.

[43] H. Murayama, A model of direct gauge mediation, *Phys. Rev. Lett.* **79** (1997) 18 [arXiv:hep-ph/9705271].

[44] S. Dimopoulos, G.R. Dvali, R. Rattazzi and G.F. Giudice, Dynamical soft terms with unbroken supersymmetry, *Nucl. Phys. B* **510** (1998) 12 [arXiv:hep-ph/9705307].

[45] M.A. Luty, Simple gauge-mediated models with local minima, *Phys. Lett. B* **414** (1997) 71 [arXiv:hep-ph/9706554].

[46] S. Dimopoulos, G.R. Dvali and R. Rattazzi, A simple complete model of gauge-mediated SUSY-breaking and dynamical relaxation mechanism for solving the mu problem, *Phys. Lett. B* **413** (1997) 336 [arXiv:hep-ph/9707537].

[47] M.A. Luty and J. Terning, New mechanisms of dynamical supersymmetry breaking and direct gauge mediation, *Phys. Rev. D* **57** (1998) 6799 [arXiv:hep-ph/9709306].

[48] Y. Shadmi and Y. Shirman, Dynamical supersymmetry breaking, *Rev. Mod. Phys.* **72** (2000) 25 [arXiv:hep-th/9907225].

[49] J. Terning, Non-perturbative supersymmetry, arXiv:hep-th/0306119.

[50] K. Intriligator and S.D. Thomas, Dynamical supersymmetry breaking on quantum moduli spaces, *Nucl. Phys. B* **473** (1996) 121 [arXiv:hep-th/9603158].

[51] K.I. Izawa and T. Yanagida, Dynamical supersymmetry breaking in vector-like gauge theories, *Prog. Theor. Phys.* **95** (1996) 829 [arXiv:hep-th/9602180].

[52] Z. Chacko, M.A. Luty and E. Ponton, Calculable dynamical supersymmetry breaking on deformed moduli spaces, *JHEP* **9812** (1998) 016 [arXiv:hep-th/9810253].

[53] N. Arkani-Hamed and H. Murayama, Renormalization group invariance of exact results in supersymmetric gauge theories, *Phys. Rev. D* **57** (1998) 6638 [arXiv:hep-th/9705189].

[54] M. Eto, K. Hashimoto and S. Terashima, Solitons in supersymmetry breaking meta-stable vacua, arXiv:hep-th/0610042.

[55] S. Franco and A.M. Uranga, Dynamical SUSY breaking at meta-stable minima from D-branes at obstructed geometries, *JHEP* **0606** (2006) 031 [arXiv:hep-th/0604136].

[56] H. Ooguri and Y. Ookouchi, Landscape of supersymmetry breaking vacua in geometrically realized gauge theories, *Nucl. Phys. B* **755** (2006) 239 [arXiv:hep-th/0606061].

[57] R. Kitano, Dynamical GUT breaking and mu-term driven supersymmetry breaking, arXiv:hep-ph/0606129.

[58] R. Kitano, Gravitational gauge mediation, *Phys. Lett. B* **641** (2006) 203 [arXiv:hep-ph/0607090].

[59] A. Amariti, L. Girardello and A. Mariotti, Non-supersymmetric meta-stable vacua in $SU(N)$ SQCD with adjoint matter, arXiv:hep-th/0608063.

[60] M. Dine and J. Mason, Gauge mediation in metastable vacua, arXiv:hep-ph/0611312.

[61] R. Kitano, H. Ooguri and Y. Ookouchi, Direct mediation of meta-stable supersymmetry breaking, arXiv:hep-ph/0612139.

[62] H. Murayama and Y. Nomura, Gauge mediation simplified, arXiv:hep-ph/0612186.

[63] O. Aharony and N. Seiberg, Naturalized and simplified gauge mediation, arXiv:hep-ph/0612308.

[64] C. Csaki, Y. Shirman and J. Terning, A simple model of low-scale direct gauge mediation, arXiv:hep-ph/0612241.

[65] S.A. Abel, C.S. Chu, J. Jaeckel and V.V. Khoze, SUSY breaking by a metastable ground state: Why the early universe preferred the non-supersymmetric vacuum, arXiv:hep-th/0610334.

[66] N.J. Craig, P.J. Fox and J.G. Wacker, Reheating metastable O'Raifeartaigh models, arXiv:hep-th/0611006.

[67] W. Fischler, V. Kaplunovsky, C. Krishnan, L. Mannelli and M. Torres, Meta-stable supersymmetry breaking in a cooling universe, arXiv:hep-th/0611018.

[68] S.A. Abel, J. Jaeckel and V.V. Khoze, Why the early universe preferred the non-supersymmetric vacuum. II, arXiv:hep-th/0611130.

[69] S. Ferrara, L. Girardello and H.P. Nilles, Breakdown of local supersymmetry through gauge fermion condensates, *Phys. Lett. B* **125** (1983) 457.

[70] J.P. Derendinger, L.E. Ibanez and H.P. Nilles, On the low-energy $D = 4$, $N = 1$ supergravity theory extracted from the $D = 10$, $N = 1$ superstring, *Phys. Lett. B* **155** (1985) 65.

[71] M. Dine, R. Rohm, N. Seiberg and E. Witten, Gluino condensation in superstring models, *Phys. Lett. B* **156** (1985) 55.

Course 4

SUPERSYMMETRY AND THE REAL WORLD

G.F. Giudice

CERN Theory Division,
Geneva, Switzerland

C. Bachas, L. Baulieu, M. Douglas, E. Kiritsis, E. Rabinovici, P. Vanhove, P. Windey
and L.F. Cugliandolo, eds.
Les Houches, Session LXXXVII, 2007
String Theory and the Real World: From Particle Physics to Astrophysics
© *2008 Published by Elsevier B.V.*

Contents

1. Summary

Low-energy supersymmetry, the topic of my lectures at the 2007 Les Houches École d'Été de Physique Théorique, is by now a well-established subject exhaustively discussed in excellent reviews and text books. For this reason, I decided to refer the students to the existing literature, rather than writing new lecture notes. I will give here just a short summary of the material I presented and a list of references to the appropriate reviews. The slides of my lectures are also available at the web page of the school.

During my lectures, assuming that students were already familiar with the formalism of $N = 1$ supersymmetric gauge theories and supergravity, I introduced the hierarchy problem as the primary motivation for low-energy supersymmetry. Next, I discussed the deep connection between supersymmetry breaking and the generation of the electroweak scale. The phenomenology of the Higgs sector was investigated. After explaining the friction between experimental data and a natural implementation of supersymmetry, I introduced the different schemes of mediation of supersymmetry breaking and studied their phenomenological consequences. I discussed gravity mediation (and the flavor problem), gauge mediation, anomaly mediation and, more briefly, gaugino and mirage mediation. Finally, I discussed how dark matter can be obtained in supersymmetric models and studied the various implementations.

The subject of supersymmetry breaking was discussed by Intriligator in his lectures at this school [1]. There are five recent books which give introductions to supersymmetric theories and their low-energy applications by Binetruy [2], by Drees, Godbole, Roy [3], by Terning [4], by Baer, Tata [5], and by Dine [6]. There is also a book, edited by Kane [7], with chapters by different authors devoted to various subjects relevant to low-energy supersymmetry. The first chapter, by Martin [8], contains a pedagogical introduction. I can also recommend some very good lecture notes [9–16]. Finally, my discussion of gauge mediation followed the review in Ref. [17].

References

[1] K. Intriligator, these proceedings.
[2] P. Binetruy, *Supersymmetry: Theory, Experiment, and Cosmology*, Oxford University Press, 2004.

[3] M. Drees, R.M. Godbole, and P. Roy, *Theory and Phenomenology of Sparticles*, World Scientific, 2004.

[4] J. Terning, *Modern Supersymmetry*, Oxford University Press, 2006.

[5] H. Baer and X. Tata, *Weak Scale Supersymmetry*, Cambridge University Press, 2006.

[6] M. Dine, *Supersymmetry and String Theory: Beyond the Standard Model*, Cambridge University Press, 2007.

[7] *Perspective on Supersymmetry*, ed. by G.L. Kane, World Scientific, 1998.

[8] S.P. Martin, A supersymmetry primer, arXiv:hep-ph/9709356.

[9] J.A. Bagger, Weak-scale supersymmetry: Theory and practice, arXiv:hep-ph/9604232.

[10] M. Drees, An introduction to supersymmetry, arXiv:hep-ph/9611409.

[11] J.D. Lykken, Introduction to supersymmetry, arXiv:hep-th/9612114.

[12] H. Murayama, Supersymmetry phenomenology, arXiv:hep-ph/0002232.

[13] G.L. Kane, Weak scale supersymmetry: A top-motivated-bottom-up approach, arXiv:hep-ph/0202185.

[14] J.R. Ellis, Supersymmetry for Alp hikers, arXiv:hep-ph/0203114.

[15] M.A. Luty, 2004 TASI lectures on supersymmetry breaking, arXiv:hep-th/0509029.

[16] M.E. Peskin, Supersymmetry in elementary particle physics, arXiv:0801.1928 [hep-ph].

[17] G.F. Giudice and R. Rattazzi, Theories with gauge-mediated supersymmetry breaking, *Phys. Rept.* **322** (1999) 419 [arXiv:hep-ph/9801271].

Course 5

LES HOUCHES LECTURES ON COSMOLOGY AND FUNDAMENTAL THEORY

Juan M. Maldacena

Institute for Advanced Study, Princeton, NJ 08540, USA

C. Bachas, L. Baulieu, M. Douglas, E. Kiritsis, E. Rabinovici, P. Vanhove, P. Windey and L.F. Cugliandolo, eds.
Les Houches, Session LXXXVII, 2007
String Theory and the Real World: From Particle Physics to Astrophysics
© 2008 Published by Elsevier B.V.

Contents

1. Review of standard cosmology

We start with a lightning review of standard cosmology. For further discussion see [1–4]. As one observes the universe at the large scales one finds that it is spatially uniform. So one can describe it using a spatially uniform metric

$$ds^2 = -dt^2 + a(t)^2 dx_i dx_i = a^2[-d\eta^2 + dx_i dx_i] \tag{1.1}$$

where t is proper time and η is called conformal time. The coordinates x^i are called "comoving" coordinates. Note that the universe is uniform and isotropic in space but it is not uniform in time, it was different in the past. In writing the metric (1.1), I have assumed that the universe is spatially flat, which is in good agreement with current observations, but one could have imagined also spatial sections with constant positive or negative curvature which would also be homogeneous and isotropic. From now on we will discuss only the flat case. We can define the expansion rate

$$H = \frac{1}{R_H} \equiv \frac{\dot{a}}{a} \tag{1.2}$$

where the dot is a derivative with respect to proper time. We have also introduced a quantity called the Hubble radius, R_H. We will later see that the Hubble radius is a length scale which characterizes the range of influence of the physics that is happening at a certain time. We will discuss this in more detail below.

The scale factor also characterizes the redshift, z, of a photon emitted at time t and observed at a later time t_0

$$(1 + Z) = \frac{\lambda_0}{\lambda_t} = \frac{a(0)}{a(t)} \tag{1.3}$$

The evolution of the universe is determined by Einstein's equations after we make some statement about the matter distribution. We assume that the matter distribution is given by a perfect fluid with a stress tensor of the form $T_\nu^\mu \sim diag(\rho, -p, -p, -p)$ characterized by the density and pressure. We include a possible cosmological constant as a contribution to the stress tensor. Einstein's equations then boil down to

$$3H^2 = \rho M_{pl}^{-2}, \quad M_{pl}^{-2} \equiv 8\pi G_N \tag{1.4}$$

183

The main contributions to the density today are those of the cosmological constant, which is about 75% and matter which is the other 25%. Dark matter contributes 21% and ordinary matter is about 4%. The rest of the components of the universe such as electromagnetic radiation, neutrinos and gravity waves give a negligible contribution the energy of the universe today. Of course, they were more important in the past since radiation goes like $\rho_{rad} \sim a^{-4}$ while matter $\rho_{mat} \sim a^{-3}$. Notice that for radiation we have that $\rho \sim T^4$. Photons redshift as $T \sim 1/a \sim (1 + Z)$ which gives the $1/a^4$ we mentioned above. Note that both for matter and radiation the number of particles per comoving volume is constant. This implies that we can talk about the ratio [5]

$$\eta_b = \frac{n_B}{n_\gamma} = (6.5 \pm 0.4)10^{-10} \tag{1.5}$$

which gives us the number of baryons per photon in the universe. This is one of the numbers that specifies the state of the universe and we would like to explain it.

Just as an aside note that for a radiation dominated universe we have the rough relation $H \sim T^2/M_{pl}$ which is saying that the scale that characterizes the time evolution of the universe is much slower than T for $T \ll M_{pl}$.

Now let us briefly recall the time-line in the evolution of the universe. Starting from the present where $T_{CMB} = 2.7$ K we can go back to $Z \sim 1000$ where the temperature is about 1 eV. Due to the large number of photons per baryon this temperature is enough to ionize hydrogen. Before this time the CMB was in thermal equilibrium with baryonic matter. At this time it ceases to be in equilibrium and the CMB is produced. The universe was very uniform at this time, the fluctuations were of the order of one part in 10^5. The universe was already matter dominated, but the cosmological constant was negligible. Going back to a redshift of about $Z \sim 3400$ we get to the matter radiation equality. Before this time the universe was radiation dominated. At a temperature between 1 MeV and 0.05 MeV primordial nucleosynthesis happened. Namely the weak interactions dropped out of equilibrium and the primordial nuclei were formed. Of course, all the heavier nuclei that form us were formed later in stars. However these primordial nuclei are very important because their relative abundances are telling us what the universe was doing at this time. Thus we know that it was radiation dominated and with no extra massless particles other than the ones we expect from the standard model. These observations also allow us to measure the baryon to photon ratio, which is also measured by the CMB at a later time in the history of the universe. According to the standard model we expect other interesting things to have happened at earlier times. One of them is the QCD phase transition, $T \sim 200$ MeV. Unfortunately this phase transition, which is now being experimentally explored at the Relativistic Heavy Ion Collider (RHIC), does not seem to lead to any observable effect because thermal equilibrium is achieved after it. At a temperature of the order of 100–1000 GeV we probably had the electroweak phase transition. Around this time or earlier WIMPs might have decoupled from the thermal bath. Baryogenesis, which gave rise to (1.5), should have happened at this time or earlier. It is generally believed that before all this the universe had a period of almost exponential expansion called inflation [6]. Inflation explains the uniformity and the flatness of the universe. In addition it gives a mechanism for the production of the primordial density fluctuations which then gave origin to galaxies and other structure in the universe. We do not have a clear understanding of the period before inflation. Quantum gravity is necessary to understand what happened before inflation. In fact we will argue later that quantum gravity is also important to explain the potential that gives rise to inflation. Gravity waves were produced during inflation and, if we could observe them, they would give us very interesting information about this period. We will discuss this in more detail later.

It is really amazing that the initial conditions for the universe, after inflation, are characterized by a very small number of parameters. One of them is (1.5), another is the ratio between ordinary matter and dark matter. We also have the amplitude of initial density fluctuations. In addition we have a number that only recently has become important which is the value of the cosmological constant (it is not yet settled whether it is a constant or whether it varies slowly). These parameters determine the Hubble constant today and the temperature of the CMB. In addition we have some other parameters which are very small, such as the spatial curvature of the universe. The spectrum of density fluctuations is characterized by a power law in wavelengths in terms of a power which is sometimes called n_s. We will later define it more carefully. For now we will just say that is close to one $n_s - 1 = -0.05 \pm 0.02$. The fluctuations are also Gaussian (though recently some non-Gaussianity was claimed [7]). In addition they are "adiabatic". This means that the fluctuations in all the fluids are proportional to their time derivatives in the unperturbed solution $\delta\rho_i/\dot\rho_i$. This holds for the photon fluid, the baryonic fluid and the dark matter fluid. More quantitatively, the fluctuations violating this relation are bounded by $\delta\rho_{iso}/\delta\rho < 0.08$. Many of these features are explained by the theory of inflation. Inflation also explains the absence of exotic objects such as magnetic monopoles, cosmic strings and domain walls. It should be noted though that strings could possibly arise after inflation [8].

We would like to have a theory of the early universe that explains the values of the parameters we observe. Inflation explains some of the features we see. However, there are other features whose origin is not clear. We do not know the nature of dark matter, the origin of the baryon asymmetry and the reason for the cosmic coincidence that the dark matter density, the baryons density, and the cosmological constant are all similar now. We do not have any explanation, other than anthropic arguments, for the value of the cosmological constant, and we do not know if it really constant.

1.1. Big questions

There are some bigger questions which definitely require quantum gravity, such as the following. What is the origin of time? Did time exist before the big bang? Why do we live in $3 + 1$ dimensions? What is the origin of dark energy? How do we get the standard model of particle physics? Does cosmology play a role in selecting the particular vacuum in which we live? How do we compute probabilities for the whole universe? What is the wavefunction of the universe? What is the correct theory of quantum gravity? And the most important questions are probably some that we have not yet thought about yet!

Of course, we would like string theory to answer all these questions. However, we do not understand string theory well enough to be able to answer these questions, or to even know which of these questions are answerable in principle. One particular area that string theory could be useful for is to provide microscopic models for inflation that are embedded in a full theory of quantum gravity.

In this lecture I will focus mainly in reviewing several aspects of inflation, since it is one promising area where we could expect that string theory might make some connection with experiment. The important topic of eternal inflation and its interpretation in a theory of quantum gravity is covered in S. Shenker's lectures on this same volume.

First, we will review several well known facts about inflationary theory and then we will discuss in some detail the mechanism for the generation of fluctuations during inflation. This is a very standard computation but recently several extensions and variations have been explored, so it is worth reviewing and understanding clearly the standard case first. Finally, we will discuss some aspects of models that produce inflation in string theory. We will also try to explain why it is necessary to have a theory of quantum gravity to explain the origin and features of the potential that drives inflation.

2. Review of standard inflationary theory

Let us first derive the Penrose diagram of the standard Big Bang theory. Writing the metric in conformal time as in (1.1) we see that conformal time is given by $d\eta = \frac{dt}{a(t)}$. So if we compute the conformal time elapsed since the size of the universe was zero $a = 0$, we find a finite answer if $a(t) \sim t^p$ for $p < 1$, which is the behavior for matter ($p = 2/3$) or radiation ($p = 1/3$). In conformal time the

Penrose diagram looks like the one of flat space, except that we have an initial singularity at some value of conformal time, which we set to zero, $\eta = 0$. When we look back we can see a portion of the initial singularity which has comoving size η_0, where η_0 is the conformal time today. On the other hand, if we think about the universe at a very early time η_{pl} then those regions can only see a small portion of the part of the initial singularity that we see today. Thus we conclude that if the initial conditions at the singularity were independent for these regions, then these regions would not have had time to equilibrate and the uniform would not be as uniform at it is observed to be. In other words, many of the distant regions of the universe that we now see were causally disconnected from each other at the time that they emitted the signals we see today. However we observe that they have similar properties and there are correlations among them. If we extrapolate all the way to the Planck time, the number of disconnected regions is given by the cube of the following number

$$\frac{\eta_0}{\eta_{pl}} \sim \left(\frac{t_0}{t_{pl}}\right)^{1/2} = \left(\frac{H_{pl}}{H_0}\right)^{1/2} \sim 10^{30} \sim e^{70} \qquad (2.1)$$

Of course the Penrose diagram could change if the form of the evolution changes at early times. The idea is to change the evolution in such a way that we give it enough conformal time so as to make sure that all the regions we see today were causally connected to some initial event that sets the initial conditions for them. We need that a decreases as fast as $a \sim t^p$ with $p > 1$. This can only occur if $\ddot{a} > 0$ or there is a period of acceleration in the early universe. In this case we get a large amount of conformal time before the ordinary radiation dominated phase. This ensures that all the regions we now see had a common origin. By writing Eintein's equation (1.4) and its time derivative, after using the equation for conservation of the stress tensor, we get

$$3H^2 M_{pl}^2 = \rho, \qquad \dot{H} M_{pl}^2 = -\frac{1}{2}(\rho + p), \qquad \frac{\ddot{a}}{a} M_{pl}^2 = -\frac{1}{6}(\rho + 3p) \quad (2.2)$$

From the last equation we see that the condition for acceleration is that $p < -\rho/3$. If we have a spatially uniform scalar field with a potential we have

$$\rho = \frac{1}{2}\dot{\phi}^2 + V, \qquad p = \frac{1}{2}\dot{\phi} - V, \qquad p + 3\rho = 2(\dot{\phi}^2 - V) \qquad (2.3)$$

So in order to have acceleration we need $\dot{\phi}^2 < V$. Typically we will demand that the kinetic energy of the scalar field is much smaller than its potential energy. The equation of motion for the scalar field is

$$\ddot{\phi} + 3H\dot{\phi} + \frac{\partial V}{\partial \phi} = 0 \qquad (2.4)$$

Now let us suppose that $\ddot{\phi} \ll \frac{\partial V}{\partial \phi}$. Then we can approximate equation (2.4) as $-3H\dot{\phi} \sim \frac{\partial V}{\partial \phi}$. Using this expression for the time derivative of the field and demanding that the potential energy dominates we get the condition

$$\epsilon \equiv \frac{1}{2} \frac{M_{pl}^2 (\partial_\phi V)^2}{V^2} \ll 1 \tag{2.5}$$

Demanding that ϵ remains small we get the additional condition

$$\eta \equiv \frac{M_{pl}^2 \partial_\phi^2 V}{V} \ll 1 \tag{2.6}$$

The number of e-folds is given by

$$N_{eff} = \int \frac{da}{a} = \int H dt = \int \frac{H}{\dot{\phi}} d\phi = M_{pl}^{-2} \int_{\phi_i}^{\phi_f} d\phi \frac{V}{(-\partial_\phi V)} \tag{2.7}$$

where ϕ_i and ϕ_f are the values of the scalar field and the begging and the end of inflation. In many models we get that the number of e-foldings is proportional to $N_{eff} \sim 1/\eta$. One requires that $N_{eff} \sim 60$ in order to put all the horizon size patches in causal contact at the time of the start of the usual radiation dominated phase (the hot big bang phase). This number could be smaller if we lower the reheating temperature or if we alter the history of the early universe (before nucleosynthesis) is some way. Typically a value of η and ϵ of the order of 0.01–0.03 is good enough. Usually inflation lasts until the slow roll conditions cease to be valid. At this point the scalar field starts to evolve more rapidly. In order to have a model with a chance of explaining nature we need that enough of the energy of the scalar field is transferred into ordinary matter and dark matter. In addition we need to reheat the universe at least up to the temperatures of big bang nucleosynthesis. Ordinarily one wants to achieve much higher temperatures in order to explain baryogenesis.

The η condition (2.6) can also be understood as follows. Suppose that we have a massive scalar field $V \sim V_0 + m^2 \phi^2$. Then we find that $\eta \sim \frac{m^2}{H^2}$. Thus we are demanding that the mass of the field is smaller than the Hubble scale. An important point to note about the slow roll conditions (2.5), (2.6) is that they involve explicitly the Planck mass. This implies that it is meaningless to try to construct inflation models within quantum field theories without gravity. It is crucial to include the gravity corrections to the potential. In particular, it is important to include possible quantum gravity corrections to the potential. For this reason the construction of an inflationary model requires quantum gravity. It is not that quantum gravity effects are important during the evolution, but quantum gravity is important to derive the effective field theory that describes inflation. We also see that for simple power law potentials $V \sim \lambda_n \phi^n$ we find that the number of e-folds is $N_{eff} = \frac{1}{2n} \frac{\Delta \phi^2}{M_{pl}^2}$. The slow roll parameters are proportional to $1/N_{eff}$ and the energy density is much smaller than the Planck scale at the time that the scales we see today were of the order of the Hubble scale during inflation. So these power law potentials are the simplest inflationary models. Note that the change of the field in Planck units is large. This is another reason for wanting to understand quantum gravity, and we will see that in string theory it is quite difficult to have such potentials. This conclusion only holds for these simple power law potentials. For other potentials it is possible to have inflation with a field that does not move much in Planck units [9], see also [10].

The last qualitative issue we would like to clarify is the behavior of perturbations in an inflationary background. Let us consider a massless field. Its equation of motion in comoving coordinates has the form

$$\ddot{\phi} + 3H\dot{\phi} + \frac{1}{a^2}\partial_x^2 \phi = 0 \tag{2.8}$$

We see that when the wavelengths of interest $a \Delta x$ are larger than the Hubble radius $1/H$, then we can neglect the last term in the equation compared to the second term. In fact we can think of $1/H$ as an instantaneous horizon radius which is telling us how far the perturbations are propagating. Two regions that are further away than $1/H$ will evolve independently. When we look at a fixed comoving distance scale Δx we see that it starts out being subplancking. As inflation proceeds it grows and becomes comparable to the size of the horizon, then it goes outside the horizon. After inflation ends the Hubble scale starts growing more rapidly and eventually the scale comes back into the horizon and the two points start being able to influence each other again. So the whole universe that we observe was at some point subplancking and thus all its regions had a common origin. This explains the uniformity of the universe and the fact that what regions that seemed to be causally disconnected in the hot big bang model have similar properties.

Inflation also predicts that the spatial curvature of the universe should be negligible. It also dilutes any object that was produced before inflation. Such objects could be domain walls, strings or monopoles. Of course, these objects could also be produced at the end of inflation when the universe is reheated, depending on the details of the model. One nice feature of inflation is that it also produces small fluctuations in the energy density. So inflation does not only produce a fairly uniform universe but it also seeds the proper energy fluctuations that will later grow into galaxies and form the structure in the universe. The computation of quantum fluctuations in inflation is a simple application of quantum field theory in time dependent backgrounds. It can also be viewed as a variant of Hawking radiation of black holes, except that now the horizon is the cosmological horizon during inflation. In the next section we will describe it in detail.

3. Generation of fluctuations during inflation

The computation of primordial fluctuations that arise in inflationary models was first discussed in [11–16] and was nicely reviewed in [17]. The discussion here is taken from [18].

The starting point is the Lagrangian of gravity and a scalar field which has the general form

$$S = \frac{1}{2} \int \sqrt{g} [R - (\nabla \phi)^2 - 2V(\phi)] \tag{3.1}$$

up to field redefinitions. We have set $M_{pl}^{-2} \equiv 8\pi G_N = 1$. Note that this definition of M_{pl} is different from the definition that some other authors use (including Planck). The dependence on G_N is easily reintroduced.

The homogeneous solution has the form

$$ds^2 = -dt^2 + e^{2\rho(t)} dx_i dx_i = e^{2\rho} (-d\eta^2 + dx_i dx_i) \tag{3.2}$$

where η is conformal time. The scalar field is a function of time only. ρ and ϕ obey the equations

$$3\dot{\rho}^2 = \frac{1}{2}\dot{\phi}^2 + V(\phi) \tag{3.3}$$

$$\ddot{\rho} = -\frac{1}{2}\dot{\phi}^2 \tag{3.4}$$

$$0 = \ddot{\phi} + 3\dot{\rho}\dot{\phi} + V'(\phi) \tag{3.5}$$

The Hubble parameter is $H \equiv \dot{\rho}$. The third equation follows from the first two. We will make frequent use of these equations.

If the slow roll parameters are small we will have a period of accelerated expansion. The slow roll parameters can also be expressed in terms of time derivatives of the fields as

$$\epsilon \equiv \frac{1}{2} \left(\frac{M_{pl}V'}{V} \right)^2 \sim \frac{1}{2}\frac{\dot{\phi}^2}{\dot{\rho}^2}\frac{1}{M_{pl}} \tag{3.6}$$

$$\eta \equiv \frac{M_{pl}^2 V''}{V} \sim -\frac{\ddot{\phi}}{\dot{\rho}\dot{\phi}} + \frac{1}{2}\frac{\dot{\phi}^2}{\dot{\rho}^2}\frac{1}{M_{pl}} \tag{3.7}$$

where the approximate relations hold when the slow roll parameters are small.

We now consider small fluctuations around the solution spatially homogeneous solution, which obeys the equations (3.3)–(3.5). Since we have translation invariance we expand the fluctuations in Fourier modes in x by writing fluctuations as $\delta = \int d^3 k \phi_{\vec{k}}(t) e^{i\vec{k}\vec{x}}$. We expect to have three physical propagating degrees of freedom, two from gravity and one from the scalar field. The scalar field mixes with other components of the metric which are also scalars under $SO(2)$ (the little group that leaves $\vec{k}$ fixed). There are four scalar modes of the metric which are δg_{00}, δg_{ii}, $\delta g_{0i} \sim \partial_i B$ and $\delta g_{ij} \sim \partial_i \partial_j H$ where B and H are arbitrary functions. Together with a small fluctuation, $\delta\phi$, in the scalar field these total five scalar modes. The action (3.1) has gauge invariances coming from reparametrization invariance. These can be linearized for small fluctuations. The scalar modes are acted upon by two gauge invariances, time reparametrizations and spatial reparametrizations of the form $x^i \rightarrow x^i + \epsilon^i(t, x)$ with $\epsilon^i = \partial_i \epsilon$. Other coordinate transformations act on the vector modes.[1] Gauge invariance removes two of the five functions. The constraints in the action remove two others so that we are left with one degree of freedom.

In order to proceed it is convenient to work in the ADM formalism [19]. We write the metric as

$$ds^2 = -N^2 dt^2 + h_{ij}(dx^i + N^i dt)(dx^j + N^j dt) \tag{3.8}$$

and the action (3.1) becomes

$$
\begin{aligned}
S &= \frac{1}{2} \int \sqrt{h} \big[N R^{(3)} - 2NV + N^{-1}(E_{ij} E^{ij} - E^2) \\
&\quad + N^{-1}(\dot{\phi} - N^i \partial_i \phi)^2 - N h^{ij} \partial_i \phi \partial_j \phi \big]
\end{aligned} \tag{3.9}
$$

where

$$E_{ij} = \frac{1}{2}(\dot{h}_{ij} - \nabla_i N_j - \nabla_j N_i), \qquad E = E_i^i \tag{3.10}$$

Note that the extrinsic curvature is $K_{ij} = N^{-1} E_{ij}$.

In the ADM formulation, spatial coordinate reparametrizations are an explicit symmetry while time reparametrizations are not so obviously a symmetry. The ADM formalism is designed so that one can think of h_{ij} and ϕ as the dynamical variables and N and N^i as Lagrange multipliers. We will choose a gauge for h_{ij} and ϕ that will fix time and spatial reparametrizations. A convenient gauge is

$$\delta\phi = 0, \qquad h_{ij} = e^{2\rho}[(1 + 2\zeta)\delta_{ij} + \gamma_{ij}], \qquad \partial_i \gamma_{ij} = 0, \qquad \gamma_{ii} = 0 \tag{3.11}$$

---

[1] There are no propagating vector modes for this Lagrangian (3.1). They are removed by gauge invariance and the constraints. Vector modes are present when more fields are included.

where ζ and γ are first order quantities. ζ and γ are the physical degrees of freedom. ζ parameterizes the scalar fluctuations and γ the tensor fluctuations. The gauge conditions (3.11) fix the gauge completely at nonzero momentum. In order to find the action for these degrees of freedom we just solve for N and N^i through their equations of motion and plug the result back in the action. This procedure gives the correct answer since N and N^i are Lagrange multipliers. The gauge (3.11) is very similar to Coulomb gauge in electrodynamics where we set $\partial_i A_i = 0$, solve for A_0 through its equation of motion and plug this back in the action. The equation of motion for N^i and N are the momentum and Hamiltonian constraints

$$\nabla_i[N^{-1}(E^i_j - \delta^i_j E)] = 0, \tag{3.12}$$

$$R^{(3)} - 2V - N^{-2}(E_{ij}E^{ij} - E^2) - \dot{\phi}^2 = 0 \tag{3.13}$$

where we have used that $\delta\phi = 0$ from (3.11). We can solve these equations to first order by setting $N^i = \partial_i \psi + N^i_T$ where $\partial_i N^i_T = 0$ and $N = 1 + N_1$. We find

$$N_1 = \frac{\dot{\zeta}}{\dot{\rho}}, \qquad N^i_T = 0, \qquad \psi = -e^{-2\rho}\frac{\zeta}{\dot{\rho}} + \chi, \qquad \partial^2\chi = \frac{\dot{\phi}^2}{2\dot{\rho}^2}\dot{\zeta} \tag{3.14}$$

In order to find the quadratic action for ζ we can replace (3.14) in the action and expand the action to second order. For this purpose it is not necessary to compute N or N^i to second order. The reason is that the second order term in N will be multiplying the Hamiltonian constraint, $\frac{\partial L}{\partial N}$ evaluated to zeroth order which vanishes since the zeroth order solution obeys the equations of motion. There is a similar argument for N^i. Direct replacement in the action gives, to second order,

$$S = \frac{1}{2}\int e^\rho(1+\zeta)\left(1+\frac{\dot{\zeta}}{\dot{\rho}}\right)\left[-4\partial^2\zeta - 2(\partial\zeta)^2 - 2V\right] \tag{3.15}$$

$$+ e^{3\rho}(1+3\zeta)\left(1-\frac{\dot{\zeta}}{\dot{\rho}}\right)\left[-6(\dot{\rho}+\dot{\zeta})^2 - \frac{2}{3}(\partial^2\psi)^2 + \dot{\phi}^2\right] \tag{3.16}$$

where we have neglected a total derivative which is linear in ψ. After integrating by parts some of the terms and using the background equations of motion (3.3)–(3.5) we find the final expression to second order

$$S = \frac{1}{2}\int dt d^3x \frac{\dot{\phi}^2}{\dot{\rho}^2}[e^{3\rho}\dot{\zeta}^2 - e^\rho(\partial\zeta)^2] \tag{3.17}$$

No slow roll approximation was made in deriving (3.17). Note that naively the action (3.15) contains terms of the order $\dot{\zeta}^2$, while the final expression contains only terms of the form $\epsilon\dot{\zeta}^2$, so that the action is suppressed by a slow roll parameter. The reason is that the ζ fluctuation would be a pure gauge mode in de-Sitter space and it gets a non-trivial action only to the extent that the slow roll parameter is non-zero. So the leading order terms in slow roll in (3.15) cancel leaving only the terms in (3.17). A simple argument for the dependence of (3.17) on the slow roll parameters is given below.

Since (3.17) is describing a free field we just have a collection of harmonic oscillators. More precisely we expand

$$\zeta(t, x) = \int \frac{d^3k}{(2\pi)^3} \zeta_k(t) e^{i\vec{k}\vec{x}} \tag{3.18}$$

Each $\zeta_k(t)$ is a harmonic oscillator with time dependent mass and spring constants. The quantization is straightforward [20]. We pick two independent classical solutions $\zeta_k^{cl}(t)$ and $\zeta_k^{cl*}(t)$ of the equations of motion of (3.17),

$$\frac{\delta L}{\delta \zeta} = -\frac{d}{dt}\left(e^{3\rho}\frac{\dot{\phi}^2}{\dot{\rho}^2}\dot{\zeta}_k\right) - \frac{\dot{\phi}^2}{\dot{\rho}^2}e^\rho k^2\zeta_k = 0 \tag{3.19}$$

Then we write

$$\zeta_{\vec{k}}(t) = \zeta_k^{cl}(t)a_{\vec{k}}^\dagger + \zeta_k^{cl*}(t)a_{-\vec{k}} \tag{3.20}$$

where a and $a^\dagger$ are some operators. Demanding that $a^\dagger$ and a obey the standard creation and annihilation commutation relations we get a normalization condition for ζ_k^{cl}. Different choices of solutions are different choices of vacua for the scalar field, where the vacuum is defined via the condition $a_{\vec{k}}|0\rangle = 0$ for all $\vec{k}$. The comoving wavelength of each mode $\lambda_c \sim 1/k$ stays constant but the physical wavelength changes in time. For early times the ratio of the physical wavelength to the Hubble scale is very small and the mode feels it is in almost flat space. We can then use the WKB approximation to solve (3.19) and choose the usual vacuum in Minkowski space. When the physical wavelength is much longer than the Hubble scale

$$\lambda_{phys}H = \frac{\dot{\rho}e^\rho}{k} \gg 1 \tag{3.21}$$

the solutions of (3.19) go rapidly to a constant.

A useful example to keep in mind is that of a massless scalar field f in de-Sitter space. In that case the action is $S = \frac{1}{2}\int H^{-2}\eta^{-2}[(\partial_\eta f)^2 - (\partial f)^2]$ and the normalized classical solution, analogous to ζ_k^{cl}, corresponding to the standard Bunch Davies vacuum is [20],

$$f_k^{cl} = \frac{H}{\sqrt{2k^3}}(1 - ik\eta)e^{ik\eta} \tag{3.22}$$

where we are using conformal time which runs from $(-\infty, 0)$. Very late times correspond to small $|\eta|$ and we clearly see from (3.22) that f^{cl} goes to a constant. Any solution, including (3.22), approaches a constant at late times as $\eta^2 \sim e^{-2\rho}$, which is exponentially fast is proper time. In de-Sitter space we can easily compute the two point function for this scalar field and obtain

$$\langle f_{\vec{k}}(\eta) f_{\vec{k}'}(\eta) \rangle = (2\pi)^3 \delta^3(\vec{k} + \vec{k}') |f_k^{cl}(\eta)|^2 \tag{3.23}$$

$$= (2\pi)^3 \delta^3(\vec{k} + \vec{k}') \frac{H^2}{2k^3}(1 + k^2\eta^2) \tag{3.24}$$

$$\sim (2\pi)^3 \delta^3(\vec{k} + \vec{k}') \frac{H^2}{2k^3} \quad \text{for } k\eta \ll 1 \tag{3.25}$$

In coordinate space the result for late times is $\langle f(x, t) f(x', t) \rangle \sim -\frac{H^2}{(2\pi)^2} \log(|x - x'|/L)$ where is an IR cutoff which is unimportant when we compute differences in f as we do in actual experiments. We now go back to the inflationary computation. If one knew the classical solution to the equation (3.19) the result for the correlation function of ζ can be simply computed as

$$\langle \zeta_{\vec{k}}(t) \zeta_{\vec{k}'}(t) \rangle = (2\pi)^3 \delta^3(\vec{k} + \vec{k}') |\zeta_k^{cl}(t)|^2 \tag{3.26}$$

If the slow roll parameters are small when the comoving scale $\vec{k}$ crosses the horizon then it is possible to estimate the late time behavior of (3.26) by the corresponding result in de-Sitter space (3.23) with a Hubble constant that is the Hubble constant at the moment of horizon crossing. The reason is that at late times ζ is constant while at early times the field is in the vacuum and its wavefunction is accurately given by the WKB approximation. Since the action (3.17) also contains a factor of $\dot{\phi}/\dot{\rho}$ we also have to set its value to the value at horizon crossing, this factor only appears in normalizing the classical solution. In other words, near horizon crossing we set $f = \frac{\dot{\phi}}{\dot{\rho}}\zeta$ where f is a canonically normalized field in de-Sitter space. This produces the well known result [11–16]

$$\langle \zeta_{\vec{k}}(t) \zeta_{\vec{k}'}(t) \rangle \sim (2\pi)^3 \delta^3(\vec{k} + \vec{k}') \frac{1}{2k^3} \frac{\dot{\rho}_*^2}{M_{pl}^2} \frac{\dot{\rho}_*^2}{\dot{\phi}_*^2} \tag{3.27}$$

where the star means that it is evaluated at the time of horizon crossing, i.e. at time t_* such that

$$\dot{\rho}(t_*) e^{\rho(t_*)} \sim k \tag{3.28}$$

The dependence of (3.27) on t_* leads to additional momentum dependence. It is conventional to parameterize this dependence by saying that the total correlation function has the form $k^{-3+(n_s-1)}$ where

$$
\begin{aligned}
n_s - 1 &= k\frac{d}{dk}\log\left(\frac{\dot{\rho}_*^4}{\dot{\phi}_*^2}\right) \sim \frac{1}{\dot{\rho}_*}\frac{d}{dt_*}\log\left(\frac{\dot{\rho}_*^4}{\dot{\phi}_*^2}\right) = -2\left(\frac{\ddot{\phi}_*}{\dot{\rho}_*\dot{\phi}_*} + \frac{\dot{\phi}_*}{\dot{\rho}_*}\right) \\
&= 2(\eta - 3\epsilon)
\end{aligned}
\tag{3.29}
$$

After horizon crossing the mode becomes classical, in the sense that the commutator $[\dot{\zeta}, \zeta] \to 0$ exponentially fast. So for measurements which only involve ζ or $\dot{\zeta}$ we can treat the mode as a classical variable.

After the end of inflation the field ϕ ceases to determine the dynamics of the universe and we eventually go over to the usual hot big bang phase. It is possible to prove [15, 17] that ζ remains constant outside the horizon as long as no entropy perturbations are generated and a certain condition on the off-diagonal components of the spatial stress tensor is obeyed. (The condition is $\partial_i\partial_j(\delta T_{ij} - \frac{1}{3}\delta_{ij}T_{ll}) = 0$.) These conditions are obeyed if the universe is described by a single fluid or by a single scalar field. We should mention that for a general fluid the variable ζ can be defined in terms of the three metric as above (3.11) in the comoving gauge where $T_i^0 = 0$. In the case of a scalar field this implies that $\delta\phi = 0$. This gauge is convenient conceptually since the variable ζ is directly a function appearing in the metric. We see that the variable ζ tells us how much the spatial directions have expanded in the comoving gauge, so that to linear order ζ determines the curvature of the spatial slices $R^{(3)} = 4k^2\zeta$ [21]. This variable ζ is very useful in order to continue through the end of inflation since it is defined throughout the evolution and it is constant outside the horizon. An intuitive way to understand why ζ is constant is the following. Two observers separated by a distance larger than the size of the horizon, R_H, at that time see the universe undergoing precisely the same history. Outside the horizon (where we can set $k = 0$ in all equations) ζ is just a rescaling of coordinates and this rescaling is a symmetry of the equations.

If we have fluctuations in more than one field, then the history of two patches at large distances need not be the same.

Other gauges can be more convenient in order to do computations in the slow roll approximation. A gauge that is particularly convenient is

$$
\delta\phi \equiv \varphi(t, x), \qquad h_{ij} = e^{2\rho}(\delta_{ij} + \gamma_{ij}), \qquad \partial_i\gamma_{ij} = 0, \qquad \gamma_{ii} = 0 \tag{3.30}
$$

where we have denoted the small fluctuation of the scalar field by φ. In order to avoid confusion, from now on ϕ will denote the background value of the scalar field and φ will be its deviation from the background value. We expect that in this gauge the action will be approximately the action of a massless scalar field φ

to leading order in slow roll. Indeed, we can check that the first order expressions for N and N^i are

$$N_{1\varphi} = \frac{\dot{\phi}}{2\dot{\rho}}\varphi, \qquad N_\varphi^i = \partial_i \chi, \qquad \partial^2 \chi = \frac{\dot{\phi}^2}{2\dot{\rho}^2}\frac{d}{dt}\left(-\frac{\dot{\rho}}{\dot{\phi}}\varphi\right) \qquad (3.31)$$

where the φ subindex reminds us that $N_{1\varphi}$, N_φ^i are computed in the gauge (3.30). We see that these expressions are subleading in slow roll compared to φ. So in order to compute the quadratic action to lowest order in slow roll it is enough to consider just the $(\nabla\varphi)^2$ term in the action (3.1) since V'' is also of higher order in slow roll. This is just the action of a massless scalar field in the zeroth order background. We can compute the fluctuations in φ in the slow roll approximation and we find a result similar to that of a scalar field in de-Sitter space (3.23) where the Hubble scale is evaluated at horizon crossing. After horizon crossing we can evaluate the gauge invariant quantity ζ. This is most easily done by changing the gauge to the gauge where $\varphi = 0$. This can be achieved by a time reparametrization of the form $\tilde{t} = t + T$ with

$$T = -\frac{\varphi}{\dot{\phi}} \qquad (3.32)$$

where t is the time in the gauge (3.11) and $\tilde{t}$ is the time in (3.30). After the gauge transformation (3.32), we find that the metric in (3.30) becomes of the form in (3.11) with

$$\zeta = \dot{\rho}T = -\frac{\dot{\rho}}{\dot{\phi}}\varphi \qquad (3.33)$$

Incidentally, this implies that χ in (3.31) is the same as χ in (3.14). So the correlation function for ζ can be computed as the correlation function for φ times the factor in (3.33). In order to get a result as accurate as possible we should perform the gauge transformation (3.33), just after crossing the horizon so that the factor in (3.33), is evaluated at horizon crossing leading finally to (3.27). In principle we could compute ζ from φ at any time. If we were to choose to do it a long time after horizon crossing we would need to take into account that φ changes outside the horizon. This would require evaluating the action (3.1) to higher order in the slow roll parameters. Of course, the dependence for φ outside the horizon is such that it precisely cancels the time dependence of the factor in (3.33) so that ζ is constant.

In summary, the computation is technically simplest if we start with the gauge (3.30) and we compute the two point function of φ after horizon exit and at that time compute the ζ variable which then remains constant. On the other hand the computation in the gauge (3.11) is conceptually simpler since the whole computation always involves the variable of interest which is ζ. In other words, the gauge (3.30) is more useful before and during horizon crossing while the gauge (3.11) is more useful after horizon crossing.

These last few paragraphs are basically simple argument presented in [13]. The computation of fluctuations of φ in de-Sitter produces fluctuations of the order $\varphi = \frac{H}{2\pi}$ and then this leads to an effective delay in the evolution by $\delta t = -\varphi/\dot{\rho}$ (see (3.32)) which in turn gives an additional expansion of the universe by a factor $\zeta = \dot{\rho}\delta t = -\frac{\dot{\rho}}{\phi}\varphi$. This additional expansion is evaluated at horizon crossing in order to minimize the error in the approximation.

We now summarize the discussion of gravitational waves [22]. Inserting (3.11) in the action and focusing on terms quadratic in γ gives

$$S = \frac{1}{8} \int [e^{3\rho}\dot{\gamma}_{ij}\dot{\gamma}_{ij} - e^{\rho}\partial_l\gamma_{ij}\partial_l\gamma_{ij}] \tag{3.34}$$

As usual we can expand γ in plane waves with definite polarization tensors

$$\gamma_{ij} = \int \frac{d^3k}{(2\pi)^3} \sum_{s=\pm} \epsilon_{ij}^s(k)\gamma_{\vec{k}}^s(t)e^{i\vec{k}\vec{x}} \tag{3.35}$$

where $\epsilon_{ii} = k^i\epsilon_{ij} = 0$ and $\epsilon_{ij}^s(k)\epsilon_{ij}^{s'}(k) = 2\delta_{ss'}$. So we see that for each polarization mode we have essentially the equation of motion of a massless scalar field. As in our previous discussion, the solutions become constant after crossing the horizon. Computing the correlator just after horizon crossing we get

$$\langle \gamma_{\vec{k}}^s \gamma_{\vec{k}'}^{s'} \rangle = (2\pi)^3\delta^3(\vec{k}+\vec{k}')\frac{1}{2k^3}\frac{2\dot{\rho}_*^2}{M_{pl}^2}\delta_{ss'} \tag{3.36}$$

where we reinstated the M_{pl} dependence. We can similarly define the tilt of the gravitational wave spectrum by saying that the correlation function scales as k^{-3+n_t} where n_t is given by

$$n_t = k\frac{d}{dk}\log\dot{\rho}_*^2 = -\frac{\dot{\phi}_*^2}{\dot{\rho}_*^2} = -2\epsilon \tag{3.37}$$

4. Models of inflation

In this section we attempt to give an overview about inflation models. We will discuss explicitly only a small subset of all proposed models. We discuss some of the difficulties encountered in building inflation models.

In order to find a suitable inflation model we need to find a potential $V(\phi)$ which is flat enough so that the slow roll parameters, (3.6), are $\epsilon, \eta < 0.05$ and the number of e-folds, (2.7) is around 60, where the detailed number depends on the details of reheating. In addition we want the amplitude of the primordial fluctuations to be such that on the largest observable scales we have

$$
\langle \zeta_k \zeta_{k'} \rangle = \frac{(2\pi)^3 \delta^3 (k + k')}{2k^3} \mathcal{A}_s(k_0) \left(\frac{k}{k_0} \right)^{n_s - 1}, \tag{4.1}
$$

$$
\mathcal{A}_s(k_0) = \frac{1}{2\epsilon} \frac{H^2}{M_p^2} = \frac{V^3}{V'^2 M_p^6} \sim 20 \times 10^{-10} \tag{4.2}
$$

In addition we want n_s close to one, perhaps a few percent below one. For a more detailed description of the constraints see [5]. In addition we should have appropriate reheating. I will not talk about this, but it is an important constraint. In principle we can also have stringy effects that arise from the details of reheating, see [8] for example.

Just to see how things work, let us consider predictions from simple looking models. Let us consider simple power law models where $V \sim \lambda_n \phi^n$, which are called "chaotic inflation". We then have

$$
N_{eff} \sim \frac{\phi_i^2}{M_p^2} \frac{1}{2n}, \tag{4.3}
$$

$$
\epsilon = \frac{n^2}{2} \frac{M_p^2}{\phi^2}, \qquad \eta = n(n-1) \frac{M_p^2}{\phi^2}, \tag{4.4}
$$

$$
n_s - 1 = -6\epsilon + 2\eta = -\frac{M_p^2}{\phi^2} n(n+1), \tag{4.5}
$$

$$
\mathcal{A}_s = \frac{\lambda_n \phi^{n+2}}{n^2 M_p^6} = \frac{\lambda_n \phi^{n-4}}{n^2} \frac{\phi^6}{M_p^6} \tag{4.6}
$$

where ϕ is the value of the field at the time that the largest scales we see today were crossing the horizon during inflation. We see that we can adjust the value of ϕ/M_p to get the desired number of e-folds. The precise number will depend on the details of the reheating mechanism, which we have not specified. We can select the right amplitude by choosing the appropriate value of λ_n. Thus all these models work in principle, but precision cosmology is already starting to rule some of them out on the basis of their predictions for gravity waves together with the spectral index, see [5]. The models with $n = 2$ seems to work well at describing current data. This seems to be a nice and simple model, just a massive scalar field. In these models the scale for the energy density during inflation is close to the GUT scale. This is an attractive feature of these models since it suggests a connection between the inflationary energy scale and the GUT scale, which, of course, is very interesting from the particle physics point of view.

On the other hand, from the perspective of a full quantum gravity theory this model looks a bit funny. It is funny because one would expect that the potential for a scalar field would have corrections of the form ϕ/M_p. In other words $V = \lambda_n \phi^n (1 + o(\phi/M_p))$. But ϕ/M_p is larger than one (it should be of order 15 for $n = 2$), so that the correction seems bigger than the original term. In some cases one might be able to argue that there is a symmetry protecting some terms, such as the case of the axion model we will discuss below. But in any case, one needs some further argument justifying this form for the potential for large values of the fields.

From the point of view of string theory this also looks funny since ϕ/M_p will correspond to the shape of the internal manifold, or to some coupling, or to some brane position. Then one needs to justify why the vacuum energy does not change by an order one amount when we change ϕ/M_p by an order one amount, which would amount to a large deformation of the geometry. Note that it is easy to find fields whose range in target space is infinite, such as the dilaton, for example. What seems difficult is to find a situation where the potential energy remains positive and roughly constant over a large range.

In fact, it is possible to consider models where the field does not change so much during inflation. But before discussing them, let us discuss an interesting connection between the change of the field during inflation and the production of gravity waves. We can consider the change of the field per e-fold and write

$$\frac{1}{M_p}\frac{d\phi}{dN} = \frac{1}{M_p}\frac{d\phi}{dt}\frac{1}{H} = M_p\frac{V'}{V} = \sqrt{2\epsilon} = \sqrt{\frac{r}{8}}, \tag{4.7}$$

$$\frac{\Delta\phi}{M_p} = \frac{1}{\sqrt{8}}\int_0^{N_{\mathrm{eff}}} dN\sqrt{r} \tag{4.8}$$

where $r = 16\epsilon$ is a measure of the gravity wave amplitude (relative to the well measured scalar amplitude). This relation is called the Lyth bound [23]. It expresses the minimum amount that the field has to change for a given level of gravity wave amplitude. The current bound on r is about 0.3 (see [5]), which, if saturated over the whole range of scales, would give $\Delta\phi/M_p \sim 10$. The bounds attainable by near future experiments will be of the order of $r \sim 0.01$ which would imply $\frac{\Delta\phi}{M_p} < 2$. Thus, if we were to have a bound in string theory on $\Delta\phi/M_p$ we get a direct bound on the gravity wave amplitude, see [24] for example.

Let us make a comment on the detection of gravitational waves. The primordial amplitude of gravity waves is huge, compared to more ordinary gravity waves, since it is of the order of 10^{-6}, while astrophysical sources produce amplitudes of the order of 10^{-22}. Unfortunately, once these gravity waves come within the horizon their amplitude starts to decrease and the amplitude behaves as $h \sim 1/a$. A second unfortunate fact is that the spectrum is red, meaning that already the primordial amplitude is getting smaller for shorter distances. However, gravity waves affect the amplitude of the scalar part of $\delta T/T$ in the CMB and also its polarization. This leads to the current bounds. Interestingly there is some information about gravity waves in the polarization of the CMB. There are particular polarization patterns, called "B-modes" that, in the linearized theory, can only be produced by gravity waves, see [25] for a review.

Let us now discuss inflation models where $\Delta\phi/M_p \ll 1$. These are typically models where $\epsilon \ll 0.01$ but with η in the percent range. In these models the Hubble constant during inflation is smaller than in the models discussed above. In these models one has to tune ϵ so as to produce the appropriate level of scalar fluctuations, which will be much larger than the amplitude of tensor fluctuations. (Some authors have argued against such models on this tuning basis [26].) Models of this type can be constructed with several fields, so that when the field ϕ reaches a critical point a second field becomes tachyonic so that inflation ends (see [27]). This allows one the freedom of making the potential very flat without increasing, at the same time, the number of e-foldings. In these models the inflationary energy scale can be very low, and some models have been proposed where it is as low as a few $(\text{TeV})^4$. Experience has shown that getting ϵ to be small is not too difficult. On the other hand, it is difficult to have a *natural* model where η is also small. By a *natural* model we mean a model where without any fine tuning we get a small value of η. Notice that we can call V'' the mass squared of the inflation and then $\eta \sim m^2/H^2$. So we are saying that it is difficult for the effective mass of the field to be smaller, in absolute value, than the Hubble scale. Of course a fine tuning a few percent is not too surprising, so it is perhaps not a problem to worry too much about. However, in any specific model one must make sure that all effects are under control so as to ensure that η is small.

Just as an example of the issues discussed here let us consider a supergravity model. Imagine that at the level of field theory we manage to have a potential which is flat enough. For example, consider a field theory with $K = \phi\bar{\phi}$ and $W = C\phi$. In this case we get a constant potential $V_{FT} = |\partial W|^2 = |C|^2$ (which could then be rendered non-constant by adding further terms ...). Once we consider a supergravity theory, the full potential then becomes

$$V = e^K (g^{\phi\bar{\phi}}|DW|^2 - 3|W|^2) \sim |C|^2\left(1 + \frac{1}{2}\phi\bar{\phi} + \cdots\right) \qquad (4.9)$$

where we expanded the potential around $\phi \sim 0$ and we have momentarily set $M_p = 1$. We see that we get an η of order one, despite the fact that the field theory potential, before coupling to gravity, was completely flat. Of course this feature depended on our choice of Kahler potential. This problem might be avoided in some cases by taking instead a Kahler potential of the form $K \sim (\phi+\bar{\phi})^2$ instead. This Kahler potential is related to the original one by a Kahler transformation $K \to K + f(\phi) + \bar{f}(\bar{\phi})$. Under a Kahler transformation we should also modify the superpotential $W \to e^{-f}W$. Thus, we have to specify the Kahler potential and the superpotential together and we need to understand them well enough in the supergravity context. This is not to say that it is impossible to find a model. In fact one can write down superpotentials and Kahler potentials that do the job (see [28] for example), but I simply want highlight the need to understand well enough the full supergravity theory.

Or course, in the context of string theory one is not allowed to postulate an arbitrary supergravity theory, one must obtain it from some string theory compactification. Thus one needs to understand the compactification well enough to be able to give the form of the supergravity Kahler potential and superpotential. It is often stated that one can consider D-term inflation which does not suffer from this problem [29]. This is a model where the potential energy comes from the D-term of a $U(1)$ field. In other words, consider a supergravity theory with a $U(1)$ gauge field and add a Fayet Iliopoulos term ξ for this $U(1)$. Then this leads to constant potential energy $V = e^2\xi^2$, even in supergravity (where e is the coupling constant). Thus, at the level of supergravity there is no problem in getting a flat potential. On the other hand, in string constructions either the FI term, ξ, or the charge e will end up depending too strongly on the inflaton field, sometimes through the modulus stabilization mechanism.

The reader should not get the impression that it is impossible to build a model. The goal of this discussion is simply to highlight the problems that a successful model should overcome. Notice that since the slow roll parameters depend explicitly on M_p one needs to understand quantum gravity corrections well enough in order to construct a viable model.

Now let us discuss an interesting class of models. These are "axion" inflation models [30]. In these models one imagines a periodic field $a \sim a + 2\pi$ similar to the QCD axion, or a pseudo Goldstone boson, but different from the QCD axion (which will have a much smaller potential). One also imagines that we get a potential which breaks the continuous shift symmetry. In addition, let us assume that this potential is given to leading order by a "one instanton effect" so that we get a cosine potential. Then the model has the form

$$S = \int \frac{1}{2}f^2(\nabla a)^2 + V_0(1 - \cos a) \tag{4.10}$$

where we have added a constant to set the minimum of the potential at zero. f is the so called "axion decay constant". Expanding around $a \sim 0$ we get a quadratic potential similar to the $m^2\phi^2$ potential discussed above. In this case, the shift symmetry of the axion, and the smallness of instanton effects, which further suppresses terms of the form $\cos na$, would ensure that corrections are under control. This would seem to give us a natural model for inflation. On further analysis one finds that, as in the ϕ^2 model, we need f/M_p to be large, about $f/M_p > 10$.

In string theory it is hard to get axions with such large decay constants [31–33]. In string theory one typically gets axions with $f < M_p$ and the ratio of M_p/f is typically the value of the action of the instantons that give corrections, $S_{inst} \sim M_p/f$ [33]. Of course we need S_{inst} to be large so that the potential is well approximated by the one instanton approximation. More precisely, one can get fields with $f > M_p$ but the instanton corrections are so large that one looses the argument controlling corrections to the potential.

One interesting suggestion is to consider a model with N axions, each with $f < M_p$ [34]. Then the axions parametrize an N dimensional hypercube and we can consider motion along the diagonal. The length of the diagonal is larger by a factor of $\sqrt{N}$. In fact, in string models one can get a large number of axions, of the order of hundreds, since this is the number of cycles of a typical CY. The most interesting aspect of this model is that it could produce observable gravity waves. On the other hand, it remains to be seen whether it can be realized in a reasonable controlled way in string theory, see [35] and reference therein for recent work on this subject.

Another class of models that was studied recently are models with moving branes. In principle one can consider various types of branes, but D3 branes are the easiest to consider because mobile D3 branes arise naturally in GKP compactifications [36] which are used in the KKLT construction [37].

Before discussing the details of these models it is useful to imagine a simple model where we have a T^6 compactification and we consider a D3 and anti-D3 pair moving in the T^6 [38]. Then the vacuum energy comes from the tension of the branes. There is a potential between the brane and the anti-brane which is attractive. The question is whether this set-up can lead to inflation. We have a Lagrangian of the form

$$S = \int \sqrt{g}\left[\frac{M_4^2}{2}R^{(4)} + 2T + T(\nabla r)^2 + V(r)\right], \quad V(r) \sim \frac{T^2}{M_{10}^8 r^4} \qquad (4.11)$$

where $M_4^2 = M_{10}^8 V_6$ where V_6 is the volume of the six torus. T is the tension of the brane. We can now compute η in this model and we find

$$\eta \sim \frac{M_4^2}{M_{10}^8 r^6} \sim \frac{V_6}{r^6} \qquad (4.12)$$

Thus we see that since the distance is bounded by the size of the space, we get η of order one. (Here we are assuming that the T^6 has all equal sides.)

An additional difficulty with this model is that since we have not stabilized the volume of the T^6 this volume will fast roll to large values. Actually, in the presence of classical energy source such a branes or classical curvature, general moduli, like the volume or the string coupling, have potentials of the form

$$S = M_p^2 \int (\nabla \sigma)^2 + e^{\alpha \sigma} \tag{4.13}$$

where α is of order one. This leads to ϵ and η of order α^2 which are then of order one.

Even though this model does not work as it is, one can modify it in order to make is more viable [39, 40]. This is done as follows. One considers a CY compactification with fluxes and one adds a mobile D3 brane. At the classical level this stabilizes the complex structure moduli of the Calabi–Yau space and breaks susy in a no scale fashion. In other words, we have a constant superpotential W_0 and a Kahler potential $\mathcal{K} = -3 \log(\rho + \bar{\rho} + k(\phi, \bar{\phi}))$ where k is the Kahler potential for the Calabi–Yau metric and ϕ parametrizes the position of the D3 brane. One can then add non-perturbative effects of the form $A(\phi)e^{-\rho}$ so that one gets an AdS_4 solution. To have a controlled large volume situation one needs to fine tune the fluxes to obtain W_0 small, because we are balancing W_0 against $Ae^{-\rho}$ and we want ρ to be somewhat large. In addition we can consider a flux configuration so that the CY space is close to a conifold singularity. In this case, in the presence of fluxes, this will lead to a highly warped region near the conifold with a geometry similar to that of the Klebanov–Strassler throat. An object inside this throat will be redshifted. We can put in an anti-D3 brane inside this throat and then raise the vacuum energy a little bit, so that we have positive vacuum energy, as in [37]. In the presence of a mobile D3 brane then the D3 brane will be attracted to the anti-D3 brane and want to annihilate with it. The direct force on the mobile D3 does not come from the Coulomb like interaction between the branes, which is very small, but rather from the effects we talked about above. Namely, the fact that the inflaton, which is the position of the D3 brane appears explicitly in the moduli stabilizing potential. This on its own would lead to η of order one. One needs to add other effects, such as the presence of a D7 brane in the throat that with sufficient freedom to fine tune the potential. In that case one can find an inflating region, see [40] and references therein. This model has not yet been embedded in the full, honest, CY compactification. One interesting aspect of these models is that when the branes and anti-branes annihilate they can produce strings that remain confined in the throat where inflation happened. These strings could be long lived and might be visible as cosmic strings [8].

Finding other inflating regions in the string theory landscape is an important problem, see [41–43] for recent reviews.

5. Alternatives to inflation

Let us briefly mention some of the alternatives that have been proposed for inflation.

Let us discuss first a scenario that uses strings (and also branes) in a crucial way [44–46]. In the simplest version, [44], one assumes that we are in a toroidal compactification with a gas of strings and momentum at a temperature near the Hagedorn temperature in such a way that they are in thermal equilibrium. The idea is the winding modes prevent the cycles from expanding and the momentum modes prevent it form contracting. Then, if the winding modes can annihilated, the cycles will expand. The winding modes will annihilate if the strings can meet. Strings only meet in less than 3 space dimensions. So the idea is the 3 of the spatial dimensions expand and the others do not. In this way the model attempts to explain why we have 3 large spatial dimensions. There are various difficulties with this model. It does not explain what fixes the values of the moduli at late times. It also does not explain what happens with the dilaton, which will be varying rapidly in during the time that we are near the Hagedorn temperature. Some attempts at describing the origins of scalar fluctuations were made in [46], but were later criticized in [47].

Another interesting model is the so called ekpyrotic model [48] or its cyclic variant [49]. In the first model the universe starts very uniform but in a rapidly collapsing phase, when scalar fluctuations are generated, then it goes through a mysterious crunch-bang transition where it emerges as the hot big bang model. The cyclic variant [49] nicely explains why the starting point was uniform, by using the exponential expansion we are starting to experience today to produce a very uniform universe before entering again into the contracting phase and repeat of the cycle. The main problem with these models lies in the lack of understanding of the crunch to bang transition. The crunch-bang transition is prevented in classical physics, for flat spatial slices, by the null energy condition, $T_{--} \geq 0$ or $\rho + p \geq 0$. In fact, we see from (2.2) that $\dot{H} \leq 0$ which prevents the transition from $H < 0$ to $H > 0$. The null (sometimes called "weak") energy condition could be violated in quantum field theory, though it is suspected that an integrated version, integrated along a null line, $\int dx^- T_{--}$ might still hold, see [50]. In fact, this averaged version would be enough to rule out wormholes and causality violation, [51]. In other words, the violation of the averaged null energy condition, which is necessary for crunch-bang transition, opens a cosmic can of worms. Of course, one might argue that since we do not understand string theory enough, perhaps these transitions are allowed. However, our ignorance about this transition has real effects on the calculability of the model. In particular, one needs to make certain assumptions in order to propagate the cosmological perturbations through the transition. Depending on how these assumptions one gets different answers, the original analysis [48], produces a nearly scale invariant spectrum while an analysis based on the constancy of the variable we called ζ above produces a non-scale invariant result that contradicts experiment ([52]). This controversy clearly highlights the fact that we do not know what the appropriate matching conditions are. There was some work on string theory orbifolds which looks like bouncing cosmologies [53]. These authors found that even if you assume that fluctuations can propagate through the singularity, then the four point function diverges, see also [54]), which at the very least, would be a strong non-Gaussian signal.

The reader should keep in mind that these spacelike singularities are not at all understood in string theory so that we cannot say whether the ekpyrotic or cyclic models are or are not consistent with string theory. What is clear, however, is that we need a quantum theory of gravity in order to answer whether the crunch into bang transition is possible.

6. Remarks on AdS/CFT and dS/CFT

6.1. AdS/CFT

The computation that we did in section three was done with inflation in mind, but the same mathematical structure arises if one considers a single scalar field with a negative potential. In the slow roll case, the background will be a slightly deformed anti-de-Sitter space. This can be understood as a slightly deformed conformal field theory [55–57]. In other words, a non-conformal field theory which is almost conformal. An incomplete list of references where situations of this sort were considered is [58–63]. Here we just mention a few results that are relevant for us, for a review see [64]. The variables γ^s that we used above are associated to the traceless components of the stress tensor while the variable ζ is associated to the trace of the stress tensor. More precisely, we have a coupling of the form $\int \frac{dk^3}{(2\pi)^3} [2\zeta_{-\vec{k}} T_i^i(\vec{k}) + 2\gamma^s_{-\vec{k}} T^s(\vec{k})]$, where T^s is defined by an expression similar to (3.35), with $\gamma \rightarrow T$. The fact that the definition of the scalar mode depends on the gauge is translated into the fact that in a field theory with a scale we can either change the dimensionfull coupling constant or we can change the overall scale in the metric. It is common to fix the coupling and change the metric, which then relates ζ to the trace of the stress tensor. Alternatively we can fix the metric and change the coupling constant. In the field theory we do not have two independent operators, we have only one operator related by the equation

$$2T_i^i = \beta_\lambda \mathcal{O} \tag{6.1}$$

where β_λ is the beta function for the coupling λ which appears in the field theory Lagrangian in front of the non-marginal operator as $\int \lambda \mathcal{O}$. The operator $\mathcal{O}$ is the one associated to the bulk scalar field ϕ and the operator $2T_i^i$ is associated to ζ, as. The factor of slow roll that relates the correlators of ζ and ϕ is precisely the factor β_λ appearing above [65].

From the computations in the previous sections we can also compute the correlation functions of the stress tensor and the trace of the stress tensor in non-conformal theories.

Two point functions of the trace of the stress tensor were considered in the AdS context in [60–63]. The derivation of the effective action for the corresponding field in AdS identical to the one in the dS context. Similarly, computations of three point functions in AdS can be done by performing minor modifications to the above formulae. We will be more explicit below.

Now we will review the AdS_4 computation (see [66] for a review) so that we can contrast it clearly to the dS_4 computation.

Let us consider a canonically normalized scalar field in Euclidean anti-Sitter space ($EAdS_4$) which is the same as hyperbolic space. The action is

$$S = R_{AdS}^2 \int \frac{dz}{z^2} \frac{1}{2} [(\partial_z f)^2 + (\partial f)^2] \tag{6.2}$$

In order to do computations it will be necessary to consider classical solutions which go to zero for large z and obey prescribed boundary conditions at $z = z_c$. In momentum space these are

$$f_{\vec{k}} = f_{\vec{k}}^0 \frac{(1 + kz)e^{-kz}}{(1 + kz_c)e^{-kz_c}}, \quad k = |\vec{k}| \tag{6.3}$$

where $f_{\vec{k}}^0$ is the boundary condition we impose at $z = z_c$. One should then compute the action for this solution as a function of the boundary conditions. Inserting (6.3 into (6.2, integrating by parts and using the equations of motion we get

$$-S = \int \frac{d^3k}{(2\pi)^3} \frac{1}{2} R_{AdS}^2 f_{-\vec{k}}^0 \frac{1}{z_c^2} \frac{df_{\vec{k}}}{dz}\bigg|_{z=z_c} \tag{6.4}$$

$$\sim -\int \frac{d^3k}{(2\pi)^3} \frac{1}{2} R_{AdS}^2 f_{-\vec{k}}^0 f_{\vec{k}}^0 \left[\frac{k^2}{z_c} - k^3 + \cdots \right] \tag{6.5}$$

where the dots indicate terms of higher order in z_c. The term divergent in z_c is local in position space[2] and it is viewed as a divergence in the CFT which should be subtracted by a local counterterm. The term independent of z_c is non-local and gives rise to the two point function

$$\langle \mathcal{O}(\vec{k})\mathcal{O}(\vec{k}')\rangle_{EAdS} = \left.\frac{\delta^2 Z}{\delta f_{\vec{k}}^0 \delta f_{\vec{k}'}^0}\right|_{f^0=0} \sim (2\pi)^3 \delta(\vec{k}+\vec{k}') R_{AdS}^2 k^3 \tag{6.6}$$

where Z is the partition function of the Euclidean CFT which is approximated by $Z \sim e^{-S_{cl}}$, with S in (6.4).

6.2. dS-CFT

The dS/CFT was proposed [67, 68] in analogy with AdS/CFT [55–57]. The dS/CFT postulates that the wavefunction of a universe which is asymptotically de-Sitter space can be computed in terms of a conformal field theory. The proposals in [68] and [67] differ by some details in how to relate the bulk and boundary observables. Our discussion here is compatible with [68]. We can think of the ds/CFT correspondence as a prescription for computing the wavefunction of the universe at late times

$$\Psi[g] = Z[g] \tag{6.7}$$

where the left hand side is the wavefunction of the universe for given three metric and the right hand side is the partition function of some dual conformal field theory. Actually the left hand side has rapidly oscillating pieces which can be expressed as local functions of the metric. We discard these pieces since they have the interpretation of local counterterms in the CFT. Here we are thinking of de-Sitter in flat slices (or Poincare coordinates) and we are imagining that all fields start in their life in the Bunch–Davies vacuum. This determines the wavefunction Ψ, at least in the context of perturbation theory. If we were considering global de-Sitter space then our discussion would be valid in a small patch in the future where it can be approximated by the Poincare patch and the memory of the particular state that could have come from the far past is lost, as long as we assume a vacuum which is locally the Bunch–Davies one. This point of view follows simply from the discussion in [68] in analogy with the standard discussion in Euclidean AdS where the same formula (6.7) is valid. In AdS/CFT formula (6.7) arises in the Euclidean context when we think of Euclidean time as the direction perpendicular to the boundary. Ψ can then be interpreted also as the Hartle–Hawking wave function [69]. See [70–72] for more on this point of

---

[2]It is proportional to $\frac{1}{z_c}\int dx^3 \frac{1}{2}(\partial f^0)^2$.

view. Nobody has found a concrete example of this duality and there is an argument that such a duality should not exist [73]. I think that the argument in [73] is not definite because it only says that we cannot describe expectation values in de-Sitter in terms of a CFT. However, the dS/CFT correspondence is a statement about the wavefunction. In particular it is a statement about the inner product between the wavefunction and the out vacuum, $\langle 0|_{out}|HH\rangle$, where $|HH\rangle$ is the Bunch–Davies or Hartle–Hawking wavefunction of the universe.

Here we will just do some computations on the gravity side in order to get some insight on the properties that this hypothetical CFT should have. We will also see the close connection with computations that are done in the AdS context. If an example were found, then it would be a more powerful way of computing the wavefunction that semiclassical physics in de-Sitter or nearly de-Sitter space. Note that an observer living in eternal de-Sitter space will not be able to measure two point correlators such as (3.36) or the wavefunction (6.7) which involves distances much larger than the Hubble scale. Only so called "metaobservers" can measure these [68]. On the other hand if the universe is approximately de-Sitter for a while and then inflation ends and we go over to a radiation or matter dominated universe then these correlation functions become observable. In fact, we are metaobservers of the early inflationary epoch [74].

In [67, 75] the relation between CFT operators and fields in the bulk was explored and various ways of defining operators were considered. It was found that given a scalar field in the bulk one could define two operators with two conformal dimensions differing by $\Delta_+ - \Delta_- = d$ where d is the dimension of the CFT. If the field we are considering in the bulk is the metric then it is clear that the corresponding operator is the stress tensor and it should have dimension d. Indeed we will see that this agrees precisely with what we expect from the prescription (6.7). Below we explain more precisely how this computation is related both to the inflationary computation (3.36) and the corresponding EAdS computation.

The first step is to compute the wavefunction as a function of a small fluctuation in a massless scalar field f. Since f is a free field, which is a collection of harmonic oscillators, all we need to do is to compute the wavefunction for these harmonic oscillators. We want to compute the Schroedinger picture wavefunction at some time η_c as a function of the amplitude of the field f. The wavefunction is given by a sum over all paths ending with amplitude f and starting at the appropriate vacuum state. Since the action is quadratic this sum reduces to evaluating the action on the appropriate classical solution. We choose the standard Euclidean (Bunch–Davies) vacuum for the fields at early times. The classical solution obeying the appropriate boundary conditions is

$$f = f_{\vec{k}}^0 \frac{(1 - ik\eta)e^{ik\eta}}{(1 - ik\eta_c)e^{ik\eta_c}} \tag{6.8}$$

The boundary conditions at large η are the ones that correspond to the statement that the oscillator is in its ground state, which can be defined adiabatically at early times. The condition is that the field should behaves as $e^{ik\eta}$ for $|\eta| \to \infty$. Note that $f_{-\vec{k}} \neq f_{\vec{k}}^*$ since the boundary condition we are imposing at early times is not a real condition on the field $f(\eta, x)$. There is nothing wrong in considering a complex solution since all we are doing is to evaluate a functional integral by a saddle point approximation. This is one of the many ways to think about the harmonic oscillator wavefunction. When we evaluate the classical action on this solution we get

$$iS = i \int \frac{d^3k}{(2\pi)^3} \frac{1}{2} R_{dS}^2 \frac{1}{\eta_c^2} f_{-\vec{k}}^0 \partial_\eta f_{\vec{k}}|_{\eta=\eta_c} \tag{6.9}$$

$$\sim \int \frac{d^3k}{(2\pi)^3} \frac{1}{2} R_{dS}^2 \left[i\frac{k^2}{\eta_c} - k^3 + \cdots \right] f_{-\vec{k}}^0 f_{\vec{k}}^0 \tag{6.10}$$

Note that we are dropping an oscillatory piece at $|\eta| \to \infty$ which is equivalent to slightly changing the contour of integration by $\eta \to \eta + i\epsilon$. This is the standard prescription for the vacuum state of a harmonic oscillator.

Notice that under

$$\eta = iz, \qquad R_{dS} = i R_{AdS} \tag{6.11}$$

the formulas (6.8) and (6.9) go into (6.3) and (6.4). The fact that (6.8) goes into (6.3) is intimately related to the statement that when the mode has short wavelength it is in the adiabatic vacuum. A consequence of this fact is that the two point function computed using dS_4 differs by a sign from the corresponding one in Euclidean AdS_4.[3] More explicitly we have

$$\langle \mathcal{O}(\vec{k})\mathcal{O}(\vec{k}') \rangle_{dS_4} \equiv \left. \frac{\delta^2 Z}{\delta f_{\vec{k}}^0 \delta f_{\vec{k}'}^0} \right|_{f^0=0} \sim (2\pi)^3 \delta(\vec{k}+\vec{k}') R_{dS}^2(-k^3) \tag{6.12}$$

We can easily check that this is the analytically continued version of (6.6) under (6.11).

Now let us understand the relation between the wavefunction computed in (6.9), which is $\Psi \sim e^{iS_{cl}}$ and the expectation values that appeared in our earlier discussion (3.23). Of course, the relation is that $\langle f^2 \rangle = \int \mathcal{D}f f^2 |\Psi(f)|^2$. We see that only the real piece in iS contributes. This has a finite limit at late times. The divergent pieces in (6.9) are all imaginary and do not contribute to the expectation value. The functional integration over f gives again (3.23). There is a crucial

---

[3] In other dimensions there are extra is that appears in the relation.

factor of 2 that comes from the square of the wavefunction, so that the relation between (3.23) and (6.12) is not a Legendre transform.

Our previous discussion focused on a scalar field and its corresponding operator $\mathcal{O}$. All that we have said above translates very simply for the traceless part of the metric and the traceless part of the stress tensor, since at the linearized level the action for the graviton in the traceless transverse gauge reduces to the action of a scalar field (3.34)–(3.35). We are defining the stress tensor operator as

$$T_{ij}(x) \equiv \frac{\delta Z[h]}{\sqrt{h}\delta h^{ij}(x)} = \frac{\delta \Psi[h]}{\sqrt{h}\delta h^{ij}(x)} \tag{6.13}$$

which is the standard definition for a Euclidean field theory.[4] In this case the divergent term in (6.9) can be rewritten as $-i\frac{1}{2\eta_c}\int d^3x \sqrt{h}R^{(3)}$. Note that there is a factor of i. We want to remove this by a counterterm in the action of the Euclidean CFT. These factors of i are related to the fact that the renormalization group transformation in the CFT should be appropriately unitary since this RG transformation corresponds, in the context of perturbation theory, to unitary evolution of the wavefunction in the bulk. If we define the central charge of the CFT in terms of the two point function of the stress tensor we get a negative answer. This negative answer has a simple qualitative explanation. We know that the wavefunction in terms of small fluctuations is bounded, in the sense that it is of the form $e^{-\alpha|f|^2}$ with α positive, since each mode is a harmonic oscillator with positive frequency. This sign implies a negative sign for the two point function of the stress tensor. Similarly the trace of the stress tensor is related to the derivative of the wavefunction with respect to ζ.

After we understood the relation between two point functions of operators and expectation values of the corresponding fluctuations we can similarly understand the relation between three point functions. The wavefunction has the form

$$\Psi = Exp\left[\frac{1}{2}\int d^3x d^3x' \langle \mathcal{O}(x)\mathcal{O}(x')\rangle f(x)f(x') \right. \tag{6.14}$$

$$\left. + \frac{1}{6}\int d^3x d^3x' d^3x'' \langle \mathcal{O}(x)\mathcal{O}(x')\mathcal{O}(x'')\rangle f(x)f(x')f(x'') \right] \tag{6.15}$$

where we emphasized that derivatives of Ψ give correlation functions for the corresponding operators. The expectation values in momentum space are related

---

[4]One might want to define it with an i so that $T_{jl} \equiv i\frac{\delta Z[h]}{\sqrt{h}\delta h^{jl}}$. This definition might be natural given that the counterterms (which represent the leading dependence of the wavefunction) are purely imaginary. In any case, it is trivial to go between both definitions.

by

$$\langle f_{\vec{k}} f_{-\vec{k}} \rangle' = -\frac{1}{2Re\langle \mathcal{O}_{\vec{k}} \mathcal{O}_{-\vec{k}} \rangle'} \tag{6.16}$$

$$\langle f_{\vec{k}_1} f_{\vec{k}_2} f_{\vec{k}_3} \rangle' = \frac{2Re\langle \mathcal{O}_{\vec{k}_1} \mathcal{O}_{\vec{k}_2} \mathcal{O}_{\vec{k}_3} \rangle'}{\prod_i (-2Re\langle \mathcal{O}_{\vec{k}_i} \mathcal{O}_{-\vec{k}_i} \rangle')} \tag{6.17}$$

where the prime means that we dropped a factor of $(2\pi)^3 \delta(\sum \vec{k})$. And Re indicates the real part. The factors of two come from the fact that we are squaring the wavefunction (6.14). Notice that this explains why $\langle TT \rangle \sim c$ while $\langle \gamma\gamma \rangle \sim 1/c$ where $c \sim -R_{dS}^2 M_{pl}^2$.

Some of the points we explained above are specific to the four dimensional dS_4 case. The situation in dS_5 is rather interesting. The computation of fluctuations for a massless scalar field gives, outside the horizon,

$$\langle f_{\vec{k}} f_{\vec{k}'} \rangle \sim H^3 (2\pi)^4 \delta(\vec{k} + \vec{k}') \frac{4}{\pi} \frac{1}{k^4}, \qquad H = R_{dS}^{-1} \tag{6.18}$$

On the other hand the wavefunction $\Psi \sim e^{iS}$ has the form

$$iS = -\frac{i}{2} R_{ds}^3 \int \frac{d^4k}{(2\pi)^4} f_{\vec{k}}^0 f_{-\vec{k}}^0 \left[\frac{k^2}{2\eta_c^2} - \frac{1}{4}k^4 \log(-\eta_c k) - i\frac{\pi}{8}k^4 + \alpha k^4 \right] \tag{6.19}$$

where α is a real number. Note that the only term contributing to (6.18 is the real term proportional to k^4. All other terms are purely imaginary. From (6.19 we can compute the non-local contribution to the two point function which gives

$$\langle \mathcal{O}(\vec{k}) \mathcal{O}(\vec{k}') \rangle_{dS_5} \sim (2\pi)^4 \delta(\vec{k} + \vec{k}') i R_{ds}^3 \frac{1}{4} k^4 \log k \tag{6.20}$$

The $EAdS_5$ answer is given by the analytic continuation (6.11). Notice that the i is due to the fact that we have an odd number of powers of R_{dS} and is consistent with the fact that the logarithmic term in the wavefunction is purely imaginary. For the stress tensor this gives an imaginary central charge and imaginary three point functions. It is rather interesting that the two point function (6.18) is related to a local term in the wavefunction, namely the term proportional to k^4, which is the only real term. In other words, the non-local piece in the wavefunction which determines the stress tensor seems unrelated to the local piece which determines the expectation value of the fluctuations. In other words, dS_5/CFT_4 would tell us how to compute the non-local piece in the wavefunction but will give us no information on the local piece. On the other hand from the inflationary point of view we would be interested in computing (6.18) which depends on the local part of the wavefunction, or the partition function of the CFT. Maybe in dS/CFT we are only allowed to use imaginary counterterms, then the field theory should be such that it allows the computation of the finite real local parts in the effective action. Note that the real term in (6.19) arises in the analytic continuation (6.11) from the term in the $EAdS_5$ wavefunction that is proportional to $k^4 \log(z_c k) \rightarrow -\frac{\pi}{2} k^4 + k^4 \log(-\eta_c k)$. So still, in some sense, the real part of the wavefunction (6.19) is intimately related to the non-local term in the wavefunction. It looks like this will be the situation in all odd dimensional dS spaces. The AdS_3 case studied in [67] seems special because there is no bulk propagating graviton. Stress tensor correlators in dS/CFT were also studied in [75, 76].

Now let us reexamine the three point functions of stress tensor operators in the limit that one of the momenta is much smaller than the other two. We can then approximate the small momentum by zero. This zero momentum insertion of the stress tensor can be viewed as coming from an infinitesimal coordinate transformation. So we then know that the three point function is going to be given by the change of the two point function by this coordinate transformation. For example, an insertion of the trace of the stress tensor at zero momentum is equivalent to performing a rescaling of the coordinates without rescaling the mass scale of the theory. Then the three point function will be given by the scale dependence of the two point function. In other words

$$\langle 2T_i^i(0)\mathcal{O}(k)\mathcal{O}(k')\rangle = -k^i \frac{\partial}{\partial k^i} \langle \mathcal{O}(k)\mathcal{O}(k')\rangle \qquad (6.21)$$

This is the reason why three point functions in this limit are proportional to the tilt of the scalar and tensor spectra respectively. There is a similar argument for the insertion of the traceless part of the stress tensor at zero momentum. Formula (6.21) is valid to all orders in slow roll. This is a consistency condition for the three point functions and it holds quite generally [18, 77].

Notice that in order to compute observable quantities from dS/CFT we will need to square the wavefunction and integrate over some range of values of the couplings and the metric of the space where the CFT is defined. In other words, in order to compute some physically interesting quantity it is not enough to consider the CFT on a fixed 3-manifold but over a range of three manifolds. This is the reason that expectation values in dS are not simply given by analytic continuation of the ones in EAdS [75] even though the wavefunction and correlation functions of operators are given by analytic continuation.[5] This makes it clear that even if dS/CFT is true there is no causality problem, one is not fixing the final state of the universe. One fixes it as an auxiliary step in order to compute the wavefunction, but in order to compute probabilities we need to sum over all final boundary

---

[5]This analytic continuation is very clear for fields with $2m R_{dS_d} < d$. For fields with mass above this bound it is not so clear what the right prescription is. In this paper we focus our attention on the easy case.

conditions. A slightly different integral over boundary conditions arises also in the $E\,AdS$ context when we consider certain relevant operators [78], or double trace operators [79]. In those cases this integration is the same as a change in the boundary condition. Note that this is *not* what happens in the dS context since we have the *square* of the wavefunction. One might conjecture that dS expectation values are given by two CFTs (one for Ψ and one for Ψ^*) coupled together in some fashion. Note that then it is not clear if we should view the resulting object as a local field theory since in the resulting object is not defined on a fixed manifold since in order to compute expectation values we need to integrate over the three metric. The two copies of the CFT that we are talking about arise just at the future boundary, so these two copies are different than the two copies talked about in [67, 68, 75, 76]. In global coordinates in addition we have the past boundary. Throughout this paper we have ignored the past boundary since we focused on distances larger than the Hubble scale but smaller than the total size of the spatial slice. In the Hartle and Hawking prescription for the wavefunction of the universe the past and future parts of the wavefunctions are complex conjugates of each other since the total wavefunction is real [69]. It is natural to suspect that these two pieces can be thought of as Ψ and Ψ^* in our discussion above.

Acknowledgements

This work was supported in part by DOE grant DE-FG02-90ER40542.

References

[1] A.D. Linde, Particle physics and inflationary cosmology, arXiv:hep-th/0503203.

[2] V. Mukhanov, *Physical Foundations of Cosmology*, Univ. Pr., Cambridge, UK (2005) 421 p; P.J.E. Peebles, Univ. Pr., Princeton, USA (1993) 718 p.

[3] P.J.E. Peebles, *Principles of Physical Cosmology*, Univ. Pr., Princeton, USA (1993) 718 p.

[4] A. Liddle, *An Introduction to Modern Cosmology*, Wiley, Chichester, UK (1998) 129 p.

[5] http://map.gsfc.nasa.gov/m_mm/pub_papers/threeyear.html, Three year WMAP results "Implications for Cosmology."

[6] A.H. Guth, *Phys. Rev. D* **23** (1981) 347; A.D. Linde, *Phys. Lett. B* **108** (1982) 389; A. Albrecht and P.J. Steinhardt, *Phys. Rev. Lett.* **48** (1982) 1220.

[7] A.P.S. Yadav and B.D. Wandelt, arXiv:0712.1148 [astro-ph].

[8] S. Sarangi and S.H.H. Tye, *Phys. Lett. B* **536** (2002) 185 [arXiv:hep-th/0204074]; E.J. Copeland, R.C. Myers and J. Polchinski, *JHEP* **0406** (2004) 013 [arXiv:hep-th/0312067].

[9] A.D. Linde, *Phys. Rev. D* **49** (1994) 748 [arXiv:astro-ph/9307002].

[10] A. Linde, *Phys. Scripta* **T117** (2005) 40 [arXiv:hep-th/0402051].

[11] V.F. Mukhanov and G.V. Chibisov, *JETP Lett.* **33** (1981) 532, [*Pisma Zh. Eksp. Teor. Fiz.* **33** (1981) 549.

[12] A.A. Starobinsky, *Phys. Lett. B* **117** (1982) 175.

[13] S.W. Hawking, *Phys. Lett. B* **115** (1982) 295.

[14] A.H. Guth and S.Y. Pi, *Phys. Rev. Lett.* **49** (1982) 1110.

[15] J.M. Bardeen, P.J. Steinhardt and M.S. Turner, *Phys. Rev. D* **28** (1983) 679.

[16] V. Mukhanov and G. Chibisov, *Zh. Eksp. Teor. Fiz.* 83 (1982) 475.

[17] V.F. Mukhanov, H.A. Feldman and R.H. Brandenberger, *Phys. Rept.* **215** (1992) 203.

[18] J.M. Maldacena, *JHEP* **0305** (2003) 013 [arXiv:astro-ph/0210603].

[19] R. Arnowitt, S. Deser and C.W. Misner, *Phys. Rev.* **117** (1960) 1595.

[20] N. Birrel and P. Davies, *Quantum Fields in Curved Space*, Cambridge Univ. Press 1982.

[21] D. Lyth, *Phys. Rev. D* 31 (1985) 1792.

[22] A.A. Starobinsky, *JETP Lett.* **30** (1979) 682, [*Pisma Zh. Eksp. Teor. Fiz.* **30** (1979) 719].

[23] D.H. Lyth, *Phys. Rev. Lett.* **78** (1997) 1861 [arXiv:hep-ph/9606387].

[24] D. Baumann and L. McAllister, *Phys. Rev. D* **75** (2007) 123508 [arXiv:hep-th/0610285].

[25] W. Hu and M.J. White, *New Astron.* **2** (1997) 323 [arXiv:astro-ph/9706147].

[26] L.A. Boyle, P.J. Steinhardt and N. Turok, *Phys. Rev. Lett.* **96** (2006) 111301 [arXiv:astro-ph/0507455].

[27] A.D. Linde, *Phys. Rev. D* **49** (1994) 748 [arXiv:astro-ph/9307002].

[28] M. Kawasaki, M. Yamaguchi and T. Yanagida, *Phys. Rev. Lett.* **85** (2000) 3572 [arXiv:hep-ph/0004243].

[29] P. Binetruy and G.R. Dvali, *Phys. Lett. B* **388** (1996) 241 [arXiv:hep-ph/9606342].

[30] K. Freese, J.A. Frieman and A.V. Olinto, *Phys. Rev. Lett.* **65** (1990) 3233.

[31] T. Banks, M. Dine, P.J. Fox and E. Gorbatov, *JCAP* **0306** (2003) 001 [arXiv:hep-th/0303252].

[32] P. Svrcek and E. Witten, *JHEP* **0606** (2006) 051 [arXiv:hep-th/0605206].

[33] P. Svrcek, arXiv:hep-th/0607086.

[34] S. Dimopoulos, S. Kachru, J. McGreevy and J.G. Wacker, arXiv:hep-th/0507205.

[35] T.W. Grimm, arXiv:0710.3883 [hep-th].

[36] S.B. Giddings, S. Kachru and J. Polchinski, *Phys. Rev. D* **66** (2002) 106006 [arXiv:hep-th/0105097].

[37] S. Kachru, R. Kallosh, A. Linde and S.P. Trivedi, *Phys. Rev. D* **68** (2003) 046005 [arXiv:hep-th/0301240].

[38] G.R. Dvali and S.H.H. Tye, *Phys. Lett. B* **450** (1999) 72 [arXiv:hep-ph/9812483].

[39] S. Kachru, R. Kallosh, A. Linde, J.M. Maldacena, L.P. McAllister and S.P. Trivedi, *JCAP* **0310** (2003) 013 [arXiv:hep-th/0308055].

[40] D. Baumann, A. Dymarsky, I.R. Klebanov, L. McAllister and P.J. Steinhardt, *Phys. Rev. Lett.* **99** (2007) 141601 [arXiv:0705.3837 [hep-th]].

[41] R. Kallosh, *Lect. Notes Phys.* **738** (2008) 119 [arXiv:hep-th/0702059].

[42] S.H. Henry Tye, arXiv:hep-th/0610221.

[43] A. Linde, eConf **C040802**, L024 (2004) [*J. Phys. Conf. Ser.* **24**, 151 (2005 PTPSA, 163,295-322.2006)] [arXiv:hep-th/0503195].

[44] R.H. Brandenberger and C. Vafa, *Nucl. Phys. B* **316** (1989) 391.

[45] R. Easther, B.R. Greene, M.G. Jackson and D. Kabat, *JCAP* **0401** (2004) 006 [arXiv:hep-th/0307233].

[46] A. Nayeri, R.H. Brandenberger and C. Vafa, *Phys. Rev. Lett.* **97** (2006) 021302, hep-th/0511140; R.H. Brandenberger, A. Nayeri, S.P. Patil and C. Vafa, hep-th/0608121; R.H. Brandenberger, A. Nayeri, S.P. Patil and C. Vafa, hep-th/0604126.

[47] N. Kaloper, L. Kofman, A. Linde and V. Mukhanov, *JCAP* **0610** (2006) 006, hep-th/0608200.

[48] J. Khoury, B.A. Ovrut, P.J. Steinhardt and N. Turok, *Phys. Rev. D* **64** (2001) 123522 [arXiv:hep-th/0103239]; J. Khoury, B.A. Ovrut, P.J. Steinhardt and N. Turok, *Phys. Rev. D* **66** (2002) 046005 [arXiv:hep-th/0109050].

[49] P.J. Steinhardt and N. Turok, *Science* **296** (2002) 1436; P.J. Steinhardt and N. Turok, *New Astron. Rev.* **49** (2005) 43 [arXiv:astro-ph/0404480].

[50] U. Yurtsever, *Class. Quant. Grav.* **7** (1990) L251.

[51] M.S. Morris, K.S. Thorne and U. Yurtsever, *Phys. Rev. Lett.* **61** (1988) 1446.

[52] P. Creminelli, A. Nicolis and M. Zaldarriaga, *Phys. Rev. D* **71** (2005) 063505 [arXiv:hep-th/0411270].

[53] H. Liu, G.W. Moore and N. Seiberg, *JHEP* **0210** (2002) 031 [arXiv:hep-th/0206182].

[54] G.T. Horowitz and J. Polchinski, *Phys. Rev. D* **66** (2002) 103512 [arXiv:hep-th/0206228].

[55] J.M. Maldacena, *Adv. Theor. Math. Phys.* **2** (1998) 231 [*Int. J. Theor. Phys.* **38** (1999) 1113], hep-th/9711200.

[56] E. Witten, *Adv. Theor. Math. Phys.* **2** (1998) 253, hep-th/9802150.

[57] S.S. Gubser, I.R. Klebanov and A.M. Polyakov, *Phys. Lett. B* **428** (1998) 105, hep-th/9802109.

[58] L. Girardello, M. Petrini, M. Porrati and A. Zaffaroni, *JHEP* **9812** (1998) 022, hep-th/9810126.

[59] D.Z. Freedman, S.S. Gubser, K. Pilch and N.P. Warner, *Adv. Theor. Math. Phys.* **3** (1999) 363, hep-th/9904017.

[60] G. Arutyunov, S. Frolov and S. Theisen, *Phys. Lett. B* **484** (2000) 295, hep-th/0003116.

[61] O. DeWolfe and D.Z. Freedman, hep-th/0002226.

[62] M. Bianchi, D.Z. Freedman and K. Skenderis, *JHEP* **0108** (2001) 041, hep-th/0105276.

[63] M. Bianchi, D.Z. Freedman and K. Skenderis, *Nucl. Phys. B* **631** (2002) 159, hep-th/0112119.

[64] K. Skenderis, hep-th/0209067.

[65] F. Larsen, J.P. van der Schaar and R.G. Leigh, *JHEP* **0204** (2002) 047, hep-th/0202127.

[66] O. Aharony, S.S. Gubser, J.M. Maldacena, H. Ooguri and Y. Oz, *Phys. Rept.* **323** (2000) 183, hep-th/9905111.

[67] A. Strominger, *JHEP* **0110** (2001) 034, hep-th/0106113.

[68] E. Witten, hep-th/0106109.

[69] J.B. Hartle and S.W. Hawking, *Phys. Rev. D* **28** (1983) 2960.

[70] E. Witten, *JHEP* **9812** (1998) 012, hep-th/9812012.

[71] G. Lifschytz and V. Periwal, *JHEP* **0004** (2000) 026, hep-th/0003179.

[72] J. de Boer, E. Verlinde and H. Verlinde, *JHEP* **0008** (2000) 003, hep-th/9912012.

[73] L. Dyson, J. Lindesay and L. Susskind, *JHEP* **0208** (2002) 045, hep-th/0202163.

[74] U.H. Danielsson, *JHEP* **0207** (2002) 040, hep-th/0205227.

[75] D. Klemm, *Nucl. Phys. B* **625** (2002) 295, hep-th/0106247. R. Bousso, A. Maloney and A. Strominger, *Phys. Rev. D* **65** (2002) 104039, hep-th/0112218. M. Spradlin and A. Volovich, *Phys. Rev. D* **65** (2002) 104037, hep-th/0112223. V. Balasubramanian, J. de Boer and D. Minic, hep-th/0207245.

[76] V. Balasubramanian, J. de Boer and D. Minic, *Phys. Rev. D* **65** (2002) 123508, hep-th/0110108.

[77] P. Creminelli and M. Zaldarriaga, *JCAP* **0410** (2004) 006 [arXiv:astro-ph/0407059].

[78] I.R. Klebanov and E. Witten, *Nucl. Phys. B* **556** (1999) 89 [arXiv:hep-th/9905104].

[79] E. Witten, arXiv:hep-th/0112258.

.

Course 6

SPONTANEOUS BREAKING OF SPACE-TIME SYMMETRIES

Eliezer Rabinovici

Racah Institute of Physics, The Hebrew University of Jerusalem, 91904, Israel
eliezer@vms.huji.ac.il

C. Bachas, L. Baulieu, M. Douglas, E. Kiritsis, E. Rabinovici, P. Vanhove, P. Windey
and L.F. Cugliandolo, eds.
Les Houches, Session LXXXVII, 2007
String Theory and the Real World: From Particle Physics to Astrophysics
© *2008 Published by Elsevier B.V.*

Contents

1. Introduction

It is not uncommon for young scientists to complain that their teachers didn't educate them appropriately. I would have liked for example to know earlier about the ideas of Kaluza and Klein. So in order to somewhat reduce the complaints that will be directed at me, I would like to use this opportunity to describe something that it is not taught extensively in particle physics courses, mechanisms to spontaneously break space-time symmetries. The world around us is actually neither explicitly invariant under translations, nor under rotations. It also is not explicitly invariant under scale and conformal symmetries. In this work we will review various mechanisms to break all these space-time symmetries. I think they may yet play an important role in particle physics as well. I will first describe attempts to break translational invariance kinematically by imposing specific boundary conditions. Then I will review the Landau theory of solidification and an attempt to apply it to generate a dynamical mechanism for compactifications. I will discuss both the success and challenges of that approach. Next, in the context of breaking time translational invariance, I will discuss various systems which are well defined but have no ground state. Following a review of the breaking of scale invariance and conformal invariance, I will also not miss this opportunity to describe in a Katoish manner that the vacuum energy in conformal/scale invariant theories is very constrained, and its zero value does not depend on the presence or absence of any spontaneously generated scales. This may eventually be recognized as an important ingredient in understanding and explaining the cosmological constant problem.

2. Spontaneous breaking of space symmetries

Space symmetries include space translations and space rotations, we address the spontaneous breaking of these space symmetries. This occurs for example when a liquid solidifies and a lattice is formed. The standard manner to identify the ground state of a system is to construct what is called the effective potential. The symmetry properties of the ground state determine whether a spontaneous breaking of symmetries which are manifest in the Lagrangian occurs.

Let's review the manner in which the effective potential is constructed. One first considers all wave functionals which have the same expectation value of the field operator $\hat{\phi}$

$$< \Psi(\phi)|\hat{\phi}|\Psi(\phi) >= \tilde{\phi}. \tag{2.1}$$

Out of these subset of wave functionals, one chooses that particular wave functional which minimizes the expectation value of the Hamiltonian. One calls it $V_{eff}(< \phi >)$

$$V_{eff}(\tilde{\phi}) = \min_{\tilde{\phi}} < \Psi(\phi)|\hat{H}|\Psi(\phi) >. \tag{2.2}$$

Eventually one draws a picture portraying V_{eff} as a function of $\tilde{\phi}$ and one searches for its minimum. The wave functional for which this energy minimum was obtained is the wave functional of the ground state of the system. However, one usually ignores the possibility that the ground state wave function would correspond to a non-constant (in x) expectation value $< \phi(x) >$. Of course it makes much easier the drawing of pictures in books, here however we will discuss cases where $< \phi(x) >$ actually does depend on x when evaluated in the ground state.

Why does one usually only consider wave functionals with constant values of $< \phi(x) >$?

The reason is expediency, when one wants to pick up the ground state of the system among various candidates, one is interested only in the winner, that is the true ground state. One does not care about missing out candidate states whose energies are just above that of the ground state of the system. As generally spontaneous breakdown of space-time symmetries in the ground state is not expected, one considers it enough to search for the ground state only among those candidates for which $< \phi(x) >$ is constant. However that need not always be the case.

2.0.1. *Kinematics: Attempts to break spatial translational invariance through boundary conditions*

I will first describe an easy way to attempt to break space symmetries. That is to break the symmetries not by the dynamics of the system but kinematically, by imposing certain boundary conditions. This easy solution, is a mirror to what is done in String Theory in several cases, including when one is considering brane sectors. To try and break translational invariance by boundary conditions one considers for example a system which depends on a scalar field ϕ. Assume the system lives in a box extending from $-L$ to L, and impose the condition of anti-periodicity, namely

$$\phi(L) = -\phi(-L), \tag{2.3}$$

where L is the spacial cutoff we put on the system.

If the system at hand is described by an effective potential that has only one minimum (see Fig. 1), where there the expectation value $< \phi >$ vanishes, then

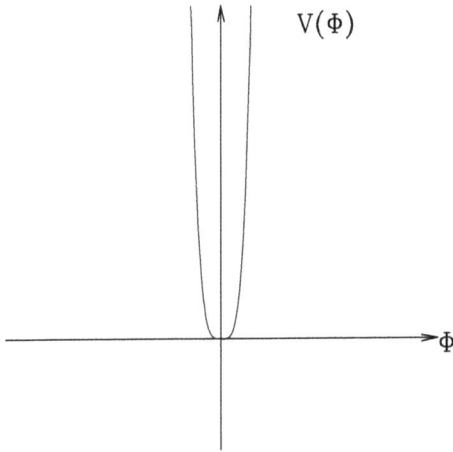

Fig. 1. Unbroken symmetry.

there is no effect resulting from imposing the boundary conditions. The ground state does fulfill the boundary condition, and it remains the one which does not break translational invariance. From the point of view of the wave functional, it is concentrated around $\phi = 0$.

However, consider the double well potential, (Fig. 2), (in circumstances where there is no tunneling).

In this case the effective potential has two minima, one at $\phi = a$ and the other at $\phi = -a$. Imposing the boundary condition removes both of the possible true vacua of the system, because neither the ground state for which $< \phi >= -a$, nor the ground state for which $< \phi >= a$ obey the boundary condition. One is driven to look for another type of ground state. We know, for example, that in a two dimensional system composed only of scalar fields there is a finite energy solution, which is a soliton that at L has a value a and at $-L$ has a value $-a$, see Fig. 3. An anti-soliton will have the opposite values. This is a stable topological configuration, and one may imagine that indeed in such a system there is no translational invariance, because the ground state will have to be such that its spacial expectation value follows the values of the soliton field, and thus is not translational invariant.

It is true that by imposing the boundary conditions one has forced the system into the soliton sector, but one has to remember that this system has a zero mode. Technically, if one solves the small fluctuations of the scalar field in the presence of a background, which is a soliton, one finds that there is a zero mode. This zero mode is a reflection of the underlying translational invariance and it actually tells that one is not able to determine, by energetic considerations, where the

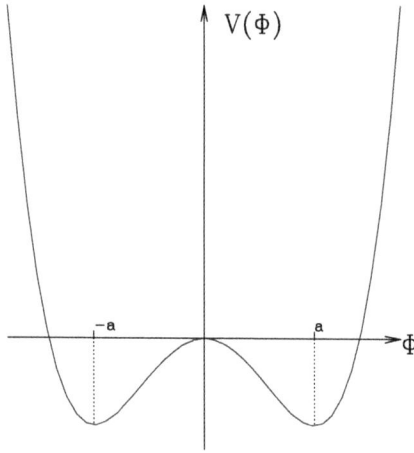

Fig. 2. Broken symmetry.

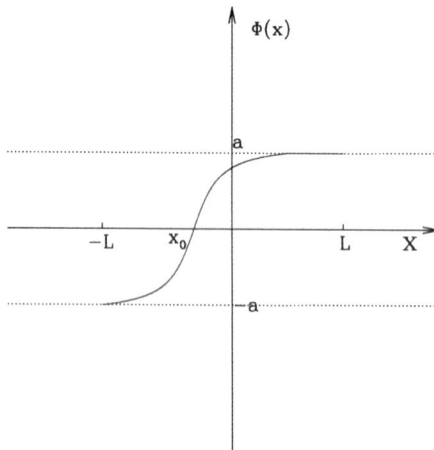

Fig. 3. A soliton attempts to break translational invariance.

inflection point x_0, the point from which one turns from one vacuum to the other, (see Fig. 3) occurs. Actually there is a valid soliton solution for each value of x_0.

Why is this important? At the case at hand, the zero mode is normalizable. This amounts to saying that the soliton mass is finite. In such a case, there is actually no bulk violation of translational invariance. What one needs to do is to construct an eigenstate configuration, which is an eigenstate of the linear momentum operation, a plane wave in terms of the center of mass coordinate of the soliton. The lowest energy state which corresponds to a momentum state has $p = 0$, one has restored in this way translational invariance. The only problem will be to fix the system very near the edges, but in the bulk, the symmetry has been restored and there is no breaking of bulk translational invariance. Could one still have a case where the boundary conditions do cause a spontaneous breaking of translational invariance?

This may occur when one drives the mass of the soliton to infinity by an appropriate choice of parameters. When the mass of the soliton is infinite, physically one cannot form a linear momentum state out of it, and technically the zero mode ceases to be normalizable. In such a case, one does indeed break translational invariance spontaneously by fixing the point where the soliton makes the transition from one vacuum to the other. This occurs for example in String Theory in a sector containing infinite branes, branes which have finite energy do not break translational invariance, and one can build out of them linear momentum states. However, branes which extend up to infinity carry infinite energy, and therefore do lead to the breakdown of translational invariance.

I will mention at this point that once upon the time, when people were considering the breakdown of extended global supersymmetries, there was a predominant common wisdom which claimed that one cannot break down extended global supersymmetry to anything but $\mathcal{N} = 0$. That is either all the supersymmetries are manifest together, or they are all broken together. The argument went in the following way: one writes the formula for the Hamiltonian

$$H = \sum_\alpha \bar{Q}_\alpha^I Q_\alpha^I, \tag{2.4}$$

where $I = 1, \ldots, \mathcal{N}$ is *not* summed, it is a non-trivial constraint to get the same Hamiltonian by summing over different supersymmetry generators. When one can do that one has an extended supersymmetry. However it is clear from this that if the Hamiltonian does not vanish on the ground state, then some of the Q^I (for each I independently) do not annihilate the ground state. Therefore the supersymmetries are either all preserved or all broken.

This type of argument assumed implicitly that Poincaré invariance is present in the system. If one now considers other systems, (see for example [1,2]), where part of the Poincaré invariance is preserved and part is broken, this exposes a loop hole in the former argument. In the absence of full translational invariance(due to the presence of infinite mass branes) one may obtain fractional BPS states, and one may break $\mathcal{N} = 4$ down to $\mathcal{N} = 2$, $\mathcal{N} = 2$ to $\mathcal{N} = 1$ and various other combinations.

This is an example where spontaneous breaking of translation invariance occurs, it has an impact also on the partial breaking of global supersymmetry and, if one wishes, this is a way to break translation invariance by forcing the system, using boundary conditions, to a certain super-selection sector.

This is not what I mainly want to discuss here. I would like to discuss a situation where the dynamics of the system drives the spontaneous breaking of translational and rotational invariance.

2.0.2. Dynamics: The Landau theory of liquid-solid phase transitions

Let us now turn to discuss the transition between a liquid and a solid. This follows the seminal work of Landau [3]. In a monumental paper, he described spontaneous symmetry breaking of both, internal and space-time symmetries. Consider a liquid, a system whose Lagrangian is either relativistic or non-relativistic, and it possesses full rotational and translational invariance. On the other hand, a solid is a system which maintains only a very small part of the translational invariance and rotational invariance, (Fig. 4).

Let us simplify the study by ignoring the point structure at each lattice point which a solid may have. That is let's not consider the atomic structure at each point. One focuses first on the question of how does the simplest lattice forms.

I will describe this following Landau and then, following [4], I am going to describe applications to String Theory. Landau starts by defining the Landau order parameter to monitor the transition between a solid and a liquid. It is a scalar order parameter $\varrho(\vec{x})$

$$\varrho(\vec{x}) = \varrho_s(\vec{x}) - \varrho_0, \tag{2.5}$$

the difference between the non-translational non-rotational invariant density of the solid $\varrho_s(\vec{x})$, and the constant density ϱ_0 of the liquid. Next, consider the Fourier decomposition of $\varrho(\vec{x})$

$$\varrho(\vec{x}) = \sum \varrho(\vec{q})e^{i\vec{q}\cdot\vec{x}} + h.c. \tag{2.6}$$

It is useful to use as order parameters the Fourier components $\varrho(\vec{q})$.

The question is thus: Does the wave functional of the ground state have support on $\vec{q} \neq \vec{0}$? If the answer is positive, then at the very least continuous translational space symmetry would be spontaneously broken. This will be determined

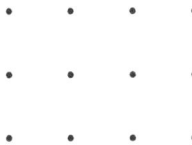

Fig. 4. The solid lattice breaks most of the translational and rotational invariances.

by studying the Landau–Ginsburg effective action as expressed in terms of the order parameter $\varrho(\vec{q})$. The first relevant term of the Landau–Ginsburg action is quadratic in the order parameter and is given by:

$$\mathcal{L}_0 = \int d\vec{q}_1 d\vec{q}_2 \varrho(\vec{q}_1)\varrho(\vec{q}_2)A(|\vec{q}_1|^2)\delta(\vec{q}_1 + \vec{q}_2). \tag{2.7}$$

The delta function $\delta(\vec{q}_1 + \vec{q}_2)$ enforces translational invariance, while rotational invariance is preserved by the dependence on $|\vec{q}|^2$ of the function $A(|\vec{q}|^2)$. The function $A(|\vec{q}|^2)$, like in any Landau–Ginsburg potential, is determined by the microscopic theory. In the particular case at hand, it will depend on the hard-core potential component in the atoms involved and on other possible potentials, as well as on the temperature of the system. In the case of neutron stars, studied in [5], the Pauli exclusion principle plays a role in determining the function $A(|\vec{q}|^2)$.

Let us treat first an example that we are familiar with, that of a free massive spin-zero particle in a relativistic field theory. In that case the function $A(|\vec{q}|^2)$ is

$$A(|\vec{q}|^2) = |\vec{q}|^2 + m^2. \tag{2.8}$$

This has a minimum at $|\vec{q}|^2 = 0$, as shown in Fig. 5, and thus the function $\varrho(\vec{q})$ should get the support only at $\vec{q} = \vec{0}$, there is no spontaneous breakdown of translational invariance in this case.

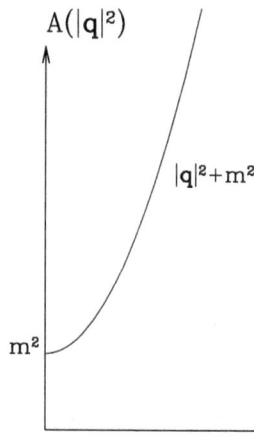

Fig. 5. The form of $A(|\vec{q}|^2)$ in a free massive relativistic field theory does not lead to spontaneous breaking of translational invariance.

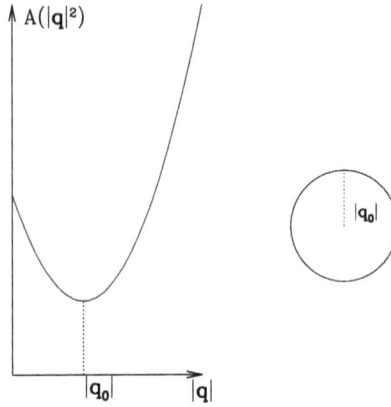

Fig. 6. Example of a function $A(|\vec{q}^2|)$ which leads to the breaking of translational invariance. An explicit microscopical realization of a such a form appears in neutron stars [5]. The wave functional is concentrated at most on the shell of a sphere of radius $|\vec{q}_0|$.

In the presence of interactions things may become more complicated, for example I am not familiar even with a proof that the Standard Model ground state does not violate space-time symmetry, (though most likely it does not). In any case the microscopic theory may allow a different function for $A(|\vec{q}|^2)$. In particular, assume that the form of $A(|\vec{q}|^2)$ is as given in Fig. 6. In this case, the function $A(|\vec{q}|^2)$ has a minimum at a value $|\vec{q}_0|^2 \neq 0$. In such a system the ground state wave functional gives rise to a density concentrated around $|\vec{q}_0|^2 \neq 0$. In particular, one would expect the support to be concentrated around a sphere in $\vec{q}$-space, whose radius is $|\vec{q}_0|$. So one is in a situation, given $A(|\vec{q}|^2)$ of that form, where one has a spontaneous breaking of translational invariance, but not yet also a breaking of rotational invariance, which is what is needed to form a solid. It is good enough to break just translational invariance.

The ground state density does depend on $\vec{x}$

$$\varrho(\vec{x}) = \int_{\mathcal{S}_{|\vec{q}_0|}} d\Omega\, \varrho(\vec{q}) e^{i\vec{q}\cdot\vec{x}} + h.c. \tag{2.9}$$

In this approximation the wave functional of the ground state is supported on a sphere $\mathcal{S}_{|\vec{q}_0|}$ whose radius is $\vec{q}_0$. In particle physics we have become rather sophisticated, and when one writes down the Landau–Ginsburg action, one usually requires that the expansion in the order parameter to be under control. For example, this means that there is a limit in which this expansion becomes exact. In the case at hand, this is not the situation, and it is actually very complicated, nevertheless one follows the usual Landau–Ginsburg expansion.

The term which follows the quadratic interaction is a cubic term

$$\mathcal{L} = \mathcal{L}_2 + \mathcal{L}_3, \tag{2.10}$$

$$\mathcal{L}_3 = \int d^3\vec{q}_1 d^3\vec{q}_2 d^3\vec{q}_3 \; \varrho(\vec{q}_1)\varrho(\vec{q}_2)\varrho(\vec{q}_3)\delta(\vec{q}_1 + \vec{q}_2 + \vec{q}_3)$$

$$\times B\left(|\vec{q}_1|^2, |\vec{q}_2|^2, |\vec{q}_3|^2, \vec{q}_1 \cdot \vec{q}_2, \vec{q}_1 \cdot \vec{q}_3, \vec{q}_2 \cdot \vec{q}_3\right). \tag{2.11}$$

I am going to assume for the purpose of illustration, as Landau did, that this is a good perturbation, namely that when one considers $\mathcal{L}_3$ one is going already to assume that the support of ϱ comes from only those values of $\vec{q}$ such that $|\vec{q}_1|^2 \cong |\vec{q}_2|^2 \cong |\vec{q}_3|^2 \cong |\vec{q}_0|^2$. This was determined by $\mathcal{L}_2$.

In (2.11), once again, the delta function $\delta(\vec{q}_1 + \vec{q}_2 + \vec{q}_3)$ enforces the explicit translational invariance, and the dependence of B on the momentum respects both translational and rotational invariance.

The integral in the $\vec{q}$'s is not over all possible values, but only over those whose lengths is determined by $|\vec{q}_0|^2$, which in turn was fixed by $\mathcal{L}_2$.

An additional structure emerges due to the effect of the delta function $\delta(\vec{q}_1 + \vec{q}_2 + \vec{q}_3)$. It restricts the candidates for the ground state, to have support on at least three different values for the $\vec{q}_i$. The three vectors appearing need to sum up to give a triangle, see Fig. 7.

Actually they are six if the field is real since one needs

$$\varrho(\vec{q}) = \varrho(-\vec{q}). \tag{2.12}$$

Thus one has at least six components of $\varrho(\vec{q})$ which do not vanish. In general, instead of $\varrho(\vec{q})$ having support on all values of a sphere, they are now broken into triplets where the $\vec{q}_i$ have to sum together to form triangles, (Fig. 7). In this manner also rotational invariance is spontaneously broken.

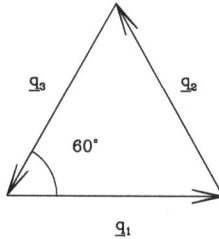

Fig. 7. The sphere $\mathcal{S}_{|\vec{q}_0|}$ is triangulated due to the presence of a cubic term in the Lagrangian. Since in this approximation all the sides of the triangles have the same length, their angles are determined to be 60°. Rotational invariance is thus spontaneously broken.

Let's be even more explicit, because we have used the approximation that all the $\vec{q}_i$ have the same length, the $\vec{q}_i$ that tessellate the sphere, have to form equilateral triangles, as in Fig. 7. Equilateral triangles single out a specific angle 60°, that is a spontaneous breaking of rotational invariance. One has obtained a non zero value for $\vec{q}$, and one has derived that the ground state is built out of objects which have to sum up to form triangles which are equilateral and thus have a 60° angle.

From energetic and combinatorial considerations, one finds that to be on an extremum, one needs all the values of $\varrho(\vec{q}_i)$ to be equal

$$|\varrho(\vec{q}_i)|^2 = |\varrho(\vec{q}_0)|^2, \tag{2.13}$$

which leads to

$$|\varrho(\vec{x})|^2 = n|\varrho(\vec{q}_0)|^2. \tag{2.14}$$

There are a couple of general way to distribute the triplets, one in which each $\vec{q}_i$ appears in only one of the triplets, and another in which each value of $\vec{q}_i$ does participate in two triplets. The number of elements is proportional to n in both cases, being either $2n/3$ or $4n/3$. When one does the analysis, and one estimates the value of $\mathcal{L}_3$, one finds that it decreases as the inverse of $\sqrt{n}$

$$\mathcal{L}_3 \sim \frac{|\varrho(\vec{q}_0)|^2}{\sqrt{n}}. \tag{2.15}$$

Thus the ground state will be obtained for some finite value of n. One needs to consider only a finite number of triplet configurations when one searches for the extrema of the free energy. Just three, i.e. six participants lead to the following density distribution

$$\varrho(x, y) = \pm\left(\frac{2}{3}\right)^{1/2} \varrho_{q_0}\left[\cos(q_0 x) + 2\cos\left(\frac{1}{2}q_0 x\right)\cos\left(\frac{\sqrt{3}}{2}q_0 y\right)\right]. \tag{2.16}$$

The corresponding free energy is

$$\mathcal{L}_3^{n=3} = \frac{2B\varrho_{q_0}^3}{3\sqrt{3}}. \tag{2.17}$$

For the case of two spatial dimensions it turns out that if $\varrho(q_0) > 0$ it is advantageous to form a triangular lattice, while if $\varrho(q_0) < 0$, the dual lattice, which is a honeycomb lattice, is formed.

This required only studying the minimal possible configuration. In three spatial dimensions, this would be a candidate for a two dimensional lattice in three dimensions, if one wishes some type of compactification.

In three dimensions one needs to consider also larger configurations to obtain the extrema. The next candidate configuration has six ($n = 6$), i.e. twelve values of $\vec{q}$. This is a more complicated configuration, whose density distribution is

$$\varrho(x, y, z) = \frac{2}{\sqrt{3}}\varrho_{q0}\left[\cos\left(\frac{\sqrt{2}}{2}q_0 x\right)\cos\left(\frac{\sqrt{2}}{2}q_0 y\right)\right.$$
$$\left. + \cos\left(\frac{\sqrt{2}}{2}q_0 x\right)\cos\left(\frac{\sqrt{2}}{2}q_0 z\right)\cos\left(\frac{\sqrt{2}}{2}q_0 y\right)\cos\left(\frac{\sqrt{2}}{2}q_0 z\right)\right].$$

(2.18)

This actually describes a BCC lattice (in real space). The value of $\mathcal{L}_3$ is larger than that one of the former configuration

$$\mathcal{L}_3^{n=6} = \frac{4B\varrho_{q0}^3}{3\sqrt{6}} > \mathcal{L}_3^{n=3},$$

(2.19)

and leads to the extrema of the free energy, being the most stable configuration.

From amazingly simple considerations, one has a prediction that solids in three dimensions are all BCC lattices, a very universal description of the system. Before confronting this claim with the data one needs to recall that the transitions between solids and liquids are not second order transitions, they are actually first order transitions. So one may question the validity of universality claims in this context. However, it turns out that in many cases one can arrange that the solidifications occur as a weak first-order transitions, in which case approximate universality properties can be present.

Returning to the data and following [6], one discovers that about 40 metals, which are on the left of the periodic table, (excluding Magnesium (Mg)), form near the solidification point a BCC configuration.

I will repeat the difficulties of the analysis and the argumentations to proceed with it. The transition is first order—the fact that in many cases it is a weak first order transition softens this problem. There is no true expansion parameter in the problem. The microscopic theory constructing A and B is very phenomenological, and therefore the real relative stability of the metal is a very delicate matter. Even taking all this into a account the result and its agreement with a large body of the experimental data is striking.

Consider what would have happened without a cubic term. In that case, the term following the quadratic term would be $\mathcal{L}_4$, which schematically would assume the form

$$\mathcal{L}_4 = \int d\vec{q}_1 d\vec{q}_2 d\vec{q}_3 d\vec{q}_4 \, \delta(\vec{q}_1 + \vec{q}_2 + \vec{q}_3 + \vec{q}_4)$$
$$\times C\left(|\vec{q}_1|^2, |\vec{q}_2|^2|\vec{q}_3|^2, |\vec{q}_4|^2, \vec{q}_1 \cdot \vec{q}_2, \vec{q}_1 \cdot \vec{q}_3, \ldots\right),$$

(2.20)

E. Rabinovici

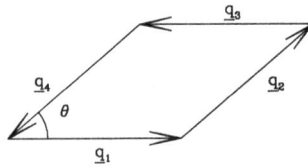

Fig. 8. In the absence of a cubic term, a quartic term would not suffice classically to induce a spontaneous breaking of rotational invariance. A rhombus does not single out a preferred angle θ.

where the delta function enforces translational invariance, and the function C should be build by such invariants that maintain both rotational and translational invariance.

This does not break rotational invariance, because unlike the case of triangles, the configurations which are enforced now, assuming perturbation theory, are those of quadrilaterals with equal sides. But for a rhombus (Fig. 8) no preferred angle is singled out. The rotational invariance is not broken. Fortunately there is no microscopic symmetry consideration that rules out the cubic term.

Another interesting type of lattices are the Abrikosov lattices formed of vortexes, which we do not discuss here.

2.1. String theory compactifications

What has been described above has a very solid basis in nature. What we will describe next is of a much more speculative nature, and it is based on work by Elitzur, Forge and myself [4], in which we try to address the issues of compactification in String Theory. There are several attitudes one might adopt regarding compactification. One, which makes a lot of sense, is to say that the Universe starts up very small, and the issue of compactification is an issue of explaining why four dimensions became very large, while the rest of the dimensions remain small. This is not what I am going to discuss here.

Here, I discuss possible dynamical aspects of compactification taking in account some of the hints learnt from the case of solid state physics. I don't have much confidence in human imagination when it is totally detached from reality, and I would hope that many of the hints available in nature will be useful to understand other phenomena. In particle physics one has learned quite a lot from the dynamics of solid state physics, and statistical mechanics systems.

Returning to the case at hand, we have just reviewed a system which has lost most of its rotational and translational invariance, and we want to see how a similar thing might happen in String Theory. One of the key ingredients driving this behavior is the presence of a bulk tachyon.

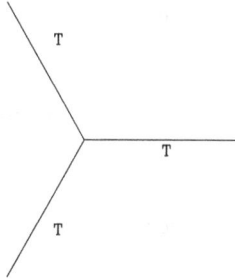

Fig. 9. Tachyonic cubic vertex.

There are actually at least three types of tachyons/instabilities with whom one is familiar right now in String theory. One is the Bosonic closed String Theory tachyon. This instability could well be an incurable one, nevertheless let's try and follow it.

The other types of instabilities, which we will discuss later, are an instability in Open String Theory, an open string tachyon, and also localized bulk tachyons.

For the moment we focus on bulk tachyons, which will be one key ingredient. Due to them it is preferable for a system in String Theory containing a tachyon to have a support on a non-zero value of q^2. One can see this from the form of the tachyon, whose vertex operator is the following

$$T(x) = e^{iq_0 x} + h.c. \tag{2.21}$$

To obtain a dimension $(1, 1)$ operator, one needs $q_0 \neq 0$. Tachyons do give us the starting point that appears in Landau theory of solidification (note that here it is not a minimum consideration). The second key ingredient that we need for Landau's theory of solidification, in order to obtain not only the breakdown of translational invariance, but also of rotational invariance, is the presence of a cubic term. We know from the OPE (operator-product-expansion) that three tachyons do couple together, (see Fig. 9). In particular, the OPE between two tachyons does contain a third tachyon. So we have in a such a theory a T^3 term. One indeed has the necessary ingredients to try to follow if tachyons could lead to the spontaneous breaking of rotational and translational invariance in String Theory, and maybe also to compactification.

In order to be more concrete, we followed the ideas of [7,8] and tried to handle in a reliable fashion almost marginal operators. Consider a tachyon which is not an exact $(1, 1)$ operator, but one which has $q_0^2 = 2 - \varepsilon$. We will also look at the subset of the full string background, a subset which contains a $c = 2$ sector. We will not deal here with the question of how the total central charge remains at the appropriate value, which is zero and how to dress operators.

As an illustration, consider the subset of the backgrounds which are string moving in flat space, where the piece of the Lagrangian on which we focus is

$$\mathcal{L} = \partial X^1 \bar{\partial} X^1 + \partial X^2 \bar{\partial} X^2 + T(X^1, X^2). \tag{2.22}$$

From Landau's theory of solidification, we know that because the system has support on a $\vec{q}_0 \neq 0$, and because the free energy of the system contains a cubic coupling, we can try to build the triplet, which again are actually six vectors, so they get a support in an appropriate way, i.e. such that they break translational and rotational invariance.

The 60° angle, discussed in the solidification case, manifests itself in a suggested tachyon configuration:

$$T(X^1, X^2) = \sum_{a=1}^{3} T_a \cos \left(\sum_{i}^{2} k_i^a X^i \right), \tag{2.23}$$

where the three momenta $\vec{k}^1$, $\vec{k}^2$, and $\vec{k}^3$ are the following

$$\vec{k}^1 = k(1, 0), \quad \vec{k}^2 = k(-1/2, \sqrt{3}/2), \quad \vec{k}^3 = k(-1/2, -\sqrt{3}/2). \tag{2.24}$$

All of them have $\vec{k}^2 = 2 - \varepsilon$, and the structure is very similar to that of the $SU(3)$ root lattice (see Fig. 10), as before for a every k_i, there is also the corresponding $-k_i$ contribution.

One can simplify the tachyon potential by taking the ansatz for the amplitudes $T_a = T$.

The Lagrangian one needs to solve is the following

$$\mathcal{L} = \partial X^1 \bar{\partial} X^1 + \partial X^2 \bar{\partial} X^2 + T(X^1, X^2), \tag{2.25}$$

and actually one can show that the beta function of the tachyon alone vanishes to order ε. So (2.23) is a solution of the approximate tachyon equations of motion. This means that had it been up to the tachyon alone one would have obtained the lattice, perhaps some honeycomb or triangular lattice, which would break both translational and rotational invariance. However, this system contains also gravity so one needs to see what is the influence of the formation of such a lattice on gravity. As shown in [4], the beta function for the graviton $\beta_{G_{\mu\nu}}$ vanishes (at leading order in α') if

$$\beta_{G_{\mu\nu}} = -R_{\mu\nu} + \nabla_\mu T \nabla_\nu T = -R_{\mu\nu} + \frac{3}{2}\varepsilon^2 \delta_{\mu\nu} = 0. \tag{2.26}$$

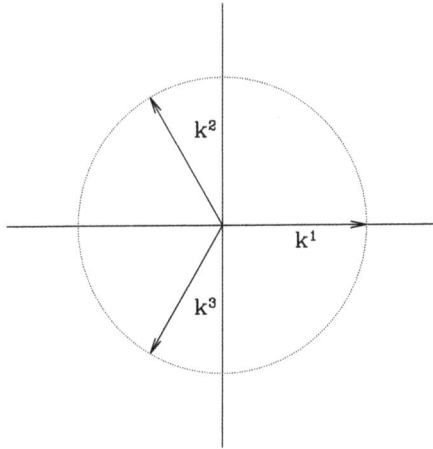

Fig. 10. $SU(2)$ roots.

For $D = 2$, due to the Liouville theorem $R_{\mu\nu}$ can be written as $R_{\mu\nu} = a\delta_{\mu\nu}$, so one can solve the equation by forming a two-dimensional sphere.

This is actually a highlight of a model for compactification. We started by having just a tachyon. The tachyon would have produced the lattice on its own, but because of the presence of the gravity the lattice of tachyons actually causes the compactification of space to a sphere.

However, it turns out, and details are presented in [4], that unfortunately this result is not obtained in a desired reliable approximation. The main problem is that in order to do reliable perturbation theory, we need to do a plane waves expansion, with the wave lengths representing a nearly marginal operator. However, the moment the sphere is formed the topology changes, and the change of topology means that one should now expand the fields in terms of spherical harmonics $Y_{l,m}$. This topological obstruction takes away the reliability of our calculation. Some defects may form in order to resolve this topological problem, and one conjecture we had at that time was that actually parafermions, which are defects, form to resolve the tension. A more complex form of compactification emerges.

Once again, recall that actually the system, when fully considered, has to be coupled to the dilaton, in order to maintain the total central charge. According to the Zamolodchikov theorem [7], once the system starts to flow, the central charge decreases from 2 and this on its own breaks the balance. In a sense, in the case of bulk tachyons we were tantalizingly close to obtain an explicit dynamical mechanism for compactification. However, due to topological obstructions, what was a solution for the beta function locally in space, it cannot be a global solution without taking into account other effects. We will return to the breaking of translational invariance in the different context of the open string tachyon.

2.2. Liquid crystals

The tachyon is a scalar order parameter, String Theory has additional fields which carry indexes. In particular, one might think that if one looks for a similarity to our universe, maybe one should would be consider the phase of liquid crystals. Such systems are translational invariant in some directions but not in other (see Fig. 11). We will give now examples of that.

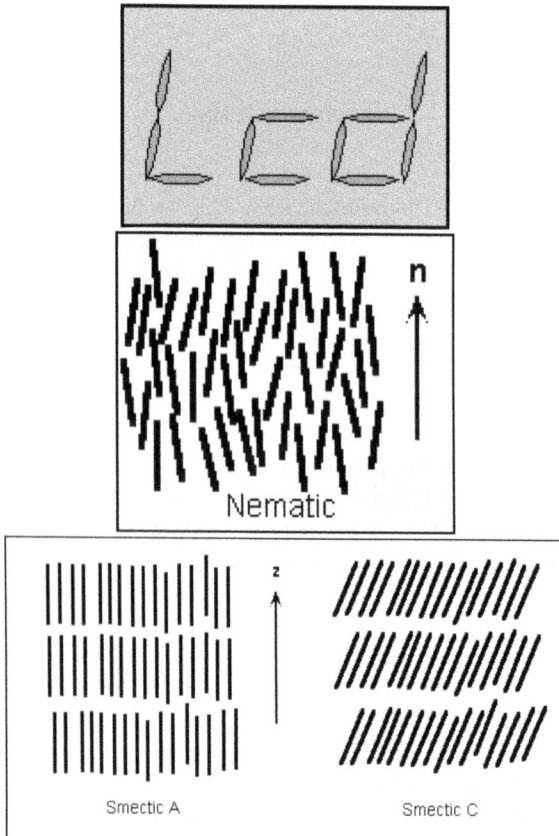

Fig. 11. Various phases of liquid crystals breaking. These systems exhibit asymmetrical breaking of translational and rotational invariance.

There are various types of liquid crystals and one can ask what is the Landau–Ginsburg theory of them. Actually, one can also ask about vector potential systems which are described, as gauge fields are, by vector order parameters. Such systems include detergents which posses a hydrophobic and a hydrophilic pole, and play a crucial role in cleaning our garments. One can try to extract from $\vec{p}(\vec{r})$ the various invariants one wants to use in order to describe this system, such as $div\ \vec{p}$, $curl\ \vec{p}$, $s_{\alpha\beta} = \partial_\alpha p_\beta + \partial_\beta p_\alpha$.

It turns out that one can write down a Landau–Ginsburg theory for detergents, which explains many of their very fascinating properties.

Consider the case of liquid crystals, these can also be written by choosing as an order parameter particular spherical harmonic functions.

For illustrative purposes, we give the dependence of the density ϕ on the angles and on the coordinates[1]

$$\phi = \sum_i \mu_i Y_2^2(\theta_i, \phi_i)e^{i\vec{k}_i \cdot \vec{r}_i} + h.c. \tag{2.27}$$

By assuming the ansatz $\mu_i = \mu$, the effective Landau–Ginsburg free energy is given below

$$F \sim (\alpha_0 + dk + ck^2)\mu^2 - \beta\mu^3 + r\mu^4, \tag{2.28}$$

from which one can extract the properties of nematic, smectic A and smectic C properties, and many other exciting things for which we refer to the literature [6].

2.3. Boundary perturbations

Next I discuss an example where a breakdown of spatial translational symmetry actually clearly occurs. As mentioned above one can formulate an intuitive theorem in the bulk, which states that under the renormalization group flow, the value of the Virasoro central charge c decreases from its UV value to a smaller IR value. This is due to the integrating out of the degrees of freedom and applies to the unitary sectors of String Theory. In String Theory with its ghosts the total central charge vanishes. One can imagine mechanisms by which the central charge of the ghosts increases [9], but basically one needs to couple the two systems maintaining a total vanishing central charge. This can be done for example with the help of a linear dilaton and leads to very interesting questions and results. Generically, the matter central charge will decrease to zero leaving one with just a $c = 0$ topological theory, but there are also other possibilities. The central charge is related to the anomaly which exists in the bulk. On the other hand, when one considers the boundary theory, there are no gravitational

---

[1] The index structure of ϕ has been omitted.

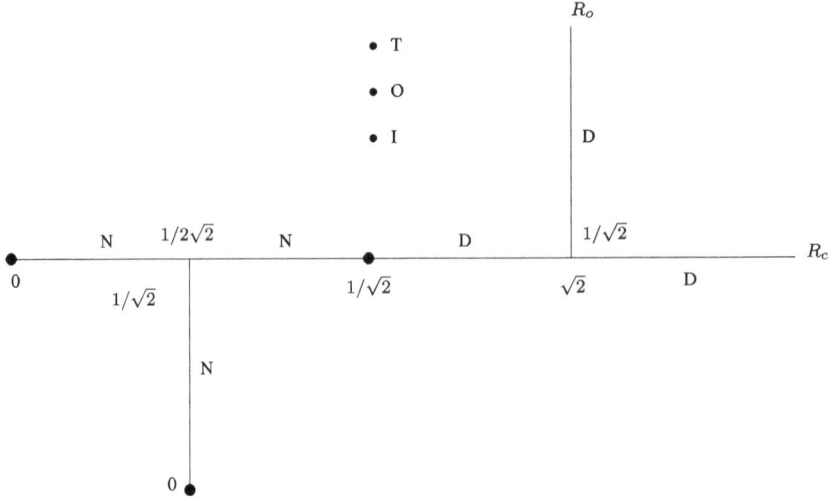

Fig. 12. Map of the preferred boundary conditions in the $c = 1$ moduli space, N stands for Neuman and D for Dirichlet boundary conditions [9].

anomalies in it. Thus in that case one can consider tachyonic open string theory perturbations. In the example given in the action below

$$ S = \int_{\Sigma} \mathcal{L}_{CFT} + \int_{\partial\Sigma} g\mathcal{O}_{Rel.}, \tag{2.29} $$

the bulk theory is defined on the surface Σ and on its boundary $\partial\Sigma$ one adds a relevant operator $\mathcal{O}_{Rel.}$. There is a boundary renormalization group flow, which does not change the bulk central charge and therefore does not lead to all the problems associated with tachyons in the bulk.

One can associate a term in the boundary which measures the effective number of degrees of freedom, this has been done by various authors [10–12].

One can prove moreover that one can define such a function whose value also decreases when the theory flows on the boundary. All this while not requiring an adjustment of the total central charge. What happens for example is that the theory flows from Neuman(N) to Dirichlet(D) boundary conditions [13], or in other words the branes may dissolve or may be created under such a flow. In the figure below (Fig. 12) we give an example of a very simple compactification for which one can identify what are the stable configurations, describing when the system chooses to obey Dirichlet and when the system chooses to obey Neuman boundary conditions [9].

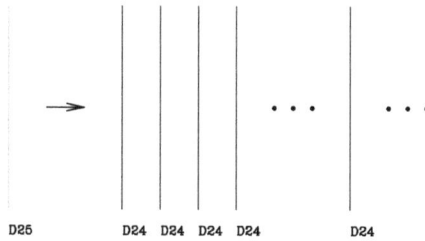

Fig. 13. A lattice of D24 branes is formed from a D25 brane in the presence of a boundary tachyon.

This can be used even further if one changes the relevant operator added on the boundary into a sine-Gordon one. In that case one can actually have situations where one breaks translational invariance in space-time by a $D - 25$ brane for example dissolving into a lattice of $D - 24$ branes [14], (Fig. 13). Again such a situation will lead also to a reduction of the original amount of supersymmetry. Thus the idea of spontaneous breaking of spatial translational invariance is demonstratively realized in String Theory by the presence of open string tachyons.

3. Spontaneous breaking of time translational invariance and of supersymmetry

Next I will discuss a somewhat different mechanism which may allow the possibility of a spontaneous breaking of time-translational invariance. For that it is useful to consider conformal and superconformal quantum mechanics. One way to motivate the interest in such systems is to recall some basic facts concerning the validity of a perturbative expansion.

Consider the Hamiltonian,

$$H = \frac{p_q^2}{2m} + \frac{1}{2}gq^n. \tag{3.1}$$

One may wonder if it is possible to make a meaningful perturbative expansion in terms of small or large g or small or large m. To answer this question one needs to find out if one can remove the g, m dependence from the operators, and relegate it to the total energy scale. This type of rescaling is used for discussing the harmonic oscillator. One attempts to define a new set of dimensionless canonical variables p_x, x that preserve the commutation relations

$$[p_q, q] = [p_x, x]\hbar, \tag{3.2}$$

and

$$H = h(m, g)\frac{1}{2}(p_x^2 + x^n). \tag{3.3}$$

The following decomposition:

$$q = f(m, g)x, \qquad p_q = \frac{1}{f(m, g)}p_x, \tag{3.4}$$

gives

$$2H = \frac{p_x^2}{mf^2(m, g)} + gf(m, g)^n x^n, \tag{3.5}$$

and so one may choose

$$gf(m, g) = \left(\frac{1}{mf(m, g)^2}\right)^{\frac{1}{n+2}}. \tag{3.6}$$

The Hamiltonian becomes:

$$H = g^{1-\frac{n}{n+2}} m^{-\frac{n}{n+2}} \frac{1}{2}(p_q^2 + q^n). \tag{3.7}$$

The role of g and m is indeed just to determine the overall energy scale. They may not serve as meaningful perturbation parameters.

This does not apply to the special case of $n = -2$, the case of conformal quantum mechanics, where g can be a real perturbative parameter.

3.1. Conformal quantum mechanics: A stable system with no ground state breaks time translational invariance

Consider the Hamiltonian

$$H = \frac{1}{2}(p^2 + gx^{-2}) \tag{3.8}$$

for a positive value of g [15].

H is part of the following algebra:

$$[H, D] = iH, \quad [K, D] = iK, \quad [H, K] = 2iD. \tag{3.9}$$

It is an $SO(2,1)$ algebra, one representation of which is:

$$D = -\frac{1}{4}(xp + px), \quad K = \frac{1}{2}x^2 \tag{3.10}$$

with H is given above. The Casimir is given by:

$$\frac{1}{2}(HK + KH) - D^2 = \frac{g}{4} - \frac{3}{16}. \tag{3.11}$$

In the Lagrangian formalism the system is described by:

$$\mathcal{L} = \frac{1}{2}\left(\dot{x}^2 - \frac{g}{x^2}\right), \quad S = \int dt \mathcal{L}. \tag{3.12}$$

Symmetries of the action S, and not of the Lagrangian $\mathcal{L}$ alone, are given by:

$$t' = \frac{at + b}{ct + d}, \quad x'(t') = \frac{1}{ct + d}x(t), \tag{3.13}$$

$$A = \begin{pmatrix} a & b \\ c & d \end{pmatrix}, \quad \det A = ad - bc = 1 \tag{3.14}$$

H acts as translation

$$A_T = \begin{pmatrix} 1 & 0 \\ \delta & 1 \end{pmatrix}, \quad t' = t + \delta. \tag{3.15}$$

D acts as dilation

$$A_D = \begin{pmatrix} \alpha & 0 \\ 0 & \frac{1}{\alpha} \end{pmatrix}, \quad t' = \alpha^2 t. \tag{3.16}$$

K acts as a special conformal transformation

$$A_K = \begin{pmatrix} 1 & \delta \\ 0 & 1 \end{pmatrix}, \quad t' = \frac{t}{\delta t + 1}. \tag{3.17}$$

The spectrum of the Hamiltonian (3.8) is the open set $(0, \infty)$, the spectrum is therefore continuous and bounded from below. The wave functions are given by:

$$\psi_E(x) = \sqrt{x} J_{\sqrt{g+\frac{1}{4}}}(\sqrt{2E}x), \quad E \neq 0. \tag{3.18}$$

The zero energy state is given by $\phi(x) = x^\alpha$:

$$H = \left(-\frac{d^2}{dx^2} + \frac{g}{x^2}\right)x^\alpha = 0. \tag{3.19}$$

This implies

$$g = -\alpha(\alpha - 1), \tag{3.20}$$

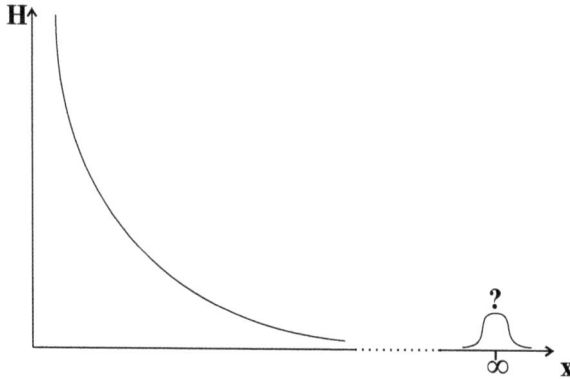

Fig. 14. Their no normalisable ground state for this potential.

solving this equation gives

$$\alpha = -\frac{1}{2} \pm \frac{\sqrt{1+4g}}{2}. \tag{3.21}$$

This gives rise to two independent solutions and by completeness these are all the solutions. The case $\alpha_+ > 0$, does not lead to a normalizable solution since the function diverges at infinity. $\alpha_- < 0$, is not normalizable either since the function diverges at the origin (a result of the scale symmetry).

Thus, there is no normalizable (not even plane wave normalizable) $E = 0$ solution!

Most of the analysis in field theory proceeds by identifying a ground state and the fluctuations around it. How do we deal with a system in the absence of a ground state?

One possibility is to accept this as a fact of life. Perhaps it is possible to view this as similar to cosmological models that also lack a ground state, such those with Quintessence. In field theory such systems have no finite energy states in the spectrum at all. Only time dependent states are allowed. In the presence of an appropriate cutoff and in quantum mechanics it is only the potential lowest energy state which is disallowed.

Another possibility is to define a new evolution operator that does have a ground state

$$G = uH + vD + wK. \tag{3.22}$$

This operator has a ground state if $v^2 - 4uw < 0$. Any choice explicitly breaks scale invariance. Take for example

$$G = \frac{1}{2}\left(\frac{1}{a}K + aH\right) \equiv R, \tag{3.23}$$

a has the dimension of a length. The eigenvalues of R are

$$r_n = r_0 + n, \qquad r_0 = \frac{1}{2}\left(1 + \sqrt{g + \frac{1}{4}}\right). \tag{3.24}$$

This is a breaking of scale invariance by a dictum and not by the dynamics of the system. Nevertheless it is very interesting to search for a physical interpretation of this. Surprisingly this question arises in the context of black hole physics [17]. Consider a particle of mass m and charge q falling into a charged black hole. The black hole is BPS, meaning that its mass M and charge Q are related, in the appropriate unites, by $M = Q$.

The blackhole metric and vector potential are given by:

$$ds^2 = -\left(1 + \frac{M}{r}\right)^{-2} dt^2 + \left(1 + \frac{M}{r}\right)^2 (dr^2 + r^2 d\Omega^2), \qquad A_t = \frac{r}{M}. \tag{3.25}$$

Now consider the near Horizon limit, i.e. $r \ll M$, which we will reach by taking $M \to \infty$ and keeping r fixed. This produces an $AdS_2 \times S^2$ geometry

$$ds^2 = -\left(\frac{r}{M}\right)^2 dt^2 + \left(\frac{M}{r}\right)^2 dr^2 + M^2 d\Omega^2. \tag{3.26}$$

We also wish to keep $M^2(m - q)$ fixed as we scale M. This means we must scale $(m - q) \to 0$, that is the particle itself becomes BPS in the limit.

The Hamiltonian for this falling in particle in this limit is given by our old friend:

$$H = \frac{p_r^2}{2m} + \frac{g}{2r^2}, \qquad g = 8M^2(m - q) + \frac{4l(l + 1)}{M}. \tag{3.27}$$

For $l = 0$, we have $g > 0$ and there is no ground state. This is associated with the coordinate singularity at the Horizon. The change in evolution operator is now associated with a change of time coordinate. One for which the world line of a static particle passes through the black hole horizon instead of remaining in the exterior of the space time. In any case, the consequence of removing the potential lowest energy state of the system from the spectrum can be described as a breaking of time translational invariance.

3.2. Superconformal quantum mechanics: A stable system with no ground state breaks also supersymmetry

The bosonic conformal mechanical system had no ground state. The absence of a $E = 0$ ground state in the supersymmetric context leads to the breaking of super-symmetry. This breaking has a different flavor from that which was discussed for the spatial translations. We next examine the supersymmetric version of confor-mal quantum mechanics [1, 16], to see if indeed supersymmetry is broken. The superpotential is chosen to be

$$W(x) = \frac{1}{2}g \, \log x^2, \tag{3.28}$$

yielding the Hamiltonian:

$$H = \frac{1}{2}\left[\left(p^2 + \left(\frac{dw}{dx}\right)^2\right)1 - \frac{d^2W}{dx^2}\sigma_3\right]. \tag{3.29}$$

Representing ψ by $\frac{1}{2}\sigma_-$ and ψ^* by $\frac{1}{2}\sigma_+$ gives the supercharges:

$$Q = \psi^+\left(-ip + \frac{dW}{dx}\right), \quad Q^+ = \psi\left(ip + \frac{dw}{dx}\right). \tag{3.30}$$

One now has a larger algebra, the superconformal algebra,

$$\{Q, Q^+\} = 2H, \quad \{Q, S^+\} = g - B + 2iD,$$
$$\{S, S^+\} = 2K, \quad \{Q^+, S\} = g - B - 2iD. \tag{3.31}$$

A realization is:

$$B = \sigma_3, \quad S = \psi^+ x, \quad S^+ = \psi x. \tag{3.32}$$

The zero energy solutions are

$$\exp(\pm W(x)) = x^{\pm g}, \tag{3.33}$$

neither solution is normalizable.

H factorizes:

$$2H = \begin{pmatrix} p^2 + \frac{g(g+1)}{x^2} & 0 \\ 0 & p^2 + \frac{g(g-1)}{x^2} \end{pmatrix}, \tag{3.34}$$

and we may solve for the full spectrum:

$$\psi_E(x) = x^{1/2}J_{\sqrt{v}}(x\sqrt{2E}), \quad E \neq 0, \tag{3.35}$$

where $v = g(g - 1) + 1/4$ for $N_F = 0$ and $v = g(g + 1) + 1/4$ for $N_F = 1$.

The spectrum is continuous and there is no normalizable zero energy state. One must interpret the absence of a normalizable ground state. It is also possible to define a new operator which has a normalizable ground state. By inspection the operator (3.23) can be used, provided one makes the following identifications:

$$N_F = 1, \quad g_B = g_{susy}(g_{susy} + 1),$$
$$N_F = 0, \quad g_B = g_{susy}(g_{susy} - 1). \tag{3.36}$$

Thus the spectrum differs between the $N_F = 1$ and $N_F = 0$ sectors and super-symmetry would be broken. One needs to define a whole new set of operators:

$$M = Q - S, \quad M^+ = Q^+ - S^+,$$
$$N = Q^+ + S^+, \quad N^+ = Q + S^+, \tag{3.37}$$

which produces the algebra:

$$
\begin{aligned}
\frac{1}{4}\{M, M^+\} &= R + \frac{1}{2}B - \frac{1}{2}g \equiv T_1, \\
\frac{1}{4}\{N, N^+\} &= R + \frac{1}{2}B + \frac{1}{2}g \equiv T_2, \\
\frac{1}{4}\{M, N\} &= L_-, \quad \frac{1}{4}\{M^+, N^+\} = L_+, \\
L_\pm &= -\frac{1}{2}(H - K \mp 2iD).
\end{aligned}
\tag{3.38}
$$

T_1, T_2, H have a doublet spectra. "Ground states" are given by:

$$T_1|0> = 0; \quad T_2|0> = 0; \quad H|0> = 0. \tag{3.39}$$

In this setup one can also exhibit [1] how the in the presence of a breaking of a space time symmetry, global $\mathcal{N} = 2$ can be broken only to $\mathcal{N} = 1$. A physical context arises when one considers a supersymmetric particle falling into a black hole [17, 18]. This is the supersymmetric analogue of the situation already discussed.

One should mention again that there is a dictum in the way one has broken scale/conformal invariance in the problem. It is amusing to mention that if one takes the dictated ground state and decomposes it in terms of the energy eigenstates, then one usually gets that the new ground state looks like a thermal distribution of the old ground states. This looks very attractive and it is related to black holes, which as mentioned above do come up.

Another example where such breakdown of time-translational invariance may occur is the Liouville model. Also there, there is no normalizable ground state.

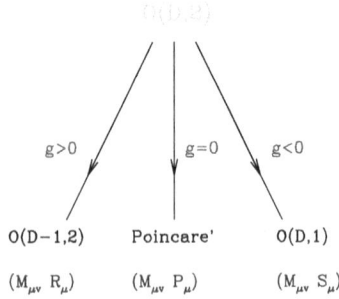

Fig. 15. The sign of the quartic coupling g determines the symmetry breaking patterns of the symmetry group $O(d, 2)$.

For works on the possible breakdown of translational invariance in the two dimensional Liouville model, see [19, 20].

Beyond $d = 2$, we can also mention that in four dimensions in $\mathcal{N} = 1$ supersymmetric theories, where the number of flavors N_F is smaller then the number of colors $0 < N_F < N_C$, one gets [21, 22]. This is another situation where the spectrum is bounded from below but there is no ground state (Fig. 15). The spectrum is open, and actually in the presence of a cutoff such systems have no finite energy states at all, which is a very interesting as far as Cosmology is concerned.

4. Spontaneous breaking of conformal invariance

Fubini also suggested to discuss such situations in a general number of dimensions [23]. He researched it in a scientific environment which did not yet fully realize that interacting finite theories might exist in various number of dimensions. Therefore much of his analysis was of a classical nature. He emphasized the conformal features of the system, and we are going to discuss the breakdown of conformal invariance. The discussion of the breakdown of time translational invariance brought us to conformal theories and now we are discussing also the breakdown of the conformal invariance.

If one considers a theory with only one scalar field, a general classic conformal invariant is given by the following Lagrangian

$$\mathcal{L} = \frac{1}{2} \partial_\mu \phi \partial^\mu \phi - g \phi^{\frac{2d}{d-2}}. \tag{4.1}$$

The symmetry of the system is the bosonic, $O(d, 2)$ symmetry, and the generator are $M_{\mu\nu}$, P_μ, of the Poincaré group, the special conformal transformation generator K_μ, and the dilatation D. The dictum of Fubini in this case is that the ground state is not translational invariant, this is not accompanied by any dynamical calculation. The vacuum expectation value $< \phi(x) >$, is x dependent, and actually it looks very much like an instanton

$$< \phi(x) >= b \left(\frac{a^2 + x^2}{2a} \right)^{-\frac{d-2}{2}}, \tag{4.2}$$

which is a solution of the equation of motion

$$\partial^2 \phi(x) - 2g \frac{d}{d-2} \phi^{\frac{d+2}{d-2}}(x) = 0. \tag{4.3}$$

By choosing this to be the vacuum, (again I emphasize, this is by dictum), one breaks down the $O(d, 2)$ symmetry, (as in Fig. 15) in the following fashion: if the coupling g of the scalar self-interaction is positive the theory breaks down to $O(d - 1, 2)$ and the resulting symmetries are $M_{\mu\nu}$, R_μ. If $g < 0$ the symmetry breaks to $O(d - 1)$, generated by $M_{\mu\nu}$, S_μ, where

$$S_\mu = \frac{1}{2} \left(aP_\mu - \frac{1}{a} K_\mu \right). \tag{4.4}$$

If $g = 0$ one remains with Poincaré invariance, (Fig. 15). In the de Sitter example, which occurs for $g > 0$, one can show again that there are signatures of temperature. A question which at the time seemed interesting was: Does a spontaneous breaking of conformal invariance require also the breakdown of translational invariance? Examples were since found where this is not the case. Counter examples to the idea that the breaking of conformal invariance must drive a breaking of Supersymmetry were discovered, We will discuss in more detail some such examples. One can break scale invariance without breaking rotational or translational invariance. We also mention briefly that conformal invariance and scale invariance are not always equivalent, and in a set of works,(see for example [24]), one can show that scale invariance leads under some certain conditions to conformal invariance.

That is the case when the spectrum of the theory is discrete, such as for a two dimensional sigma model description in which the target space is compact. But for non-compact target spaces one can find counter examples [25], in which scale invariance does not lead to conformal invariance. In recent years it has been fully realized that theory which are quantum mechanically scalar invariant and finite may exist in $d = 2, 3, 4, 5, 6$ dimensions. Such theories can exhibit spontaneous breaking, for example the $d = 4$, $\mathcal{N} = 4$ Super Yang–Mils with $SU(N)$ gauge group is characterized by the following spectrum

$$(A_\mu^a, \lambda^a, \phi^a + i\varrho^a).$$

The theory is parameterized by the complex parameter $ig + \theta$, where g is the coupling constant and θ is the θ angle. Such a theory has flat directions which allow phases where either $< \phi >$ vanishes and the theory is realized in a conformal manner, or a phase in which $< \phi > \neq 0$ along flat directions. This is the Coulomb phase, in which the gauge group $SU(N)$ may be reduced all the way to $U(1)^N$, where N is the rank of the gauge group. This is the maximum possible breaking of the gauge group when the fields are in the adjoint representation. In such a case, scale invariance is broken spontaneously and the vacuum energy remains zero, and there is no breakdown of either translational invariance or Supersymmetry. Such a theory will have a Goldstone boson, associated with the spontaneous breaking of scale invariance, which is called the dilaton. This is a *true* dilaton worthy of his name. It is interesting to note that in such a system the vacuum energy is not influenced by the value of $< \phi >$, and it vanishes in all the phases.

5. $O(N)$ vector models in $d = 3$: Spontaneous breaking of scale invariance and the vacuum energy

The next example that we have is related to the spontaneous breaking of scale invariance in a three dimensional bosonic theory. Such a theory describes the mixing of He_3 and He_4, (see [26] and references there in).

The most general Lagrangian describing such a system is

$$\mathcal{L} = \frac{1}{2}\partial_\mu \vec{\phi}\,\partial^\mu \vec{\phi} - \frac{1}{2}\lambda_2(\vec{\phi})^2 + \frac{\lambda_4}{4N}(\vec{\phi})^4 + \frac{\lambda_6}{N^2}(\vec{\phi})^6, \tag{5.1}$$

and it can be treated at $d = 3 - \varepsilon$. The system has two order parameters $< (\vec{\phi})^2 >$, and $< \vec{\phi} >$.

In a classical analysis performed for $d = 3 - \varepsilon$, when the sign of λ_2 changes then $< \vec{\phi} >$ is produced. However, $< (\vec{\phi})^2 > \neq 0$ even for $\lambda_2 > 0$, which is exemplified by the diagram shown in Fig. 16.

When one goes to three dimensions, the point which is denoted by CP, which is a critical point, and the point CEP which is the critical end point, do actually meet together and lead to a very interesting structure. Going directly to $d = 3$, one can write down the $O(N)$ vector model written below

$$\mathcal{L} = \frac{1}{2}\partial_\mu \vec{\phi}\,\partial^\mu \vec{\phi} - \frac{1}{2}\lambda_2(\vec{\phi})^2 + \frac{\lambda_4}{4!N}(\vec{\phi})^4 + \frac{\lambda_6}{6!N^2}(\vec{\phi})^6. \tag{5.2}$$

It should be emphasized that everything said depends on the very specific manner of taking the limit. One first keeps the cutoff Λ fixed and takes $N \to \infty$, by performing a functional integral or selecting a subset of diagrams, and only

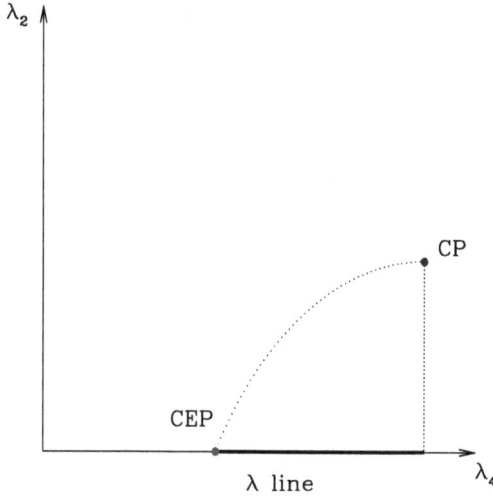

Fig. 16. The Phase Diagram of a $d = 3 - \varepsilon$ Conformal Theory, in three dimensions the *CP* and *CEP* points coincide to produce a flat direction, see [26] for more details.

then does one remove the cutoff, sending it to infinity, setting the renormalized quadratic and quartic couplings to zero. Such a system turns out to be not only classically conformally invariant, but also quantum mechanically, having a vanishing beta function [27]. We next elaborate on such systems.

Let us now review some more known facts about the three dimensional theory once a classically marginal operator, $(\vec{\phi}^2)^3$, is added [27]. For any finite value of N, the coupling g_6 of this operator is infrared free quantum mechanically, as the marginal operator gets a positive anomalous dimension already at one loop. This implies that the theory is only well defined for zero value of the coupling of this operator. In the presence of a cutoff interacting particles have mass of the order of the cutoff. At its tri-critical point the $O(N)$ model in three dimensions is described by the Lagrangian

$$\mathcal{L} = \frac{1}{2} \partial_\mu \vec{\phi} \, \partial^\mu \vec{\phi} + \frac{1}{6N^2} g_6 (\vec{\phi}^2)^3 \,, \tag{5.3}$$

where the fields $\vec{\phi}$ are in the vector representation of $O(N)$.

In the limit $N \to \infty$ [27]

$$\beta_{g_6} = 0. \tag{5.4}$$

$1/N$ corrections break conformality. In the large N limit, then, g_6 is a modulus. It turns out that there is no spontaneous breaking of the $O(N)$ symmetry and it is instructive to write the effective potential in terms of an $O(N)$ invariant field,

$$\sigma = \vec{\phi}^2. \tag{5.5}$$

The effective potential is [26]:

$$V(\sigma) = f(g_6)|\sigma|^3, \tag{5.6}$$

where:

$$f(g_6) = g_c - g_6 \tag{5.7}$$

with

$$g_c = (4\pi)^2. \tag{5.8}$$

The system has various phases. For values of g_6 smaller than g_c, i.e. when $f(g_6)$ is positive, the system consists of N massless non-interacting ϕ particles. These particles do not interact in the infinite N limit; thus, correlation functions do not depend on g_6. For the special value $g_6 = g_c$, $f(g_6)$ vanishes and a flat direction in σ opens up: the expectation value of σ becomes a modulus. For a zero value of this expectation value, the theory continues to consist of N massless ϕ fields. For any non-zero value of the expectation value the system has N massive ϕ particles. All have the same mass due to the unbroken $O(N)$ symmetry. Scale invariance is broken spontaneously though the vacuum energy still vanishes. The Goldstone boson associated with the spontaneous breaking of scale invariance, the dilaton, is massless and identified as the $O(N)$ singlet field $\delta\sigma \equiv \sigma - \langle\sigma\rangle$. All the particles are non-interacting in the infinite N limit. This theory is not conformal: in the infrared limit, it flows to another theory containing a single, massless, $O(N)$-singlet particle. For larger values of g_6 the exact potential is unbounded from below. The system is unstable (in the supersymmetric case the potential is bounded from below and the larger g_6 structure is similar to the smaller g_6 structure [28]). This case is useful to illustrate the fate of some gravitational instabilities [29]. Actually this instability is an artifact of the dimensional regularization used above, which does not respect the positivity of the renormalized field σ. In any case a more careful analysis [27] shows that the apparent instability reflects the inability to define a renormalizable interacting theory. All the masses are of the order of the cutoff, and there is no mechanism to scale them down to low mass values. In other words, the theory depends strongly on its UV completion.

This is summarized in Table 1.

Table 1

Marginal perturbations of the $O(N)$ model

$f(g_6)$	$|\langle\sigma\rangle|$	S.B.	masses	V
$f(g_6) > 0$	0	No	0	0
$f(g_6) = 0$	0	No	0	0
$f(g_6) = 0$	$\neq 0$	Yes	Massless dilaton, N particles of equal mass	0
$f(g_6) < 0$	∞	Yes, but ill defined	Tachyons or masses of order the cutoff	$-\infty$

There, S.B. denotes spontaneous symmetry breaking of scale invariance and V is the vacuum energy. For $f(g_6) < 0$ the theory is unstable. Note that the vacuum energy always vanishes whenever the theory is well-defined.

When $\langle\sigma\rangle \neq 0$ and the scale invariance is spontaneously broken, one can write down the effective theory for energy scales below $\langle\sigma\rangle$, and integrate out the degrees of freedom above that scale. The vacuum energy remains zero however, and it is not proportional to $\langle\sigma\rangle^3$, [26, 30–32, 39], as might be expected naively.

For completeness we note that by adding more vector fields one has also phases in which the internal global $O(N)$ symmetry is spontaneously broken.

An example is the $O(N) \times O(N)$ model [32] with two fields in the vector representation of $O(N)$, with Lagrangian:

$$
\begin{aligned}
\mathcal{L} &= \partial_\mu\vec{\phi}_1 \cdot \partial^\mu\vec{\phi}_1 + \partial_\mu\vec{\phi}_2 \cdot \partial^\mu\vec{\phi}_2 + \lambda_{6,0}(\vec{\phi}_1^2)^3 \\
&\quad + \lambda_{4,2}(\vec{\phi}_1^2)^2(\vec{\phi}_2^2) + \lambda_{2,4}(\vec{\phi}_1^2)(\vec{\phi}_2^2)^2 + \lambda_{0,6}(\vec{\phi}_2^2)^3.
\end{aligned}
\tag{5.9}
$$

Again, the β functions vanish in the strict $N \to \infty$ limit. There are now two possible scales, one associated with the breakdown of a global symmetry and another with the breakdown of scale invariance. The possibilities are summarized by the table below:

$O(N)$	$O(N)$	scale	massless	massive	V	
+	+	+	all	none	0	
−	+	−	$(N-1)\pi's, D$	N, σ	0	(5.10)
+	−	−	$(N-1)\pi's, D$	N, σ	0	
−	−	−	$2(N-1)\pi's, D$	σ	0	

Again in all cases the vacuum energy vanishes. Assume a hierarchy of scales where the scale invariance is broken at a scale much above the scale at which the $O(N)$ symmetries are broken. One would have argued that one would have had a low energy effective Lagrangian for the massless pions and dilaton with a vacuum energy given by the scale at which the global symmetry is broken. This is not true, the vacuum energy remains zero. This system has a critical surface, on one patch the deep infrared theory contains only one massless particle: an $O(N) \times O(N)$ singlet. For the other patches the deep infrared theory is described by $O(N)$ massless particles, most of which are not $O(N)$ singlets. The dimension of the surface for which spontaneous symmetry breakings occur is smaller then that of the full space of parameters. I will not consider spontaneous symmetry breaking as fine tuning.

In general, effective field theories should have all possible symmetries of the underlying theory, whether they are realized linearly or non-linearly. In finite scale invariant theories the vacuum energy E_{vac} should be determined by all scales and symmetries involved. It should have the same value, (zero in this case), in all phases of the system whether or expectation values are formed. This punches a hole in Zaldowitch like arguments [33], and offers a different view on the gravity of the Cosmological Constant problem [34]. If the theory has a global scale invariance, which is spontaneously broken, it will produce a dilaton. The question is: Where is the dilaton? The dilaton should be a massless field. Several authors, [35, 36] tried to check the possibilities that the dilatons might exist, noting that the dilaton must be a massless Goldstone boson. Under certain assumptions, one finds out that actually in certain models having a massless dilaton would not violate experimental data. Perhaps it even predicts deviations of the equivalence principle from Galileo famous experiment just below the present experimental sensitivity $\delta a / a \sim 10^{-12}$.

This is done under the assumption that the dilaton couples in the following universal fashion

$$\mathcal{L} = F(\Phi)\big(R - F^2 + 2[\nabla^2 \Phi - (\nabla \Phi)^2]\big). \tag{5.11}$$

It could also happen that the dilaton gets swallowed in some Higgs like mechanism. One should also mention that if kinematically a finite scale-invariant is forced by some super-selection rule (such as having a non trivial monopole number [40]) into a certain solitonic sector, then the rest energy of the system should be accounted for and the vacuum energy will be slightly lifted from zero. Let us finish this section by noticing an amusing thing, there are various solutions that go under the name of Randall and Sundrum [37]. One of the constructions contains two types of branes, near the boundary of the space there is a Planck brane with tension T_1, which is fine tuned so to have zero Cosmological Constant. Then at a certain distance, very deep inside the bulk theory, one places the TeV brane, it has negative tension and the tension is again fine tuned, so that the Cosmological Constant vanishes also on that brane.

The two branes are separated by some distance which in [38] is associated to massless particle, which is the dilaton or the radion, (see Fig. 17).

Fig. 17. Planck brane and TeV brane.

In principle, there are circumstances where this distance is not fixed, and there are several possible situations whose outcome is very similar to that one discussed in the $d = 3$ conformal theory. If the sum of the tensions $T_1 + T_2$ is arranged to vanish, then the system behaves as a spontaneously broken system, the magnitude of the vev of the field is the distance between the two branes.

If $T_1 + T_2 > 0$ the two branes actually are attracted to sit one on top of the other, and when $T_1 + T_2 < 0$ the branes repel, the system is unstable and as a result one of the branes is exiled to infinity.

These three examples are in full correspondence with the conditions on the coefficients of the $(\vec{\phi})^6$ theory that we discussed above. The difference between the two theories, and an important difference is that in case of the $(\vec{\phi})^6$ theory we are certain that in the large N limit, the theory is indeed finite quantum mechanically.

For the case of $(\vec{\phi})^4$ we don't have such an assurance, and it would be nice to find a system for which we are guaranteed to be finite also quantum mechanically, which exhibits the same type of behavior.

5.1. Conclusions

• Spontaneous breaking of translational and rotational symmetry are possible. It fits data for many phases of matter, and it may have a manifestation in the dynamics of compactification.

• Conformal/Scale invariant theories which are stable but have no ground states indicate a new mechanism of breaking time translational invariance as well as supersymmetry.

• A finite scale invariant theory has the same (vanishing) vacuum energy in all its phases.

Acknowledgements

The author thanks Matteo Cardella for various discussions on this manuscript. The author wishes to thank his various collaborators on these subjects, especially

W. Bardeen, S. Elitzur, M. Einhorn, A. Forge, A. Giveon, M. Porrati, A. Schwimmer, and G. Veneziano. A very similar but not identical version of this work was presented as a contribution to "String Theory and Fundamental Interactions. Celebrating Gabriele Veneziano on his 65th Birthday", edited by M. Gasperini and J. Maharana.

References

[1]   S. Fubini and E. Rabinovici, *Nucl. Phys. B* **245** (1984) 17.

[2]   J. Hughes and J. Polchinski, *Nucl. Phys. B* **278** (1986) 147.

[3]   L.D. Landau, *Phys. Z. Soviet II* **26** (1937) 545.

[4]   S. Elitzur, A. Forge and E. Rabinovici, Some global aspects of string compactifications, *Nucl. Phys. B* **359** (1991) 581.

[5]   C. Baym, H.A. Bethe and C.J. Pethick, *Nucl. Phys. A* **175** (1971) 25.

[6]   S. Alexander, *Symmetries and Broken Symmetries in Condensed Matter Physics*, ed. N. Boccara (IDSET, Paris, 1981) p. 141, and references therein.

[7]   A.B. Zamolodchikov, *Sov. Phys. JETP Lett.* **43** (1986) 730.

[8]   A.W.W. Ludwing and J.L. Cardy, *Nucl. Phys. B* **285** [FS19] (1987) 687.

[9]   S. Elitzur, E. Rabinovici and G. Sarkissian, *Nucl. Phys. B* **541** (1999) 246.

[10] I. Affleck and A.W.W. Ludwig, *Phys. Rev. Lett.* **67** (1991) 161.

[11] S.L. Shatashvili, *Alg. Anal.* **6** (1994) 215.

[12] D. Friedan and A. Konechny, *Phys. Rev. Lett.* **93** (2004) 030402 [arXiv:hep-th/0312197].

[13] E. Witten, *Phys. Rev. D* **47** (1993) 3405.

[14] J.A. Harvey, D. Kutasov and E.J. Martinec, On the relevance of tachyons, arXiv:hep-th/0003101.

[15] V. de Alfaro, S. Fubini and G. Furlan, *Nuovo Cim. A* **34** (1976) 569.

[16] V.P. Akulov and A.I. Pashnev, *Theor. Math. Phys.* **56** (1983) 862 [*Teor. Mat. Fiz.* **56** (1983) 344].

[17] P. Claus, M. Derix, R. Kallosh, J. Kumar, P.K. Townsend and A. Van Proeyen, *Phys. Rev. Lett.* **81** (1998) 4553.

[18] R. Kallosh, Black holes and quantum mechanics, arXiv:hep-th/9902007.

[19] E. D'Hoker and R. Jackiw, *Phys. Rev. Lett.* **50** (1983) 1719.

[20] C.W. Bernard, B. Lautrup and E. Rabinovici, *Phys. Lett. B* **134** (1984) 335.

[21] T.R. Taylor, G. Veneziano and S. Yankielowicz, *Nucl. Phys. B* **218** (1983) 493.

[22] I. Affleck, M. Dine and N. Seiberg, *Nucl. Phys. B* **241** (1984) 493.

[23] S. Fubini, A new approach to conformal invariant field theories, CERN-TH-2129, Feb 1976. 46 pp. Published in *Nuovo Cim.* **A34** (1976) 521.

[24] J. Polchinski, *Nucl. Phys.* **303** (1988) 226.

[25] S. Elitzur, A. Giveon, E. Rabinovici, A. Schwimmer and G. Veneziano, *Nucl. Phys. B* **435** (1995) 147.

[26] D.J. Amit and E. Rabinovici, *Nucl. Phys. B* **257** (1985) 371.

[27] W.A. Bardeen, M. Moshe and M. Bander, *Phys. Rev. Lett.* **52** (1984) 1188.

[28] W.A. Bardeen, K. Higashijima and M. Moshe, *Nucl. Phys. B* **250** (1985) 437.

[29] S. Elitzur, A. Giveon, M. Porrati and E. Rabinovici, *JHEP* **0602** (2006) 006.

[30] D.S. Berman and E. Rabinovici, Supersymmetric gauge theories, arXiv:hep-th/0210044.

[31] M.B. Einhorn, G. Goldberg and E. Rabinovici, *Nucl. Phys. B* **256** (1985) 499.

[32] E. Rabinovici, B. Saering and W.A. Bardeen, *Phys. Rev. D* **36** (1987) 562.

[33] Ya.B. Zeldovich, *Sov. Phys. Uspekhi* **11** (1968) 381.

[34] S. Weinberg, *Rev. Mod. Phys.* **61** (1989).

[35] T. Damour and A.M. Polyakov, *Nucl. Phys. B* **423** (1994) 532.

[36] T. Damour, F. Piazza and G. Veneziano, *Phys. Rev. D* **66** (2002) 046007.

[37] L. Randall and R. Sundrum, *Phys. Rev. Lett.* **83** (1999) 4690.

[38] R. Rattazzi and A. Zaffaroni, *JHEP* **0104** (2001) 021.

[39] E. Rabinovici, In: W.M. Alberico and S. Sciuto, *Symmetry and Simplicity in Physics. Proceedings, Symposium on the Occasion of Sergio Fubini's 65th Birthday*, Turin, Italy, February 24–26, 1994, Singapore (World Scientific 1994) p. 220.

[40] E. Rabinovici, unpublished.

Course 7

THE STANDARD MODEL FROM D-BRANES IN STRING THEORY

Angel M. Uranga

PH-TH CERN, CH-1211 Geneva 23, Switzerland
and
Instituto de Física Teórica UAM/CSIC, Facultad de Ciencias C-XVI, 28049 Madrid

C. Bachas, L. Baulieu, M. Douglas, E. Kiritsis, E. Rabinovici, P. Vanhove, P. Windey and L.F. Cugliandolo, eds.
Les Houches, Session LXXXVII, 2007
String Theory and the Real World: From Particle Physics to Astrophysics
© *2008 Published by Elsevier B.V.*

257

Contents

1. Introduction

String theory has the remarkable property that it provides a description of gauge and gravitational interactions in a unified framework consistently at the quantum level. It is this general feature (beyond other beautiful properties of *particular* string models) that makes this theory interesting as a possible candidate to unify our description of the different particles and interactions in Nature.

Now if string theory is indeed realized in Nature, it should be able to lead not just to 'gauge interactions' in general, but rather to gauge sectors as rich and intricate as the gauge theory we know as the Standard Model of Particle Physics. In these lecture we describe compactifications of string theory where sets of D-branes lead to gauge sectors close to the Standard Model. We furthermore discuss the interplay of such D-brane systems with flux compactifications, recently introduced to address the issues of moduli stabilization and supersymmetry breaking.

Before starting, it is important to emphasize that there are other constructions in string theory which are candidates to reproduce the physics of the Standard Model at low energies, which do not involve D-branes. For instance, compactifications of heterotic string on Calabi–Yau threefolds, M-theory compactifications on G_2-holonomy spaces, etc. We emphasize D-brane models because of their simplicity, and also because they are often related to these other compactifications via string dualities. Hence, they provide a simple introduction from which the interested reader may jump onto the big picture.

This first lecture introduces D-branes and their properties, and deals with model building using intersecting D-branes. Useful reviews for this lecture are for example [1].

These lectures are organized as follows. In Section 2 we quickly review properties of D-branes and their world-volume dynamics. In Section 3 we describe that configurations of intersecting D6-branes naturally lead to four-dimensional chiral fermions, and discuss their spectrum and supersymmetry. In Section 4 we construct compactifications of type IIA string theory to four dimensions, including configurations of intersecting D6-branes. We provide explicit descriptions of toroidal compactifications of this kind, and generalizations to more general Calabi–Yau compactifications. In Section 5 we introduce further ingredients to improve these models, namely orientifold 6-planes. We describe their properties, discuss configurations of D6-branes and O6-planes, and describe how to include

261

them in compactifications in Section 5.3. These techniques are exploited in Section 6 to construct models whose chiral spectrum is that of the standard model, and in Section 7 to describe supersymmetric chiral compactifications with intersecting branes. Appendix Appendix A provides some details on the computation of open string spectra for parallel and intersecting D-branes.

2. Overview of D-branes

2.1. Properties of D-branes

The study of string theory beyond perturbation theory has led to the introduction of new objects in string theory, D-branes. For a complementary description of D-branes and their properties see [2, 3].

Type II string theories contains certain 'soliton-like' states in their spectrum, with $p+1$ extended dimensions, the p-branes. They were originally found as solutions of the low-energy supergravity equations of motion. This is schematically shown in Fig. 1. Subsequently, it was realized [4] that certain of these objects (known as Dp-branes) admit a fully stringy description, as $(p+1)$-dimensional subspaces on which open strings can end. Notice that these open strings are not present in the vacuum of the underlying string theory, but rather represent the fluctuations of the theory around the topological defect background. Namely, the closed string sector still describes the dynamics of the vacuum (gravitational interactions, etc), while open strings rather describe the dynamics of the object. The situation is shown in Fig. 2.

The basic properties of Dp-branes for our purposes in these lecture are:

• Dp-branes are dynamical, and for instance have non-trivial interactions with closed string modes. Due to these couplings, they carry tension (they interact with the 10d graviton) and charge under a RR $(p+1)$-form potential C_{p+1}, see

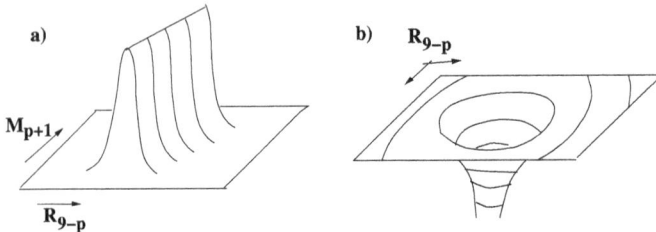

Fig. 1. Two pictures of the p-brane as a lump of energy. The second picture shows only the transverse directions, where the p-brane looks like point-like.

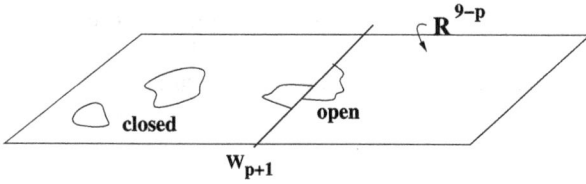

Fig. 2. String theory in the presence of a Dp-brane. The closed string sector describes the fluctuations of the theory around the vacuum (gravitons, dilaton modes, etc), while the sector of open strings describes the spectrum of fluctuations of the soliton.

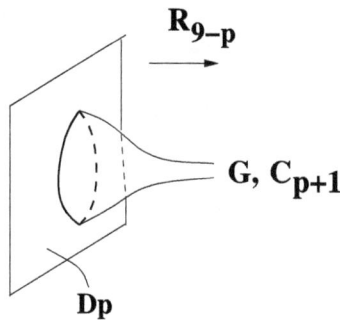

Fig. 3. Disk diagram describing the interaction of a Dp brane with closed string modes.

Fig. 3. Hence, type IIA (resp. IIB) string theory contains Dp-branes with p even (resp. odd).

• A flat Dp-brane in flat spacetime preserves half the supersymmetries of the theory. Denoting Q_L, Q_R the two 16-component spinor supercharges of type II string theories, arising from the left- or right-moving world-sheet degrees of freedom, a Dp-brane with world-volume spanning the directions $012\ldots p$ preserves the linear combination

$$Q = \epsilon_R Q_R + \epsilon_L Q_L \tag{2.1}$$

where $\epsilon_{L,R}$ are spinor coefficients satisfying

$$\epsilon_L = \Gamma^{01\ldots p}\epsilon_R \tag{2.2}$$

Thus Dp-branes are BPS states, and their charge and tension are equal.

• Dp-branes may have curved world-volumes. However, they tend to minimize the volume of the submanifold they span, hence in flat space Dp-branes tend to span flat world-volumes. In curved spaces, arising e.g. in compactifications, they may however wrap curved non-trivial homology cycles.

• As mentioned already, open string modes in the presence of D-branes are localized on the world-volume of the latter. This implies that such open strings represent the collective coordinates of the non-perturbative object, and thus their dynamics controls the dynamics of the object. In next section we will center on the zero modes, corresponding to the massless open string sector.

2.2. World-volume fields

The spectrum of fluctuations of the theory in the presence of the Dp-brane is obtained by quantizing closed strings and open strings ending on the Dp-brane. Since the open string endpoints are fixed on the D-brane, the massless modes in the latter sector yield fields propagating on the $(p + 1)$-dimensional D-brane world-volume W_{p+1}.

A simplified calculation of the quantization of open strings for a configuration of a single type II Dp-brane in flat 10d is carried out in appendix Appendix A.1. The resulting set of massless particles on the Dp-brane world-volume is given by a $U(1)$ gauge boson, $9 - p$ real scalars and some fermions (transforming under Lorentz as the decomposition of the $\mathbf{8}_C$ of $SO(8)$ under the $(p + 1)$-dimensional little group $SO(p - 1)$). The scalars (resp. fermions) can be regarded as Goldstone bosons (resp. Goldstinos) of the translational symmetries (resp. supersymmetries) of the vacuum broken by the presence of the D-brane. The open string sector fills out a $U(1)$ vector multiplet with respect to the 16 supersymmetries unbroken by the D-brane.

As mentioned above, Dp-branes are charged under the corresponding RR $(p + 1)$-form C_{p+1} of type II string theory, via the minimal coupling $\int_{W_{p+1}} C_{p+1}$. Since flat Dp-branes in flat space preserve 1/2 of the 32 supercharges of the type II vacuum, such D-branes are BPS states, and their RR charge is related to their tension. This implies that there is no net force among parallel branes (roughly, gravitational attraction cancels against 'Coulomb' repulsion due to their RR charge). Hence one can consider dynamically stable configurations of several parallel Dp-branes, labeled by a so-called Chan-Paton index a, at locations x_a^i in the transverse coordinates, $i = p + 1, \ldots, 9$.

We would like to consider the situation with n coincident Dp-branes, located at the same position in transverse space. In such situation there are n^2 open string sectors, labeled ab for an open string starting at the ath D-brane and ending at the bth D-brane. The computation for each sector ab is similar to the single brane case. Hence the spectrum of physical states contains, at the massless level, n^2 gauge bosons, $n^2 \times (9 - p)$ scalars, and n^2 sets of $(p + 1)$-dimensional fermions (in representations obtained from decomposing the $\mathbf{8}_C$ of $SO(8)$).

This multiplicity renders interactions between open strings non-Abelian. It is possible to see that the gauge bosons in the aa sector correspond to a $U(1)^n$ gauge

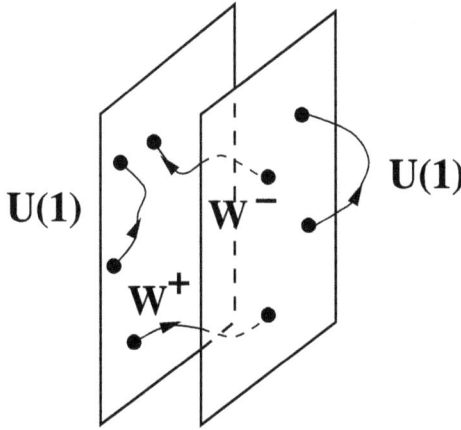

Fig. 4. Non-Abelian gauge bosons in a configuration of coincident D-branes.

symmetry, and that states in the ab sector have charges $+1$ and -1 under the ath and bth $U(1)$, respectively. This enhances the gauge symmetry to $U(n)$, and makes the different fields transform in the adjoint representation. The complete massless open string spectrum is given by $U(n)$ gauge bosons, $9 - p$ adjoint scalars and adjoint fermions, filling out a $U(n)$ vector multiplet with respect to the 16 unbroken supersymmetries. The structure of gauge bosons for $n = 2$ is shown in Fig. 4.

D-branes provide a nice and simple realization of non-Abelian gauge symmetry in string theory. The low-energy effective action for the massless open string modes has several pieces. One of them is the Dirac-Born-Infeld action, which has the form

$$S_{\text{DBI}} = -T_p \int_{W_{p+1}} \mathrm{d}^{p+1}x^\mu \left[-\det(G + B + 2\pi\alpha' F) \right]^{1/2} \tag{2.3}$$

where T_p is the Dp-brane tension, and $G_{\mu\nu} = \partial_\mu \phi^i \partial_\nu \phi^j \, G_{ij}$ is the metric induced on the D-brane worldvolume, and similarly $B_{\mu\nu}$ is the induced 2-form. These terms introduce the dependence of the action on the world-volume scalars $\phi^i(x^\mu)$. Finally $F_{\mu\nu}$ is the field strength of the worldvolume gauge field.

Neglecting the dependence on the field strength, it reduces to the D-brane tension times the D-brane volume $\int (\det G)^{1/2}$. At low energies, i.e. neglecting the α' corrections, it reduces to a kinetic term for the scalars plus the $(p + 1)$-dimensional Yang–Mills action for the worldvolume gauge fields, with gauge coupling given by $g^2_{U(n)} = g_s$. Of course the above action should include superpartner fermions, etc, but we skip their discussion.

A second piece of the effective action is the Wess–Zumino terms, of the form

$$S_{WZ} = -Q_p \int_{W_{p+1}} \mathcal{C} \wedge \mathrm{ch}(F) \, \hat{A}(R) \tag{2.4}$$

where $\mathcal{C} = C_{p+1} + C_{p-1} + C_{p-3} + \cdots$ is a formal sum of the RR forms of the theory, and $\mathrm{ch}(F)$ is the Chern character of the worldvolume gauge bundle on the D-brane volume

$$\mathrm{ch}(F) = \exp\left(\frac{F}{2\pi}\right) = 1 + \frac{1}{2\pi}\mathrm{tr}\,F + \frac{1}{8\pi^2}\mathrm{tr}\,F^2 + \cdots \tag{2.5}$$

and $\hat{A}(R)$ is the A-roof genus, characterizing the tangent bundle of the D-brane world-volume $\hat{A}(R) = 1 - \mathrm{tr}\,R^2/(2\pi^2) + \cdots$. Integration is implicitly defined to pick up the degree $(p + 1)$ pieces in the formal expansion in wedge products. Hence we get terms like

$$S_{WZ} = -Q_p \left(\int_{W_{p+1}} C_{p+1} + \frac{1}{2\pi} \int_{W_{p+1}} C_{p-1} \wedge \mathrm{tr}\,F \right.$$
$$\left. + \frac{1}{8\pi^2} \int_{W_{p+1}} C_{p-3} \wedge (\mathrm{tr}\,F^2 - \mathrm{tr}\,R^2) + \cdots \right) \tag{2.6}$$

A very important property of this term is that it is topological, independent of the metric or on the particular field representatives in a given topological sector. This is related to the fact that these terms carry the information about the RR charges of the D-brane configuration.

2.3. Chirality and D-branes

We have obtained simple configurations of D-branes leading to non-Abelian gauge symmetries on their world-volume. It is interesting to wonder if such configurations could be exploited to reproduce the gauge sector describing high energy particle physics, so as to embed it into a string theory model. Clearly, the main obstruction is that the standard model of particle physics is chiral in four dimensions. This property is incompatible with the large amount of supersymmetry preserved by the D-brane configurations considered.

There is an alternative heuristic way to intuitively understand the lack of chirality in our D-brane configuration. Four-dimensional chirality is a violation of four-dimensional parity. In the spectrum of open strings there is a correlation (implied by the GSO projection) between the 4d chirality and the chirality in the six extra dimensions. Hence to achieve 4d parity violation the configuration must violate 6d parity. However, the above configurations of D-branes do not violate 6d parity, do not introduce a preferred six-dimensional orientation.

The latter remark indeed suggest how to proceed to construct configurations of D-branes leading to four-dimensional chiral fermions. The requirement is that the configuration introduces a preferred orientation in the six transverse dimensions. There are several ways to achieve this, as we discuss now.

• D-branes sitting at *singular* (rather than smooth) points in transverse space can lead to chiral open string spectra. The prototypical example is given by stacks of D3-branes sitting at the singular point of orbifolds of flat space, e.g. orbifold singularities $\mathbf{C}^3/\mathbf{Z_N}$, as studied in [5]. A particularly simple and interesting case is the $\mathbf{C}^3/\mathbf{Z_3}$ orbifold, which will be studied in our second lecture. The key idea is that the discrete rotation implied by the $\mathbf{Z_3}$ action defined a preferred orientation in the 6d space, and allows for chirality on the D-branes.

• Consider a stack of D9-branes in flat 10d spacetime, split as $M_4 \times \mathbf{R}^2 \times \mathbf{R}^2 \times \mathbf{R}^2$. For simplicity we ignore for the moment the issue of RR tadpole cancellation. Otherwise, to make the configuration consistent it suffices to introduce orientifold 9-planes, namely consider the configuration in type I string theory. Now introduce non-trivial field strength background for the world-volume $U(1)_a$ gauge fields, F_a^i in the ith $\mathbf{R}^2$, with $i = 1, 2, 3$, (see [8,9] for early discussions, and [10–12] for more recent ones). The magnetic fields introduce a preferred orientation in the transverse six dimensions (obtained by using $F \wedge F \wedge F$ as the volume form, where F is the 2-form associated to the field strength). Hence the configurations lead naturally to 4d chiral fermions, as we describe in our second lecture.

• Sets of intersecting D-branes can also lead to chiral fermions in the sector of open strings stretched between different kinds of D-brane [13], and are the topic of our lecture today.

3. Intersecting D6-branes

3.1. Local geometry and spectrum

The basic configuration of intersecting D-branes leading to chiral 4d fermions at their intersection is two stacks of D6-branes in flat 10d intersecting over a 4d subspace of their volumes. Consider flat 10d space $M_4 \times \mathbf{R}^2 \times \mathbf{R}^2 \times \mathbf{R}^2$, and two stacks of D6-branes, spanning M_4 times a line in each of the three 2-planes. Figures 5, 6 provide two pictorial representations of the configurations. The local geometry is fully specified by the three angles θ_i which define the rotation between the two stacks of D6-branes. As we discuss below, the chiral fermions are localized at the intersection of the brane volumes.

The appearance of chirality can be understood from the fact that the geometry of the two D-brane introduces a preferred orientation in the transverse 6d space,

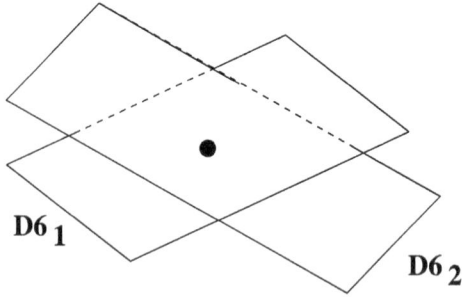

Fig. 5. Picture of D6-branes intersecting over a 4d subspace of their volumes.

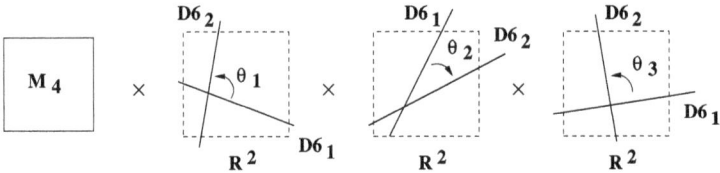

Fig. 6. A more concrete picture of the configuration of two D6-branes intersecting over a 4d subspace of their volumes.

namely by considering the relative rotation of the second D6-brane with respect to the first. This also explains why one should choose configurations of D6-branes. For example, two sets of D5-branes intersecting over 4d do not lead to 4d chiral fermions, since they do not have enough dimensions to define an orientation in the transverse 6d space.

A more detailed computation of the spectrum of open string models on systems of intersecting branes is provided in appendix Appendix A.2. Here it will suffice to mention the results of the spectrum for this configuration. The open string spectrum in a configuration of two stacks of n_1 and n_2 coincident D6-branes in flat 10d intersecting over a 4d subspace of their volumes consists of three open string sectors:

$6_1 6_1$ Strings stretching between D6$_1$-branes provide $U(n_1)$ gauge bosons, three real adjoint scalars and fermion superpartners, propagating over the 7d world-volume of the D6$_1$-branes.

$6_2 6_2$ Similarly, strings stretching between D6$_2$-branes provide $U(n_2)$ gauge bosons, three real adjoint scalars and fermion superpartners, propagating over the D6$_2$-brane 7d world-volume.

$6_1 6_2 + 6_2 6_1$ Strings stretching between both kinds of D6-brane lead to a 4d chiral fermion, transforming in the representation $(n_1, \bar{n}_2)$ of $U(n_1) \times U(n_2)$, and localized at the intersection. The chirality of the fermion is encoded in the orientation defined by the intersection; this will be implicitly taken into account in our discussion.

So we have succeeded in constructing a configuration of D-branes leading to 4d chiral fermions in the open string sector. Again, let us emphasize that the appearance of chiral fermions in the present system is the angles between the branes (technically, leading to the reduction of the Clifford algebra of fermion zero modes in the open strings between branes). Notice that the 4d chiral fermions lead to a localized anomaly at the intersection of the D6-branes. This anomaly is however canceled by the anomaly inflow mechanism, see [14].

In addition to the chiral fermions at intersections, there are several potentially light complex scalars at the intersection, transforming in bifundamental representations, and with masses (in α' units) given by

$$
\begin{array}{ll}
\dfrac{1}{2\pi}(-\theta_1 + \theta_2 + \theta_3) & \dfrac{1}{2\pi}(\theta_1 - \theta_2 + \theta_3) \\[3mm]
\dfrac{1}{2\pi}(\theta_1 + \theta_2 - \theta_3) & 1 - \dfrac{1}{2\pi}(-\theta_1 - \theta_2 - \theta_3)
\end{array}
\tag{3.1}
$$

These scalars, as we further discuss in Section 3.2), can be massless, massive or tachyonic.

3.2. Supersymmetry for intersecting D6-branes

It is interesting to consider if the above configurations preserve some supersymmetry. This can be analyzed following [13]. The condition that there is some supersymmetry preserved by the combined system of two D6-brane stacks is that there exist spinors ϵ_L, ϵ_R that satisfy

$$
\begin{aligned}
\epsilon_L &= \Gamma_6 \epsilon_R; & \Gamma_6 &= \Gamma^0 \ldots \Gamma^3 \Gamma^4 \Gamma^6 \Gamma^8 \\
\epsilon_L &= \Gamma_{6'} \epsilon_R; & \Gamma_{6'} &= \Gamma^0 \ldots \Gamma^3 \Gamma^{4'} \Gamma^{6'} \Gamma^{8'}
\end{aligned}
\tag{3.2}
$$

where 468 and $4'6'8'$ denote the directions along the two D6-branes in the six dimensions 456789. The above is simply the condition (2.3) for each of the branes.

Let R denote the $SO(6)$ rotation that takes the first D6-brane into the second, acting on the spinor representation. Then we have $\Gamma_{6'} = R\Gamma_6 R^{-1}$. A preserved spinor exists if and only if there is a 6d spinor which is invariant under R. This implies that R must belong to an $SU(3)$ subgroup of $SO(6)$. This can be more explicitly stated by rewriting R in the vector representation as

$$
R = \text{diag}\,(e^{i\theta_1}, e^{-i\theta_1}, e^{i\theta_2}, e^{-i\theta_2}, e^{i\theta_3}, e^{-i\theta_3})
\tag{3.3}
$$

The condition that the rotation is within $SU(3)$ is

$$\theta_1 \pm \theta_2 \pm \theta_3 = 0 \quad \mathrm{mod} \quad 2\pi \qquad \text{for some choice of signs} \qquad (3.4)$$

Indeed, one can check that the open string spectrum computed above is boson-fermion degenerate in such cases. In the generic case, there is no supersymmetry invariant under the two stacks of branes, and the open string sector at the intersection is non-supersymmetric. However, if $\theta_1 \pm \theta_2 \pm \theta_3 = 0$ for some choice of signs, one of the scalars becomes massless, reflecting that the configuration is $\mathcal{N} = 1$ supersymmetric. $\mathcal{N} = 2$ supersymmetry arises if e.g. $\theta_3 = 0$ and $\theta_1 \pm \theta_2 = 0$, while $\mathcal{N} = 4$ arises only for parallel stacks $\theta_i = 0$.

As described above, the light scalars at intersections may be massless, massive or tachyonic. The massless case corresponds to a situation with some unbroken supersymmetry. The massless scalar is a modulus, whose vacuum expectation value (vev) parametrizes the possibility of recombining the two intersecting D-branes into a single smooth one, as pictorially shown in Fig. 7. That is, the intersecting geometry belongs to a one- (complex) parameter family of supersymmetry preserving configurations of D-branes. Mathematically, there is a one-parameter family of supersymmetric 3-cycles, i.e. special Lagrangian submanifolds of $\mathbf{R}^6$, with the same asymptotic behaviour as the intersecting D-brane configuration. In the simpler situation of D-branes intersecting at $SU(2)$ angles (i.e. $\mathcal{N} = 2$ supersymmetry), the recombination is very explicit. It is given by deforming two intersecting 2-planes, described by the complex curve $uv = 0$, to the smooth 2-cycle $uv = \epsilon$, with ϵ corresponding to the vev of the scalar at the intersection.

The configuration with tachyonic scalars corresponds to situations where this recombination is triggered dynamically. Namely, the recombination process correspond to condensation of the tachyon at the intersection. It is interesting to point out that in the degenerated case where the intersecting brane system becomes a brane-antibrane system (e.g. $\theta_1 = \theta_2 = 0$, $\theta_3 = 1$), the tachyons are mapped to the well-studied tachyon of brane-antibrane systems. The situation

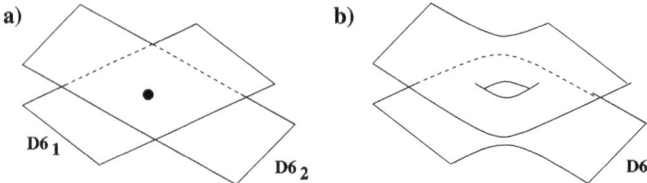

Fig. 7. Recombination of two intersecting D6-branes into a single smooth one, corresponding to a vev for an scalar at the intersection.

where all light scalars have positive squared masses correspond to a non-super-symmetric intersection, which is nevertheless dynamically stable against recombination. Namely, the recombined 3-cycle has volume larger than the sum of the volumes of the intersecting 3-cycles.

Indeed, the different regimes of dynamics of scalars at intersections have a one-to-one mapping with the different relations between the volumes of intersecting and recombined 3-cycles [15]. Namely, the conditions to have or not tachyons are related to the angle criterion [16] determining which the particular 3-cycle having smaller volume. The supersymmetric situation corresponds to both the intersecting and recombined configurations having the same volume; the tachyonic situation corresponds to the recombined 3-cycle having smaller volume; the massive situation corresponds to the intersecting 3-cycle having smaller volume.

4. Compact four-dimensional models

Once we have succeeded in describing configurations of D-branes leading to charged chiral fermions, in this section we employ them in building models with 4d gravity and gauge interactions. Although intersecting D6-branes provide 4d chiral fermions already in flat 10d space, gauge interactions remain 7d and gravity interactions remain 10d unless we consider compactification of spacetime.

The general kind of configurations we are to consider (see Fig. 8) is type IIA string theory on a spacetime of the form $M_4 \times X_6$ with compact X_6, and with stacks of N_a D6$_a$-branes with volumes of the form $M_4 \times \Pi_a$, with $\Pi_a \subset X_6$ a 3-cycles. It is important to realize that generically 3-cycles in a 6d compact space intersect at points, so the corresponding wrapped D6-branes will intersect at M_4 subspaces of their volumes. Hence, compactification reduces the 10d and

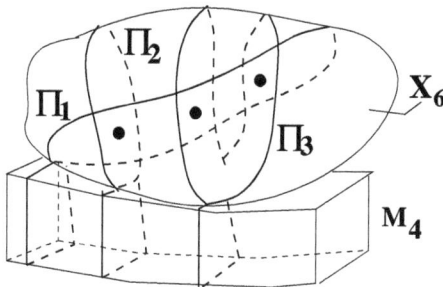

Fig. 8. Compactification with intersecting D6-branes wrapped on 3-cycles.

7d gravitational and gauge interactions to 4d, and intersections lead to charged 4d chiral fermions. Also, generically two 3-cycles in a 6d space intersect several times, therefore leading to a replicated sector of opens strings at intersections. This is a natural mechanism to explain/reproduce the appearance of replicated families of chiral fermions in Nature!

4.1. Toroidal models

4.1.1. Construction

In this section we mainly follow [12], see also [11]. To start with the simplest configurations, consider compactifying on a six-torus factorized as $\mathbf{T^6} = \mathbf{T^2} \times \mathbf{T^2} \times \mathbf{T^2}$. Now we consider stacks of D6$_a$-branes (with a an index labeling the stack), spanning M_4 and wrapping a 1-cycle (n_a^i, m_a^i) in the ith 2-torus. Namely, the ath D6-brane wraps n_a^i, m_a^i times along the horizontal and vertical directions in the ith two-torus, see Fig. 9 for examples.[1]

The general kind of configurations we are to consider (see Fig. 8) is thus type IIA string theory on a spacetime of the form $\mathbf{M_4} \times \mathbf{T^6}$, and with stacks of N_a D6$_a$-branes with volumes of the form $M_4 \times \Pi_a$, with $\Pi_a \subset \mathbf{X_6}$ a 3-cycle as described above. It is important to realize that generically 3-cycles in a 6d compact space intersect at points, so the corresponding wrapped D6-branes will intersect at M_4 subspaces of their volumes. Hence, compactification reduces the 10d and 7d gravitational and gauge interactions to 4d, and intersections lead to charged 4d chiral fermions.

Also, generically two 3-cycles in a 6d space intersect several times. Locally, each intersection is exactly of the form studied in Section 3.1, therefore the con-

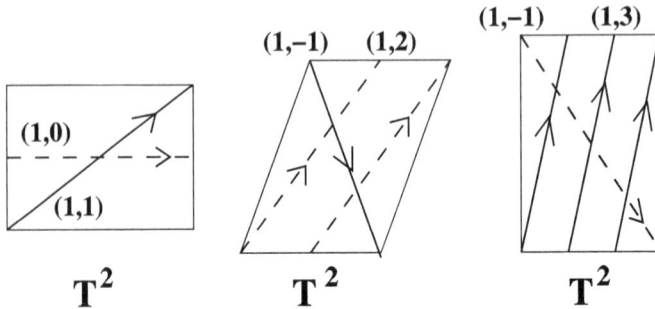

Fig. 9. Examples of intersecting 3-cycles in $\mathbf{T^6}$.

---

[1]These factorizable branes are not the most general possibility. Branes wrapped on non-factorizable cycles exist, and can be obtained e.g. by recombination of factorized branes. For simplicity, we will not use them in these lectures.

struction leads to a replicated sector of open strings at intersections. This is a natural mechanism to explain/reproduce the appearance of replicated families of chiral fermions in Nature, as we show below. It is also important to notice that in compactifications, the angles between branes are derived quantities, and depend on the closed string moduli controlling the torus geometry. For instance, for a rectangular torus of radii R_1, R_2 along the horizontal and vertical directions, the angle between the 1-cycle $(1, 0)$ and (n, m) is

$$\tan \theta = \frac{m R_2}{n R_1} \tag{4.1}$$

In this toroidal case, the intersection number is given by the product of the number of intersections in each 2-torus, and reads

$$I_{ab} = (n_a^1 m_b^1 - m_a^1 n_b^1) \times (n_a^2 m_b^2 - m_a^2 n_b^2) \times (n_a^3 m_b^3 - m_a^3 n_b^3) \tag{4.2}$$

It is useful to introduce the 3-homology class $[\Pi_a]$ of the 3-cycle Π_a, which can be thought of as a vector of RR charges of the corresponding D6-brane. The 1-homology class of an (n, m) 1-cycle in a 2-torus is $n[a] + m[b]$, with $[a]$, $[b]$ the basic homology cycles in $\mathbf{T}^2$. For a 3-cycle with wrapping numbers (n_a^i, m_a^i) we have

$$[\Pi_a] = \otimes_{i=1}^3 (n_a^i [a_i] + m_a^i [b_i]) \tag{4.3}$$

The intersection number (4.2) is intersection number in homology, denoted $I_{ab} = [\Pi_a] \cdot [\Pi_b]$. This is easily shown using $[a_i] \cdot [b_j] = \delta_{ij}$ and linearity and antisymmetry of the intersection pairing.

With the basic data defining the configuration, namely $\mathbf{N_a}$ D6$_a$-branes wrapped on 3-cycles $[\Pi_\mathbf{a}]$, with wrapping numbers $(\mathbf{n_a^i}, \mathbf{m_a^i})$ on each $\mathbf{T}^2$ and intersection numbers $\mathbf{I_{ab}}$, we can compute the spectrum of the model.

The closed string sector produces 4d $\mathcal{N} = 8$ supergravity. There exist different open string sectors:

$\mathbf{6_a 6_a}$ String stretched among D6-branes in the ath stack produce 4d $U(N_a)$ gauge bosons, 6 real adjoint scalars and 4 adjoint Majorana fermions, filling out a vector multiplet of the 4d $\mathcal{N} = 4$ supersymmetry preserved by the corresponding brane.

$\mathbf{6_a 6_b + 6_b 6_a}$ Strings stretched between the ath and bth stack lead to I_{ab} replicated chiral left-handed fermions in the bifundamental representation $(N_a, \overline{N}_b)$. Negative intersection numbers lead to a positive number of chiral fermions with right-handed chirality. Additional light scalars may be present, with masses determined by the wrapping numbers and the $\mathbf{T}^2$ moduli.

Generalization for compact spaces more general than the 6-torus will be discussed in Section 4.2. We have therefore obtained a large class of four-dimensional theories with interesting non-Abelian gauge symmetries and replicated charged chiral fermions. Hence compactifications with intersecting D6-branes provide a natural setup in which string theory can produce gauge sectors with the same rough features of the Standard Model. In coming sections we explore them further as possible phenomenological models, and construct explicit examples with spectrum as close as possible to the Standard Model.

4.1.2. RR tadpole cancellation

String theories with open string sectors must satisfy a crucial consistency condition, known as cancellation of RR tadpoles. As mentioned above, D-branes act as sources for RR p-forms via the disk coupling $\int_{W_{p+1}} C_p$, see Fig. 3. The consistency condition amounts to requiring the total RR charge of D-branes to vanish, as implied by Gauss law in a compact space (since RR field fluxlines cannot escape, Fig. 10). In our setup, the 3-cycle homology classes are vectors of RR charges, hence the condition reads

$$[\Pi_{\text{tot}}] = \sum_a N_a [\Pi_a] = 0 \tag{4.4}$$

Equivalently, the condition of RR tadpole cancellation can be expressed as the requirement of consistency of the equations of motion for RR fields. In our situation, the terms of the spacetime action depending on the RR 7-form C_7 are

$$\begin{aligned}
S_{C7} &= \int_{\mathbf{M_4} \times \mathbf{X_6}} H_8 \wedge *H_8 + \sum_a N_a \int_{\mathbf{M_4} \times \Pi_\mathbf{a}} C_7 \\
&= \int_{\mathbf{M_4} \times \mathbf{X_6}} C_7 \wedge d H_2 + \sum_a N_a \int_{\mathbf{M_4} \times \mathbf{X_6}} C_7 \wedge \delta(\Pi_a)
\end{aligned} \tag{4.5}$$

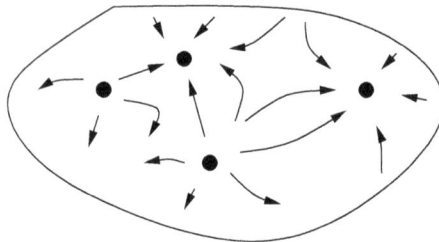

Fig. 10. In a compact space, fluxlines cannot escape and the total charge must vanish.

where H_8 is the 8-form field strength, H_2 its Hodge dual, and $\delta(\Pi_a)$ is a bump 3-form localized on Π_a in $\mathbf{X_6}$. The equations of motion read

$$dH_2 = \sum_a N_a \, \delta(\Pi_a) \tag{4.6}$$

The integrability condition (4.4) is obtained by taking this equation in homology.

In the toroidal setup the RR tadpole conditions provide a set of constraints, given by

$$\sum_a N_a n_a^1 n_a^2 n_a^3 = 0$$

$$\sum_a N_a n_a^1 n_a^2 m_a^3 = 0 \text{ and permutations}$$

$$\sum_a N_a n_a^1 m_a^2 m_a^3 = 0 \text{ and permutations} \tag{4.7}$$

$$\sum_a N_a m_a^1 m_a^2 m_a^3 = 0$$

4.1.3. Anomaly cancellation

Cancellation of RR tadpoles in the underlying string theory configuration implies cancellation of four-dimensional chiral anomalies in the effective field theory in our configurations. Recall that the chiral piece of the spectrum is given by I_{ab} chiral fermions in the representation $(N_a, \overline{N}_b)$ of the gauge group $\prod_a U(N_a)$.

Cubic non-Abelian anomalies

The $SU(N_a)^3$ cubic anomaly is proportional to the number of fundamental minus antifundamental representations of $SU(N_a)$, hence it is proportional to

$$A_a = \sum_b I_{ab} N_b. \tag{4.8}$$

It is easy to check this vanishes due to RR tadpole cancellation: Starting with (4.4), we consider the intersection of $[\Pi_{\text{tot}}]$ with any $[\Pi]$ to get

$$0 = [\Pi_a] \cdot \sum_b N_b \, [\Pi_b] = \sum_b N_b I_{ab} \tag{4.9}$$

as claimed.[2]

---

[2]It is interesting to notice that RR tadpole cancellation is slightly stronger than cancellation of cubic non-Abelian anomalies. In fact, the former requires that the number of fundamental minus antifundamentals vanishes even for the cases $N_a = 1, 2$, where no gauge theory anomaly exists. This observation will turn out relevant in phenomenological model building in Section 6.

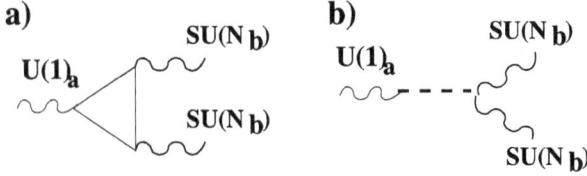

Fig. 11. Triangle and Green–Schwarz diagrams contributing to the mixed $U(1)$-non-Abelian anomalies.

Cancellation of mixed anomalies

The $U(1)_a$-$SU(N_b)^2$ mixed anomalies also cancel as a consequence of RR tadpole cancellation. They do so in a trickier way, namely the anomaly receives two non-zero contributions which cancel each other, see Fig. 11. Mixed gravitational triangle anomalies cancel automatically, without Green–Schwarz contributions.

The familiar field theory triangle diagrams give a contribution which, even after using RR tadpole conditions, is non-zero and reads

$$A_{ab} \simeq N_a \, I_{ab} \tag{4.10}$$

On the other hand, the theory contains contributions from Green–Schwarz diagrams, where the gauge boson of $U(1)_a$ mixes with a 2-form which subsequently couples to two gauge bosons of $SU(N_b)$, see Fig. 11. These couplings arise in the KK reduction of the D6-brane world-volume couplings $N_a \int_{D6_a} C_5 \wedge \mathrm{tr}\, F_a$ and $\int_{D6_b} C_3 \wedge \mathrm{tr}\, F_b^2$, as follows.

Introducing a basis $[\Lambda_k]$ and its dual $[\Lambda_{\tilde{l}}]$, we can define the KK reduced 4d fields

$$(B_2)_k = \int_{[\Lambda_k]} C_5, \quad \phi_{\tilde{l}} = \int_{[\Lambda_{\tilde{l}}]} C_3 \quad \text{with } d\phi_{\tilde{l}} = -\delta_{k\tilde{l}} *_{4d} (B_2)_k \tag{4.11}$$

The KK reduced 4d couplings read

$$N_a q_{ak} \int_{4d} (B_2)_k \mathrm{tr}\, F_a, \quad q_{b\tilde{l}} \int_{4d} \phi_{\tilde{l}} \mathrm{tr}\, F_b^2 \tag{4.12}$$

with $q_{ak} = [\Pi_a] \cdot [\Lambda_k]$, and similarly for $q_{b\tilde{l}}$. The total amplitude is proportional to

$$A_{ab}^{GS} = -N_a \sum_k q_{ak} q_{b\tilde{l}} \delta_{k\tilde{l}} = \cdots = -N_a I_{ab} \tag{4.13}$$

leading to a cancellation between both kinds of contributions.

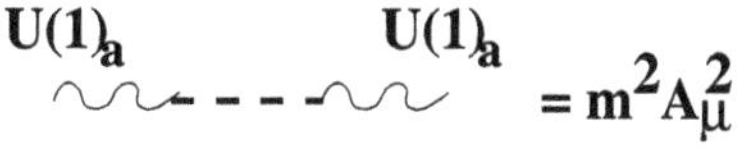

$$\text{U(1)}_a \qquad \text{U(1)}_a$$
$$\sim\!\sim\!\cdot - - - \sim\!\sim = \text{m}^2\text{A}_\mu^2$$

Fig. 12. The $B \wedge F$ couplings lead to a $U(1)$ gauge boson mass term.

An important observation is that any $U(1)$ gauge boson with $B \wedge F$ couplings gets massive, with mass roughly of the order of the string scale, see Fig. 12. Such $U(1)$'s disappear as gauge symmetries from the low-energy effective field theory, but remain as global symmetries, unbroken in perturbation theory. Introducing the generators Q_a of the $U(1)$ inside $U(N_a)$, the condition that a $U(1)$ with generator $\sum_a c_a Q_a$ remains massless is

$$\sum_a N_a q_{ak} c_a = 0 \quad \text{for all } k \tag{4.14}$$

Such $U(1)$ factors remain as gauge symmmetries of the low energy theory.

4.2. Generalization beyond torus: Model building with A-type branes

Clearly the above setup is not restricted to toroidal compactifications. Indeed one may take any compact 6-manifold as internal space, for instance a Calabi–Yau threefold, which would lead to 4d $\mathcal{N} = 2$ supersymmetry in the closed string sector. In this situation we should pick a set of 3-cycles Π_a on which we wrap N_a D6-branes (for instance special Lagrangian 3-cycles of $\mathbf{X_6}$ if we are interested in preserving supersymmetry), making sure they satisfy the RR tadpole cancellation condition $\sum_a N_a[\Pi_a] = 0$.

The final open string spectrum (for instance, in the case of supersymmetric wrapped D6-branes) arises in two kinds of sectors

$\mathbf{6}_a\text{-}\mathbf{6}_a$ Leads to $U(N_a)$ gauge bosons ($\mathcal{N} = 1$ vector multiplets in the supersymmetric case) and $b_1(\Pi_a)$ real adjoint scalars (chiral multiplets in susy case).

$\mathbf{6}_a\text{-}\mathbf{6}_b+\mathbf{6}_b\text{-}\mathbf{6}_a$ We obtain I_{ab} chiral fermions in the representation $(N_a, \overline{N}_b)$ (plus light scalars, massless in supersymmetry preserving intersections). Here $I_{ab} = [\Pi_a] \cdot [\Pi_b]$.

Notice that the chiral spectrum is obtained in terms of purely topological information of the configuration, as should be the case.

Our whole discussion up to this point has simply been a pedagogical way of describing a general class of string compactifications. Namely, compactifications of type IIA string theory on Calabi–Yau threefolds with A-type D-branes. In the geometrical large volume regime, these are described as D-branes wrapped on special Lagrangian 3-cycles, and reproduce the structures we have been discussing. A-type branes are extensively studied from the point of view of topological strings, with results of immediate application to our models. To name a few, the fact that such D-brane states do not have lines of marginal stability in Kahler moduli space, that their world-volume superpotential arises exclusively from worldsheet instantons, and their nice relation via mirror symmetry with type IIB compactifications with B-type D-branes, see next lecture. It is very satisfactory that a phenomenological motivation has driven us to consider a kind of configurations so interesting from the theoretical viewpoint as well.

The phenomenology of toroidal and non-toroidal models is quite similar to that of toroidal compactifications with D-branes, see next subsection. Thus, the later are in any event good toy model for many features of general compactifications with intersecting branes. This is particularly interesting since it is relatively difficult to construct explicit configurations of intersecting D6-branes in Calabi–Yau models (although some explicit examples have been discussed in [17, 18]).

4.3. Phenomenological features

We now turn to a brief discussion of the phenomenological properties natural in this setup [12].

• Most models constructed in the literature are non-supersymmetric. It is however possible to construct fully $\mathcal{N} = 1$ supersymmetric models, see Section 7. For non-supersymmetric models, unless alternative solutions to the hierarchy model are provided, the best proposal low string scale $M_s \simeq$ TeV to avoid hierarchy, along the lines of [19].

• The proton is stable in these models, since the $U(1)$ within the $U(3)$ color factor plays the role of baryon number, and is preserved as a global symmetry, exactly unbroken in perturbation theory. Non-perturbative effects breaking it arise from euclidean D2-branes wrapped on 3-cycles, and have the interpretation of spacetime gauge theory instantons, hence reproducing the non-perturbative breaking of baryon number in the Standard Model.

• These models do *not* have a natural gauge coupling unification, even at the string scale. Each gauge factor has a gauge coupling controlled by the volume of the wrapped 3-cycle. Gauge couplings are related to geometric volumes, hence their experimental values can be adjusted/reproduced in concrete models, rather than predicted by the general setup.

• There exists a geometric interpretation for the spontaneous electroweak symmetry breaking. In explicit models, the Higgs scalar multiplet arises from the light scalars at intersections, and parametrizes the possibility of recombining two intersecting cycles into a single smooth one, as shown in Fig. 13. In the process, the gauge symmetry is reduced, corresponding to a Higgs mechanism in the effective field theory. See [20] for further discussion.

• There is a natural exponential hierarchy of the Yukawa couplings. Yukawa couplings among the scalar Higgs and chiral fermions at intersections arise at tree

a)   b)

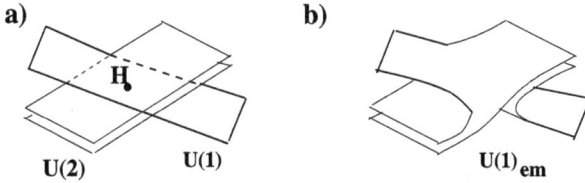

Fig. 13.

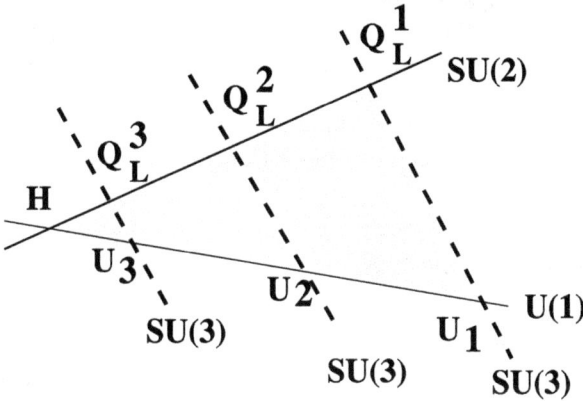

Fig. 14. Geometric origin of the hierarchy of Yukawa couplings for different generations.

level in the string coupling from open string worldsheet instantons; namely from string worldsheets spanning the triangle with vertices at the intersections and sides on the D-branes. Their value is roughly given by e^{-A}, with A the triangle area in string units. Since different families are located at different intersections, their triangles have areas increasing linearly with the family index, leading to an exponential Yukawa hierarchy, see Fig. 14. See e.g. [21] for further analysis of Yukawa couplings in explicit models.

5. Orientifold models

The above constructed models are non-supersymmetric. One simple way to see it is that we start with type IIA string theory compactified on X_6, and introduce D6-branes. Since RR tadpole cancellation requires that the total RR charge vanishes, we are forced to introduce objects with opposite RR charges, in a sense branes and antibranes, a notoriously non-supersymmetric combination.

An equivalent derivation of the result is as follows: If we would succeed in constructing a supersymmetric configuration of D6-branes, the system as a whole would be a supersymmetric BPS state of type IIA on $\mathbf{X_6}$. Since for a BPS state the tension is proportional to the RR charge, and the latter vanishes due to RR tadpole cancellation, the tension of the state must vanish. The only D6-brane configuration with zero tension is having no D6-brane at all. Hence the only supersymmetric configuration would be just type IIA on $\mathbf{X_6}$, with no brane at all.

These arguments suggest a way out of the impasse. In order to obtain $\mathcal{N} = 1$ supersymmetric compactifications we need to introduce objects with negative tension and negative RR charge, and which preserve the same supersymmetry as the D6-branes. Such objects exist in string theory and are orientifold 6-planes, O6-planes. Introduction of these objects leads to an interesting extension of the configurations above constructed, and will be studied in Section 5.3. In particular we will use them to construct supersymmetric compactifications with intersecting D6-branes.

5.1. Properties of O6-planes

To start, consider type IIA string theory on 10d flat space M_{10}, and mod it out by the so-called orientifold action $\Omega R(-)^{F_L}$. Here Ω is world-sheet parity, which flips the orientation of the fundamental strings; R is a $\mathbf{Z_2}$ geometric action, acting locally as $(x^5, x^7, x^9) \rightarrow (-x^5, -x^7, -x^9)$; finally $(-)^{F_L}$ is left-moving world-sheet fermion number, introduced for technical reasons.

The quotient theory contains a special subspace in spacetime, fixed under the geometric part R of the above action. Namely, it is a 7d plane defined by $x^5 = x^7 = x^9 = 0$, and spanned by the coordinates 0123468. This set of points fixed under the orientifold action is called an orientifold 6-plane (O6-plane), since it has six spatial dimensions (in general one can define other orientifold quotients of type II string theories, containing Op-planes of p spatial dimensions). Physically, it corresponds to a region of spacetime where the orientation of a string can flip (since a string at the O6-plane is identified, by the orientifold action, with itself with the opposite orientation). The description of string theory is the presence of orientifold planes is modified only by the inclusion of unoriented world-sheets, for instance with the topology of the Klein bottle.

Orientifold planes have some features similar to D-branes of the same dimension. For instance, Op-planes carry tension and are charged under the RR $(p + 1)$-form C_{p+1}. The diagram responsible for these couplings is shown in Fig. 15. For instance, and O6-plane is charged under the RR 7-form, and its charge is given by $Q_{O6} = \pm 4$, in units where the D6-brane charge is $+1$. Here the two possible signs correspond to two different kinds of O6-planes; we will center on the negatively charged O6-plane in what follows. Also, O-planes pre-

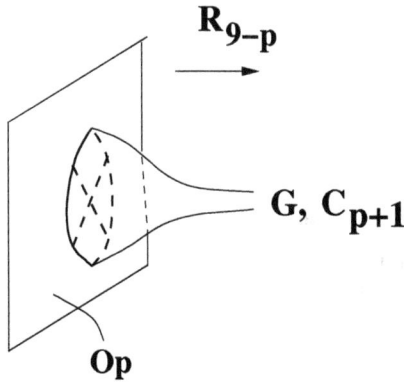

Fig. 15. Diagram describing the interaction of an Op brane with closed string modes. The dashed cross denotes a crosscap, namely a disk with an identification of antipodal points in the boundary, so that the world-sheet is closed and unoriented.

serve the same supersymmetry as a D-brane. This implies that there is a relation between the tension and charge of O-planes.

There are however some important differences between O-planes and D-branes, the main one being that O-planes do not carry world-volume degrees of freedom. Hence, they are better regarded as part of the spacetime geometrical data, rather than dynamical objects.

5.2. O6-planes and D6-branes

It is interesting to include orientifold planes in compactifications or configurations with D-branes. These configurations are most simply described in the covering space of the orientifold quotient. Here we must include the images of the D-branes under the orientifold action, denoted by primed indices. The spectrum of open strings in the orientifold quotient theory is obtained by simply computing the spectrum in the covering space, and then imposing the identifications implied by the orientifold action (taking into account the flip in the open string orientation implied by the latter).

• Let us start by considering the simple situation of configurations of parallel D6-branes and O6-planes. Consider first a stack of n D6-branes on top of an O6-plane, see Fig. 16a. The open string spectrum before the orientifold action is given by the n^2 open string sectors, giving rise to an $U(n)$ vector multiplet. The orientifold action implies the following identification among the ab open strings

$$|ab\rangle \leftrightarrow \pm|ba\rangle \tag{5.1}$$

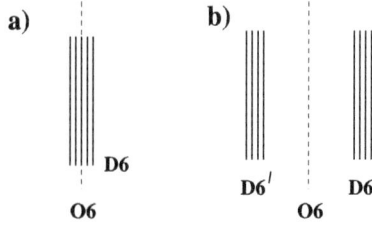

Fig. 16. Configurations of D6-brane stacks parallel to an O6-plane. Figure a) show the situation where the branes are on top of the O6-plane, while figure b) corresponds to branes separated from it. Although the branes within a stack are coincident, they are shown slightly separated, for clarity.

with the negative (positive) sign corresponding to the choice of negatively (positively) charged O6-plane. Centering on the former case, the physical states in the quotient correspond to the $n(n-1)/2$ antisymmetric linear combinations $(|ab\rangle - |ba\rangle)/2$. The massless modes correspond to an $SO(n)$ vector multiplet with respect to the 16 supercharges unbroken by the O6/D6 configuration.

• Consider now a configuration of n coincident D6-branes, parallel but separated from the O6-plane. The configuration must include an orientifold image of the D-brane stack, namely a set of n D6'-branes, see Fig. 16b. The massless open string spectrum before the orientifold projection is given by a $U(n) \times U(n)'$ gauge group plus superpartners. The orientifold action implies and identification of the degrees of freedom in both $U(n)$ factors, so that only a linear combination survives. In the quotient, we just obtain an $U(n)$ vector multiplet (which agrees with the intuition that massless modes on D-branes are not sensitive to distant objects, hence the n D6-branes in the quotient do not notice, at the level of the massless spectrum, the distant O6-plane).

An important observation in the identification of the $U(n)$ factors is that, due to the orientation reversal, and open string starting on the D6-brane stack is mapped to an open string ending on the D6'-brane stack, and vice versa, see Fig. 17. This implies that the $U(n)$ is identified with $U(n')$ with the fundamental $\square$ mapping to the anti-fundamental $\overline{\square}'$, and vice versa. This will be important in the computation of open string massless spectra in more involved configurations (or in these simple ones, if one is interested in computing the massive spectrum).

Let us consider another local geometry similar to the above. Let us consider configurations of D6-branes orthogonal to the O6-plane in some of the directions, so that the D6-brane stack is still mapped to itself under the orientifold action. A configuration which appears often (since it preserves 8 supercharges) is when there are four dimensions not commonly along or commonly transverse to the objects. For instance, consider an O6-plane along 0123456 and a D6-brane along

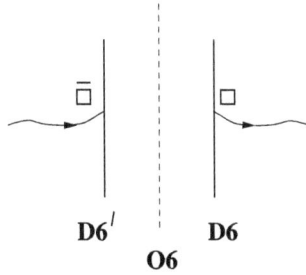

Fig. 17. The orientifold projection relates the gauge groups on D-branes and images such that open string endpoint in the fundamental representation of one map to endpoints in the antifundamental of the other.

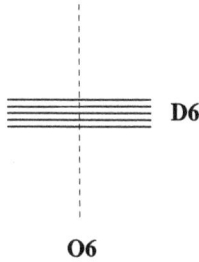

Fig. 18. Configurations of D6-brane stacks with some directions orthogonal to an O6-plane.

0123478, see Fig. 18. For one such stack of n D6-branes, the final gauge group (for a negatively charged O6-plane) is $USp(n)$ (hence n must be even) and fills out a vector multiplet with respect to the eight unbroken supersymmetries. In addition there is a hypermultiplet in the two-index symmetric (reducible) representation. The change of gauge group with respect to the case of the parallel O6/D6 system is due to an additional sign in the orientifold action [23].

Let us now consider situations with intersecting D6-branes (and their images) in the presence of O6-planes. All the D6-branes and O6-planes are taken to be parallel in four of their common dimensions, so that the intersections are geometrically of the kind studied above. There are several different situations to be considered, depending on the relative geometry of the intersection and the O6-plane. To simplify the discussion, we center on describing the gauge group and chiral fermions at intersections.

• Consider two stacks of D6-branes, labeled a and b, intersecting away from the O6-plane. The configuration also includes the image D6′-branes, labeled a', b', see Fig. 19a. Before the orientifold projection, the gauge group is $U(N_a) \times$

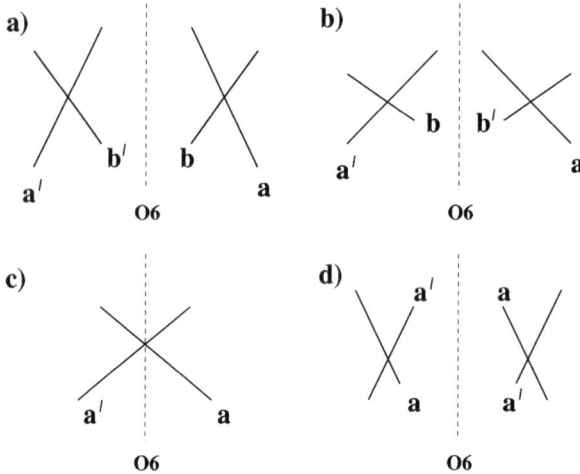

Fig. 19. Configurations of intersecting D6-brane stacks in the presence of an O6-plane. Figure a) shows the intersections of two stacks a and b, away from the O6-plane. Figure b) shows the intersection of a stack a with the image of another b'. Figures c) and d) show the intersection of the stacks a and its image a' on top of the O6-plane and away from it, respectively.

$U(N_b) \times U(N_a)' \times U(N_b)'$. Also, the intersections in the figure (ignoring other possible intersections of the branes) provide 4d chiral fermions in the representation $(\square_a, \overline{\square}_b)$ and $(\square_{b'}, \overline{\square}'_a)$, due to the different relative orientation of the branes and their images. After the identification implied by the orientifold action (recalling the effect on fundamental representations and their images), we are left with a gauge group $U(N_a) \times U(N_b)$ and a 4d chiral fermion in the $(\square_a, \overline{\square}_b)$.

• Consider now the intersection of a stack of D6$_a$-branes with D6$'_b$-branes, namely the orientifold image of a stack of D6$_b$-branes, see Fig. 19b. Before the orientifold projection, the gauge group is $U(N_a) \times U(N_b)' \times U(N_a)' \times U(N_b)$, with 4d chiral fermions in the representation $(\square_a, \overline{\square}'_b)$ and $(\square_b, \overline{\square}'_a)$. After the orientifold action, we have a gauge group $U(N_a) \times U(N_b)$ and a 4d chiral fermion in the $(\square_a, \square_b)$.

• Consider the intersection of a stack of D6$_a$-branes, with its own image, on top of the O6-plane, see Fig. 19c. Before the orientifold action, the gauge group is $U(N_a) \times U(N_a)'$ and there is a 4d chiral fermion in the $(\square_a, \overline{\square}'_a)$. The orientifold action reduces the gauge group to $U(N_a)$. The initial 4d chiral fermions thus transform under this as the tensor product of $\square_a$ and $\overline{\square}'_a = \square_a$, namely $\boxminus_a + \square\square_a$. After the orientifold projection (for a negatively charged O6-plane), however, only 4d fermions in the $\boxminus_a$ component survive.

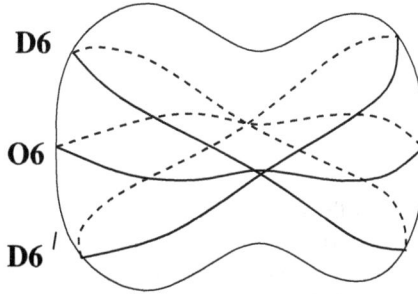

Fig. 20. D6-branes and their images in an orientifold compactification.

• Consider finally the intersection of a stack of D6$_a$-branes, with D6$'_a$-branes, away from the O6-plane, see Fig. 19d. Before the orientifold action, the gauge group is $U(N_a) \times U(N_a)'$ and there are 4d chiral fermions in the $2(\Box_a, \overline{\Box}'_a)$, due to the two intersections. The orientifold action reduces the gauge group to $U(N_a)$, and identifies both intersections. Thus, the 4d chiral fermions in the quotient transform in the representation $\boxminus_a + \Box\Box_a$.

It is easy to derive the spectra for intersections of generic D6-brane stack with stacks overlapping or orthogonal to the O6-plane. With these ingredients, we have enough information to describe compactifications with O6-planes and intersecting D6-branes.

5.3. Orientifold compactifications with intersecting D6-branes

5.3.1. Construction

Consider type IIA theory on e.g. a Calabi–Yau $\mathbf{X_6}$, and mod out the configuration by $\Omega R(-)^{F_L}$, where R is an antiholomorphic $\mathbf{Z_2}$ symmetry of $\mathbf{X_6}$. Hence it locally acts as $(z_1, z_2, z_3) \rightarrow (\bar{z}_1, \bar{z}_2, \bar{z}_3)$ on the CY complex coordinates, or as $(x^5, x^7, x^9) \rightarrow (-x^5, -x^7, -x^9)$ in suitable real ones. The set of fixed points of R are O6-planes, similar to those introduced above, with the difference that they are not flat in general, but rather wrap a (special Lagrangian) 3-cycle in $\mathbf{X_6}$. Let us denote Π_{O6} the total 3-cycle spanned by the set of O6-planes in the configuration.

We now introduce stacks of N_a D6$_a$-branes, and their image D6$'_a$-branes, in the above orientifold quotient, see Fig. 20. They are wrapped on 3-cycles, denoted Π_a and Π'_a, respectively. The model is $\mathcal{N} = 1$ supersymmetric if all the D6-branes are wrapped on special Lagrangian 3-cycles, see Section 7 for concrete examples.

Taking into account the different sources of RR 7-form in the configuration, the RR tadpole cancellation conditions read[3]

$$\sum_a N_a [\Pi_a] + \sum_a N_a [\Pi_{a'}] - 4 \times [\Pi_{O6}] = 0 \qquad (5.2)$$

The open string spectrum in orientifolded models can be easily computed. It only requires computing the relevant numbers of intersections, and if required how many lie on top of the orientifold planes. The results for the different sectors and the corresponding chiral spectra, assuming that no D6-branes are mapped to themselves under the orientifold action, are as follows

aa+a'a' Contains $U(N_a)$ gauge bosons and superpartners

ab+ba+b'a'+a'b' Contains I_{ab} chiral fermions in the representation $(N_a, \overline{N}_b)$, plus light scalars.

ab'+b'a+ba'+a'b Contains $I_{ab'}$ chiral fermions in the representation (N_a, N_b), plus light scalars.

aa'+a'a Contains $n_{\Box\Box_a}$ 4d chiral fermions in the representation $\Box\Box_a$ and $n_{\Box}{}_a$

in the $\Box_a$, with

$$n_{\Box\Box_a} = \frac{1}{2}(I_{aa'} - I_{a,O6}), \quad n_{\Box}{}_a = \frac{1}{2}(I_{aa'} + I_{a,O6}) \qquad (5.3)$$

where $I_{a,O6} = [\Pi_a] \cdot [\Pi_{O6}]$ is the number of aa' intersections on top of O6-planes.

As expected, the new RR tadpole conditions in the presence of O6-planes guarantee the cancellation of 4d anomalies of the new chiral spectrum, in analogy with the toroidal case. (In the orientifold case, mixed gravitational anomalies may receive Green–Schwarz contributions, see appendix in the first reference in [24]). The condition that a $U(1)$ remains massless is given by the orientifold version of (4.14)

$$\sum_a N_a(q_{ak} - q_{a'k})c_a = 0 \quad \text{for all } k \qquad (5.4)$$

5.3.2. Toroidal orientifold models

A simple class of examples is provided by compactifications on $\mathbf{X_6} = \mathbf{T^6}$, with factorized $\mathbf{T^6}$, and with R given by the action $y_i \to -y_i$, where y_i are the vertical direction on each $\mathbf{T^2}$. This is a symmetry for rectangular two-tori or for two-tori tilted by a specific angle [25], see Fig. 22. Let us introduce a quantity $\beta = 0, \frac{1}{2}$, corresponding to the rectangular and tilted cases.

---

[3]There are additional discrete constraints arising from cancellation of $\mathbf{Z_2}$-valued K-theory charges. We skip their discussion for the moment.

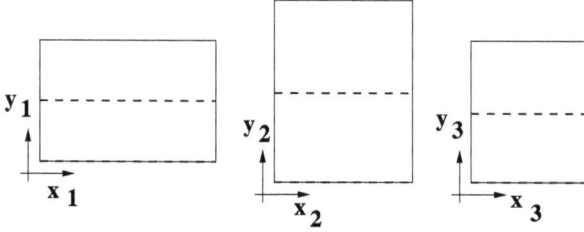

Fig. 21. Orientifold 6-planes in the orientifold quotient of IIA on $\mathbf{T^6}$ by $\Omega R(-)^{FL}$, with $R : y_i \to -y_i$.

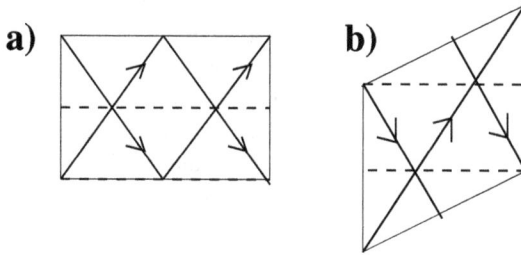

Fig. 22. Cycles and their orientifold images in a rectangular and tilted 2-tori.

For a geometry with rectangular two-tori, as in Fig. 21, the set of fixed points is given by x_i arbitrary, $y_i = 0$, $R_{y_i}/2$, hence it has 8 components. They correspond to O6-planes wrapped on the 3-cycle with wrapping numbers $(n_i, m_i) = (1, 0)$, so that $[\Pi_{O6}] = 8[a_1][a_2][a_3]$.

We now introduce D6$_a$-branes, with multiplicities N_a and wrapping numbers (n_a^i, m_a^i) of the D6-brane stacks. We also introduce their orientifold images, with wrapping numbers $(n_a^i, -m_a^i)$ for rectangular 2-tori, or $(n_a^i, -n_a^i - m_a^i)$ for tilted tori, see Fig. 22. To unify their description, we introduce $\tilde{m}_a = m_a + \beta n_a$, so that branes and images have wrapping numbers $(n_a, \tilde{m}_a)$ and $(n_a, -\tilde{m}_a)$ respectively.

The RR tadpole conditions are simple to obtain. In the case of rectangular two-tori, they are explicitly given by

$$\sum_a N_a n_a^1 n_a^2 n_a^3 = 16,$$

$$\sum_a N_a n_a^1 m_a^2 m_a^3 = 0 \text{ and permutations} \tag{5.5}$$

The spectrum is as discussed above, with the specific intersection numbers computed using (4.2). Namely, explicit examples are discussed in further sections.

6. Getting just the standard model

In this section we consider some of the phenomenologically most interesting constructions, where the chiral part of the low-energy spectrum is given by that of the Standard Model (SM). The models are based on brane configurations first discussed in [21, 26].

As we have discussed above, models without orientifold planes do not lead to the chiral spectrum of the Standard Model. In fact, there is a general argument [26] showing that any such construction always contains additional chiral fermions in $SU(2)$ doublets, beyond those in the SM, as follows. First notice that in such models, the gauge group is a product of unitary factors, so the electroweak $SU(2)$ must belong to a $U(2)$ factor in the gauge group. As mentioned in Section 4.1.3, the RR tadpole cancellation conditions imply that the number of fundamentals and antifundamentals for each $U(N)$ factor must be equal, even for $U(2)$ (where the 2 and the $\bar{2}$ are distinguished by their $U(1)$ charge). Now in any such model with SM gauge group containing $SU(3) \times SU(2)$, the left-handed quarks must belong to a representation $3(3, \bar{2})$, contributing nine anti-fundamentals of $SU(2)$. The complete spectrum must necessarily contain nine fundamentals of $SU(2)$, three of which may be interpreted as left-handed leptons; the remaining six doublets are however exotic chiral fermions, beyond the spectrum of the SM.

The introduction of orientifold planes in the construction allows to avoid this issue in several ways, as we describe in this section. In fact, as a consequence, they allow to construct string compactifications with the chiral spectrum of just the SM. This is a remarkable achievement.

6.1. The $U(2)$ class

One possibility [26] is to exploit the fact that in orientifold models there are two different kinds of bifundamental representations that arise in the spectrum, namely $(\square, \bar{\square})$ and $(\square, \square)$. This allows an alternative construction of the SM chiral fermion spectrum, satisfying the RR tadpole constraint on the spectrum without exotics, as follows. Consider realizing the three families of left-handed quarks as $(3, \bar{2}) + 2(3, 2)$. This contributes three net $SU(2)$ doublets, hence the three $SU(2)$ doublets required in the model correspond simply to the three left-handed leptons.

Indeed, it is possible to propose a set of intersection numbers, such that any configuration of D6-branes wrapped on 3-cycles with those intersections numbers reproduces the chiral spectrum of the SM. Consider [26] four stacks of D6-branes, denoted a, b, c, d (and their images), giving rise to a gauge group

$U(3)_a \times U(2)_b \times U(2)_c \times U(1)_d$. If the intersections numbers of the corresponding 3-cycles are given by

$$I_{ab} = 1; \quad I_{ab'} = 2; \quad I_{ac} = -3; \quad I_{ac'} = -3;$$
$$I_{bd} = 0; \quad I_{bd'} = -3; \quad I_{cd} = -3; \quad I_{cd'} = 3 \tag{6.1}$$

then the chiral spectrum of the model has the non-Abelian quantum numbers of the chiral fermions in the SM (plus right-handed neutrinos). In order to reproduce exactly the SM spectrum, one also needs to require that the linear combination of $U(1)$'s

$$Q_y = \frac{1}{6}Q_a - \frac{1}{2}Q_c + \frac{1}{2}Q_d \tag{6.2}$$

which reproduces the hypercharge quantum numbers, remains as the only massless $U(1)$ in the model.

It is important to emphasize that at this level, we have not constructed any explicit model. Rather, we have made a general proposal of what kind of structure one must implement in concrete examples to lead to the SM chiral spectrum. This is however a very useful step.

In [26] there is a large class of examples of models of this kind, constructed explicitly in terms of D6-branes on factorized 3-cycles in the orientifold of $\mathbf{T^6}$ discussed in Section 5.3.2. To illustrate the discussion with an example, consider the model in [26] corresponding to the parameters

$$\beta^1 = \beta^2 = 1; \quad \epsilon = \rho = 1; \quad n_a^2 = 4;$$
$$n_b^1 = 1; \quad n_c^1 = 5; \quad n_d^2 = 2 \tag{6.3}$$

The D6-brane configuration (without specifying the images) is given by

	N	(n^1, m^1)	(n^2, m^2)	$(n^3, \tilde{m}^3)$
a	3	$(1, 0)$	$(4, 1)$	$(1, \frac{1}{2})$
b	2	$(1, 1)$	$(1, 0)$	$(1, \frac{3}{2})$
c	1	$(5, 3)$	$(1, 0)$	$(0, 1)$
d	1	$(1, 0)$	$(2, -1)$	$(1, \frac{3}{2})$

Let us emphasize again that the proposal to obtain the SM from models with the intersection numbers above is not restricted to the toroidal orientifold setup. Indeed, they have been discussed in [17, 18] in the large volume regime of geometric compactifications, and in [27] a large class of models has been constructed in Gepner constructions. The latter models are fully supersymmetric, leading to almost MSSM spectra (differing from it in the structure of the non-chiral Higgs sector), showing that the proposal can be exploited to construct supersymmetric models as well.

6.2. The USp(2) class

Another possible way to avoid the problem of the extra $SU(2)$ doublets, is to exploit the fact that D6-branes in the presence of orientifold planes may contain $USp(N)$ gauge factors (see Section 5.2). For the latter, all representations are real, and RR tadpole conditions do not impose any constraint on the matter content. Since $USp(2) \equiv SU(2)$, it is possible to realize the electroweak $SU(2)$ in terms of such D6-brane with $USp(2)$ gauge group, and thus circumvent the constraints on the number of doublets.

Indeed such a construction is proposed in [21]. The SM spectrum would arise in terms of a configurations of four stacks of D6-branes, leading to a gauge group $U(3)_a \times USp(2)_b \times U(1)_c \times U(1)_d$, with intersection numbers

$$I_{ab} = 3; \quad I_{ab'} = 3; \quad I_{ac} = -3; \quad I_{ac'} = -3; \quad I_{db} = 3;$$
$$I_{db'} = 3; \quad I_{dc} = -3; \quad I_{dc'} = 3; \quad I_{bc} = -1; \quad I_{bc'} = 1 \tag{6.4}$$

the $U(1)$ that needs to be massless in order to reproduce the SM hypercharge is

$$Q_Y = \frac{1}{6}Q_a - \frac{1}{2}Q_c - \frac{1}{2}Q_d \tag{6.5}$$

Moreover, explicit realizations of D6-branes on 3-cycles with those intersection numbers (and with massless hypercharge) have been constructed in toroidal orientifolds [21]. Let us consider an illustrative example, corresponding to $\rho = 1$ in that reference. The set of D6-branes (to which we should add the images) is specified by

	N	(n^1, m^1)	(n^2, m^2)	(n^3, m^3)
a	3	$(1, 0)$	$(1, 3)$	$(1, -3)$
b	1	$(0, 1)$	$(1, 0)$	$(0, -1)$
c	1	$(0, 1)$	$(0, -1)$	$(1, 0)$
d	1	$(1, 0)$	$(1, 3)$	$(1, -3)$

One needs to add additional branes to satisfy the RR tadpole condition, but this may be done with the latter having no intersection with the above one. Hence the additional D6-branes are decoupled, and we do not discuss them for simplicity. The above D6-brane configuration can preserve supersymmetry locally, but with supersymmetry is eventually broken by the additional decoupled D6-brane sector.

Notice that in this realization, the $USp(2)$ factor arises from the D6-brane b and its image, when they are coincident. Notice also that the D6-brane d and its image, and the a and c stacks, can be taken to coincide. Thus the above standard model configuration can be considered a spontaneously broken Pati-Salam theory, with original gauge group $U(4) \times USp(2)_L \times USp(2)_R$.

As emphasized above, the proposed intersection numbers may be realized in other contexts, also with or without supersymmetry. Explicit constructions of models with those intersection numbers, with supersymmetry will be studied in next section (see also [28], and [27] for Gepner model constructions).

7. Supersymmetric models

In this section we review some simple supersymmetric 4d chiral models of intersecting D6-branes, in [24] to which we refer the reader for additional details (see e.g. [29–31] for additional models in other orbifolds, see also [32] for early work on diverse non-chiral supersymmetric orbifolds with intersecting branes). For a more geometric description of the model, adapting the recipe in Section 4.2, see [17].

7.1. Orientifold of $T^6/Z_2 \times Z_2$

In order to obtain supersymmetric models, one needs a sufficient number of O6-planes in the construction. One of the simplest possibilities is the $\Omega R(-)^{F_L}$ orientifold of the $\mathbf{T^6}/(\mathbf{Z_2} \times \mathbf{Z_2})$ orbifold.

We consider type IIA theory on $\mathbf{T^6}/(\mathbf{Z_2} \times \mathbf{Z_2})$, with generators θ, ω associated to the twists $v = (\frac{1}{2}, -\frac{1}{2}, 0)$ and $w = (0, \frac{1}{2}, -\frac{1}{2})$, hence acting as

$$\theta : (z_1, z_2, z_3) \rightarrow (-z_1, -z_2, z_3),$$
$$\omega : (z_1, z_2, z_3) \rightarrow (z_1, -z_2, -z_3) \tag{7.1}$$

where z_i are complex coordinates in the $\mathbf{T^6}$. The action projects out some of the moduli, in particular implies the $\mathbf{T^6}$ is factorizable. We mod out this theory by $\Omega R(-)^{F_L}$, where

$$R : (z_1, z_2, z_3) \rightarrow (\bar{z}_1, \bar{z}_2, \bar{z}_3) \tag{7.2}$$

The model contains four kinds of O6-planes, associated to the actions of ΩR, $\Omega R\theta$, $\Omega R\omega$, $\Omega R\theta\omega$, as shown in Fig. 23 (for rectangular 2-tori). For simplicity we henceforth center on rectangular two-tori.

In order to cancel the corresponding RR tadpoles, we introduce D6-branes wrapped on three-cycles as in previous discussions. Also for simplicity we assume that each stack of D6-branes is passing through $\mathbf{Z_2} \times \mathbf{Z_2}$ fixed points. These

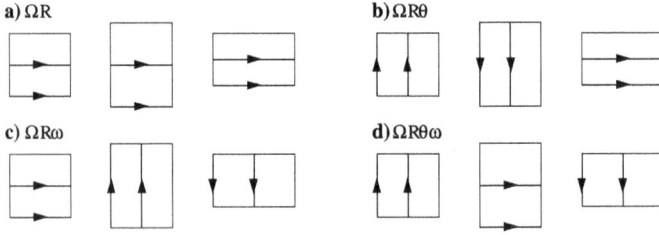

Fig. 23. O6-planes in the orientifold of $\mathbf{T}^6/(\mathbf{Z}_2 \times \mathbf{Z}_2)$.

extra projections are responsible for the fact that N D6-branes lead to an $U(N/2)$ gauge symmetry. The RR tadpole conditions have the familiar form

$$\sum_a N_a \, [\Pi_a] + \sum_a N_a \, [\Pi_{a'}] - 4[\Pi_{O6}] = 0 \qquad (7.3)$$

where $[\Pi_{O6}]$ is the homology charge of the complete set of O6-planes. More explicitly, for instance for rectangular tori we have

$$\sum_a N_a n_a^1 n_a^2 n_a^3 - 16 = 0,$$

$$\sum_a N_a n_a^1 m_a^2 m_a^3 + 16 = 0,$$

$$\sum_a N_a m_a^1 n_a^2 m_a^3 + 16 = 0, \qquad (7.4)$$

$$\sum_a N_a m_a^1 m_a^2 n_a^3 + 16 = 0$$

Skipping the details, the chiral spectrum is

Sector	Representation
aa	$U(N_a/2)$ vector multiplet
	3 Adj. chiral multiplets
$ab + ba$	I_{ab} ($\square_a, \overline{\square}_b$) fermions
$ab' + b'a$	$I_{ab'}$ ($\square_a, \square_b$) fermions
$aa' + a'a$	$-\frac{1}{2}(I_{aa'} - 4I_{a,O6})$ $\square\square$ fermions
	$-\frac{1}{2}(I_{aa'} + 4I_{a,O6})$ $\square\!\square$ fermions

The condition that the system of branes preserves $N = 1$ supersymmetry is that each stack of D6-branes is related to the O6-planes by a rotation in $SU(3)$,

Table 1

D-brane magnetic numbers giving rise to an $\mathcal{N} = 1$
MSSM like model, in the $\mathbf{T^6}/(\mathbf{Z_2} \times \mathbf{Z_2})$ orientifold

N_a	(n_a^1, m_a^1)	(n_a^2, m_a^2)	(n_a^3, m_a^3)
$N_a = 6$	$(1, 0)$	$(3, 1)$	$(3, -1)$
$N_b = 2$	$(0, 1)$	$(1, 0)$	$(0, -1)$
$N_c = 2$	$(0, 1)$	$(0, -1)$	$(1, 0)$
$N_d = 2$	$(1, 0)$	$(3, 1)$	$(3, -1)$
$N_{h_1} = 2$	$(-2, 1)$	$(-3, 1)$	$(-4, 1)$
$N_{h_2} = 2$	$(-2, 1)$	$(-4, 1)$	$(-3, 1)$
40	$(1, 0)$	$(1, 0)$	$(1, 0)$

Table 2

$\mathcal{N} = 1$ spectrum derived from the D-brane content of Table 1 after D-brane recombination. There is no chiral matter arising from ah, ah', hh' or charged under $USp(40)$. The generator of $U(1)'$ is now given by $Q' = \frac{1}{3}Q_a - 2Q_h$

Sector	Matter	$SU(4) \times SU(2) \times SU(2) \times [USp(40)]$	Q_a	Q_h	Q'
(ab)	F_L	$3(4, 2, 1)$	1	0	$1/3$
(ac)	F_R	$3(\bar{4}, 1, 2)$	-1	0	$-1/3$
(bc)	H	$(1, 2, 2)$	0	0	0
(bh)		$2(1, 2, 1)$	0	-1	2
(ch)		$2(1, 1, 2)$	0	$+1$	-2

see Section 3.2. More specifically, denoting by θ_i the angles the D6-brane forms with the horizontal direction in the ith two-torus, supersymmetry preserving configurations must satisfy

$$\theta_1 + \theta_2 + \theta_3 = 0 \qquad (7.5)$$

For fixed wrapping numbers (n^i, m^i), the condition translates into a constraint on the ratio of the two radii on each torus. For rectangular tori, denoting $\chi_i = (R_2/R_1)_i$, with R_2, R_1 the vertical resp. horizontal directions, the constraint is

$$\arctan\left(\chi_1 \frac{m_1}{n_1}\right) + \arctan\left(\chi_2 \frac{m_2}{n_2}\right) + \arctan\left(\chi_3 \frac{m_3}{n_3}\right) = 0 \qquad (7.6)$$

To provide an illustrative example, we consider a model [30] containing a sector of branes leading to the SM fields (belonging to the $USp(2)$ class in Section 6.2, plus an additional set of branes required for RR tadpole cancellation (and contributing vector-like exotic matter in the spectrum). The set of branes is given in Table 1.

The spectrum is fairly complicated. However, under recombination of the branes h_1, h_2 and images, it has the simpler form given in Table 2.

We hope that these examples suffice to illustrate the flexibility of the techniques we have discussed and allows the reader to safely jump into the literature for further details.

Appendix A. Spectrum of open strings

Appendix A.1. Single D-brane in flat 10d space

In this appendix we describe a simplified calculation of the spectrum of open strings for a configuration of a single type II Dp-brane in flat 10d space.

In string theory the physical degrees of freedom for the string oscillation (in the light-cone gauge) are described by a set of functions $X^m(\sigma, t)$, which, at each (world-sheet) time t, define the graph of the string oscillation in the ith transverse dimension, with $m = 2, \ldots, 9$. The coordinate σ parametrizes the length of the string, and runs from 0 to $\ell = 4\pi\alpha' p^+$, where $\alpha' = M_s^{-2}$ is the inverse of the string tension, and p^+ is the light cone momentum. This is shown in Fig. 24. For each such transverse direction, denoted generically $X(\sigma, t)$, one can perform a general mode expansion

$$
\begin{aligned}
X(\sigma, t) &= x + \frac{p}{p^+}t + i\sqrt{\frac{\alpha'}{2}} \sum_\nu \frac{\alpha_\nu}{\nu} \exp[-\pi i \, \nu \, (\sigma + t)/\ell] \\
&\quad + i\sqrt{\frac{\alpha'}{2}} \sum_{\tilde{\nu}} \frac{\tilde{\alpha}_{\tilde{\nu}}^i}{\tilde{\nu}} \exp[-\pi i \, \tilde{\nu} \, (\sigma - t)/\ell]
\end{aligned}
$$

We need to impose that the open string endpoints can move freely along the coordinates spanned by the D-brane world-volume, denoted X^μ, with $\mu = 2, \ldots, p$, but are fixed at the D-brane location in the transverse coordinates, denoted X^i, with $i = p + 1, \ldots, 9$. This is implemented by the boundary conditions (of Neumann and Dirichlet type)

$$
\begin{array}{llll}
\mu = 2, \ldots, p & \partial_\sigma X^\mu(\sigma, t) = 0 & \text{at } \sigma = 0, \ell & NN \\
i = p + 1, \ldots, 9 & \partial_t X^i(\sigma, t) = 0 & \text{at } \sigma = 0, \ell & DD
\end{array}
\tag{A.1}
$$

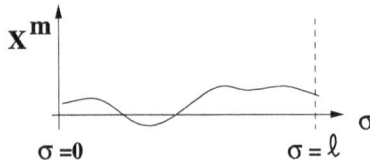

Fig. 24. A string configuration is specified (in the light-cone gauge) by the position $X^i(\sigma)$ in transverse space for the point at the coordinate value σ along the string.

Imposing these constraints, the expansions become

$$
\begin{aligned}
X^\mu(\sigma, t) &= x^\mu + \frac{p^\mu}{p^+}t + i\sqrt{2\alpha'}\sum_{n\neq 0}\frac{\alpha_n^\mu}{n}\cos[\pi n\sigma/\ell]\exp[-\pi int/\ell]\\
X^i(\sigma, t) &= x^i + \sqrt{2\alpha'}\sum_{n\neq 0}\frac{\alpha_n^i}{n}\sin[\pi n\sigma/\ell]\exp[-\pi int/\ell]
\end{aligned}
\tag{A.2}
$$

The parameters x^μ are arbitrary, while x^i must correspond to the coordinates of the D-brane in the corresponding dimension. Hence the string centre of mass is localized on the D-brane world-volume, as announced.

The oscillation modes α_n^μ, α_n^i satisfy the commutation relations

$$
[\alpha_n^\mu, \alpha_{-m}^\nu] = n\delta_{n,m}\delta^{\mu\nu}; \quad [\alpha_n^i, \alpha_{-m}^j] = n\delta_{n,m}\delta^{ij}
\tag{A.3}
$$

corresponding to one infinite set of decoupled harmonic oscillators, for each spacetime dimension.

Recall that type II superstrings also have fermionic oscillation degrees of freedom. They can be similarly described in terms of an infinite set of decoupled fermionic harmonic oscillators, for each spacetime dimension. Hence we have an additional set of operators, Ψ_{n+v}^μ, Ψ_{n+r}^i, with $r = \frac{1}{2}, 0$ for fermions with NS or R boundary (or periodicity) conditions, obeying the anticommutation relations

$$
\{\Psi_{n+r}^\mu, \alpha_{-(m+r)}^\nu\} = \delta_{n,m}\delta^{\mu\nu}; \quad \{\Psi_{n+r}^i, \Psi_{-(m+r)}^j\} = \delta_{n,m}\delta^{ij}
\tag{A.4}
$$

To construct the Hilbert space of string oscillation states, one first defines a vacuum state given by the product of groundstates of the infinite (bosonic and fermionic) harmonic oscillators, namely annihilated by all positive modding oscillators. Next one builds physical states by applying raising operators, corresponding to negative modding oscillators. In string theory, each string oscillation quantum state corresponds to a particle in spacetime. Its spacetime mass is given by

$$
\alpha'M^2 = N_B + N_F - \frac{1}{2}r(1 - r)
\tag{A.5}
$$

where N_B and N_F are the bosonic and fermionic oscillator number operators

$$
N_B = \sum_{n=1}^{\infty}\sum_{\mu,i}\alpha_{-n}\alpha_n; \quad N_F = \sum_{n=0}^{\infty}\sum_{\mu,i}(n+r)\Psi_{-(n+r)}\Psi_{n+r}
\tag{A.6}
$$

We will be interested in the lightest (in fact, massless, states).

In the NS sector (where world-sheet fermions satisfy NS periodicity conditions) there are no fermion zero modes and the vacuum is non-degenerate. The lightest states, along with their mass and their interpretation as particles is the $(p + 1)$-dimensional D-brane world-volume, are

State	$\alpha' M^2$	GSO proj.	$(p + 1)$-dim.field
$|0\rangle$	$-\frac{1}{2}$	Out	–
$\psi^{\mu}_{-\frac{1}{2}}|0\rangle$	0	OK	A_{μ}
$\psi^{i}_{-\frac{1}{2}}|0\rangle$	0	OK	ϕ^i

We have also indicated the effect of the GSO projection (required from open-closed duality and the GSO projection for closed strings) on these fields. Notice that the tachyonic mode in the first line is projected out and removed from the physical spectrum, while the massless modes survive.

In the R sector, there are fermion zero modes Ψ^{μ}_0, Ψ^i_0. Since their application has zero cost in energy, the vacuum state is degenerate. Denoting zero modes collectively by Ψ^k_0, they satisfy the algebra $\{\Psi^k_0, \Psi^l_0\} = \delta^{kl}$. This implies that the degenerated vacuum states form a representation of this Clifford algebra, namely they transform as spinors of its $SO(8)$ invariance group. Hence the groundstates can be labeled by the two $SO(8)$ spinor representations of opposite chiralities, denoted $\mathbf{8}_S$, $\mathbf{8}_C$. The corresponding particles in the $(p + 1)$-dimensional world-volume transform in the representations of the Lorentz group obtained by decomposing the representations of $SO(8)$ under $SO(p-1)$ (the subgroup of $SO(p+1)$ manifest in light-cone gauge). This corresponds to a set of $(p + 1)$-dimensional fermions, whose detailed structure is dimension-dependent, but straightforward to determine in each specific case. The set of light states, along with their masses, behaviour under GSO, and $(p + 1)$-dimensional interpretation, are

State	$\alpha' M^2$	GSO proj.	$(p + 1)$-dim.field
$\mathbf{8}_S$	0	Out	–
$\mathbf{8}_C$	0	OK	λ_{α}

The final result is that the set of massless particles on the Dp-brane world-volume is given by a $U(1)$ gauge boson, $9-p$ real scalars and some fermions. The scalars (resp. fermions) can be regarded as Goldstone bosons (resp. Goldstinos) of the translational symmetries (resp. supersymmetries) of the vacuum broken by the presence of the D-brane. The open string sector fills out a $U(1)$ vector multiplet with respect to the 16 supersymmetries unbroken by the D-brane.

As described in the main text, when n parallel D-branes overlap, the world-volume gauge theory is enhanced to an $U(n)$ gauge group, and matter fields transform in the adjoint representation.

Appendix A.2. Open string spectrum for intersecting D6-branes

In this section we carry out the computation of the spectrum of open strings in the configuration of two stacks of intersecting D6-branes [13]. In particular, we explicitly show the appearance of 4d chiral fermions from the sector of open strings stretching between the different D6-brane stacks. The key point in getting chiral fermions is that the non-trivial angles between the branes removes fermion zero modes in the R sector, and leads to a smaller Clifford algebra.

As discussed above, the spectrum of states for open strings stretched between branes in the same stack is exactly as in Section 2.2. It yields an $U(n_a)$ vector multiplet in the 7d world-volume of the ath D6-brane stack.

We thus center in the computation of the spectrum of states for open strings stretched between two stacks a, b. The open string boundary conditions for the coordinates along M_4 are of the NN kind, and lead to the oscillators α_n^μ, Ψ_{n+r}^μ. For the directions where the branes form non-trivial angles, for instance in the 45 2-plane, we have boundary conditions

$$\partial_\sigma X^4\big|_{\sigma=0} = 0$$
$$\partial_t X^5\big|_{\sigma=0} = 0$$
$$\left(\cos \pi\theta^1 \, \partial_\sigma X^4 + \sin \pi\theta^1 \, \partial_\sigma X^5\right)\big|_{\sigma=\ell} = 0 \tag{A.7}$$
$$\left(-\sin \pi\theta^1 \, \partial_t X^4 + \cos \pi\theta^1 \, \partial_t X^5\right)\big|_{\sigma=\ell} = 0$$

where $\pi(\theta_1)_{ab}$ is the angle from the ath to the bth D6-brane, written $\pi\theta_1$ for short. One has similar expression for the coordinates associated to the remaining two-planes.

It is convenient to define complex coordinates $Z^i = X^{2i+2} + iX^{2i+3}$, $i = 1, 2, 3$. The boundary conditions for ab open strings thus read

$$\partial_\sigma (\text{Re } Z^i)\big|_{\sigma=0} = 0; \qquad \partial_t (\text{Im } Z^i)\big|_{\sigma=0} = 0$$
$$\partial_\sigma [\text{Re } (e^{i\theta_i} Z^i)]\big|_{\sigma=0} = \ell; \qquad \partial_t [\text{Im } (e^{i\theta_i} Z^i)]\big|_{\sigma=\ell} = 0 \tag{A.8}$$

These boundary conditions shift the oscillator moddings by an amount $\pm\theta^i$. The oscillator operators (which are now associated to complex coordinates) are $\alpha_{n+\theta_i}^i$, $\alpha_{n-\theta_i}^{\bar{i}}$, $\Psi_{n+r+\theta_i}^i$, $\Psi_{n+r-\theta_i}^{\bar{i}}$. It is important to point out that the centre of mass degrees of freedom are frozen in these directions, so that the open strings are localized at the intersection between the D6-branes.

All these operators satisfy decoupled harmonic oscillator (anti)commutation relations. As before, the Hilbert space of string oscillation modes is obtained by first constructing a vacuum (annihilated by all positive modding oscillators) and then applying creation operators to it (corresponding to negative modding oscillators). Each oscillation state corresponds to a particle that propagates on the 4d intersection of the D6-brane world-volumes. Its spacetime mass is given by

$$\alpha' M^2 = N_B + N_F + E_0 \tag{A.9}$$

where N_B and N_F are the oscillator numbers, and $E_0 = -\frac{1}{2}(1 + \sum_i \theta_i)$ in the NS sector and $E_0 = 0$ in the R sector (in the normal ordering here and in what follows, we have assumed that $0 \le \theta_i \le 1$).

In the NS sector, the groundstate is non-degenerate. The lightest states surviving the GSO projection are (for the above range of θ_i)

State	$\alpha' M^2$	4d field
$\Psi^1_{-\frac{1}{2}+\theta_1}\|0\rangle$	$\frac{1}{2}(-\theta_1 + \theta_2 + \theta_3)$	Scalar
$\Psi^2_{-\frac{1}{2}+\theta_2}\|0\rangle$	$\frac{1}{2}(\theta_1 - \theta_2 + \theta_3)$	Scalar
$\Psi^3_{-\frac{1}{2}+\theta_3}\|0\rangle$	$\frac{1}{2}(\theta_1 + \theta_2 - \theta_3)$	Scalar
$\Psi^1_{-\frac{1}{2}+\theta_1}\Psi^2_{-\frac{1}{2}+\theta_2}\Psi^3_{-\frac{1}{2}+\theta_3}\|0\rangle$	$1 - \frac{1}{2}(-\theta_1 - \theta_2 - \theta_3)$	Scalar

These scalars are complexified by similar scalars in the ba open string sector. Hence we obtain complex scalars, with masses given above, in the bifundamental representation $(n_a, \bar{n}_b)$ of the $U(n_a) \times U(n_b)$ gauge factor.

In the R sector, there are two fermion zero modes associated to the M_4 directions, hence the vacuum is degenerate. They satisfy a Clifford algebra, so the vacuum fills out two opposite-chirality spinor representations of $SO(2)$. Denoting them by $\pm\frac{1}{2}$, where the label corresponds to the 4d chirality, the $+\frac{1}{2}$ state is projected out by the GSO projection, while the $-\frac{1}{2}$ state survives. Taking into account a similar state surviving in the ba sector, in total we have a 4d left-handed chiral fermion in the bi-fundamental representation $(n_a, \bar{n}_b)$. The chirality of the fermion is determined by the orientation of the intersection.

References

[1] A.M. Uranga, Chiral four-dimensional string compactifications with intersecting D-branes, *Class. Q. Grav.* **20** (2003) S373;
    F. Marchesano, Intersecting D-brane models, hep-th/0307252;
    R. Blumenhagen, M. Cvetic, P. Langacker, G. Shiu, Toward realistic intersecting D-brane models, hep-th/0502005.
[2] J. Polchinski, Tasi lectures on D-branes, *Fields, Strings and Duality* (Boulder 1996) p. 293, hep-th/9611050.

[3] C.V. Johnson, *D-branes*, Cambridge, USA, Univ. Pr. (2003) 548.

[4] J. Polchinski, Dirichlet Branes and Ramond-Ramond charges, *Phys. Rev. Lett.* **75** (1995) 4724.

[5] M.R. Douglas and G.W. Moore, D-branes, quivers, and ALE instantons, hep-th/9603167;
M.R. Douglas, B.R. Greene and D.R. Morrison, Orbifold resolution by D-branes, *Nucl. Phys. B* **506** 84.

[6] G. Aldazabal, L.E. Ibáñez, F. Quevedo and A.M. Uranga, D-branes at singularities: A Bottom up approach to the string embedding of the standard model, *JHEP* **08** (2000) 002.

[7] D. Berenstein, V. Jejjala and R.G. Leigh, The Standard model on a D-brane, *Phys. Rev. Lett.* **88** (2002) 071602;
L.F. Alday and G. Aldazabal, In quest of just the standard model on D-branes at a singularity, *JHEP* **05** (2002) 022.

[8] E. Witten, Some properties of $O(32)$ superstrings, *Phys. Lett. B* **149** (1984) 351.

[9] C. Bachas, A way to break supersymmetry, hep-th/9503030.

[10] C. Angelantonj, I. Antoniadis, E. Dudas and A. Sagnotti, Type I strings on magnetized orbifolds and brane transmutation, *Phys. Lett. B* **489** (2000) 223.

[11] R. Blumenhagen, L. Görlich, B. Körs and D. Lüst, Noncommutative compactifications of type I strings on tori with magnetic background flux, *JHEP* **10** (2000) 006.

[12] G. Aldazabal, S. Franco, L.E. Ibanez, R. Rabadan and A.M. Uranga, $D = 4$ chiral string compactifications from intersecting branes, *J. Math. Phys.* **42** (2001) 3103, hep-th/0011073;
Intersecting brane worlds, *JHEP* **0102** (2001) 047, hep-ph/0011132.

[13] M. Berkooz, M.R. Douglas and R.G. Leigh, Branes intersecting at angles, *Nucl. Phys. B* **480** (1996) 265, hep-th/9606139.

[14] M.B. Green, J.A. Harvey and G.W. Moore, I-brane inflow and anomalous couplings on d-branes, *Class. Q. Grav.* **14** (1996) 47.

[15] M.R. Douglas, Topics in D geometry, *Class. Q. Grav.* **17** (2000) 1057.

[16] G. Lawlor, The angle criterion, *Invent. Math.* **95** (1989) 437.

[17] R. Blumenhagen, V. Braun, B. Körs and D. Lüst, Orientifolds of K3 and Calabi–Yau manifolds with intersecting D-branes, *JHEP* **0207** (2002) 026, hep-th/0206038.

[18] A.M. Uranga, Local models for intersecting brane worlds, hep-th/0208014.

[19] N. Arkani-Hamed, S. Dimopoulos and G.R. Dvali, The hierarchy problem and new dimensions at a millimeter, *Phys. Lett. B* **429** (1998) 263–272, hep-ph/9803315;
I. Antoniadis, N. Arkani-Hamed, S. Dimopoulos and G.R. Dvali, New dimensions at a millimeter to a Fermi and superstrings at a TeV, *Phys. Lett. B* **436** (1998) 257–263, hep-ph/9804398.

[20] D. Cremades, L.E. Ibanez and F. Marchesano, Intersecting brane models of particle physics and the Higgs mechanism, *JHEP* **0207** (2002) 022, hep-th/0203160.

[21] D. Cremades, L.E. Ibanez and F. Marchesano, Towards a theory of quark masses, mixings and CP violation, hep-ph/0212064.

[22] M. Cvetic and I. Papadimitriou, Conformal field theory couplings for intersecting D-branes on orientifolds, *Phys. Rev. D* **68** (2004) 046001, Erratum-ibid. **D70** (2004) 029903;
D. Lüst, P. Mayr, R. Richter and S. Stieberger, Scattering of gauge, matter, and moduli fields from intersecting branes, *Nucl. Phys. B* **696** (2004) 205, hep-th/0404134.

[23] E.G. Gimon and J. Polchinski, Consistency conditions for orientifolds and D-manifolds, *Phys. Rev. D* **54** (1996) 1667.

[24] M. Cvetic, G. Shiu and A.M. Uranga, Chiral four-dimensional $N = 1$ supersymmetric type 2A orientifolds from intersecting D6 branes, *Nucl. Phys. B* **615** (2001) 3, hep-th/0107166;
Three family supersymmetric standard—like models from intersecting brane worlds, *Phys. Rev. Lett.* **87** (2001) 201801, hep-th/0107143.

[25] R. Blumenhagen, B. Kors and D. Lust, Type I strings with F flux and B flux, *JHEP* **0102** (2001) 030, hep-th/0012156.

[26] L.E. Ibanez, F. Marchesano and R. Rabadan, Getting just the standard model at intersecting branes, *JHEP* **0111** (2001) 002, hep-th/0105155.

[27] T.P.T. Dijkstra, L.R. Huiszoon and A.N. Schellekens, Chiral supersymmetric standard model spectra from orientifolds of Gepner models, *Phys. Lett. B* **609** (2005) 408, hep-th/0403196; Supersymmetric standard model spectra from RCFT orientifolds, *Nucl. Phys. B* **710** (2005) 3, hep-th/0411129.

[28] G. Honecker and T. Ott, Getting just the supersymmetric standard model at intersecting branes on the Z(6) orientifold, *Phys. Rev. D* **70** (2004) 126010, Erratum-ibid. *D* **71** (2005) 069902.

[29] R. Blumenhagen, L. Gorlich and T. Ott, Supersymmetric intersecting branes on the type 2A T6/Z(4) orientifold, hep-th/0211059; G. Honecker, Chiral supersymmetric models on an orientifold of Z(4) × Z(2) with intersecting D6-branes, *Nucl. Phys. B* **666** (2003) 175; R. Blumenhagen, J.P. Conlon and K. Suruliz, Type IIA orientifolds on general supersymmetric Z(N) orbifolds, *JHEP* **07** (2004) 022.

[30] F. Marchesano and G. Shiu, MSSM vacua from flux compactifications, *Phys. Rev. D* **71** (2005) 011701, hep-th/0408059; Building MSSM flux vacua, *JHEP* **0411** (2004) 041, hep-th/0409132.

[31] M. Cvetic, P. Langacker, T. Li and T. Liu, Supersymmetric Pati–Salam models from intersecting D6-branes: A road to the standard model, *Nucl. Phys. B* **698** (2004) 163, hep-th/0403061; D6-brane splitting on type IIA orientifolds, *Nucl. Phys. B* **709** (2005) 241, hep-th/0407178.

[32] R. Blumenhagen, L. Gorlich and B. Kors, Supersymmetric 4-D orientifolds of type IIA with D6-branes at angles, *JHEP* **0001** (2000) 040, hep-th/9912204; S. Forste, G. Honecker and R. Schreyer, Supersymmetric Z(N) × Z(M) orientifolds in 4-D with D branes at angles, *Nucl. Phys. B* **593** (2001) 127, hep-th/0008250.

Course 8

NON-RENORMALISATION THEOREMS IN SUPERSTRING AND SUPERGRAVITY THEORIES

Pierre Vanhove

CEA, DSM, Institut de Physique Théorique,
IPhT, CNRS, MPPU, URA2306,
and
Niels Bohr Institute, University of Copenhagen,
Blegdamsvej 17, DK–2100 Copenhagen Ø, Denmark
pierre.vanhove@cea.fr

C. Bachas, L. Baulieu, M. Douglas, E. Kiritsis, E. Rabinovici, P. Vanhove, P. Windey
and L.F. Cugliandolo, eds.
Les Houches, Session LXXXVII, 2007
String Theory and the Real World: From Particle Physics to Astrophysics
© *2008 Published by Elsevier B.V.*

Contents

1. Introduction

The theoretical construction of unification models for particle physics has led to remarkable progress in the understanding of the fundamental interactions involved in particle physics phenomena at accelerator energy scale, or in cosmological phenomena responsible for the formation of the visible matter of our universe. However a good understanding of quantum gravity effects at either short distances or large (cosmological) scales is still lacking. It is expected that subtle quantum gravity effects could be at work behind some of the outstanding fundamental problems of modern cosmology and particle physics models and their ultra-violet completion (or may be the absence of constraints from the ultra-violet completion). For instance, because of our poor understanding of the rules for a correct quantisation of the gravitational forces one gets a landscape of vacua for unification models coupled to gravity, as was explained by Nima Arkani-Hamed, Frederik Denef and Michael Douglas at this school. Hopefully, the difficulties of charting the physically relevant vacua of string theory (or any other consistent theory of quantum gravity) would be resolved once the correct boundary conditions and quantization rules for quantum gravity has been better understood.

It is a remarkable feature of the string setup to provide a consistent theory for quantum gravity and its supersymmetric extensions [1–3]. String theory provides a consistent framework for analysing perturbative and non-perturbative aspects of quantum gravity. The low energy approximation of various compactifications of string theory leads to the supergravity theories and their quantum corrections. In string theory based models the various coupling constant depends on the moduli of the theory which are acted on by the perturbative and non-perturbative symmetries of string theory. These symmetries are the U-dualities [4] connecting all the different corners of the M-theory moduli space [5].

A typical amplitude computed within string theory is given by an integral over the moduli space of the punctured Riemann surface used for the definition of the amplitude [1,2,6]. This moduli space resums in a very compact expression the contributions from the many Feynman graphs one has to sum in the usual field theoretical analysis [7–10]. This compact formulation makes explicit some cancellations that are not a priori obvious using the traditional Feynman rules for constructing the amplitudes. Even if the maximal $N = 8$ supergravity is perturbatively ultra-violet finite, it will not be complete in the ultra-violet and the inclusion of extra non-perturbative states will be needed for getting a consistent theory. These extra states are charged under the U-dualities of M-theory and their decoupling from the supergravity massless states is singular [11].

In these lecture notes we will describe the various lessons than one can draw about the role of linearised on-shell supersymmetry and the string dualities in the analysis of amplitude computations in the gravitational sector of string theory and supergravity theories in various dimensions. We will review, in Section 2 the role of the on-shell linearised extended supersymmetry on the ultraviolet behaviour of multi-graviton amplitudes. The pure spinor formalism of N. Berkovits [12–15], has provided a new understanding of the role of supersymmetry for the case of $N = 8$ supergravity in diverse dimensions, and allowed to construct a new set of higher-derivative gravitational F-terms [16]. We will then discuss, in Section 3, some conditions for expecting that multi-graviton amplitude in $N = 8$ supergravity in flat space could have a much better ultraviolet behaviour and its relation with $N = 4$ super-Yang–Mills will be reviewed. Then in Section 4 we will discuss the construction of higher-loop amplitudes in supergravity theory using the on-shell unitarity method [17]. In Section 5 we will apply the relation between the supergravity in eleven dimensions and perturbative string theory provided by M-theory to get constraints on the low-energy behaviour of multi-graviton amplitudes. In Section 4.2 and 4.3 we analyse the structure of the four-graviton one-loop and two-loop amplitudes in $N = 8$ supergravity, and in Section 8 we describe various non-renormalisation conditions on some gravitational couplings to the low-energy effective action of string theory. In Appendix A we describe some of the structure of graviton amplitudes at the tree-level, genus-one, genus two and higher-order in string theory.

2. Ultra-violet divergences ...

A perturbative treatment of gravity [7, 8] in the background field method [9, 10], by linearization of the Einstein–Hilbert actions

$$\mathcal{L} = \frac{1}{\kappa_{(D)}^2} \int d^D x \sqrt{-g^{(D)}} \, \mathcal{R}^4[g] \tag{2.1}$$

around a specific background $g_{\mu\nu} = g_{\mu\nu}^{(0)} + \kappa_{(D)} h_{\mu\nu}$ give rise to an infinite set of effective vertices of two derivative nature. The Newton's constant in D-dimension has dimension $\kappa_{(D)}^2 = (length)^{D-2}$ and an L-loop, n-graviton amplitude in D-dimension as mass dimension

$$[M_{n;L}] \sim \kappa_{(D)}^{2(L-1)+n} \, (mass)^{(D-2)L+2}. \tag{2.2}$$

One notices that the mass dimension of the amplitude is independent of the number of external legs n (except for the dependence on $\kappa_{(D)}$ from the normalisation of the external states) due to the two derivative coupling nature of the interactions.

2.1. ... in pure gravity

In pure gravity the one-loop four-graviton amplitude has dimension

$$[M_{4;1}^{(4)}] \sim \kappa_{(4)}^4 (mass)^4 \tag{2.3}$$

and has a logarithmically ultra-violet divergence which requires the introduction of a counter-term of dimension four

$$\delta_1 M_{4;1}^{(4)} = \alpha \, (\kappa_{(4)}^2 \, R_{mnpq})^2 + \beta \, (\kappa_{(4)}^2 \, R_{mn})^2 + \gamma \, (\kappa_{(4)}^2 \, R)^2 \tag{2.4}$$

given by a precise linear combination of the square of the Riemann tensor, the Ricci tensor and the Ricci scalar. But for pure gravity this quantity vanishes on-shell and the divergence is accidentally zero [9]. This not true when the theory is coupled to matter and there a divergence at one-loop. At two-loop order the four-graviton amplitude has dimension

$$[M_{4;2}^{(4)}] \sim \kappa_{(4)}^6 \, (mass)^6 \tag{2.5}$$

which requires the introduction of a counter-term of dimension six constructed from three powers of the Riemann tensor R_{mnpq}, the Ricci tensor R_{mn} and the Ricci scalar R, symbolically represented as

$$\delta_2 M_{4;2}^{(4)} = (\kappa_{(4)}^2 \, R_{mnpq})^3 \tag{2.6}$$

and the theory of pure gravity is divergent at two-loop [18–20].

From the formula (2.2) one can read off the critical dimension for the appearance of ultra-violet divergences

$$D \geq D_c = 2 + \frac{2}{L}. \tag{2.7}$$

This formula indicates that pure gravity is finite only in two dimensions.

2.2. ... in extended supergravity

For at least $N = 1$ linearly realised on-shell supersymmetry in four dimensions R^3_{mnpq} cannot be supersymmetrised [21, 22]. Therefore the two-loop four-graviton amplitude in four dimensions is finite for $N \geq 1$ supergravity.

In a four dimensional N extended supersymmetric theory the *on-shell* counter-terms have to correspond to an integral over full superspace (a D-term) of the form [23–26]

$$\delta L = \kappa_{(4)}^{d+2N-4} \int d^4x \, d^{4N}\theta \, \det(E) \, \mathcal{L}(\mathbf{R}, \mathbf{T}), \tag{2.8}$$

where $\det(E)$ is the determinant of the super-vielbein and $\mathcal{L}(\mathbf{R}, \mathbf{T})$ is a superspace density of length dimension $-d$ expressed in terms of the super-curvature $\mathbf{R}$ and the supertorsion $\mathbf{T}$. The superspace variables have the following dimensions $[x] = length$ and $[\theta] = (length)^{1/2}$.

The Bianchi identities for the superspace formalism with N on-shell linearly realised supersymmetry are expressed in term of a scalar superfield φ and in terms of the chiral superfield of spin $2 - N/2$ for $N \leq 4$ [24–28].

• For $1 \leq N \leq 3$ the superfield φ has the Weyl tensor[1] $C_{\alpha\beta\gamma\delta}$ appearing at the order θ^N

$$\varphi_{(\beta_{N+1}\cdots\beta_4)} = \phi_{(\beta_{N+1}\cdots\beta_4)} + \cdots + \frac{1}{N!}\theta^{\beta_1}_{a_1} \cdots \theta^{\beta_N}_{a_N} \epsilon^{a_1\cdots a_N} C_{\beta_1\cdots\beta_4} + \cdots, \tag{2.9}$$

where a_i are indices for the $SU(N)$ R-symmetry. This superfield has length dimension $-(2 - N/2)$ and the *first* possible counter-term allowed by supersymmetry is

$$\delta L = \kappa_{(4)}^4 \int d^4x \, d^{4N}\theta \, (\varphi_{\alpha_1\cdots\alpha_4}\bar{\varphi}^{\dot\alpha_1\cdots\dot\alpha_4})^2 \sim \frac{1}{\kappa_{(4)}^4} \int d^4x \, (\kappa_{(4)}^2 \, C)^4 \tag{2.10}$$

which is a *three-loop* contribution to the four-graviton amplitude. In this case because the dimension of the superfield depends on the number of linearly realised supersymmetries, its dimension balances the one from the fermionic measure and the order of the appearance of the counter-term is always three-loop.

• For $4 \leq N \leq 8$ the Bianchi identities are solved in terms of the scalar superfield of dimension 0 where the Weyl tensor appears at the order θ^4

$$\varphi_{[a_1\cdots a_4]} = \phi_{[a_1\cdots a_4]} + \cdots + \theta^{\beta_1}_{a_1} \cdots \theta^{\beta_4}_{a_4} C_{\beta_1\cdots\beta_4} + \cdots \tag{2.11}$$

and the *first* possible counter-term allowed by linearized supersymmetry is

$$\delta L = \kappa_{(4)}^{2N-4} \int d^4x \, d^{4N}\theta \, (\varphi_{a_1\cdots a_4}\bar{\varphi}^{a_1\cdots a_4})^{\frac{N}{2}} \sim \frac{1}{\kappa_{(4)}^4} \int d^2x \, (\kappa_{(4)}^2 \, C)^N. \tag{2.12}$$

---

[1] In four dimensions we use the spinorial notation for the Riemann and Weyl tensors

In this case, because the superfield has dimension 0, the order at which the counter-term can appear is controlled by the dimension of the superspace integration. The linearized superspace integrals correspond to a three loop contribution to the four-graviton amplitude for $N = 4$, a four loop contribution to the four-graviton amplitude for $N = 5$, a five-loop contribution to the six-graviton amplitude for $N = 6$, a six-loop contribution to the seven-graviton amplitude for $N = 7$ and a seven-loop contribution to the eight-graviton amplitude for $N = 8$.

The absence of a three-loop divergence in four-graviton amplitude in four dimensions for $N = 8$ supergravity [29] indicates that more than sixteen on-shell linearized supersymmetries are controlling the perturbative computations of $N = 8$ supergravity. The precise number of supersymmetries is still unknown and we will describe below various constraints from string theory and dualities that can shed some light on this issue.

For the particular case of $N = 8$ supergravity the three-loop counter-term was constructed using the scalar superfield of the superspace formalism in four dimensions [27], but this superfield is not invariant under the U-duality symmetries of the theory which are expected to play an important role [11]. So we will be considering the construction of on-shell counter-terms that are invariant under the local $SU(8)$ R-symmetry and global E_7 of $N = 8$ supergravity [30].

• With increasing mass dimensions the first superfield invariant under the local $SU(8)$ R-symmetry and global E_7 of $N = 8$ supergravity is the dilatino superfield of length dimension $-1/2$ given in Table 1

$$\chi_\alpha^{ijk} = D_{\alpha,l}\varphi^{ijkl} \tag{2.13}$$

from which one can construct an *eight loop* counter-term [23]

$$\delta L = \kappa_{(4)}^{14} \int d^4x d^{32}\theta \, \det(E) \, (\chi_{ijk}^\alpha \bar{\chi}_\alpha^{ijk})^2 \sim \frac{1}{\kappa_{(4)}^4} \int d^4x \, \kappa_{(4)}^{18} \, D^{10}C^4. \tag{2.14}$$

This superfield is invariant under the local $SU(8)$ R-symmetry of $N = 8$ supergravity [23] unlike the scalar superfield discussed. All the Bianchi identities

Table 1

Basic superfields of linearized $N = 8$ supergravity. The lowest-component of these superfields are respectively given by the Weyl tensor $C_{\alpha\beta\gamma\delta}$, the 8 gravitino curvatures, 28 (Ramond-Ramond) field strengths, 56 Weyl spinors and 70 scalars

spin/dimension	2	3/2	1	1/2	0
superfield	$W_{\alpha\beta\gamma\delta}$	$W_{\alpha\beta\gamma}^i$	$W_{\alpha\beta}^{ij}$	χ_α^{ijk}	φ^{ijkl}
$SU(8)_R$ rep.	1	8	28	56	70

of $N = 8$ supergravity can be expressed in terms of this spinor superfield [27]. But bearing in mind the properties of the maximal supergravity theories with 32 supercharges in every dimension where they can be defined, the dimension 1/2 superfield does not appear in the $N = 1$, $D = 11$ formulation. We are considering gravity amplitudes, so one wants to construct counter-term using only Lorentz invariant quantities expressed in terms of the curvature of the (super)graviton. In ten dimensions the dimension 1/2 superfield starts with the dilatino field $\chi_\alpha = \lambda_\alpha + \theta^\beta F_{\alpha\beta} + \cdots$ (from which the four dimensional χ_α^{ijk} is the dimensional reduction) and $SO(1, 9)$ Lorentz invariance.

• The integrated graviton vertex operators in the pure spinor formalism for perturbative string theory were constructed by Berkovits [12–15]

$$V = \int d^2z \left(G_{MN} \partial x^M \bar{\partial} x^N + W_{\alpha\beta,+-} d_+^\alpha d_-^\beta + \cdots \right) \tag{2.15}$$

from the superfield $W_{\alpha\beta,+-}$ with dimension $(length)^{-1}$ has the Weyl tensor at order θ^2

$$W_{\alpha\beta,a_1a_2} = F_{\alpha\beta,a_1a_2} + \cdots + \theta_{a_1}^\gamma \theta_{a_2}^\delta C_{\alpha\beta\gamma\delta} + \cdots . \tag{2.16}$$

The lowest component of this superfield are the Ramond-Ramond field-strengths. In ten-dimensional type IIb supergravity this superfield arises by taking two fermionic derivatives on the scalar superfield dilaton $\Phi = \tau + \cdots + \theta^2(R + \partial F_5) + \cdots$ [31], and in type IIa supergravity this superfield is readily obtained by a dimensional reduction from the mass dimension one, four-form superfield of eleven dimensional supergravity [32]

$$(\Gamma^{m_1\cdots m_3} D)_\alpha W_{m_1\cdots m_4} = 0 \tag{2.17}$$

which has the following (schematic) expansion

$$W_{m_1\cdots m_4} = F_{m_1\cdots m_4} + \cdots + \theta^2 C + \cdots \tag{2.18}$$

starting with the field-strength for the three-form potential $F_4 = dC_3$ and having the Weyl tensor at θ^2 order. By dimensional reduction of this superfield gives the ten dimensional and four dimensional superfields discussed above.

• In four dimensions this superfield is obtained by dimensional reduction of the superfields of $N = 8$ supergravity. Using this "string favoured" building block in four dimensions one can construct the following *nine loop* counter-term [33]

$$\delta L = \kappa_{(4)}^{16} \int d^4x d^{32}\theta \, \det(E) \, (W_{\alpha\beta})^4 \sim \frac{1}{\kappa_{(4)}^4} \int d^4x \, \kappa_{(4)}^{20} \, D^{12}C^4. \tag{2.19}$$

3. Critical dimension for (logarithmic) ultra-violet divergences

In order to determine the critical dimension for logarithmic ultra-violet divergences of gravity amplitudes one needs to know precisely how the loop integral diverges. On general grounds a L-loop n-point gravity amplitude in D-dimension will be decomposed as

$$M_{n;L} = \sum_i t_i \, I^{(D)}_{n;L\,(i)}[\ell^\nu], \tag{3.1}$$

where t_i is some tensor constructed from the polarisations and the momenta of the external states, and $I^{(D)}_{n;L}[\ell^\nu]$ is an n-point L-loop integral defined in dimension D with ν powers of loop momenta in the numerators. Each individual integral can have a worse ultra-violet behaviour than the total amplitude $M_{n;L}$ where subtle cancellations are expected to occur.

If Λ is momentum cut-off, the low-energy expansion of the four-graviton amplitude at L loops and D dimensions will behaves as

$$M_{4;L} \sim \Lambda^{\delta_L} \mathcal{O}_{k_L} + \cdots , \tag{3.2}$$

where $\mathcal{O}_{k_L}$ is an operator of dimension $(length)^{-k_L}$. The dimension of this operator can change with the loop order. The leading degree of ultra-violet divergence is

$$\delta_L = (D-2)L + 2 + k_L \tag{3.3}$$

so that the total dimension of the loop amplitude is $(D-2)L + 2$. The ellipsis in the eq. (3.2) are for the sub-leading ultra-violet divergences. Ultra-violet divergences occur when $\delta_L \geq 0$

$$D \geq D_c = 2 + \frac{2 + k_L}{L}. \tag{3.4}$$

For $N = 8$ supergravity all the four-graviton amplitudes have at least a factor of $\hat{R}^4$ (see eq. (4.9)) but more derivatives can be factorised and the low energy expansion starts contributing with an operator of dimension $8 + 2\beta_L$

$$\mathcal{O}_{k_L} = D^{2\beta_L} R^4. \tag{3.5}$$

In this case the critical dimension for the appearance of ultra-violet logarithmic divergences is

$$D_c = 2 + \frac{6 + 2\beta_L}{L}. \tag{3.6}$$

Explicit results for the four-graviton amplitude give that $\beta_1 = 0$ at one-loop [34], and $\beta_2 = 2$ at two loops [12, 35–39], and $\beta_L = 3$ at three loops[2] [29]. Assuming that $\beta_L = 3$ for $L \geq 3$ one concludes that the critical dimension for ultra-violet divergence is

$$D_c = 2 + \frac{12}{L} \quad \text{for } L \geq 3. \tag{3.7}$$

Predicting a first divergence for $N = 8$ supergravity in four dimension at $L = 6$ loop. This formula predicts as well that the four-loop four-graviton amplitude diverges logarithmically in $D = 5$.

In ten dimensions using the non-minimal pure spinor formalism Berkovits showed in [16] up to $L = 6$ the four-graviton amplitudes are F-term satisfying the rule $\beta_L = L$. This result which only makes use of the fermionic zero mode saturation does not involve any massive string excitations and makes only use of the fact that the vertex operators are constructed using the mass dimension one superfield $W_{\alpha\beta}$ of eq. (2.17).

Since this superfield exists for all the formulations of the maximal ($N = 8$) supergravity in every dimensions, it was argued in [41] that this leads to the following critical dimension

$$D \geq D_c = 2 + \frac{18}{L} \quad \text{for } L \geq 6. \tag{3.8}$$

And the first divergence of $N = 8$ supergravity in four dimensions is expected at nine loops.

All these formulæ give that only $D = 2$ supergravity is finite (for any number of supersymmetries). As long $2 + k_L$ in eq. (3.4) or $6 + 2\beta_L$ in eq. (3.6) is bounded when the loop order L increases the equation $D_c = 4$ will always have a solution at some loop order and an ultra-violet divergence will occurs in four dimensions. It was proposed in [33, 41] that at each loop order two extra powers of the external momenta factors out the low energy limit of the L-loop amplitude, so that

$$\beta_L = L \quad \text{for } L \geq 2 \tag{3.9}$$

giving the critical dimension

$$D \geq D_c = 4 + \frac{6}{L} \quad \text{for } L \geq 2. \tag{3.10}$$

---

[2]Notice that β_L is the leading power of the amplitude in the low-energy expansion, and not all the various diagrams composing the $L = 3$ have an overall power of $D^6 R^4$ in the solution presented in [29]. The string based analysis presented in [16] assures that there is representation of the $L = 3$ amplitude with an explicit power of $D^6 R^4$. The field theory result of [29] can be rewritten in such a form as well [40].

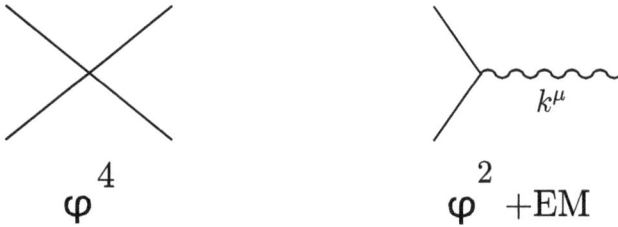

Fig. 1. Effective interactions that could be giving an ultra-violet behaviour with the critical dimension $D_c = 4 + 6/L$.

Since the critical dimension is always bigger than 4, then $N = 8$ supergravity would be finite in four dimensions.

When the condition $\beta_L = L$ holds the constant piece in D_c has changed its value from 2 to 4. This means that at each loop order the mass dimension of the loop integral increases by a factor of $(mass)^{D-2}$ typical of φ^4 scalar interactions or cubic derivative interactions of Fig. 1. These vertices are the elementary one of $N = 4$ super-Yang–Mills. The φ^4 vertex are needed from four-loop order [42,43]. These vertex appear in the construction of the multiloop amplitudes so that no triangle sub-graph are generated. Supersymmetry cannot be responsible for such reduction which are due to extra cancellations from pure gravity interactions [44–46] accounted for the general coordinate gauge invariance in one-loop amplitudes and the sum over all the permutation of the external legs in a theory without color ordering [47].

In four dimensions the loop amplitude has negative mass dimension -6

$$[M_{4;L}^{(4)}] \sim (mass)^{-6} D^{2L} R^4, \tag{3.11}$$

which means that the amplitude has no ultra-violet divergences but only IR divergences. At one-loop order, the amplitude is given by R^4 times the scalar box amplitude I_4 which has dimension $(mass)^{-4}$ in four dimension [34], at two-loop order the amplitude is given by $D^4 R^4$ times the planar and non-planar double-box [35] of dimension $(mass)^{-6}$ [48,49]. For more details on the structure of four-graviton loop amplitude we refer to Section 4.

3.1. Is $N = 8$ equal to $(N = 4)^2$?

The formula (3.8) is the same critical formula as $N = 4$ super-Yang–Mills indicating a very close relation between $N = 8$ supergravity and $N = 4$ super-Yang–Mills. The relation is that the basis of integrals on which $N = 8$ supergravity amplitudes are expressed is the same as the one of $N = 4$ super-Yang–Mills amplitudes once all the planar and non-planar contributions are included. We will see this in practice at the one-loop $L = 1$ and two-loop order $L = 2$ in Section 4.

The n-point L-loop amplitude in D dimensions for $N = 4$ super-Yang–Mills has the following mass dimension

$$[A_{n;L}] \sim g_{YM}^{2(L-1)+n} (mass)^{(D-4)L} \tag{3.12}$$

with a coupling constant g_{YM}^2 of dimensions $(length)^{D-4}$. In four dimensions the theory is logarithmic divergent. For $N = 4$ super-Yang–Mills supersymmetric the finiteness in four dimensions is only a consequence of supersymmetry [50] and the supersymmetric cancellations give that the low-energy expansion of the amplitude is given by (for $L \geq 2$) [50]

$$A_{4;L} \sim \Lambda^{(D-4)L-6} D^2 F^4 + \cdots \tag{3.13}$$

which is enough for the perturbative ultra-violet finiteness of the theory in four dimensions. This leads to the critical dimension for super-Yang–Mills amplitudes

$$D \geq D_c = 4 + \frac{6}{L}. \tag{3.14}$$

This formula indicates that the dimension six operator $D^2 F^4$ factorises at higher loops which is enough for assuring that the theory is perturbatively finite and does not have logarithmic divergences. It is important to remark here that the dimension of the operator that is factorised does not depend on the loop order.

The KLT [51] relation express the supergravity tree-level amplitudes as sum of square of Yang–Mills amplitudes

$$M_n^{tree} = \sum_{\sigma \in S_n} p_{n-3}^{(i)}(s_{ij}) A_n^{tree}(1, 2, \ldots, n) A_n^{tree}(\sigma(1), \sigma(2), \ldots, \sigma(n)). \tag{3.15}$$

In this expression σ is a permutation of the external legs and $p_{n-3}^{(i)}(s_{ij})$ are polynomials of order $n - 3$ in the kinematic invariants $s_{ij} = (k_i + k_j)^2$ needed to cancel the spurious poles appearing when multiplying the two gauge theory amplitudes [52–54]. For instance the tree-level four gauge boson (colour stripped) amplitude is

$$A_4^{tree}(s, t) = \frac{1}{s\,t} t_8 F^4 \tag{3.16}$$

which squared gives the tree-level four-graviton amplitude

$$M_4^{tree}(s, t) = s\, A_4^{tree}(s, t)\, A_4^{tree}(s, u) = \frac{1}{stu} t_8 t_8 R^4. \tag{3.17}$$

These tree-level relations have some important implications at the loop level structure where the amplitude is developed on the same basis [17, 35, 47, 55] of integrals as for $N = 4$ super-Yang–Mills with for coefficients tensorial structure that are 'square' of the $N = 4$ super-Yang–Mills tree coefficients from the KLT relation [51]. These relations are specific to the on-shell amplitude computations and are not properties of the effective action of supergravity theories but are nevertheless a useful guide for constructing some higher derivative superinvariants for the ten- and eleven-dimensional supergravity effective action in [56] under linearized supersymmetry.

One important difference between the gravity and Yang–Mills is the absence of colours. At one-loop order the sum over all the external legs for a colorless theory played was needed for cancelling the triangle contributions [47]. At higher-loop order this requires that one sums over all the planar and non-planar contributions to a specific amplitude, and any good ultra-violet properties of $N = 8$ supergravity must rely on subtle cancellations between the planar and non planar sector. We will see examples of this when discussing the explicit example of higher-loop contributions in the next section.

In the context of $N = 4$ supergravity one expects $\beta_L = L/2$ giving a critical dimension

$$D_c = 3 + \frac{6}{L} \tag{3.18}$$

and a first divergence at $L = 6$ in four dimensions, which is higher than the three loop divergence prediction based on the linearized $N = 4$ supersymmetry.

In the following we will describe a construction of the higher-loop amplitude four-graviton amplitude in $N = 8$ supergravity and their regularisation using the string theory induced scheme [57, 58]. This construction will give a set of non renormalisation theorems for the higher derivative gravitational interactions. Since all the possible counter-term from the superspace formalism have a contribution to the four-graviton amplitude, we will be able to confront the predictions from supersymmetry and dualities for the ultra-violet behaviour of the multiloop amplitudes in supergravity in various dimensions.

4. Higher loop amplitudes in supergravity theories

For computing higher-loop amplitude in eleven dimensions one could envisage using background field method linearizing the eleven dimensions supergravity action of [59] around a specific background and making use of Feynman rules. This has been used to extract information about one- and two-loop amplitudes in gravity [10, 18–20] and some of the structure of the higher order corrections to M-theory [60, 61]. The method is very cumbersome an obscures many of the properties of the gravitational interactions (see [17] and a presentation of these points).

In the following we will describe how to construct the four-graviton L-loop amplitude of $N = 1$ supergravity in eleven dimensions using on the on-shell unitarity method for supergravity developed by Zvi Bern and his collaborators [62].

4.1. Cut construction of the loop amplitudes

We construct the scattering S_{if}-matrix between initial state i and final state f from its discontinuities across the branch cuts in the complex energy plane (the Mandelstam plane). Because in an unitary local quantum field theory the S-matrix $S_{ij} = \delta_{if} + i\,T_{if}$ the S-matrix satisfies $SS^{\dagger} = \mathbb{I}$ the transition matrix T satisfies the relation

$$2i\,(T - T^{\dagger})_{if} = \sum_{k} T_{ik} T^{\dagger}_{kf}, \tag{4.1}$$

where the sum on the right hand side is over all possible intermediate states. Perturbatively the states k are the one running in the loops. In the context of supergravity theories the sum will be over the massless supergravity multiplet.

This relation is valid for any unitary quantum field theory independently of any perturbative expansion.

A perturbative expansion relation (4.1) relates the value of the discontinuity of the scatting matrix across a branch cuts at a given loop order L to the integration over the intermediate states exchanged between lower loops order amplitude.

Unfortunately there are various ambiguities in reconstructing the real part of the S-matrix associated with the fact that the dispersion relations give rise to diverging expressions which need to be regularised and introduce some ambiguities in the rational part of the amplitude, and one has to use the dispersion relation for obtaining the real part of the amplitude [63]. The only ambiguity with a physical meaning are the one associated with the usual ultra-violet and infra-red divergences of the amplitude. A traditional way of dealing with this problem is to consider the unitarity method in the context of the dimensional regularisation [64, 65]. The application of this method in the context of the helicity formalism proved to be a very powerful tool for constructing higher-loop amplitude in gauge and supergravity theories [17, 66–68].

In some particular cases the amplitudes do not contain rational terms and can therefore be completely reconstructed by considering the behaviour of the S-matrix across the branch cuts.

The appearance of rational terms in the amplitude is connected to the number of loop momenta in the Feynman integrals (and therefore the ultra-violet behaviour) of the amplitude by various steps of the Passarino–Veltman reduction [69].

Consider an n-point one-loop amplitude in D dimensions with φ^3 vertices

$$I_{n;\nu}^{(D)}(k_1, \ldots, k_n) = \int d^D \ell \, \frac{\mathcal{P}_\nu(\ell)}{\ell_1^2 \cdots \ell_n^2}, \tag{4.2}$$

where $\ell_i^2 = (\ell - k_1 - \cdots - k_i)^2$ are the various propagators along the loop and $\mathcal{P}_\nu(\ell)$ in the numerator is a polynomial of degree ν in the loop momentum ℓ.

The dimension of the integral in eq. (4.2) is

$$[I_{n;\nu}^{(D)}] \sim (mass)^{D+\nu-2n} \, \mathcal{O}_\nu(h, k) \tag{4.3}$$

the superficial degree of divergence $\delta_{1,n} = D + \nu - 2n$ is a function of the dimension D, the number of external states and the order ν of the polynomial. The amplitude is ultra-violet finite when $2n \geq D + \nu$, therefore by analysing the ultra-violet behaviour of the integral in arbitrary dimensions and with an increasing number of external legs one can determine the power of ν of loop momenta in the numerators of the amplitudes.

For on-shell massless external states using the identity

$$\frac{2\ell \cdot k_1}{\ell^2 (\ell - k_1)^2} = \frac{1}{(\ell - k_1)^2} - \frac{1}{\ell^2} \tag{4.4}$$

one reduces [69] the integral I_n in eq. (4.2) into the difference of two $n - 1$-point loop amplitude with a numerator of degree $\nu - 1$

$$\begin{aligned}
I_{n;\nu}^{(D)}(k_1, \ldots, k_n) &= I_{n-1,\nu-1}^{(D)}(k_1 + k_2, \ldots, k_n) \\
&\quad - I_{n-1;\nu-1}^{(D)}(k_2, \ldots, k_n + k_1).
\end{aligned} \tag{4.5}$$

On the right hand side one sees the two massive external legs with momentum $k_1 + k_n$ and $k_1 + k_2$ due to the cancellation of the propagators with the identity (4.4).

For gravity or gauge theories the higher power of loop momenta in $\mathcal{P}_\nu(\ell)$ is given by the cubic vertex, which means that $\mathcal{P}_\nu(\ell) \leq n$ for gauge theories and $\mathcal{P}_\nu(\ell) \leq 2n$ for gravity. For $N = 4$ super-Yang–Mills a vertex brings a power of the loop momenta, and the saturation of eight fermionic zero modes (needed by the $N = 4$ supersymmetry) cancels four loops momenta leading to $\nu \leq n - 4$. In gravity each vertices bring two powers of loop momenta, and for supergravity theories with N supersymmetries it is conjectured that [45]

$$\nu \leq 2n - N - (n - 4), \tag{4.6}$$

where supersymmetry only guarantees that $\nu \leq 2n - N$. Therefore $n - 4$ extra cancellations of loop momentum are needed.

One remarks that there is a special case where the extra $n - 4$ cancellations are not needed. This is the case of the amplitudes with more than four external states the one-loop amplitude which contains at most $n - 4$ powers of propagators ℓ_i^2 in the numerator and reduce directly to a massive boxes

$$\int d^D\ell \, \frac{\ell_5^2 \cdots \ell_n^4}{\ell_1^2 \cdots \ell_n^2} = \int d^D\ell \, \frac{1}{\ell_1^2 \cdots \ell_4^2}. \tag{4.7}$$

In this case one concludes that
- Clearly theories with $\nu < n$ are one-loop cut constructible since no rational terms can be obtained by a succession of Passarino–Veltman reductions.
- $N = 4$ super-Yang–Mills does not have rational term and is cut-constructible.
- For $N = 8$ supergravity it is conjectured that $\nu \leq n - 4 < n$, which means that one-loop n-graviton amplitudes under a Passarino–Veltman reduction but do not contain integral functions more singular than (massive) boxes and in particular do not contain triangles or bubble functions [45, 70]. At one-loop order this fact has been linked to gauge invariance and the summation over all permutations of the external legs for a colorless theory [47]. Because of the absence of colors in gravity on has to sum over all planar and non-planar contributions at higher-loop order. This plays an essential role in the cancellation of various unwanted contributions that would bring a worst ultra-violet behaviour.

When using the on-shell unitarity method for constructing the higher-loop amplitude in $N = 8$ supergravity, one has to sum over all the 256 massless states of the graviton supermultiplet. These states are the same in every dimensions and the construction of the amplitudes are valid in every dimensions where $N = 8$ supergravity can be defined.

4.2. The one-loop amplitude

The amplitude between four massless state of the supergravity multiplet of $N = 8$ supergravity in D dimensions can be constructed completely from its 2-particle cut with the result [62]

$$M_{4;1}(k_1, \ldots, k_4) = \frac{\kappa_{(D)}^4}{(2\pi)^D} \, \hat{R}^4 \left[I_4^{(D)}(S, T) + I_4^{(D)}(S, U) + I_4^{(D)}(T, U) \right], \tag{4.8}$$

where $\hat{R}^4$ is kinematic factor for four massless external states defined in [1]

$$\hat{R}^4 = h_1^{AA'} h_2^{BB'} h_3^{CC'} h_4^{DD'} \, K_{ABCD} \tilde{K}_{A'B'C'D'}, \tag{4.9}$$

the indices A, B on the superhelicity run over both vector and spinor values and span the 256 states of the massless $N = 8$ gravity supermultiplet. $I_4^{(D)}$ is the

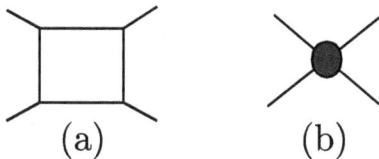

Fig. 2. The one-loop four-graviton amplitudes in $N = 8$ supergravity is given by the box diagram in figure (a). In $D \leq 8$ the amplitude has a ultra-violet divergence which is subtracted by the counter-term represented in figure (b).

D-dimension massless box integral ($I_{4;0}^{(D)}$ in the notations of eq. (4.2)) given by

$$I_4^{(D)}(S, T) = \int_{\Lambda^{-2}}^{\infty} \frac{dt}{t} \, t^{4 - \frac{D}{2}} \int_0^1 \prod_{i=1}^4 dv_i \, e^{\pi t Q_4(k_i)}, \tag{4.10}$$

where

$$Q_n(k_1, \ldots, k_n) = \sum_{1 \leq i < j \leq n} k_i \cdot k_j \, (v_{ij}^2 - |v_{ij}|). \tag{4.11}$$

The absolute value in Q_n forces the breaking of the integral into three different physical regions where the integral converges [71,72]. The (s, t)-region for $\mathcal{T}_{ST} = \{0 \leq v_1 \leq v_2 \leq v_3 \leq v_4 \leq 1\}$, the (s, u)-region for $\mathcal{T}_{SU} = \{0 \leq v_2 \leq v_1 \leq v_3 \leq v_4 \leq 1\}$ and the (t, u)-region for $\mathcal{T}_{TU} = \{0 \leq v_1 \leq v_3 \leq v_2 \leq v_4 \leq 1\}$. The integral is to be evaluated with $s, t < 0$ where it converges and then analytically continued to the physical region.

The expression (4.8) for the amplitude for four massless $N = 8$ states of the supergravity multiplet has been constructed using the on-shell unitarity method and is valid in any dimensions $D \leq 11$. In dimension $D \leq 10$ this expression has been obtained by Green et al. [34] by compactifying the genus-one string loop amplitude and decoupling the massive string modes, Kaluza–Klein and winding modes.[3]

The lower bound on the integral is an ultra-violet cut-off[4] and the amplitude $M_{1;4}$ has the leading ultra-violet behaviour

$$[M_{4;1}] \sim \frac{\kappa_{(D)}^4}{(2\pi)^D} \, \hat{R}^4 \, (mass)^{D-8} \tag{4.12}$$

---

[3]The string one-loop amplitude is of course ultra-violet finite, and gives a resulting total amplitude which also ultra-violet finite because the residue of the $1/\epsilon$ in $5 \leq D \leq 10$ and $\frac{1}{\epsilon^2}$ pole in $D = 4$ is proportional to $s + t + u = 0$ and vanishes on-shell.

[4]In $D \leq 4$ the amplitude develops infra-red (IR) divergences which will not be discussed in here.

and the superficial degree of divergence of the 4 graviton amplitude is given by the behaviour of the scalar box in D dimensions

$$\delta_{4;1} = D - 8. \tag{4.13}$$

In $D = 11$ this amplitude has a cubic divergence which will be regulated by the addition of a local counter-term

$$\delta_1 M_{4;1} = c_4 \frac{\kappa_{(11)}^4}{(2\pi)^{11}} \frac{\pi^3}{2} \ell_P^3 \hat{R}^4. \tag{4.14}$$

The precise renormalisation scheme will be described in Section 6.

4.3. Two-loop amplitude

The on-shell unitarity method [62] gives that the two-loop four-graviton amplitude for $N = 8$ supergravity in D-dimension is expressed as a the sum of scalar double-box amplitude represented in Fig. 3 and given by

$$M_{4;2} = i \frac{\kappa_{(D)}^6}{(2\pi)^{2D}} \hat{R}^4 \left[S^2 I^{(S)} + T^2 I^{(T)} + U^2 I^{(U)} \right] \tag{4.15}$$

which is given by a sum of contributions from the S, T and U-channel with

$$I^{(S)} = \frac{1}{2} \left(I^P(S, T) + I^P(S, U) + I^{NP}(S, T) + I^{NP}(S, U) \right), \tag{4.16}$$

with analogous expressions for $I^{(T)}$ and $I^{(U)}$. The loop integral $I^P(S, T)$ is the planar two-loop φ^3 contribution

$$I^P(S, T) = \int \frac{d^D p \, d^D q}{p^2 (p - k_1)^2 (p - k_{12})^2 (q - k_{12})^2 (q - k_4)^2 q^2 (p + q)^2} \tag{4.17}$$

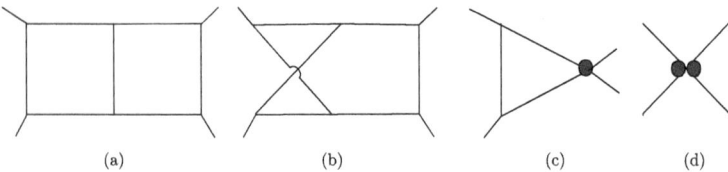

(a)   (b)   (c)   (d)

Fig. 3. The two-loop four-graviton amplitudes in $N = 8$ supergravity is given by the double box diagrams in figure (a) and (b). Figure (c) represents the contribution induces by the one-loop counter-term of Fig. 2(b). Figure (d) represents the new primitives ultra-violet divergences arising for $D \geq 7$.

and $I^{NP}(S, T)$ is the non-planar two-loop φ^3 contribution

$$I^{NP}(S, T) = \int \frac{d^D p\, d^D q}{p^2 (p - k_1)^2 (p - k_{12})^2 (q - k_4)^2 q^2 (p + q)^2 (p + q - k_3)^2}.$$
(4.18)

At this order we see the appearance of a non-planar contribution to the amplitude. In the case of gravity there is no colour and the planar and non-planar contributions have to be summed in the total amplitude. The two-loop string amplitude is given by a single expression in eq. (A.17) expressed as the integral over the moduli space genus two Riemann surfaces [36, 37]. At this order the field theory limit has a remaining 'modular' symmetry putting together planar and the non-planar contribution in a single contribution [58, 73].

Finally because there are no diagrams with three external legs on the same loop proper-time, the two loop amplitude does not contain any triangle. This is a consequence of the vanishing of the factor $|\mathcal{Y}_s|^2$ in eq. (A.19) when three external legs are on the same loop proper-time in the integrand of the field theory limit of the genus two [73].

In eleven dimensions the one-loop counter-term in eq. (4.14) induces the following contribution at two-loop order

$$\delta_{2a} M_{4;2} = i c_4 \frac{\pi^3}{2} \ell_P^3 \frac{\kappa_{(D)}^6}{(2\pi)^{2D}} \hat{R}^4 I_{3;1}^{(D)},$$
(4.19)

regulating the one-loop ultraviolet sub-divergence of the amplitude.

In eleven dimension a new set of primitive divergences arise at two-loop order

$$M_{4;2} \sim \Lambda^8 \mathcal{D}^4 \hat{R}^4 + \Lambda^6 \mathcal{D}^6 \hat{R}^4 + \Lambda^4 \mathcal{D}^8 \hat{R}^4 + \Lambda^2 \mathcal{D}^{10} \hat{R}^4 + \log(\Lambda) \mathcal{D}^{12} \hat{R}^4$$
(4.20)

these divergences are subtracted by local counter-term

$$\delta_{2b} M_{4;2} = \frac{\kappa_{(11)}^6}{(2\pi)^{22}} \sum_{k=2}^{6} c_k \mathcal{D}^{2k} \hat{R}^4,$$
(4.21)

the value of this counter-term will be determined in Section 7 in the context of the duality compatible renormalisation scheme we use for regulating the ultraviolet divergences of the supergravity amplitudes.

4.4. Higher-loop amplitudes

The four-graviton three-loop amplitude in $N = 8$ supergravity has been constructed in ref. [29] and has the two classes of vacuum diagram represented in

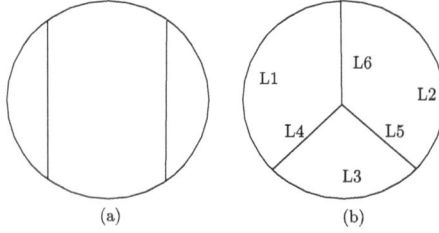

Fig. 4. The vacuum diagrams for the three-loop four-graviton amplitudes in $N = 8$ supergravity has two different topologies depicted in figure (a) and (b). The vacuum diagram in figure (a) leads to the class of scalar φ^3 triple ladder and cross-ladder boxes. They have a prefactor of $D^8 R^4$. The vacuum diagram in figure (b) leads to various diagrams that are no purely scalar φ^3.

Fig. 4. It was shown in ref. [29] that the low-energy limit of this amplitude is given by

$$M_{4;3} \sim \Lambda^{3(D-4)-6} D^6 \hat{R}^4 + \cdots .$$ (4.22)

One class of diagrams is given by all the scalar φ^3 planar and non-planar triple-box diagrams one obtains by putting the legs external legs on the vacuum diagram of Fig. 4(a) without generating diagrams containing triangles. These diagrams have an explicit factor of $D^8 R^4$ and are much too ultraviolet convergent to be the leading contribution to the low-energy limit of the $L = 3$ amplitude. At this order appears a new class of diagrams which are not purely scalar φ^3 diagrams but with some powers of the loop momenta in the numerator of the integrals. They arise from the vacuum diagram of Fig. 4(b). Although the solution presented in [29] did not have an explicit factor of $D^6 R^4$ (some of the diagrams only have an explicit $D^4 \hat{R}^4$ factor in front of the loop integrals) the structure of the genus 3 amplitude derived in [16] guarantees that there is an expression of the three loop result with an explicit $D^6 R^4$ in front the loop integrals, with loop integrals obtained by putting the external legs on the diagram in Fig. 4(b). The resulting four-point loop integrals must have a maximun of two powers of loop propagators in the numerators to have the dimension $[I_{4;3}^{(D)}] = (mass)^{15}$.

The construction of the gravitational F-terms by Berkovits in [16] guarantees that up to six loops the four-graviton amplitude will start contributing from $D^{2L} R^4$ in the low energy limit.

The construction of the higher loop amplitude in supergravity is a difficult task essentially because one has to sum over all the permutations of the external states and include both planar and non-planar contributions. It was found in [58] that the $L = 2$ amplitude has an hidden modular symmetry inherited from the symmetries of the string theory genus two moduli space. This symmetry puts together the planar and non-planar contributions in a single an compact expression [58, 73]. One may wonder about the existence of some remnant of the higher-genus string amplitude modular symmetries organising the field theory contributions by putting together planar and non-planar contributions and facilitating the construction of the higher loop contributions.

5. Duality constraints

5.1. The M-theory conjecture

The eleven dimensional supergravity [59] has a Lagrangian of the form

$$
\mathcal{S}_{CJS}^{(11)} = \frac{1}{2\kappa_{(11)}^2} \int d^{11}x \sqrt{-G} \left[\mathcal{R}_{(11)} + \frac{1}{2}|G_4|^2 \right.
$$

$$
\left. + \frac{1}{6} C_3 \wedge G_4 \wedge G_4 + \text{fermions} \right], \tag{5.1}
$$

where $\mathcal{R}_{(11)}$ is the Ricci scalar in eleven dimensions and $G_4 = dC_3$ the four form-field strength. The coupling constant $\kappa_{(11)}^2 = (2\pi)^8 \ell_P^9$.

This Lagrangian can be seen as a consequence of the κ-symmetry invariance of the M2-brane [74] and has an on-shell superspace description in eleven dimensions [32].

The M-theory conjecture states that this Lagrangian is the kinetic part of the effective action a fundamental theory. The microscopic degrees of freedom of M-theory are not known so far.

Even if the cancellations described in the previous section occur to all orders the eleven dimensional supergravity will have ultraviolet divergences, and needs to be regulated. We will regulate the theory by adding *local counter-terms* to the eleven dimension supergravity effective action

$$
\delta\mathcal{S}_{CJS} = \frac{1}{2\kappa_{(11)}^2} \int d^{11}x \sqrt{-G} \sum_k c_k \, \ell_P^{2(k-1)} \, \mathcal{R}^k + \cdots, \tag{5.2}
$$

where $\mathcal{R}^i$ represents an higher dimensional operators composed by powers of Riemann tensors, or derivatives on the Riemann tensor $\mathcal{D}^m \mathcal{R}^n$ with $n + 2m = k$. (There are of course counter-terms depending on the four-form field-strength but they will not be discussed here.) The knowledge of the microscopic degrees of freedom of M-theory and its symmetries would dictates the infinite series of counter-terms to add to the theory for making it ultraviolet finite and determine the values of the constants c_k in eq. (5.2). This cutoff should be determined by the microscopic degree of freedom of M-theory and related to the tension of the M2-brane $T_{M2} \sim 1/\ell_P^3$ or the M5-brane $T_{M5} \sim 1/\ell_P^6$. The $N = 8$ supersymmetric cancellations of loop momenta in one-loop amplitudes assure that the higher power of one-loop sub-divergences is given by Λ^3. Therefore one only expects in eleven dimensions primitive divergences of the type Λ^{3n}. These divergences are subtracted by the following infinite set of counter-term to the M-theory action

$$S_{M\text{-}theory} = \frac{1}{\ell_P^9} \int d^{11}x \left[\mathcal{R}^4_{(11)} + \sum_{k \geq 0} c_k \, \ell_P^{6k+6} \, R^{3k+4} \right]. \tag{5.3}$$

The infinite set of coefficients is constraints by the duality symmetries of M-theory, and have been determined up to order $k = 2$ in [57, 58, 73, 75]

$$S_{M\text{-}theory} = \frac{1}{\ell_P^9} \int d^{11}x \left[\mathcal{R}^4_{(11)} + 4\zeta(2) R^4 + 2\zeta(4) D^6 R^4 \right.$$
$$\left. + \frac{196}{142} \zeta(6) D^{12} R^4 + \cdots \right]. \tag{5.4}$$

For instance the first non-zero correction, the $\ell_P^6 \mathcal{R}^4$ term in (5.2) can be seen as originating from membrane effects [56, 76–80] induced for instance by world-sheet higher-loop effects on the world-volume of the M2-brane. It would be interesting to test if the pattern of the higher-derivative corrections appearing in δS determined in the references [57, 58, 73] and reviewed in the Sections 6 and 7 can be reproduced from the infinite dimensional group of symmetries of M-theory considered in [81–83].

In the following we will use a renormalisation scheme for regulating the loop amplitudes by matching the result to perturbative string data[5] which will allow us to derive a set of counter-terms to the effective action.

Taking this theory on a circle of radius[6] $R_{11} \ell_P$ the theory is identified with type IIa string theory if the masses are measured with the following metric [5, 84]

$$\frac{G_{MN}}{\ell_P^2} dx^M dx^N = \frac{1}{R_{11}} \frac{g^A_{\mu\nu}}{\ell_s^2} dx^\mu dx^\nu + R_{11}^2 (dx^{11} - C_\mu dx^\mu)^2, \tag{5.5}$$

where $M = 0, \ldots, 10$, $\mu = 0, \ldots, 9$, $g^A_{\mu\nu}$ is the type IIa sigma model metric, and C_μ is the 1-form RR-potential carrying the D0-brane charge. The radius R_{11} is related to the string coupling constant by the relation

$$R_{11}^3 = (g_s^A)^2 \tag{5.6}$$

---

[5]Considering this theory on a circle of finite value of R_{11} the S-matrix $\mathcal{S}(R_{11}, \ell_P)$ can be expanded in powers of R_{11} (there are as well $\exp(-1/R_{11})$ effects which are D-branes which will be commented on below) which will be matched with corresponding quantities from string perturbation.

[6]We always take the radii as dimensionless quantities.

which together with the following relation between the eleven dimensional Planck length and the string scale

$$\ell_P = (g_s^A)^{\frac{1}{3}} \ell_s, \tag{5.7}$$

gives the dictionary between M-theory variables and string variables. These relations relate the strong coupling limit $g_s \to \infty$ of type IIa string to the eleven dimensional theory $R_{11} \to \infty$.

Plugging these relations into eq. (5.1) the Einstein–Hilbert term transform into [5]

$$S^{(IIa)} = \frac{1}{2\,\kappa_{(10)}^2} \int d^{10}x\,\sqrt{-g^A}\,\frac{1}{(g_s^A)^2}\,[\mathcal{R}_{(10)}^4 + \cdots] \tag{5.8}$$

the Einstein–Hilbert term in ten dimensions in the string frame, and the ellipsis are for the various non-gravitational contributions arising from the reduction of the action (5.1) leading to the type IIa supergravity action in ten dimensions. We have $2\kappa_{(10)}^2 = (2\pi)^7 \alpha'^4$.

Because of the specific dependence on R_{11} in the metric (5.5) the massless supergraviton multiplet in eleven dimensions gives the ten dimensional supergraviton multiplets as well as Ramond 1-form carrying $D0$-brane charges.

The type IIb theory in ten dimensions can be obtained as well by considering the M-theory on a torus of vanishing volume $\mathcal{V} \to 0$ and fixed complex structure Ω [85, 86]. The complex structure $\Omega = \Omega_1 + i\Omega_2$ with $\Omega_2 = R_{10}/R_{11}$ becomes the complexified coupling constant of the type IIb superstring $\tau = C^{(0)} + i/g_s^B$ where $C^{(0)}$ is the RR 0-form potential which couples to D-instantons. For external states without momenta in the internal directions the resulting amplitude will be invariant under the $Sl(2, \mathbb{Z})$ symmetry inherited by large diffeomorphism of the compactification torus. This (geometric) symmetry becomes in the type IIb limit the non-perturbative S-duality symmetry of the theory.

Under this reduction the higher derivative corrections in eq. (5.2) to the effective action in eleven dimensions transforms as

$$\frac{\ell_P^{2k-2}}{2\,\kappa_{(11)}^2} \int d^{11}x\,\sqrt{-G}\,c_k\,\mathcal{R}^k \to \frac{\ell_s^{2k-2}}{2\,\kappa_{(10)}^2} \int d^{10}x\,\sqrt{-g^A}\,c_k\,(g_s^A)^{\frac{2(k-4)}{3}}\,\mathcal{R}^k \tag{5.9}$$

giving contributions to be interpreted as string perturbative contributions to the type IIa string effective action only if [87]

$$k = 4 + 3m \quad \text{with } m \geq 0. \tag{5.10}$$

When the condition (5.10) is not satisfied the coefficient c_k is *set to zero* and give is the same condition as in (5.3). And when this condition is satisfied an higher derivative correction to the classical action of M-theory is expected. The precise value for this coefficient will be determined by matching the value of the perturbative string genus $m + 1$ contribution for this operator.

The contributions in eq. (5.2) will contribute to higher-loop amplitude computations as counter-term to the ultraviolet divergences of the graviton amplitudes.

For instance the $\mathcal{R}^4$ term

$$\delta^{(4)}S = \frac{1}{2\kappa_{(11)}^2} \int d^{11}x \sqrt{-G} \, c_4 \, \frac{\kappa_{(11)}^2}{\ell_P^3} \, \mathcal{R}^4 \tag{5.11}$$

will be a one-loop $L = 1$ counter-term for the $\Lambda^3 \sim 1/\ell_P^3$ divergence in the four-graviton amplitude. The contributions that do not match the condition in eq. (5.10) have a zero coefficient $c_k = 0$ which means that a primitive ultraviolet divergence of higher-loops amplitudes in eleven dimensions is subtracted with a zero remainder. One interesting case that occurs in the dimension 20 operators $\mathcal{D}^{12}\mathcal{R}^4$.[7] Its contribution to the effective action of M-theory can be understood as a counter-term to the logarithmic divergence of the two-loop $L = 2$ four-graviton amplitude in eleven dimensions

$$\delta^{(10)}S = \frac{1}{2\kappa_{(11)}^2} \int d^{11}x \sqrt{-G} \, c_{10} \, \kappa_{(11)}^4 \, \mathcal{D}^{12}\mathcal{R}^4, \tag{5.12}$$

or a counter-term to a superficial $\Lambda^9 \sim 1/\ell_P^9$ divergence of the three-loop $L = 3$ four-graviton amplitude in eleven dimensions

$$\delta^{(10)}S = \frac{1}{2\kappa_{(11)}^2} \int d^{11}x \sqrt{-G} \, c_{10} \, \frac{\kappa_{(11)}^6}{\ell_P^9} \, \mathcal{D}^{12}\mathcal{R}^4. \tag{5.13}$$

Both of these points of view will be discussed in the Section 7.

5.2. An all order argument for $\beta_L = L$

The general structure of an L-loop amplitude in eleven dimensions is

$$M_{4;L} = \sum_i \mathcal{D}^{2n_i} \hat{R}^4 \, I_{4;L(i)}^{(11)}, \tag{5.14}$$

where $I_{4;L(i)}^{(11)}$ is a Feynman loop integral (not necessarily of scalar type for $L \geq 3$) so that the low-energy expansion is given by

$$M_{4;L} \sim \mathcal{D}^{2\beta_L} \hat{R}^4 \, \Lambda^{9L-6-2\beta_L} + \cdots, \tag{5.15}$$

---

[7] On-shell there are two types of couplings [72, 73] given by $(S^2 + T^2 + U^2)^3 \mathcal{R}^4$ and $(S^3 + T^3 + U^3)^2 \mathcal{R}^4$. Each coupling receives distinct contribution in string theory at tree-level order (A.6) and at loop order (A.12).

where the ellipsis are for sub-leading ultraviolet divergences. After compact-ification on a circle of radius $R_{11} \ell_P$, there are Kaluza–Klein states of mass $n^2/(R_{11} \ell_P)^2$ running in the loops, and the low-energy expansion of the loop amplitude becomes [33]

$$M_{4;L} \sim \sum_{n \geq 0} \sum_{\nu=0}^{9L-6-2\beta_L} \Lambda^{9L-6-2\beta_L-\nu} (R_{11} \ell_P)^{-\nu} (R_{11} \ell_P \mathcal{D})^{2n} \mathcal{D}^{2\beta_L} \hat{R}^4 + \cdots$$
(5.16)

converting this expression to the string frame using the metric of eq. (5.5) and the relation $R_{11}^3 = (g_s^A)^2$ we get that the amplitude contributes in ten dimensions to

$$M_{4;L}^{(11 \to 10)} \sim \sum_{n \geq 0} \sum_{\nu=0}^{9L-6-2\beta_L} (\Lambda \ell_P)^{9L-6-2\beta_L-\nu} (g_s^A)^{2(h-1)} \mathcal{D}^{2(\beta_L+n)} R^4 + \cdots$$
(5.17)

with

$$h = n + \frac{\beta_L - \nu}{3}.$$
(5.18)

The quantity h is the highest genus order at which the higher derivative contribu-tion $\mathcal{D}^{2(\beta_L+n)} R^4$ can occur in string perturbation in ten dimensions.

Since $\nu \geq 0$ we have that $h \leq n + \beta_L/3 \leq n + \beta_L$. We have seen in Section 4 that the four-graviton one-loop amplitude has $\beta_1 = 0$ and that $\beta_L \geq 2$ for $L \geq 2$, therefore only the Kaluza–Klein contributions from the $L = 1$ loop contribution gives the maximal genus contribution $h = k$ for the operator $\mathcal{D}^{2k} R^4$ which implies that the low-energy limit of the supergravity and string theory amplitudes satisfy the rule[8]

$$\beta_L = L.$$
(5.19)

This argument indicates that the operator $\mathcal{D}^{2k} R^4$ to string effective action in ten dimensions receives a perturbative contributions until genus k.

The same argument implies that the operators $\mathcal{D}^{2k} R^4 r_A^{-2n}$ to the type IIa string effective action in nine dimensions receives perturbative contributions until genus $k + n$.

We will see in the next section how the duality symmetries of M-theory relate the higher-order Kaluza–Klein contributions of the $L = 1$ loop amplitude to the contributions from higher-loop orders.

---

[8]Infra-red singularities could reduce the derivative order to which the amplitude contributes in the low-energy limit. We are assuming that no infra-red singularities are encountered when taking the low-energy limit of the loop amplitude. The issue of infra-red singularities in the low-energy of the amplitude does not arise in ten dimensions but could be a problem for the compactified cases [11].

6. The contributions from the one-loop amplitude

We consider the reduction on a d dimensional torus $\mathbb{T}^d$ of the four-graviton $L = 1$ loop amplitude $M_{1;4}$ given in eq. (4.8). The result is given by the sum of the scalar integrals for each channels in eq. (4.10)

$$I_4^{(11-d)}(S, T) = \frac{\pi^{\frac{11-d}{2}}}{\ell_P^d \, \mathcal{V}} \int_{\Lambda^{-2}}^{\infty} \frac{dt}{t} t^{\frac{d-3}{2}} \int_{\mathcal{T}_{ST}} \prod_{r=1}^{4} dv_r \sum_{\{m\} \in \mathbb{Z}^d} e^{-\pi t \, G^{IJ} m_I m_J + \pi t \, Q_4}.$$

(6.1)

The masses of the Kaluza–Klein state running in the loop is denoted $G^{IJ} m_I m_J$ and the volume of the torus is $\ell_P^d \, \mathcal{V}_d$. This expression contains a non-analytic contribution from the massless supergravity states in dimensions $11 - d$, and analytic terms. The non-analytic part is the usual field theory contribution from the massless states given by

$$I_{4,\text{nonana}}^{(11-d)}(S, T) \sim \int_0^1 \prod_{r=1}^{3} dv_r \, (Q_4)^{\frac{d-3}{2}}.$$

(6.2)

For $d = 0$ this is the eleven dimensional supergravity contribution $M_{4;1} \sim (-S)^{3/2}$, for $d = 1$ this is the ten dimension supergravity contribution $M_{4;1}^{(10)} \sim S \log(-S)$ of eq. (A.13), and for $d = 2$ this is the nine dimensional contribution $M_{4;1}^{(9)} \sim (-S)^{-1/2}$ (see [33,57,58] for a detailed discussion on the relation between these various contributions by considering the decompactification limits $d = 2 \to d = 1$ and $d = 1 \to d = 0$).

The expression for $I_4^{(11-d)}$ has the same leading ultraviolet divergence Λ^3 as the parent integral $M_{4;1}$, with Λ an ultraviolet cut-off measured in eleven dimensional Planck units. The ultraviolet divergences arise from the momentum independent part for small values of the proper time $t = \Lambda^{-2} \sim 0$, and correspond to a local ultraviolet divergence in eleven dimensions.

In order to isolate the divergences one must perform a Poisson resummation over the Kaluza–Klein modes m_I to get [57,58]

$$I_o^{(11-d)}(S, T) = \pi^{\frac{11-d}{2}} \int_0^{\Lambda^2} d\hat{t} \, \hat{t}^{\frac{1}{2}} \sum_{\{\hat{m}\} \in \mathbb{Z}^d} e^{-\pi \hat{t} \, G_{IJ} \hat{m}^I \hat{m}^J}$$

$$= \Lambda^3 + \frac{1}{\mathcal{V}^3} \sum_{\substack{\{\hat{m}\} \in \mathbb{Z}^d \\ \{\hat{m}\} \neq (0,\dots,0)}} \frac{1}{(\hat{m}^I \, \hat{G}_{IJ} \, \hat{m}^J)^{\frac{3}{2}}},$$

(6.3)

where $G_{IJ} = \mathcal{V}\,\hat{G}_{IJ}$ is the metric of the d-torus and $\det \hat{G}_{IJ} = 1$. The ultraviolet divergence is now localised in the zero winding sector $\hat{m}_I = 0$. The finite part is the contribution from the non zero winding modes given by

$$\mathcal{E}_s(G_{IJ}) \equiv \sum_{\substack{\{\hat{m}\}\in\mathbb{Z}^d \\ \{\hat{m}\}\neq(0,\dots,0)}} \frac{1}{(\hat{m}^I\,\hat{G}_{IJ}\,\hat{m}^J)^s}. \tag{6.4}$$

This expression is invariant under the large diffeomorphism $Sl(d,\mathbb{Z})$ of the d dimensional torus. The higher order terms in the external momenta expansion gives

$$\tilde{I}_4^{(11-d)}(S,T) = 2\pi^{7-d-n}\,(\ell_P^d\mathcal{V})^{n+\frac{d-5}{2}}\,\frac{\mathcal{G}_{ST}^n}{n!}\,\Gamma\left(\frac{d-3}{2}+n\right)\,\mathcal{E}_{\frac{d-3}{2}+n}(\hat{G}^{IJ}), \tag{6.5}$$

where

$$\mathcal{G}_{ST}^n \equiv \int_{T_{ST}} \prod_{i=1}^4 dv_i\,(Q_4)^n. \tag{6.6}$$

The superficial divergence of the one-loop amplitude is subtracted by adding to the eleven dimensional action the *local* counter-term of equation (5.11) which contributes to the one-loop amplitude by the following contribution

$$\delta_1 M_{4;1}^{(11)} = c_4\,\frac{\kappa_{(11)}^4}{(2\pi)^{11}}\,\frac{\pi^3}{2}\,\ell_P^3\,\hat{R}^4. \tag{6.7}$$

We now determine the value of c_4 by matching the total one-loop amplitude with corresponding string theory expressions reviewed in section Appendix A.

6.1. The circle compactification

For the case of a circle compactification $\mathbb{T}^1 = S^1$ of radius R_{11} one gets [57,58]

$$M_{4;1} + \delta_1 M_{4;1} = \frac{\kappa_{(11)}^4}{\ell_P^3}\,\hat{R}^4\left[(\Lambda\,\ell_P)^3 + c_4 + \frac{2\zeta_3}{R_{11}^3}\right] + \cdots, \tag{6.8}$$

where the ellipsis are for higher derivative contributions discussed in eq. (6.13) and eq. (6.15) that are independent of the cut-off. Using the dictionary in eq. (5.5) the amplitude translates into the ten dimensional expression

$$M_{4;1} + \delta_1 M_{4;1} \sim \kappa_{(10)}^2\,\hat{R}^4\left[(\ell_P\Lambda)^3 + c_4 + \frac{2\zeta_3}{(g_s^A)^2}\right] + \cdots, \tag{6.9}$$

where we recognise tree-level and one-loop string contributions. Comparing with the value of the $\hat{R}^4$ contribution at genus-one in string theory in eq. (A.12) we deduce that

$$(\ell_P \Lambda)^3 + c_4 = \frac{2\pi^2}{3}. \tag{6.10}$$

Another equivalent way of formulating the same regularisation scheme is to use the T-duality properties of the perturbative string amplitudes, which we discuss now.

6.2. The torus compactification

The $\mathbb{T}^2$ torus compactification of the one-loop four-point amplitude $M_{1;4}$ gives the following perturbative contributions [57,58]

$$M_{4;1} + \delta_1 M_{4;1} \sim \kappa_{(10)}^2 \, \hat{R}^4 \left[\frac{1}{r_A} \left((\ell_P \Lambda)^3 + c_4 \right) + r_A \left(\frac{2\zeta_3}{(g_s^A)^2} + 4\zeta_2 \right) + \cdots \right], \tag{6.11}$$

where we used that $r_A = R_{10}\sqrt{R_{11}}$. The T-duality invariance of the four-graviton amplitude at tree-level and genus-one in string theory requires that the above expression is invariant under the transformation $r_B \to r_A = 1/r_B$. Although the tree-level contribution is invariant thanks to the transformation rules of the ten dimensional dilaton or as consequence of the M-theory dictionary given in eq. (5.5)

$$\frac{r_A}{(g_s^A)^2} = \frac{R_{10}\sqrt{R_{11}}}{R_{11}^3} = \left(\frac{R_{10}}{R_{11}} \right)^2 \frac{1}{R_{10}\sqrt{R_{11}}} = \frac{r_B}{(g_s^B)^2}, \tag{6.12}$$

the one-loop contribution is invariant only if the condition (6.10) is satisfied.

The rest of the higher order derivative contributions to string genus-one amplitude in (6.9) give ultraviolet finite contributions which read in the type IIa frame

$$M_{4;1} + \delta_1 M_{4;1} \sim \cdots + \kappa_{(10)}^2 8\pi^{\frac{3}{2}} \sum_{n=2}^{\infty} \frac{\Gamma(n - \frac{1}{2})\zeta_{2n-1}}{n!} r_A^{2n-1} (\ell_s \mathcal{D}^{2n} \hat{R}^4) + \cdots, \tag{6.13}$$

and in the type IIb frame

$$M_{4;1} + \delta_1 M_{4;1} \sim \cdots + \kappa_{(10)}^2 8\pi^{\frac{3}{2}} \sum_{n=2}^{\infty} \frac{\Gamma(n - \frac{1}{2})\zeta_{2n-1}}{n!} r_B^{-2n+1} (\ell_s \mathcal{D}^{2n} \hat{R}^4) + \cdots, \tag{6.14}$$

where $\mathcal{D}^{2n} = \mathcal{G}^n_{ST} + \mathcal{G}^n_{TU} + \mathcal{G}^n_{SU}$. The ellipsis represent the contributions given in eq. (6.9) and in eq. (6.15). These expressions match the corresponding contributions to the derivative expansion of the genus-one amplitude compactified to nine dimensions on a circle of radius r_A or r_B derived in [72], but they are not invariant under the T-duality symmetry $r_A \to r_B = 1/r_A$. We will see below that the missing contributions are provided by the higher loop $L \geq 2$ contributions in eleven dimensions.

Finally the last piece from $L = 1$ amplitude are the higher-derivative contributions that give higher string genus contributions in the type IIa frame

$$M_{4;1} + \delta_1 M_{4;1} \sim \cdots + \kappa^2_{(10)} 8\pi^2 r_A \sum_{n=2}^{\infty} \frac{\Gamma(n-1)\zeta_{2n-2}}{n!} (g_s^A)^{2n-2} (\ell_s \mathcal{D}^{2n} \hat{R}^4),$$

(6.15)

and in the type IIb frame

$$M_{4;1} + \delta_1 M_{4;1} \sim \cdots + \kappa^2_{(10)} 8\pi^2 \sum_{n=2}^{\infty} \frac{\Gamma(n-1)\zeta_{2n-2}}{n!} \frac{(g_s^B)^{2n-2}}{r_B^{2n+1}} (\ell_s \mathcal{D}^{2n} \hat{R}^4).$$

(6.16)

These contributions are genus n contributions to the operator $\mathcal{D}^{2n} \mathcal{R}^4$, satisfying the relation $\beta_n = n$ derived in Section 5.2. The value of the coefficient for the $\mathcal{D}^4 \hat{R}^4$ term matches the genus two contributions derived from string theory in eq. (A.17) [58, 88].

We will return to these contributions in Section 8 when we will discuss non renormalisation theorems in string theory and supergravity.

6.3. The three-torus compactification

Taking the amplitude on a three-torus $\mathbb{T}^3$ one gets for the $\hat{R}^4$ term in the type IIa variables

$$M_{4;1} + \delta_1 M_{4;1} \sim \kappa^2_{(10)} \left([(\ell_P \Lambda)^3 + c_1] U_2 + \mathcal{E}_{\frac{3}{2}}(\hat{G}_{IJ}) \right) \hat{R}^4,$$

(6.17)

where $\mathcal{E}_{3/2}(\hat{G})$ is the $Sl(3, \mathbb{Z})$ modular form defined in eq. (6.4), which has the perturbative expansion [89, 90]

$$\mathcal{E}_{\frac{3}{2}}(\hat{G}) = 2\zeta_3 \frac{T_2}{(g_s^A)^2} - \pi \log(T_2 |\eta(T)|^2) - \frac{4\pi}{3} \log(T_2^2/g_s^A) + n.p.,$$

(6.18)

where the first term corresponds to the tree-level contribution for the type IIa string on two torus of volume T_2, the second term gives the contributions from the wrapped F-string on the two torus and contains the genus-one perturbative contributions, the third term is a logarithmic term expressed in terms of the eight dimensional dilaton $\exp(-2\phi^{(8)}) = T_2 \exp(-2\phi^{(10)})$. This contribution arises from the massless threshold $\hat{R}^4 \log(s)$ in eight dimensions after performing a Weyl rescaling to get to the Einstein frame. The non perturbative effects are the D-instanton effects and the (p, q)-string wrapped around the two-torus.

The $Sl(3, \mathbb{Z})$ invariance of the coupling, inherited from the large diffeomorphism of the torus, is part of the eight dimensional U-duality group $Sl(3, \mathbb{Z}) \times Sl(2, \mathbb{Z})_U$ where U is the complex structure of the two torus on which the type IIa string is compactified. As for the case of the compactification to nine dimensions the string perturbative answer has to be invariant under the T-duality sub-group $Sl(2, \mathbb{Z})_T \times Sl(2, \mathbb{Z})_U$, and symmetric under the exchange between T and U. The perturbative part of eq. (6.17) is invariant only if the condition (6.10) is satisfied and the dependence on the complex structure U is given by $\log(U_2|\eta(U)|^2)$. These requirements allow the determination of the U-duality invariant coupling uniquely and reproduce the result given by [89]

$$\int d^9x \sqrt{-g^{(8)}} \left(\mathcal{E}_{\frac{3}{2}}(M) - \pi \log(U_2|\eta(U)|^2)\right) \hat{R}^4, \tag{6.19}$$

with the $Sl(3, \mathbb{Z})$ modular forms defined by

$$\mathcal{E}_s(M) = \sum_{(m_1,m_2,m_3)\neq(0,0,0)} \frac{v_8^{-\frac{s}{3}}}{\left(\frac{|m_1+m_2\Omega+m_3B|^2}{\Omega_2} + \frac{m_3^2}{v_8}\right)^s}, \tag{6.20}$$

where $1/\sqrt{v_8} = V^2/\ell_P^2 = g_S^{-1/2} T_2$ is the dimensionless compactification volume measured in Planck length unit, and $B = B_R + \Omega B_{NS}$ is the combination of the RR and NS B-field. The dependence on the B-field is needed in order to consider wrapped M2-branes along the internal directions which are not included in the construction from the multiloop amplitude we have described. The total eight dimensional coupling

$$\mathcal{E}_{(0,0)}^{(8d)} = \mathcal{E}_{\frac{3}{2}}(M) - 2\pi \log(U_2|\eta(U)|^2) \tag{6.21}$$

is a zero mode of the $SO(3)\backslash Sl(3, \mathbb{Z}) \times SO(2)\backslash Sl(2, \mathbb{Z})$ Laplacian associated with the U-duality group in eight dimensions [89]

$$\Delta_{Sl(3)\times Sl(2)} \mathcal{E}_{(0,0)}^{(8d)} = 8\pi. \tag{6.22}$$

The $D^4 \hat{R}^4$ contribution from eq. (6.5) gives the coupling [90]

$$(T_2 (g_s^A)^2)^{\frac{1}{3}} \mathcal{E}_2(\hat{G}^{-1}) D^4 \hat{R}^4 \sim \left[\zeta_4 (g_s^A)^2 + 90\pi T_2 E_2(U) + n.p.\right] \hat{R}^4, \tag{6.23}$$

where $E_s(U)$ are the usual $Sl(2, \mathbb{Z})$ Eisenstein series depending on the complex structure of the torus on which type IIa string is compactified. We recognise the genus two contributions given in eq. (6.15) and the genus-one contribution depending on the moduli T and U of the torus on which type IIa string is compactified. This expression fails to be invariant under T-duality but this symmetry will be recovered once the $L = 2$ contribution has been added.

7. The contributions from the two-loop amplitude

In this section we describe the contributions from the $L = 2$ amplitude compactified on a circle and on a torus. The analysis of the two loop amplitude brings new technical difficulties which we will not comment on and refer to the papers [58, 73, 75] for details. We only describe its consequences on the M-theory and string theory low energy effective action.

• The $\mathcal{D}^4 \hat{R}^4$ term is the leading contribution to the low-energy limit of the $L = 2$ amplitude of eq. (4.15).

Taking the two-loop amplitude on a circle of radius $R_{11} \ell_P$ one gets at this order

$$I_1 \sim \left((\ell_P \Lambda)^8 + c_{2;0} + \pi^5 \zeta_5 [(\ell_P \Lambda)^3 + c_4] \frac{1}{R_{11}^5} \right) \mathcal{D}^4 \hat{R}^4. \tag{7.1}$$

We have included the contribution from the new counter-term $c_{2;0}$ of eq. (4.21) needed to regulate the new primitive divergence at this order, and the contributions from the triangle diagram with the one-loop counter-term of Fig. 3(c). Converted into string variables using the relation between the M-theory parameters and the string variables given in eq. (5.5) one gets

$$I_1 \sim \left([(\Lambda \ell_P)^8 + c_{2;0}] (g_s^A)^{\frac{10}{3}} + \pi^5 \zeta_5 [(\Lambda \ell_P)^3 + c_4] \frac{1}{(g_s^A)^2} \right) \mathcal{D}^4 \hat{R}^4. \tag{7.2}$$

The first term does not have any meaning in string perturbation theory, therefore the leading divergence has to be subtracted with no finite remainder

$$(\Lambda \ell_P)^8 + c_{2;0} = 0. \tag{7.3}$$

The second term is a string three loop contribution, which contribution which is determined by the regulated one-loop divergence in eq. (6.10) giving the correct value of the tree-level contribution to the $\mathcal{D}^4 \hat{R}^4$ term [58].

Taking the amplitude on a torus one gets

$$I_1 \sim \left([(\Lambda\, \ell_P)^8 + c_{2;0}] + \frac{\pi^5}{2}[(\Lambda, \ell_P)^3 + c_4]\, \frac{E_{\frac{5}{2}}(\Omega)}{\mathcal{V}^{\frac{5}{2}}} + \frac{2\pi^4 \zeta_3 \zeta_4}{\mathcal{V}^4} \right) \mathcal{D}^4 \hat{R}^4. \tag{7.4}$$

The second term in this expression decompactifies to a finite $E_{5/2}(\Omega)\,\mathcal{D}^4 \hat{R}^4$ contribution to the ten dimensional type IIb limit $\mathcal{V} \to 0$ with Ω kept constant. The last term in this expression gives a genus 1 contribution to the effective action of type II string in nine dimensions. This contribution together with the $\mathcal{D}^4 \hat{R}^4$ from the $L = 1$ amplitude in eq. (6.13) gives the complete contribution to the string genus-one amplitude at this order [71,72]

$$\delta I_1 \sim \frac{\zeta_3}{15}\left(r_A^3 + \frac{1}{r_A^3} \right) \mathcal{D}^4 \hat{R}^4. \tag{7.5}$$

One notices that the $L = 2$ contribution provided by the genus-one $1/r_A^3\, \mathcal{D}^4 \hat{R}^4$ missing in eq. (6.13) for the T-duality symmetry $r_A \to r_B = 1/r_A$ of the nine dimensional contribution to the low energy expansion of the genus-one string amplitude on a circle.

This is the manifestation of a generic phenomena where the compactification of the higher-loop four-graviton amplitudes will gives genus one contributions to the higher-derivative operators $\mathcal{D}^{2k} \hat{R}^4$ completing the expression in eq. (6.13) in a T-duality invariant expression.

The complete $\mathcal{D}^4 \hat{R}^4$ in type IIb variables is given in the Einstein frame by [73]

$$\int d^9x \sqrt{-g^{(9)}}\, \mathcal{E}^{(9d)}_{(1,0)}\, \mathcal{D}^4 \hat{R}^4 \tag{7.6}$$

with

$$\mathcal{E}^{(9d)}_{(1,0)} = v_9^{-\frac{3}{7}} E_{\frac{3}{2}}(\Omega) + \frac{2\pi^2}{3}\, v_9^{\frac{4}{7}}, \tag{7.7}$$

where $v_9^{1/2} = r_B/g_s^{1/4}$ is the radius of the circle measures Planck length unit. This coupling satisfies the differential equation [73]

$$\left[\Delta_\Omega + \frac{7}{4} v_9(\partial_{v_9} v_9 \partial_{v_9}) + \frac{1}{2} v_9 \partial_{v_9} \right] \mathcal{E}^{(9d)}_{(1,0)} = \frac{30}{7}\, \mathcal{E}^{(9d)}_{(1,0)}. \tag{7.8}$$

Considering the amplitude on $\mathbb{T}^3$ we obtain

$$\mathcal{E}_{\frac{5}{2}}(G) = 2 \left(\frac{T_2}{(g_s^A)^2} \right)^{\frac{5}{3}} \zeta_5 + \frac{4}{3} \left(\frac{(g_s^A)^2}{T_2} \right)^{\frac{1}{3}} E_2(T)$$

which added to the $L = 1$ contribution in eq. (6.18) gives an expression invariant under the exchange of T and U moduli. The complete U-duality $Sl(3, \mathbb{Z}) \times Sl(2, \mathbb{Z})$ invariant expression has been derived in [90] with the result in the string frame being

$$\int d^8x \sqrt{-g^{(8)}} \, g_s^{-\frac{4}{3}} T_2^{-\frac{2}{3}} \left(\mathcal{E}_{\frac{5}{2}}(M) - 8 E_{-\frac{1}{2}}(M) E_2(U) \right) \mathcal{D}^4 \mathcal{R}^4, \tag{7.9}$$

where $\mathcal{E}_s(M)$ are the $Sl(3, \mathbb{Z})$ modular forms defined in eq. (6.20). The total eight dimensional coupling

$$\mathcal{E}_{(1,0)}^{(8d)} = \mathcal{E}_{\frac{5}{2}}(M) - 8 E_{-\frac{1}{2}}(M) E_2(U), \tag{7.10}$$

is an eigenfunction of the $Sl(3, \mathbb{Z}) \times Sl(2, \mathbb{Z})$ Laplacian [73]

$$\Delta_{Sl(3) \times Sl(2)} \, \mathcal{E}_{(1,0)}^{(8d)} = \frac{10}{3} \, \mathcal{E}_{(1,0)}^{(8d)}. \tag{7.11}$$

- Expanding the $L = 2$ loop to the next order, we obtain for the $\mathcal{D}^6 \hat{R}^4$ term on a circle [73, 75]

$$I_2 \sim \left([(\Lambda \, \ell_P)^6 + c_{2;1}] + [(\Lambda \, \ell_P)^3 + c_{2;1}] \frac{\pi \zeta_3}{3 R_{11}^3} + \frac{\zeta_3^2}{2 R_{11}^6} \right) \mathcal{D}^6 \hat{R}^4. \tag{7.12}$$

Converted to string variables and using the relation (6.10) we have

$$I_2 \sim \left([(\Lambda \, \ell_P)^6 + c_{2;1}] (g_s^A)^2 + \zeta_2 \zeta_3 + \frac{\zeta_3^2}{2(g_s^A)^2} \right) \mathcal{D}^6 \hat{R}^4, \tag{7.13}$$

where we recognise a genus two, genus-one and tree-level string contribution. In order to determine the precise value for the new counter-term $c_{2;1}$ one would need to know the value of the genus 2 contribution to the $\mathcal{D}^6 \hat{R}^4$ in ten dimensions. Since this value is not known it will be determined later using the duality relations. The finite remainder after subtracting the divergence will provide a new corrections to the M-theory effective action of eq. (5.2) in agreement with the considerations of Section 5.2 and the reference [87].

Taking the amplitude on a torus one gets [73, 75]

$$I_2 \sim \left([(\Lambda \, \ell_P)^6 + c_{2;1}] + [(\Lambda \, \ell_P)^3 + c_{2;1}] \frac{E_{\frac{3}{2}}(\Omega)}{V^{\frac{3}{2}}} + \frac{\mathcal{E}_{(\frac{3}{2}, \frac{3}{2})}(\Omega)}{V^3} \right) \mathcal{D}^6 \hat{R}^4. \tag{7.14}$$

The contribution to the $\mathcal{D}^6 \hat{R}^4$ to the ten dimensional effective action for the type IIb string $\mathcal{E}_{(3/2,3/2)}$ has been determined in [75] and has the following weak-coupling expansion

$$
\mathcal{E}_{(\frac{3}{2},\frac{3}{2})}(\Omega) = 4\zeta_3^2 \Omega_2^3 + 8\zeta_2\zeta_3\,\Omega_2 + \frac{48}{5}\zeta_2^2\Omega_2^{-1} + \frac{8}{9}\zeta_6\Omega_2^{-3}
$$
$$
+ O(e^{-2\pi\Omega_2}). \tag{7.15}
$$

Converting eq. (7.14) into type IIa string variables and using the relation (6.10) we have

$$
I_2 \sim \left(4\zeta_3^2\,\frac{r_A}{(g_s^A)^2} + 8\zeta_2\zeta_3\left(r_A + \frac{1}{r_A}\right) + 16\zeta_2^2\,\frac{(g_s^A)^2}{r_A}\right.
$$
$$
\left. + [(\Lambda\,\ell_P)^6 + c_{2;1}]\,r_A\,(g_s^A)^2 + \frac{48}{5}\,\zeta_2^2\,\frac{(g_s^A)^2}{r_A^3} + \frac{8}{9}\,\zeta_6\,\frac{(g_s^A)^4}{r_A^5}\right)\mathcal{D}^6\hat{R}^4. \tag{7.16}
$$

The first line gives the string tree-level and the genus-one contributions in nine dimensions to $\mathcal{D}^6\hat{R}^4$ couplings. The value of these couplings match the results extracted from string perturbation theory (see the Appendix A and [71, 72]). Again we stress the fact that the genus-one contribution is invariant under the T-duality transformation $r_A \to r_B = 1/r_A$ thanks to the relation in eq. (6.10) determining the one-loop counter-term.

The second line of this expression gives string genus two contributions. Counting the number of fermionic zero modes involved in the gravity amplitude, the genus two type IIa and type IIb string contributions are the same at the $\mathcal{D}^6\hat{R}^4$ order[9] and the second line in eq. (7.16) should be invariant under the T-duality symmetry $r_A \to r_B = 1/r_A$. The first term in this expression is invariant thanks to the transformation rules of the eight-dimensional dilaton in eq. (6.12). The second and the third terms are exchanged only if

$$
(\Lambda\,\ell_P)^6 + c_{2;1} = \frac{48}{5}\,\zeta_2^2, \tag{7.17}
$$

which determines the value of the $\mathcal{D}^6\hat{R}^4$ in the M-theory effective action (5.2).

The third line in eq. (7.16) is a genus three contribution. This expression is invariant under T-duality once summed with the three loop contribution from the

---

[9]In the RNS formalism, the chirality dependence of the gravity amplitudes enters from the *odd/odd* spin structure contributions. Because string perturbations has $2(g-1)$ odd moduli at genus g, the four-graviton amplitude in type II superstring can get contributions from the *odd/odd* spin structure sector from genus three. Using the non-minimal pure spinor formalism Berkovits showed that the chirality dependences arises from genus five [16].

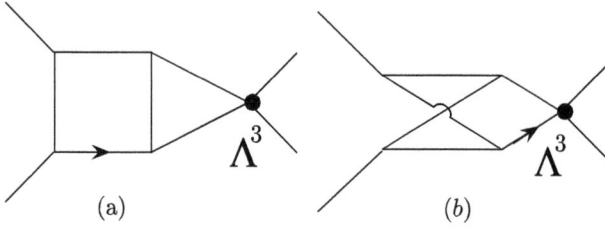

Fig. 5. Sub-divergences of the $L = 3$ amplitude constructed from Fig. 4(b). Diagram (a) and (b) contribute to a $\Lambda^3/\mathcal{V}^6 \mathcal{D}^6 \hat{R}^4$. The arrow indicates the quadratic dependence on the loop momenta in the numerator of the expression for the two-loop integrals.

$L = 1$ contribution in eq. (6.15) to give

$$\frac{8}{9} \zeta_6 \left(r_A + \frac{1}{r_A^5} \right) (g_s^A)^4 \, \mathcal{D}^6 \hat{R}^4. \tag{7.18}$$

The $L = 3$ three-loop supergravity amplitude in eleven dimensions will contribute to the $\mathcal{D}^6 \hat{R}^4$ coupling only from the class of diagrams constructed from the vacuum diagram given in Fig. 4(b) because the diagrams obtained from Fig. 4(a) all have a prefactor of $\mathcal{D}^8 \hat{R}^4$ [29]. This amplitude satisfies the $\beta_3 = 3$ rule and has mass dimension

$$[M_{4,3}] \sim (mass)^{15} \, \mathcal{D}^6 \hat{R}^4. \tag{7.19}$$

The sub-divergence in Fig. 5(a,b) can respectively contribute to a term of the type

$$\frac{\Lambda^3}{\mathcal{V}^6} \, \mathcal{D}^6 \, \hat{R}^4, \tag{7.20}$$

contributing to a genus-one term completing the genus-one term $r_A^5 \zeta_5 \, \mathcal{D}^6 \hat{R}^4$ from the $L = 1$ amplitude given in eq. (6.13) into a T-duality invariant expression

$$\zeta(5) \left(r_A^5 + \frac{1}{r_A^5} \right) \mathcal{D}^6 \hat{R}^4. \tag{7.21}$$

This amplitude can only contribute to the ten dimensional type IIb effective action from a contribution of the type [73]

$$I_3 \sim \frac{\Lambda^9}{\mathcal{V}^3} \, f(\Omega), \tag{7.22}$$

but a simple dimensional analysis on the various diagrams entering the $L = 3$ amplitude shows that no such contributions can be found in the solution given in [29] and the coupling to the ten dimensional type IIb effective action given by the eq. (7.15) is not renormalised.

The complete $\mathcal{D}^6 \hat{R}^4$ coupling in nine dimensions in the Einstein frame is given by [73]

$$\int d^9 x \sqrt{-g^{(9)}} \, \mathcal{E}^{(9d)}_{(0,1)} \, \mathcal{D}^6 \, \hat{R}^4, \tag{7.23}$$

with

$$\mathcal{E}^{(9d)}_{(0,1)} = v_9^{-\frac{6}{7}} \mathcal{E}_{(0,1)} + 4\zeta(2) \, v_9^{\frac{1}{7}} \, E_{\frac{3}{2}} + \frac{12\zeta(2)}{63} \, v_9^{\frac{15}{7}} \, E_{\frac{5}{2}}$$

$$+ \frac{24\zeta(2)\zeta(5)}{63} \, v_9^{-\frac{20}{7}} + \frac{48\zeta(2)^2}{5} \, v_9^{\frac{8}{7}}, \tag{7.24}$$

satisfying the Laplace equation

$$\left[\Delta_\Omega + \frac{7}{4} v_9 (\partial_{v_9} v_9 \partial_{v_9}) + \frac{1}{2} v_9 \partial_{v_9} \right] \mathcal{E}^{(9d)}_{(0,1)} = 12 \, \mathcal{E}^{(9d)}_{(0,1)} - 6\left(\mathcal{E}^{(9d)}_{(0,0)}\right)^2, \tag{7.25}$$

where $\mathcal{E}^{(9d)}_{(0,0)}$ is the nine dimensional $\hat{R}^4$ coupling

$$\mathcal{E}^{(9d)}_{(0,0)} = v_9^{-\frac{3}{7}} \, E_{\frac{3}{2}}(\Omega) + 4\zeta_2 \, v_9^{\frac{4}{7}}. \tag{7.26}$$

The $\mathcal{D}^6 \hat{R}^4$ coupling in eight dimensions is given in the string frame [91]

$$\int d^8 x \sqrt{-g^{(8)}} \, \frac{g_s^2}{T_2} \, \mathcal{E}^{(8d)}_{(0,1)} \, \mathcal{D}^6 \, \hat{R}^4 \tag{7.27}$$

with

$$\mathcal{E}^{(8d)}_{(0,1)} = \mathcal{E}_{(\frac{3}{2}, \frac{3}{2})}(M) + \frac{20}{3} E_{-\frac{3}{2}}(M) E_3(U)$$

$$+ \frac{1}{2} E_{\frac{3}{2}}(M) E_1(U) + f(U, \bar{U}), \tag{7.28}$$

where the $Sl(3, \mathbb{Z})$ modular forms $\mathcal{E}_{(3/2, 3/2)}(M)$ is defined by

$$\Delta_{Sl(3)} \mathcal{E}_{(\frac{3}{2}, \frac{3}{2})}(M) = 12 \, \mathcal{E}_{(\frac{3}{2}, \frac{3}{2})}(M) - \frac{3}{2} \, E_{\frac{3}{2}}^2(M) \tag{7.29}$$

and the function $f(U, \bar{U})$ is defined by the differential equation with a source term [91]

$$\Delta_U f(U, \bar{U}) = 12 f(U, \bar{U}) - 6 E_1^2(U). \tag{7.30}$$

The total eight dimensional coupling satisfies the Laplace equation with for source term the $\hat{R}^4$ the eight dimensional coupling given in eq. (6.21)

$$\Delta_{Sl(3) \times Sl(2)} \mathcal{E}_{(0,1)}^{(8d)} = 12 \mathcal{E}_{(0,1)}^{(8d)} + \frac{3}{2} (\mathcal{E}_{(0,0)}^{(8d)})^2. \tag{7.31}$$

8. Non-renormalisation theorems

We have described the structure of gravity amplitudes in supergravity and superstring theory, and in particular formulated various constraints that these amplitudes should satisfy in maximal supergravity.

For the case of maximal supergravity theories with $N = 8$ supersymmetry one expects that the low-energy limit of an L-loop four-graviton amplitude will start contributing from $\mathcal{D}^{2L} R^4$, summarised by the rule $\beta_L = L$. In Section 4.4 we gave an argument based on supersymmetry for the validity of this rule up-to $L = 6$ and in Section 5.2 we gave a argument based on dimensional analysis and the duality relations of M-theory for an all order confirmation of this rule.

When one applies the rule $\beta_L = L$ one deduces important non renormalisation theorems that the couplings to the low-energy effective action of superstring theories have to satisfy. At the level of amplitude computations in supergravity such conditions imply that the ratio of the four-graviton L-loop amplitude to the tree-amplitude would behave as

$$\left[\frac{M_{4;L}}{M_{4;tree}} \right] = \mathcal{D}^{2L} stu \, (mass)^{(D-4)L-6}, \tag{8.1}$$

indicating that in four dimensions, the L-loop amplitude would have to be the sum of an L-loop dimensionless Feynman integral times a power of the external momenta $\mathcal{D}^{2L}$ increasing with the loop order. Up to genus three it has been shown that the polarisation dependence of the four-graviton amplitude is the same as the tree amplitude [1, 6, 36, 37, 92] and that the structure of the amplitude has the above mentioned structure.

The conjectured S-duality invariance of the type IIb effective action implies that the higher-derivative terms of the type $\mathcal{D}^{2k} \hat{R}^4$ must have coefficients that are modular functions $f_k(\Omega)$ under the action of $Sl(2, \mathbb{Z})$ on the complexified coupling constant $\Omega = C^{(0)} + i/g_s$ where $C^{(0)}$ is the Ramond-Ramond 0-form potential. This conjecture can hold only if the polarisation dependence of the four-graviton amplitude at all orders in the genus expansion is the same.

The non renormalisation conditions for the R^4 couplings to the ten dimensional type IIa and IIb effective action [57, 93] beyond one-loop is guaranteed by the fact that higher-loop amplitudes in $N = 8$ supergravity have $\beta_L \geq 2$ for $L \geq 2$.

The exactness of the type IIb $E_{5/2}(\Omega) \mathcal{D}^4 R^4$ coupling extracted from the two-loop amplitude in eleven dimensions in [58] (see Section 7 of this text as well) is assured by the fact that the rule $\beta_L \geq 3$ for $L \geq 3$ is satisfied by the four-graviton amplitudes in $N = 8$ supergravity [29] and the higher genus (F-terms) computation by Berkovits in [16].[10]

The dimensional analysis argument of Section 5.2 implies that the higher-derivative operators $\mathcal{D}^{2k} R^4$ do not receive perturbative corrections beyond genus k in string perturbation theory. This fact has been confirmed up to $k \leq 5$ for the analysis of superstring amplitudes in [16] and supergravity computation in [33, 73].

One important consequence of the structure of the $L = 1$ and the $L = 2$ loop amplitudes and the relation between the eleven dimensional M-theory and string theory is that the higher-order couplings satisfy some differential equations.

The R^4 and $\mathcal{D}^4 R^4$ couplings to the ten dimensional type IIb effective action in ten dimensions satisfy the differential equation

$$\Delta\, f_{\frac{3}{2}+p}(\Omega) = \frac{(3+2p)(1+2p)}{4}\, f_{\frac{3}{2}+p}(\Omega), \tag{8.2}$$

where $\Delta = 4\Omega_2^2 \partial_\Omega \bar{\partial}_{\bar\Omega}$ is the $Sl(2, \mathbb{Z})$ Laplacian and with respectively $p = 0$ and $p = 1$ as a trivial consequence of the structure[11] of the $L = 1$ [57] and the $L = 2$ loop amplitudes [58, 73]. This equation together with the boundary condition

$$\lim_{\Omega_2 \to \infty} \Omega_2^{\frac{1}{2}-p}\, f_{\frac{3}{2}+p}(\Omega) = \text{Cste} \tag{8.3}$$

with the constant given by the expansion of the tree-level amplitude given in eq. (A.6), determine uniquely the couplings.

At higher-order in the derivative expansion the supersymmetry constraints on the couplings will change their structure with the appearance of source terms [75]. The $\mathcal{D}^6 R^4$ coupling to the type IIb effective action in ten dimensions satis-

---

[10]The issue of infra-red singularities in the low-energy of the amplitude does not arise in ten dimensions but could be a problem for the compactified cases [11].

[11]These equations have been shown to be a consequence of the on-shell supersymmetry of the type IIb supergravity in ten dimensions in [94, 95].

fies the equation

$$\Delta\, f_3(\Omega) = 12\, f_3(\Omega) - 6E_{\frac{3}{2}}(\Omega)^2 \tag{8.4}$$

with the boundary condition

$$\lim_{\Omega_2\to\infty}\, \Omega_2^{-1}\, f_3(\Omega) = 4\zeta_3^2, \tag{8.5}$$

has a unique solution the function $f_3(\Omega) = \mathcal{E}_{(3/2,3/2)}(\Omega)$ found in [75] which has the weak coupling expansion given in eq. (7.15). For compactification of string theory to lower dimensions these couplings satisfy equivalent differential equations for the Laplacian associated with the U-duality group. In eight dimensions the U-duality group is $Sl(3, \mathbb{Z}) \times Sl(2, \mathbb{Z})$ and the action of the $SO(3)\backslash Sl(3) \times SO(2)\backslash Sl(2)$ Laplacian on the previous coupling is given by [73, 89–91]

$$\Delta_{Sl(3)\times Sl(2)}\, \mathcal{E}_{(p,0)}^{(8d)} = \frac{(3+2p)2p}{3}\mathcal{E}_{(p,0)}^{(8d)} \tag{8.6}$$

for the R^4 ($p = 0$) and the $D^4 R^4$ ($p = 1$) and

$$\Delta_{Sl(3)\times Sl(2)}\, \mathcal{E}_{(0,1)}^{(8d)} = 12\, \mathcal{E}_{(0,1)}^{(8d)} - \frac{3}{2}\big(\mathcal{E}_{(0,0)}^{(8d)}(M)\big)^2, \tag{8.7}$$

for the $D^6 \hat{R}^4$ term.

The structure of the differential equations for the $\hat{R}^4$ coupling in eq. (6.22), for the $D^4\, \hat{R}^4$ in eqs. (7.8) and (7.11) and for the $D^6\, \hat{R}^4$ in eqs. (7.25) and (7.31) have the same structure as the one for the ten dimensional type IIb theory. These equations that are expected to be a consequence of $N = 8$ supersymmetry which is preserved by the torus compactification we have considered.

The presence of the source term in (8.4) can be motivated from the structure of the on-shell supersymmetry for the type IIb effective actions as follows. The on-shell supersymmetry variations of the α' corrections to the effective action of the type IIb superstring

$$S = S^{(0)} + \alpha'^3\, S^{(3)} + \alpha'^5\, S^{(5)} + \alpha'^6\, S^{(6)} + \cdots, \tag{8.8}$$

requires the modification of the on-shell supersymmetry transformations [56] at each order

$$\delta_\epsilon = \delta_\epsilon^{(0)} + \alpha'^3\, \delta_\epsilon^{(3)} + \alpha'^5\, \delta_\epsilon^{(5)} + \alpha'^6\, \delta_\epsilon^{(6)} + \cdots, \tag{8.9}$$

according to the following pattern

$$
\begin{aligned}
\delta_\epsilon^{(0)} S^{(0)} &= 0, \\
\delta_\epsilon^{(0)} S^{(3)} &= \delta_\epsilon^{(3)} S^{(0)}, \\
\delta_\epsilon^{(0)} S^{(5)} &= \delta_\epsilon^{(5)} S^{(0)}, \\
\delta_\epsilon^{(0)} S^{(6)} + \delta_\epsilon^{(3)} S^{(3)} &= \delta_\epsilon^{(6)} S^{(0)}.
\end{aligned}
\tag{8.10}
$$

At the order α'^6 for the first time, the lowest order modifications to the on-shell supersymmetry transformations enter in the equations giving rise to the source term in eq. (8.4). The structure of the differential equations at higher order is expected to follow a similar pattern where coupling are given by a finite sum of modular functions $\mathcal{E}_s^{(i)}(\Omega) = \sum_{i=1}^{n_s} e_s^{(i)}(\Omega)$ with each function $e_s^{(i)}(\Omega)$ satisfying a differential equation with a source as in eq. (8.4) [73].

Acknowledgements

I would like to thank M.B. Green, and J. Russo for many discussions and collaborations on various computations that lead to (published and non published) results helping to shape the general picture described in this text. I would like to thank N. Berkovits, Z. Bern, P. Howe and K. Stelle for many useful discussions. And Paul Cook for useful feedback on the manuscript.

   This lecture note is based on the dissertation submitted for the degree of 'habilitation à diriger les recherches' granted by the University of Paris VI. I would like to thank the referees of this thesis Jean-Pierre Derendinger, Paolo di Vecchia and Kelly Stelle.

   The author would like to thank the Niels Bohr institute for their hospitality when most of these lecture notes have been written down. This work has been partially supported by the following RTN contracts MRTN-CT-2004-503369, MRTN-CT-2004-005104, the ANR project ANR-06-BLAN-3_137168.

Appendix A. The string S-matrix

In this appendix we collect various data from perturbative string calculations that are needed for comparison with the predictions from the $L = 1$ and $L = 2$ amplitudes of eleven dimensional supergravity on a circle and a torus.

   The unitary string theory S-matrix for four-graviton scattering has the following perturbative expansion [1,71,72,88]

$$
S(\hat{\sigma}_2, \hat{\sigma}_3) = \kappa_{(10)}^2 g_s^4 \left(\frac{1}{g_s^2} A^{\text{tree}}(\hat{\sigma}_2, \hat{\sigma}_3) + 2\pi A^{g=1} + \pi g_s^2 A^{g=2} + \cdots \right), \tag{A.1}
$$

where A^{tree}, $A^{g=1}$ and $A^{g=2}$ are respectively the tree-level, genus-one and genus two amplitudes for four massless states ($s+t+u=0$) described in the following section.

Appendix A.1. The tree-amplitude

The tree-level amplitude is given by

$$A^{\text{tree}} = \frac{\Gamma(-\alpha's)\Gamma(-\alpha't)\Gamma(-\alpha'u)}{\Gamma(1+\alpha's)\Gamma(1+\alpha't)\Gamma(1+\alpha'u)} \hat{R}^4, \tag{A.2}$$

where $\hat{R}^4$ defined in eq. (4.9).

Separating the supergravity contribution from the effect of massive string modes

$$A^{\text{tree}} = \left(\frac{1}{\hat{\sigma}_3} + T\right) \hat{R}^4. \tag{A.3}$$

We introduce the symmetric polynomials of the Mandelstam variables

$$\hat{\sigma}_n \equiv \left(\frac{\alpha'}{4}\right)^n (s^n + t^n + u^n) \tag{A.4}$$

$$= n \sum_{2p+3q=n} \frac{(p+q-1)!}{p!q!} \left(\frac{\hat{\sigma}_2}{2}\right)^p \left(\frac{\hat{\sigma}_3}{3}\right)^q,$$

which for $n \geq 4$ are all expressible as polynomials in $\hat{\sigma}_2$ and $\hat{\sigma}_3$ because of the on-shell condition $s+t+u=0$ [71]. The α' expansion of the dynamical factor T takes the form

$$T = \sum_{p,q=0}^{\infty} T_{(p,q)} \hat{\sigma}_2^p \hat{\sigma}_3^q, \tag{A.5}$$

so that [1,71,72]

$$T = 2\zeta_3 + \zeta_5 \hat{\sigma}_2 + \frac{2}{3}\zeta_3^2 \hat{\sigma}_3 + \frac{1}{2}\zeta_7 \hat{\sigma}_2^2 + \frac{2}{3}\zeta_3\zeta_5 \hat{\sigma}_2\hat{\sigma}_3$$

$$+ \frac{1}{4}\zeta_9 \hat{\sigma}_2^3 + \frac{2}{27}\left(2\zeta_3^3 + \zeta_9\right) \hat{\sigma}_3^2 + \cdots . \tag{A.6}$$

One remark here is that this expansion satisfies a transcendentality principle by giving a weight n to the Riemann zeta value ζ_n, the tree-level coefficient of the operator of order $\alpha'^{n+3} \mathcal{D}^{2n} \hat{R}^4$ is given by a polynomials in the odd zeta value of total weight $n+3$.

Appendix A.2. The genus-one amplitude

The genus-one amplitude is given by

$$A^{g=1} = \hat{R}^4 \int_{\mathcal{F}_{(1)}} \frac{d^2\tau}{\tau_2^2} \int_{\mathcal{T}_{(1)}} \prod_{i=1}^{3} \frac{d^2 v^{(i)}}{\tau_2} e^{D_{(1)}}, \tag{A.7}$$

where the integrations are performed over the domains

$$\begin{aligned}
\mathcal{F}_{(1)} &= \{|\tau_1| \leq 1/2, |\tau|^2 \geq 1\}, \\
\mathcal{T}_{(1)} &= \left\{ -\frac{1}{2} \leq v_1 < \frac{1}{2}, \ 0 \leq v_2 < \tau_2 \right\}
\end{aligned} \tag{A.8}$$

and

$$D_{(1)} = \frac{\alpha'}{2} \sum_{1 \leq i < j \leq 4} k_i \cdot k_j \, \mathcal{P}_{(1)}(v^{(ij)}|\tau), \tag{A.9}$$

where $\mathcal{P}_{(1)}(v^{(ij)}|\tau)$ is the two dimensional propagator which can be written as [71]

$$\mathcal{P}_{(1)}(v|\tau) = \frac{1}{4\pi} \sum_{(m,n)\neq(0,0)} \frac{\tau_2}{|m\tau + n|^2} \exp\left[\frac{2\pi i}{\tau_2} \Im((m\tau + n)\bar{v}) \right] + C(\tau, \bar{\tau}). \tag{A.10}$$

The piece $C(\tau, \bar{\tau})$ cancels out of the $Sl(2, \mathbb{Z})$-invariant combination in (A.9) due to the on-shell condition $s + t + u = 0$.

The low-energy expansion of $A^{g=1}$ is complicated by the presence of massless thresholds giving rise to non-analytic contributions. Separating the analytic and the non-analytic pieces as

$$A^{g=1} = (A_{\text{ana}}^{g=1} + A_{\text{nonana}}^{g=1}) \, \hat{R}^4, \tag{A.11}$$

one finds for the analytic contributions [71, 72]

$$\begin{aligned}
A_{\text{ana}}^{g=1}(\hat{\sigma}_2, \hat{\sigma}_3) = \frac{2\zeta_2}{\pi} \Bigg(1 &+ \frac{\zeta_3}{3}\hat{\sigma}_3 + 0\,\hat{\sigma}_2^2 + \frac{97}{7776} \zeta_5 \,\hat{\sigma}_2\hat{\sigma}_3 \\
&+ \frac{1}{30} \zeta_3^2\hat{\sigma}_2^3 + \frac{61}{10801}\zeta_3^2\hat{\sigma}_3^2 + \cdots \Bigg).
\end{aligned} \tag{A.12}$$

The non-analytic contributions take the following form (see [71,72] for details)

$$A_{\text{nonana}}^{g=1}(\hat{\sigma}_2, \hat{\sigma}_3) = \frac{\pi}{240} \frac{(\alpha' s)(s + 3t)}{u} \log(-\alpha' s) + \frac{4\pi \zeta_3}{45} (\alpha' s/4)^4 \log\left(-\frac{\alpha' s}{\mu_4}\right)$$

$$+ \frac{\pi \zeta_5}{2520 \cdot 4^6} (87 (\alpha' s)^6$$

$$+ (\alpha' s)^4 (\alpha' t - \alpha' u)^2) \log\left(-\frac{\alpha' s}{\mu_6}\right) + \cdots . \tag{A.13}$$

The unitarity relation of the string S-matrix implies that the 2-particle s-channel discontinuity takes the form

$$\text{Disc}_s A^{g=1}(p_1, p_2, p_3, p_4)$$

$$= -i \frac{\kappa_{(10)}^2}{\alpha'} \frac{\pi}{2} \int \frac{d^{10}k}{(2\pi)^{10}} \delta^{(+)}(k^2) \delta^{(+)}((q - k)^2)$$

$$\times \sum_{\{h_r, h_s\}} A^{\text{tree}}(p_1, p_2, (-k)^{h_r}, (k - q)^{h_s}) A^{\text{tree}}(p_3, p_4, (k)^{-h_r}, (q - k)^{-h_s}),$$

$$\tag{A.14}$$

where the sum inside the cut is over all states within the supergraviton multiplet and $\delta^{(+)}(p^2) = \delta^{(10)}(p^2)\theta(p^0)$ imposes the mass-shell condition on each intermediate state. The sum over all the helicities is performed easily thanks to the recycling identity for the $\hat{R}^4$ factor of eq. (4.9) derived in [62]

$$\sum_{\{h_r, h_s\}} \hat{R}^4((p_1)^{h_1}, (p_2)^{h_2}, (k - q)^{h_r}, (-k)^{h_s})$$

$$\times \hat{R}^4((k)^{-h_r}, (q - k)^{-h_s}, (p_3)^{h_3}, (p_4)^{h_4})$$

$$= s^4 \hat{R}^4((p_1)^{h_1}, (p_2)^{h_2}, (p_3)^{h_3}, (p_4)^{h_4}). \tag{A.15}$$

The first contribution in eq. (A.13) corresponds to the supergravity massless threshold obtained by injecting in the unitarity relation the $\hat{R}^4/\hat{\sigma}_3$ of eq. (A.3) for each tree-level factor, the higher-order α' corrections arise by the α' corrections from the factor T in eq. (A.13).

Giving a transcendentality weight 1 to π and $\log(x)$, the analytic contribution to the operators $\alpha'^{n+3} D^{2n} \hat{R}^4$ in eq. (A.12) have a total weight $n + 1$. The coefficients are of the form of the volume of the fundamental domain [96] $\text{vol}(\mathcal{F}_{(1)}) = 2\zeta_2/\pi$ defined in eq. (A.8), times a polynomials in the odd zeta values of total weight n. The non-analytic contributions $\alpha'^{n+3} s^n \log(-\alpha' s) \hat{R}^4$ in eq. (A.13) have a total weight coefficient n and are of the form $\text{vol}(\mathcal{F}_{(1)})$ times a polynomial in the odd zeta values of total weight $n - 1$.

At each order in the α' expansion of the genus one amplitude one brings down an extra factor of the two dimensional propagator $\mathcal{P}_{(1)}$ of eq. (A.10) which is a modular form of weight one. The fact that the coefficients are only polynomials in odd zeta values is a consequence of the unitarity relation (A.14) and the structure of the tree-level amplitude in eq. (A.6).

The analytic contribution to the genus-one four-graviton amplitude in type II superstring compactified on a circle of radius $\ell_s\, r$ is given by [72]

$$
I^{(d=9)}_{an}(r; s, t) = \frac{\pi}{3}\Bigg[r + r^{-1} + \hat{\sigma}_2\left(\frac{\zeta(3)}{15}r^3 + \frac{\zeta(3)}{15}r^{-3}\right)
$$

$$
+ \hat{\sigma}_3\left(\frac{\zeta(5)}{63}r^5 + \frac{\zeta(3)}{3}r + \frac{\zeta(3)}{3}r^{-1} + \frac{\zeta(5)}{63}r^{-5}\right)
$$

$$
+ \hat{\sigma}_2^2\left(\frac{\zeta(7)}{315}r^7 + \frac{2\zeta(3)}{15}r\log(r^2\lambda_4) + \frac{\zeta(5)}{36}r^{-3} + \frac{\zeta(3)^2}{315}r^{-5} + \frac{\zeta(7)}{1050}r^{-7}\right)
$$

$$
+ O(r^{-3}) + O(e^{-r})\Bigg].
$$

$$\tag{A.16}$$

Appendix A.3. The genus two amplitude

The genus two amplitude is given by [36–39, 88]

$$
A^{g=2} = \hat{R}^4 \int_{\mathcal{F}_{(2)}} \frac{d^3\Omega \wedge d^3\bar{\Omega}}{(\det \Im m\Omega)^3} \int_{\mathcal{T}_{(2)}} \frac{|\mathcal{Y}_S|^2}{(\det \Im m\Omega)^2}\, e^{D_{(2)}}
\tag{A.17}
$$

with

$$
D_{(2)} = \frac{\alpha'}{2} \sum_{1 \le i < j \le 4} k_i \cdot k_j\, \mathcal{P}_{(2)}(v^{(ij)}|\Omega),
\tag{A.18}
$$

where $\mathcal{P}_{(2)}(v^{(ij)}|\Omega)$ is the genus two propagator given in [36, 37, 88] and $\mathcal{Y}_S$ is a $(2, 0)$-form given by

$$
\mathcal{Y}_s = (t - u)\,\Delta(1, 2)\Delta(3, 4) + (s - t)\,\Delta(1, 3)\Delta(4, 2)
$$

$$
+ (u - s)\,\Delta(1, 4)\Delta(2, 3)
\tag{A.19}
$$

expressed in term of anti-symmetric combination of the Abelian differentials $\Delta(i, j) = \omega_1(z_i)\omega_2(z_j) - \omega_1(z_j)\omega_2(z_i)$. Finally Ω is the genus two period matrix, where the domains of integration $\mathcal{F}_{(2)}$ and $\mathcal{T}_{(2)}$ are defined in [36, 37, 88]. The leading term in the low energy expansion of this genus two amplitude is given by

$$
A^{g=2} = \frac{4}{3\pi}\,\zeta_4\,\hat{\sigma}_2^2\,\hat{R}^4 + O(\alpha'),
\tag{A.20}
$$

where we used the fact that the value of the volume of the genus two moduli space domain is given by $\mathrm{vol}(\mathcal{F}_{(2)}) = \zeta_4/(3\pi)$ [96].

Appendix A.4. Higher genus contributions

At higher genus order using the non-minimal pure spinor formalism [13, 14, 16] Berkovits showed that the leading behaviour of the low-energy expansion of the four graviton amplitude is given by F-terms of the schematic form

$$A^g = \int d^{16}\theta_L d^{16}\theta_R \, \theta_L^{12-2g} \, \theta_R^{12-2g} \, W_{\alpha\beta}^4 \int_{\Sigma_g} (\cdots) + O(\alpha') \tag{A.21}$$

$$\propto \mathcal{D}^{2g} \, \hat{R}^4 + O(\alpha'), \tag{A.22}$$

where $W_{\alpha\beta}$ is the dimension 1 superfield introduced in eq. (2.17) appearing in the graviton vertex operators, and the ellipsis $(\cdots)$ is for the contributions from the other fields but do not contain any dependence on the superspace fermionic coordinates θ_L and θ_R integrated over the genus g Riemann surface.

This expression is valid up to genus 6, and at genus 6 the zero mode factor gives the D-term contribution discussed in Section 2.2.

References

[1] M.B. Green, J.H. Schwarz and E. Witten, *Superstring Theory*, Cambridge, UK: Univ. Pr. (1987) (Cambridge Monographs On Mathematical Physics).

[2] J. Polchinski, *String Theory. Vol. 1: An Introduction to the Bosonic String*, Cambridge, UK: Univ. Pr. (1998) 402 p.; *String Theory. Vol. 2: Superstring Theory and Beyond*, Cambridge, UK: Univ. Pr. (1998) 531 p.

[3] E. Kiritsis, *String Theory in a Nutshell*, Princeton, USA: Univ. Pr. (2007) 588 p.

[4] C.M. Hull and P.K. Townsend, Unity of superstring dualities, *Nucl. Phys. B* **438** (1995) 109 [arXiv:hep-th/9410167].

[5] E. Witten, String theory dynamics in various dimensions, *Nucl. Phys. B* **443** (1995) 85 [arXiv:hep-th/9503124].

[6] E. D'Hoker and D.H. Phong, The geometry of string perturbation theory, *Rev. Mod. Phys.* **60** (1988) 917.

[7] R.P. Feynman, Quantum theory of gravitation, *Acta Phys. Polon.* **24** (1963) 697.

[8] B.S. DeWitt, Quantum theory of gravity. 1. The canonical theory, *Phys. Rev.* **160** (1967) 1113; Quantum theory of gravity. II. The manifestly covariant theory, *Phys. Rev.* **162** (1967) 1195; Quantum theory of gravity. III. Applications of the covariant theory, *Phys. Rev.* **162** (1967) 1239.

[9] G. 't Hooft and M.J.G. Veltman, One loop divergencies in the theory of gravitation, *Annales Poincare Phys. Theor. A* **20** (1974) 69.

[10] M.J.G. Veltman, *Quantum Theory Of Gravitation*, In *Les Houches 1975, Proceedings, Methods In Field Theory*, Amsterdam 1976, 265–327.

[11] M.B. Green, H. Ooguri and J.H. Schwarz, Nondecoupling of maximal supergravity from the superstring, *Phys. Rev. Lett.* **99** (2007) 041601.

[12] N. Berkovits, Multiloop amplitudes and vanishing theorems using the pure spinor formalism for the superstring, *JHEP* **0409** (2004) 047 [arXiv:hep-th/0406055].

[13] N. Berkovits, Pure spinor formalism as an $N = 2$ topological string, *JHEP* **0510** (2005) 089 [arXiv:hep-th/0509120].

[14] N. Berkovits and N. Nekrasov, Multiloop superstring amplitudes from non-minimal pure spinor formalism, *JHEP* **0612** (2006) 029 [arXiv:hep-th/0609012].

[15] N. Berkovits, Explaining the pure spinor formalism for the superstring, arXiv:0712.0324 [hep-th].

[16] N. Berkovits, New higher-derivative R^4 theorems, *Phys. Rev. Lett.* **98** (2007) 211601 [arXiv:hep-th/0609006].

[17] Z. Bern, Perturbative quantum gravity and its relation to gauge theory, *Living Rev. Rel.* **5** (2002) 5 [arXiv:gr-qc/0206071].

[18] M.H. Goroff and A. Sagnotti, Quantum gravity at two loops, *Phys. Lett.* B **160** (1985) 81.

[19] M.H. Goroff and A. Sagnotti, The ultraviolet behavior Of Einstein gravity, *Nucl. Phys.* B **266** (1986) 709.

[20] A.E.M. van de Ven, Two loop quantum gravity, *Nucl. Phys.* B **378** (1992) 309.

[21] M.T. Grisaru, Two loop renormalizability of supergravity, *Phys. Lett.* B **66** (1977) 75.

[22] E. Tomboulis, On the two loop divergences of supersymmetric gravitation, *Phys. Lett.* B **67** (1977) 417.

[23] R.E. Kallosh, Counterterms in extended supergravities, *Phys. Lett.* B **99** (1981) 122.

[24] P.S. Howe and U. Lindstrom, Higher order invariants in extended supergravity, *Nucl. Phys.* B **181** (1981) 487.

[25] S. Deser, J.H. Kay and K.S. Stelle, Renormalizability properties of supergravity, *Phys. Rev. Lett.* **38** (1977) 527.

[26] S. Deser and J.H. Kay, Three loop counterterms for extended supergravity, *Phys. Lett.* B **76** (1978) 400.

[27] L. Brink and P.S. Howe, The $N = 8$ supergravity in superspace, *Phys. Lett.* B **88** (1979) 268; Eleven-dimensional supergravity on the mass-shell in superspace, *Phys. Lett.* B **91** (1980) 384.

[28] P.S. Howe, R^4 terms in supergravity and M-theory, arXiv:hep-th/0408177.

[29] Z. Bern, J.J. Carrasco, L.J. Dixon, H. Johansson, D.A. Kosower and R. Roiban, Three-loop superfiniteness of $N = 8$ supergravity, *Phys. Rev. Lett.* **98** (2007) 161303 [arXiv:hep-th/0702112].

[30] E. Cremmer and B. Julia, The $N = 8$ supergravity theory. 1. The Lagrangian, *Phys. Lett.* B **80** (1978) 48.

[31] P.S. Howe and P.C. West, The complete $N = 2$, $D = 10$ supergravity, *Nucl. Phys.* B **238** (1984) 181.

[32] E. Cremmer and S. Ferrara, Formulation of eleven-dimensional supergravity in superspace, *Phys. Lett.* B **91** (1980) 61.

[33] M.B. Green, J.G. Russo and P. Vanhove, Non-renormalisation conditions in type II string theory and maximal supergravity, *JHEP* **0702** (2007) 099

[34] M.B. Green, J.H. Schwarz and L. Brink, $N = 4$ Yang–Mills and $N = 8$ supergravity as limits of string theories, *Nucl. Phys.* B **198** (1982) 474.

[35] Z. Bern, L.J. Dixon, M. Perelstein and J.S. Rozowsky, Multi-leg one-loop gravity amplitudes from Gauge theory, *Nucl. Phys.* B **546** (1999) 423 [arXiv:hep-th/9811140].

[36] E. D'Hoker and D.H. Phong, Lectures on two-loop superstrings, arXiv:hep-th/0211111.

[37] E. D'Hoker and D.H. Phong, Two-loop superstrings VI: Non-renormalization theorems and the 4-point function, *Nucl. Phys. B* **715** (2005) 3 [arXiv:hep-th/0501197]; Two-loop superstrings V: Gauge slice independence of the N-point function, *Nucl. Phys. B* **715** (2005) 91 [arXiv:hep-th/0501196]; Two-loop superstrings IV: The cosmological constant and modular forms, *Nucl. Phys. B* **639** (2002) 129 [arXiv:hep-th/0111040]; Two-loop superstrings III, slice independence and absence of ambiguities, *Nucl. Phys. B* **636** (2002) 61 [arXiv:hep-th/0111016]; Two-loop superstrings II, the chiral measure on moduli space, *Nucl. Phys. B* **636** (2002) 3 [arXiv:hep-th/0110283]; Two-loop superstrings I, main formulas, *Phys. Lett. B* **529** (2002) 241 [arXiv:hep-th/0110247].

[38] N. Berkovits, Super-Poincare covariant two-loop superstring amplitudes, *JHEP* **0601** (2006) 005 [arXiv:hep-th/0503197].

[39] N. Berkovits and C.R. Mafra, Equivalence of two-loop superstring amplitudes in the pure spinor and RNS formalisms, *Phys. Rev. Lett.* **96** (2006) 011602 [arXiv:hep-th/0509234].

[40] Z. Bern, *private discussion*

[41] M.B. Green, J.G. Russo and P. Vanhove, Ultraviolet properties of maximal supergravity, *Phys. Rev. Lett.* **98** (2007) 131602 [arXiv:hep-th/0611273].

[42] Z. Bern, L.J. Dixon and V.A. Smirnov, Iteration of planar amplitudes in maximally supersymmetric Yang–Mills theory at three loops and beyond, *Phys. Rev. D* **72** (2005) 085001 [arXiv:hep-th/0505205].

[43] F. Cachazo and D. Skinner, On the structure of scattering amplitudes in $N = 4$ super Yang–Mills and $N = 8$ supergravity, arXiv:0801.4574 [hep-th].

[44] P. Benincasa, C. Boucher-Veronneau and F. Cachazo, Taming tree amplitudes in general relativity, arXiv:hep-th/0702032.

[45] Z. Bern, J.J. Carrasco, D. Forde, H. Ita and H. Johansson, Unexpected cancellations in gravity theories, arXiv:0707.1035 [hep-th].

[46] N. Arkani-Hamed and J. Kaplan, On tree amplitudes in Gauge theory and gravity, arXiv:0801.2385 [hep-th].

[47] N.E.J. Bjerrum-Bohr and P. Vanhove, Explicit cancellation of triangles in one-loop gravity amplitudes, arXiv:0802.0868 [hep-th].

[48] V.A. Smirnov, Analytical result for dimensionally regularized massless on-shell double box, *Phys. Lett. B* **460** (1999) 397 [arXiv:hep-ph/9905323].

[49] J.B. Tausk, Non-planar massless two-loop Feynman diagrams with four on-shell legs, *Phys. Lett. B* **469** (1999) 225 [arXiv:hep-ph/9909506].

[50] P.S. Howe, K.S. Stelle and P.K. Townsend, Miraculous ultraviolet cancellations in supersymmetry made manifest, *Nucl. Phys. B* **236** (1984) 125.

[51] H. Kawai, D.C. Lewellen and S.H.H. Tye, A relation between tree amplitudes of closed and open strings, *Nucl. Phys. B* **269** (1986) 1.

[52] F.A. Berends, W.T. Giele and H. Kuijf, On relations between multi- gluon and multigraviton scattering, *Phys. Lett. B* **211** (1988) 91.

[53] J. Bedford, A. Brandhuber, B.J. Spence and G. Travaglini, A recursion relation for gravity amplitudes, *Nucl. Phys. B* **721** (2005) 98 [arXiv:hep-th/0502146].

[54] H. Elvang and D.Z. Freedman, Note on graviton MHV amplitudes, arXiv:0710.1270 [hep-th].

[55] A. Brandhuber, S. McNamara, B. Spence and G. Travaglini, Recursion relations for one-loop gravity amplitudes, *JHEP* **0703** (2007) 029 [arXiv:hep-th/0701187].

[56] K. Peeters, P. Vanhove and A. Westerberg, Supersymmetric higher-derivative actions in ten and eleven dimensions, the associated superalgebras and their formulation in superspace, *Class. Quant. Grav.* **18** (2001) 843 [arXiv:hep-th/0010167].

[57] M.B. Green, M. Gutperle and P. Vanhove, One loop in eleven dimensions, *Phys. Lett. B* **409** (1997) 177 [arXiv:hep-th/9706175].

[58] M.B. Green, H.h. Kwon and P. Vanhove, Two loops in eleven dimensions, *Phys. Rev. D* **61** (2000) 104010 [arXiv:hep-th/9910055].

[59] E. Cremmer, B. Julia and J. Scherk, Supergravity theory in 11 dimensions, *Phys. Lett. B* **76** (1978) 409.

[60] S. Deser and D. Seminara, Tree amplitudes and two-loop counterterms in $D = 11$ supergravity, *Phys. Rev. D* **62** (2000) 084010 [arXiv:hep-th/0002241].

[61] S. Deser and D. Seminara, Graviton-form invariants in $D = 11$ supergravity, *Phys. Rev. D* **72** (2005) 027701 [arXiv:hep-th/0506073].

[62] Z. Bern, L.J. Dixon, D.C. Dunbar, M. Perelstein and J.S. Rozowsky, On the relationship between Yang–Mills theory and gravity and its implication for ultraviolet divergences, *Nucl. Phys. B* **530** (1998) 401

[63] R.J. Eden, P.V. Landshoff, D.I. Olive and J.C. Polkinghorne (eds.), *The ANalytic S-matrix*, ISBN 0521523362, pp. 295, Cambridge, UK: Cambridge University Press, April 2002.

[64] G. 't Hooft and M.J.G. Veltman, Regularization and renormalization of Gauge fields, *Nucl. Phys. B* **44** (1972) 189.

[65] W.L. van Neerven, Dimensional regularization of mass and infrared singularities in two loop on-shell vertex functions, *Nucl. Phys. B* **268** (1986) 453.

[66] Z. Bern, L.J. Dixon and D.A. Kosower, Progress in one-loop QCD computations, *Ann. Rev. Nucl. Part. Sci.* **46** (1996) 109 [arXiv:hep-ph/9602280].

[67] Z. Bern, L.J. Dixon and D.A. Kosower, $N = 4$ super-Yang–Mills theory, QCD and collider physics, *Comptes Rendus Physique* **5** (2004) 955 [arXiv:hep-th/0410021].

[68] Z. Bern, L.J. Dixon and D.A. Kosower, On-shell methods in perturbative QCD, *Annals Phys.* **322** (2007) 1587 [arXiv:0704.2798 [hep-ph]].

[69] G. Passarino and M.J.G. Veltman, One loop corrections for e^+e^- annihilation into $\mu^+\mu^-$ in the Weinberg model, *Nucl. Phys. B* **160** (1979) 151.

[70] N.E.J. Bjerrum-Bohr, D.C. Dunbar, H. Ita, W.B. Perkins and K. Risager, The no-triangle hypothesis for $N = 8$ supergravity, *JHEP* **0612** (2006) 072 [arXiv:hep-th/0610043].

[71] M.B. Green and P. Vanhove, The low energy expansion of the one-loop type II superstring amplitude, *Phys. Rev. D* **61** (2000) 104011 [arXiv:hep-th/9910056].

[72] M.B. Green, J.G. Russo and P. Vanhove, Low energy expansion of the four-particle genus-one amplitude in type II superstring theory, arXiv:0801.0322 [hep-th]

[73] M.B. Green, J. Russo and P. Vanhove, Connections between four-graviton scattering in two-loop maximal supergravity and superstring theory, in preparation.

[74] E. Bergshoeff, E. Sezgin and P.K. Townsend, Properties of the eleven-dimensional super membrane theory, *Annals Phys.* **185** (1988) 330.

[75] M.B. Green and P. Vanhove, Duality and higher derivative terms in M theory, *JHEP* **0601** (2006) 093 [arXiv:hep-th/0510027].

[76] M. Cederwall, U. Gran, M. Nielsen and B.E.W. Nilsson, Manifestly supersymmetric M-theory, *JHEP* **0010** (2000) 041 [arXiv:hep-th/0007035].

[77] M. Cederwall, U. Gran, B.E.W. Nilsson and D. Tsimpis, Supersymmetric corrections to eleven-dimensional supergravity, *JHEP* **0505** (2005) 052 [arXiv:hep-th/0409107].

[78] P.S. Howe and D. Tsimpis, On higher-order corrections in M theory, *JHEP* **0309** (2003) 038 [arXiv:hep-th/0305129].

[79] N. Lambert and P. West, Duality groups, automorphic forms and higher derivative corrections, *Phys. Rev. D* **75** (2007) 066002 [arXiv:hep-th/0611318].

[80] N. Lambert and P. West, Enhanced coset symmetries and higher derivative corrections, *Phys. Rev. D* **74** (2006) 065002 [arXiv:hep-th/0603255].

[81] T. Damour, M. Henneaux and H. Nicolai, E_{10} and a 'small tension expansion' of M theory, *Phys. Rev. Lett.* **89** (2002) 221601 [arXiv:hep-th/0207267].

[82] P.C. West, E_{11} and M theory, *Class. Quant. Grav.* **18** (2001) 4443 [arXiv:hep-th/0104081].

[83] F. Riccioni and P. West, The E_{11} origin of all maximal supergravities, *JHEP* **0707** (2007) 063 [arXiv:0705.0752 [hep-th]].

[84] P.K. Townsend, The eleven-dimensional supermembrane revisited, *Phys. Lett. B* **350** (1995) 184 [arXiv:hep-th/9501068].

[85] P.S. Aspinwall, Some relationships between dualities in string theory, *Nucl. Phys. Proc. Suppl.* **46** (1996) 30 [arXiv:hep-th/9508154].

[86] J.H. Schwarz, An $SL(2, Z)$ multiplet of type IIB superstrings, *Phys. Lett. B* **360** (1995) 13 [Erratum-ibid. B **364** (1995) 252] [arXiv:hep-th/9508143].

[87] J.G. Russo and A.A. Tseytlin, One-loop four-graviton amplitude in eleven-dimensional supergravity, *Nucl. Phys. B* **508** (1997) 245 [arXiv:hep-th/9707134].

[88] E. D'Hoker, M. Gutperle and D.H. Phong, Two-loop superstrings and S-duality, *Nucl. Phys. B* **722** (2005) 81 [arXiv:hep-th/0503180].

[89] E. Kiritsis and B. Pioline, On R^4 threshold corrections in type IIB string theory and (p,q) string instantons, *Nucl. Phys. B* **508** (1997) 509 [arXiv:hep-th/9707018].

[90] A. Basu, The $D^4 R^4$ term in type IIB string theory on T^2 and U-duality, arXiv:0708.2950 [hep-th].

[91] A. Basu, The $D^6 R^4$ term in type IIB string theory on T^2 and U-duality, arXiv:0712.1252 [hep-th].

[92] C.R. Mafra, Pure spinor superspace identities for massless four-point kinematic factors, arXiv:0801.0580 [hep-th].

[93] M.B. Green and M. Gutperle, Effects of D-instantons, *Nucl. Phys. B* **498** (1997) 195 [arXiv:hep-th/9701093].

[94] M.B. Green and S. Sethi, Supersymmetry constraints on type IIB supergravity, *Phys. Rev. D* **59** (1999) 046006 [arXiv:hep-th/9808061].

[95] A. Sinha, The $\hat{G}^4 \lambda^{16}$ term in IIB supergravity, *JHEP* **0208** (2002) 017 [arXiv:hep-th/0207070].

[96] Carl Ludwig Siegel, Symplectic geometry, *American Journal of Mathematics* **65**(1) (1943) 1–86.

Course 9

PREPARING THE LHC EXPERIMENTS FOR FIRST DATA

F. Gianotti

CERN, Physics Department
Genève, Switzerland

C. Bachas, L. Baulieu, M. Douglas, E. Kiritsis, E. Rabinovici, P. Vanhove, P. Windey
and L.F. Cugliandolo, eds.
Les Houches, Session LXXXVII, 2007
String Theory and the Real World: From Particle Physics to Astrophysics
© *2008 Published by Elsevier B.V.*

Contents

355

1. Introduction

This write-up is a short summary of the lectures given at the Summer 2007 Les Houches School "String theory and the real world." The purpose of this brief report is to recall the main topics discussed there and to point to additional reference material.

The LHC [1] will provide pp collisions at the unprecedented centre-of-mass energy $\sqrt{s} = 14$ TeV. According to the present plans, the design luminosity $L = 10^{34}$ cm^{-2} s^{-1} should be achieved after 2010, while in the first years of operation the luminosity should increase gradually from $L \sim 10^{29}$ cm^{-2} s^{-1} to $L \sim 2 \times 10^{33}$ cm^{-2} s^{-1}. The machine will also deliver heavy ion collisions, for instance lead-lead collisions at the colossal centre-of-mass energy of about 1000 TeV. The accelerator has been installed in the 27 km underground ring previously used for the LEP e^+e^- collider, and is presently being commissioned. First collisions are expected in Summer 2008.

Four main experiments will take data at the LHC: two general-purpose detectors, ATLAS [2] and CMS [3], which have a very broad physics programme; one experiment, LHCb [4], dedicated to the study of B-hadrons and CP violation; one experiment, ALICE [5], which will study ion-ion and p-ion physics. Only the ATLAS and CMS experiments and their physics programs are discussed here in some detail, since their physics goals address directly the exploration of the energy frontier.

The lectures covered the following points: the status of the LHC machine (Section 2); the main experimental challenges (Section 3); the ATLAS and CMS experiments (Section 4); the LHC main physics goals, and the physics potential with emphasis on first data (Section 5).

2. Machine status

A high-technology accelerator is needed to reach a beam energy of 7 TeV in a 27 km ring, in particular the bending power of 1232 superconducting dipole magnets providing a 8.3 T magnetic field. The latter is far higher than the confining fields of the HERA proton ring (4.7 T) and of the Tevatron (4.2 T). The main challenges and parameters of the machine can be found in Refs. [1,6].

The present (February 2008) machine status can be summarised as follows:

• All magnets have been installed in the underground tunnel and the ring is closed and pressure-tested.

• Two of the eight machine sectors[1] have been cooled down to the operation temperature of 1.9 K, and two additional sectors are being cooled down.

• One sector has undergone extensive electrical tests, and has been powered up to a magnet current of 10 kA (the nominal value is 12 kA).

• The two injection lines from the SPS to the LHC have been successfully tested with beams.

According to the present official schedule, first beam injection is expected in June 2008 and first collisions a couple of months later.

3. The main experimental challenges

Operation at a high-energy and high-luminosity hadron collider like the LHC brings many advantages and some difficulties. The main advantage is that the event rate will be huge, so that the LHC will become a factory of all particles with masses up to a few TeV which have reasonable couplings to Standard Model (SM) particles. At the same time, both the high energy and high luminosity entail several experimental challenges that set stringent requirements on the trigger and detector performance.

The main difficulty related to the high centre-of-mass energy is illustrated in Fig. 1, which shows the production cross-sections for several channels at hadron colliders, as a function of $\sqrt{s}$. One can notice two features. First at the LHC, just as at previous hadron colliders, hard-scattering processes will be dominated by QCD jet production, a strong process with a huge cross-section. In contrast, the most interesting physics channels are usually much rarer, either because they involve the production of heavy particles, or because they arise from electroweak interactions (e.g. W or Higgs production). It can be seen, for example, that at 14 TeV the cross-section for jets with $p_T > 100$ GeV (where p_T is the momentum component perpendicular to the beam line) is five orders of magnitude larger than the cross-section for a Higgs boson of mass 150 GeV. As a consequence, in contrast to e^+e^- machines, there is no hope for experiments at the LHC to detect a Higgs boson decaying into jets, unless it is produced in association with other particles giving a cleaner signature, since such final states will be swamped by the much larger QCD background. Decays into leptons or photons have to be used instead, so that in general only part of the available cross-section is *de facto* usable. Similar arguments apply to any other relatively light (mass in the

---

[1] An LHC sector is 3.3 km long and contains 154 dipoles.

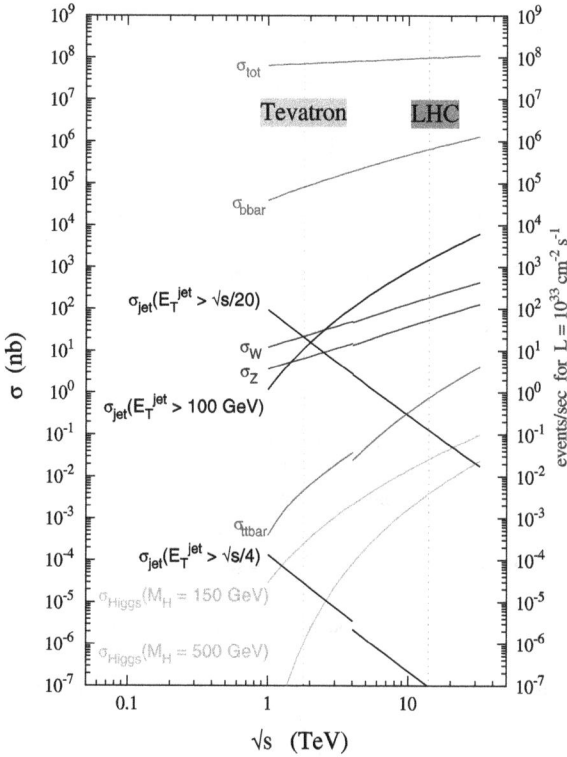

Fig. 1. Production cross-sections for various processes at hadron colliders (pp and $p\bar{p}$), as a function of the machine centre-of-mass energy. The discontinuities in some of the curves are due to the transition from $p\bar{p}$ to pp collisions.

few hundred GeV range) object, whereas the situation improves for very massive particles (e.g. an excited quark in the TeV range decaying as $q^* \to qg$) since the QCD background decreases fast with the invariant mass of the final-state products. In addition, because the probability for a jet to fake an electron or photon is small but non-vanishing, and because of the above-mentioned huge difference between the signal and background cross-sections, excellent detector (and trigger) performance in terms of particle identification capability and energy resolution are needed in most cases in order to extract a clean signal above the various reducible and irreducible backgrounds. For example, mass resolutions of $\sim 1\%$ for objects decaying into leptons or photons, and rejection factors against jets faking photons larger than 10^3 are required.

The main difficulty related to operating at $L = 10^{34}$ cm^{-2} s^{-1}, which will be a factor of about 40 higher than the instantaneous luminosity achieved so far at the Tevatron, is that the pp interaction rate will be as large as 10^9 Hz. Hence, at each crossing between two proton bunches, i.e. every 25 ns, an average of about 20 low-p_T events (so-called "minimum-bias") will be produced simultaneously in the ATLAS and CMS detectors. The impact of this event "pile-up" on the design of the LHC detectors in terms of response time, radiation hardness and granularity, as well as on their performance, is discussed in [7]. It should be noted that fast and highly-selective trigger systems are also needed, with the ability to reduce the initial event rate of 10^9 Hz to a rate-to-storage of $\sim$100 Hz while preserving high efficiency for the interesting physics processes. This is also discussed in [7].

4. ATLAS and CMS

Since we don't know how new physics will manifest itself, the LHC experiments must be able to detect as many particles and signatures as possible. Therefore ATLAS (A Toroidal Lhc ApparatuS, left panel in Fig. 2) and CMS (Compact Muon Solenoid, right panel in Fig. 2) are multi-purpose detectors that will provide efficient and precise measurements of e.g. electrons, muons, taus, neutrinos, photons, jets, b-jets.

The main features of the two experiments, which are complementary in several aspects, are presented in Table 1.

CMS has only one magnet, a big solenoid that surrounds the inner detector and the calorimeters and provides a magnetic field of 4 T in the inner detector volume. ATLAS has four magnets: a solenoid sitting in front of the electromagnetic calorimeter and producing a field of 2 T in the inner cavity, and external barrel and end-cap air-core toroids.

The CMS inner detector consists of layers of Pixels and Silicon strips. Thanks mainly to the high magnetic field, excellent momentum resolution is expected (see Table 1). The ATLAS inner detector also contains Pixel and Silicon strip layers close to the interaction region and, in addition, a Transition Radiation Detector (TRT) at larger radii. Due to the lower magnetic field and somewhat smaller cavity, the expected momentum resolution is a factor of about three worse than that of CMS. However, the Transition Radiation Detector provides electron/pion separation capabilities.

The CMS electromagnetic calorimeter is a high-resolution crystal detector. The ATLAS calorimeter is a lead-liquid argon sampling calorimeter, therefore with a worse intrinsic energy resolution. However, thanks to a very fine lateral and good longitudinal segmentation, the ATLAS calorimeter provides more robust particle identification capabilities than the CMS calorimeter.

In both experiments the hadronic calorimeters are sampling detectors with scintillator or liquid-argon as active medium. The ATLAS calorimeter offers

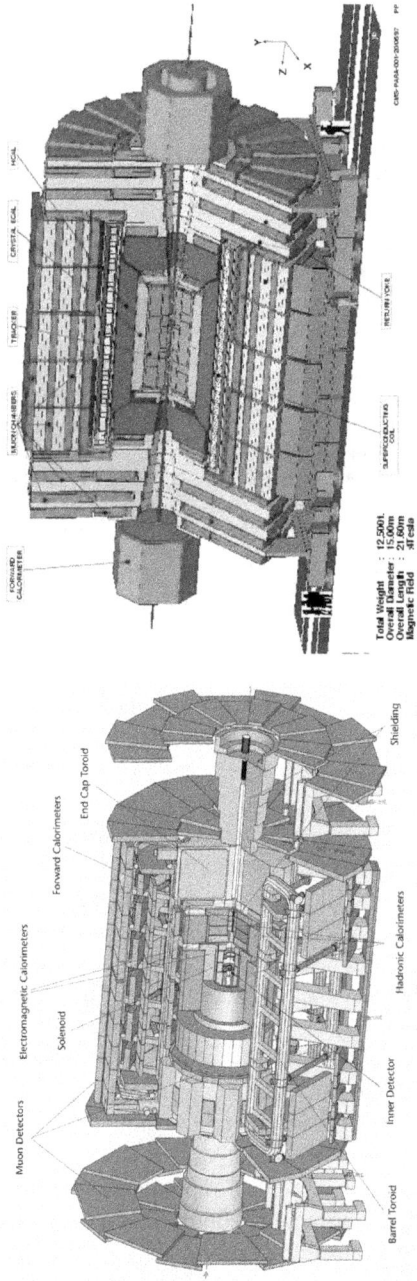

Fig. 2. Layout of the ATLAS (left) and CMS (right) detector.

Table 1

Main features of the ATLAS and CMS detectors

	ATLAS	CMS
Magnet(s)	Air-core toroids + solenoid in inner cavity	Solenoid
	Calorimeters in field-free region	Calorimeters inside field
	4 magnets	1 magnet
Inner detector	Si pixels and strips	Si pixels and strips
	TRT $\rightarrow$ particle identification	No particle identification
	$B = 2$ T	$B = 4$ T
	$\sigma/p_T \sim 5 \times 10^{-4} p_T\,(\text{GeV}) \oplus 0.01$	$\sigma/p_T \sim 1.5 \times 10^{-4} p_T\,(\text{GeV}) \oplus 0.005$
EM calorimeter	Lead-liquid argon	PbWO$_4$ crystals
	$\sigma/E \sim 10\%/\sqrt{E\,(\text{GeV})}$	$\sigma/E \sim 2 - 5\%/\sqrt{E\,(\text{GeV})}$
	Longitudinal segmentation	No longitudinal segmentation
HAD calorimeter	Fe-scintillator + Cu-liquid argon	Brass-scintillator
	$\geq 10\lambda$	$\geq 5.8\lambda$ + tail catcher
	$\sigma/E \sim 50\%/\sqrt{E\,(\text{GeV})} \oplus 0.03$	$\sigma/E \sim 100\%/\sqrt{E\,(\text{GeV})} \oplus 0.05$
Muon spectrometer	Chambers in air	Chambers in solenoid return yoke (Fe)
	$\sigma/p_T \sim 7\%$ at 1 TeV	$\sigma/p_T \sim 5\%$ at 1 TeV
	with spectrometer alone	combining spectrometer and inner detector

a better energy resolution because it is thicker (the CMS hadronic calorimeter suffers from space constraints dictated by the external solenoid) and has a finer sampling frequency.

Finally, the external Muon spectrometer of CMS consists of chamber stations embedded into the iron of the solenoid return yoke, where multiple scattering is not negligible. ATLAS has a spectrometer in air, where multiple scattering is minimised, and therefore offers the possibility of good standalone (i.e. without the inner detector contribution) measurements. The expected momentum resolution is better than 10% for muons of $p_T = 1$ TeV in both experiments. This performance is achieved by the Muon spectrometer alone in ATLAS, and by combining the information from the Muon spectrometer and the inner detector in CMS.

At the moment (February 2008), the installation of both experiments in the underground cavern is almost completed. ATLAS is missing some forward muon chambers, and CMS has still to install the end-cap electromagnetic calorimeter and the pixel detector. A spectacular picture is shown in Fig. 3.

The experiments commissioning with cosmics muon data started more than two years ago (see e.g. Fig. 4). This important phase for the preparation to physics has allowed increasingly more complete and integrated detector components to be tested in the surface assembly halls (in the past) and in the underground pits (more recently).

Fig. 3. View of the ATLAS detector in the underground cavern, with one end-cap calorimeter being moved to its position inside the barrel toroid system (CERN courtesy).

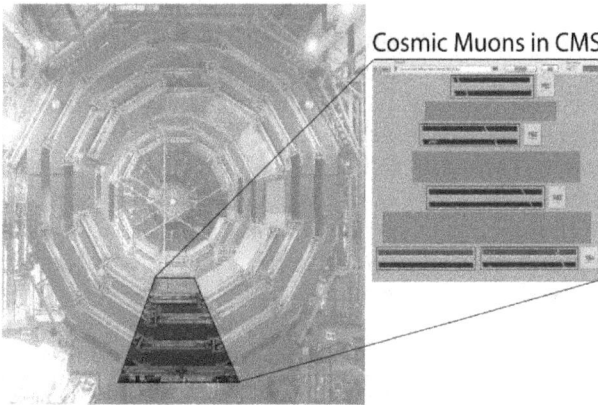

Cosmic Muons in CMS

Fig. 4. View of the CMS detector in the surface assembly hall, showing the iron return yoke of the solenoid with several muon stations installed, and a cosmic muon traversing four chambers located at the bottom of the detector (CERN courtesy).

Comprehensive discussions of the expected detector performance, as obtained from test-beam measurements and extensive simulations of both experiments, can be found in [2, 3].

5. Physics goals and physics potential

The main LHC goals are related to the exploration of the highly-motivated TeV scale. They include, first and foremost, the elucidation of the mechanism responsible for the electroweak symmetry breaking (in particular looking for the Higgs boson), the search for new physics able to fix the SM problems (e.g. the divergent radiative corrections to the Higgs mass), and the quest for answers to outstanding questions (e.g. what is the dark matter of the Universe). Precision measurements (e.g. top mass, B-mesons and CP-violation), as well as the study of quark-gluon plasma formation, are also in the agenda. A review of the LHC motivations and goals can be found in Ref. [8].

A few examples of the LHC physics potential are discussed shortly below. For a more complete picture the reader could look at Refs. [2, 3].

5.1. First data and first measurements

Thanks to the huge LHC centre-of-mass energy, large samples of interesting data should become available after only a few weeks of data taking, i.e. by the end of 2008. This is illustrated in Table 2, which shows the numbers of events expected to be recorded by each of the two experiments, ATLAS or CMS, for some representative physics processes and for an integrated luminosity of only 100 pb^{-1}. The latter corresponds to a few weeks of data taking at an instantaneous luminosity of only 10^{31} cm^{-2} s^{-1}. It can be seen that huge event samples (similar in size to those collected at LEP and Tevatron over the whole life of these colliders) should be produced, over such a short time period, by a variety of Standard Model processes. These data will be used to understand the experiments in all details, to calibrate the detectors *in situ* using well-known final states, such as $Z \rightarrow \ell\ell$ and $t\bar{t}$ events, and to perform extensive measurements of as many SM processes as possible, e.g. W, Z, $t\bar{t}$ production. The latter are important on their own, but also as potential backgrounds to new physics.

As an example of initial measurement of SM physics, it has been demonstrated [9] that a $t\bar{t}$ signal can be observed with few pb^{-1} of data, with a very simple analysis and a detector still in the commissioning phase. In turn such a signal can be used to improve the knowledge of the detector performance and

Table 2

For some physics processes, and for each experiment (ATLAS or CMS), the expected numbers of events to tape for an integrated luminosity of 100 pb^{-1}. The total event samples (expected to be) collected at colliders operating before the LHC start-up are given in the last column

Channel	Number of events to tape for 100 pb^{-1}	Total numbers of events from previous colliders
$W \rightarrow \mu\nu$	$\sim 10^6$	$\sim 10^4$ LEP, $\sim 10^6$ Tevatron
$Z \rightarrow \mu\mu$	$\sim 10^5$	10^6 LEP, $\sim 10^5$ Tevatron
$t\bar{t} \rightarrow \mu + X$	$\sim 10^4$	$\sim 10^4$ Tevatron
QCD jets $p_T > 1$ TeV	$> 10^3$	–
$\tilde{g}\tilde{g}$, m($\tilde{g}$)=1 TeV	~ 50	–

physics. The feasibility of this early measurement is due to the large cross-section (∼250 pb) for the gold-plated semileptonic $t\bar{t} \rightarrow b\ell v b jj$ channel (where $\ell = e, \mu$) and the clear signature of these events. A simple analysis required an isolated electron or muon with $p_T > 20$ GeV, large missing transverse energy, and four jets with $p_T > 40$ GeV in the final state. The additional constraint that two of the jets have an invariant mass compatible with the W mass was imposed. The resulting mass spectrum of the three jets giving the highest p_T of the top quark is presented in Fig. 5. It should be noted that no b-tagging of two of the jets was required, assuming (conservatively) that the performance of the vertex detector would not be well understood at this early stage. Figure 5 shows that, even in these pessimistic conditions, a clear top signal should be observed above the background. An integrated luminosity of less than 30 pb^{-1}, which should be collected in 2008, would be sufficient to this purpose. The production cross-section could be initially measured with a precision of ∼20% and the top mass to 7–10 GeV (whereas the ultimate LHC precision is expected to be ∼1 GeV).

In addition, such a top sample will be very useful to understand several aspects of the detector performance. For example, the two b-jets in the final state can be used to study the efficiency of the b-tagging procedure, and the jet en-

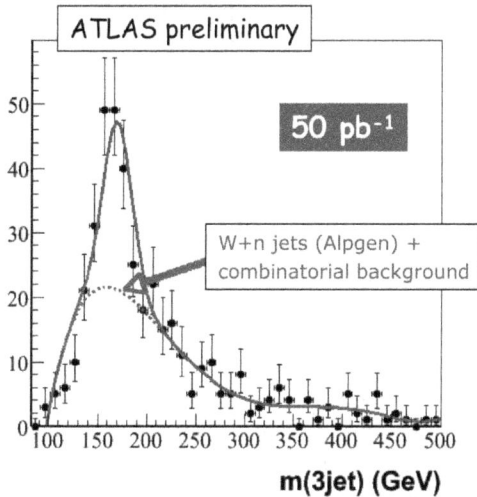

Fig. 5. Three-jet invariant mass distribution for events selected as described in the text, as obtained from a full simulation of the ATLAS detector. The dots with error bars show the expected signal from $t\bar{t}$ events plus the background, the dashed curve indicates the background alone. The number of events corresponds to an integrated luminosity of 50 pb^{-1}. Courtesy of W. Verkerke.

ergy scale of the experiment can be established in a preliminary way from the reconstructed $W \rightarrow jj$ mass peak. Furthermore, the (reconstructed) p_T spectrum of the top-quark is very sensitive to higher-order QCD corrections, and this feature can be exploited to test the theory and tune the Monte Carlo generators.

5.2. Early discoveries

Only after the detectors and SM processes will have been well understood can the LHC experiments hope to extract convincing discovery signals from their data. Three examples are discussed briefly below, ranked by increasing difficulty: an easy case, namely a possible $Z' \rightarrow e^+e^-$ signal, an intermediate case, Supersymmetry, and a difficult case, a light Standard Model Higgs boson.

5.2.1. $Z' \rightarrow e^+e^-$

A particle of mass 1–2 TeV decaying into e^+e^- pairs, such as a possible new gauge boson Z' or a graviton in Randall–Sundrum extra-dimension theories, is probably the easiest object to discover at the LHC, for three main reasons. First, if the branching ratios into leptons are at least at the percent level, the expected number of events after all experimental cuts is relatively large, e.g. about ten for an integrated luminosity of only 300 pb^{-1} and a Z' mass as large as 1.5 TeV. Second, the dominant background, di-lepton Drell–Yan production, is small in the TeV region, and even if it were to be a factor of two-three larger than expected today (which is unlikely for such a theoretically well-known process), it would still be negligible compared to the signal. Finally, the signal will be indisputable, since it will appear as a resonant peak on top of a smooth background, and not just as an overall excess in the total number of events. These expectations are not based on ultimate detector performance, since they hold also if the calorimeter response is understood to a conservative level of a few percent.

5.2.2. Supersymmetry

If Supersymmetry has something to do with stabilizing the Higgs mass, new particles are expected at the TeV scale, and some of them (e.g. squarks and gluinos) could be discovered rather quickly. This is because of the huge production cross-section at 14 TeV for squarks and gluinos, which are strongly-interacting particles, with about ten events per day expected in each experiment at instantaneous luminosities of only 10^{32} cm^{-2} s^{-1} and for squark/gluino masses as large as $\sim$1 TeV. In addition, cascade decays of (heavy) squarks and gluinos should give rise to clear-signature final states, containing several high-p_T jets, leptons and, in R-parity conserving models, large missing transverse energy coming from the escaping stable neutralinos (χ_1^0). With less than 1 fb^{-1} of data, and provided the detectors and the numerous backgrounds are well understood (see below), AT-LAS and CMS should be able to discover gluinos up to masses beyond 1 TeV, whereas the ultimate LHC reach extends up to masses of 2.7 TeV. Rather quickly therefore (sometimes in 2009?), the LHC could provide indications about the energy scale and mass spectrum of new physics. It may take considerably longer, however, to unravel the nature of new physics.

5.2.3. Standard Model Higgs boson

Figure 6 shows the integrated luminosity per experiment, as a function of the Higgs boson mass, needed to discover a possible Higgs signal (5σ excess required) or to exclude it at the 95% C.L. by combining both experiments. Two conclusions can be drawn from these projections. First, with a few fb^{-1} of well-understood data the LHC can say the final word about the SM Higgs mechanism, i.e. discover the Higgs boson or exclude it over the full allowed mass range.

Second, ignoring masses much larger than 200 GeV, which are disfavoured by the present electroweak data, two regions can be identified. If the Higgs mass is

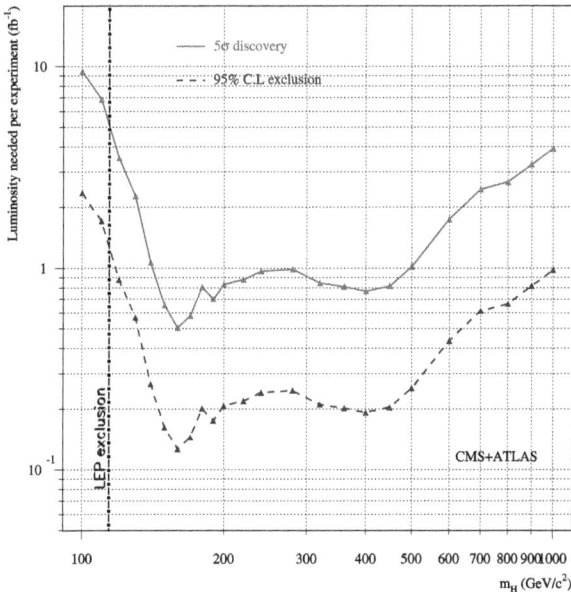

Fig. 6. Integrated luminosity per experiment, as a function of the Higgs boson mass, needed for a 5σ discovery (upper curve) and for a 95% C.L. exclusion (lower curve) of a SM Higgs boson signal at the LHC, combining ATLAS and CMS [10].

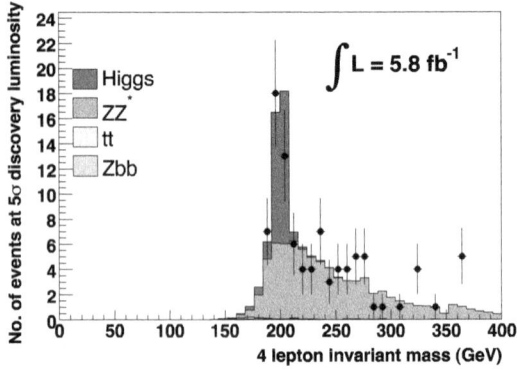

Fig. 7. The expected $H \to ZZ \to e^+e^-\mu^+\mu^-$ signal in CMS [3] for a Higgs mass of 200 GeV on top of the backgrounds, for an integrated luminosity of 5.8 fb^{-1}. The results of one simulated experiment are shown as dots.

around 150 GeV or larger, discovery should be easier thanks to the gold-plated $H \to 4\ell$ channel. As shown in Fig. 7, the expected signal sample is tiny for integrated luminosities of a few fb^{-1}, but these events are very pure (since the background is small) and should cluster in a narrow mass peak. About 1 fb^{-1} of data could be enough for discovery, combining both experiments and provided that the detectors are well understood. This may well happen by end 2009.

If, on the other hand, the Higgs mass is around 115–120 GeV, i.e. just above the experimental lower limit set by LEP, more luminosity is needed and observation of a possible signal is less straightforward. This is because in this mass region the experimental sensitivity is equally shared by three different channels ($H \to \gamma\gamma$, $t\bar{t}H$ production with $H \to b\bar{b}$, and Higgs production in vector-boson fusion followed by $H \to \tau\tau$) which all require close-to-ultimate detector performance and a control of the huge backgrounds at the few percent level. Discovery in this region is not likely to happen before 2010.

6. Conclusions

LHC operation will start in Summer 2008, and the accelerator and the experiments are progressing at full speed toward this goal. In the next months, all efforts will continue to focus on the commissioning of a machine and detectors of unprecedented complexity, technology and performance.

With the first 14 TeV pp data (1–100 pb^{-1}), the most urgent tasks will be to understand the detectors in detail and perform first measurements of SM physics.

If Nature is kind enough, new physics could be discovered up to the 1 TeV range with less than ~ 1 fb^{-1}.

With more time and more data, the LHC will be able to explore the highly-motivated TeV scale in detail, with a direct discovery potential up to particle masses of ~ 5–6 TeV. Hence, if new physics is there the LHC should find it. It will also provide definitive answers about the SM Higgs mechanism, Supersymmetry, and other TeV-scale predictions that have resisted experimental verification for decades. Finally, and perhaps more importantly, the LHC will likely tell us which are the right questions to ask and how to continue.

References

[1] L. Evans, *New. J. Phys.* **9** (2007) 335.

[2] ATLAS Collaboration, *Detector and Physics Performance Technical Design Report*, CERN/LHCC/99-15.

[3] CMS Collaboration, *Physics Technical Design Report*, CERN/LHCC/06-01 and CERN/LHCC/06-021.

[4] LHCb Collaboration, *Technical Proposal*, CERN/LHCC/98-004; LHCb Collaboration, *Reoptimized detector Technical Design Report*, CERN/LHCC/2003-030.

[5] ALICE Collaboration, *Technical Proposal*, CERN/LHCC/95-71.

[6] J. Wenninger, *LHC Accelerator Complex*, in: Second CERN-Fermilab Hadron Collider Physics Summer School, 2007, http://indico.cern.ch/conferenceOtherViews.py?view=cdsagenda&confId=6238.

[7] T.S. Virdee, *Phys. Rep.* **403–404** (2004).

[8] C. Quigg, *Rept. Prog. Phys.* **70** (2007) 1019.

[9] M. Cobal and S. Bentvelsen, *Top studies for the Atlas detector commissioning*, ATLAS Note ATL-PHYS-PUB-2005-024, http://documents.cern.ch/cgi-bin/setlink?base=atlnot&categ=PUB&id=phys-pub-2005-024.

[10] J.-J Blaising et al., *Potential LHC contributions to Europe's future strategy at the high-energy frontier*, input n. 54 to the CERN Council Strategy group, http://council-strategygroup.web.cern.ch/council-strategygroup/.

Course 10

STRING THEORY, GRAVITY AND EXPERIMENT

Thibault Damour[1] and Marc Lilley[2]

[1] *Institut des Hautes Etudes Scientifiques, 35 route de Chartres, F-91440 Bures-sur-Yvette, France*
[2] *Institut d'Astrophysique de Paris, 98, bis Blvd. Arago, F-75014 Paris, France*

C. Bachas, L. Baulieu, M. Douglas, E. Kiritsis, E. Rabinovici, P. Vanhove, P. Windey
and L.F. Cugliandolo, eds.
Les Houches, Session LXXXVII, 2007
String Theory and the Real World: From Particle Physics to Astrophysics
© *2008 Published by Elsevier B.V.*

Contents

1. Introduction

The common theme of these lectures is *gravity*, and their aim is to discuss a few cases where string theory might have an interesting interplay either with gravity theory, or with gravity phenomenology. We shall discuss the following topics:

- *Classical black holes as dissipative branes.* The idea here is to review the "classic" work on black holes of the seventies which led to the picture of black holes as being analog to dissipative branes endowed with finite electrical resistivity, and finite surface viscosity. In particular, we shall review the derivation of the (classical) surface viscosity of black holes, which has recently acquired a new (quantum) interest in view of AdS/CFT duality.

- *Hawking radiation from black holes.* To complete our classical account of irreversible properties of black holes, we shall also give a direct derivation of the phenomenon of Hawking radiation, because of its crucial importance in fixing the coefficient between the area of the horizon and black hole "entropy".

- *Experimental tests of gravity.* Before discussing possible phenomenological consequences of string theory in the gravitational sector, we find useful to summarize the present status of experimental tests of gravity, as well as the theoretical frameworks used to interpret them. In particular, we emphasize that binary pulsar experiments have already given us accurate tests of some aspects of strong-field (and radiative) relativistic gravity.

- *String-inspired phenomenology of the gravitational sector.* In this section we shall discuss (without any attempt at completeness) some of the ideas that have been suggested about observable signals possibly connected to string theory. In particular, we shall discuss the *cosmological attractor mechanism* which leads to a rather rich gravitational phenomenology that will be probed soon by various gravitational experiments.

- *String-related signals in cosmology.* After discussing a few alternatives to slow-roll inflation (and the possible relaxation of the Lyth bound when using non-linear kinetic terms for the inflaton), we discuss in some detail *cosmic superstrings*. We explain, in particular, how one computes the *gravitational wave burst signal* emitted by the cusps that periodically form during the dynamical evolution of generic string loops.

A final warning: by lack of time (and energy), no attempt has been made to give exhaustive and fair references to original and/or relevant work. The given

375

references are indicative, and should be viewed as entry points into the relevant literature. With the modern, web-based, easy access to the scientific literature it is hoped that the reader will have no difficulty in using the few given references as starting points for an instructive navigation on the vast sea of the physics literature.

2. Classical black holes as dissipative branes

Early work on (Schwarzschild, Reissner–Nordström, or Kerr–Newman) black holes (BHs) in the 1950's and 1960's treated them as *passive objects*, i.e., as given geometrical backgrounds (and potential wells). This viewpoint changed in the early 1970's when the study of the *dynamics* of BHs was initiated by Penrose [1], Christodoulou and Ruffini [2,3], Hawking [4], and Bardeen, Carter and Hawking [5]. In the works [1–5], only the *global dynamics* of BHs was considered, i.e., their total mass, their total angular momentum, their total irreducible mass, and the variation of these quantities. This viewpoint further evolved in the works of Hartle and Hawking [6], Hanni and Ruffini [7], Damour [8–10], and Znajek [11], in which the *local dynamics of BH horizons* was studied. In this new approach (which was later called the "membrane paradigm" [12]) a BH horizon is interpreted as a brane with dissipative properties, such as, for instance, an electrical resistivity ρ, equal to 377 Ohms [8,11] independently of the type of BH, and a surface (shear) viscosity, equal to $\eta = \frac{1}{16\pi}$ [9,10]. When divided by the entropy density found by Hawking ($S/A = \frac{1}{4}$), the shear viscosity yields the ratio $\frac{1}{4\pi}$, a result which has recently raised a renewed interest in connection with AdS/CFT, through the work of Kovtun, Son, and Starinets [13,14].

2.1. Global properties of black holes

Let us start by reviewing the study of the *global dynamics* of BHs. Initially, BHs were thought of as given geometrical backgrounds. In the case of a spherically symmetric object of mass M without any additional attribute, Schwarzschild derived the first exact solution of Einstein's equations only a few weeks after Einstein had obtained them in their final form. Schwarzschild's solution is as follows. In $3 + 1$ dimensions, setting $G = c = 1$, the metric for a spherically symmetric background can be written in the form

$$\mathrm{d}s^2 = -A(r)\mathrm{d}T^2 + B(r)\mathrm{d}r^2 + r^2\left(\mathrm{d}\theta^2 + \sin^2\theta\,\mathrm{d}\varphi^2\right), \tag{2.1}$$

where T denotes the usual Schwarzschild-type time coordinate, and where the coefficients $A(r)$ and $B(r)$ read

$$A(r) = 1 - \frac{2GM}{r},$$

$$B(r) = \frac{1}{A(r)}. \tag{2.2}$$

This result was generalized in independent works by Reissner, and by Nordström (1918) for electrically charged spherically symmetric objects, in which case $A(r)$ and $B(r)$ are given by

$$A(r) = 1 - \frac{2M}{r} + \frac{Q^2}{r^2},$$

$$B(r) = \frac{1}{A(r)}. \tag{2.3}$$

We shall not review here the long historical path which led to interpreting the above solutions, as well as their later generalizations due to Kerr (who added to the mass M the spin J[1]), and Newman *et al.* (mass, spin and charge), as BHs.

Up to the 1960's BHs were viewed only as *passive* gravitational wells. For instance, one could think of adiabatically lowering a small mass m at the end of a string until it disappears into the BH, thereby converting its mass-energy mc^2 into work. More realistically, one was thinking of matter orbiting a BH and radiating away its potential energy (up to a maximum, given by the binding energy of the last stable circular orbit around a BH). This viewpoint changed in the 1970's, when BHs started being considered as *dynamical* objects, able to exchange mass, angular momentum and charge with the external world. Whereas in the simplest case above, one uses the attractive potential well created by the mass M without extracting energy from the BH, Penrose [1] showed that energy could in principle be *extracted* from a BH itself by means of what is now called a (gedanken) "Penrose process" (see Fig. 1). Namely, if one considers a time-independent background and a BH that is more complicated than Schwarzschild's, say a Kerr BH, one may extract energy using a test particle 1 coming in from infinity with energy E_1, angular momentum p_{φ_1}, and electric charge e_1. By Noether's theorem, the time-translation, axial and $U(1)$ gauge symmetries of the background guarantee the conservation of E, p_φ and e during the "fall" of the test particle. Moreover, if, in a quantum process, the test particle 1 splits, near the BH, into two particles 2 and 3, with E_2, p_{φ_2}, e_2, and E_3, p_{φ_3}, e_3 respectively, then, under certain conditions, one finds that particle 3 can be absorbed by the BH, and that particle 2 *may come out at infinity with more energy than the incoming particle 1*. A detailed analysis of the efficiency of such gedanken Penrose processes by Christodoulou and Ruffini [2,3] then led to the understanding of the existence

---

[1]Note that in the case of a spinning BH, one often introduces the useful quantity $a = J/M$, i.e., the ratio of the total angular momentum to the mass of the BH, which has the dimension of length.

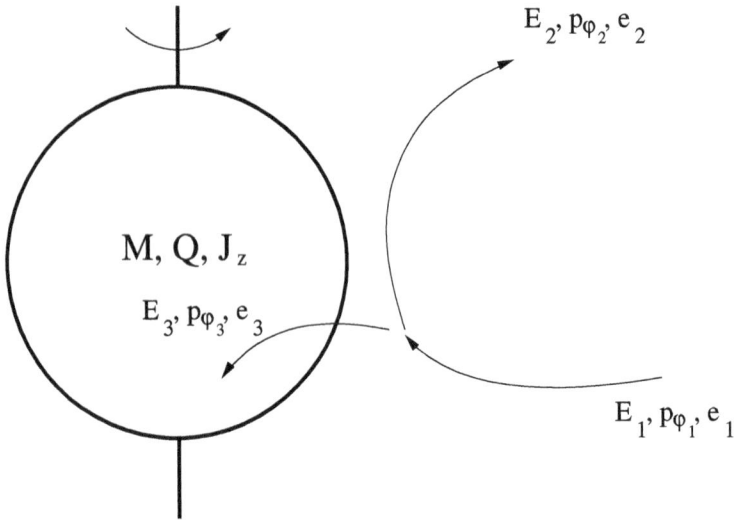

Fig. 1. In this figure, we schematically illustrate the "Penrose process", i.e., the splitting of an ingoing particle into one that falls into the BH and another that exits at infinity.

of a fundamental *irreversibility* in BH dynamics, and to the discovery of the BH *mass formula*. Let us explain these results.

The basic idea is to explore the physics of BHs through a sequence of infinitesimal changes of their state obtained by injecting in them some test particles. One starts by writing that the total mass-energy, spin and charge of the BH change, by absorption of particle 3, as

$$\begin{aligned} \delta M &= E_3 = E_1 - E_2, \\ \delta J &= J_3 = J_1 - J_2, \\ \delta Q &= e_3 = e_1 - e_2. \end{aligned} \qquad (2.4)$$

This preliminary result can be further exploited by making use of the Hamilton-Jacobi equation. Considering an on-shell particle of mass μ, and adopting the $(-+++)$ signature, the Hamilton-Jacobi equation reads

$$g^{\mu\nu}(p_\mu - eA_\mu)(p_\nu - eA_\nu) = -\mu^2, \qquad (2.5)$$

in which $p_\mu = \partial S/\partial x^\mu$, S is the action and the partial derivatives are taken w.r.t. the coordinates x^μ. The details of the splitting process will be irrelevant, as only particle 3 matters in the calculation. In an axially symmetric and time-independent background, S can be taken as a linear function of T and φ,

$$S = -ET + p_\varphi \varphi + S(r, \theta), \qquad (2.6)$$

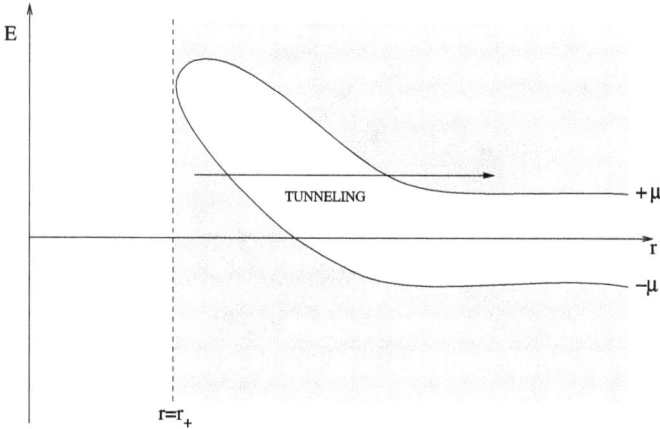

Fig. 2. This figure depicts the classically allowed energy levels (shaded region) as a function of radius, for test particles in the neighborhood of a BH. There exist positive- and negative-energy solutions, corresponding (after second quantization) to particles and anti-particles. Classically (as in the Penrose process) one should consider only the "positive-square-root" energy levels, located in the upper shaded region. The white region is classically forbidden. Note the possibility of tunneling (this corresponds to particle creation via the "super-radiant", non-thermal mechanism briefly mentioned below).

where $E = -p_T = -p_0$ is the conserved energy, p_φ is the conserved φ-component of angular momentum and the last term is the contribution from terms that depend on the angle θ and on the radial distance r. Let us consider the case of a Reissner–Nordström BH, where calculations are easier: the inverse metric is easily computed and (2.5) can then be written explicitly as

$$-\frac{1}{A(r)}(p_0 - eA_0)^2 + A(r)p_r^2 + \frac{1}{r^2}\left(p_\theta^2 + \frac{1}{\sin^2\theta}p_\varphi^2\right) = -\mu^2, \qquad (2.7)$$

which we re-write as

$$(p_0 - eA_0)^2 = A(r)^2 p_r^2 + A(r)\left(\mu^2 + \frac{L^2}{r^2}\right). \qquad (2.8)$$

The electric potential is $-A_0 = +V = +Q/r$. The above expression is quadratic in E (it is the generalization of the famous flat-spacetime $E^2 = \mu^2 + \mathbf{p}^2$) and one finds two possible solutions for the energy as a function of momenta and charge (see Fig. 2):

$$E = \frac{eQ}{r} \pm \sqrt{A(r)^2 p_r^2 + A(r)\left(\mu^2 + \frac{L^2}{r^2}\right)}. \qquad (2.9)$$

In flat space, $A(r) = 1$, so that, if we ignore charge, we recover the usual Dirac dichotomy on the choice of the $+$ or $-$ sign between particle and antiparticle: $E = \pm\sqrt{\mu^2 + \mathbf{p}^2}$. This shows that one should take the *plus sign* in the equation above. We remind the reader that for a charged BH, there exists a regular horizon only if $Q < M$ (which can be interpreted as a BPS bound). [We have set $G = 1$]. Remembering that $A(r) = 1 - 2M/r + Q^2/r^2$, there exists both an outer and an inner horizon defined by $r_\pm = M \pm \sqrt{M^2 - Q^2}$ (which are the two roots of $A(r) = 0$). The horizon of relevance for BH physics is the outer one $r_+ = M + \sqrt{M^2 - Q^2}$ (it gives the usual result $2M$ when $Q = 0$). As particle 3 is absorbed by the BH, we can compute its (conserved) energy when it crosses the horizon, i.e., in the limit where the radial coordinate r is equal to r_+. This simplifies the expression of E_3 to

$$E_3 = \frac{e_3\, Q}{r_+} + |p^r|, \tag{2.10}$$

where we have introduced the contravariant component $p^r = g^{rr} p_r = A(r) p_r$, which has a finite limit on the horizon. Note the presence of the *absolute value* of p^r (coming from the limit of a positive square-root). The change in the mass of the BH is equal to the energy E_3 of the particle absorbed, i.e., particle 3. Using $e_3 = \delta Q$, this yields

$$\delta M = \frac{Q\delta Q}{r_+(M, Q)} + |p^r|. \tag{2.11}$$

From the positivity of $|p^r|$ we deduce that

$$\delta M \geq \frac{Q\delta Q}{r_+(M, Q)}. \tag{2.12}$$

We have derived an *inequality* and have thereby demonstrated (by following Christodoulou and Ruffini) the irreversibility property of BH energetics. There exist two types of processes, the *reversible* ones with an '=' sign in (2.12), and the *irreversible* ones with an '>' sign. The former ones are reversible because if a BH first absorbs a particle of charge $+e$ with vanishing $|p^r|$ (thereby changing its mass by $\delta'M = eQ/r_+(M, Q)$ and its charge by $\delta'Q = e$), and then a particle of charge $-e$ with vanishing $|p^r|$ (thereby changing its mass by $\delta''M = -eQ/r_+(M, Q)$ and its charge by $\delta''Q = -e$), it will be left, at the end, in the same state as the original one (with mass $M + \delta'M + \delta''M = M$ and charge $Q + \delta'Q + \delta''Q = Q$). Evidently, such reversible transformations are delicate to perform, and one expects that irreversibility will occur in most BH processes. The situation here is clearly similar to the relation between reversible and irreversible processes in thermodynamics.

The same computation as for the Reissner–Nordström BH can be performed for the Kerr–Newman BH. One obtains in that case, by a slightly more complicated calculation,

$$\delta M - \frac{a\delta J + r_+ Q\delta Q}{r_+^2 + a^2} = \frac{r_+^2 + a^2\cos^2\theta}{r_+^2 + a^2}|p^r|, \tag{2.13}$$

in which $r_+(M, J, Q) = M + \sqrt{M^2 - Q^2 - a^2}$. We recall that $a = J/M$, and that one has the bound $Q^2 + (J/M)^2 \leq M^2$.

The idea now is to consider an infinite sequence of infinitesimal reversible changes (i.e., $p^r \to 0$), and to study the BH states which are reversibly connected to some initial BH state with given mass M, angular momentum J and charge Q. This leads to a partial differential equation for δM,

$$\delta M = \frac{a\delta J + r_+ Q\delta Q}{r_+^2 + a^2}, \tag{2.14}$$

which is found to be integrable. Integrating it, one finds the Christodoulou–Ruffini mass formula [3]

$$M^2 = \left(M_{\text{irr}} + \frac{Q^2}{4M_{\text{irr}}}\right)^2 + \frac{J^2}{4M_{\text{irr}}^2}. \tag{2.15}$$

Here the *irreducible mass* $M_{\text{irr}} = \frac{1}{2}\sqrt{r_+^2 + a^2}$ appears as an integration constant. The mass squared thus appears as a function of three contributions, with one term containing the square of the sum of the irreducible mass and of the Coulomb energy, and the other one containing the rotational energy. Inserting this expression into Eq. (2.13), one finds

$$\delta M_{\text{irr}} \geq 0 \tag{2.16}$$

with $\delta M_{\text{irr}} = 0$ under reversible transformations and $\delta M_{\text{irr}} > 0$ under irreversible transformations. The irreducible mass M_{irr} can only increase or stay constant. This behaviour is certainly reminiscent of the second law of thermodynamics. The free energy of a BH is therefore $M - M_{\text{irr}}$, i.e., this is the maximum extractable energy. In this view, BHs are no longer passive geometrical backgrounds but contain stored energy that can be extracted. Actually, the stored energy can be enormous because a BH can store up to 29% of its mass as rotational energy, and up to 50% as Coulomb energy!

The irreducible mass is related to the area of the horizon of the BH, by $A = 16\pi M_{\text{irr}}^2$ so that in a reversible process $\delta A = 0$, while in an irreversible one $\delta A > 0$. Hawking showed [4] that this irreversible evolution of the area of the horizon was a general consequence of Einstein's equations, when assuming the weak energy condition. He also showed that in the merging of two BHs of area A_1 and A_2, the total final area satisfied $A_{tot} \geq A_1 + A_2$.

While such results evoque the second law of thermodynamics, the analog of the first law $[dE(S, \text{extensive parameters}) = dW + dQ$, where the work dW is linked to the variation of extensive parameters (volume, etc.) and where $dQ = TdS$ is the heat exchange] reads, for BH processes,

$$dM(Q, J, A) = VdQ + \Omega dJ + \frac{g}{8\pi}dA. \tag{2.17}$$

Comparing this result with expression (2.13), one has

$$V = \frac{Qr_+}{r_+^2 + a^2},$$
$$\Omega = \frac{a}{r_+^2 + a^2}, \tag{2.18}$$

and

$$g = \frac{1}{2}\frac{r_+ - r_-}{r_+^2 + a^2}, \tag{2.19}$$

which, in the Kerr–Newman case, is given by

$$g = \frac{\sqrt{M^2 - a^2 - Q^2}}{r_+^2 + a^2}. \tag{2.20}$$

V is interpreted as the electric potential of the BH, and Ω as its angular velocity. Expression (2.17) resembles the usual form of the first law of thermodynamics in which the area term has to be interpreted as some kind of entropy. The parameter g is called the "surface gravity". [In the Schwarzschild case, it reduces to M/r_+^2 (in $G = 1$ units), i.e., the usual formula for the surface gravitational acceleration $g = GM/R^2$.] In the Les Houches Summer School of 1972, a more general version of the first law was derived, that included the presence of matter around the BH, and energy exchange [5]. An analog of the zeroth law was also derived [15], in the sense that the surface gravity g (which is analog to the temperature) was found to be uniform on the surface of a BH in equilibrium.

In 1974, Bekenstein went further in taking seriously (and no longer as a simple analogy) the thermodynamics of BHs. First, note that one can write the formal BH "heat exchange" term in various ways

$$dQ = TdS = \frac{gdA}{8\pi} = 4gM_{irr}dM_{irr}. \tag{2.21}$$

In light of this, is the appropriate physical analog of the entropy the irreducible mass or the area of a BH? Is the analog of temperature proportional to the surface gravity g or to the product $M_{irr}\,g$? Can one give a physical meaning to the temperature and entropy of a BH? To address such questions, Bekenstein used several different approaches.

In particular, he used Carnot-cycle-type arguments. For instance, one may extract work from a BH by slowly lowering into it a box of radiation of infinitesimal size. In fact, in this ideal case, one can theoretically convert all the energy of the box of radiation, mc^2, into work. The efficiency of Carnot cycles is defined in terms of both a hot and a cold source as

$$\eta = 1 - \frac{T_{cold}}{T_{hot}}. \tag{2.22}$$

From what we just said, it would seem that the efficiency of classical BHs as thermodynamic engines is 100%, $\eta = 1$. This would then correspond to a BH temperature (= the cold source) $T_{BH} = T_{cold} = 0$. The point made by Bekenstein was that this classical result will be modified by quantum effects. Indeed, one expects (because of the uncertainty principle) that a box of thermal radiation at temperature T (made of typical wavelengths $\lambda \sim 1/T$) cannot be made infinitesimally small, but will have a minimum finite size $\sim \lambda$. From this limit on the size of the box, Bekenstein then deduced an upper bound on the efficiency η, and therefore a lower bound on the BH temperature $T_{BH} \neq 0$.

Let us indicate another reasoning (of Bekenstein) which suggests that the absorption of a single particle by a BH augments its surface by a finite amount proportional to $\hbar$. As we said above the change of BH energy as it absorbs a particle is (when $a = 0$, for simplicity)

$$E_3 = \frac{eQ}{r_+} + \lim_{r \to r_+} |p^r|. \tag{2.23}$$

We also showed that the transformation will be reversible (i.e., will *not* increase the surface area of the BH) only if $\lim_{r \to r_+} |p^r| = 0$. However, for this to be true *both* the (radial) position and the (radial) momentum of the particle must be exactly fixed: namely, $r = r_+$ and $p^r = 0$. This would clearly be in contradiction with the Heisenberg uncertainty principle.

Technically, we must consider the conjugate momentum to the position r which is the *covariant* component p_r of the radial momentum (instead of the *contravariant* component p^r used in the equation above). The uncertainty relation therefore reads

$$\delta r \delta p_r \geq \frac{1}{2}\hbar. \tag{2.24}$$

Near the horizon (i.e. when $\delta r \equiv r - r_+$ is small), the contravariant radial momentum reads (using $g^{rr} = 1/g_{rr} = A(r)$)

$$
\begin{aligned}
p^r &= A(r)p_r \\
&= \frac{(r - r_+)(r - r_-)}{r^2} p_r \\
&\simeq \delta r \frac{(r_+ - r_-)}{r_+^2} p_r \\
&\simeq \left(\frac{\partial A}{\partial r}\right)_{r_+} \delta r p_r,
\end{aligned}
\tag{2.25}
$$

so that Heisenberg's uncertainty relation yields a *lower bound* for p^r. We can reexpress this lower bound in terms of the BH surface gravity g introduced above by noting that the partial derivative of A w.r.t. r, $\left(\frac{\partial A}{\partial r}\right)_{r_+}$, entering the last equation, is proportional to g:

$$
\left(\frac{\partial A}{\partial r}\right)_{r_+} = 2g.
\tag{2.26}
$$

This then gives

$$
p^r \simeq 2g\delta r \delta p_r \geq g\hbar.
\tag{2.27}
$$

From the relation $\delta M = \frac{Q\delta Q}{r_+(M,Q)} + |p^r|_{r_+}$, we finally obtain

$$
\delta M - \frac{Q\delta Q}{r_+} = |p^r| \geq g\hbar,
\tag{2.28}
$$

which can be rewritten as

$$
\delta A \geq 8\pi \hbar.
\tag{2.29}
$$

In other words, quantum mechanics tells us that when one lets a particle fall into a BH, one cannot do so in a perfectly reversible way. The area must increase by a quantity of order $\hbar$. If (still following Bekenstein) one considers that the irreversible absorption of a particle by a BH corresponds to the loss of one bit of information (for the outside world), we are led to the idea of attributing to a BH an entropy (in the sense of "negentropy") equal (after re-introducing the constants c and G) to [16]

$$
S_{\mathrm{BH}} = \hat{\alpha}\frac{c^3}{\hbar G}A,
\tag{2.30}
$$

with a dimensionless numerical coefficient equal to $\hat{\alpha} = \ln 2/8\pi$ according to the reasoning just made. More generally, Bekenstein suggested that the above formula should hold with a dimensionless coefficient $\hat{\alpha} \approx \mathcal{O}(1)$, without being able to fix in a unique, and convincing, manner the value of $\hat{\alpha}$. This result in turn implies (by applying the law of thermodynamics) that one should attribute to a BH a temperature equal to

$$T_{\text{BH}} = \frac{1}{8\pi\hat{\alpha}} \frac{\hbar}{c} g. \tag{2.31}$$

This attribution of a finite temperature to a BH looked rather strange in view of the definition of a BH has being "black", i.e., as allowing no radiation to come out of it. In particular, Stephen Hawking resisted this idea, and tried to prove it wrong by studying quantum field theory in a BH background. However, much to his own surprise, he so discovered (in 1974) the phenomenon of quantum radiation from BH horizons (see below) which remarkably vindicated the physical correctness of Bekenstein's suggestion. Hawking's calculation also unambiguously fixed the numerical value of $\hat{\alpha}$ to be $\hat{\alpha} = \frac{1}{4}$ [17]. [We shall give below a simple derivation (from Ref. [18]) of Hawking's radiation.]

Summarizing so far: The results on BH dynamics and thermodynamics of the early 1970's modified the early view of BHs as passive potential wells by endowing them with *global* dynamical and thermodynamical quantities, such as mass, charge, irreducible mass, entropy, and temperature. In the following section, we shall review the further changes in viewpoint brought by work in the mid and late 1970's ([6, 8–11]) which attributed *local* dynamical and thermodynamical quantities to BHs, and led to considering BH horizons as some kind of *dissipative branes*. Note that, in the following section, we shall no longer consider only Kerr–Newman BHs (i.e., stationary BHs in equilibrium, which are not distorted by sources at infinity). We shall consider more general non-stationary BHs distorted by outside forces.

2.2. Black hole electrodynamics

The description of BHs we give from here on is essentially "holographic" in nature since it will consist of excising the interior of a BH, and replacing the description of the interior BH physics by quantities and phenomena taking place entirely on the "surface of the BH" (i.e., the horizon). The surface of the BH is defined as being a null hypersurface, i.e., a surface everywhere tangent to the lightcone, separating the region inside the BH from the region outside. As just said, we ignore the region inside, including the spacetime singularity, and consider the physics in the outside region, completing it with suitable "boundary effects" on the horizon. These boundary effects are fictitious, and do not really exist on the BH surface but play the role of representing, in a holographic sense, the physics that goes on inside. In the end, we shall have a horizon, a set of surface quantities on the horizon and a set of bulk properties outside the horizon. We first consider Maxwell's equations, namely $F_{\mu\nu} = \partial_\mu A_\nu - \partial_\nu A_\mu$, and

$$\nabla_\nu F^{\mu\nu} = 4\pi J^\mu,$$
$$\nabla_\mu J^\mu = 0.$$
$$(2.32)$$

A priori, the electromagnetic field $F_{\mu\nu}$ permeates the full space time, existing both inside and outside the horizon, and the current, i.e., the source term of $F_{\mu\nu}$ that carries charge, is also distributed both outside and inside the BH. In order to replace the internal electrodynamics of the BH by surface effects, we replace the real $F_{\mu\nu}(x)$ by $F_{\mu\nu}(x)\Theta_H$, where Θ_H is a Heaviside-like step function, equal to 1 outside the BH and 0 inside. Then we consider what equations are satisfied by this Θ_H-modified electromagnetic field. The corresponding modified Maxwell equations contain two types of source terms,

$$\nabla_\nu (F^{\mu\nu}\Theta) = (\nabla_\nu F^{\mu\nu})\Theta + F^{\mu\nu}\nabla_\nu\Theta$$
$$= 4\pi \left(J^\mu\Theta + j_v^\mu \right),$$
$$(2.33)$$

where we have introduced a *BH surface current* j_H^μ as

$$j_H^\mu = \frac{1}{4\pi} F^{\mu\nu}\nabla_\nu\Theta.$$
$$(2.34)$$

This surface current contains a Dirac δ-function which restricts it to the horizon. Indeed, let us consider a scalar function $\varphi(x)$ such that $\varphi(x) = 0$ on the horizon, with $\varphi(x) < 0$ inside the BH, and $\varphi(x) > 0$ outside it. The BH Θ-function introduced above is simply equal to $\Theta_H = \theta(\varphi(x))$, where θ denotes the standard step function of one real variable. Therefore, the gradient of Θ_H reads

$$\partial_\mu\Theta_H = \partial_\mu\theta(\varphi(x)) = \delta(\varphi(x))\partial_\mu\varphi,$$
$$(2.35)$$

where δ is the (one dimensional) usual Dirac δ, so that $\delta(\varphi(x))$ is a δ function with support on the horizon. Morally, the gradient $\partial_\mu\varphi$ yields a vector "normal to the horizon". In the case of a BH (by contrast to the usual case of a hypersurface in Euclidean space), there exists an extra subtlety in the exact definition of the normal to the horizon. The horizon is a null hypersurface which by definition is normal to a null covariant vector ℓ_μ satisfying both $\ell_\mu\ell^\mu = 0$ and $\ell_\mu dx^\mu = 0$ for any infinitesimal displacement dx^μ within the hypersurface. Since ℓ_μ is null, it cannot be normalized in the same way as in Euclidean space. This leads to an ambiguity in the physical observables related to ℓ_μ. In stationary-axisymmetric spacetimes, one uniquely normalizes ℓ_μ by demanding that the corresponding directional gradient $\ell^\mu \partial_\mu$ be of the form $\partial/\partial t + \Omega \partial/\partial \phi$ (with a coefficient one in front of the time-derivative term). We shall assume (in the general non-stationary case) that ℓ_μ is normalized so that its normalization is compatible with the usual normalization when considering the limiting case of stationary-axisymmetric spacetimes. Anyway, given any normalization, there exists a scalar ω such that

$$\ell_\mu = \omega \partial_\mu \varphi, \tag{2.36}$$

and we can then define an "horizon δ-function"

$$\delta_H = \frac{1}{\omega} \delta(\varphi), \tag{2.37}$$

such that

$$\partial_\mu \Theta_H = \ell_\mu \delta_H. \tag{2.38}$$

One can then define a "BH surface current density"

$$K^\mu = \frac{1}{4\pi} F^{\mu\nu} \ell_\nu. \tag{2.39}$$

With this definition, the BH current j_H^μ reads

$$j_H^\mu = K^\mu \delta_H, \tag{2.40}$$

and satisfies

$$\nabla_\mu \left(\Theta_H J^\mu + K^\mu \delta_H \right) = 0, \tag{2.41}$$

which is a conservation law for the sum of the outside bulk current $\Theta_H J^\mu$ and of the boundary current $K^\mu \delta_H$. In picturesque terms, the surface current $K^\mu \delta_H$ effectively "closes" the external current lines penetrating the BH (analogously to the case of external currents being injected in a perfect conductor and leading to currents flowing on its surface). In addition, Eq. (2.39) shows that this surface current is linked to the electromagnetic fields which are on the horizon. We have thus endowed the horizon with surface quantities, defined uniquely and locally on the horizon.

Before we proceed, we introduce a convenient coordinate system to describe the physics on the horizon of a general BH. We assume some regular "slicing" of the horizon and its neighbourhood by some (advanced) Eddington–Finkelstein-like time coordinate $t = x^0$. Then we assume that the first coordinate x^1 is such that it is equal to zero on the horizon (like $r - r_+$ in the Kerr–Newman case). Finally x^A for $A = 2, 3$ denote some angular-like coordinates on the two-dimensional spatial slice S_t ($x^0 = t$) of the horizon. In this coordinate system, we normalize ℓ^μ such that

$$\ell^\mu \partial_\mu = \frac{\partial}{\partial t} + v^A \frac{\partial}{\partial x^A}. \tag{2.42}$$

Here, we have used the fact that the "normal" vector ℓ^μ, being null, is also *tangent* to the horizon, so that $\ell^\mu \partial_\mu$ is a general combination of $\partial/\partial t$ and $\partial/\partial x^A$ but has no component along the "radial" (or "transverse") coordinate x^1. Because ℓ^μ is a vector tangent to the hypersurface, we can consider its integral lines $\ell^\mu = dx^\mu/dt$, which lie within the horizon. These integral curves are called the *generators* of the horizon. They are null geodesic curves, lying entirely within the horizon.

Expression (2.42) for the directional gradient along ℓ^μ suggests that v^A be interpreted as the velocity of some "fluid particles" on the horizon, which are the "constituents" of a null membrane. Similarly to the usual description of the motion of a fluid, one has to keep track of the changes in the distance between two fluid particles as the fluid expands and shears. For a usual fluid, one considers the gradient of the velocity field, splitting it into its symmetric and anti-symmetric parts. The antisymmetric part is simply a local rotation which has no incidence on the physics and can be ignored. The symmetric part is further split into its trace and tracefree parts, namely

$$\frac{1}{2}(\partial_i v_j + \partial_j v_i) = \sigma_{ij} + \frac{1}{d}\partial \cdot v \delta_{ij}, \tag{2.43}$$

where d is the spatial dimension of the considered fluid (which will be $d = 2$ in our case). Here the first term describes the shear, and the second describes the rate of expansion. We will see later how the BH analogs of these quantities are defined. For the moment let us consider the distances on the horizon. They are measured by considering the restriction to the horizon of the spacetime metric (which is assumed to satisfy Einstein's equations). As we are considering a null hypersurface, we have

$$ds^2|_{x^1=0} = \gamma_{AB}(t, x^C)(dx^A - v^A dt)(dx^B - v^B dt), \tag{2.44}$$

where $v^A = \frac{dx^A}{dt}$. Note that ds^2 is a degenerate metric: indeed, on a (three-dimensional) null hypersurface, there is no real time direction (ds^2 vanishes along the generators). One has only two positive-definite space dimensions along, e.g., the spatial slices S_t. This metric describes the geometry on the horizon from which one can compute the area element of the spatial sections S_t

$$dA = \sqrt{\det \gamma_{AB}}\, dx^2 \wedge dx^3. \tag{2.45}$$

One can decompose the current density K^μ into a time component $\sigma_H = K^0$, and two spatial components K^A tangent to the spatial slices S_t (t = const.) of the horizon,

$$K^\mu \partial_\mu = \sigma_H \partial_t + K^A \partial_A \tag{2.46}$$

in which $\partial_t = \ell^\mu \partial_\mu - v^A \partial_A$ so that

$$K^\mu \partial_\mu = \sigma_H \ell^\mu + (K^A - \sigma_H v^A) \partial_A. \tag{2.47}$$

The total electric charge of the spacetime is defined by a surface integral at ∞, say

$$Q_{\text{tot}} = \frac{1}{4\pi} \oint_{S_\infty} \frac{1}{2} F^{\mu\nu} dS_{\mu\nu}. \tag{2.48}$$

This result can be re-written as the sum of a surface integral on the horizon and a volume integral in between the horizon and ∞. The volume integral is simply the usual charge contained in space, so that we can define the BH charge Q_H as

$$Q_H = \frac{1}{4\pi} \oint_H \frac{1}{2} F^{\mu\nu} dS_{\mu\nu}, \tag{2.49}$$

where the tensorial horizon surface element reads $dS_{\mu\nu} = \frac{1}{2}\varepsilon_{\mu\nu\rho\sigma} dx^\rho \wedge dx^\sigma = (n_\mu \ell_\nu - n_\nu \ell_\mu) dA$. Here, n^μ is a second null vector, which is transverse to the horizon, and which is orthogonal to the spatial sections S_t. It is normalized such that $n^\mu \ell_\mu = +1$. Using the above definitions for the BH surface current, one easily finds that the total BH charge can be rewritten as

$$Q_H = \oint_H \sigma_H dA, \tag{2.50}$$

where σ_H is the time component of the BH surface current introduced above. Though it is a priori only the integrated BH charge which has a clear physical meaning, it is natural to consider the density σ_H appearing in the above surface integral as defining a charge distribution on the horizon. Then the link

$$\sigma_H = K^\mu n_\mu = \frac{1}{4\pi} F^{\mu\nu} n_\mu l_\nu \tag{2.51}$$

can be thought of as being analog to the result $\sigma = \frac{1}{4\pi} E^i n_i$ giving the electric charge distribution on a metallic object. This can again be viewed as part of a holographic approach in which the interior of the BH is replaced by boundary effects. This analogy extends to the (spatial) currents flowing along the surface of the BH. Indeed, using the conservation law $\nabla_\mu (\Theta_H J^\mu + K^\mu \delta_H) = 0$, which is just a Bianchi identity, one has

$$\frac{1}{\sqrt{\gamma}} \frac{\partial}{\partial t} \left(\sqrt{\gamma} \sigma_H \right) + \frac{1}{\sqrt{\gamma}} \frac{\partial}{\partial x^A} \left(\sqrt{\gamma} K^A \right) = -J^\mu \ell_\mu. \qquad (2.52)$$

This shows, in a mathematically precise way, how an external current injected "normally" to the horizon "closes" onto a combination of currents flowing along the horizon, and/or of an increase in the local horizon charge density. One can also introduce the electromagnetic 2-form and restrict it on the horizon. It then defines the electric and magnetic fields on the horizon according to

$$\frac{1}{2} F_{\mu\nu} dx^\mu \wedge dx^\nu |_H = E_A dx^A \wedge dt + B_\perp dA. \qquad (2.53)$$

Taking the exterior derivative of the left-hand-side then gives

$$\nabla \times \vec{E} = -\frac{1}{\sqrt{\gamma}} \partial_t \left(\sqrt{\gamma} B_\perp \right). \qquad (2.54)$$

which relates the electric and magnetic fields on the horizon.

From the various formal definitions above, one also gets the following relation

$$E_A + \epsilon_{AB} B_\perp v^B = 4\pi \gamma_{AB} (K^B - \sigma_H v^B), \qquad (2.55)$$

or

$$\vec{E} + \vec{v} \times \vec{B}_\perp = 4\pi (\vec{K} - \sigma_H \vec{v}). \qquad (2.56)$$

We recognize here a BH analog of the usual Ohm's law relating the electric field to the current (especially in the case where $v \to 0$, i.e., in the absence of the various "convection effects" linked to the horizon "velocity" $\vec{v}$). From this form of Ohm's law, we can read off that BHs have a *surface electric resistivity* equal to $\rho = 4\pi = 377$ Ohm [8, 11].

Let us give an example in which this BH Ohm's law can be "applied" to a specific system. We consider for simplicity the case of a Schwarzschild BH and set up an electric circuit "on the surface of the BH" by injecting on the North pole (through an electrode penetrating the horizon under a polar angle θ_1, with, say, $\theta_1 \ll 1$) an electric current I, and letting it escape[2] from the South pole (via an electrode penetrating the horizon under a polar angle θ_2, with, say, $\pi - \theta_2 \ll 1$). When viewing the BH as a membrane with surface resistivity ρ, this set up will give rise to a fictitious current flow on the horizon, closing the circuit between

---

[2] Actually, as (classical) charges cannot escape from a BH, we need to inject in the South electrode a flow of negative charges (while injecting a flow of positive charges down the North pole).

the North and the South poles. Associated to the current flow on the horizon, there will be a potential drop V between the poles. This potential drop is simply given by the usual Ohm's law, $V = RI$, i.e., the product of the current I by a "resistance" R:

$$V = -A_0(\theta_1) + A_0(\theta_2) = RI. \tag{2.57}$$

The BH resistance R can be computed in two different ways, either by solving Maxwell's equations in a Schwarzschild background, or by computing, in usual Euclidean space, the total resistance of a spherical metallic shell with a uniform surface resistivity $\rho = 4\pi$ (by decomposing the problem in many elementary resistances, some being in parallel, and others in series). Both methods give the same answer, namely

$$R = 2\ln\frac{\tan\frac{\theta_2}{2}}{\tan\frac{\theta_1}{2}}, \tag{2.58}$$

expressed in units of $30\,\Omega$.[3] This result is saying that the typical total resistivity of a BH is of the order of $30\,\Omega$. In addition, if one considers a rotating BH placed in a magnetic field out of alignment with its axis of rotation (a field uniform at ∞, but distorted on the horizon), one expects to find eddy currents on the horizon, currents which dissipate the energy. These currents exist, can be computed and do indeed brake the rotation of the BH. In such a situation, one also finds a torque which acts to restore the alignment of the BH with the field [8].

2.3. Black hole viscosity

In the previous section, we introduced the electromagnetic dissipative properties of a BH, using a holographic approach which kept the physics outside up to infinity, and replaced the physics inside the BH by defining suitable quantities on the horizon, and then showed that they satisfied equations similar to ones well-known (such as Ohm's law). We now turn to the viscous properties of BHs and show how suitably defined "surface hydrodynamical" quantities satisfy a sort of Navier–Stokes equation. Technically, we would like to do, for the gravitational surface properties, something similar to what we did for the electrodynamic properties. Namely, we would like to replace the spacetime connection, say ω, by some sort of "screened connection" $\Theta_H\omega$, and see what kind of quantities and physics will be so induced on the surface of the BH. However, Einstein's equations being nonlinear, one cannot simply use a BH step function Θ_H as was done

---

[3]Indeed, in CGS-Gaussian units (as used, say, in the treatise of Landau and Lifshitz) 30 ohms is equal to the velocity of light (or its inverse, depending on whether one uses esu or emu). Then, when using (as we do here) units where $c = 1$, $30\,\Omega = 1$.

for BH electrodynamics. We shall therefore motivate the definition of suitable "surface quantities" related to ω in a slightly different way and then study the evolution of these surface quantities and their connection to the physics outside the horizon, up to ∞. Our presentation will be sketchy; for technical details, see [9, 10, 19, 20].

Let us start by considering an axisymmetric spacetime. Then there exists a Killing vector $\vec{m} = m^\mu \partial/\partial x^\mu = \partial/\partial \varphi$, to which, by Noether's theorem, one can associate a conserved total angular momentum, which can be written as a surface integral at ∞. The total angular momentum J_z w.r.t. φ reads

$$J_\infty = -\frac{1}{8\pi} \int_{S_\infty} \frac{1}{2} \nabla^\nu m^\mu dS_{\mu\nu}, \tag{2.59}$$

where $dS_{\mu\nu} = \frac{1}{2} \varepsilon_{\mu\nu\rho\sigma} dx^\rho \wedge dx^\sigma$, ∇^ν denotes a covariant derivative, and the surface integral is performed over the 2-sphere, S_∞. This starting point is the analog of the surface-integral expression for the total electric charge used above to motivate the definition of a BH surface charge distribution.

In a way similar to what was done in the electromagnetic case, we can use Gauss' theorem to rewrite this integral as the sum of two contributions: (i) a volume integral (over the 3-volume contained between the horizon and infinity) measuring the angular momentum of the matter present outside the horizon, and (ii) a surface integral over a (topological) 2-sphere S_H at the horizon, representing what we can call the BH angular momentum J_H, i.e.,

$$J = J_{\text{matter}} + J_H, \tag{2.60}$$

where J_H is given by the same surface-integral formula as J_∞, except for the replacement of S_∞ by S_H as the integration domain.

The horizon being tangent to the lightcone, one defines on the horizon, as above, a null vector ℓ_μ both normal and tangent to it. ℓ_μ can in turn be complemented by another null vector n_μ such that $\ell^\mu n_\mu = 1$ and such that the surface element $dS_{\mu\nu}$ can then be re-expressed as $(n_\mu \ell_\nu - n_\nu \ell_\mu) dA$. Remembering that the Killing symmetry preserves the generators of the horizon i.e., the commutator $[\vec{\ell}, \vec{m}] = 0$, one has $\ell^\nu \nabla_\nu m^\mu = m^\nu \nabla_\nu \ell^\mu$, so that we can re-express the BH angular momentum J_H as the following surface integral

$$J_H = -\frac{1}{8\pi} \int_{S_H} n_\mu m^\nu \nabla_\nu \ell^\mu dA. \tag{2.61}$$

This result involves the directional (covariant) derivative of the horizon *normal* vector $\vec{\ell}$ along a vector $\vec{m}$ which is *tangent* to the horizon. The crucial point now is to realize that, very generally, given any hypersurface, the parallel transport along some *tangent* direction, say $\vec{t}$, of the (normalized) vector $\vec{\ell}$ normal to the hypersurface yields *another tangent vector*. The technical proof of this fact consists of starting from the fact that $\vec{\ell} \cdot \vec{\ell} = \epsilon$, where ϵ is a *constant* which is equal to ± 1 in the case of a time-like or spacelike hypersurface, and to 0 in the case (of interest here) of a null hypersurface. Then, taking the directional gradient of this starting equality along an arbitrary tangent vector $\vec{t}$ yields $(\nabla_{\vec{t}}\vec{\ell}).\vec{\ell} = 0$. From this result, one deduces that the vector $(\nabla_{\vec{t}}\vec{\ell})$ must be *tangent* to the hypersurface. Therefore, there exists a certain linear map K, acting in the tangent plane to the hypersurface, such that $\nabla_{\vec{t}}\vec{\ell} = K(\vec{t})$. For a usual (time-like or space-like) hypersurface, the linear map K is called the "Weingarten map" and is simply the mixed-component K^{i}_{j} version of the extrinsic curvature of the hypersurface (usually thought of a being a symmetric covariant tensor K_{ij}). On the other hand, in the case of a null hypersurface, there is no unique way to define the analog of the covariant tensor K_{ij} (where the indices i, j are "tangent" to the hypersurface), but it is natural, and useful, to consider the mixed-component tensor K^{i}_{j}, intrinsically defined as the Weingarten map K in $\nabla_{\vec{t}}\vec{\ell} = K(\vec{t})$.

To explicitly write out the various components of the linear map K (acting on the hypersurface tangent plane), we need to define a basis of vectors tangent to the horizon. This basis contains the null vector $\vec{\ell}$ (which is both normal and tangent to the horizon), and two spacelike vectors. Using a coordinate system x^0, x^1, x^A ($A = 2, 3$) of the type already introduced (with the horizon being located at $x^1 = 0$), we can choose, as two spacelike horizon tangent vectors, the vectors $\vec{e}_A = \partial_A$. Then one finds that the Weingarten map K is fully described by the set of equations

$$\begin{aligned}
\nabla_{\vec{\ell}}\vec{\ell} &= g\,\vec{\ell}, \\
\nabla_A\vec{\ell} &= \Omega_A\vec{\ell} + D^{B}_{A}\vec{e}_B.
\end{aligned} \tag{2.62}$$

The first equation follows from the fact that $\vec{\ell}$ is tangent to a null geodesic lying within the null hypersurface. [In turn, this follows from the fact that $\vec{\ell}$ is proportional to the gradient of some scalar, say φ (satisfying the eikonal equation $(\nabla\varphi)^2 = 0$).] The coefficient g entering the first equation defines (in the most general manner) the *surface gravity* of the BH. We see that it represents one component of the Weingarten map K. The other components are the two-vector Ω_A, and the mixed two-tensor D^{B}_{A}. One can show that the components D^{B}_{A} are the mixed components of a symmetric two-tensor D_{AB}, which measures the "deformation", in time, of the geometry of the horizon. We remind the reader of the expression of the horizon metric, introduced above, $ds^2|_H = \gamma_{AB}(t,\vec{x})(dx^A - v^A dt)(dx^B - v^B dt)$. Here, $\gamma_{AB}(t,\vec{x})$ is a symmetric rank 2 tensor i.e., a time-dependent 2-metric such that the horizon may by viewed as a 2-dimensional brane. In addition, we have the generators, which are the vectors tangent to $\vec{\ell}$. When decomposing $\vec{\ell} = \partial_t + v^A \partial_A$ w.r.t. our coordinate system, they appear to have a "velocity" v^A which can also be viewed as the velocity of a fluid particle on the horizon. D_{AB} is then defined as the deformation tensor of the horizon geometry, namely $D_{AB} = \gamma_{BC} D_A^C = \frac{1}{2} \frac{D\gamma_{AB}}{dt}$, where D/dt denotes the *Lie derivative* along $\vec{\ell} = \partial_t + v^A \partial_A$. It is explicitly given by

$$
\begin{aligned}
D_{AB} &= \frac{1}{2} \left(\partial_t \gamma_{AB} + v^C \partial_C \gamma_{AB} + \partial_A v^C \gamma_{CB} + \partial_B v^C \gamma_{AC} \right) \\
&= \frac{1}{2} \left(\partial_t \gamma_{AB} + v_{A|B} + v_{B|A} \right),
\end{aligned}
\tag{2.63}
$$

where '$|$' denotes a covariant derivative w.r.t. the Christoffel symbols of the 2-geometry γ_{AB}. Note the contribution from the ordinary time derivative of γ_{AB}, and that from the variation of the generators of velocity v^A along the horizon. It is then convenient to split the deformation tensor D_{AB} into a tracefree part and a trace, i.e., $D_{AB} = \sigma_{AB} + \frac{1}{2}\theta\gamma_{AB}$, where the tracefree part σ_{AB} is the "shear tensor" and the trace, $\theta = D_A^A = \frac{1}{2}\gamma^{AB}\partial_t\gamma_{AB} + v_{|A}^A$, the "expansion". The remaining component of the Weingarten map, namely the 2-vector Ω_A, is defined as $\Omega_A = \vec{n}.\nabla_A\vec{\ell}$ with $\vec{\ell}.\vec{n} = 1$. Its physical meaning can be seen from looking at the BH angular momentum J_H.

Indeed, from the definition above of J_H, one finds that the total BH angular momentum is the projection of Ω_A on the direction of the rotational Killing vector $\vec{m} = \partial_\varphi$ introduced at the beginning of this section, so that we have

$$
J_H = -\frac{1}{8\pi} \oint_S m^A \Omega_A dA,
\tag{2.64}
$$

where $m^A \Omega_A$ is the φ-component of Ω_A. It is therefore natural to define, for a BH, a "surface density of linear momentum" as $\pi_A = -\frac{1}{8\pi}\Omega_A = -\frac{1}{8\pi}\vec{n} \cdot \nabla_A\vec{\ell}$. With this definition, one has

$$
J_H = \int_S \pi_\varphi dA,
\tag{2.65}
$$

which is similar to the result above giving the BH electric charge as the surface integral of the "charge surface density" σ_H.

Having so defined some (fictitious) "hydrodynamical" quantities on the surface of a BH (fluid velocity, linear momentum density, shear tensor, expansion rate, etc.), let us now see what evolution equations they satisfy as a consequence of Einstein's equations. By contracting Einstein's equations with the normal to the horizon, we can relate the quantities just defined to the flux of the energy-momentum tensor $T_{\mu\nu}$ into the horizon. For instance, by projecting Einstein's equations along $\ell^\mu e_A^\nu$, one finds

$$\frac{D\pi_A}{dt} = -\frac{\partial}{\partial x^A}\left(\frac{g}{8\pi}\right) + \frac{1}{8\pi}\sigma_{A\,|B}^B - \frac{1}{16\pi}\partial_A\theta - \ell^\mu T_{\mu A}, \qquad (2.66)$$

where

$$\frac{D\pi_A}{dt} = (\partial_t + \theta)\,\pi_A + v^B \pi_{A|B} + v_{|A}^B \pi_B,$$

$$\sigma_{AB} = \frac{1}{2}\left(\partial_t \gamma_{AB} + v_{A|B} + v_{B|A}\right) - \frac{1}{2}\theta\gamma_{AB}, \qquad (2.67)$$

$$\theta = \frac{\partial_t \sqrt{\gamma}}{\sqrt{\gamma}} + v_{|A}^A,$$

correspond to a convective derivative, a shear and an expansion rate respectively. Let us recall that the Navier–Stokes equation for a viscous fluid reads

$$(\partial_t + \theta)\,\pi_i + v^k \pi_{i,k} = -\frac{\partial}{\partial x^i}p + 2\eta\sigma_{i,k}^k + \zeta\theta_{,i} + f_i, \qquad (2.68)$$

where π_i is the momentum density, p the pressure, η the shear viscosity, $\sigma_{ij} = \frac{1}{2}(v_{i,j} + v_{j,i}) - $ Tr. the shear tensor, ζ the bulk viscosity, $\theta = v_{,i}^i$ the expansion rate, and f_i the external force density. The two equations are remarkably similar. This suggests that a BH can be viewed as a (non-relativistic[4]) brane with (positive[5]) surface pressure $p = +\frac{g}{8\pi}$, external force-density $f_A = -\ell^\mu T_{\mu A}$ which corresponds to the flow of external linear momentum, surface shear viscosity $\eta = +\frac{1}{16\pi}$, and surface bulk viscosity $\zeta = -\frac{1}{16\pi}$ (in units where $G = 1$). Note, finally, that both the surface shear viscosity and the surface bulk viscosity apply to any type of deformed non-stationary BH.

2.4. Irreversible thermodynamics of black holes

In previous sections, we have introduced some electrodynamic and fluid dynamical quantities associated to a kind of dissipative dynamics of BH horizons. In addition, following Bekenstein, we would like to endow a BH with a surface density of entropy equal to a dimensionless constant $\hat{\alpha}$ (in units where $\hbar = G = c = 1$). Any dissipative system verifying Ohm's law and the Navier–Stokes equation is also expected to satisfy corresponding thermodynamic dissipative equations, namely Joule's law and the usual expression of the viscous heat rate proportional

---

[4]The non-relativistic character of the BH hydrodynamical-like equations may seem surprising in view of the "ultra-relativistic" nature of a BH. This non-relativistic-looking character is due to our use of an adapted "light-cone frame" (ℓ, n, e_A). It is well-known that light-cone-gauge results have a distinct "non-relativistic" flavour.

[5]This is consistent with the idea that the BH surface pressure must counteract the self-gravity.

to the sum of the squares of the shear tensor and of the expansion rate. More precisely, we would expect to have a "heat production rate" in each surface element dA of the form

$$\dot{q} = dA\left[2\eta\sigma_{AB}\sigma^{AB} + \zeta\theta^2 + \rho(\vec{\mathcal{K}} - \sigma_H\vec{v})^2\right], \tag{2.69}$$

where ρ is the surface resistivity, and η and ζ the shear and bulk viscosities. In addition, one expects that this heat production rate should be associated with a corresponding local increase of the entropy $s = \hat{\alpha}dA$ contained in any local surface element of the form

$$\frac{ds}{dt} = \frac{\dot{q}}{T} \tag{2.70}$$

with a local temperature T expected to be equal to $T = \frac{g}{8\pi\hat{\alpha}}$.

Remarkably, the "scalar" $(\ell^\mu\ell^\nu)$ projection of Einstein's equations, (i.e., , the Raychauduri equation) yields an evolution law for the entropy $s = \hat{\alpha}dA$ of a local surface element which is very analogous to what one would expect. Indeed, it yields

$$\frac{ds}{dt} - \tau\frac{d^2s}{dt^2} = \frac{dA}{T}\left[2\eta\sigma_{AB}\sigma^{AB} + \zeta\theta^2 + \rho(\vec{\mathcal{K}} - \sigma_H\vec{v})^2\right], \tag{2.71}$$

where $T = \frac{g}{8\pi\hat{\alpha}}$, where $2\eta\sigma_{AB}\sigma^{AB} + \zeta\theta^2$ are exactly the expected viscous contributions, and where $\rho(\vec{\mathcal{K}} - \sigma_H\vec{v})^2$ is Joule's law.

The only unexpected term in this result is the second term on the l.h.s.; this term goes beyond usual near-equilibrium thermodynamics (which involves only the first order time derivative of the entropy), and is proportional to the second time derivative of the entropy density and to a time scale $\tau = \frac{1}{g}$. It is interesting to note that, for the value $\hat{\alpha} = 1/4$, corresponding to the Bekenstein–Hawking entropy density, τ is equal to $\frac{1}{2\pi T}$, i.e., the inverse of the lowest "Matsubara frequency" associated to the temperature T (one also notes that $\tau = D = 2\mathcal{D}$, where D, $\mathcal{D}$ are the diffusion constants of [14]). The minus sign in front of this new term is also a particularity of BH physics. In the approximation of a constant τ, and in solving for the non-equilibrium second law of thermodynamics, one finds that the rate of increase of entropy is given by

$$\frac{ds}{dt} = \int_t^\infty \frac{dt'}{\tau}\exp\left(-\frac{(t'-t)}{\tau}\right)\left(\frac{\dot{q}}{T}\right)(t'), \tag{2.72}$$

i.e., it is defined not as the value of the heat dissipated instantaneously, nor as an integral over the past heat dissipation, but as an integral over the *future*. This highlights the acausal nature of BHs, i.e., a BH is defined as a null hypersurface which *will become stationary in the far future*. As such, it has to anticipate any external perturbation. Failing this, the null hypersurface would generically tend either to collapse, or blow up toward ∞.

We also note that the ratio of the shear viscosity $\eta = 1/(16\pi)$ to the entropy density $\hat{s} = s/dA = \hat{\alpha}$ is given by

$$\frac{\eta}{\hat{s}} = \frac{1}{\hat{\alpha} \, 16\pi} = \frac{1}{4\pi}, \tag{2.73}$$

where, in the second equality, we have used the Bekenstein–Hawking value for $\hat{s} = s/dA = \hat{\alpha} = \frac{1}{4}$. It is interesting to note that the result $\frac{\eta}{\hat{s}} = \frac{1}{4\pi}$ is indeed the ratio found by Kovtun, Son and Starinets in the gravity duals of strongly coupled gauge theories [13, 14].

Finally, let us note another remarkable agreement between BH dissipative dynamics and a rather general property of ordinary (near-equilibrium) irreversible thermodynamics. This agreement concerns what Prigogine has called the "minimum entropy production principle". Let us consider the total "dissipation function"

$$D = \oint_{S} \dot{q} \tag{2.74}$$

as a functional of the velocity field $v^A(x^2, x^3)$ and of the electric potential $\phi(x^2, x^3)$ in the presence of given external influences such as an external magnetic field or tidal forces acting on the rotating BH. Then, $D[\phi]$ or $D[v^A]$ (imposing $v^A_{|A} = 0$ as a constraint), reach a *minimum* when (and only when) the lowest order (Einstein–Maxwell) dynamical equations for ϕ or v^A are satisfied.

2.5. Hawking radiation

In this section we discuss the phenomenon of Hawking radiation, first obtained in Ref. [17] (we shall follow here the derivation of Ref. [18]). For simplicity, we will consider a $3 + 1$ dimensional spherically symmetric BH. We remind the reader that the coefficient $A(r)$, associated to the time coordinate (here denoted as T), goes to zero on the horizon, so that the horizon is an *infinite redshift surface*. It is also a *Killing* horizon, i.e., the (suitably normalized) normal vector $\vec{\ell} = \partial/\partial t + \Omega \partial/\partial \phi$ is a Killing vector. These points will be crucial in the following. Since the coefficient of the radial coordinate is defined by $B(r) = \frac{1}{A(r)}$ (see Section 2.1), it is singular on the horizon. To get a good coordinate system on the horizon, we first factorize $A(r)$,

$$ds^2 = -A(r)\,dT^2 + \frac{dr^2}{A(r)} + r^2 d\Omega^2$$
$$= -A(r)\left(dT^2 - \left(\frac{dr}{A(r)}\right)^2\right) + r^2 d\Omega^2, \tag{2.75}$$

and then introduce a new radial coordinate, the so-called *tortoise* coordinate, defined by

$$r_* = \int \frac{dr}{A(r)}. \tag{2.76}$$

Note that as $r \to r_+$, $A(r) \simeq \left(\frac{\partial A}{\partial r}\right)_{r_+} (r - r_+)$, where $\left(\frac{\partial A}{\partial r}\right)_{r_+}$ is (as mentioned above) equal to twice the surface gravity g. This implies

$$r_* \simeq \int \frac{dr}{(r - r_+)\left(\frac{\partial A}{\partial r}\right)_{r_+}}$$
$$\simeq \frac{\ln(r - r_+)}{2g} \tag{2.77}$$

such that as $r \to +\infty$, $r_* \simeq r + 2M \ln r$ and as $r \to r_+$, $r_* \simeq \frac{\ln(r-r_+)}{2g}$. The line element can thus be re-written as

$$ds^2 = -A\,(dT - dr_*)(dT + dr_*) + r^2 d\Omega^2. \tag{2.78}$$

We now switch to the so-called *Eddington–Finkenstein* coordinates, (t, r, θ, φ), where the combination $t = T + r_*$ of T and r_* remains regular across the (future) horizon[6] and define $t = T + r_*$. The time translation Killing vector $\partial/\partial t$ coïncides with the usual one $\partial/\partial T$. In terms of these new coordinates the metric reads

$$ds^2 = -A(r)dt^2 + 2dt\,dr + r^2 d\Omega^2. \tag{2.79}$$

This metric is now regular (with a non-vanishing determinant) as the radial coordinate r penetrates within the horizon, i.e., becomes smaller than r_+.

Having defined a regular coordinate system, we now consider a massless scalar field, the dynamics of which is given by the massless Klein–Gordon equation

$$0 = \Box_g \varphi = \frac{1}{\sqrt{g}} \partial_\mu \left(\sqrt{g}\, g^{\mu\nu} \partial_\nu \varphi\right) \tag{2.80}$$

---

[6]Given that the horizon is an infinite redshift surface, it takes an infinite time T to fall into it, while $r_* \simeq \ln(r - r_+)/2g$ goes to $-\infty$ at the horizon. The sum of the two remains, however, finite.

The solutions to this equation in a spherically symmetric and time-independent background are given by mode functions which are themselves given simply by products of a Fourier decomposition into frequencies, spherical harmonics and a radial dependence and thus read

$$\varphi_{\omega\ell m}(T, r, \theta, \varphi) = \frac{e^{-i\omega T}}{\sqrt{2\pi|\omega|}} \frac{u_{\omega\ell m}(r)}{r} Y_{\ell m}(\theta, \varphi). \tag{2.81}$$

The problem is then reduced to solving a radial equation with the radial coordinate r_*,

$$\frac{\partial^2 u}{\partial r_*^2} + \left[\omega^2 - V_\ell(r(r_*))\right]u = 0. \tag{2.82}$$

In the case of the Schwarzschild metric (i.e., when $A(r) = 1 - 2M/r$) the effective radial potential $V_\ell(r)$ is given by

$$V_\ell(r) = \left(1 - \frac{2M}{r}\right)\left(\frac{\ell(\ell+1)}{r^2}\right). \tag{2.83}$$

An essential point to note is that the effective potential vanishes both at ∞ (like a massless centrifugal potential $\ell(\ell+1)/r^2$, *and* at the horizon (where it is proportional to $A(r)$). Therefore, in these two regimes (which correspond to $r_* \to \pm\infty$), the solution of the wave equation behaves essentially as in flat space (see Fig. 3). The effect of the coupling to curvature is non-negligible only in an intermediate region, where the combined effect of curvature and centrifugal effects yield a *positive* potential barrier. In turn, this potential barrier yields a *grey body factor* which diminishes the amplitude of the quantum modes considered below, i.e., those generated near the horizon and which must penetrate through the potential barrier on their way towards ∞. Far from the potential barrier, the general solution for φ is

$$\varphi_{\omega\ell m} \sim \frac{e^{-i\omega(T\pm r_*)}}{\sqrt{2\pi|\omega|}} \frac{1}{r} Y_{\ell m}(\theta, \varphi). \tag{2.84}$$

The quantification of the scalar field φ is rather standard (and similar to what one does when studying the amplification of quantum fluctuations during cosmological inflation). The quantum operator for the scalar field is decomposed into mode functions, i.e., the eigenfunctions of the Klein–Gordon equation, with coefficients given by creation and annihilation operators. The subtlety lies, however, in the definition of positive and negative frequencies.

Let us start by formally considering the simpler case of a quantum scalar field $\hat{\varphi}(x)$ in a background spacetime which becomes stationary *both* in the infinite

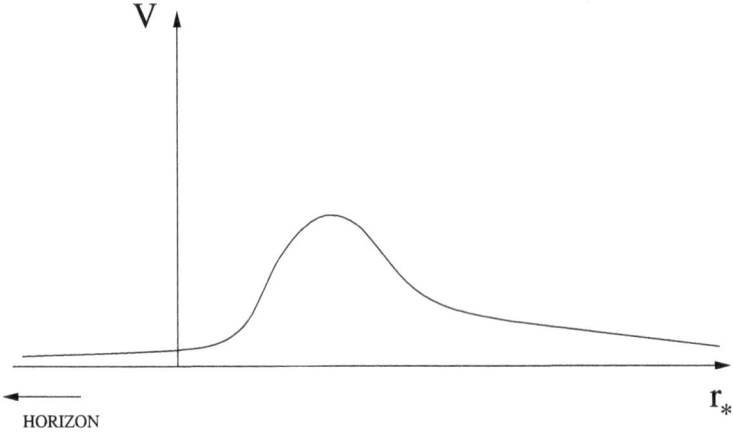

Fig. 3. This figure is a schematic representation of the effective gravitational potential in the neighborhood of a BH. Note that as far as the particles are concerned, the spacetime is essentially flat both at infinity and near the horizon. The tidal-centrifugal barrier that separates the horizon from infinity gives rise to the grey body factor.

past, and in the infinite future. In that case, one can define positive and negative frequencies in the usual way, in the two asymptotic regions $t \to \pm\infty$. The coefficient of the positive frequency modes, say $p(x)$, then defines an annihilation operator $\hat{a}$. However, there are two sorts of positive-frequency $(p(x))$, and negative-frequency $(n(x))$ modes. The *in* ones p_i^{in}, n_i^{in} (defined in the asymptotic region $t \to -\infty$, and then extended everywhere by solving the Klein–Gordon equation), and the *out* ones p_i^{out}, n_i^{out} (defined in the asymptotic region $t \to +\infty$). The operator-valued coefficients of these modes define some corresponding *in* and *out* annihilation or creation operators, so that we can write the field operator $\hat{\varphi}(x)$ as (here i is a label that runs over a basis of modes)

$$
\begin{aligned}
\hat{\varphi}(x) &= \sum_i \hat{a}_i^{in} p_i^{in}(x) + \left(\hat{a}_i^{in}\right)^\dagger n_i^{in}(x), \\
&= \sum_i \hat{a}_i^{out} p_i^{out}(x) + \left(\hat{a}_i^{out}\right)^\dagger n_i^{out}(x).
\end{aligned}
\tag{2.85}
$$

Here, the modes are normalized as $(p_i, p_j) = \delta_{ij}$ and $(n_i, n_j) = -\delta_{ij}$, where '$(\ ,\ )$' is the Klein–Gordon scalar product, i.e., $\sim i \int d\sigma^\mu (\varphi_1^* \partial_\mu \varphi_2 - \partial_\mu \varphi_1^* \varphi_2)$. The operators a_i and $a_j^\dagger$ are correspondingly normalized in the usual way as $[a_i, a_j^\dagger] = \delta_{ij}$. One defines both an *in vacuum* $|in\rangle$ and an *out vacuum* $|out\rangle$ as the states that are respectively annihilated by a_i^{in} or a_i^{out}. Then, the phenomenon of particle creation corresponds to the fact that the *out* vacuum differs from the *in* one. More quantitatively, the expectation value of the number of *out* particles, in the mode labelled by i, which will be observed when the quantum field is in the *in* vacuum state is given by

$$\langle N_i \rangle = \langle in | \left(a_i^{out} \right)^\dagger a_i^{out} | in \rangle$$
$$= \sum_j |T_{ij}|^2, \tag{2.86}$$

where we have introduced the transition amplitude $T_{ij} = (p_i^{out}, n_j^{in})$ from an initial negative frequency mode n_j^{in} into a final positive frequency one p_i^{out}. These transition amplitudes (also called Bogoliubov coefficients) enter the calculation because, as is easily deduced from the double expansion of the field $\hat{\varphi}(x)$ above, they give the part of a_i^{out} which is proportional to $(a_j^{in})^\dagger$. The application of the previous general formalism to the BH case is delicate since a BH background is not asymptotically stationary in the infinite future (because of the BH interior where the Killing vector $\partial/\partial t$ is spacelike), and is asymptotically stationary in the infinite past only if we do not consider explicitly the collapse leading to the formation of a BH from an initially stationary star. However, Hawking showed how to essentially bypass these difficulties by focussing on two types of modes:

• The high-frequency modes coming from the infinite past, which reach the horizon with practically no changes (because of their high-frequency nature) and,

• the outgoing modes, viewed in the asymptotically flat region and in the far future.

Concerning the outgoing modes, they can be unambiguously decomposed in positive- and negative-frequency parts, because, as explained above, their asymptotic behaviour is given by a sum of essentially flat-spacetime modes, (2.84). One then defines the outgoing p_i^{out}'s as being proportional to $e^{-i\omega(T-r_*)}$ with a *positive ω*.

Let us now focus on the definition of positive- and negative-frequency modes near the horizon. We recall that, as mentioned at the beginning of this section, there is a physically infinite redshift between the surface of the horizon and asymptotically flat space at infinity. If one is interested in particle creation with a finite given frequency, as observed at infinity, the corresponding wave packets will have very high frequency near the horizon and can therefore be approximated by very localized wave packets. Given that the spacetime geometry in the vicinity of the horizon is regular, with a finite radius of curvature, it can be regarded as a piece of flat spacetime locally if one looks in a small enough region. In this approximation, the calculation can be performed in a single step.

We wish to compute the average number of final outgoing particles[7] seen in the *in* vacuum. Then the average number of outgoing particles of type p_i^{out} is given by $\sum_j |T_{ij}|^2$, where $T_{ij} = (p_i^{out}, n_j^{in})$ is the transition amplitude from an initial negative frequency mode n_j^{in} into a final *outgoing* positive frequency one p_i^{out} (recorded at spatial infinity). To compute this transition amplitude, we need to describe what is an initial negative frequency mode n_j^{in}. As said above, Hawking suggested that only high-frequency initial modes are important, and that they essentially look the same (some kind of WKB wave) in the real *in* region (in the far past, before the formation of the BH) as in the vicinity of the horizon. Our technical problem is then reduced to characterizing the negative frequency modes n_j^{in} as seen in a small neighborhood of the horizon, which looks like the Minkowski vacuum.

To do this, it is convenient to have a technical criterion for characterizing positive and negative frequency modes in (a local) Minkowski spacetime. Locally, one can perform a Fourier decomposition of the wave packet and use the mathematical fact that the Fourier space properties are mapped onto analytic continuation properties in x-space. This relation can then be used to define positive and negative frequency modes. This is easy to see. Consider a general *negative frequency* wave packet in Minkowski spacetime. It has the form $\varphi_-(x) = \oint_{C^-} d^4k \tilde{\varphi}(k) e^{ik_\mu x^\mu}$ where k_μ is *timelike-or-null* and *past-directed*, i.e., $k^\mu \in C^-$. We now perform a complex shift of the spacetime coordinate, $x_\mu \to x^\mu + i y^\mu$, where y^μ is timelike-or-null and *future-directed* (i.e., $y^\mu \in C^+$), then, the $e^{ik_\mu x^\mu}$ term will be suppressed by a $e^{-k_\mu y^\mu}$ term, where the scalar product $k_\mu y^\mu$ is *positive* because it involves two timelike vectors that point in opposite directions (we use the "mostly plus" signature). This ensures that a negative-frequency wavepacket can indeed be analytically continued to complex spacetime points of the form $x^\mu + i y^\mu$, with $y^\mu \in C^+$.

The strategy for applying this criterion to characterizing negative-frequency modes n_j^{in} in the vicinity of the horizon is then the following. One starts from a wavepacket which is not purely a "negative-frequency" one near the horizon, but which has the property of evolving into an *outgoing* positive-frequency wave packet (so that it will have a non-zero transition amplitude to some p_i^{out}).

---

[7]Note that the "out" label, in the general discussion of particle creation above, referred to "final" particles (as defined in the final, asymptotic, stationary spacetime background). In the case of a BH background, the "final" spacetime is made of two separate asymptotic regions: (i) the outgoing wave region at spatial infinity, and (ii) the vicinity of the (spacelike?) singularity within the BH. The definition of positive- and negative-frequency modes in the latter region is ill-defined. However, luckily, the calculation of the physically relevant flux of final, outgoing modes can be performed without worrying about the physics near the BH interior singularity. In other words, it is enough to consider as "out" positive-frequency modes p_i^{out} only the ones outgoing at spatial infinity (i.e., on "scri$^+$"), though they do not constitute a complete basis of final modes.

Then, one modifies the initial wavepacket so that it becomes a purely negative-frequency mode n_j^{in} near the horizon.

When looking at a wavepacket of the form of $\varphi_\omega^{out}(t, r) \propto e^{-i\omega(T-r_*)}$ (with positive ω) just outside the horizon, one must first switch to well-defined coordinates to examine its physical content. We therefore replace the Schwarzschild-type time coordinate T by the Eddington–Finkenstein time coordinate, $t = T+r_*$ which is regular on the horizon. After rearranging terms according to $T - r_* = (T + r_*) - 2r_*$ the previous (outgoing, positive-frequency) wave reads

$$
\begin{aligned}
\left[\varphi_\omega^{out}(t, r)\right]_{r_+} \propto e^{-i\omega(T-r_*)} &= e^{-i\omega t} e^{2i\omega r_*} \\
&= e^{-i\omega t} e^{i\frac{\omega}{g}\ln(r-r_+)} \\
&= e^{-i\omega t} (r - r_+)^{\frac{i\omega}{g}}.
\end{aligned}
\tag{2.87}
$$

This describes the behaviour, just outside the horizon, of a wavepacket which will become (modulo some grey-body factor) an outgoing positive frequency wavepacket at ∞. However, locally on the horizon, it is neither a positive nor a negative frequency wavepacket because, at this stage, it is defined only outside the horizon, but not inside. Let us now show how one must continue this wavepacket inside the horizon, so that it becomes a genuine negative-frequency wavepacket "straddling" the horizon. Using the criterion explained above, we can "continue" the wavepacket inside the horizon by a suitable *analytic continuation*. More precisely, we need an analytic continuation of the form $x^\mu \to x^\mu + iy^\mu$, where y^μ belongs to the future lightcone to ensure that we will then be dealing with a local negative frequency wavepacket. It is easy to see, from a spacetime diagram of the lightcone on the horizon, that the vector $\partial/\partial r$ is everywhere null and past-directed, such that $r \to r - \varepsilon$, where $\varepsilon > 0$, is everywhere null and future directed. The analytic continuation of $\varphi_\omega^{out}(t, r)$ to $r \to r - i\varepsilon$ will therefore define a good local negative-frequency mode $n_j^{in} = n_{\omega\ell m}^{in}$. One easily sees that this analytic continuation in r generates a new component to the wavepacket which is located inside the BH (i.e., for $r < r_+$). More precisely, a short calculation yields

$$
\begin{aligned}
n_{\omega\ell m}^{in}(r, t) &= N_\omega \varphi_\omega^{out}(t, r - i\varepsilon) \\
&= N_\omega \left[\theta(r - r_+) \varphi_\omega^{out}(r - r_+)\right. \\
&\quad \left. + e^{\frac{\pi\omega}{g}} \theta(r_+ - r) \varphi_\omega^{out}(r_+ - r)\right],
\end{aligned}
\tag{2.88}
$$

where the second term is the wavefunction inside the horizon that has acquired an additional exponential factor due to the rotation $e^{-i\pi}$ in the complex plane from $r > r_+$ to $r < r_+$. The overall factor N_ω is a normalization factor (needed because we have extended the mode inside the BH), such that

$$\langle n^{in}_{\omega\ell m}(r,t)\, n^{in}_{\omega'\ell'm'}(r,t)\rangle = \delta(\omega-\omega')\delta_{\ell\ell'}\delta_{mm'}, \tag{2.89}$$

from which we obtain (when remembering that φ^{out}_ω was correctly normalized)

$$|N_{\omega\ell m}|^2 = \frac{1}{e^{2\pi\omega/g}-1}. \tag{2.90}$$

The physical meaning of equation (2.88) is the description of the splitting of the in mode $n^{in}_j = n^{in}_{\omega\ell m}$ into a positive-frequency wave outgoing from the horizon and a wave falling from the horizon towards the singularity. One can read off from it the needed transition amplitude $T_{ij} = (p^{out}_i, n^{in}_j)$. It is essentially given by the factor N_ω, which must, however, be corrected by a grey-body factor $\sqrt{\Gamma_\ell(\omega)}$ taking into account the attenuation of the outgoing wave $e^{-i\omega(\bar{r}-r_*)}$ as it crosses the curvature + centrifugal potential barrier $V_\ell(r)$ on its way from the horizon to ∞. Then (using Fermi's golden rule), one easily finds that the general result (2.86) yields a rate of particle creation given by

$$\frac{d\langle N\rangle}{dt} = \sum_{\ell,m}\int \frac{d\omega}{2\pi}\frac{\Gamma_\ell(\omega)}{e^{\frac{2\pi\omega}{g}}-1}. \tag{2.91}$$

One recognizes here a thermal (Planck) spectrum (corrected by a grey-body factor). From the Planck factor, one reads off the Hawking temperature, $T = \hbar\frac{g}{2\pi}$. This result fixes the dimensionless coefficient $\hat{\alpha}$ in the Bekenstein entropy to the famous result $\hat{\alpha} = \frac{1}{4}$, i.e.,

$$S_{BH} = \frac{A}{4G\hbar}. \tag{2.92}$$

Let us end by two final comments. First, the generalisation of the Hawking radiation to a more general rotating and/or charged BH is given essentially by replacing in the result above the frequency ω by $\omega - \omega_0$ where $\omega_0 = m\Omega + eV$ exhibits the couplings of the created particles to the angular velocity Ω and the electric potential V of the BH. Then, in astrophysically realistic conditions, the "Hawking" part of the particle creation (i.e., the thermal aspect) is too small to be relevant, while the combined effect of the grey body factor and of the zero-temperature limit of $(e^{\frac{2\pi(\omega-\omega_0)}{g}} - 1)^{-1}$ yield potentially relevant particle creation phenomena in Kerr–Newman BHs, associated to the "superradiance" of modes with frequencies $\mu < \omega < \omega_0$, where μ is the mass of the created particle (see, e.g., [18] for more details and references). We conclude by noting that the situation just described is not only technically similar to the one in the inflationary scenario for cosmological perturbations, but also physically similar in that, in both cases, *transplanckian frequency modes* in the ultraviolet are redshifted to a finite, observable frequency.

3. Experimental tests of gravity

Before discussing various possibilities of string-inspired phenomenology (and of possible string-inspired deviations from Einstein's theory of General Relativity) we give an overview of what is known experimentally about the gravitational sector.

3.1. Universal coupling of matter to gravity

The standard model of gravity is Einstein's General Relativity (GR). In GR, all fields of the standard model of particle physics (SM) are universally coupled to gravity by replacing the flat spacetime metric $\eta_{\mu\nu}$ by a curved spacetime one $g_{\mu\nu}$. In "standard GR" one also assumes that gravity is the only long range coupling (apart from electromagnetism). We shall see below, how the presence of other long range interactions (coupled to bulk matter) modify the usual "pure GR" phenomenology. The action for the matter sector, S_{SM}, has the structure

$$
\begin{aligned}
S_{SM} = \int d^4x \bigg[& -\frac{1}{4} \sum \sqrt{g}\, g^{\mu\alpha} g^{\nu\beta} F^a_{\mu\nu} F^a_{\alpha\beta} - \sum \sqrt{g}\, \bar{\psi} \gamma^\mu D_\mu \psi \\
& -\frac{1}{2}\sqrt{g}\, g^{\mu\nu} \overline{D_\mu H} D_\nu H - \sqrt{g}\, V(H) \\
& -\sum \lambda \sqrt{g}\, \bar{\psi} H \psi - \sqrt{g}\, \rho_{\text{vac}} \bigg],
\end{aligned}
\tag{3.1}
$$

where D denotes a (gauge and gravity) covariant derivative, while the dynamics of $g_{\mu\nu}$ is described by the Einstein–Hilbert action, S_{EH},

$$
S_{EH} = \int d^4x \frac{c^4}{16\pi G} \sqrt{g}\, g^{\mu\nu} R_{\mu\nu}(g).
\tag{3.2}
$$

The total action is therefore given by

$$
S = S_{EH}[g_{\mu\nu}] + S_{SM}[\psi, A_\mu, H; g_{\mu\nu}],
\tag{3.3}
$$

and its variation w.r.t. $g_{\mu\nu}$ yields the well-known Einstein field equations

$$
R_{\mu\nu} - \frac{1}{2} R\, g_{\mu\nu} = \frac{8\pi G}{c^4} T_{\mu\nu},
\tag{3.4}
$$

where $T^{\mu\nu} = \frac{2}{\sqrt{g}} \frac{\delta \mathcal{L}_{SM}}{\delta g_{\mu\nu}}$. The universal coupling of any type of particle to $g_{\mu\nu}$ is made manifest in S_{SM} while S_{EH} contains all the information on the propagation of gravity. For instance, expanding S_{EH} in powers of $h_{\mu\nu}$ (where $g_{\mu\nu} \equiv \eta_{\mu\nu} + h_{\mu\nu}$), one obtains, at quadratic order in $h_{\mu\nu}$, the spin-two Pauli–Fierz Lagrangian. Higher orders in $h_{\mu\nu}$ contain an infinite series of nonlinear self-couplings of gravity: $\partial\partial hhh$, $\partial\partial hhhh$, etc. As we shall see, this nonlinear structure has been verified experimentally to high accuracy (both in the weak-field regime, where the cubic vertex $\partial\partial hhh$ has been checked, and in the strong-field regime of binary pulsars, where the fully nonlinear GR dynamics has been confirmed). In the following, we discuss, successively, (i) the experimental tests of the coupling of matter to gravity, and (ii) the tests of the dynamics of the gravitational field: kinetic terms (describing the propagation of gravity), and cubic and higher gravitational vertices.

The universal nature of matter's coupling to gravity, i.e., the coupling of matter to a universal deformation of spacetime, has many experimental consequences. These experimental consequences can be derived by using a simple theorem by Fermi and Cartan. Given any pseudo-Riemannian manifold, for instance a curved spacetime endowed with a metric $g_{\mu\nu}$, and given any worldline $\mathcal{L}$ in this spacetime (not assumed to be a geodesic), there always exists a coordinate system such that, all along $\mathcal{L}$, $g_{\mu\nu}(x^\lambda) = \eta_{\mu\nu} + \mathcal{O}(\vec{x}^2)$, where $\vec{x}$ denotes the *spatial* deviation away from the central worldline $\mathcal{L}$. It is important to note that there is no linear term in $\vec{x}$, but only $\vec{x}^2$ effects, i.e., tidal effects. There exists a very simple and intuitive demonstration of this Fermi–Cartan theorem. Let us view the curved manifold as being some "brane" embedded within a *flat* ambient auxiliary manifold. For instance, consider an ordinary 2-surface Σ within a three-dimensional flat euclidean space. Given any (smooth) curve $\mathcal{L}$ traced on Σ, we can take a flat sheet of paper and progressively "apply" (or "fit") this sheet on Σ along the curve $\mathcal{L}$. The orthogonal projection of Σ onto this applied flat sheet defines a map from Σ to a coordinatized flat manifold which has the property enunciated above. Note that, in this "development" of the neighbourhood of $\mathcal{L}$ within Σ onto a flat sheet, the shape (as seen on the flat sheet) of the "developped" curve $\mathcal{L}$ is generically not a straight line. It is only when $\mathcal{L}$ was a *geodesic* line on Σ, that its development will be a straight line. This proof, and its consequences, are valid in any dimension and signature.

Here, we have in mind applying this result to the "center of mass" worldline $\mathcal{L}$ of an arbitrary body moving in a background spacetime, or more generally of any sufficiently small laboratory (containing several bodies, between which we can neglect gravitational effects). We assume that we can neglect the backreaction of the body (or bodies) on the spacetime. In the approximation where we can neglect the tidal effects (linked to the $\mathcal{O}(\vec{x}^2)$ terms in $g_{\mu\nu}$ in Fermi coordinates), we can consider that we have a body, or a small lab, moving in a flat spacetime. In other words, the theorem of Fermi and Cartan tells us that we can essentially "efface" (modulo small, controllable tidal effects) the background gravitational field $g_{\mu\nu}$ all along the history of a small lab, or a body. This "effacement property" is telling us, for instance, that the physical properties we can measure in a small lab will be independent of where the lab is, and when the measurements are made. In particular, all the (dimensionless) coupling constants[8] that enter the interpretation of local experiments (such as various mass ratios, the fine-structure constant, etc.) must be independent of where and when they are (locally) measured (*constancy of the constants*). A second consequence of this effacement property is that local physics should be Lorentz SO(3,1) invariant, because this is a symmetry of the (approximate) flat spacetime appearing after one has effaced the tidal effects (*local Lorentz invariance*).

Moreover, in absence of coupling to other long-range fields (such as electromagnetism for a charged body), the center of mass of an isolated body (viewed as moving in a flat spacetime) must follow a straight worldline (principle of inertia). We therefore conclude (by the theorem above) that $\mathcal{L}$ has to be a geodesic in the original curved spacetime. This is true independently of the internal properties of the object. One may thus conclude that isolated neutral bodies fall along geodesics independently of the internal properties of the object, since at no point in the demonstration had we to rely on any internal properties of the object. This is therefore a proof of the *weak equivalence principle*, i.e., all bodies in a gravitational field fall with the same acceleration. Note, once again, that the absence of other long range fields besides $g_{\mu\nu}$ that could influence the object considered is crucial.

Finally, another universality property, that of the gravitational redshift, may be shown by a comparison of the GR formulation with the Newtonian one. In lowest order approximation, the deviation of the g_{00} component of the metric from $\eta_{00} = -1$ is twice the Newtonian potential $U(x)$. Indeed, comparing the action for a geodesic,

$$S_E = -m \int dt \sqrt{-g_{\mu\nu}\dot{x}^\mu\dot{x}^\nu} \tag{3.5}$$

with

$$S_{\text{Newton}} = \int dt \left[\frac{1}{2}m\dot{x}^2 + mU(x) \right], \tag{3.6}$$

one finds $g_{00} = -1 + \frac{2}{c^2}U(x) + \mathcal{O}(\frac{1}{c^4})$ where $U = \sum_a \frac{Gm_a}{|\vec{x}-\vec{x}_a|}$.

---

[8] We assume here that the cutoff length scale $\epsilon = 1/\Lambda$ of any low-energy effective QFT description of the physics in a small lab is fixed, when measured in units of $ds = \sqrt{g_{\mu\nu}dx^\mu dx^\nu}$.

Experimentally, one may transfer electromagnetic signals from one clock to another identical clock located in a gravitational field. If we are in a stationary situation (i.e., if there exists a coordinate system w.r.t. which the physics is independent of time $x^0 = ct$), the time translation invariance of the background shows that electromagnetic signals will take a *constant coordinate time* to propagate from clock 1 to clock 2. We can then use the link $d\tau_i = \sqrt{-g_{00}(\vec{x}_i)}dt_i$ between the proper time (at the location of clock i ($i = 1, 2$)) and the corresponding coordinate time, as well as the (approximate) result above for g_{00}. Finally, we conclude that two identically constructed clocks located at two different positions in a static external Newtonian potential exhibit, when intercompared by electromagnetic signals, the (apparent) difference in clock rate

$$\frac{\tau_1}{\tau_2} = \frac{\nu_2}{\nu_1} = 1 + \frac{1}{c^2}\left[U(\vec{x}_1) - U(\vec{x}_2)\right] + \mathcal{O}\left(\frac{1}{c^4}\right). \qquad (3.7)$$

This gravitational redshift effect is proportional to the difference in the Newtonian potential between the two locations, independently of the constitution of the clocks (say Hydrogen maser, or Cesium clock, etc.). This is a property known as the *universality of the gravitational redshift*.

The various consequences, discussed above, of the universal character of the coupling of matter to gravity are usually summarized under the generic name of *equivalence principle*. In the next section, we discuss the experimental tests of the equivalence principle and their accuracy.

3.2. Experimental tests of the coupling of matter to gravity

3.2.1. How constant are the constants?

The best tests of the "constancy of the constants" concern the fine structure constant $\alpha = e^2/\hbar c \simeq 1/137.037$, and the ratio of the electron mass to that of the proton $\frac{m_e}{m_p}$ (see Ref. [21] for a review). There exist several types of tests, based, for instance, on geological data (e.g., measurements made on the nuclear decay products of old meteorites), or on measurements (of astronomical origin) of the fine structure of absorption and emission spectra of distant atoms, as, e.g., the absorption lines of atoms on the line-of-sight of quasars at high redshift. Such kinds of tests all depend on the value of α. There exist, in addition, several laboratory tests such as, for example, comparisons made between several different high-stability clocks. However, the best measurement of the constancy of α to date is the Oklo phenomenon.[9] It sets the following (conservative) limits on the

---

[9]The Oklo phenomenon was discovered by scientists at the *Commissariat à l'énergie atomique* (CEA) in France. A study of the uranium ore in a Gabonese mine revealed an unusual depletion in U^{235} (used in fission reactors) w.r.t. the usual proportion. Uranium ore is a mix of two isotopes, with, in usual samples, 99.28% U^{238} and 0.72% U^{235}. By contrast, the Oklo ore had only $\ll 0.72\%$

variation of α over a period of two billion years [22–24]

$$-0.9 \times 10^{-7} < \frac{\alpha^{\text{Oklo}} - \alpha^{\text{today}}}{\alpha^{\text{today}}} < 1.2 \times 10^{-7}. \tag{3.8}$$

Converting this result into an average time variation, one finds

$$-6.7 \times 10^{-17}\,\text{yr}^{-1} < \frac{\dot{\alpha}}{\alpha} < 5 \times 10^{-17}\,\text{yr}^{-1}. \tag{3.9}$$

Note that this variation is a factor of $\sim 10^7$ smaller than the Hubble scale, which is itself $\sim 10^{-10}\,\text{yr}^{-1}$. Comparably stringent limits were obtained using the Rhenium 187 to Osmium 187 ratio in meteorites [25] yielding an upper bound $\frac{\Delta\alpha}{\alpha} = (8 \pm 8) \times 10^{-7}$ over 4.6×10^9 years. Laboratory limits were also obtained from the comparison, over time, of stable atomic clocks. More precisely, given that $\frac{v}{c} \sim \alpha$ for electrons in the first Bohr orbit, direct measurements of the variation of α over time can be made by comparing the frequencies of atomic clocks that rely on different atomic transitions. The upper bound on the variation of α using such methods is $\frac{\dot{\alpha}}{\alpha} = (-0.9 \pm 2.9) \times 10^{-15}\,\text{yr}^{-1}$ [26]. It should be mentioned that a few years ago claims were made concerning observational evidence of non-zero time variations of α and $\frac{m_e}{m_p}$ from analyses of some astronomical spectra (see Ref. [21]). Other recent astronomical data indicate no variability of these constants (see Ref. [21] and chapter 18 of the Review of Particle Physics[10] for references).

3.2.2. *Tests of local Lorentz invariance*

We should first mention that the Michelson-Morley experiment[11] has been repeated (with high accuracy) and strong limits have been obtained on a possible anisotropy of the propagation of light. In its modern realizations (Brillet and Hall, 1979), it has been performed with laser technology on rotating platforms. This experiment is now viewed as a test of the isotropy of space on the moving Earth, and thereby as a test of local Lorentz invariance. There also exists another idea for testing the isotropy of space, and although its interpretation is not totally

---

of U^{235}. It was realized that a natural fission process took place, prompted by the presence of ground water, in Oklo some two billion years ago, and lasted for about two million years. Scientists analysed in detail the 2 billion year-old fission decay products. One can then infer from these measurements the scattering cross-sections of slow neutrons on various isotopes. Then, modulo some further assumptions about the dependence of various nuclear quantities on α, one could constrain the variation of α between the time of the fission reaction (roughly two billion years ago) and now. For details about the analysis and interpretation of Oklo data see [22] and references therein.

[10] Available on http://pdg.lbl.gov/

[11] First performed as part of a series of experiments, beginning in Potsdam in 1881 (by Michelson alone) and then in the US until 1887 (by both Michelson and Morley) to test the existence of the *aether*.

clear, it is a conceptually interesting idea. This is why we choose to outline it in these lectures.

For simplicity, consider the hydrogen atom. Assuming the isotropy of space, i.e., the existence of a SO(3) symmetry, we know that there should exist a degeneracy in the energy levels, given by the magnetic quantum number m. However, it is interesting to understand how the SO(3) symmetry comes about dynamically (and therefore, how it might be dynamically violated). The Hamiltonian for the electron is given by

$$\hat{H} = -\frac{\hbar^2}{2m}\Delta - \frac{e^2}{r}, \tag{3.10}$$

where the first term is the kinetic term (Δ being the Laplacian), and the second term is the Coulomb potential. Note that in fact, $\Delta = \delta^{ij}\partial_{ij}$ and $r^2 = \delta_{ij}x^i x^j$, such that both terms depend on the same spatial structure δ_{ij}, the flat metric, thereby ensuring the SO(3) symmetry. However, both terms also come from an underlying field theoretic formulation: (i) the non-relativistic electron kinetic energy term $\propto \Delta = \delta^{ij}\partial_{ij}$ comes from the kinetic term in the Dirac action, $\bar{\psi}\gamma^\mu\partial_\mu\psi - m\bar{\psi}\psi$, with $\{\gamma^\mu, \gamma^\nu\} = \eta_{\mu\nu}$, while (ii) the e^2/r term is the static Green's function of the electromagnetic field, which comes from inverting the kinetic term of the photon $\eta^{\alpha\mu}\eta^{\beta\nu}F_{\alpha\beta}F_{\mu\nu}$, which manifestly depends, by assumption, on the same spacetime metric $\eta_{\mu\nu}$. Einstein assumed that, in order to take into account the coupling to gravity, it was sufficient to replace $\eta_{\mu\nu}$ by the same $g_{\mu\nu}$ both for the electron and the photon. By contrast, let us consider the possibility that electrons ("matter") and photons ("electromagnetism") have a different coupling to gravity, e.g., described by saying that they couple to two different (spatial) metrics, say

$$g_{ij}^{\text{matter}} = \delta_{ij},$$
$$g_{ij}^{\text{em}} = \delta_{ij} + h_{ij}. \tag{3.11}$$

Then, computing the new propagators for the electron and the photon in their respective metrics, one finds that the SO(3) symmetry would be violated by tensor terms, appearing in the Hamiltonian, of the form $\delta H \sim \frac{e^2}{2}h_{ij}\frac{x^i x^j}{r^3}$. This is a violation, at a deep level, of the universality discussed in the previous section. The usual SO(3) symmetry implies that all energy levels with magnetic quantum number m are degenerate. But if tensor terms violating SO(3) were to exist, then, observables effects would include potentially measurable quadrupole-type splittings in the energy levels, which, applied to the atomic nucleus (whose energy levels are a more sensitive probe of anisotropy), are $\propto \langle I\,M|\hat{Q}_{ij}|I\,M\rangle$, where I and M are the nuclear spin quantum numbers, and where $\hat{Q}_{ij}$ is a symmetric tracefree tensor operator that couples to the tracefree part of h_{ij}. Such types of measurements have been performed on the energy levels of nuclei with impressively high accuracy, the current upper bound being

$$\left| h_{ij} - \frac{h_{kk}}{3} \delta_{ij} \right| \le 10^{-27}. \tag{3.12}$$

The universality of space is thus valid to one part in 10^{27}, showing how delicate Einstein's postulate is.

3.2.3. Universality of free fall

The most recent limits on the deviation from the universality of free fall have been obtained by Eric's Adelberger's group [27]. In particular, they compared the acceleration of a Beryllium mass and a Copper one in the Earth's gravitational field and found

$$\left(\frac{\Delta a}{a} \right)_{Be-Cu} = (-1.9 \pm 2.5) \times 10^{-12}, \tag{3.13}$$

where $\Delta a = a_{Be} - a_{Cu}$. Other limits exist, such as, for instance, the fractional difference in acceleration of earth-core-like ($\sim$ iron) and moon-mantle-like (silica) bodies,

$$\left(\frac{\Delta a}{a} \right)_{Earth-core-Moon-mantle} = (3.6 \pm 5.0) \times 10^{-13}. \tag{3.14}$$

There are also excellent limits concerning celestial bodies. In particular the possible difference in the accelerations of the Earth and the Moon towards the Sun have been measured using laser ranging (with 5 mm accuracy) with retro-reflectors (corner cubes) placed on the Moon, giving the result [28]

$$\left(\frac{\Delta a}{a} \right)_{Earth-Moon} = (-1.0 \pm 1.4) \times 10^{-13}. \tag{3.15}$$

One should, however, remember that only a fraction ($\sim 1/3$) of the Earth mass is made of iron, while the rest is mostly silica (which is the main material the Moon is made of). As, independently of the equivalence principle, silica must fall like silica, one looses a factor 3, so that the resulting bound on a possible violation of the equivalence principle is only around the 5×10^{-13} level, which is comparable to laboratory bounds.

3.2.4. Universality of the gravitational redshift

We conclude the section on the tests of the coupling of matter to gravity by just mentioning that the universality of the gravitational redshift, namely the apparent change in the frequencies of two similar clocks in a gravitational field, has been tested by comparing the frequencies of hydrogen masers at the Earth surface and in a rocket. Vessot and Levine (1979) in Ref. [29] verified that the fractional change in the measured frequencies is consistent with GR to the 10^{-4} level:

$$\frac{\Delta \nu}{\nu} = \left(1 \pm 10^{-4}\right) \frac{\Delta U}{c^2}. \tag{3.16}$$

The universality of this redshift has also been verified by measurements involving other types of clocks.

3.3. Tests of the dynamics of the gravitational field

3.3.1. Brief review of the theoretical background

Until now we have only considered the coupling between matter and gravity, and various tests of its universality. We now discuss the tests of the *dynamics of the gravitational field*, i.e., tests probing either the propagator of the gravitational field, or the cubic or higher order gravitational vertices (for more detailed reviews see Refs. [30, 31]). We first consider the weak field regime, regime in which we can write $g_{\mu\nu} = \eta_{\mu\nu} + h_{\mu\nu}$, where $h_{\mu\nu}$ is numerically much smaller than one. For instance, in the solar system, $h_{\mu\nu} \sim 10^{-6}$ on the surface of the Sun, $\sim 10^{-8}$ on the Earth orbit around the Sun, or $\sim 10^{-9}$ on the Earth surface. With values so small, it is clear that the solar system will not allow one to test many nonlinear terms in the perturbative expansion of $g_{\mu\nu}$.

We start by considering the gravitational interaction between two particles of masses m_A and m_B. At linear order in $h_{\mu\nu}$, we will have an interaction corresponding to the following (classical Feynman-like) graph

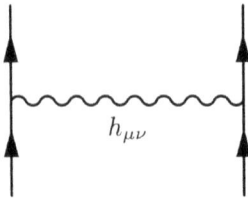

To compute explicitly what the preceding graph means, we must start from the full action describing two gravitationally interacting bodies A and B:

$$S = -m_A \int \mathrm{d}s_A - m_B \int \mathrm{d}s_B + \int \frac{\sqrt{g}\,R}{16\pi\,G}. \tag{3.17}$$

Expanding S in the deviations of $g_{\mu\nu}$ away from $\eta_{\mu\nu}$, one obtains (denoting $h \equiv \eta^{\mu\nu} h_{\mu\nu}$)

$$
\begin{aligned}
S = &-m_A \int \sqrt{-\eta_{\mu\nu} dx_A^\mu dx_A^\nu} - m_B \int \sqrt{-\eta_{\mu\nu} dx_B^\mu dx_B^\nu} \\
&+ \frac{1}{2} \int h_{\mu\nu} T_A^{\mu\nu} + \frac{1}{2} \int h_{\mu\nu} T_B^{\mu\nu} \\
&+ \int \frac{1}{32\pi G} h^{\mu\nu} \Box \left(h_{\mu\nu} - \frac{1}{2} h \eta_{\mu\nu} \right) + \mathcal{O}(h^2 T) + \mathcal{O}(h^3),
\end{aligned}
\tag{3.18}
$$

where $T_A^{\mu\nu}$ is the (flat-space limit) of the stress-energy tensor of particle A (given by a δ-function localized on the worldline of A), and where the kinetic term of $h_{\mu\nu}$ is the one corresponding to the harmonic gauge (i.e., $\partial_\nu(\sqrt{g} g^{\mu\nu}) = 0$). Inverting this kinetic term yields for $h_{\mu\nu}$ the following lowest-order equation (corresponding to Einstein's equations at linearized order)

$$
\Box h_{\mu\nu} = -\frac{16\pi G}{c^4} \left(T_{\mu\nu} - \frac{1}{D-2} T \eta_{\mu\nu} \right)
\tag{3.19}
$$

with $T_{\mu\nu} = \eta_{\mu\alpha} \eta_{\nu\beta} (T_A^{\alpha\beta} + T_B^{\alpha\beta})$. We can then "integrate out" $h_{\mu\nu}$ by solving the latter field equation for h, and by replacing the result in the original action. Modulo self-interaction terms $\propto T_A^{\mu\nu} \Box^{-1} P_{\mu\nu\rho\sigma} T_A^{\rho\sigma}$, the action then splits into the sum of three terms, a term $-m_A \int \sqrt{-\eta_{\mu\nu} dx_A^\mu dx_A^\nu}$, describing the free propagation of body A, a similar term for B, and an interaction term,

$$
S^{\text{int}} = -\frac{8\pi G}{c^4} \int T_A^{\mu\nu} \Box^{-1} \left(T_{\mu\nu}^B - \frac{1}{D-2} T^B \eta_{\mu\nu} \right).
\tag{3.20}
$$

More explicitly, if we introduce the scalar Green's function $G(x)$, such that $\Box G(x) = -4\pi \delta^D(x)$, this lowest-order interaction term reads

$$
\begin{aligned}
S^{\text{int}} = 2G \int \int & ds_A ds_B m_A u_A^\mu u_A^\nu P_{\mu\nu}^{\rho\sigma} \\
& G\left(x_A(s_A) - x_B(s_B)\right) m_B u_{B\rho} u_{B\sigma},
\end{aligned}
\tag{3.21}
$$

in which one easily identifies the usual structure of a Feynman graph (namely the one depicted above), with the coupling constant G in front, and a graviton propagator (comprising the scalar Green's function, together with the spin-two projection operator $P_{\mu\nu}^{\rho\sigma}$, which can be read off the previous explicit result) sandwiched between two source terms.

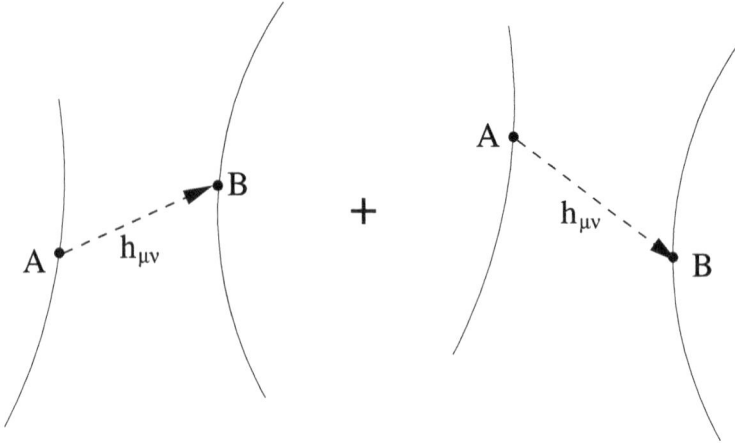

Fig. 4. Time-symmetric half-advanced half-retarded contributions to the gravitational interaction between particles A and B.

In the stationary approximation, the scalar Green's function reduces to the usual Newtonian propagator $1/r$. If one further neglects the relative velocity of the two worldlines one can replace the spacetime velocities u_A^μ and u_B^μ by $(1, 0, 0, 0, \cdots)$. This yields the usual Newtonian interaction term $G \int dt m_A m_B/r$. However, the "one-graviton exchange" diagram above contains many Einsteinian effects that go beyond the Newtonian approximation. To compute them explicitly, we first need the explicit expression of the relativistic scalar Green's function $G(x)$. As we are deriving here the part of the gravitational interaction which is "conservative" (i.e., energy conserving), we must use the *time-symmetric* (half-advanced half-retarded) Green's function. In four dimensions, it is given by $\delta((x_A - x_B)^2)$. It is the sum of two terms (a retarded and an advanced one), as depicted in Fig. 4. Note in passing that this classical time-symmetric propagator corresponds to the real part of the Feynman propagator. Indeed, for a massless scalar particle in $D = 4$ the Feynman propagator (in x space) is proportional to

$$\frac{i}{x^2 + i\varepsilon} = iPP\frac{1}{x^2} + \pi\delta(x^2), \tag{3.22}$$

where the first term on the r.h.s. is a distributional "principal part" (it is pure imaginary and "quantum"), while the second (real) term is the classical contribution (classically the interaction propagates along the light cone, see Fig. 4). Note that, contrary to the Newtonian picture where the interaction is instantaneous, we have here an interaction which depends both on the future and on the past.[12] When considering the case (of most importance in applications) where A and B move slowly relative to c, the time-symmetric propagator can be expanded in powers of $1/c$. The first term in this expansion yields the usual Newtonian instantaneous interaction, while all the higher-order terms can be expressed in terms of successive derivatives of the positions of A and B, (so that the acausality formally disappears, and is replaced by a dependence of the Lagrangian on derivatives higher than the velocities, i.e., accelerations, derivatives of accelerations, and so forth. Actually, such higher-derivative terms start appearing only at the so-called "second post-Newtonian" (2PN) order, i.e., the order $\mathcal{O}\left(\frac{1}{c^4}\right)$. Such higher-order post-Newtonian (PN) contributions are important for some applications (binary pulsars, coalescing black holes), and have been computed up to the 3PN ($\mathcal{O}\left(\frac{1}{c^6}\right)$), as well as 3.5 PN ($\mathcal{O}\left(\frac{1}{c^7}\right)$) levels. Here we shall consider only the first post-Newtonian, 1PN, level, i.e., $\mathcal{O}\left(\frac{1}{c^2}\right)$. At this level, the action can be written entirely in terms of the velocities of A and B (taken at the same instant t in some Lorentz frame). By expanding the time-symmetric acausal one-graviton-exchange action written above to order $1/c^2$ one finds the following explicit 1PN Lagrangian (now considered for an N-body system made of masses labelled $A, B = 1, 2, \ldots, N$.)

$$
\mathcal{L}^{2-\text{body}} = \frac{1}{2} \sum_{A \neq B} \frac{G m_A m_B}{r_{AB}} \left[1 + \frac{3}{2c^2} \left(\vec{v_A}^2 + \vec{v_B}^2 \right) \right.
$$
$$
\left. - \frac{7}{2c^2} \vec{v_A} . \vec{v_B} - \frac{1}{2c^2} \left(\vec{n}_{AB} . \vec{v_A} \right) \left(\vec{n}_{AB} . \vec{v_B} \right) + \mathcal{O} \left(\frac{1}{c^4} \right) \right].
$$

(3.23)

Note that the coefficients $3/2$, $7/2$, $1/2$, etc., arise from the spin 2 nature of the graviton, i.e., they are uniquely fixed by Einstein's propagator.

When considering a gravitationally bound N-body system, we must remember that there is a link $v^2 \sim \frac{GM}{r}$, due to the "virial theorem". This link says that the $\frac{v^2}{c^2}$ contributions in the one-graviton-exchange graph considered above must be completed by computing non-linear interaction graphs containing more gravitons, namely the ones of order G^2/c^2 involving two powers of the coupling constant G. These contributions correspond to the terms $\mathcal{O}(h^2 T)$ or $\mathcal{O}(h^3)$ in the h-expansion of the exact Einstein action. In terms of Feynman-like diagrams,

---

[12]This "acausal" behaviour is due to our considering the conservative ("Fokker") action. If we were computing the "real" classical equations of motion of the two particles, we would use only a *retarded* Green's function. The equations of motion so obtained would then be "causal" and would automatically contain some (physically needed), time-asymmetric "radiation reaction" terms. The trick, used here, to employ an acausal time-symmetric Green's function is a technical shortcut allowing one to derive the action yielding the conservative part of the equations of motion.

this means the following graphs:

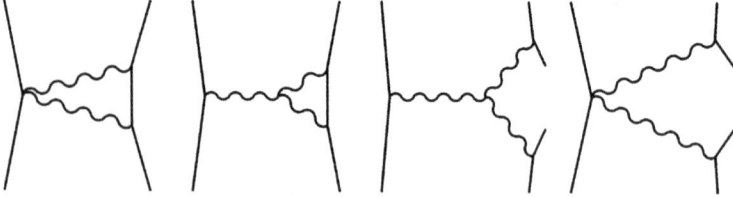

Note, in particular, that these terms involve the graviton cubic vertex (whose structure will therefore be probed by solar-system experiments). Note also that some of these terms involve only two bodies (being proportional, say, to $m_A m_B^2$), while others can involve three distinct bodies ($\propto m_A m_B m_C$). The full G^2/c^2 result (containing both two-body and three-body terms) is found to be equal to

$$\mathcal{L}^{3-\text{body}} = -\frac{1}{2} \sum_{B \neq A \neq C} \frac{G^2 m_A m_B m_C}{r_{AB} r_{AC} c^2}, \tag{3.24}$$

where the factor of $1/2$ is a prediction from Einstein's theory. Note that the summation is restricted by $B \neq A \neq C$ which allows for the two-body terms where $B = C$.

When looking at the nonlinear diagrams above one sees some "loops" made by graviton propagator lines closing up on a matter worldline. This may seem paradoxical because we are considering here classical gravitational effects, and classical theory is usually thought of as involving only tree diagrams. Indeed, if we replace all our "source worldlines" (drawn above as continuous worldlines) by separate external sources (i.e., by replacing the line describing $T_A^{\mu\nu}(x)$ by separate "blobs" on which graviton propagators start or end), we see that all the diagrams above open up and become tree diagrams. However, the presence of loops in the diagrams used here do correspond to essentially some of the same physical effects that "quantum loops" describe. This is particularly clear for the diagrams below (which are included in the classical calculations)

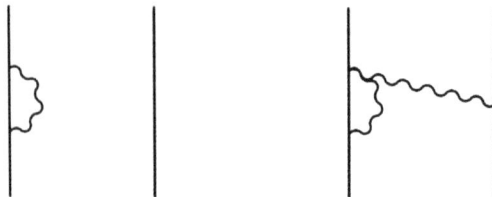

It is clear that these diagrams describe the self-gravity effects of a mass m_A on itself. As such, they do describe the classical limit of quantum loops such as the simplest one-loop diagram

which describes the back action of the emission and reabsorption of a graviton on a quantum particle. Another similarity between "classical loops" and quantum ones, is that, in practice, multi-loops are associated to the presence of multiple integrals which are increasingly difficult to compute.[13] In addition, the loop diagrams depicted in the penultimate graph lead (like quantum loops) to formally divergent integrals. These divergences arise from the fact that we have been describing the gravitationally interacting bodies as "pointlike", i.e., mathematically described by a δ-function (on a worldline). There are several ways of dealing with this technical problem. One can complete the formal perturbative calculations done with point-like bodies by another approximation scheme in which each body is locally viewed (in its own rest frame) as a weakly perturbed isolated body. The development of such a dual perturbation method [32, 33] shows that the non-point-like, internal structure of non-rotating *compact* bodies (neutron stars or black holes) will enter their translational dynamics only at the 5PN level ($\sim G^6/c^{10}$), corresponding to 5 loops! Knowing this, one expects that the use of a gauge-invariant regularization method for treating gravitationally-interacting point masses should give a physically unique answer up to 5 loops (excluded). Using *dimensional regularization*, one finds that all self-gravity effects are unambiguous and finite at 1PN and 2PN [33]. Recent work has pushed the calculation to the 3PN (i.e., 3 loop) level. Again, one finds, either (when using a convenient gauge) a finite answer [34], or, when using the harmonic gauge, an equivalent answer after renormalizing the position of the worldline used in the δ-function source [35].

3.3.2. *Experimental tests in the solar system*
The 1PN-level results described above are accurate enough for describing the gravitational dynamics in the solar system. Testing the validity of GR's description of the gravitational field's dynamics is then achieved by verifying the agreement of the coefficients introduced above with experimental measurements. In this section, we will see that these GR-predicted coefficients agree with their experimentally measured value to better than the 10^{-5} level for the coefficients entering the one-graviton-exchange term, and to about the 10^{-3} level for the additional 1PN multi-graviton term. There are many observables that can be used to test relativistic gravity in the solar system. One may use the advance of the perihelion of planets, the deflection, by the local curvature, of light reaching the

---

[13]Note that classical diagrams must all be computed in x-space. An increasing number of loops signals the presence of intermediate vertices in x-space on which one must integrate.

Earth from distant stars, the additional time delays suffered by electromagnetic signals compared to their flat spacetime counterparts, or also, general relativistic corrections to the Moon's motion using the laser ranging technique mentioned before.

When testing Einstein's predictions it is convenient to embed GR within a class of alternative gravity theories. For instance, one could consider not only the interaction of matter with the usual Einstein (pure spin-2) graviton but also an interaction with a long-range scalar field φ, i.e., a spin-zero massless field, coupled to the trace of $T^{\mu\nu}$ with strength $\sqrt{G}\alpha(\varphi)$. This leads to an additional attractive force, so that the effective gravitational constant measured in a Cavendish experiment is $G_{\text{eff}} = G(1 + \alpha^2)$. This also modifies the v^2/c^2 terms in the two-body action introduced above by terms proportional to α^2. These modifications are often summarized by writing the 1PN-level metric generated by an N-body system in the form

$$ds^2 = -\left(1 - \frac{2U}{c^2} + 2\left(1 + \bar{\beta}\right)\frac{U^2}{c^4}\right)c^2 dt^2 + \left(1 + 2\left(1 + \bar{\gamma}\right)\frac{U}{c^2}\right)\delta_{ij}dx^i dx^j,$$

where $U = G_{\text{eff}}\sum_A m_A/r_A$ is the (effective) Newtonian potential and where the two dimensionless coefficients $\bar{\beta}$ and $\bar{\gamma}$ encode the two possible (Lorentz-invariant) deviations from GR which enter the 1PN level. For an additional coupling to a scalar with φ-dependent coupling strength $\alpha(\varphi)$, one finds that the "post-Einstein" parameter $\bar{\gamma}$ is given by $\bar{\gamma} = -2\alpha^2/(1 + \alpha^2)$. As for the other "post-Einstein" parameter $\bar{\beta}$ it measures a possible modification of the (cubic-vertex related) three-body action $\mathcal{L}^{3-\text{body}}$ written above, and it is given by $\bar{\beta} = +\frac{1}{2}\beta\alpha^2/(1 + \alpha^2)^2$ where β denotes the derivative of the scalar coupling α w.r.t. the field φ.

The most accurate test of GR in the solar system is the one made using the Cassini spacecraft by the authors of Ref. [36]. This test is (essentially) only sensitive to the post-Einstein parameter $\bar{\gamma}$ (i.e., it depends only on the graviton propagator and the coupling to matter but not on nonlinear terms). The experiment used electromagnetic signals sent from the Earth to the Cassini spacecraft and transponded back to Earth, and monitored the ratio between the electromagnetic frequency $\nu + \Delta\nu$ recorded back on Earth to the initial frequency ν. This ratio was used to probe the change in the geometry of spacetime in the vicinity of the Sun as the line of sight moved, especially when it was nearly grazing the Sun. The theoretical prediction for the experimental quantity measured in this experiment is

$$\left(\frac{\Delta\nu}{\nu}\right)^{2-\text{way}} = -4\left(2 + \bar{\gamma}\right)\frac{GM_{\text{sun}}}{c^3 b}\frac{db}{dt}, \tag{3.25}$$

where b is the impact parameter, i.e., the distance of closest approach of the signal's trajectory to the center of the Sun. The experimental data gave the following result for the parameter $\bar{\gamma}$

$$\bar{\gamma} = (2.1 \pm 2.3) \times 10^{-5}. \tag{3.26}$$

This confirms GR (namely $\bar{\gamma}^{GR} = 0$) to the $\mathcal{O}(1) \times 10^{-5}$ level.

The three-graviton vertex can be probed by considering a body, having a non-negligible gravitational self-binding energy, in an external gravitational field. Indeed, as emphasized by Nordtvedt [37, 38], the free fall acceleration of a self-gravitating body is, in most gravity theories (except GR) sensitive to its gravitational binding energy. For instance, the Earth and the Moon will have, in a general theory, a slightly different acceleration of free fall towards the Sun. The effect is proportional to the combination of post-Einstein parameters $4\bar{\beta} - \bar{\gamma}$. Lunar laser ranging data have allowed one to put a stringent upper limit on such a possibility [28], namely

$$4\bar{\beta} - \bar{\gamma} = (4.4 \pm 4.5) \times 10^{-4}. \tag{3.27}$$

Thus, to date, predictions of Einstein's theory in the linear (one-graviton-exchange) approximation have been verified to the 10^{-5} level, while some of the cubically nonlinear aspects have been verified to the 10^{-3} level.

The tests discussed up to now concern the quasi-stationary, weak-field regime, as it can be probed in the solar system. We shall now discuss the tests obtained from binary pulsar data, which go beyond the solar-system tests in probing part of the strongly nonlinear regime of gravity.

3.3.3. Objects with strong self-gravity: binary pulsars

Binary pulsars were discovered by Hulse and Taylor in 1974. Such systems are made of two objects going around each other in very elliptical orbits. Both objects are neutron stars,[14] of which one is a pulsar, i.e., a rotating, magnetized object that emits a beam of electromagnetic noise (which includes radio waves, as well as other parts of the electromagnetic spectrum). When one looks at the geometry generated by a neutron star, and computes deviations from flat space of the metric components, one finds (on the surface of the star)

$$g_{00} = -1 + \frac{2GM}{c^2 R} \simeq -1 + 0.4 \tag{3.28}$$

for a star of (typical) mass $1.4M_\odot$, and radius $R = 10$ km. This is a 40% deviation from flat space. By contrast, we recall that, in the solar system, the

---

[14]Except in a few cases where the companion is another compact (though less compact) star remnant, a white dwarf.

largest metric deviation from flat space occurs on the Sun's surface and is of order $GM/(c^2R) \sim 10^{-6}$. We should therefore a priori expect that such objects might provide tests that go beyond the solar system ones in probing some of the strong-field aspects of relativistic gravity. In addition, in the solar system, the time-irreversible radiative aspects of gravity (i.e., radiation reaction) are negligible (this is why we focussed above on time-symmetric interactions). Here, not only are strongly nonlinear effects relevant, but one must also take into account the time-dissymmetric effects linked to using a *retarded* propagator $\propto \delta\left(t - \frac{|x_A - x_B|}{c}\right)/|x_A - x_B|$. The corresponding time delay $\frac{|x_A - x_B|}{c}$ is typically $\simeq 1$ sec (since typical separations between the two objects are of order $300,000$ km) and plays an essential role in the equations of motion of a binary pulsar. As a consequence, binary pulsars have given us firm experimental evidence for the reality of gravitational radiation, and for the fact that on-shell gravitational radiation is described by two transverse tensorial degrees of freedom travelling at the velocity of light.

In practice, only a subset of the known binary pulsars can be used for testing the strong nonlinear regime of GR, and/or its radiative regime. Among these, the very best ones are PSR1913+16 (where the numbers 19h13m and +16 deg measure angles on the sky), PSR1534+12, PSRJ1141-6545, and PSRJ0737-3039, the first *double* binary pulsar (made of two radio pulsars, simultaneously emitting toward the Earth).

From the theoretical point of view, methods have been developed to deal with strongly self-gravitating objects, both in Einstein's theory and in alternative theories (see [39] for a recent review, and references). To adequately discuss the observations of binary pulsars, one has had to push the post-Newtonian perturbative calculation to the 2.5 PN level, i.e., to order $(v/c)^5$. This odd power of the ratio v/c is linked to the time-dissymmetric, retarded nature of the propagator (together with some nonlinear effects). It is the first PN level where radiation reaction effects arise. Any experimental test of the presence of such $(v/c)^5$ terms in the equations of motion is a probe of the reality of gravitational radiation. In addition, as mentioned above, one must carefully treat, and disentangle, the various strong-field effects that are linked to the self-gravity of each neutron star in the system.

The existing experimental tests are based on the *timing* of binary pulsars. Each time the beam of radio waves sweeps across the Earth, one observes a pulse of electromagnetic radiation. The data consists in recording the successive arrival times, say t_N (with $N = 1, 2, 3, \ldots$), of these pulses. Were the pulsar fixed in space, these arrival times would be equally spaced in time, i.e., $t_N = t_0 + NP$, where P would be the (fixed) period of the pulsar. However, in a binary pulsar, the sequence of arrival times is a more complicated function of the integer N than such a simple linear dependence. Indeed, one must take into account many effects: the fact that the pulsar moves on an approximately elliptical orbit, the deviation of this orbit from a usual Keplerian ellipse, the deviation of the orbital velocity from the usual Kepler areal velocity law, the existence of various additional relativistic effects: gravitational redshift, second-order (relativistic) Doppler effect, time-delay when the electromagnetic pulse passes near the companion, radiation reaction effects in the orbital motion, etc. To compute all these effects, one needs to solve Einstein's equations of motion with high ($\sim (v/c)^5$) accuracy.

The final result of these theoretical calculations is to derive the so-called *DD timing formula*, which gives the Nth pulse arrival time t_N as an explicit function of various "Keplerian" (p^{K}), and "post-Keplerian" (p^{PK}) parameters, say

$$t_N - t_0 = F\left[N; p^{\text{K}}; p^{\text{PK}}\right]. \tag{3.29}$$

Here, the Keplerian parameters (p^{K}) comprise parameters that would exist in a purely Keplerian description of the timing: the orbital period P_b, the eccentricity of the orbit e, the time of passage at some initial periastron T_0, and some corresponding angular position of the periastron ω_0, and, finally, the projected semi-major axis $x = \frac{a_1 \sin i}{c}$, where a_1 is the semi-major axis of the orbit of the observable pulsar[15] and i is the inclination angle w.r.t. the plane of the sky. The post-Keplerian parameters (p^{PK}) then correspond to many relativistic effects that go beyond a Keplerian description, namely: a dimensionless parameter k measuring the progressive advance of the periastron $k = \langle \dot{\omega} \rangle P_b / 2\pi$, a parameter γ_t measuring the combined second-order Doppler and gravitational redshift effects, possible secular variations in Keplerian parameters $\dot{e}$, $\dot{x}$, $\dot{P}_b$, two parameters r, s measuring the "range" and the "shape" of the additional time delay that appears when the radio waves pass near the companion, and finally a parameter δ_θ measuring the distortion of the orbit w.r.t an ellipse. By least-squares fitting the observed arrival times t_N^{obs} to the above general theoretical timing formula one can accurately determine the numerical values of all the Keplerian parameters, as well as some of the post-Keplerian ones. At this stage, the determination of these phenomenological parameters is (in great part) independent of the choice of a theory of gravity. On the other hand, in any specific theory of gravity, each post-Keplerian parameter is predicted to be some well-defined function of the Keplerian parameters and of the two masses, m_1 and m_2, of the pulsar and its companion. For instance, *within GR* the advance of the periastron is given by

$$k^{\text{GR}}\left(p^{\text{K}}, m_1, m_2\right) = \frac{3}{c^2} \frac{(GMn)^{2/3}}{1 - e^2}, \tag{3.30}$$

---

[15] In general, only one of the two objects, here labelled as 1, is a pulsar.

where $n = 2\pi/P_b$ and $M = m_1 + m_2$, while the secular variation of the orbital period (caused by radiation reaction effects) is given by

$$\dot{P}_b^{\mathrm{GR}}\left(p^{\mathrm{K}}, m_1, m_2\right) = -\frac{192\pi}{5c^5} \frac{1 + \frac{73}{24}e^2 + \frac{37}{96}e^4}{(1 - e^2)^{7/2}} (GMn)^{5/3} \frac{m_1 m_2}{M^2}. \qquad (3.31)$$

Note that while k is proportional to $1/c^2$ (1PN level), the secular variation of the orbital period is proportional to $1/c^5$ (and is indeed numerically of order $(v/c)^5$). The GR-predicted value for $\dot{P}_b$ is a direct reflection of the presence of $\mathcal{O}((v/c)^5)$ time-asymmetric radiation damping terms in the equations of motion. Numerically, $\dot{P}_b$ (which is dimensionless) is predicted to be of typical order of magnitude $\dot{P}_b \sim 10^{-12}$, which seems very small, but happens to be large enough to be measured with good accuracy in several binary pulsars.

The crucial point to notice is that the GR predictions (of which two are given here as examples) for the link between the post-Keplerian parameters and the masses are specific to the structure of GR, and will be replaced, in other theories of gravity, by different functions $k^{\mathrm{theory}}(p^{\mathrm{K}}, m_1, m_2)$, $\dot{P}_b^{\mathrm{theory}}(p^{\mathrm{K}}, m_1, m_2)$, etc. In particular, it has been explicitly shown in various cases (and notably in the case of generic tensor-scalar theories where gravity is mediated both by a spin-2 field and a spin-0 one) that the large self-gravity of neutron stars would generically enter these functions, and drastically modify the usual prediction of GR, see [40].

To see which theory of gravity is in agreement with pulsar timing data, one can proceed as follows. Within each theory of gravity, the measurement of each post-Keplerian parameter defines a corresponding curve in the m_1, m_2 plane. Therefore, in general, the measurement of two post-Keplerian parameters is sufficient to determine the (a priori unknown) numerical values of the two masses m_1 and m_2 (as the location where the two curves intersect). Then, the measurement of any additional post-Keplerian parameter yields a clear test of the validity of the theory considered: the corresponding third curve should pass precisely through the intersection point of the first two curves. If it does not, the theory is invalidated by the binary pulsar data considered. By the same reasoning, the measurement of n different post-Keplerian parameters yields $n - 2$ tests of the underlying theory of gravity. Many such stringent tests have been obtained in binary pulsar observations (more precisely, nine different tests in all have been obtained when considering the data from four binary pulsars). Remarkably, *GR has been found to be consistent with all these tests*. Many alternative gravity theories have fallen by the wayside, or their parameters have been constrained so as to make the theory extremely close to GR in all circumstances (including strong-field ones).

Let us just give two impressive examples of the beautiful agreement between GR and pulsar data. In the case of the original Hulse-Taylor pulsar PSR1913+16 the ratio between the observed value of $\dot{P}_b$ to that predicted by GR is given by

$$\left[\frac{\dot{P}_b^{\text{obs}} - \dot{P}_b^{\text{gal}}}{\dot{P}_b^{\text{GR}}[k^{\text{obs}}, \gamma^{\text{obs}}]} \right] = 1.0026 \pm 0.0022, \tag{3.32}$$

where $\dot{P}_b^{\text{gal}}$ is a Galactic correction. The fact that this ratio is close to one corresponds to a confirmation of the relativistic force law acting on the pulsar, of the symbolic form $F = \frac{GM}{r^2}\left(1 + \cdots + \left(\frac{v}{c}\right)^5\right)$, where the crucial last term $\sim (v/c)^5$ (i.e., an effect of order 10^{-12}) has been verified with a fractional accuracy of order 10^{-3}. Note that this corresponds to an absolute accuracy of order 10^{-15} compared to the leading Newtonian term $\sim GM/r^2$!

The timing data from the recently discovered double binary pulsar PSRJ0737-3039 led to the following ratio between the observed, and GR-predicted, values of the post-Keplerian parameter s

$$\left[\frac{s^{\text{obs}}}{s^{\text{GR}}[k^{\text{obs}}, R^{\text{obs}}]} \right] = 0.99987 \pm 0.00050. \tag{3.33}$$

The agreement for this parameter is at the 5×10^{-4} level.

Summarizing: binary pulsar timing data have led to accurate confirmations of the strong-field and radiative structure of GR. Roughly speaking, these confirmations exclude any alternative theory containing long-range fields[16] coupled to bulk (hadronic) matter.

3.3.4. Tests of gravity on very large scales

So far, we have mainly focussed on tests of GR on spatial scales of several astronomical units (the size of the solar system), and on scales of 300,000 km (the typical separation between two neutron stars). We conclude this section on experimental tests of gravity by mentioning the existence of tests made on very large spatial and temporal scales. *Gravitational lensing effects* by galaxy clusters allow one to probe some aspects of relativistic gravity on scales ~ 100 kpc. Here, one is talking of the effect of the curved spacetime metric generated by the cluster on light emitted by very distant quasars and passing near a galaxy cluster containing (in addition to visible galaxies) a lot of dark matter, as well as some X-ray gas. Data on the temperature distribution of the X-ray gas allows one to directly probe the Newtonian gravitational potential $U(x)$ of the cluster (without having

---

[16]By which, one really means here fields with range larger than the distance between the two pulsars, i.e., $\sim 300,000$ km.

to assume much about the (dark) matter distribution). In turn, the potential $U(x)$ determines the relativistic lensing of light, via the spacetime metric predicted by Einstein's theory, i.e., $-g_{00} = 1 - \frac{2U}{c^2}$, $g_{ij} = \left(1 + \frac{2U}{c^2}\right)\delta_{ij}$. According to Ref. [41], the agreement is of the order of 30%. This confirms the validity of GR on scales ~ 100 kpc.

Primordial nucleosynthesis of light elements (e.g., Helium, Lithium, Deuterium) in the early universe depends on both the expansion rate and on the weak-interaction reaction rate for the conversion between neutrons and protons. Given that the Hubble parameter $H^2 \propto G\rho \propto GT^4$, the creation of light elements at early times (and high temperatures T) depends on Newton's constant. The comparison between theoretical predictions and observations of the abundance of light elements typically constrains the value of G at the time of Big Bang nucleosynthesis, say G^{BB} to differ by less than $\mathcal{O}(10\%)$ from its current value G^{now} (see e.g., Chapter 18 of the Review of Particle Physics, http://pdg.lbl.gov/).

4. String-inspired phenomenology of the gravitational sector

4.1. Overview

From the previous sections, one can conclude that GR is a very well confirmed theory so that one might be tempted to require of any future theory (and especially string theory) that it lead to essentially no observable deviations from usual 4-dimensional GR. For instance, one might require that all the a priori massless scalar fields that abound in (tree level, compactified) string theory acquire large masses. However, as there is yet no clear understanding of how to fit our world within string theory, it is phenomenologically interesting to keep an open mind and explore whether there exist possibilities for deviations of GR that have naturally escaped detection so far.

String theory predicts the existence of an extended mass spectrum ($g_{\mu\nu}(x)$, $\Phi(x)$, $B_{\mu\nu}$, moduli fields, etc.) from which there could result some long range or short range modification of gravity. The existence of branes and large extra dimensions could also be sources of modified gravity (e.g., KK gravity). There could exist short distance effects at scales of order the string scale ℓ_s which are observable in cosmology or in high energy astrophysics. We shall also consider possible gravitational wave signals from string-cosmology models. Finally, we refer the reader to the lectures by Juan Maldacena for a discussion of non-Gaussianities in CMB data.

A phenomenologically interesting idea (though it is not supported by precise theoretical arguments) is a possible breakdown of Lorentz invariance, on large scale physics, linked to string-scale cutoff-related effects. An example of this is a modified dispersion relation of the type

$$E^2 = m^2 + \vec{p}^2 + \beta_1 \frac{E^3}{m_P} + \beta_2 \frac{E^4}{m_P^2} + \cdots, \tag{4.1}$$

where m_P denotes the Planck mass. One could think that because of the large value of the Planck mass, any such corrections to the usual dispersion relation are unobservable. However, there exist astrophysical phenomena, such as high energy cosmic rays, for instance high energy γ-rays, for which such a small change in this relation could be observed. For example, by comparing the times-of-arrival of γ-rays of different energies, one has been able to place strong limits on the parameter β_1. Such modifications of the dispersion relation have also been used in the analysis of the CMB, since, in the standard inflationary model, initial quantum fluctuations (the seeds of today's large scale structures) arise in the deep ultraviolet i.e., at transplanckian scales. Note that there exist theoretical difficulties[17] with the inclusion of the $\beta_1 \frac{E^3}{m_P}$ term (the one which is severely constrained experimentally), while the more conventional fourth order term would be too small to be observed. Note that in the case of the photon, a modification on short scales could imply a birefringence of the vacuum as $\omega_\pm = |k| \left(1 \pm \beta \frac{|k|}{m_P}\right)$. For references on these issues see [42, 43]. Speaking of string-inspired astrophysical effects, let us mention the suggestion of Ref. [44] that string theory might imply a violation of the usual Kerr bound on the spin of rotating black holes: $J \leq GM^2$.

Other possible predictions of string theory arise from the picture in which one considers the existence of branes on which (open string) SM particles are confined, while (closed string) gravitons are free to propagate in the bulk (Fig. 5). The extra dimensions of the bulk can then be compactified, on a Calabi-Yau or simply on a torus (thereby "localizing" gravity around the SM brane). Constraints on the size of the compactified dimensions then come either from the gravitational phenomenology, or from effects on SM particles. This is the "large" extra dimensions idea [45] which could be tested at the LHC, and so is of interest today. Other realizations include models with "very large" extra dimensions [46], but it is less clear how they are realized in string theory. In the Randall–Sundrum model [46], a brane can be like a defect in a bulk with a negative cosmological constant, in which case the zero mode of bulk gravitational waves behaves as a surface wave localized on the brane due to the discontinuity located at the interface of the brane with the bulk. In the DGP model [47], the approximate localization of bulk gravity on the SM brane is achieved through the interplay of two dynamics for the gravitational sector: a 5D Einstein action, plus a 4D "induced" Einstein action, with a different value of Newton's constant, on the

---

[17]Linked to its proportionality to $1/m_P$, while most theoretical models suggest a proportionality to $1/m_P^2$.

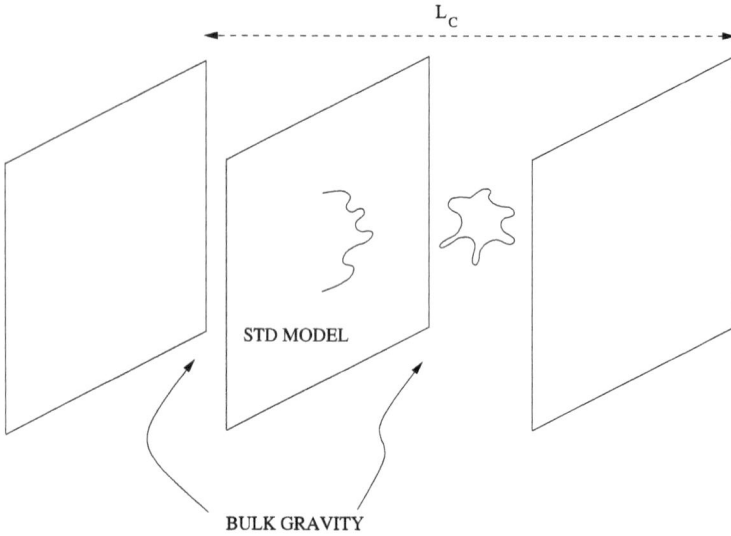

Fig. 5. The ends of open strings are attached to a brane, giving rise to SM particles, while closed strings are free to propagate in the bulk.

brane. Combining the two inverse propagators, the global propagator drastically modifies gravity on large length scales r:

$$r \geq L = \frac{G_5}{G_4}. \tag{4.2}$$

In addition, even on length scales $r \leq L$ there exist modifications of usual gravity. Indeed, the claim is that Newton's potential is modified as [48]

$$U \simeq \frac{GM}{r} \left[1 - \frac{1}{L} \sqrt{\frac{r^3 c^2}{GM}} \right]. \tag{4.3}$$

At the phenomenological level, it is interesting that (*Newtonian*) gravity be modified in this way. Estimates indicate that effects are small enough to have escaped detection so far, but could be seen in refined solar system experiments (e.g., Lunar Laser Ranging). Some authors have argued that such models may have acausal behaviours, with, for instance, the appearance of closed timelike curves [49].

Another conceptually interesting idea involves the possible existence of several (parallel) Randall–Sundrum branes. The confining mechanism of gravity in the Randall–Sundrum model is such that the wavefunction of surface gravitons is exponentially decaying away from the brane. If two branes are nearby, such quasi-confined gravitational effects can tunnel from one brane to the other via exponentially small effects. As a consequence, the effective Lagrangian would contain two metric tensors with two gravitons, one massless, the other massive [50]. There are, however, theoretical difficulties with any massive gravity theory, in relation with the van Dam–Veltman–Zakharov discontinuity (see, e.g., [51] and references therein).

4.2. Long range modifications of gravity

It is well known that, at tree level in string theory, there exist many massless scalar fields with gravitational strength coupling, the so-called moduli fields. Phenomenologically, one would expect that having massless scalar fields at low energies is undesirable (Would a theory containing such massless fields not immediately fail the GR tests discussed in the section above?). General arguments suggest that such scalar fields should not be expected to remain massless after supersymmetry breaking [52]. Recently, a large "industry" has been devoted to try to construct explicit compactification models where all moduli are fixed, and actually acquire very heavy masses, which is needed if inflation is to happen in the usual way. Here however, in the spirit of keeping an open mind, we will instead assume that a scalar field remains massless in the low-energy effective theory, and discuss ways in which it might not disagree with existing tests of general relativity. In other words, we suppose there exists a flat or almost flat direction in the total scalar potential $V(\varphi)$, such that there remains a massless field after supersymmetry breaking. Let us mention in this respect the idea suggested, in particular, by Eliezer Rabinovici [53] that the ultimate explanation for the smallness of the cosmological constant might be a mechanism of spontaneous breaking of an underlying scale invariance. In that case, we would expect to have an associated massless Goldstone boson (the "dilaton", in the original sense of the word).

4.2.1. The cosmological attractor mechanism
Let us discuss here the idea of the *least coupling principle*, realized via a cosmological attractor mechanism (see e.g., Refs [54, 55]), which can reconcile the existence of a massless scalar field in the low energy world with existing tests of GR (and with cosmological inflation). Note that, to date, it is not known whether this mechanism can be realized in string theory. We assume the existence of a massless scalar field Φ (i.e., of a flat direction in the potential), with gravitational-strength coupling to matter. *A priori*, this looks phenomenologically forbidden but we are going to see that the cosmological attractor mechanism (CAM) tends to drive Φ towards a value where its coupling to matter becomes naturally $\ll 1$.

In the string frame, we start with an effective action of the generic form

$$
S_{\text{eff}} = \int d^4x \sqrt{\hat{g}} \left[B_g\left(\Phi\right) \frac{\hat{R}}{\alpha'} + \frac{B_\Phi\left(\Phi\right)}{\alpha'} \left(4\Box\Phi - 4\left(\nabla\Phi\right)^2\right) - B_F\left(\Phi\right) \frac{k}{4} F_{\mu\nu}^2 \right.
$$
$$
\left. - B_\Psi\left(\Phi\right) \Psi D\bar{\Psi} - \frac{1}{2} B_\chi\left(\Phi\right) \left(\hat{\nabla}\hat{\chi}\right)^2 - \frac{1}{2} m_\chi^2\left(\Phi\right) \chi^2 \right],
$$

where Φ is the massless dilaton field, χ is the inflaton, to which has been associated a simple chaotic-inflation-type potential term, with the exception that here m_χ is a function of Φ. In heterotic string theory for instance, B_g, B_Φ, B_F, B_Ψ and B_χ are given by expansions in powers of the string coupling $g_s = e^\Phi$, as

$$
B_i = e^{-2\Phi} + c_0^{(i)} + c_1^{(i)} e^{2\Phi} + \cdots, \tag{4.4}
$$

where the first term is the tree level term, followed by an infinite series of correction terms involving positive powers of g_s (or non-perturbative functions of g_s). Switching to the Einstein frame, and redefining $\hat{g}_{\mu\nu}$ and the nonstandard Φ kinetic terms according to

$$
\hat{g}_{\mu\nu} \rightarrow g_{\mu\nu} = C B_g\left(\Phi\right) \hat{g}_{\mu\nu},
$$
$$
\varphi = \int d\Phi \left[\frac{3}{4}\left(\frac{B_g'}{B_g}\right)^2 + 2\frac{B_\Phi'}{B_g} + 2\frac{B_\Phi}{B_g} \right]^{1/2}, \tag{4.5}
$$

(where a prime denotes $d/d\Phi$), the effective action turns into

$$
S_{\text{eff}} = \int d^4x \sqrt{g} \left[\frac{\tilde{m}_P^2}{4} R - \frac{\tilde{m}_P^2}{2}\left(\nabla\varphi\right)^2 - \frac{\tilde{m}_P^2}{2} F\left(\varphi\right)\left(\nabla\chi\right)^2 \right.
$$
$$
\left. - \frac{1}{2} m_\varphi^2\left(\chi\right) \chi^2 \right] + \cdots, \tag{4.6}
$$

with $\tilde{m}_P^2 \equiv \frac{1}{4\pi G}$, and in which the χ terms are important during inflation while additional terms that include the gauge fields and ordinary matter such as

$$
-\frac{1}{4} B_F\left(\varphi\right) F_{\mu\nu}^2 - \sum_A \int m_A \left[B_F\left(\varphi\left(x_A\right)\right)\right] \sqrt{-g_{\mu\nu}\left(x_A\right) dx_A^\mu dx_A^\nu} - V_{\text{vac}}\left(\varphi\right)
$$
$$
\tag{4.7}
$$

are relevant in the matter dominated era.

As we shall see, the CAM leads to some generic predictions even without knowing the specific structure of the various coupling functions, such as e.g., $m_\chi(\varphi), m_A(B_F(\varphi)), \ldots$. The basic assumption one has to make is that the string-loop corrections are such that there exists a *minimum* in (some of) the functions

$m(\varphi)$ at some (finite or infinite) value, φ_m. During inflation, the dynamics is governed by a set of coupled differential equations for the scale factor, χ and φ. In particular, the equation of motion for φ contains a term $\propto -\frac{\partial}{\partial\varphi}m_\chi^2(\varphi)\chi^2$. During inflation (i.e., when χ has a large vacuum expectation value) this coupling drives φ towards the special point φ_m where $m_\chi(\varphi)$ reaches a minimum. Once φ has been so attracted near φ_m, φ essentially (classically) decouples from χ (so that inflation proceeds as if φ was not there). A similar attractor mechanism exists during the other phases of cosmological evolution, and tends to decouple φ from the dominant cosmological matter. For this mechanism to efficiently decouple φ from all types of matter, one needs the special point φ_m to approximately minimize all the important coupling functions. This can be naturally realized by assuming that φ_m is a special point in field space: for instance it could be the fixed point of some Z_2 symmetry of the T- or S-duality type (so that one could say that "symmetry is attractive"). An alternative way of having such a special point in field space is to assume that $\varphi_m = +\infty$[18] is a limiting point where all coupling functions have finite limits. This leads to the so-called *runaway dilaton* scenario [55]. In that case the mere assumption that $B_i(\Phi) \simeq c_1^i + \mathcal{O}(e^{-2\Phi})$ as $\Phi \to +\infty$ implies that $\varphi_m = +\infty$ is an attractor where all couplings vanish.

4.2.2. *Observable consequences of the cosmological attractor mechanism*
Before discussing the observational predictions of the CAM, let us remind the reader of a few facts that are relevant for studying the possible effects of a string-inspired modification of gravity. The main source of modification of gravity comes from the fact that the "moduli" field φ will influence the values of the masses of the (low-energy) particles and nuclei. This means that the classical action of, say an atom A, will be

$$-\int m_A(\varphi)\mathrm{d}s_A = -\int m_A(\varphi)\sqrt{-g_{\mu\nu}\mathrm{d}x_A^\mu\mathrm{d}x_A^\nu}, \tag{4.8}$$

where $g_{\mu\nu}$ is the Einstein-frame metric. Then, one finds that the scalar field φ will be coupled to the atom A with the strength $\alpha_A\sqrt{G}$, where the dimensionless coupling strength α_A (with the same normalization as the one discussed above for usual tensor-scalar theories[19]) is simply given by

$$\alpha_A = \frac{\partial}{\partial\varphi}\ln m_A(\varphi). \tag{4.9}$$

---

[18]This is viewed as a strong-(bare-)coupling limit, by contrast to the usual weak-coupling limit $\varphi \to -\infty$ and $\Phi \to -\infty$.

[19]In particular, the effective Newton constant for a Cavendish experiment between a body made of atoms A and another one made of atoms B is $G_{AB}^{\mathrm{eff}} = G(1 + \alpha_A\alpha_B)$.

To see better the various ways in which φ might enter into m_A, let us consider for instance the various parts constituting the mass of an atom:

$$m_A(\varphi) = Zm_p + Nm_n + Zm_e + E_{\text{SU3}}^{\text{nucleus}} + E_{\text{U1}}^{\text{nucleus}}, \qquad (4.10)$$

where Z is the atomic number, m_p the mass of the proton, N the neutron number, m_n the mass of the neutron, m_e the mass of the electron, $E_{\text{SU3}}^{\text{nucleus}}$ and $E_{\text{U1}}^{\text{nucleus}}$ the nuclear and Coulomb interaction energies of the nucleus, respectively. In addition, one must note that the mass of the proton is given by

$$\begin{aligned} m_p(\varphi) &= a\Lambda_{\text{QCD}}\big(g_3^2(\varphi)\big) + b_u m_u(\varphi) \\ &\quad + b_d m_d(\varphi) + c_p \Lambda_{\text{QCD}}\alpha_{\text{em}}(\varphi), \end{aligned} \qquad (4.11)$$

and the main scale that determines the mass of the proton is Λ_{QCD}. It depends on all the moduli including the massless field φ and is roughly of the form

$$\Lambda_{\text{QCD}}(\varphi) = C_g^{1/2}(\varphi)\, B_g^{-1/2}(\varphi) \exp\left[-\frac{8\pi B_F(\varphi)}{b_3}\right] \tilde{M}_{\text{string}}. \qquad (4.12)$$

Here, C_g is the conformal factor from the string to the Einstein frame. The most important contribution to the φ dependence of Λ_{QCD} is that given by the φ dependence of the exponential term. This dependence comes from the well-known running (via the β-function of SU(3)) of some (unified) gauge coupling constant between its value $1/g_3^2 \propto B_F(\varphi)$ considered at a GUT-scale cut-off (here approximately related to $\tilde{M}_{\text{string}}$), to a value of order unity at the confining scale Λ_{QCD}. The other contributions to the mass of the proton are the quark masses, which are determined by the vev of the Higgs boson and by the Yukawa coupling constants, which, again, are expected to be functions of φ at high energy. There also exists a contribution from the electromagnetic sector since part of the mass of the proton is a function of the fine structure constant $\alpha_{\text{em}}(\varphi)$. Finally the nuclear binding energy of a nucleus is quite important and must also be expressed as a function of basic scales. In an approximate form it reads

$$E_{\text{SU3}}^{\text{nucleus}} \simeq (N+Z)\,a_3 + (N+Z)^{2/3}\,b_3, \qquad (4.13)$$

where

$$a_3 \simeq a_3^{\text{chiral limit}} + \frac{\partial a_3}{\partial m_\pi^2}\,m_\pi^2(\varphi). \qquad (4.14)$$

In the chiral limit (i.e., taking the quark masses to zero) one gets a non-zero limit $a_3^{\text{chiral limit}}$ to which must be added a term approximately proportional to the squared pion mass. In turn, m_π^2 is proportional to the product of Λ_{QCD} and $m_u + m_d$, both of which are expected to be functions of φ. Incidentally, let us note

that there exists a delicate balance between attractive and repulsive nuclear interactions [56], which implies a strong sensitivity of the binding energy of nuclei to
the value of the quark masses [57]. A recent result shows that if the quark masses
were to increase by 50% (at one 1σ, or 64% at 2σ), all heavy nuclei would fall
apart because there would be no nuclear binding [58].

At leading order, the mass of any nucleus is a pure number times Λ_{QCD}. In this
approximation, m_A would depend universally on φ (via $\Lambda_{QCD}(\varphi)$), and the scalar
coupling strength α_A would be independent of the atomic species A considered.
As a consequence, there would be no violation of the universality of free fall. This
shows that the violations of the universality of free fall will depend on the
small fractional corrections in m_A proportional to the ratios

$$\frac{m_u}{\Lambda_{QCD}}, \quad \frac{m_d}{\Lambda_{QCD}}, \quad \text{and} \quad \alpha_{em}. \tag{4.15}$$

When differentiating the mass of an atom w.r.t. φ, say

$$m_A(\varphi) = \mathcal{N}\Lambda_{QCD}\left(1 + \varepsilon_A^\sigma \frac{m_u + m_d}{\Lambda_{QCD}} + \varepsilon_A^\delta \frac{m_d - m_u}{\Lambda_{QCD}} + \varepsilon_A^{em}\alpha_{em}\right), \tag{4.16}$$

where $\mathcal{N}$ is a pure number (which depends on N and Z), one obtains for the
scalar coupling strength $\alpha_A(\varphi) = \frac{\partial}{\partial\varphi} \ln m_A(\varphi)$ an (approximate) expression of
the form

$$\alpha_A(\varphi) \simeq \alpha_{had}(\varphi) + \varepsilon_A^\sigma \frac{\partial}{\partial\varphi}\left(\frac{m_u + m_d}{\Lambda_{QCD}}\right) + \varepsilon_A^\delta \frac{\partial}{\partial\varphi}\left(\frac{m_d - m_u}{\Lambda_{QCD}}\right)$$
$$+ \varepsilon_A^{em}\frac{\partial}{\partial\varphi}\alpha_{em}, \tag{4.17}$$

where $\alpha_{had} \equiv \frac{\partial}{\partial\varphi} \ln \Lambda_{QCD}(\varphi)$. When the CAM has attracted φ near a value φ_m
which minimizes all the separate coupling functions entering the various ingre-
dients of $m_A(\varphi)$, each term in the above expression for $\alpha_A(\varphi)$ will be (approximately) proportional to the small difference $\varphi - \varphi_m$. As a consequence all the
contributions to $\alpha_A(\varphi)$ will be small, so that all the observable deviations from
GR will be naturally small.

Let us describe more precisely the possible observable consequences of the
CAM. In this mechanism, the couplings of the massless scalar field to the various physical sectors are not assumed to be initially small (they are given by the
various coupling functions $B_i(\varphi)$ entering the Lagrangian, and these functions
are "of order unity"). However, via its coupling to cosmological evolution, the
scalar field is driven towards a point where the couplings to matter become small,
but not exactly zero. Indeed, one can analytically estimate the "efficiency" of the

cosmological evolution in driving φ towards φ_m, and one finds some expression for the difference[20] $\delta\varphi \equiv \varphi - \varphi_m$ [54, 55]. The deviations from GR are all proportional to the small quantity $\delta\varphi^2$ because the scalar coupling strengths α_A, α_B are proportional to $\delta\varphi$, and all "post-Einstein" observables contain two scalar couplings, say $\alpha_A\alpha_B$ when talking about the scalar exchange between A and B (for instance the modified gravitational constant for a Cavendish experiment involving two bodies made of atoms A and B is $G_{AB} = G(1 + \alpha_A\alpha_B)$). In addition to predicting small values for the (approximately composition-independent) "post-Einstein" parameters $\bar{\gamma}$ and $\bar{\beta}$ this mechanism also predicts various (small) violations of the equivalence principle.

For instance, the above expressions for the ingredients entering m_A and α_A lead to generic predictions about the type of violation of the universality of free fall that one might expect in string theory. Indeed, one finds that the fractional difference in the free fall acceleration of two bodies (made of atoms A and B) takes the form

$$
\frac{a_A - a_B}{\langle a \rangle} \simeq 2 \times 10^{-5} \alpha_{\text{had}}^2 \left[\Delta\left(\frac{E}{M}\right)_{AB} + c_B \Delta\left(\frac{N+Z}{M}\right)_{AB} + c_D \Delta\left(\frac{N-Z}{M}\right)_{AB} \right],
\tag{4.18}
$$

with

$$
\frac{E}{M} = \frac{Z(Z-1)}{(N+Z)^{1/3}},
\tag{4.19}
$$

where $(\Delta Q)_{AB} \equiv Q_A - Q_B$, and where the first, second and third terms in the brackets are contributions from the Coulomb energy of the nucleus ($\propto \alpha_{\text{em}}$), and from the φ-dependence of the sum and difference of the quarks masses, i.e., $m_u + m_d$ and $m_u - m_d$.

This mechanism also predicts (approximately composition-independent) values for the post-Einstein parameters $\bar{\gamma}$ and $\bar{\beta}$ parametrizing 1PN-level deviations from GR. They are of the form

$$
\bar{\gamma} = -2\frac{\alpha_{\text{had}}^2}{1 + \alpha_{\text{had}}^2} \simeq -2\alpha_{\text{had}}^2,
\tag{4.20}
$$

and

$$
\bar{\beta} = \frac{1}{2}\frac{\alpha_{\text{had}}^2 \frac{\partial\alpha_{\text{had}}}{\partial\varphi}}{(1 + \alpha_{\text{had}}^2)^2} \simeq \frac{1}{2}\alpha_{\text{had}}^2 \frac{\partial\alpha_{\text{had}}}{\partial\varphi}.
\tag{4.21}
$$

---

[20]When φ_m is infinite, $\delta\varphi \equiv \varphi - \varphi_m$ is replaced, e.g., by $e^{-c\varphi}$.

In this model, one in fact violates all tests of GR. However, all these violations are correlated. For instance, using the numerical value $\Delta\left(\frac{E}{M}\right) \simeq 2.6$ (which applies both to the pair Cu–Be and to the pair Pt–Ti), one finds the following link between equivalence-principle violations and solar-system deviations

$$\left(\frac{\Delta a}{a}\right) \simeq -2.6 \times 10^{-5}\bar{\gamma}. \tag{4.22}$$

Given that present tests of the equivalence principle place a limit on the ratio $\Delta a/a$ of the order of 10^{-12}, one finds $|\bar{\gamma}| \leq 4 \times 10^{-8}$. Note that the upper limit given on $\bar{\gamma}$ by the Cassini experiment was 10^{-5}, so that in this case the necessary sensitivity has not yet been reached to test the CAM.

As another example, one can compute the evolution of the fine structure constant w.r.t. time. Given that it is a function of φ, and that φ evolves as a function of cosmological evolution due to its coupling to matter, $\alpha_{\rm em}$ is indeed a function of time, and its time derivative can be written as

$$\frac{\rm d}{{\rm d}t}\ln\alpha_{\rm em} \sim \pm 10^{-16}\sqrt{1 + q_0 - \frac{3\Omega_m}{2}}\sqrt{10^{12}\frac{\Delta a}{a}}\ {\rm yr}^{-1}. \tag{4.23}$$

The first square root on the r.h.s. of this equation can also be written as

$$\frac{\Omega_m\alpha_m + 4\Omega_v\alpha_v}{\Omega_m + 2\Omega_v}, \tag{4.24}$$

where Ω_m denotes the fraction of the cosmological closure density due to dark matter, and α_m the scalar coupling to dark matter, while Ω_v and α_v denote the corresponding quantities for "dark energy" (or "vacuum energy"). For instance, if we assume $\alpha_v \sim 1$ (so that φ is a kind of "quintessence") while $\alpha_m \ll 1$, we see from the result above that the current experimental limit $\frac{\Delta a}{a} < 10^{-12}$, implies the following upper bound on a possible time variation of the fine-structure constant: $\frac{\rm d}{{\rm d}t}(\ln\alpha_{\rm em}) \leq 10^{-16}$ yr^{-1}. This upper bound is below the current laboratory limits on $\dot{\alpha}/\alpha$, but comparable to the Oklo limit mentioned above. When working out the generic predictions of the *runaway dilaton* version of the cosmological attractor mechanism, one finds that it naturally predicts (when assuming an inflationary potential $\propto \chi^2$) a level of deviation from GR of order $-\bar{\gamma} \sim 4C \times 10^{-8}$, corresponding, for instance, to a violation of the equivalence principle at the level $\Delta a/a \sim C \times 10^{-12}$. Here, C is a combination of model-dependent dimensionless parameters, which are generically expected to be "of order unity". This suggests (if C is smaller, but not much smaller than 1) that the current sensitivity of equivalence principle experiments may be close to what is needed to test the deviations from GR predicted by such a runaway dilaton. Let us note in this respect that ongoing improved lunar laser ranging experiments will probe $\Delta a/a$ to better than the $\sim 10^{-13}$ level, and that the CNES satellite mission MICROSCOPE (to be launched in the coming years) will reach $\Delta a/a \sim 10^{-15}$. Another more ambitious satellite mission (which is not yet approved), STEP (Satellite Test of the Equivalence Principle), plans to probe violations of the equivalence principle down to the 10^{-18} level.[21] In addition, post-Newtonian solar system experiments at the 10^{-7} level would be of interest. The approved micro-arcsecond global astrometry experiment GAIA will probe $\bar{\gamma} \sim 10^{-7}$, while the planned laser experiment LATOR might reach $\bar{\gamma} \sim 10^{-9}$. In addition, the comparison of cold-atom clocks might soon reach the interesting level $\frac{d}{dt}(\ln \alpha_{em}) \sim 10^{-16} \text{ yr}^{-1}$.

Finally, let us mention that one can combine the basic mechanism of the CAM (which consists in using the coupling of φ to matter, i.e., the presence of a term of the form $a(\varphi)\rho_{matter}$ in the action) with the presence of a "quintessence"-like potential $V(\varphi) \propto 1/\varphi^p$. This yields the "chameleon" mechanism [60] in which both the value φ_m towards which φ is attracted, and the effective mass (or inverse range) of φ, depend on the local matter density ρ_{matter}. Whatever be one's opinion concerning the a priori plausibility of having some nearly massless moduli field surviving in the low-energy physics of string theory, it is clear that such experiments are important and could teach us something new about reality.[22]

5. String-related signals in cosmology

5.1. Alternatives to slow-roll inflation

In the usual inflationary scenario, the period of exponential expansion is based on the slow roll mechanism, i.e., one has to assume a sufficiently flat potential so that the scalar field, the inflaton, slowly rolls down to its minimum in such a way that the approximate equality $p_\varphi \simeq -\rho_\varphi$ lasts sufficiently long, say for a minimum of 60–70 e-folds. The simplest inflationary Lagrangian reads

$$\mathcal{L} = -\frac{1}{2}(\partial\varphi)^2 - V(\varphi) \tag{5.1}$$

with a usual kinetic term and a potential $V(\varphi)$. This simple inflationary framework leads to specific predictions such as a relation between the ratio of tensor to scalar primordial perturbations and the "distance" in field space over which

---

[21] Let us also mention the suggestion [59] that atom interferometry might be used for testing the equivalence principle down to the $\Delta a/a \sim 10^{-17}$ level.

[22] For instance, if one considers it very unlikely that such a field can exist, these experiments are important because they can *falsify* string theory. By contrast, if one finds a violation of the equivalence principle which is nicely consistent with the prediction (4.18) for the composition dependence of a moduli field, this might be viewed as a *confirmation* of string theory.

φ runs during inflation (the so-called Lyth bound, see the lectures by Juan Maldacena in these proceedings). Let us, however, emphasize that these predictions (which lead to constraints on the model) do depend on the assumption that inflation is realized by the simple action (5.1) with a slow-roll potential. There are, however, other ways of realizing inflation, in which these constraints might be relaxed. Let us note in this respect that inflation can be realized even if the potential $V(\varphi)$ in (5.1) is *not* of the slow-roll type [61]. Moreover, one may have inflation without a potential at all if the Lagrangian is a complicated enough function of $X \equiv -(\partial\varphi)^2$. Indeed, if one has an action of the type $\mathcal{L} = p(X)$, one finds that there can exist attractors toward a de Sitter expansion phase, corresponding to a line where the effective equation of state deduced from $\mathcal{L} = p(X)$ is $p = -\rho$ (e.g., k-inflation [62]; ghost inflation [63]). To have a "graceful exit" from this de Sitter phase one needs, for instance, to introduce some additional φ dependence in $\mathcal{L}$. It has been suggested in [64] that such a mechanism might be realized in string theory, via a Dirac–Born–Infeld-type action, say

$$p(X, \varphi) = -\frac{\varphi^4}{\lambda^2}\left(\sqrt{1 - \frac{\lambda\dot{\varphi}^2}{\varphi^4}} - 1\right) - V(\varphi). \tag{5.2}$$

In such a "DBI inflation", the use of non-standard kinetic terms greatly relaxes the restrictions imposed on the flatness of the potential $V(\varphi)$ which must be imposed in the usual case of (5.1). It also tends to produce larger non-gaussianities in the CMB [65]. Let us also point out that the use of a non-linear kinetic term might significantly affect the Lyth bound. For instance, if one considers the action

$$\mathcal{L} = p(X, \varphi) = K(X) - V(\varphi), \tag{5.3}$$

where $K(X)$ is a non-linear function of the kinetic term $X \equiv -(\partial\varphi)^2$, one finds the following modified form of the relation between the ratio r of tensor to scalar primordial perturbations and the derivative of φ w.r.t. the number of efolds N:

$$\frac{r}{8} = \left(\frac{d\varphi}{dN}\right)^2 a. \tag{5.4}$$

Here a is an additional amplification factor, which is given by the following expression in terms of the kinetic function $K(X)$

$$a = \frac{2K'}{\sqrt{1 + 2X\frac{K''}{K'}}} = \begin{cases} 1 \text{ for } K = \dfrac{X}{2}, \\ 1 \text{ for DBI type: } -\sqrt{1 - X}, \\ \gg 1 \text{ for, e.g., } -\dfrac{1}{2\alpha}(1 - X)^\alpha, \text{ with } \alpha < \dfrac{1}{2}. \end{cases}$$

A large amplification factor $a \gg 1$ would (formally) correspond to a relaxed Lyth bound on the excursion of φ, given a minimum number N of efolds. It is interesting to note that $a = 1$ (unchanged Lyth bound) in the DBI-type square root model. However, we note that a more general power α, with $\alpha < 1/2$, would formally relax the Lyth bound. [A more detailed study is, however, necessary for seeing whether the bound is *physically* relaxed.]

The present section presented only a very partial and sketchy picture. It was only intended as an illustration that folklore results and constraints on inflationary models do depend on using the standard slow-roll action (5.1), and that there exist other mechanisms in which those results and constraints might be different and possibly relaxed.

5.2. Cosmic superstrings

5.2.1. Phenomenological origin

The existence and detection of *cosmic superstrings* is an exciting possibility that was first suggested in Ref. [66], then kept alive for a number of years, and recently revived dramatically notably in Ref. [67] and in other papers [68–70]. They arise in brane–antibrane scenarios where the inflaton is the brane–antibrane separation (see e.g., [71]). In such scenarios, for large enough brane–antibrane separations, the potential behaves like $c_1 - c_2/\varphi^4$ such that it satisfies the slow roll conditions (Fig. 6). When the branes are near, some of the modes connecting the two branes become tachyonic, i.e., a complex field (with kinetic term $-\partial T \partial \bar{T}$) having a potential $V(|T|^2)$ with wrong-sign curvature near $T = 0$. This instability can generate topological defects since the phase of the vev of T need not be uniformly the same all over space. Contrary to the situation in which strings are created at the beginning of inflation and then diluted away, this scenario naturally produces strings at the end of inflation so that they are not diluted by the

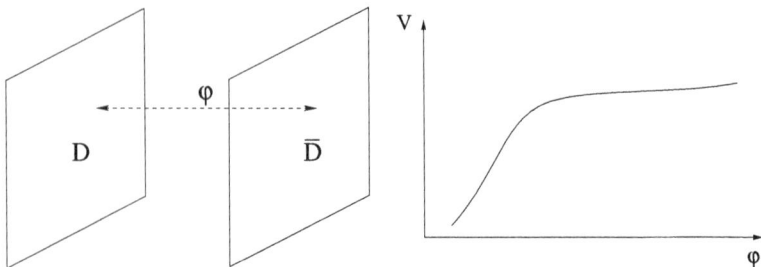

Fig. 6. Left: The brane–antibrane distance as a scalar field φ. Right: $V(\varphi)$ behaves as $c_1 - c_2\varphi^{-4}$ for large brane–antibrane separations.

expansion. For causality reasons, the value of the field's vev in a given Hubble patch should be uncorrelated to that in other Hubble patches. This is what creates a network of strings and one can then compute the initial density and correlation length of the string network. The string tension μ in Planck units, i.e., the dimensionless parameter $G\mu$, where G is Newton's constant, was initially thought to be high, of the order of 10^{-3} at best, because of the string theoretic origin of these objects and of the then expected relation between α' and the Planck length. However, in models with warping factors and large fluxes, the string tension can be lowered to much smaller values. In practice, the string tension is tuned to fit current CMB data. Tye and collaborators [68] find a window of the type $10^{-12} < G\mu < 10^{-6}$, while in the more detailed KKLMMT model [67] one finds $G\mu \sim 10^{-10}$.

In trying to gain insight into the observational predictions that can be made from cosmic superstring models, one must consider not only the stretching by the cosmological expansion of an initial network of cosmic strings with a correlation length of the order of the Hubble scale, but also string interactions. A string can for instance self-intersect or two strings can intersect and reconnect. The Hubble expansion tends to locally straighten out the strings while interconnections tend to produce loops and small-scale structure. Given an initial correlation length and reconnection probability p, working out the time evolution of a string network is essentially a classical problem. Two types of strings develop, long strings with correlation length of the order of the time scale t, and small loops that loose energy by gravitational radiation [72–74].

In order to define the typical size of string loops (at the time they are formed) we introduce a dimensionless parameter α such that $\ell_{\text{loop}}(t) = \alpha\, c\, t$. It was initially thought that $\alpha = 50G\mu$, an estimate linked to the idea that gravitational damping is the essential mechanism which determines the lifetime of loops. More recently, is has been suggested that α might be significantly different from $50G\mu$. There is, however, no consensus on the "correct" value of α. Estimates vary between $\alpha \sim (50G\mu)^\beta$, with $\beta > 1$ (leading to "small loops") and $\alpha \sim 0.1$ (leading to large loops). [For an introduction to this problem, and references, see e.g., the talk of Joe Polchinski at the 2007 String meeting in Madrid.] Happily, some of the predictions we shall discuss below (notably those concerning the observability of gravitational waves from a cosmic string network) are rather insensitive to the value of α.

Several numerical simulations confirm the tendency of string networks to display a scale-invariant behavior [75, 76]. There have also been recent attempts at refining the theoretical description of string networks [77, 78]. However, there is, to date, no consensus among experts as to the typical size distribution of loops (i.e., the dominant gravitational wave-emitting string type). In several simulations, the distribution of the size of loops is bimodal, with one peak at $\alpha \sim 0.1$

and another peak at the UV cutoff, but it has been argued by Vilenkin and collaborators that only the "large loop" part, i.e., $\alpha \sim 0.1$, will survive.

In the following, assuming KKLMMT-type brane inflation and the stability of strings over cosmologically interesting time scales, we discuss the phenomenological predictions made by treating p and α as free parameters and their possible observable signals.

5.2.2. Observational signatures

Partly for historical reasons, the phenomenology of cosmic strings has been studied mostly in the context of CMB observations. Slow-roll inflation generates a random $\frac{\delta T}{T}$ angular distribution on the sky that fits well the observations. Adding a random network of cosmic strings generates additional (non-Gaussian) fluctuations in the CMB which have less angular structure (the string has a lensing effect proportional to its velocity v over the sky, $\delta T/T \sim 8\pi G\mu v\gamma$). CMB observations can then be used to place an upper bound on $G\mu$, of order 5×10^{-7}. Much smaller values of the string tension will not lead to any observable signature in the CMB. Let us also mention that cosmic (super)strings might be detected via their gravitational lensing of galaxies, or microlensing of stars.

By contrast to the CMB (or lensing) observations, which are only sensitive to string tensions $G\mu > 10^{-7}$, existing or planned gravitational wave interferometers could detect cosmic (super)strings with tensions in the much wider range $10^{-15} < G\mu < 10^{-6}$. Let us recall that a gravitational wave (GW) detector is actually measuring tidal forces, and more precisely a component of the Riemann tensor projected "along" a detector having a quadrupolar structure.[23] In other words, a GW antenna measures the second time derivative $\ddot{h}(t)$ of a projection of the metric fluctuation $h_{\mu v}$. Current detectors are sensitive down to the level $h \sim 10^{-22}$, for frequencies $f \sim 100$ Hz.

The GW signal from a string network is an incoherent background of GWs made of the superposition of all GWs ever emitted by string loops (from zero to very large redshifts). This signal is distributed over a very large spectrum of frequencies (including wavelengths of the size of the universe, as well as very short ones). In order to determine the frequency distribution, the number of loops and how they evolve, one needs to know the evolution of the universe during the inflationary, radiation and matter dominated eras.

Besides detecting the GWs from a string network in a man-made interferometer, another observational possibility lies in the timing of isolated millisecond pulsars. In a stationary spacetime, the pulses emitted by an isolated pulsar would be observed on Earth (after correcting for the Earth motion) at very regular in-

---

[23]This is the spin-2 analog of saying that electromagnetic antennas are sensitive to the projection of the electric field along the direction of a dipolar antenna.

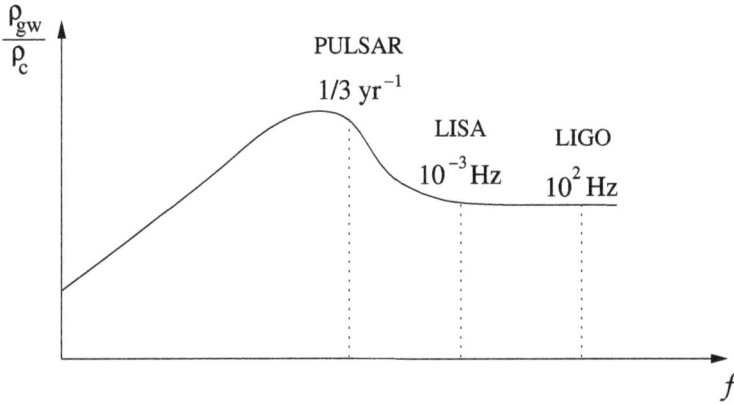

Fig. 7. Expected frequency distribution of the ratio $\Omega_{GW} = \frac{\rho_{GW}}{\rho_c}$ for the stochastic gravitational wave background of cosmic strings.

tervals. By contrast, in presence of a fluctuating background of GWs, the times of arrival of successive pulses would fluctuate, and exhibit some red noise. Pulsar timing over some time interval T (which is typically several years) is most sensitive to the part of the GW frequency spectrum with frequencies $f \sim 1/T$. Therefore, pulsar timing is most likely to detect long wavelength GWs (several light years long).

Along with LIGO-type ground based interferometers, a space-based one, the Laser Interferometer Space Antenna (LISA) has been conceived, with arm lengths of the order of 10^6 km instead of the 3 or 4 km ones constructed on the ground. LISA can therefore explore much smaller frequencies. The best achievable sensitivity for LIGO-type instruments is reached for frequencies $f \sim 100$ Hz, i.e., rather fast events lasting $\sim 10^{-2}$ seconds, while space experiments may probe events with periods ~ 1000 secs, which are quite slow events (see Fig. 7). In Ref. [74], the possible existence of sharp gravitational wave bursts above the background caused by string cusps was pointed out. For a typical oscillating loop, there occurs a cusp once or twice per oscillation with the extremity of the cusp going at the velocity of light and emitting a strong gravitational wave signal in the direction in which the cusp is moving. Statistically, these events are random. A GW burst will be detected if it happens to be emitted towards the detector. Under some conditions, these cusps can create signals which stand much above the quasi-Gaussian random mean square background "GW noise". This raises the exciting possibility that LIGO/VIRGO/GEO or LISA might detect GW signals emitted by giant superstrings at cosmological distances.

Let us now give an introduction to the physics behind the occurrence of those cusps, and the associated emission of GW bursts.

5.2.3. String dynamics

We consider the string position X^μ as a function of the worldsheet coordinates τ and σ. We treat the string dynamics in a locally flat spacetime. Introducing the lightcone coordinates in conformal gauge,

$$\sigma_\pm = \tau \pm \sigma, \tag{5.5}$$

$X^\mu(\tau, \sigma)$ satisfies

$$\frac{\partial}{\partial \sigma_+} \frac{\partial}{\partial \sigma_-} X^\mu(\tau, \sigma) = 0 \tag{5.6}$$

such that the generic string solution is the sum of left and right movers

$$X^\mu(\tau, \sigma) = \frac{1}{2} \left[X_+^\mu(\sigma_+) + X_-^\mu(\sigma_-) \right] \tag{5.7}$$

in which the factor $1/2$ is introduced for convenience. The Virasoro constraints read

$$\begin{aligned}
\left(\partial_+ X_+^\mu \right)^2 &= 0, \\
\left(\partial_- X_-^\mu \right)^2 &= 0.
\end{aligned} \tag{5.8}$$

In the time gauge, the worldsheet is sliced by constant coordinate time hyperplanes $X^0 = x^0 = \tau$, so that

$$X^0(\tau, \sigma) = \tau = \frac{1}{2}(\sigma_+ + \sigma_-). \tag{5.9}$$

We thus have $X_+^0 = \sigma_+$ and $X_-^0 = \sigma^-$. Then $\partial_\pm X^0 = 1$ contributes a -1 in the Virasoro constraints, so that

$$\left(\partial_+ X_+^\mu \right)^2 = -1 + \left(\partial_+ X^i \right)^2, \tag{5.10}$$

and similarly for the $-$ equation. This means that the derivatives (w.r.t. their argument) of the spatial components $X_\pm^i(\sigma_\pm)$ of the left and right modes are constrained to be *unit euclidean vectors*:

$$\left(\dot{X}_\pm^i \right)^2 = 1. \tag{5.11}$$

The $X_\pm^i$ are periodic and the time derivative of the spatial component of X are unit vectors. We may now use a representation first introduced in Ref. [79]. The derivatives $\dot{X}_\pm^i$ can be seen as drawing two curves on the unit sphere (the Turok-Kibble sphere). In addition, as $X_\pm^i$ is periodic in three-dimensional space (there is no winding), we have $\int d\sigma_\pm \dot{X}_\pm^i = 0$. As a result, the "center of mass" of both

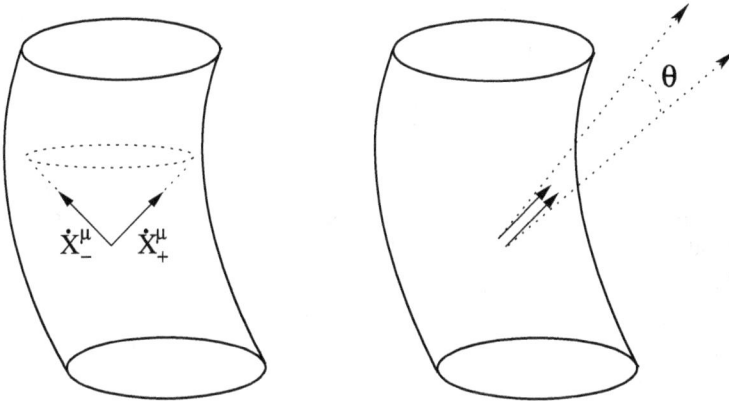

Fig. 8. Left: The lightcone generically intersects the worldsheet into two separate null directions corresponding to the velocity of the left and right movers. Right: When a cusp occurs, the two null vectors are parallel and the worldsheet is tangent to the lightcone. There results a large burst of outgoing gravitational radiation.

left and right moving curves must be at the center of the sphere. This implies that the two curves generically[24] intersect twice [80]. Now, the main point is that an intersection between the two curves represents a *cusp*. More technically, such an intersection corresponds to particular points on the string worldsheet at which the two null (see (5.8)) tangent vectors $\dot{X}^{\mu}_{+}$ and $\dot{X}^{\mu}_{-}$ are parallel in spacetime. In general, the string worldsheet intersects locally the light cone along the two distinct directions $\dot{X}^{\mu}_{+}$ and $\dot{X}^{\mu}_{-}$. The cusps are special points where the worldsheet is *tangent to the lightcone* (see Fig. 8). This is a singularity of the classical worldsheet at which a strong gravitational wave signal is emitted along the common null vector. Let us now indicate how one computes the emission of gravitational wave bursts from cuspy strings. We consider Einstein's theory in the linearized approximation,

$$g_{\mu\nu}(x) = \eta_{\mu\nu} + h_{\mu\nu}(x). \tag{5.12}$$

We use the harmonic gauge, $\partial^{\nu}\bar{h}_{\mu\nu} = 0$, so that Einstein's equations simplify to

$$\Box\bar{h}_{\mu\nu} = -16\pi\, G\, T_{\mu\nu}(x), \tag{5.13}$$

where

$$\bar{h}_{\mu\nu} = h_{\mu\nu} - \frac{1}{2}h\eta_{\mu\nu}. \tag{5.14}$$

---

[24]There exist, however, specially contrived curves that can avoid intersecting.

The stress-energy tensor is obtained by differentiating the Nambu action w.r.t. $g_{\mu\nu}$. Taking its Fourier transform, one finds

$$T^{\mu\nu}\left(k^\lambda\right) = \frac{\mu}{T_\ell} \int_{\Sigma_\ell} d\tau d\sigma \, \dot{X}_+^{(\mu} \dot{X}_-^{\nu)} e^{-\frac{i}{2}k.(X_+ + X_-)}, \tag{5.15}$$

where $(\mu\nu)$ indicates symmetrization over the indices μ, ν, and where, in the exponential, we have replaced X^μ by the half-sum of the left and right movers. The fundamental period of a loop of length ℓ is $T_\ell = \frac{\ell}{2}$. Note that ℓ is the invariant total length $\frac{E_0}{\mu}$, where μ is the string tension. In string theory, one usually uses a worldsheet gauge where ℓ is either 1 or 2π but here one finds it more convenient to use a gauge where σ and τ are connected to an external definition of time (namely $X^0 = x^0 = \tau$).

We wish to compute the integral giving $T^{\mu\nu}(k^\lambda)$ over a periodic domain Σ_ℓ in the τ, σ plane. We can rewrite the integral as an integral over $d\sigma_+ d\sigma_-$. This yields the famous left-right factorization of closed string amplitudes[25] and the Fourier transform of the string stress-energy tensor reads

$$T^{\mu\nu}(k) = \frac{\mu}{\ell} I_+^{(\mu} I_-^{\nu)}, \tag{5.16}$$

where

$$I_\pm^\mu = \int_0^\ell d\sigma_\pm \dot{X}_\pm^\mu (\sigma_\pm) \, e^{-\frac{i}{2}k.X_\pm}. \tag{5.17}$$

By solving Einstein's equation (5.13), one finds that the spacetime-Fourier transform of the source on the r.h.s. actually gives the time-Fourier transform of the asymptotic GW amplitude (emitted in the direction $n^i = k^i/k^0$), i.e., the time-Fourier transform of the quantity $\kappa_{\mu\nu}(t - r, \vec{n})$ appearing in the asymptotic expansion

$$\bar{h}_{\mu\nu}(t, \vec{x}) = \frac{\kappa_{\mu\nu}(t - r, \vec{n})}{r} + \mathcal{O}\left(\frac{1}{r^2}\right). \tag{5.18}$$

Here the $1/r$ decrease in amplitude as a function of distance away from the string is caused by the retarded Green's function in $3 + 1$ dimensions. $\kappa_{\mu\nu}$ is a function of both the time variable and the angle of emission. As we just said, the time-Fourier transform of $\kappa^{\mu\nu}$ is proportional to the spacetime-Fourier transform $T^{\mu\nu}(k)$ of the source, and is explicitly given by

---

[25]Though we are doing here a classical calculation, one recognizes that the result is given by the graviton vertex operator.

$$\kappa_{\mu\nu}(f, \vec{n}) = |f| \int dt\, e^{2\pi i f(t-r)} \kappa^{\mu\nu}(t-r, \vec{n})$$
$$= 2G\mu |f| I_+^{(\mu}(\omega, \omega\vec{n}) I_-^{\nu)}(\omega, \omega\vec{n}). \tag{5.19}$$

This formula shows that we can compute what is observed in a GW detector as a function of string tension, frequency, and the product of two integrals involving left and right moving modes.

We can then estimate the generic features of the GW burst emitted by a cusp by noticing that, in the Fourier domain, each integral $I_\pm^\mu$ is dominated (when considering large frequencies: $f \gg T_\ell^{-1}$) by the singular behaviour of the two integrands $\dot{X}_\pm^\mu(\sigma_\pm) e^{-\frac{i}{2}k.X_\pm}$ near a cusp. The calculation proceeds by (Taylor) expanding the vectors $X_\pm^\mu$ and $\dot{X}_\pm^\mu$ in powers of $\sigma_\pm$. One finds that the first few leading terms in this expansion can be gauged away, so that the signal amplitude is much smaller than what could have (and had) been initially thought. After Fourier transforming back to the time domain, it is finally found that [74]

$$\kappa(t) \propto |t - t_c|^{1/3},$$
$$\ddot{\kappa}(t) \propto |t - t_c|^{-5/3}. \tag{5.20}$$

As this result seems to crucially depend on the presence of a mathematically singular behaviour of the *classical* string worldsheet at a cusp, one might worry that quantum effects could blur away the sharp cusp, and make the above classical burst signal disappear. It was checked that this is not the case [81] (the basic reason being that, finally, the strong GW burst signal is emitted by a large segment of the string around the cusp).

5.2.4. Gravitational waves from a cosmological string network

In order to understand the observational signature of a cosmic string network and not just a single string, one must combine the analysis of the previous section with the cosmological expansion of a Friedman–Lemaître universe and with an integration over redshift. A crucial point is then to estimate the number density of string loops. This density can be analytically estimated as a function of the string parameters, such as the reconnection probability p, and the string tension $G\mu$. Note that the reconnection probability is expected to be quite different for cosmic superstrings compared to the traditionally considered field-theory strings. Field theory strings are expected to reconnect, when they cross, with essentially unit probability ($p \simeq 1$), while fundamental or D-strings are expected to reconnect with a smallish probability, $10^{-3} < p < 1$ [82] (because of the presence of the string coupling, and other factors).

The loop number density can be approximately estimated as [83]

$$n_\ell \sim \frac{1}{p\,50G\mu\,t^3} + \cdots , \qquad (5.21)$$

where the first term on the r.h.s. comes from loops that were created at redshifts ≤ 1, while the '...' denotes a possible additional contribution from high-redshift strings. When the loop-size parameter α is smaller or equal to the "tradition- ally expected" value $50G\mu$, the contribution from high-redshift strings is neg- ligible (because strings decay in less than a Hubble time). By contrast, when $\alpha \gg 50G\mu$, the strings survive over many Hubble times, and the contribution of high-redshift strings starts to dominate the loop density. Note the somewhat unexpected feature displayed by the first term in n_ℓ, namely that it increases both as $G\mu$ and/or p are decreased. This feature is one of the features which allow GW signals from strings to be detected down to very small values of the string tension (contrary to CMB effects). Indeed, as $G\mu$ is decreased, though each individual string signal will decrease proportionally to $G\mu$, there will be more emitting loops. After integrating over redshifts, one finds that the observable sig- nal is a complicated, *non monotonic* function of $G\mu$. The numerical estimates of Ref. [83] considered the case in which the loop size parameter $\alpha < 50G\mu$, in which case the first term in (5.21) is dominant. If, on the other hand, one assumes $\alpha \sim 0.1$ (as is suggested by some numerical simulations [75]), strings survive longer so that higher redshift contributions are non-negligible. It has been found that in such cases these contributions increase the number of loops (which increases the GW signal) but tend to drown the cusp signal within the quasi-Gaussian random-mean-square GW background [84].

Based on current detector capabilities and on the sensitivity estimates for fu- ture detectors, one finds that if $\alpha \leq 50G\mu$, LIGO could detect $G\mu \geq 10^{-12}$ while LISA could detect $G\mu \geq 10^{-14}$. On the other hand, if $\alpha \gg 50G\mu$ LISA could reach $G\mu \geq 10^{-16}$. One has looked in the current LIGO data for the pos- sible presence of a background of GW's, but without success so far [85]. The best current bound on $G\mu$ comes from pulsar timing [86] and is roughly at the $G\mu \leq 10^{-9}$ level (which is about three orders of magnitude more stringent than the limits than can be obtained from CMB data).

Gravitational wave detectors are thus excellent probes of cosmic (super)strings. There is therefore the possibility that they could confirm or refute KKLMMT- type scenarios in a large domain of parameter space. However, there are large uncertainties in string network dynamics which prevent one from being able to make reliable analytical estimates. If one is in a region of parameter space where the rather specific cusp-related signals are well above the r.m.s. background one might find rather direct experimental evidence for the existence of cosmic strings. There would however remain the task of discriminating between string theoretic strings and field theoretic ones. One way would be (assuming one could strongly reduce the string network uncertainties) to determine the reconnection probability p from its influence on the loop number density, and, thereby, on the recurrence rate of observed signals. Another more ambitious possibility would be to exploit the presence of two populations of strings, namely D and F strings, in D-brane anti D-brane annihilation, and attempt to measure two different values of $G\mu$, the ratio of which satisfies $\mu_D = \mu_F/g_s$.

6. Conclusion

We hope that these lectures have shown that gravity phenomenology is a potentially interesting arena for eventually confronting string theory to reality.

Acknowledgement

TD wishes to thank the organizers of this Les Houches session for putting together a timely and interesting programme. He is especially grateful to Nima Arkani-Hamed, Michael Douglas, Igor Klebanov and Eliezer Rabinovici for informative discussions. ML wishes to thank the organizers of the school and the lecturers for a very stimulating time in les Houches.

Bibliography

[1] R. Penrose, *Riv. Nuov. Cimento.* **1** (1969) 252.

[2] D. Christodoulou, *Phys. Rev. Lett.* **25** (1970) 1596.

[3] D. Christodoulou and R. Ruffini *Phys. Rev. D* **4** (1971) 3552.

[4] S.W. Hawking, *Phys. Rev. Lett.* **26** (1971) 1344.

[5] J. Bardeen, B. Carter and S.W. Hawking, *Comm. Math. Phys.* **31** (1973) 161.

[6] S.W. Hawking and J.B. Hartle, *Comm. Math. Phys.* **27** (1972) 283.

[7] R.S. Hanni and R. Ruffini, *Phys. Rev. D* **8** (1973) 3259.

[8] T. Damour, *Phys. Rev. D* **18** (1978) 3598.

[9] T. Damour, in: *Quelques propriétés mécaniques, électromagnétiques, thermodynamiques et quantiques des trous noirs*; Thèse de Doctorat d'Etat, Université Pierre et Marie Curie, Paris VI, 1979. available (see files these1.pdf to these6.pdf) on http://www.ihes.fr/~damour/Articles/

[10] T. Damour, in: *Surface Effects in Black Hole Physics*; Proceedings of the Second Marcel Grossmann Meeting on General Relativity, (edited by R. Ruffini, North Holland, 1982) pp. 587–608; available (see file surfaceeffects.pdf) on http://www.ihes.fr/~damour/Articles/

[11] R.L. Znajek, *MNRAS* **185** (1978) 833.

[12] K.S. Thorne, R.H. Price and D.A. Macdonald, *Black Holes: The Membrane Paradigm*, New Haven, USA: Yale Univ. Press (1986) 367 p.

[13] P.K. Kovtun, D.T. Son and A.O. Starinets, *Phys. Rev. Lett.* **94** (2005) 111601.

[14] D.T. Son and A.O. Starinets, *Ann. Rev. Nucl. Part. Sci.* **57** (2007) 95 [arXiv:0704.0240 [hep-th]].

[15] B. Carter, in: *Black Holes*, Proceedings of 1972 Les Houches Summer School, (edited by C. DeWitt and B.S. DeWitt, Gordon and Breach, NY, 1973).

[16] J. Bekenstein, *Phys. Rev. D* **7** (1973) 2333.

[17] S.W. Hawking, *Comm. Math. Phys.* **43** (1975) 199.

[18] T. Damour and R. Ruffini, *Phys. Rev. D* **14** (1976), 332.

[19] E. Gourgoulhon, *Phys. Rev. D* **72** (2005) 104007 [arXiv:gr-qc/0508003].

[20] E. Gourgoulhon and J.L. Jaramillo, *Phys. Rept.* **423** (2006) 159 [arXiv:gr-qc/0503113].

[21] J.P. Uzan, *Rev. Mod. Phys.* **75** (2003) 403.

[22] T. Damour and F. Dyson, *Nucl. Phys. B* **480** (1996) 37.

[23] A.I. Shlyakhter, in: ATOMKI Report A/1 (Debrecen, Hungary) 1983.

[24] Y. Fujii, *Nucl. Phys. B* **573** (2000) 337.

[25] K.A. Olive, M. Pospelov, Y.-Z. Qian, G. Manhes, E. Vangioni-Flam, A. Coc and M. Casse, *Phys. Rev. D* **69** (2004) 027701.

[26] S. Bize et al., *Phys. Rev. Lett.* **90** (2003) 150802; H. Marion et al., *Phys. Rev. Lett.* **90** (2003) 150801; M. Fischer et al., *Phys. Rev. Lett.* **92** (2004) 230802.

[27] C.D. Hoyle, U. Schmidt, B.R. Heckel, E.G. Adelberger, J.H. Gundlach, D.J. Kapner and H.E. Swanson, *Phys. Rev. Lett.* **86** (2001) 1418; E.G. Adelberger, B.R. Heckel and A.E. Nelson, *Ann. Rev. Nucl. Part. Sci.* **53** (2003) 77; C.D. Hoyle, D.J. Kapner, E.G. Adelberger, J.H. Gundlach, U. Schmidt and H.E. Swanson, *Phys. Rev. D* **70** (2004) 042004; D.J. Kapner, T.S. Cook, E.G. Adelberger, J.H. Gundlach, B.R. Heckel, C.D. Hoyle and H.E. Swanson, *Phys. Rev. Lett.* **98** (2007) 021101.

[28] J.G. Williams, S.G. Turyshev and D.H. Boggs, *Phys. Rev. Lett.* **93** (2004) 261101 [arXiv:gr-qc/0411113].

[29] R.F.C. Vessot and M.W.A. Levine, *General Relativity and Gravitation* **10** (1979) 181.

[30] C. Will, *Living Rev. Relativity* **4** (2001).

[31] See chapter 18 (*Experimental tests of gravitational theory*, by T. Damour) in W.-M. Yao et al. (Particle Data Group), *J. Phys. G* **33** (2006) 1; a 2007 partial update for the 2008 edition is available on the PDG WWW pages (URL: http://pdg.lbl.gov/).

[32] P.D. D'Eath, *Phys. Rev. D* **12** (1975) 2183.

[33] T. Damour, Gravitational radiation and the motion of compact bodies, in: *Gravitational Radiation*, edited by N. Deruelle and T. Piran, North-Holland, Amsterdam, 1983, pp. 59–144.

[34] T. Damour, P. Jaranowski and G. Schafer, *Phys. Lett. B* **513** (2001) 147 [arXiv:gr-qc/0105038].

[35] L. Blanchet, T. Damour and G. Esposito-Farese, *Phys. Rev. D* **69** (2004) 124007 [arXiv:gr-qc/0311052].

[36] B. Bertotti, L. Iess and P. Tortora, *Nature* **425** (2003) 374.

[37] K. Nordtvedt, *Phys. Rev.* **169** (1968) 1014.

[38] K. Nordtvedt, *Phys. Rev.* **169** (1968) 1017.

[39] T. Damour, arXiv:0705.3109 [gr-qc].

[40] T. Damour and G. Esposito-Farese, *Phys. Rev. D* **54** (1996) 54.

[41] A. Dar, *Nucl. Phys. (Proc. Supp.)* **B28** (1992) 321.

[42] T. Jacobson, S. Liberati and D. Mattingly, *Annals Phys.* **321** (2006) 150.

[43] G. Amelino-Camelia, *Lect. Notes Phys.* **669** (2005) 59 [arXiv:gr-qc/0412136].

[44] E.G. Gimon and P. Horava, arXiv:0706.2873 [hep-th].

[45] I. Antoniadis, N. Arkani-Hamed, S. Dimopoulos and G. Dvali, *Phys. Lett. B* **436** (1998) 257.

[46] L. Randall and R. Sundrum, *Phys. Rev. Lett.* **83** (1999) 3370.

[47] G. Dvali, G. Gabadadze and M. Porrati, *Mod. Phys. Lett.* **83** (2000) 1717.

[48] G. Dvali, A. Gruzinov and M. Zaldarriaga, *Mod. Phys. D* **68** (2003) 024012.

[49] A. Adams, N. Arkani-Hamed, S. Dubovsky, A. Nicolis and R. Rattazzi, *JHEP* **0610** (2006) 014 [arXiv:hep-th/0602178].

[50] I.I. Kogan, S. Mouslopoulos, A. Papazoglou, G.G. Ross and J. Santiago, *Nucl. Phys. B* **584** (2000) 313; R. Gregory, V.A. Rubakov and S.M. Sibiryakov, *Phys. Rev. Lett* **84** (2000) 5928; T. Damour and I.I. Kogan, *Phys. Rev. D* **66** (2002) 104024 [arXiv:hep-th/0206042].

[51] T. Damour, I.I. Kogan and A. Papazoglou, *Phys. Rev. D* **67** (2003) 064009 [arXiv:hep-th/0212155].

[52] M.R. Douglas and S. Kachru, *Rev. Mod. Phys.* **79** (2007) 733 [arXiv:hep-th/0610102].

[53] E. Rabinovici, arXiv:0708.1952 [hep-th].

[54] T. Damour and A.M. Polyakov, *Nucl. Phys. B* **423** (1994) 532.

[55] T. Damour, F. Piazza and G. Veneziano, *Phys. Rev. D* **66** (2002) 046007; T. Damour, F. Piazza and G. Veneziano, *Phys. Rev. Lett.* **89** (2002) 081601.

[56] B.D. Serot and J.D. Walecka, *Int. J. Mod. Phys. E* **6** (1997) 515.

[57] J.F. Donoghue, *Phys. Rev. C* **74** (2006) 515.

[58] T. Damour and J.F. Donoghue, arXiv:0712.2968 [hep-ph].

[59] S. Dimopoulos, P.W. Graham, J.M. Hogan and M.A. Kasevich, *Phys. Rev. Lett.* **98** (2007) 111102 [arXiv:gr-qc/0610047].

[60] J. Khoury and A. Weltman, *Phys. Rev. Lett.* **93** (2004) 171104 [arXiv:astro-ph/0309300].

[61] T. Damour and V.F. Mukhanov, *Phys. Rev. Lett.* **80** (1998) 3440.

[62] C. Armendariz-Picon, T. Damour and V.F. Mukhanov, *Phys. Lett. B* **458** (1999) 209.

[63] N. Arkani-Hamed, P. Creminelli, S. Mukohyama and M. Zaldarriaga, *JCAP* **0404** (2004) 001.

[64] E. Silverstein and D. Tong, *Phys. Rev. D* **70** (2004) 103505.

[65] M. Alishahiha, E. Silverstein and D. Tong, *Phys. Rev. D* **70** (2004) 123505.

[66] E. Witten, *Nucl. Phys. B* **249** (1985) 557.

[67] S. Kachru, R. Kallosh and A. Linde, J. Maldacena, L. McAllister and S.P. Trivedi, *JCAP* **0310** (2003) 013.

[68] S. Sarangi and S.-H. Tye, *Phys. Lett. B* **536** (2002) 185.

[69] G. Dvali and A. Vilenkin, *JCAP* **0403** (2004) 010.

[70] E.J. Copeland, R.C. Myers and J. Polchinski, *JHEP* **0406** (2004) 013.

[71] G. Dvali and S.-H Tye, *Phys. Lett. B* **450** (1999) 72.

[72] A. Vilenkin, *Phys. Rev. D* **23** (1981) 852.

[73] A. Vilenkin and E.P.S. Shellard, *Cosmic Strings and Other Topological Defects*, Cambridge University Press, Cambridge, 2000.

[74] T. Damour and A. Vilenkin, *Phys. Rev. Lett.* **85** (2000) 3761; T. Damour and A. Vilenkin, *Phys. Rev. D* **64** (2001) 064008.

[75] V. Vanchurin, K. Olum and A. Vilenkin, *Phys. Rev. D* **72** (2005) 063514; V. Vanchurin, K. Olum and A. Vilenkin, *Phys. Rev. D* **74** (2006) 063527.

[76] C. Martins and E. Shellard, *Phys. Rev. D* **73** (2006) 043515.

[77] J. Polchinski and J.V. Rocha, *Phys. Rev. D* **75** (2007) 123503.

[78] F. Dubath and J.V. Rocha, *Phys. Rev. D* **76** (2007) 024001.

[79] T.W.B. Kibble and N. Turok, *Phys. Lett. B* **116** (1982) 141.

[80] N. Turok, *Nucl. Phys. B* **242** (1984) 520.

[81] D. Chialva and T. Damour, *JCAP* **0608** (2006) 003.

[82] M.G. Jackson, N.T. Jones and J. Polchinski, *JHEP* **0510** (2005) 013 [arXiv:hep-th/0405229].

[83] T. Damour and A. Vilenkin, *Phys. Rev. D* **71** (2005) 063510.

[84] C.J. Hogan, *Phys. Rev. D* **74** (2006) 043526.

[85] B. Abbott et al. [LIGO Collaboration], *Astrophys. J.* **659** (2007) 918 [arXiv:astro-ph/0608606].

[86] F.A. Jenet et al., *Astrophys. J.* **653** (2006) 1571 [arXiv:astro-ph/0609013].

Course 11

ASPECTS OF HAGEDORN HOLOGRAPHY

J.L.F. Barbón[1] and E. Rabinovici[2]

[1] *Instituto de Física Teórica IFT UAM/CSIC*
Madrid 28049, Spain
[2] *Racah Institute of Physics, The Hebrew University*
Jerusalem 91904, Israel

C. Bachas, L. Baulieu, M. Douglas, E. Kiritsis, E. Rabinovici, P. Vanhove, P. Windey
and L.F. Cugliandolo, eds.
Les Houches, Session LXXXVII, 2007
String Theory and the Real World: From Particle Physics to Astrophysics
© *2008 Published by Elsevier B.V.*

Contents

451

1. Preamble

This review develops the idea that known systems with a Hagedorn density of states have natural high-energy completions with asymptotic behavior of field-theoretical type. In these regularized systems, the 'Hagedorn phase' becomes a superheated transient, typically hiding behind a first-order phase transition.

A collection of early references on the subject is [1, 2]. A more recent review with a more complete list of references is [3]. In these lectures, we base our discussion on our own work on the subject, in Refs [3–8], where more technical explanations may be found. Here, we adopt a deliberately intuitive style aimed at putting forward the bare physical ideas.

2. Introduction

In general a system described by a local field theory in $d + 1$ spacetime dimensions containing a *finite* number of mass thresholds has an asymptotic density of states $\Omega(E)$ that increases according to the entropic law $S(E) = \log \Omega(E) \propto E^{\frac{d}{d+1}}$, where the energy E is assumed to be well above all the thresholds. The fact that the power of the energy is positive, but less than unity, for any finite dimension ensures that the temperature of the system can be increased at will and the derived specific heat is always positive.

Free string theory does not have a finite number of mass thresholds and indeed its entropy for energies well above the string scale, m_s, is larger than that of a field theory at a given energy, it is $S(E) \propto E/m_s$. This results in a bound on the largest possible temperature, $T_s \sim m_s = 1/\ell_s$, called the Hagedorn temperature as well as in zero specific heat [1]. Even for the free string such a behavior would lead to the emergence of tachyons unless the system is non generic, essentially supersymmetric [9]. The behavior of the system at a finite value of the string coupling g_s it at question.

Schwarzschild black holes in flat space have an even larger density of states: the entropy of a black hole of mass E grows as $S(E) \propto E^{\frac{d-1}{d-2}}$. This large density of states leads not only to a maximal temperature but to a system which has negative specific heat and eventually such an asymptotic behavior does not allow to discuss a thermodynamical limit with non-zero energy density states. It also leads to other types of highly non-standard behavior [10].

One may well need to eventually readjust one's thinking to accommodate these features. After all, if our universe turns out to be metastable many issues need to be reevaluated. However another possibility is that the excessive number of degrees of freedom is a cry for help. Perhaps above a certain energy the nature of the weakly coupled—and thus reliable—degrees of freedom describing the system is very different. Once the system is diagnosed in terms of the new degrees of freedom the peculiarities may disappear.

Historically this was the case for strings. At the onset of string theory it was thought that the spectrum of strings could be related to the hadronic one. The hadrons had a Hagedorn spectrum but it was realized that, once they interact, they are no longer the relevant degrees of freedom at high energies. It is the hadronic constituents, the gluons and quarks, which describe better the system. Their entropy is field theoretic, thus removing the upper bound on the temperature. The asymptotic density of states was field theoretic but there is a range of energies over which the Hagedorn spectrum is the appropriate description of the system. Also for finite g_s—that is in the presence of gravity—it was suggested [14] that the long excited strings cease to be the relevant degrees of freedom above a certain energy and that the black hole description takes over. This however only made the behavior of the system more puzzling from the point of view discussed above.

For black holes the situation was emeliorated once the black holes were embedded in an AdS space [15]. Such black holes have asymptotically a positive specific heat and no limiting temperature. In fact their entropy is of the field-theoretic class, something that was only fully appreciated with the advent of the AdS/CFT correspondence [16].

For strings propagating in an $AdS_5 \times S^5$ background one can identify in the density of states intermediate energy ranges for which there exist effectively both a Hagedorn type and black-hole type entropy function reflecting unstable states. But above a certain energy scale the system has a field-theoretic type entropy, i.e. the system has chosen field-theory degrees of freedom as its UV exit strategy. This is reviewed in detail in Section 3.

The AdS/CFT correspondence was derived in the context of a $1/N$ expansion. The leading order relates a small curvature, classical, ten-dimensional gravitational background to a strongly coupled four dimensional gauge theory. This relation is an example of the property of holography attributed to some gravitational systems [11]. While this leading order was tested successfully from many points of view, new issues come up when one examines the next to leading order. One component of the $1/N$ corrections is the emergence of quantum radiation in the given gravitational background. One needs to understand if and how also the radiation becomes holographic, this is done in Section 4 with an emphasis on the physical qualitative arguments.

It was suggested [24] that there exists a special type of theory which is in some sense more than a field theory but less than a full fledged string theory called Little String Theory (LST). The system is supposed to have a Hagedorn-type entropy at asymptotic energies. In Section 5 we review the construction of the theory and the emergence of the Hagedorn behavior in the classical gravitational approximation and then we go on to discuss the holographic properties of the radiation correction. We find that they are very different from the AdS case. One has to enforce an infrared volume cutoff on the gravitational theory and test the various limits: large energy, large volume and large N. We find the limits do not commute in the bulk and thus holography is difficult to impose.

In Section 7 we discuss the construction of a system which is in a sense a UV completion of LST. At intermediate scales the Hagedorn LST behavior is demonstrated. The system is described by field-theoretical fixed points at both high and low energies. We discuss both the classical and quantum aspects from the point of view of the bulk theory and discover a first-order phase transition similar to the Hawking–Page one, however it occurs only after quantum corrections are identified. In the process we analyse the way holography is imposed or not on the radiation of various systems including those having noncommutative dimensions, which emerge as a byproduct of some types of UV completion.

All in all we find that for all systems studied up till now the Hagedorn behavior is allowed at intermediate but not asymptotic energies. We find that a negative specific heat can be an indication of a first-order phase transition that can be actualized once an explicit UV completion is found.

3. The density of states of weakly coupled strings

Weakly coupled strings show two main universal properties at the level of the physical spectrum. One is the existence of a finite number of massless modes, often associated to gauge and gravitational interactions which in turn provide the basic link of string models to the 'real' low energy world. The second universal property is a characteristic asymptotics of the high-energy spectrum, usually referred to as 'Hagedorn behavior'. In the approximation of free strings, the density of mass levels is exponential:

$$\Omega(M) = \exp\left(\frac{M}{T_s} + \cdots\right),\tag{3.1}$$

where the dots stand for subleading corrections depending on the interactions or on finite-size effects. The universality of the Hagedorn spectrum stems from the geometrical interpretation of (3.1) as the natural degeneracy of a highly excited string, depicted as a random walk: the number of random walks of length ℓ scales exponentially with ℓ, whereas their mass scales linearly in the length of the walk: $M(\ell) \sim \ell/\alpha'$, where $(2\pi\alpha')^{-1}$ is the tension of the string. In this way we obtain (3.1) with $T_s \sim 1/\sqrt{\alpha'}$.

The associated entropy $\log \Omega(M) \approx M/T_s$ assigns effective 'temperature' equal to T_s to the highly excited string, hence our choice of notation in (3.1). This 'Hagedorn temperature' is effectively maximal in the sense that the naive canonical partition function

$$\int^{M} dM' \, \Omega(M') \, e^{-\beta M'} \tag{3.2}$$

becomes ill defined for $\beta T_s < 1$.

Before we proceed further, let us fix some notation. We shall consider free strings on spacetime backgrounds of the form $K \times \mathbb{R}^{d+1}$, where K is some compact manifold of characteristic size R. When $R \sim \ell_s$ we may loosely speak of 'compact manifold' despite the fact that, properly speaking, K stands for a particular worldsheet conformal field theory. We will also implicitly take $R \geq \ell_s$, after appropriate T-duality symmetries are applied. If no explicit reference to the compact manifold is made, we shall assume that $R \sim \ell_s$. When a finite-volume regularization of 'noncompact' space is needed, we will replace $\mathbb{R}^d$ by a square torus $\mathbb{T}^d$ of size L and volume $V = L^d$. Later in this review we will find it useful to introduce a negative-curvature regularization instead, i.e. Anti-de Sitter space AdS_{d+1} with radius of curvature L.

The Hagedorn temperature appearing in (3.1) is in the order of magnitude of the string mass scale, $T_s \sim m_s \equiv 1/\ell_s$, where ℓ_s is the string length scale $\ell_s \equiv \sqrt{\alpha'}$. The inverse Hagedorn temperature $\beta_s = 1/T_s$ is therefore of the order of the string length. The ten-dimensional Planck scale is given by $\ell_p^4 = g_s \, \ell_s^4$, where g_s is the string coupling constant. The ten-dimensional Newton's constant is $16\pi G_{10} = (2\pi)^7 \ell_p^8$, whereas the effective coupling in $\mathbb{R}^d$ will be defined through the standard Kaluza–Klein relation $G_{d+1} = G_{10}/\text{Vol}(K)$.

The physical question of whether T_s is really a maximal temperature turns out to have a rather complicated answer, even in the strict limit of free strings. It depends in a detailed way on the subleading terms left out in (3.1), and is very sensitive to dimensionality and finite-size effects, which enter when considering the kinetic degrees of freedom of the strings' center of mass, as well as the equipartition of energy between multistring states [6, 12]. When these effects are taken into account, the effective entropy of the string ensemble is still Hagedorn-like to leading order in a large-energy expansion, i.e.

$$\log \Omega(E) \approx \beta_s E. \tag{3.3}$$

A useful parametrization of the density of states uses the 'microcanonical temperature function', given by

$$T(E) \equiv \Omega(E) \left(\frac{\partial \Omega(E)}{\partial E} \right)^{-1}. \tag{3.4}$$

For energy densities below Hagedorn,

$$E < E_s \equiv V T_s^{d+1}, \tag{3.5}$$

the system is dominated by the massless modes in $d + 1$ dimensions, leading to a field-theoretical dependence

$$T(E) \propto E^{\frac{1}{d+1}}. \tag{3.6}$$

As E raises above the E_s threshold we enter the 'Hagedorn regime' dominated by 'long' strings and an effective entropy linear in the energy. This regime is very apparent in the plot of the function $T(E)$, corresponding to an approximate 'plateau' at $T(E) \approx T_s$.

The question of whether the ideal string gas has T_s as a maximal temperature is equivalent to the question of whether the finite-size corrections make the function $T(E)$ approach the 'plateau value' T_s from above or from below. If the 'Hagedorn temperature', T_s is approached from below as $E/E_s \to \infty$, then T_s is a maximal temperature and the thermodynamics is stable in the sense that the fixed-volume heat capacity is positive. On the other hand, should the function $T(E)$ approach T_s 'from above' as $E/E_s \to \infty$, the heat capacity is negative,

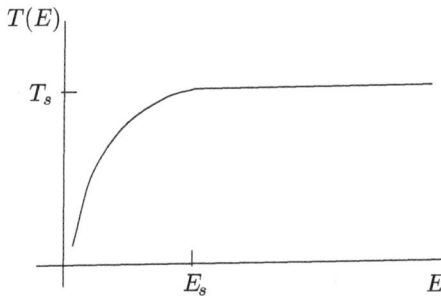

Fig. 1. Picture of the microcanonical temperature function $T(E)$, showing the field-theoretical law up to the string energy threshold E_s, followed by the Hagedorn plateau $T(E) \approx T_s$. Corrections that make $T(E)$ approach T_s from below as $E \to \infty$ correspond to stable thermodynamical states. When $T(E)$ is monotonically decreasing we have a system with negative specific heat, i.e. locally thermodynamically unstable.

suggesting that this Hagedorn phase of long strings is not thermodynamically stable.

Considering the many possibilities posed by different compactification scales, number of non-compact dimensions, or the possibility of having open strings stuck on D-brane defects of varying dimensionalities, there is a fairly complicated casuistics regarding the behavior of the microcanonical temperature function. There is, however, a good rule of thumb on these issues. If we have a total energy E available, should it be assigned to a single, very long string, the size of this state would be of order

$$\ell(E) \sim \ell_s \, (\ell_s E)^{1/2}, \tag{3.7}$$

as corresponds to a random walk of length $\ell_s^2 E$. If the available length scale of the 'container' is L, we can say that the random walk is 'well contained' for $\ell(E) \ll L$. In the opposite case, $\ell(E) \gg L$, we say that the random walk saturates on the available volume. In some sense, saturation is a form of interaction, since the random walk is 'forced to fit' on a scale smaller than its natural (free) size. The rule of thumb is that systems dominated by well-contained strings tend to be thermodynamical unstable, with $T(E)$ approaching T_s from above, whereas systems where the random walks saturate by finite-size effects tend to be stable, with T_s having the standard interpretation of a maximal temperature, and $T(E)$ being well approximated by the canonical-ensemble temperature. The rule has some exceptions when the random walks are forced to live in a sufficiently low effective dimensionality, a condition that implies stability independently of the finite-size effects.

The prototypical examples of ideal string gases with unstable thermodynamics are open strings on Dp-branes with $p < 5$ and *noncompact* transverse dimensions. Closed-string gases are rather subtle. The naive critical dimension, obtained discarding finite-size effects, is $d = 2$. This means that strings on more than two non-compact spatial dimensions would be unstable. On the other hand, if a finite-box regulator is introduced at length L and E is scaled up extensively, i.e. $E \propto L^d$, then the maximal size of random walks grows as $\ell(E) \propto L^{d/2}$, which is larger than L for $d > 2$. Hence, in practice the ideal gases of closed strings are *always* stable, with a maximal temperature T_s, unless the energy is scaled *subextensively* as we approach the infinite-volume limit.

What are we to make of the 'unstable' cases? The curve $T(E)$ raises slightly above T_s around the scale $E \sim E_s$, reaching a maximum $T_{\max}$ still of order T_s, and then approaches T_s from above as $E \to \infty$. Let us suppose that we use T as a control parameter instead of the total energy. In this case, there are two configurations for any T in the interval $T_s < T < T_{\max}$, with the entropy being largest for the high-energy one. Hence, the canonical system has a first-order phase transition to a state of formally infinite energy, exactly at temperature T_s. Less dramatically, we may regularize the system's energy to remain below some large value E_Λ, then the system jumps at temperature $T(E_\Lambda)$ with latent heat of order $E_\Lambda - E_s$.

3.1. Beyond the ideal gas

The introduction of string interactions at a perturbative level, $g_s \ll 1$, is not bound to change the qualitative features of the density of states until new collective effects set in. A good criterion for those is the formation of black holes in the string thermal gas. A Schwarzschild black hole in $\mathbb{R}^{d+1}$ has metric

$$ds^2 = -dt^2 \left(1 - c_d \frac{G_{d+1} M}{r^{d-2}} \right) + \frac{dr^2}{1 - c_d G_{d+1} M / r^{d-2}} + r^2 d\Omega_{d-1}^2, \quad (3.8)$$

with c_d a d-dependent $O(1)$ numerical constant. The Bekenstein–Hawking entropy scales as

$$S_{\rm bh} \sim \frac{(R_S)^{d-1}}{G_{d+1}}, \tag{3.9}$$

where R_S is the horizon radius. In terms of the string coupling $g_s^2 \sim G_{d+1}/\ell_s^{d-1}$ we have a contribution to the density of states

$$\log \Omega(E)_{\rm bh} \sim g_s^{\frac{2}{d-2}} (\ell_s E)^{\frac{d-1}{d-2}}. \tag{3.10}$$

If we have intermediate length scales associated to the compact manifolds, $R(\mathrm{K}) \gg \ell_s$, then these formulae will be modified according to the different dimensionalities when the Schwarzschild radius of the black hole crosses the compactification threshold, $R_S \sim R$.

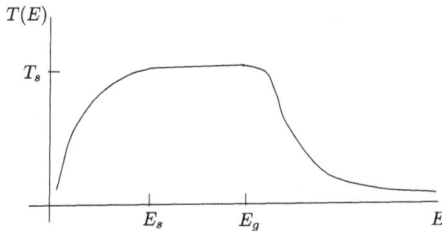

Fig. 2. Microcanonical entropy function beyond the black-hole nucleation threshold $E_g = m_s/g_s^2$.

Comparing (3.10) to the Hagedorn law, we see that as soon as E raises significantly over the threshold $E_g = m_s/g_s^2$ the system gains entropy by nucleating a black hole [14]. Hence, at $E \gg m_s/g_s^2$ we find the microcanonical temperature

$$T(E)_{bh} \sim \frac{1}{R_S(E)} \sim \frac{m_s}{g_s^{\frac{2}{d-2}}} \left(\frac{m_s}{E}\right)^{\frac{1}{d-2}}, \qquad (3.11)$$

a decreasing function, indicating that the black-hole dominated regime is thermodynamically unstable. In particular, (3.11) approaches zero at infinite energy.

It is interesting to study in somewhat more detail the threshold region, in the vicinity of $E \approx m_s/g_s^2$, as it turns out that there are no stable equilibria between the Hagedorn string gas and the black holes. To see this, let us maximize the total entropy of a system of energy E, partitioned into a black hole of mass M, and a gas of long strings of energy $E - M$. The total entropy is approximately given by

$$S(E, M) \approx S_{bh}(M) + S_{Hag}(E - M) \approx g_s^{\frac{2}{d-2}} (\ell_s M)^{\frac{d-1}{d-2}} + \beta_s(E - M). \tag{3.12}$$

This is a concave function of M in the interval $[0, E]$, with a local *minimum* at $M \sim m_s/g_s^2$. The configuration that maximizes the entropy is either a pure string gas (or $M = 0$) for E below threshold, or a pure black hole, i.e. $E = M$, for large energies. There are no *stable* equilibrium configurations with a mixture of the two phases.

Using the temperature as control parameter, the system in infinite volume is unstable towards the nucleation of ever larger black holes. A very convenient infrared regulator to study this black hole instability was introduced in [15], in the form of negative asymptotic curvature, an issue that we discuss in the next section.

4. AdS as a finite-volume regulator

Anti de Sitter space (AdS) has peculiar thermodynamical properties. In particular, black holes behave very differently once the horizon radius grows beyond the curvature radius of AdS, which we shall denote by L. The metric of a Schwarzschild black hole in AdS is obtained by the replacement

$$1 - c_d \frac{G_{d+1}M}{r^{d-2}} \longrightarrow 1 - c_d \frac{G_{d+1}M}{r^{d-2}} + \frac{r^2}{L^2} \qquad (4.1)$$

in (3.8). The main consequence of the new term proportional to the AdS curvature is that black holes with large horizon radius $R_S \gg L$ have a mass that scales as

$$G_{d+1} M \sim \frac{(R_S)^d}{L^2},$$
(4.2)

while the temperature

$$T(R_S) \sim \frac{R_S}{L^2}$$
(4.3)

now grows in a thermodynamically stable branch. All together, we find a microcanonical entropy function

$$T(E)_{\text{large bh}} \sim \frac{1}{L} \left(\frac{EL}{N_{\text{eff}}} \right)^{1/d},$$
(4.4)

where

$$N_{\text{eff}} \equiv \frac{L^{d-1}}{G_{d+1}}$$
(4.5)

is a measure of the AdS curvature radius in Planck units. Formula (4.4) has the remarkable property of scaling like the temperature of a conformal field theory in d spacetime dimensions, and with N_{eff} effective degrees of freedom. Indeed, this is one of the crucial facts supporting the AdS/CFT correspondence [16], which states that the equivalence of gravity on AdS and CFT in one dimension less is an exact statement in the quantum theory. Thus, the strongest form of the AdS/CFT correspondence gives a *definition* of quantum gravity on AdS spaces.

4.1. Holography and radiation effects

Another very important property of AdS space is its good 'containment' of radiation states. As far as radiation thermodynamics is concerned, AdS behaves like a box of size L. A radiation state in equilibrium has a local temperature that *vanishes* at large radius as

$$T(r) = \frac{T}{\sqrt{-g_{tt}(r)}} \longrightarrow \frac{T}{\sqrt{1 + r^2/L^2}},$$
(4.6)

where T is a reference temperature, corresponding to $r = 0$ in the chosen coordinate system. The whole contribution of the asymptotic region $L < r < \infty$ to the entropy of a radiation state is finite and proportional to

$$S_{\text{rad}}(r \geq L) \sim \int_{r \geq L} d\vec{V} \, T(r)^d \sim \int_L^\infty T(r)^d \frac{r^{d-1} \, dr}{\sqrt{1 + r^2/L^2}} \sim (L \, T)^d, \quad (4.7)$$

the same as the radiation inside a box of size L and temperature T. Equation (4.7) shows standard extensive behavior in $d + 1$ spacetime dimensions, with the curvature scale L effectively taking the role of the volume scale. This result brings the question of how this $(d + 1)$-dimensional scaling can be compatible with the d-dimensional scaling of the 'classical' black hole entropy that follows from (4.4),

$$S_{bh} \sim (N_{eff})^{1/d} \, (E \, L)^{\frac{d-1}{d}} \sim N_{eff} \, (L \, T)^{d-1}. \tag{4.8}$$

In fact, there is no contradiction here, as was pointed out in Ref. [4]. The large black hole has a horizon radius of order $R_S \sim L^2 T$. Hence, the contribution to the entropy by all the radiation that sits far away from the black hole can be estimated by the integral (4.7), with the lower limit replaced by $r_{min} \sim R_S$. The result is $S_{rad} \sim (L \, T)^{d-1}$, which scales exactly like (4.8) and can be interpreted as a contribution to the leading correction in an expansion in powers of $1/N_{eff}$.[1]

In more general situations, where a compact manifold K may introduce intermediate length scales in the problem, one must always compare the local temperature of the radiation to the local size of K, i.e. if $T(r)R_K(r) > 1$ the dimension of the compact manifold, d_K, will feature in the extensivity scaling of the radiation. Conversely, in regions where $T(r)R_K(r) < 1$, the translational modes on K are not excited and the full resulting radiation entropy is approximately independent of d_K.

These effects can have neat consequences. For example, in the well known $AdS_5 \times S^5$ background, the radiation never samples the whole ten dimensionality of the system, at least within maximally stable configurations. The reason is that any radiation state hot enough to reveal the five-sphere will always collapse into a black hole larger than the curvature radius of AdS [19], whereas the radiation that is left out 'far above' the horizon is too cold to 'see' the five-sphere [4].

There is an alternative way of arriving at the same 'box' picture of AdS thermodynamics, which is quite illuminating in other contexts, such as the discussion of the next section (see [5, 20] for more detailed discussions). Consider any field in AdS with effective action

$$S_{eff} = \int_{AdS_{d+1}} \Psi \, \mathcal{K} \, \Psi + \cdots, \tag{4.9}$$

---

[1] Converting this estimate into a systematic calculation procedure is not free from subtleties. The red-shift ansatz (4.7) yields a formally infinite entropy for the radiation sitting right on top of the horizon (cf. [17]). One can either cut-off this integral at Planck distance from the horizon and regard the divergent term as a contribution to the 'black hole' part (4.8) or, more systematically, one can work in Euclidean space and use a zeta-function regularization of the radiation determinants (cf. [18] for more details).

where $\mathcal{K}$ is an appropriate kinetic operator of first or second order in the covariant derivative ∇. The associated field equation is, neglecting interactions

$$\mathcal{K} \, \Psi = 0. \tag{4.10}$$

By appropriate local rescalings of the fields, $\Psi \to \tilde{\Psi}$, and making use of the isometries of the spacetime, one can reduce this equation to 'canonical form'

$$\left(\partial_t^2 - \partial_z^2 + V_{\mathrm{eff}}^{(\alpha)}(z) \right) \tilde{\Psi}_\alpha(t, z) = 0, \tag{4.11}$$

where z is an appropriate radial variable known as the Regge–Wheeler or 'tortoise' coordinate. The α index stands for those quantum numbers of 'internal' nature as well as momenta and/or angular momenta referring to spacetime directions other than the (t, z) subspace. The normal frequency modes are the solutions with $\omega \Psi_\omega = -i \partial_t \Psi_\omega$ and $\omega^2 \in \mathbb{R}$. They arise as the real positive spectrum of the Schrödinger operator

$$\omega_\alpha^2 = -\partial_z^2 + V_{\mathrm{eff}}^{(\alpha)}(z). \tag{4.12}$$

In this formalism, the radiation containment of AdS spaces is literally seen as a radial barrier of infinite height in the effective potential, i.e. one finds a second order pole at some finite value of the new radial variable, z_∞:

$$V_{\mathrm{eff}}(\mathrm{AdS}) \propto \frac{L^2}{(z - z_\infty)^2}, \tag{4.13}$$

where $z \to z_\infty$ corresponds to $r \to \infty$ in the usual variables. If AdS is written in global coordinates, so that the fixed radius submanifolds have topology $\mathbb{R} \times S^{d-1}$, the effective potential also shows a barrier at low z, just like any Schrödinger problem at the origin of polar coordinates. The result is a box of size $\Delta z \sim L$ with a mass gap of order $1/L$. Thus, the metaphor of AdS 'as a box' is quite literal when working with the Regge–Wheeler coordinate. For a more detailed discussion of Hagedorn behavior and the 'AdS box' see [3, 13].

4.2. The physical picture

Putting together all these elements, we can now give a qualitative view of the density of states, painted in broad brush strokes (cf. [3, 7, 21, 22]). Since AdS space has a mass gap of order $1/L$, the density of states is vacuum-dominated for $EL < 1$. Above this threshold, we have a region of graviton dominance in $d + 1$ dimensions, $1/L < E < E_s$, with $E_s \sim m_s(Lm_s)^d$, until we reach the string-scale density. Beyond this point, the Hagedorn plateau extends for $E_s < E < m_s/g_s^2$, ending at the onset of black hole nucleation. Black holes

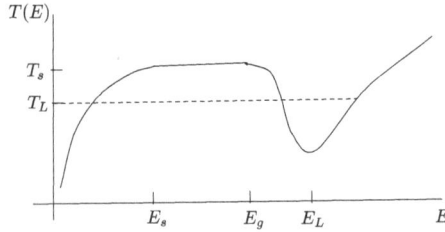

Fig. 3. Microcanonical temperature function showing the stable branch of large black holes beyond $E_L \sim N_{\text{eff}}/L$. A Maxwell-type construction shows that the system undergoes a first-order phase transition at the Hawking–Page temperature $T_L \sim 1/L$.

grow large and *cold* until we reach energies of order $E_L = (Lm_s)^{d-1}m_s/g_s^2$. At this point, the black holes have grown up to the size of AdS's curvature radius and, beyond this energy they continue to grow but at the same time get *hotter*, following a curve of positive heat capacity and a density of states characteristic of a CFT plasma in one lower dimension.

In typical examples of AdS/CFT duals, one finds spaces of the form $\text{AdS}_{d+1} \times K_{d_K}$ with K a compact Einstein manifold of dimension d_K and positive curvature of order $1/L^2$. In this case there is no genuine phase that we could call $d + 1$ dimensional. In all phases dominated by objects smaller than the curvature radius L, i.e. gravitons, Hagedorn strings and small black holes, the system is effectively $d + d_K + 1$ dimensional.

Finally, we may comment that some of the 'windows' depicted above may shut down when we take the parameters of the problem to extreme values. Namely the graviton gas phase only exists for spaces with sufficiently low curvature in string units, $L \gg \ell_s$. Similarly, the Hagedorn plateau disappears completely if the coupling is too strong compared to the 'size' of the AdS space, i.e. we require $g_s^2 \ll (m_s L)^d$ to clearly see a phase dominated by weakly coupled, long strings. On the other hand, the small black hole transient may be absent if AdS is very small, i.e. when $L \sim \ell_s$, which typically corresponds to a weakly coupled dual CFT, although suitable 'precursors' of these small black holes have been identified in the details of planar perturbation theory in the gauge theory side (see e.g. [23]).

By forcing a high-energy exit on the Hagedorn phase, in the form of a dual CFT in the AdS/CFT context, we have been able to sharpen the intuition that Hagedorn strings match onto black holes, both of them representing just transient configurations that can be considered metastable at best. Technically speaking the Hagedorn phase is a superheated state that decays into a large AdS black hole blocking all the 'box' through a holographic picture in terms of a CFT plasma. In the formal limit in which we remove the finite-volume regularization, $Lm_s \rightarrow$

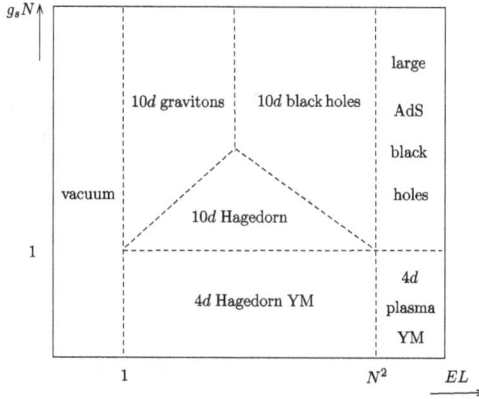

Fig. 4. Complete phase diagram of quantum gravity in AdS$_5$ × S^5 with N units of RR flux on the five-sphere, curvature radius L, and string coupling g_s, according to the AdS/CFT correspondence. The triple point that merges the phases of gravitons, Schwarzschild black holes and Hagedorn strings in ten dimensions corresponds to $g_s N \sim N^{8/17}$ and $EL \sim N^{20/17}$. At large string coupling, the diagram is reflected off the line $g_s N = N$ by the S-duality of the type IIB string theory. The combination $g_s N$ corresponds to the 't Hooft coupling of the dual $SU(N)$ Yang–Mills theory on a 3-sphere of radius L. For $g_s N < 1$ there are no good descriptions of the system as a weakly coupled gravitational theory, and we must use the dual picture of a four-dimensional Yang–Mills theory in either the 'hadron' picture (Hagedorn) or 'parton' picture (plasma).

∞, the whole portion that we may approximate by $\mathbb{R}^{d+1}$ spacetime is literally projected on the boundary, at a black hole horizon [8].

5. Little Hagedorn holography

In the previous sections we have studied the fate of the Hagedorn behavior in critical string theories. We have found that the Hagedorn phase is naturally superseded by black hole dynamics. In some sense, this was always a foregone conclusion, since critical string theories are fundamentally gravitational theories, and we have accumulated a large body of evidence over the years pointing to the smooth transition between highly excited string states and small black holes [14].

In the remainder of this lectures we shall turn to a more exotic instance of Hagedorn spectrum, i.e. that of Little String Theory (LST), a somewhat mysterious string theory that may be implicitly defined in terms of the world-volume dynamics of maximally supersymmetric fivebranes [24, 25], and which admits a dual gravitational description in the spirit of the AdS/CFT correspondence [26].

5.1. The model

We shall adopt the following technical definition of LSTs: start from the world-volume theory on a stack of N NS5-branes of IIA string theory and take the decoupling limit $g_s \to 0$ at fixed N and fixed α'. The result is a $5+1$ dimensional model decoupled from bulk ten-dimensional gravity. The more explicit definition is obtained in the geometrical description. The supergravity solution of the full charge-N NS5 background [27] degenerates in the near-horizon region to the 'tube geometry' $\mathbb{R}^{5+1} \times \mathbb{R} \times S^3$,

$$ds^2 = -dt^2 + dy_5^2 + dz^2 + R^2 \, d\Omega_3^2, \tag{5.1}$$

where $R = \sqrt{N\alpha'}$. The $(t, y_5) \in \mathbb{R}^{1+5}$ coordinates correspond to the world-volume of the NS5-brane, whereas the Regge–Wheeler coordinate, z, takes the role of radial holographic coordinate, in the language of the AdS/CFT correspondence. Finally, there is a fixed angular sphere of radius R. In addition to this metric, there is also a linear dilaton profile of the form

$$\phi(z) = -\frac{z}{R}, \tag{5.2}$$

so that the theory on the tube is asymptotically free at large z, corresponding to the 'entrance' to the throat in the original NS5-brane spacetime. The curvature of the tube background is controlled by the radius R, which is large in string units whenever $N \gg 1$. Hence, for large N we expect the low-energy supergravity description to be accurate. The dual model on the tube with a linear dilaton gives an explicit definition of the LST theory, one that we shall adopt in this lecture.

Further down the throat, at $z \leq 0$, we hit a region of strong string coupling. We can find weakly coupled variables for this regime by uplifting the solution to eleven dimensions [30]. This is the so-called smeared M5-brane solution of eleven-dimensional supergravity, which can be viewed as a long-distance approximation to the fully localized solution of N M5-branes transverse to the compact circle. In turn, the latter solution is asymptotic to AdS$_7 \times$ S^4 as we go further down to negative z values. Hence, we have the following physical picture. At very low energies we have the six-dimensional $(2, 0)$ CFT described by the AdS$_7 \times$ S^4 background with a compact circle. When the size of this eleventh circle becomes of the order of the S^4 radius, around $z \sim z_H = -\frac{1}{2}R \log N$, the metric merges into that of the smeared M5-branes. Further up, at $z \sim z_s = 0$, we enter the regime of the ten-dimensional tube, with geometry $\mathbb{R}^{5+1} \times \mathbb{R} \times S^3$, length scale R and a linear dilaton with slope $1/R$.

The model on the 'tube' has a mass gap of order $1/R$. In addition to the gap associated to the fixed sphere of radius R, less apparent is the fact that excitations propagating in the tube along the z direction have an effective mass of order $1/R$, as a result of the linear dilaton coupling [31]. To see this at an intuitive level, notice that in the NS sector we have an effective action

$$S_\Psi \sim - \int e^{-2\phi} |\partial \Psi|^2 = - \int e^{2z/R} |\partial \Psi|^2 \tag{5.3}$$

for a nominally massless field Ψ. The field redefinition $\Psi = e^{-z/R} \widetilde{\Psi}$ yields the alternative (canonical) form of the action

$$S_{\widetilde{\Psi}} \sim - \int \left(|\partial \widetilde{\Psi}|^2 + \frac{1}{R^2} \right), \tag{5.4}$$

up to boundary terms. Hence, the linear dilaton induces an effective dispersion relation

$$p^2 + p_z^2 + \frac{1}{R^2} = 0 \tag{5.5}$$

for such superficially massless modes, where p is the momentum along the $\mathbb{R}^{5+1}$ of the world-volume and p_z is the momentum along the z coordinate. By supersymmetry, the same gaps must be present in the Ramond and fermionic sectors.

If the field Ψ had a mass term $m^2 \Psi^2$ in (5.3), then the two contributions would add in quadrature. In the language of the effective potential (4.11) we would have $V_{\text{eff}} = m^2 + 1/R^2$ in the tube. In fact, it is quite interesting to draw the full effective potential, even beyond the tube region. One must match the effective potential of an AdS$_7$ excitation at large negative z, with the flat potential in the tube, as in Fig. 5.

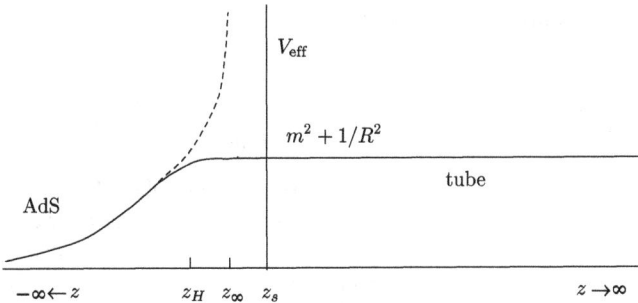

Fig. 5. Effective potential for a field of mass m in the LST background. When the field has no momentum in the eleventh direction, the plateau in the ten-dimensional tube region continues through the eleven-dimensional region between $z_H = -R \log \sqrt{N}$ and $z_s = 0$. At large negative z we have the $(2, 0)$ region of the spectrum with an approximate AdS potential (we show the continuation of the pure AdS potential past z_H). The true spectrum of the system is continuous above zero, despite the apparent mass gap for excitations that propagate in the tube $z \gg z_H$.

The potential (Fig. 5) is very instructive in studying the physics of excitations in the tube. For example, states that look formally non-normalizable as $z \to -\infty$ in the type IIA variables of the tube, become *bona fide* normalizable states in the full effective potential, with wave functions that are localized on the low-energy $(2, 0)$ CFT degrees of freedom. For further relevant work on the peculiarities of LST holography, see [36].

5.2. The classical thermodynamics

The connection with a stringy Hagedorn behavior stems from the consideration of classical black hole solutions in this model. Following standard AdS/CFT logic, these should provide the leading asymptotics of the density of states in a $1/N$ expansion.

At very low energies, the density of states corresponds to that of a $(2, 0)$ CFT in six dimensions, i.e. we have

$$T(E) \sim \left(\frac{E}{N^3 L^5} \right)^{1/6}, \tag{5.6}$$

where $N^3 = N_{\text{eff}}$ is the number of effective degrees of freedom at the deep infrared of the $(2, 0)$ conformal field theory, and L^5 is the spatial world-volume of the NS5-brane. In the gravitational description, this law is obtained from the Hawking temperature of large planar black holes in AdS$_7$. At the merger between AdS$_7 \times$ S^4 with the smeared M5-brane solution, we reach temperature $T_H = 1/2\pi R$. As the black hole grows to larger radii, its classical thermodynamics changes, with the Hawking temperature remaining constant at this value, even beyond the S-duality transition at $z = R$. We can see this by considering the ten-dimensional tube metric with a black hole horizon at $z = z_0$,

$$ds^2 = -dt^2 \left(1 - e^{2(z_0-z)/R} \right) + dy_5^2 + \frac{dz^2}{1 - e^{2(z_0-z)/R}} + R^2 \, d\Omega_3^2, \tag{5.7}$$

where z_0 denotes the position of the horizon in the tube. By a standard calculation, we can see that the Hawking temperature of this metric is constant $T_H = (2\pi R)^{-1}$ and, since such temperature is invariant under string dualities, it is also constant for the eleven-dimensional uplifted solution, the smeared M5-brane solution. Hence, the thermodynamics of this system is identical to that of a string gas with massless sector given by the six-dimensional $(2, 0)$ CFT and with a string scale $M_{\text{eff}} = 1/R$ (see [29]). The 'Hagedorn plateau' characteristic of free string theories arises here as the black hole entropy in the particular holographic geometry that we are using, hence the term 'Hagedorn holography' in the title of these notes.

The transition between the six-dimensional field-theory behavior (5.6) and the Hagedorn plateau takes place at the point $z_H = -\frac{1}{2} R \log N$. The threshold energy at this point is $E = E_H$ with

$$E_H \equiv N^3 \, 2\pi L^5 \, T_H^6 \,, \tag{5.8}$$

corresponding to a six-dimensional gas of N^3 degrees of freedom at temperature T_H. Beyond this energy, the relation between the black-hole mass and the horizon location is given by

$$E(z_0)_{\text{bh}} = N \, E_H \, \exp(2z_0/R). \tag{5.9}$$

On the other hand, the free energy is degenerate with respect to the horizon location, a natural behavior of a fixed-temperature plateau,

$$F(z_0)_{\text{bh}} = E(z_0)_{\text{bh}} - T_H \, S(z_0)_{\text{bh}} = 0. \tag{5.10}$$

5.3. Stability

The Hagedorn plateau is a situation of marginal thermodynamical stability. A small correction can effect a tilt upwards or downwards, thus determining a stable or unstable (respectively) thermal ensemble. In the case at hand, we must then identify the leading corrections to the microcanonical temperature function. We shall address this question in the ten-dimensional set up; we assume that E is sufficiently deep into the Hagedorn regime, $E \gg E_H$, so that the horizon position is well inside the ten-dimensional regime, i.e. $z_0 \gg R$, where the type IIA string theory is weakly coupled. In this case there are potential α' corrections to the function $T(E)$ and of course string loop corrections.

It can be argued that the tube background admits an exact conformal field theory description, featuring the coset $Sl(2, \mathbb{R})/U(1)$ as the factor representing the (t, z) section [32]. The S^3 factor in turn is given by a $SU(2)$ WZW model at level $N - 2$. This worldsheet conformal field theory admits a marginal deformation that can be interpreted as moving the position of the horizon, with a fixed time identification at infinity, i.e. the constancy of $T(E)$ should hold to all orders in the α' expansion [33].

There remains the estimation of the string loop corrections. For the purposes of this lecture, we will split the estimate into the contribution 'near the horizon' and the contribution 'far along the tube'. Furthermore, we shall work in the case that N is parametrically large, so that the natural temperature of the system, $T_H \sim R^{-1} \sim (N\alpha')^{-1/2}$ is much smaller than the ten-dimensional string mass scale. In this case, we can summarize the effect of massive string modes in the low-energy effective action by the addition of effective local operators weighed by a power of the local string coupling e^ϕ, while any local operator depending on the curvature 3-form flux or derivatives of the dilaton will scale dimensionally with the appropriate power of $T_H = (2\pi R)^{-1}$, the only dimensionful parameter of the background. Hence, loop effects visible as local contributions in the low energy effective action will give corrections of order (cf. [34])

$$\frac{\delta T(E)}{T_H} \sim \exp(2\phi(z_0)) \sim N \frac{E_H}{E}, \tag{5.11}$$

so that, far up in the tube, the corrections are always small. The second contribution that can be estimated in simple terms is that of low-lying modes of the strings in the tube. This contribution is extensive in the z-direction and thus is always dominant at the level of the free energy or entropy. In the low-energy approximation we can just keep the modes that are massless in ten dimensions, and which exhibit a mass gap for z-propagation as we have argued in the previous section. The free energy of such gas at temperature T_H is given by

$$F_{\text{tube}} = -C_I\, L^5\, \Delta z\, T_H^7, \tag{5.12}$$

where the numerical coefficient in front is of $O(1)$ in the large N limit, although it will not coincide exactly with the standard coefficient for 256 massless modes (the precise coefficient was computed in the full string theory in [33]). The reason is the existence of a mass gap at scale $1/R$, which renders the tube gas at temperature $T_H = (2\pi R)^{-1} < 1/R$ effectively non-relativistic. The standard scaling of (5.12) still persists though, since T_H is the only dimensional scale of the background. In view of the potential in Fig. 5 it is interesting to see that, despite the infinitely long barrier of finite height, states of the $(2, 0)$ theory still 'leak out' because of the gain in phase space that corresponds to the infinite volume of the z direction.

The most notable property of (5.12) is its extensivity in seven dimensions, in particular along the length of the tube. When added to (5.10), the system at fixed temperature equal to T_H minimizes free energy by increasing Δz. Since the tube is infinite for the fully decoupled LST theory, we see that the black hole is unstable towards complete evaporation into radiation in the tube (see also the early work [35]). This result calls into question the very occurrence of a Hagedorn phase in this system.

5.4. Balancing the act

In order to regularize the problem of radiation propagation in the tube, let us make it finite by imposing a cutoff at z_Λ. The largest black hole horizon that fits into this cutoff tube has energy $E_\Lambda = N E_H \exp(2z_\Lambda/R)$. More general configurations of total energy $E \leq E_\Lambda$ can be approximated by a black hole with energy E_{bh} and horizon at z_0, and a homogeneous state of radiation with energy $E_{\text{rad}} = E - E_{\text{bh}}$, filling the rest of the tube from z_0 up to z_Λ.

We can ask what is the maximally entropic configuration as a function of z_0. If $z_0 < z_H$, we have $E_{bh} < E_H$ and the horizon contribution is well approximated by the AdS$_7$ black hole,

$$S(E_{bh})_{IR} \sim N^{1/2} (L\, E_{bh})^{5/6}, \quad \text{for } E_{bh} < E_H, \tag{5.13}$$

whereas we have the standard linear Hagedorn law

$$S(E_{bh})_{Hag} \sim E_{bh}/T_H, \quad \text{for } E_H < E_{bh} < E_\Lambda. \tag{5.14}$$

As for the radiation component, we can neglect the portion located in AdS$_7$, always small compared to the classical one (5.13) provided $N \gg 1$. On the other hand, the radiation in the tube scales extensively with its length in the z direction, which can be as large as

$$\Delta z_{max} = z_\Lambda - z_H \sim L_z \equiv \frac{R}{2}\, \log(E_\Lambda/E_H). \tag{5.15}$$

Hence, its effect can be very important for a long tube, as we have seen in the previous section.

The radiation in the tube will be modeled as a maximally entropic homogeneous state in the available volume. The entropy has a qualitatively different scaling depending on whether the effective radiation temperature is above or below the gap $1/R$. For $T_{rad} > 1/R$, the radiation is effectively ten-dimensional, with an effective 'radiation volume' $V_{rad} \sim R^3 L^5 \Delta z$, where $\Delta z = z_H - z_0$ is the length of the tube that is not occupied by the black hole. Conversely, for $T_{rad} < 1/R$ the radiation is effectively seven-dimensional, with $V_{rad} \sim L^5 \Delta z$, but at the same time it is non-relativistic, as a result of the tube gap for propagation with non-zero radial momentum p_z. Hence, we can summarize the entropy law in tube radiation by the following asymptotic forms. First, in the high temperature regime

$$S_{rad} \sim V_{rad}\, T_{rad}^9 \sim R^3\, L^5\, \Delta z\, T_{rad}^9, \tag{5.16}$$

for $T_{rad} > 1/R$. In microcanonical variables, we have

$$S_{rad} \sim V_{rad}^{1/10}\, E_{rad}^{9/10} \quad \text{for } E_{rad} \gg V_{rad} T_H^{10} \sim \Delta z\, L^5\, T_H^6. \tag{5.17}$$

Second, in the low-temperature regime,

$$S_{rad} \sim V_{rad}\, T_{rad}^6\, (T_{rad} R)^{-3}\, \exp\left(-1/T_{rad} R\right), \tag{5.18}$$

with $V_{rad} = \Delta z\, L^5$. Again, in microcanonical variables we have

$$S_{rad} \sim \frac{E_{rad}}{T_H} \left(\log\left(\Delta z L^5 T_H^7 / E_{rad} \right) + c \right) \quad \text{for } E_{rad} \ll V_{rad} T_H^7 \sim \Delta z\, L^5\, T_H^7, \tag{5.19}$$

where c is an $O(1)$ constant that is small compared to the logarithm in the regime of interest. We include it in order to properly match (5.17) and (5.19) across the temperature threshold $T_{rad} \sim 1/R$.

We must maximize the total entropy as a function of the horizon position. The broad features are determined by the thresholds that mark the dominance of radiation over the black hole, i.e. the solutions of $S_{rad} \sim S_{bh}$. It turns out that the radiation in the tube starts dominating over the black hole before the end of the $(2, 0)$ CFT regime, i.e. at energies larger than

$$E_t = \frac{E_H}{\left(\log(L_z/N^3 R)\right)^6}.$$ (5.20)

On the other hand, for E sufficiently close to E_Λ, the radiation in the tube loses out eventually to the black hole in the tube. The threshold is

$$E_* = \frac{E_H L_z}{N^3 R}.$$ (5.21)

At energies in the interval $E_* < E < E_\Lambda$ the black hole dominates, but the small perturbation of radiation in the tube can be seen to produce a downwards tilt in the microcanonical temperature function $T(E)$, i.e. these configurations are actually unstable (c.f. [5]). The final picture is depicted in Fig. 6, showing that the Hagedorn plateau is reduced to this branch of unstable states in the band $[E_*, E_\Lambda]$. The naive Hagedorn threshold, E_H, loses any practical significance, as the entire interval $[E_t, E_*]$ is dominated by a nonrelativistic radiation bath in the tube.

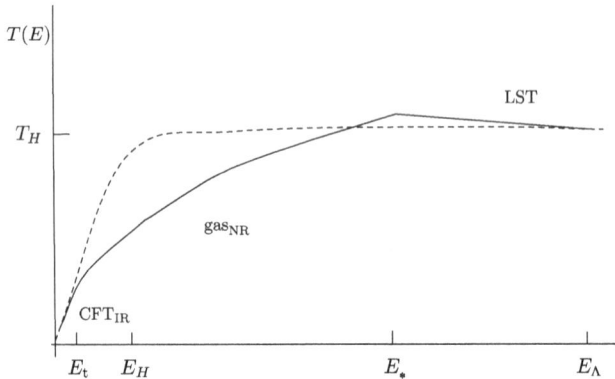

Fig. 6. Picture of the microcanonical temperature function in the cutoff LST model, after radiation effects are taken into account. The Hagedorn plateau is partially replaced by a phase dominated by seven-dimensional non-relativistic gas. If the UV regularization is removed, $E_\Lambda/E_H \to \infty$, no vestiges of the Hagedorn behavior or even the $(2, 0)$ CFT remain, since $E_t/E_H \to 0$ and $E_*/E_H \to \infty$ in this limit.

6. Hagedorn holography deconstructed

It would be certainly desirable to find a more physical interpretation of the tube cutoff. In other words, we would like to interpret the LST as an approximate low-energy description below the E_Λ threshold. One natural possibility is to embed the LST into some well defined UV fixed point, a CFT in D dimensions that arises in the very high energy limit. If this UV CFT is strongly coupled, we may be able to describe it in terms of gravity in some low-curvature AdS_{D+1} background, i.e. the density of states will be approximated by the exponential of the entropy of large black holes in such AdS space. The original purpose of the cutoff procedure was to limit the extensivity of the radiation in the tube, and we know that an asymptotic AdS space does the job, behaving as a finite box as far as radiation 'containment' is concerned.

Independently of the details at intermediate thresholds, the broad features of the system can be geometrically summarized by the following cartoon-type background. At low energies we have $AdS_7 \times S^4$ glued to a tube $\mathbb{R}^{5+1} \times [z_H, z_\Lambda] \times S^3$ of length $L_z = z_\Lambda - z_H$, which is then glued to $AdS_{D+1} \times K_{UV}$. The physical interpretation is that of a UV CFT fixed point, slightly perturbed in a way that gives rise to an intermediate transient below energy E_Λ, which is well approximated by a six-dimensional LST theory. Then, below the energy scale E_H this LST theory flows to the six-dimensional $(2, 0)$ CFT represented by the AdS_7 background at low radii.

Of course, a particular model has many details around the merging points that are just not visible in this simple cartoon. For example, we have seen that the ten-dimensional tube merges with the eleven-dimensional $AdS_7 \times S^4$ via an intermediate transient with the smeared M5-brane metric, growing an extra circle as z decreases. Similarly, the detailed structure of the high-energy threshold will

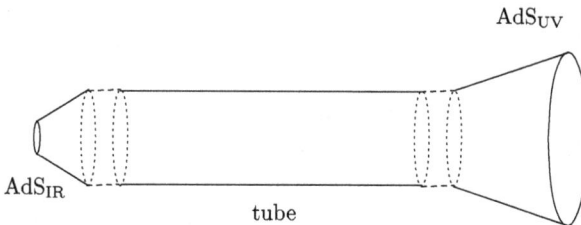

Fig. 7. Schematic representation of the 'cartoon background'. We have the $(2, 0)$ CFT at the IR fixed point, represented by the AdS_{IR} region, glued to the ten-dimensional tube of the Hagedorn plateau. The tube has finite length L_z and is subsequently glued to another AdS region corresponding to the UV CFT fixed point. The transition regions at the endpoints of the tube are not specified in detail, their effects being subleading in the large L_z limit.

determine the appropriate values of D and the properties of K_{UV}, as well as the intermediate geometry around z_Λ, such as the growth of new dimensions.

Still, we can obtain some parameters of the UV completion by simply matching the LST theory at the threshold E_Λ to the CFT, i.e. the AdS$_{UV}$ background. A black hole that fills the tube completely up to energies of order E_Λ has temperature of order T_H. As the black hole exits the tube and grows in AdS$_{D+1}$ it follows the law

$$T(E) \sim \left(\frac{E}{N_\infty L^{D-1}} \right)^{1/D}, \quad \text{for } E > E_\Lambda. \tag{6.1}$$

Then, the threshold matching $T(E_\Lambda) = T_H$ determines the effective number of degrees of freedom of the UV CFT as

$$N_\infty \sim \frac{E_\Lambda}{L^{D-1} T_H^D}. \tag{6.2}$$

These matching relations survive the radiation corrections. Since

$$\frac{E_\Lambda}{E_*} = \frac{RN^3}{L_z} \exp(2L_z/R), \tag{6.3}$$

we see that a sufficiently long tube ensures that $E_* \ll E_\Lambda$, and the matching to the UV manifold is always done under black-hole dominance. The remaining radiation in the AdS$_{UV}$ space is then bounded by $(LT_H)^{D-1}$, according to our previous results. Hence, the matching of the density of states around E_Λ is rather insensitive to radiation corrections.

As remarked above, the details of the merger between the LST regime represented by the 'tube' and the AdS$_{D+1}$ regime corresponding to the UV CFT depend on the particular LST embedding that we envisage. Some standard embeddings of LST into CFTs exist in the literature, using ideas of 'deconstruction' [37], namely we consider some peculiar vacua on the Higgs–Coulomb branch of the UV CFTs, in such a way that the low energy spectrum resulting from the Higgsing emulates a Kaluza–Klein tower within some approximations. In this way one can deconstruct six-dimensional fivebrane theories, such as the $(2, 0)$ CFT or the full LST, in terms of four-dimensional CFTs. There are two main examples in the literature, using a maximally supersymmetric gauge theory with group G_{UV} as the UV CFT. In one case $G_{UV} = SU(N)^{n^2}$ and the Higgs vacua simulate a rank N LST compactified on an $n \times n$ lattice [38]. The other method uses 'noncommutative vacua' on the Higgs branch of an $G_{UV} = SU(Nn)$ theory [39]. In this latter case the $n \times n$ lattice is noncommutative, and the resulting LST is also noncommutative below the scale E_Λ, with the usual commutative LST only arising below a second threshold $E_\theta < E_\Lambda$.

6.1. The noncommutative Hagedorn model

Even if the noncommutative deconstruction is more complicated, it is perhaps easier to implement in the dual geometrical description, since the relevant background metrics have been written down in Ref. [39]. This case is also interesting from a more general point of view, as the intermediate threshold that performs the gluing between the tube and the asymptotically AdS space is the metric of a so-called Noncommutative Little String Theory (NCLST) (see for example [40]). In fact, the full NCLST, with no reference to the UV fixed point, is very interesting on its own, since these theories have gravity duals that naturally quench the effect of radiation, *without* affecting the leading density of states, obtained from the Bekenstein–Hawking formula. Hence, in the model of Ref. [39] we can study at once two interesting possible cures for the infrared instability posed by formula (5.12).

Following Ref. [5] we can write down the metric of the noncommutative LST background as the following deformation of (5.1):

$$ds^2 = \sqrt{F(z, z_\theta)} \left(-dt^2 + dx_4^2 \right) + \frac{dw^2}{\sqrt{F(z, z_\theta)}} + \sqrt{F(z, z_\theta)}(dz^2 + R^2 d\Omega_3^2),$$

$$(6.4)$$

where we have split the original world-volume coordinates of the fivebrane as $y = (x_4, w)$, with $x_4 \in \mathbb{T}^4$, a four-torus of size L, and $w \in S^1$, parametrizing a compact circle of length L_θ. The model has an extra parameter z_θ, determining the location along the tube where 'noncommutative effects' are felt, [2] with the profile function given by

$$F(z, z_\theta) = 1 + e^{2(z-z_\theta)/R}.$$

$$(6.5)$$

The model is further specified by a Wilson line of a Ramond–Ramond one-form along the circle of length L_θ as well as a dilaton profile

$$\phi(z) = -\frac{z}{R} + \frac{1}{2} \log F(z, z_\theta).$$

$$(6.6)$$

The location z_θ, i.e. the gate to the 'noncommutative deformed' tube also determines the threshold energy E_θ, through the usual formula

$$E_\theta = N E_H \exp(2z_\theta/R).$$

$$(6.7)$$

---

[2]The noncommutative character of this background is only manifest after an S-duality transformation to a type IIB background. In that case we have a standard D5-brane background with a Neveu–Schwarz magnetic field.

On the other hand, the compactification length L_θ controls the scale at which the metric (6.4) matches approximately to a background of smeared D4-branes. The metric (6.4) determines a compact w-circle of size

$$L_\theta(z) = L_\theta \, \exp\left(\frac{z_\theta - z}{2R}\right) \tag{6.8}$$

and an S^3 of local size

$$R(z) = R \, \exp\left(\frac{z - z_\theta}{2R}\right). \tag{6.9}$$

The matching occurs when $R(z_\Lambda) \sim L_\theta(z_\Lambda)$, determining $z_\Lambda - z_\theta \sim R \log(L_\theta/R)$, or

$$\frac{E_\Lambda}{E_\theta} \sim \left(\frac{L_\theta}{R}\right)^2, \tag{6.10}$$

in energy variables. Around the matching point the topology $S^1 \times S^3$ 'flops' into S^4 by a topology-changing transition of the type studied in Ref. [28], corresponding to a background of localized D4-branes. This can be completed in the UV by yet another $(2, 0)$ CFT in six compact dimensions.

Alternatively, we may take (6.4) at face vale as a well-defined background and extend it all the way to $z = \infty$ without any approximate matching to a UV CFT. This will be the purpose of this subsection, keeping in mind that the noncommutative tube will only be a finite transient in we are ultimately interested in an embedding into some UV fixed point.

The deformed metric (6.4) has the peculiar property of supporting the same density of states as the undeformed one, at least in the leading (classical) approximation. This follows from consideration of black hole solutions with horizon at coordinate z_0. These have metrics of the form (6.4) after the usual replacement $dt^2 \to h(z, z_0)dt^2$ and $dr^2 \to dr^2/h(z, z_0)$, with

$$h(z, z_0) = 1 - e^{2(z_0 - z)/R}. \tag{6.11}$$

A characteristic property of all noncommutative backgrounds is the independence of the Hawking temperature *and* the Bekenstein–Hawking entropy on the value of the deformation parameter, z_θ. Therefore, the leading approximation to the density of states is the standard Hagedorn law, with $T_H = (2\pi R)^{-1}$ as before. As pointed out in Ref. [41], things work very differently when it comes to the radiation corrections. The 'warp' factor $F^{1/2}$ in (6.4) affects the time coordinate, so that the local temperature $T(z) = T_H/\sqrt{-g_{tt}(z)}$ is exponentially redshifted as $z \to \infty$, rapidly quenching the radiation contribution to the entropy coming from the noncommutative region of the tube, $z > z_\theta$.

Thus, it appears that we have a viable model of a Hagedorn plateau that can be defined at arbitrarily high energies, and be free of the infrared problems posed by the radiation in the tube. In order to bring this issue into further scrutiny, let us consider the effective potential (4.12) for the case at hand. Choosing momenta (p_x, p_w) dual to translations along $\mathbb{T}^4 \times S^1$ and eigenvalues Δ_{S^3} of the Laplacian on S^3 one finds a potential of the form (see [5])

$$V_{\text{eff}}(z) = p_x^2 + \frac{\Delta_{S^3}}{R^2} + \left(\frac{a(z, z_\theta)}{R}\right)^2 + F(z, z_\theta)\, p_w^2, \tag{6.12}$$

where the function $a(z, z_0)$ interpolates between unity at $z \ll z_\theta$ and 9/4 at $z \gg z_\theta$. The large-z asymptotics of this potential is very sensitive to the value of p_w. In the compact theory, at finite L_θ, we have $p_w = n_w/L_\theta$, with $n_w \in \mathbb{Z}$. Then, the potential (6.12) has an asymptotic plateau at finite height in the zero-mode sector $n_w = 0$, whereas it grows exponentially as $\exp(2z/R)$ for $z \gg z_\theta$ in the non-zero momentum sectors.

Hence, the states of non-zero momenta on the circle S^1 of length L_θ are confined within the region $z \leq z_\theta + R$. The radiation in these modes, which scales extensively with $L_\theta L^4$, is cut off and shows no infrared divergence at large z. Conversely, the radiation in the zero mode $n_w = 0$, whose entropy only scales extensively with the volume L^4 of $\mathbb{T}^4$, is *not* confined by the potential and suffers from the large-z infrared problem.

We arrive at a very interesting crossroads. Even if the 'local' non-zero modes along the w-direction are efficiently cutoff in the length of the tube, the zero mode is insensitive to the 'noncommutative' quenching and still suffers from the infrared instability. For the purposes of defining the thermodynamics of the system

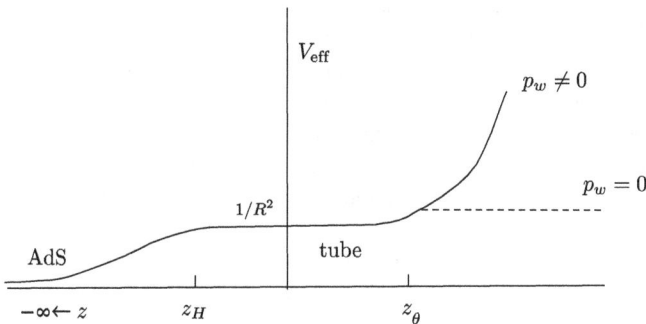

Fig. 8. Effective radiation potential for a massless field in the noncommutative LST background. The tube plateau evolves into an exponential wall for non-zero modes, $p_w \neq 0$ beyond z_θ. Zero modes still see a finite-height stepped plateau.

we must always start with a finite volume, which is scaled to infinity in defining thermodynamic densities. Hence, in order to define the thermodynamics of the noncommutative model we still need a UV completion of the type discussed in the previous section.

7. Conclusion

We have discussed the survival conditions for Hagedorn phases in the context of known holographic dualities. In the standard example of AdS/CFT duality with a weakly curved AdS background we can localize an energy transient dominated by highly excited strings with an approximate Hagedorn spectrum. In the whole phase diagram this transient is only accessible as a band of superheated states that decay quickly into the dual of a CFT plasma phase—the large AdS black hole.

In these lectures we have focused on more recent results regarding a different Hagedorn system with a holographic avatar, namely Little String Theory (LST). In this case, we are looking at an exotic string theory without gravity, which admits a gravitational dual description. There are also strings in the 'bulk' description, but these are standard critical ten-dimensional strings in a particular background where the low-energy supergravity approximation is reliable. Hence, this time the Hagedorn spectrum refers to 'long strings' in the boundary theory. In the bulk description the Hagedorn spectrum is not realized in terms of long critical strings. Instead, we have black hole configurations whose Bekenstein–Hawking entropy is linear in the energy, a result of the peculiar properties of the geometry and the presence of a linear dilaton profile. Therefore, the Hagedorn behavior of LST models is just the dynamics that 'decodes' the hologram at the horizon of these particular black holes.

We have studied the impact of quantum effects in the form of radiation corrections to the entropy function and we have found that these are significative. In fact, they are formally divergent and, when regulated by a high-energy cutoff, the whole band of six-dimensional Hagedorn states is pushed back to a purely threshold effect near the UV cutoff. In one removes the cutoff, there goes the Hagedorn behavior with it, leaving behind a non-holographic ideal gas with seven-dimensional thermodynamics.

This basic result also holds if one gives a physical entity to the UV cutoff, interpreting it as the door into a high-energy asymptotic region with an UV fixed point. In that case, the 'Hagedorn transient' follows the same fate as in the previously discussed AdS/CFT setup. Namely it is relegated to a band of unstable superheated states sitting behind a first-order phase transition between the plasma phase of the infrared (2, 0) CFT in six dimensions and the plasma phase of whatever CFT we use as an UV regulator, which in turn becomes the stable high-energy phase of the system.

Hence, we confirm that Hagedorn phases are quite elusive as asymptotic regimes in solvable models. They tend to become destabilized and survive only in the form of fine-tuned transients.

Acknowledgements

We thank C. A. Fuertes for many discussions and collaboration in the more recent results reported here. The work of J.L.F.B. was partially supported by MEC and FEDER under grant FPA2006-05485, the Spanish Consolider-Ingenio 2010 Programme CPAN (CSD2007-00042), Comunidad Autónoma de Madrid under grant HEPHACOS P-ESP-00346 and the European Union Marie Curie RTN network under contract MRTN-CT-2004-005104. The work of E.R. was partially supported by European Union Marie Curie RTN network under contract MRTN-CT-2004-512194, the American-Israel Bi-National Science Foundation, the Israel Science Foundation, the Einstein Center in the Hebrew University and by a grant of DIP (H.52).

References

[1] R. Hagedorn, *Suppl. Nuovo Cimento* **3** (1965) 147.

[2] S. Fubini and G.Veneziano, *Nuovo Cimento A* **64** (1969) 1640; K. Huang and S. Weinberg, *Phys. Rev. D* **25** (1970) 895; S. Frautschi, *Phys. Rev. D* **3** (1971) 2821; R.D. Carlitz, *Phys. Rev. D* **5** (1972) 3231; E. Alvarez, *Phys. Rev. D* **31** (1985) 418, *Nucl. Phys. B* **269** (1986) 596; M. Bowick and L.C.R. Wijewardhana, *Phys. Rev. Lett.* **54** (1985) 2485; B. Sundborg, *Nucl. Phys. B* **254** (1985) 883; S.N. Tye, *Phys. Lett. B* **158** (1985) 388; E. Alvarez and M.A.R. Osorio, *Phys. Rev. D* **36** (1987) 1175; P. Salomonson and B. Skagerstam, *Nucl. Phys. B* **268** (1986) 349, *Physica A* **158** (1989) 499; D. Mitchell and N. Turok, *Phys. Rev. D* **58** (1987) 1577, *Nucl. Phys. B* **294** (1987) 1138; I.I. Kogan, *JETP Lett.* **45** (1987) 709; B. Sathiapalan, *Phys. Rev. D.* **35** (1987) 3277; M. Axenides, S.D. Ellis and C. Kounnas, *Phys. Rev. D* **37** (1988) 2964; A.A. Abrikosov Jr. and Ya.I. Kogan, *Sov. Phys. JETP* **69** (1989) 235; M.J. Bowick and S.B. Giddings, *Nucl. Phys. B* **325** (1989) 631; S.B. Giddings, *Phys. Lett. B* **226** (1989) 55; J.J. Atick and E. Witten, *Nucl. Phys. B* **310** (1988) 291.

[3] J.L.F. Barbon and E. Rabinovici, arXiv:hep-th/0407236.

[4] J.L.F. Barbon and E. Rabinovici, *Nucl. Phys. B* **545** (1999) 371, arXiv:hep-th/9805143.

[5] J.L.F. Barbon, C.A. Fuertes and E. Rabinovici, *JHEP* **0709** (2007) 055, arXiv:0707.1158 [hep-th].

[6] S. Abel, J.L.F. Barbon, I.I. Kogan and E. Rabinovici, *JHEP* **9904** (1999) 015, arXiv:hep-th/9902058.

[7] J.L.F. Barbon, I.I. Kogan and E. Rabinovici, *Nucl. Phys. B* **544** (1999) 104, arXiv:hep-th/9809033.

[8] J.L.F. Barbon and E. Rabinovici, *JHEP* **0203** (2002) 057, arXiv:hep-th/0112173; *Found. Phys.* **33** (2003) 145, arXiv:hep-th/0211212.

[9] D. Kutasov and N. Seiberg, *Nucl. Phys. B* **358** (1991) 600.

[10] O. Aharony and T. Banks, *JHEP* **9903** (1999) 016, arXiv:hep-th/9812237.

[11] G. 't Hooft, arXiv:gr-qc/9310026. J. Bekenstein, *Phys. Rev. D* **49** (1994) 1912, arXiv:gr-qc/9307035; L. Susskind, *J. Math. Phys.* **36** (1995) 6377, arXiv:hepth/9409089.

[12] R. Brandenberger and C. Vafa, *Nucl. Phys. B* **316** (1989) 391; N. Deo, S. Jain and C.-I. Tan, *Phys. Lett. B* **220** (1989) 125, *Phys. Rev. D* **40** (1989) 2646; N. Deo, S. Jain, O. Narayan and C.-I. Tan, *Phys. Rev. D* **45** (1992) 3641.

[13] M. Kruczenski and A. Lawrence, *JHEP* **0607** (2006) 031, arXiv:hep-th/0508148.

[14] G.Veneziano, *Europhys. Lett.* **2** (1986) 199; G. Veneziano, *NATO-ASI Series B: Physics* **346** (1995) 63; Hagedorn Festschrift, J. Letessier, H. Gutbrod and J. Rafelski (eds.) (Plenum Press, New York); L. Susskind, arXiv:hep-th/9309145; G.T. Horowitz and J. Polchinski, *Phys. Rev. D* **57** (1998) 2557, arXiv:hep-th/9707170; T. Damour and G. Veneziano, *Nucl. Phys. B* **568** (2000) 93, arXiv:hep-th/9907030.

[15] S.W. Hawking and D.N. Page, *Commun. Math. Phys.* **87** (1983) 577.

[16] J.M. Maldacena, *Adv. Theor. Math. Phys.* **2** (1998) 231, arXiv:hep-th/9711200; S.S. Gubser, I.R. Klebanov and A.M. Polyakov, *Phys. Lett. B* **428** (1998) 105, arXiv:hep-th/9802109; E. Witten, *Adv. Theor. Math. Phys.* **2** (1998) 253, arXiv:hep-th/9802150.

[17] G. 't Hooft, *Nucl. Phys. B* **256** (1985) 727.

[18] J.L.F. Barbon and R. Emparan, *Phys. Rev. D* **52** (1995) 4527, arXiv:hep-th/9502155.

[19] E. Witten, *Adv. Theor. Math. Phys.* **2** (1998) 505, arXiv:hep-th/9803131.

[20] J.S. Dowker and G. Kennedy, *J. Phys. A* **11** (1978) 895; G.W. Gibbons and M. J. Perry, *Proc. Roy. Soc. Lond. A* **358** (1978) 467; J.L.F. Barbon, *Phys. Lett. B* **339** (1994) 41, arXiv:hep-th/9406209; *Phys. Rev. D* **50** (1994) 2712, arXiv:hep-th/9402004.

[21] O. Aharony, S. Gubser, J.M. Maldacena, H. Ooguri and Y. Oz, *Phys Rept.* **323** (2000) 183, arXiv:hep-th/9905111.

[22] M. Li, E. Martinec and V. Sahakyan, *Phys. Rev. D* **59** (1999) 044035, arXiv:hep-th/9809061; E. Martinec and V. Sahakyan, *Phys. Rev. D* **59** (1999) 124005, arXiv:hep-th/9810224; *Phys. Rev. D* **60** (1999) 064002, arXiv:hep-th/9901135.

[23] B. Sundborg, *Nucl. Phys. B* **573** (2000) 349, arXiv:hep-th/9908001; O. Aharony, J. Marsano, S. Minwalla, K. Papadodimas and M. Van Raamsdonk, *Adv. Theor. Math. Phys.* **8** (2004) 603, arXiv:hep-th/0310285; L. Alvarez-Gaume, C. Gomez, H. Liu and S. Wadia, *Phys. Rev. D* **71** (2005) 124023, arXiv:hep-th/0502227.

[24] N. Seiberg, *Phys. Lett. B* **408** (1997) 98, arXiv:hep-th/9705221; O. Aharony, M. Berkooz, S. Kachru, N. Seiberg and E. Silverstein, *Adv. Theor. Math. Phys.* **1** (1998) 148, arXiv:hepth/9707079; M. Berkooz, M. Rozali and N. Seiberg, *Phys. Lett. B* **408** (1997) 105, arXiv:hepth/9704089.

[25] O. Aharony, *Class. Quant. Grav.* **17** (2000) 929, arXiv:hep-th/9911147.

[26] O. Aharony, M. Berkooz, D. Kutasov and N. Seiberg, *JHEP* **9810** (1998) 004, arXiv:hepth/9808149.

[27] C.G. Callan, J.A. Harvey and A. Strominger, arXiv:hep-th/9112030.

[28] R. Gregory and R. Laflamme, *Phys. Rev. Lett.* **70** (1993) 2837, arXiv:hep-th/9301052.

[29] J.M. Maldacena, *Nucl. Phys. B* **477** (1996) 168, arXiv:hep-th/9605016.

[30] N. Itzhaki, J.M. Maldacena, J. Sonnenschein and S. Yankielowicz, *Phys. Rev. D* **58** (1998) 046004, arXiv:hep-th/9802042.

[31] S. Minwalla and N. Seiberg, *JHEP* **9906** (1999) 007, arXiv:hep-th/9904142.

[32] S. Elitzur, A. Forge and E. Rabinovici, *Nucl. Phys. B* **359** (1991) 581; G. Mandal, A.M. Sengupta and S.R. Wadia, *Mod. Phys. Lett. A* **6** (1991) 1685; E. Witten, *Phys. Rev. D* **44** (1991) 314.

[33] D. Kutasov and D.A. Sahakyan, *JHEP* **0102** (2001) 021, arXiv:hep-th/0012258.

[34] M. Berkooz and M. Rozali, *JHEP* **0005** (2000) 040, arXiv:hep-th/0005047; T. Harmark and N.A. Obers, arXiv:hep-th/0010169; *Phys. Lett. B* **485** (2000) 285, arXiv:hep-th/0005021; *Nucl. Phys. B* **742** (2006) 41, arXiv:hep-th/0510098; *Fortsch. Phys.* **53** (2005) 536, arXiv:hep-th/0503021; T. Harmark, V. Niarchos and N.A. Obers, *JHEP* **0510** (2005) 045, arXiv:hep-th/0509011.

[35] J.M. Maldacena and A. Strominger, *JHEP* **9712** (1997) 008, arXiv:hep-th/9710014.

[36] O. Aharony, A. Giveon and D. Kutasov, *Nucl. Phys. B* **691** (2004) 3, arXiv:hep-th/0404016; A.W. Peet and J. Polchinski, *Phys. Rev. D* **59** (1999) 065011, arXiv:hep-th/9809022; O. Aharony, M. Berkooz and N. Seiberg, *Adv. Theor. Math. Phys.* **2** (1998) 119, arXiv:hep-th/9712117; O. Aharony and M. Berkooz, *JHEP* **9910** (1999) 030, arXiv:hep-th/9909101; M. Rangamani, *JHEP* **0106** (2001) 042, arXiv:hep-th/0104125; A. Giveon, D. Kutasov, E. Rabinovici and A. Sever, *Nucl. Phys. B* **719** (2005) 3, arXiv:hep-th/0503121; D. Marolf, JHEP **0703** (2007) 122, arXiv:hep-th/0612012; D. Marolf and A. Virmani, arXiv:hep-th/0703251.

[37] N. Arkani-Hamed, A.G. Cohen and H. Georgi, *Phys. Rev. Lett.* **86** (2001) 4757, arXiv:hep-th/0104005.

[38] N. Arkani-Hamed, A.G. Cohen, D.B. Kaplan, A. Karch and L. Motl, *JHEP* **0301** (2003) 083, arXiv:hep-th/0110146.

[39] N. Dorey, *JHEP* **0407** (2004) 016, arXiv:hep-th/0406104.

[40] A. Hashimoto and N. Itzhaki, *Phys. Lett. B* **465** (1999) 142, arXiv:hep-th/9907166; J.M. Maldacena and J.G. Russo, *JHEP* **9909** (1999) 025, arXiv:hep-th/9908134; M. Alishahiha, Y. Oz and M.M. Sheikh-Jabbari, *JHEP* **9911** (1999) 007, arXiv:hep-th/9909215.

[41] J.L.F. Barbon and E. Rabinovici, *JHEP* **9912** (1999) 017, arXiv:hep-th/9910019.

Course 12

LECTURES ON CONSTRUCTING STRING VACUA

Frederik Denef

Jefferson Physical Laboratory, Harvard University,
Cambridge, MA 02138, USA
and
Instituut voor Theoretische Fysica, KU Leuven,
Celestijnenlaan 200D, B-3001 Leuven, Belgium

C. Bachas, L. Baulieu, M. Douglas, E. Kiritsis, E. Rabinovici, P. Vanhove, P. Windey
and L.F. Cugliandolo, eds.
Les Houches, Session LXXXVII, 2007
String Theory and the Real World: From Particle Physics to Astrophysics
© *2008 Published by Elsevier B.V.*

Contents

485

1. Introduction

The real world as we know it happens at energies well below the Planck scale, so it is very well described by effective field theory. There is a continuous infinity of consistent effective field theories. Remarkably, only a measure zero fraction of those seems to be obtainable from string theory. These effective field theories arise as low energy descriptions of certain "vacua" of string theory, which in some approximations schemes can be thought of as solutions to the equations of motion for the compactification space. Constraints on low energy effective particle spectra and interactions then typically arise from topological constraints on the internal degrees of freedom. Discreteness of the allowed values of parameters in the low energy effective action often arises from quantization effects, such as quantization of internal magnetic fluxes.

It is this remarkable selectivity of string theory which fuels the field known as string phenomenology. If the constraints from requiring the existence of a string theory UV completion are strong enough, then this can take us a significant step beyond the theorizing we can do based on field theory alone. This prompts the question: Are UV completion constraints merely academically fascinating, or are they strong enough to lead to experimental predictions?

In part triggered by developments in constructing string vacua meeting a number of rough observational constraints, a hypothetical picture has emerged in which the real world as we know it is just a tiny patch in a vast, eternally inflating multiverse, which effectively samples an gargantuan "landscape" of string vacua [1]. If correct, this has profound implications for a number of paradigms in physics, including the notion of naturalness and how we should read some of the stunning fine tunings of parameters in nature, as explained by Nima Arkani-Hamed at this school. The string theory landscape picture is not uncontested [2], and prompts the question: Is it correct?

We are far from a systematic understanding of the Hilbert space(s) of string theory, or even the space of its approximate, semiclassical vacua with four large dimensions. In view of this, one might consider the above questions premature. Nevertheless, we do understand parts. This includes in particular AdS vacua which have a known dual CFT description, and, to some extent at least, approximate semiclassical vacua which can be constructed as solutions to the classical equations of motion in regimes where quantum corrections are small. The former class is not immediately useful yet as a description of the universe as we know it, as observations indicate our vacuum has a positive effective cosmological constant. We can however try to construct controlled vacua of the latter kind, and address within this class the general questions raised above, even if our vacuum might not be accessible in this way.

The main goal of these lectures is to provide a detailed introduction to the best studied and for phenomenological applications richest set of such approximate semiclassical vacua: type IIB flux compactifications. I will almost exclusively focus on the formal construction of these vacua in string theory and the development of general methods for their analysis, leaving out specific phenomenological applications. I felt this would best complement the existing literature, and be most likely to be useful in the long run. The path from string theory to the real world is long and twisted, and if we want to get beyond loosely string-inspired but further unconstrained effective field theories, there is no choice but to dive deep into the bowels of string theory itself.

I have tried to make the lectures more or less self-contained. In particular all of the geometrical tools needed to build actual models are introduced from scratch, assuming only basic knowledge of the differential geometry contained in Section 2 of [3]. The framework in which I will be working is the F-theory description of IIB theory, because for many applications this is the most powerful and versatile approach, including for constructing semi-realistic string vacua. Nevertheless, as far as I know, no extensive elementary introduction bringing together all the basic ideas needed for applications of constructing F-theory vacua is available in the literature. I have therefore devoted a significant part of these lectures to explaining what F-theory is, how precisely it relates to M-theory and to the weak coupling limit of type IIB string theory, and how flux vacua and their number distributions over parameter space are obtained in this framework.

By the end of these lectures, you should be able to construct and analyze your own string vacuum.

The outline of this extended write-up is as follows:

1. In Section 2, I outline some of the main challenges that arise when trying to construct controlled string vacua (of any kind, not just IIB) and give a brief overview of several of the scenarios that have been proposed. In particular I will explain why successful explicit constructions must be "dirty", involving many ingredients which might seem unnecessarily contrived at first sight.

2. In Section 3, I explain what F-theory is, give its detailed construction in M-theory, discuss ways to think about branes, fluxes and tadpoles in this framework, explain how in general perturbative type IIB orientifold compactifica-

tions arise as a particular weak coupling limit, and how in this limit localized D7-branes and their worldvolume gauge fields emerge. I have tried to be as explicit as possible, following an elementary physical approach rather than an algebraic geometrical one.

3. In Section 4 we turn to the construction of F-theory flux vacua. First the four dimensional low energy effective action is given, both in the general F-theory setting and in its perturbative IIB weak coupling limit. Next the effect of turning on fluxes is considered, how they induce an effective superpotential stabilizing the shape and 7-brane moduli and how they can produce strong warping effects. Fluxes leave the size moduli massless at tree level. The latter can get lifted by various quantum effects, which are discussed next. Finally, two concrete scenarios to achieve fully moduli stabilized vacua with small positive cosmological constant are reviewed, the KKLT [4] and large volume [5] scenarios.

4. To find and analyze actual interesting concrete models of either of these two scenarios, a number of geometrical tools is needed. These are introduced from scratch in the hands-on geometrical toolbox which makes up Section 5.

5. Another indispensable ingredient in constructing and analyzing these vacua are techniques for computing approximate distributions of flux vacua over parameter space. These techniques are introduced and explained in quite a bit of detail in Section 6, including a general abstract derivation of the continuum index approximation to counting zeros of ensembles of vector fields, which can then be applied to various flux vacua counting problems, including F-theory flux vacua. A summary is given of various results, and the section concludes with a short discussion of metastability, landscape population and probabilities in the context of flux vacua.

6. Finally, in Section 7, we put all of these basic results together and outline how explicit models of moduli stabilized F-theory flux vacua can be obtained.

The different sections can to a large extent be read independently, and readers only interested in certain aspects of the constructions can probably safely skip sections.

Finally, let me emphasize that these lecture notes are in no way supposed to be a comprehensive review of the subject. The references are not meant to reflect proper historical attribution and are seriously incomplete. They are merely intended to provide pointers to articles which can be used as a starting point for further reading.

2. Basics of string vacua

2.1. Why string vacua are dirty

Even if you have never gotten your hands dirty constructing semi-realistic string vacua yourself, you probably have heard or read [6] that they tend to have a certain Rube Goldberg [7] flavor to them. In the following I will sketch what the challenges are to construct such vacua in a reasonably controlled way, and how meeting these challenges unavoidably requires adding several layers of complications.

2.1.1. The Dine–Seiberg problem
Perturbative supersymmetric string theories in flat Minkowski space and weakly curved deformations thereof only exist in ten dimensions. Observations on the other hand suggest only four large, weakly curved dimensions. The most obvious way out of this conundrum is to consider string theory on a space of the form

$$M_{10} = M_4 \times X, \tag{2.1}$$

where M_4 corresponds to visible space and X a compact manifold sufficiently small to have escaped detection so far. The size of the space X could be of the order of the fundamental scale, it could be highly curved and even defy classical notions of geometry, and it could break supersymmetry at a very high scale. Currently available techniques to analyze such situations are limited. Therefore the sensible thing to do is to consider well controlled cases and hope that these either will be favored by nature too, or that at least we can draw valuable lessons from them.

The most obvious well controlled cases are provided by compactifications (2.1) for which X_6 is large compared to the fundamental scale, and for which supersymmetry is broken at a scale well below the compactification scale. The latter is most easily achieved by first constructing a compactification which preserves some supersymmetry, and then perturb this in a controlled way to break supersymmetry.

In this regime, we can use the long distance, low energy approximation to string theory, that is, ten dimensional supergravity, described by an effective action for the massless fields. We can also consider eleven dimensional supergravity, the low energy approximation to M-theory [8], of which perturbative string theory is believed to be a particular weak coupling limit. Formally we can even go up one more dimension and imagine twelve dimensional F-theory [9], although this can be thought of more conservatively as a convenient geometrized description of type IIB string theory.

The following compactifications of string/M/F theory give rise to $\mathcal{N} = 1$ supersymmetry in four dimensional Minkowski space:

1. Heterotic or type I string theory on a three complex dimensional Calabi–Yau manifold.

2. Type II string theory on a 3d Calabi–Yau orientifold.

3. M-theory on a G_2 holonomy manifold.

4. F-theory on a Calabi–Yau fourfold.

All of these compactification manifolds are Ricci flat to leading order at large volume.

However, such compactifications immediately present a problem: at tree level, i.e. classically, they always come with moduli. Moduli are deformations of the compactification which do not change the 4d effective energy and therefore correspond to massless scalars in four dimensions. For example the size of the compactification manifold X is always a modulus at tree level, due to the scale invariance $g_{\mu\nu} \rightarrow rg_{\mu\nu}$ of vacuum Einstein equations. Other possible moduli are

1. The dilaton e^ϕ, for all string theories. This is already a modulus of the ten dimensional theory in Minkowski space. It is the parameter controlling the worldsheet perturbative expansion of the theory.

2. Axions: These arise when the supergravity theory under consideration has p-form potentials C_p and X has nontrivial harmonic p-forms (or equivalently nontrivial p-cycles Σ_p), as adding such a harmonic p-form to C_p will not affect the field strength $F_{p+1} = dC_p$ and hence not affect the energy. On the other hand adding a generic harmonic form is not a gauge transformation either, so these are physical, massless modes in the 4d theory.

3. Metric moduli:

   (a) CY complex structure (or shape) moduli, analogous to the complex structure modulus $\tau = \omega_2/\omega_1$ of the two-torus $T^2 = \mathbb{C}/(\mathbb{Z}\omega_1 \oplus \mathbb{Z}\omega_2)$.

   (b) CY Kähler (or size) moduli, analogous to the overall size of the T^2.

   (c) G_2 structure (shape and size) moduli.

4. Brane deformation and/or vector bundle moduli. Including branes or bundles is often forced upon us, both by tadpole cancelation constraints and by the phenomenological desire to have gauge bosons and charged matter in the theory that could reproduce the Standard Model.

In generic compactifications, there are thousands of these moduli. This is not good. Massless or very light scalars, if they couple at least with gravitational strength to matter, would be observed as long range "fifth" forces. No such forces have been observed. Moreover, it is difficult to allow for light scalars while preserving the successful predictions of standard cosmology.

Now, this would not seem to be such a big deal, since including quantum corrections, at least after breaking supersymmetry, are virtually guaranteed to give masses to the moduli, since there is in general nothing that protects scalars from becoming massive after supersymmetry is broken. However, one then runs into a universal problem of theoretical physics, which can be sloganized by saying that when corrections can be computed, they are not important, and when they are important, they cannot be computed. More concretely, the problem here is what is usually referred to as the Dine–Seiberg problem [10]. The argument is very simple. Let ρ be a modulus such as the volume V_X or the inverse string coupling $e^{-\phi}$, with the property that $\rho \to \infty$ corresponds to the weakly coupled region where we trust our tree level low energy effective action. Then if, as expected, quantum corrections generate a potential $V(\rho)$ in the 4d effective theory, this potential will satisfy

$$\lim_{\rho \to \infty} V(\rho) = 0, \tag{2.2}$$

precisely because of our assumption that at $\rho \to \infty$, we can trust the tree level low energy effective action, which has zero potential for ρ, by definition, since ρ is a modulus at tree level.

There are then two possibilities, as shown in Fig. 1 on the left: either $V > 0$ at large ρ, in which case the scalar has a runaway direction to $\rho = \infty$, or $V < 0$ at large ρ, in which case the scalar is pulled to the strong coupling region. A local minimum can only arise if higher order corrections are included; one needs

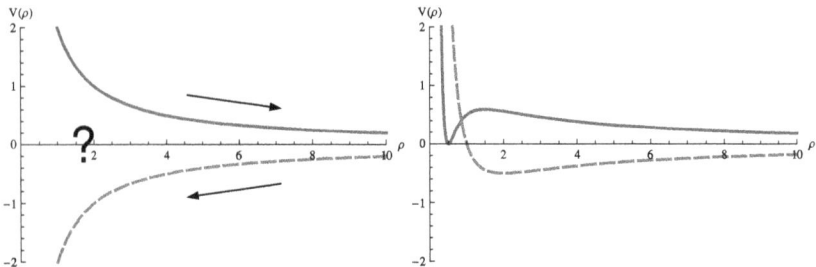

Fig. 1. On the left: two possible behaviors for the effective potential to first order. On the right: including higher order corrections.

two more corrections for the first case and one more for the second, as illustrated in Fig. 1 on the right. But, tautologically, when these corrections are important enough to cause a significant departure from the first order shape of V, as is necessary to get a local minimum, one is no longer in the weakly coupled region, and in principle all higher order corrections might be important too. In the absence of extended supersymmetry ($\mathcal{N} \geq 2$ in 4d), we generally lack the tools to compute more than a few orders in perturbation theory, so unavoidably we lose control.

On these grounds, Dine and Seiberg concluded in 1985 that the string vacuum we live in is probably strongly coupled. They may very well be right.

2.1.2. Flux vacua and no-go theorems

But new developments since then have changed the outlook somewhat. A crucial ingredient in these developments was the idea that by turning on p-form magnetic fluxes F in the internal manifold X, many new string vacua could be designed [11]. Dirac quantization requires these fluxes F to be integrally quantized, that is, integrals $\int_\Sigma F$ over closed p-cycles Σ must be integral (for a suitable normalization of F). So fluxes are discrete degrees of freedom. The type of flux and the possible values of p depend on the theory under consideration, but the basic idea is always the same. The crucial point is that turning on flux generates a moduli potential at *tree level*, of the form[1]

$$V_F = \frac{m_p^4}{V_X^2} \int_X \sqrt{g}\, g^{mr} \cdots g^{ns} F_{m \cdots n} F_{r \cdots s}, \qquad (2.3)$$

where m_p is the four dimensional Planck mass and V_X is the volume of X. The prefactor appears in the effective potential after rescaling the 4d metric as $g_{\mu\nu} \to \frac{m_p^2}{V_X} g_{\mu\nu}$ to remove the V_X-dependence of the Einstein–Hilbert term in the 4d effective action.

Since the internal metric depends on the moduli we originally had, we thus generate a tree level potential for the moduli. This is great, but unfortunately does not solve our problem. Under a rescaling $g_{mn} \to \lambda^2 g_{mn}$ of the internal metric, the potential scales as $V \to \lambda^{-2d+d-2p} V$, so if we parametrize the internal metric by $g_{mn} = r^2 g_{mn}^0$ where we normalize g_{mn}^0 such that $\int_X \sqrt{g^0} = 1$, we get

$$V_F/m_p^4 = r^{-d-2p} \int_X \sqrt{g^0}\, (g^0)^{mr} \cdots (g^0)^{ns} F_{m \cdots n} F_{r \cdots s}. \qquad (2.4)$$

This is manifestly positive definite, so there is a runaway direction towards large r, except if $\int F^2 = 0$. This conclusion remains unchanged if we have different sets

---

[1] For simplicity of exposition we suppress for now the additional dilaton dependence which we get in string theory (which is different for the two types of flux one can turn on, RR and NSNS). We work in units in which the 10/11 dimensional Planck scale is set to one.

of fluxes with different values of p, since each term will be positive definite. Now, if $\int F^2 = 0$ and the geometry is nonsingular, then positive definiteness of the internal metric implies $F = 0$, so there was no potential to begin with. Thus we conclude that in the regime in which classical geometry can be trusted, there can be no such flux vacua.

This analysis is a little too naive though. We neglected the possible backreaction of the fluxes on our originally Ricci-flat metric on X. If the Einstein–Hilbert action of X becomes nonzero, this will give an additional effective potential term in four dimensions, which will also scale under $g_{mn} \rightarrow r^2 g_{mn}$:

$$V/m_p^4 = \sum_p r^{-2p-d} \int_X \sqrt{g^0} F_p^2|_0 - r^{-2-d} \int_X \sqrt{g^0} R^0. \tag{2.5}$$

If the internal curvature is negative, the runaway gets only worse. So let us take it to be positive, as is the case for example for a sphere. If F_p is nonzero only for $p \geq 2$, the curvature will in fact give the dominant contribution at large r. We are then in the case of the dashed line in Fig. 1, and we see that flux vacua with negative cosmological constant might be possible. Indeed, such flux vacua are abundant. A subset are the so-called Freund–Rubin vacua [12], where $p = d$ and X is taken to be an Einstein space; the simplest example is M-theory on $AdS_4 \times S^7$ with 7-form flux on the S^7. We will discuss these in more detail in the next subsection. However, such compactifications cannot be viewed as deformations of the compactifications to flat Minkowski space we started off with: The flux is what supports the internal manifold X; if we send F to zero, X collapses to zero size. You can see this from (2.5) from the fact that the minimum $r_* \rightarrow 0$ when scaling F to zero keeping g_{mn}^0 fixed, or explicitly for example for $AdS_4 \times S^7$, where the size of the S^7 (and, pegged to it, the curvature radius of AdS_4) is proportional to a positive power of $N := \int_X F_d$. Moreover, in examples studied so far, the KK scale is of order of the AdS scale. If this is so, then to get to our observed near-zero value of the cosmological constant, some quantum effect has to provide an additional contribution to the effective potential of the order of the KK scale, which tautologically means quantum corrections are not small compared to our leading order potential, and we lose control again. To see this correlation between the scales, note that

$$M_{\text{AdS}}^2 = \frac{V}{m_p^2} \sim r_*^{-2} \int_X \sqrt{g^0} R^0, \tag{2.6}$$

$$M_{\text{KK}}^2 \sim \frac{1}{D^2} = r_*^{-2} \frac{1}{D_0^2}, \tag{2.7}$$

$$\frac{M_{\text{AdS}}^2}{M_{\text{KK}}^2} \sim D_0^2 \int_X \sqrt{g^0} R^0. \tag{2.8}$$

Here r_* is the local minimum of $V(r)$, D is the diameter[2] of X for the metric g, and D_0 is the diameter for the metric g^0. For the round sphere, (2.8) is of order 1, and it appears difficult to find examples where this scale ratio can be made very small. I don't known of a proof that this cannot be done though. If you looked a little harder, you might well be able to find constructions with a large scale hierarchy. We will briefly return to these compactifications in the next subsection.

So far we discussed the cases with $F_1 = F_0 = 0$. When only $F_1 \neq 0$ (i.e. a nonzero scalar gradient), things are qualitatively different. Now the two terms in (2.5) scale in the same way, so if the curvature can adjust itself so the two terms cancel at a given r, they will cancel at all r, and the potential will be zero. Such solutions do exist in type IIB theory. A simple example is $X = S^2 \times T^4$ with 24 7-branes transversal to the S^2, which source RR 1-form flux F_1. More generally, F-theory compactifications can be thought of as being of this kind. We will discuss these in detail further on. Still, such compactifications do not solve our problem. By construction, the overall scale modulus r remains massless. On top of that, there will be other nonstabilized geometric moduli—a typical F-theory compactification for example has thousands of moduli.

When $F_0 \neq 0$, the dominant term in (2.5) at $r \to \infty$ is the corresponding flux term. Together with other flux terms, this could in principle lead to the situation corresponding to the solid line in Fig. 1, and could thus possibly even lead to metastable Minkowski or de Sitter vacua. We will return to this case shortly.

The above arguments are a baby version of a very general no-go theorem, proven by Maldacena and Nuñez [14] (see also [15] and recently [16]). The theorem can be stated as follows. We start with *any* D dimensional gravity theory whose gravitational dynamics is given by the standard Einstein–Hilbert action (without higher curvature corrections), coupled to arbitrary massless fields (scalars, p-forms, nonAbelian gauge fields, ...) with positive kinetic terms, and with zero or negative potential (which could depend on the scalars). We then compactify this theory on a manifold X, with coordinates y^m, in the most general way, including a possible y-dependent "warp factor" $\Omega(y)$, to obtain a vacuum solution in $a < D$ dimensions:

$$ds_D^2 = \Omega(y)^2(ds_a^2(x) + ds_X^2(y)). \tag{2.9}$$

By vacuum solution we mean a metric $ds_a^2(x) = \eta_{\mu\nu}(x)dx^\mu dx^\nu$ which is either Anti-de Sitter, Minkowski, or de Sitter. We assume that X is compact and that the warp factor does not diverge anywhere.[3] The theorem now says that under these

---

[2] The diameter is the largest distance between two points. Its relation to the KK scale, i.e. the bottom of the eigenvalue spectrum of the Laplacian, is intuitively plausible but mathematically not trivial. See [13] for further discussion.

[3] This is somewhat stronger than the conditions under which the theorem was proven in [14].

conditions, *there are no compactifications down to de Sitter space, and none to Minkowski space except if $p = 1$ or $p = D - 1$, in which case we get Minkowski with $\Omega = constant$.*

This is consistent with our simple analysis above. (The case $p = D-1$, which is related to $p = 1$ by $F_1 = \star F_{D-1}$ was not considered in our analysis because we restricted to magnetic fluxes, which have all legs in the internal space.)

As promised above, we now return to the case $F_0 \neq 0$. This was studied separately in [14] for the particular string theory where it could occur, namely "massive" type IIA, with the conclusion that no compactifications to Minkowski or de Sitter can exist, dashing the hope left open by our simple scaling analysis earlier. A few AdS solutions with $F_0 \neq 0$ are known [17], but unfortunately they all have KK scales of the order of the AdS scale.

2.1.3. Orientifold planes and type II flux vacua

The Maldacena–Nuñez no-go theorem sounds like bad news, and in fact it is. Not because the assumptions of the theorem cannot be violated—string theory violates them immediately, because its low energy effective action does have higher order curvature corrections, and because the theory contains singular negative tension objects (O-planes)—but because it forces us to depart from the clean world of actions to second order in derivatives and smooth geometries, and to migrate to the dirty world of higher order corrections and orientifold singularities. As a result, control problems creep back in.

Still, let us proceed. We begin by returning to our simple scaling analysis, and add to (2.5) some of the extra contributions D-branes and O-planes provide. Space filling $D(3 + k)$-branes wrapping a k-cycle in X will give a contribution $\sim r^{-2d+k}$, while orientifold planes on a k-cycle give a similar contribution but (possibly) negative. Curvature of Dp-branes typically gives a negative contribution to the energy density scaling like the energy of a $D(p - 4)$-brane. Worldvolume fluxes on Dp-branes give positive energy contributions scaling like those of lower dimensional branes.

Clearly now there are many more terms in the effective potential, include some more with negative signs, so we can expect to get minima more easily. We consider two special cases of interest. The first one is type IIB string theory on a $d = 6$ Calabi–Yau orientifold with O3 and O7 planes, D3 and D7 branes, and NSNS (H) and RR (F) 3-form fluxes. Schematically this gives the potential

$$
\begin{aligned}
V(r, \phi)/m_p^4 \quad = \quad & e^{4\phi} \big[r^{-12} \, (e^{-\phi} \, T_3|_0 + \int F^2|_0 + e^{-2\phi} \int H^2|_0) \\
& + r^{-8} \, (e^{-\phi} \, T_7|_0 - e^{-2\phi} \int R|_0) \big].
\end{aligned} \tag{2.10}
$$

Here we reinstated the dependence on the dilaton e^{ϕ}, and $T_p|_0$ denotes the total tension from Dp and Op branes in the metric g_{mn}^0 and with $\phi \equiv 0$. The structure of the potential suggests we can find nontrivial Minkowski flux vacua with $R = 0$, provided the D7 and O7 tensions cancel, so $T_7 = 0$, and provided the contributions from fluxes, O3 and D3-branes cancel as well. Indeed such vacua turn out to exist [18–20], and we will discuss them in great detail in Section 4.

Of course, by construction now, r is still a modulus. In fact it turns out that all Kähler moduli remain unfixed by the flux potential. To stabilize those in this setup, one must resort to quantum corrections again, and the Dine–Seiberg problem kicks back in. Nevertheless an ingenious scenario for how this could be made to work in a reasonably controlled way, and moreover how a small positive cosmological constant could be achieved, was proposed by Kachru, Kallosh, Linde and Trivedi (KKLT) [4]. We will return to this in Section 4.

The second case we consider is type IIA on a Calabi–Yau orientifold with O6 planes, order 1 units of RR flux F_0 and NSNS flux H, and N units of RR flux F_4. Setting O(1) quantities to 1, this generates a potential of the form

$$V(r, \phi)/m_p^4 \sim e^{4\phi}\left[N^2 r^{-14} + r^{-12}e^{-2\phi} - r^{-9}e^{-\phi} - r^{-8}e^{-2\phi}\int R + r^{-6}\right].$$
(2.11)

The first negative term is the O6 contribution. Setting $R = 0$, this has minima for large N at

$$r \sim N^{1/4}, \qquad e^{\phi} \sim N^{-3/4},$$
(2.12)

that is, large volume and weak string coupling. In [23], a more refined analysis was done and it was shown that such flux vacua indeed exist in type IIA string theory, at least at the level of the 4d effective theory, with all geometric moduli fixed. In [24] this was promoted to full ten dimensional solutions in the approximation of smeared O6 charge. Although you cannot see this from the simple considerations we made, the minima turn out to be always AdS minima, but of a different nature than those of Freund–Rubin type we mentioned earlier. From the scaling (2.12), we see that $m_p^2 = e^{-2\phi}r^6 \sim N^3$, $V/m_p^4 \sim -N^{-9/2}$, and consequently

$$M_{\text{AdS}}^2 = \frac{V}{m_p^2} \sim N^{-3/2}, \qquad M_{\text{KK}}^2 = \frac{1}{r(N)^2 D_0(N)^2} \sim \frac{N^{-1/2}}{D_0(N)^2}.$$
(2.13)

Hence, unlike in the Freund–Rubin case, provided the diameter $D_0(N)$ of the unit volume normalized metric does not grow with N (or grows less fast than $N^{1/2}$), we automatically get a hierarchy of KK and AdS scales in the large N limit. This removes the immediate obstruction to controlled lifting to positive cosmological constant we mentioned for Freund–Rubin type vacua. Which is not to say that lifting is now straightforward or that there are no other control issues with these compactifications. We will come back to this in the next subsection.

In conclusion, we arrived at classical moduli stabilization scenarios in type IIB and IIA string theory which might have a chance of producing some reasonably controlled vacua. But they are not the simple smooth exact classical solutions we might have hoped for. The constructions we have at this point need many different ingredients. Calling them Rube Goldberg contraptions would be excusable. But as I hope I have made clear, it is the failures of simpler ideas[4] that has driven us this far.

2.2. A brief overview of some existing scenarios

We now turn to a brief overview of some of the constructions that have been proposed, and of their virtues and drawbacks. I will not try to be complete, far from it; the idea is to just give a flavor of what has been done and what the issues are. Several concepts mentioned may be foreign to you; some of the material will become more clear further on in the lectures. Much more can be found in the reviews [25, 26]. The references below are very incomplete and only meant to give you some pointers to the relevant clusters of papers.

2.2.1. IIB orientifolds/F-theory

These are variants of the KKLT scenario mentioned in the previous subsection. We will deeply get into the details of these scenarios in the next sections. Some key references for the basic setup are [4,5,18–20,27]. The virtues of this scenario, when it works, are:

+ Complex structure moduli, dilaton, D7 moduli can be stabilized classically at high mass scales by RR, NSNS and D7 worldvolume $U(1)$ fluxes.

+ Because there are always many more fluxes than moduli, there is a very high degree of discrete tunability of physical parameters, which helps in producing controlled models. In particular the cosmological constant can in principle be discretely tuned to become extremely small, easily of the order of the measured cosmological constant or less [28].

+ The classical geometry of the compactification manifold remains Calabi–Yau after turning on fluxes, up to warping [20]. This means in particular that many of the powerful techniques from algebraic geometry, which were invaluable to get a handle on Calabi–Yau compactifications without fluxes, can still be used to describe these vacua.

+ Strongly warped throats of Klebanov–Strassler type [29] can be achieved through the warping of the Calabi–Yau geometry. This can generate large

---

[4]and perhaps the strategy to start from the highly supersymmetric string vacua we do know and control.

scale hierarchies, useful for e.g. controlled supersymmetry breaking by adding anti-D3-branes at the bottom of the throat, or for embedding Randall–Sundrum [30] type scenarios in string theory.

+ These vacua can (with fine tuning of initial conditions) accommodate slow roll inflation, at least in local models [31–33].

+ There is a rich set of explicit D-brane constructions possible in these models, useful for particle physics model building; for a review see [34]. More general F-theory model building is also possible and provides probably the most extensive class of particle physics models in string theory, allowing in particular unification to arise naturally [21, 22].

+ The F-theory description provides g_s corrections to the geometry which smooth out the O7 singularities [35]. This is needed if one wants a large radius geometrical description of the background because O-plane singularities tend to be of a very bad kind, ripping up space at finite distance due to the negative tension of O-planes.

The main drawbacks are

– One needs quantum corrections to stabilize the Kähler (size) moduli, making the Dine–Seiberg problem something to worry about.

– Generic F-theory compactifications, away from special (orientifold) limits, do not have a globally well defined weakly coupled worldsheet description, even in the infinite volume limit, because the string coupling undergoes S-duality transformations when circling around generic (p, q) 7-branes. Hence for generic compactifications, it is unclear how to systematically compute e.g. α' corrections even in principle.

– Similarly, it is not known how to systematically compute such corrections in the presence of RR flux. This is a universal problem of any flux compactification involving RR fluxes.

2.2.2. IIA orientifolds
This is the IIA flux model mentioned at the end of the previous subsection. A key reference for the basic setup is [23]. The virtues of this scenario are

+ Classical RR and NSNS fluxes are sufficient to stabilize all geometrical moduli, in what from the low energy effective action point of view at least appears to be a parametrically controlled regime. Axions are not lifted, but these can get masses by quantum effects without triggering control issues.

+ Intersecting D-brane models can be embedded (although they are special Lagrangian, and very few explicit constructions of special Lagrangians in compact manifolds are known).

The drawbacks are:

– Because there are about as many moduli as fluxes, there is only limited discrete tunability. Warped throats cannot be generated classically, so controlled
supersymmetry breaking by adding anti-D-branes is not possible in this way.

– The presence of the localized O6 makes the solutions geometrically incomplete, since the metric and string coupling blow up at finite radius from the
O6. In flat space, this can be regularized by lifting to M-theory (the analog
of lifting type IIB O7-planes to F-theory), where the O6 turns into the smooth
Atiyah–Hitchin manifold [36]. Unfortunately, there is no direct M-theory lift
in the case at hand, due to the presence of F_0 flux. The control issues this
implies are further discussed in [37].

– There is a no-go theorem [38] excluding slow roll inflation without adding
more ingredients than those considered in [23]. (Quite a bit more ingredients
were considered in [39] however, showing how the no-go theorem could be
evaded.)

2.2.3. M on G_2

G_2 flux vacua can be viewed as M-theory uplifts of IIA orientifold flux vacua,
but with $F_0 = 0$. The latter restriction must be made because, although type
IIA with $F_0 \neq 0$ (also known as massive type IIA or Romans theory) can be
viewed as a limit of M-theory via reduction on a 2-torus and a twisted version
of T-duality [40], it cannot be directly be obtained by compactification of eleven
dimensional supergravity.

The good news is that everything is geometrical in this setup; in particular
there are no orientifold planes to worry about. The bad news is that, in accordance
with our general scaling arguments, the flux potential does not have local minima.
(This agrees with the fact that in IIA, when $F_0 = 0$, there are no vacua.)

Thus, a new ingredient is needed. In [41], a proposal was made for such a new
ingredient, making use of constraints from supersymmetry, leading to a negative
contribution to the potential from certain nonAbelian excitations around a locus
of singularities in the G_2 manifold. Although unfortunately no explicit example
is known, and its physical origin remains to be elucidated, the virtue of such a
model would be that it freezes all geometric moduli at once. The drawback, besides the fact that it is not known conclusively if such compactifications actually
exist, are similar to those of type IIA models.

2.2.4. Pure flux

This includes all flux compactifications which exist without the addition of "extra" elements such as orientifold planes. The simplest class of examples are the supersymmetric Freund–Rubin vacua [12] of M-theory of the form $AdS_4 \times X_7$ where X_7 is a Sasaki–Einstein 7-manifold (for reviews see [42–44]), supported by N units of flux F_7 on X. Sasaki–Einstein manifold are obtained as the base of Calabi–Yau fourfold cones. The simplest example is $X = S^7$, obtained by considering eight dimensional flat space as a Calabi–Yau cone over a sphere.

Such vacua are very well controlled in the large N limit, and have $2 + 1$ dimensional superconformal field theory duals obtained from placing N M2 branes at the tip of the cone, so in principle they are even defined nonperturbatively as quantum gravity theory. Their disadvantage as far as realistic model building is concerned is that known examples do not have a large hierarchy between KK and AdS scales, as mentioned earlier. Quite a few examples also have residual moduli, descending from the moduli of the Calabi–Yau cone.

In IIA, Freund–Rubin compactifications with just one flux are not possible because of a dilaton runaway. More involved IIA pure flux compactifications on non-Calabi–Yau manifolds with multiple fluxes do exist however; for a recent examples and a nice overview, see [17]. The examples include $\mathbb{CP}^3$ with all kinds of fluxes turned on, carrying a non-Kähler, non-nearly-Kähler, non-Einstein metric. Again though, the KK and AdS scales are observed to be of the same order.

Finally, the IIA Calabi–Yau flux compactifications of [11, 45] are also orientifold-free. Their low energy effective action is gauged $\mathcal{N} = 2$ supergravity. In principle, these could exhibit AdS, Minkowski, or dS vacua in four dimensions without violating the Maldacena–Nuñez no-go theorem, once quantum corrections to the scalar metric are taken into account. Because of the $\mathcal{N} = 2$ supersymmetry, an infinite series of such corrections is known. I am not aware however of examples of controlled dS or Minkowski vacua with all moduli stabilized in this setup. Also, after supersymmetry breaking, control over quantum corrections will become problematic again.

2.2.5. Heterotic

Heterotic string or heterotic M theory [46] have the important advantage that it naturally gives rise to grand unified models, something which is apparently not natural in weakly coupled type II intersecting D-brane models.[5] Moduli stabilization has been more challenging in this setting, due to the absence of RR fluxes, limited tunability, and technical difficulties in working with the holomor-

---

[5]However, as noted in Section 2.2.1, it is natural in more general F-theory model building.

phic vector bundles which are the core of these compactifications. Significant progress has been made in recent years however, see e.g. [47].

2.2.6. Nongeometric

Not all compactifications of string theory are geometric. Some recent considerations of nongeometric compactifications include [48] based on "over"-T-dualization of toroidal flux compactifications involving H-flux, and [49] based on Landau–Ginzburg models. The latter approach in particular allows to study IIB orientifolds on the mirror of rigid CY manifolds, which do not have any Kähler moduli and are therefore necessarily nongeometric. The main advantage is that since there are no Kähler moduli, we no longer need to invoke quantum corrections, which was the main issue with geometric IIB compactifications. The disadvantage, perhaps, is that one can no longer directly use geometric notions such as fluxes, warping and so on; instead CFT equivalents have to be found, which is more challenging.

2.2.7. Noncritical

Finally, string theories also do not need to be critical; the dimension of the target space can exceed $d = 10$. There is a whole landscape of supercritical string theories, connected to the more familiar critical landscape [50]. Noncritical string theories do not have conventional time-independent vacuum solutions, but in a cosmological setting, this is not necessarily a problem. Relatively little has been explored in this arena.

3. F-theory and type IIB orientifold compactifications

We now turn to the details of constructing string vacua. We will stay on the more conservative and best understood end of the spectrum of possibilities outlined in the previous section, namely type IIB (F-theory) flux vacua. As outlined in Section 2.2.1, this class of models also provides a very rich and interesting phenomenology.

For many purposes, including moduli stabilization, F-theory provides the most elegant and powerful framework to analyze questions in type IIB string theory. I will therefore spend some time first to explain what F-theory is, and how exactly it relates to IIB orientifold compactifications.

3.1. What is F-theory?

Type IIB supergravity has $\mathcal{N} = 2$ supersymmetry in 10 dimensions (32 supersymmetry generators). To write down the action, it is convenient (and it makes

S-duality manifest) to define

$$\tau \; := \; C_0 + i e^{-\phi}, \tag{3.1}$$

$$G_3 \; := \; F_3 - \tau H_3, \tag{3.2}$$

$$\tilde{F}_5 \; := \; F_5 - \frac{1}{2} C_2 \wedge H_3 + \frac{1}{2} B_2 \wedge F_3. \tag{3.3}$$

$$F_p \; := \; dC_{p-1} \quad (p = 1, 3, 5), \qquad H_3 := dB_2. \tag{3.4}$$

We will work with the 10d Einstein frame metric, which has canonical Einstein–Hilbert term in ten dimensions and is related to the string frame metric by

$$g_{MN}^E = e^{-\phi/2} g_{MN}^S, \tag{3.5}$$

where ϕ is the dilaton ($g_{\text{IIB}} = e^{\phi}$ is the string coupling constant). Due to the presence of the self-dual 5-form field strength, there is no standard manifestly covariant action for this theory,[6] but the following gives formally the correct equations of motion:

$$S_{\text{IIB}} \; = \; \frac{2\pi}{\ell_s^8} \left[\int d^{10}x \sqrt{-g}\, R - \frac{1}{2} \int \frac{1}{(\text{Im}\,\tau)^2}\, d\tau \wedge *d\bar{\tau} \right.$$
$$\left. + \frac{1}{\text{Im}\,\tau} G_3 \wedge *\overline{G_3} + \frac{1}{2} \tilde{F}_5 \wedge *\tilde{F}_5 + C_4 \wedge H_3 \wedge F_3 \right]. \tag{3.6}$$

This has to be supplemented with the selfduality constraint $*\tilde{F}_5 = \tilde{F}_5$, *after* varying the action, to get the complete equations of motion. The string length ℓ_s is related to α' by $\ell_s = 2\pi\sqrt{\alpha'}$. In this notation D$p$-brane tensions are $T_{Dp} = \frac{2\pi}{\ell_s^{p+1}}$.

The action (3.6) is manifestly invariant under $SL(2, \mathbb{Z})$ *S*-duality:

$$\tau \; \rightarrow \; \frac{a\tau + b}{c\tau + d}, \tag{3.7}$$

$$\begin{pmatrix} H \\ F \end{pmatrix} \; \rightarrow \; \begin{pmatrix} d & c \\ b & a \end{pmatrix} \begin{pmatrix} H \\ F \end{pmatrix}, \tag{3.8}$$

$$\tilde{F}_5 \; \rightarrow \; \tilde{F}_5, \tag{3.9}$$

$$g_{MN} \; \rightarrow \; g_{MN}. \tag{3.10}$$

The action (3.6) looks uncannily like something obtained by compactification of a *twelve* dimensional theory on a torus with modulus τ, with F_3 and H_3 the components of some twelve dimensional $\widehat{F}_4$ reduced along the two 1-cycles of

---

[6]There is a non-manifestly covariant action for selfdual p-forms [51]. The quantum field theory framework for such fields was developed in [52].

the torus. Moreover, the $SL(2, \mathbb{Z})$ gauge symmetry then simply becomes the geometrical $SL(2, \mathbb{Z})$ reparametrization gauge symmetry of the torus. This and other uncanniness has led to the proposal of F-theory [9], a putative twelve dimensional Father of all theories. However, this does not work as straightforwardly as one might wish. To begin with, there is no twelve dimensional supergravity with metric signature $(1, 11)$. Also, if there were actually a twelve dimensional theory with some $\widehat{F}_4$, then we would have to explain why redacting $\widehat{F}_4$ along the full T^2 and a point do not show up as 2- resp. 4-form field strengths in type IIB. Even more directly, why would the complex structure modulus τ of the torus appear in (3.6), but not the overall size modulus?

Proposals have been made to circumvent these problems, but there is actually an alternative geometrical interpretation in M-theory, which works perfectly in the most conservative way. The rough idea is as follows. We start with M-theory on a small T^2 with modulus τ. Taking one of the small T^2 circles to be the M-theory circle gives weakly coupled IIA on the other small circle. T-dualizing along this circle gives IIB on a large circle. In the limit of vanishing M-theory T^2, this becomes uncompactified IIB. This can be extended to T^2 fibrations by performing this procedure fiberwise, resulting in type IIB compactifications with varying dilaton-axion given by the geometric τ-modulus of the T^2, effectively realizing the F-theory idea through this chain of dualities.

One might worry though that this fiberwise duality procedure might not give rise to a four dimensional Lorentz invariant solution, given the very different origin of one of the spatial directions in the IIB theory. But in fact, somewhat miraculously, it turns out that the result is fully Lorentz invariant in the limit.

Let us make this more precise. We start with M-theory, whose eleven dimensional low energy effective action is

$$S_{\mathrm{M}} = \frac{2\pi}{\ell_M^9} \left[\int d^{11}x \sqrt{-g}\, R - \frac{1}{2} \int G_4 \wedge *G_4 - \frac{1}{6} C_3 \wedge G_4 \wedge G_4 \right.$$
$$\left. + \ell_M^6 \int C_3 \wedge I_8(R) + \cdots \right], \qquad (3.11)$$

where $G_4 := dC_3$, $I_8(R)$ is a polynomial of degree 4 in the curvature [53], and ℓ_M is the eleven dimensional Planck length. In this notation the tension of an Mp brane is $T_{Mp} = \frac{2\pi}{\ell_M^{p+1}}$, in analogy to the string case. Although the I_8 correction is higher derivative, we include it here because it plays a crucial role in anomaly/tadpole cancelation. Moreover further terms related to it by supersymmetry allow to evade the Maldacena–Nuñez no-go argument against flux compactifications to Minkowski space, providing negative energy balancing the decompactification pressure of flux.

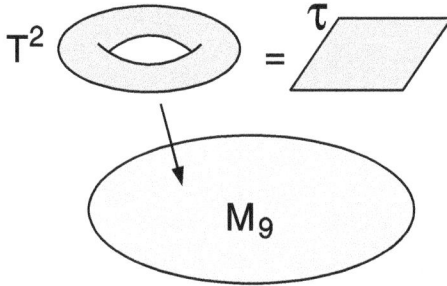

Fig. 2. F-theory from M-theory. Starting point: T^2 fibration over M_9.

Now, as illustrated in Fig. 2, we compactify this theory on $T^2 \times M_9$, or more generally a T^2 fibration over M_9, with metric

$$ds_M^2 = \frac{v}{\tau_2}\big((dx + \tau_1 dy)^2 + \tau_2^2 dy^2\big) + ds_9^2, \tag{3.12}$$

where x and y are periodic coordinates with periodicity 1. This corresponds to a T^2 with complex structure modulus $\tau = \tau_1 + i\tau_2$ and total area v. We can allow v and τ to depend on the coordinates of M_9; in this case we get a T^2 fibration rather than a direct product. We call the 1-cycle along the x-direction the A-cycle, and the one along the y-direction the B-cycle. We will reduce from M to IIA along the A-cycle, and then T-dualize to IIB along the B-cycle.

The relation between the circle compactified M-theory and type IIA metrics is in general given by

$$ds_M^2 = L^2 e^{4\chi/3}(dx + C_1)^2 + e^{-2\chi/3} ds_{\text{IIA}}^2, \tag{3.13}$$

where x is a coordinate on the M-theory circle with periodicity 1 and L is a conventional length which sets the scale of the M-theory circle and which we can choose at our convenience (since rescaling L can be absorbed in shifting χ by a constant). The circle bundle connection C_1 is the type IIA RR 1-form potential. This immediately gives

$$C_1 = \tau_1 dy, \qquad e^{4\chi/3} = \frac{v}{L^2\tau_2}, \qquad ds_{\text{IIA}}^2 = \frac{\sqrt{v}}{L\sqrt{\tau_2}}(v\tau_2 dy^2 + ds_9^2). \tag{3.14}$$

Now we want to T-dualize this geometry along the y-circle. T-duality maps IIA to IIB, the circle length L_A to $L_B = \ell_s/L_A$, the RR axion becomes $C_0 = (C_1)_y$ and the string coupling $g_{\text{IIB}} = \frac{\ell_s}{L_A} g_{\text{IIA}}$. To compute L_B and g_{IIB}, we thus need to know ℓ_s and g_{IIA}. Reducing the M2 probe action to F1 resp. D2 probe actions on the metric (3.13), we get the relations

$$\frac{1}{\ell_s^2} = \frac{L}{\ell_M^3}, \qquad \frac{1}{g_{IIA}\ell_s^3} = \frac{1}{e^{\chi}\,\ell_M^3}. \tag{3.15}$$

This and the above allows us to express ℓ_s and g_{IIA} as a function of v, τ_2, L and ℓ_M, and hence to compute the IIB metric and coupling in terms of these quantities. The final result is

$$C_0 + \frac{i}{g_{IIB}} = \tau, \qquad ds_{IIB,S}^2 = \frac{\sqrt{v\,g_{IIB}}}{L}\left(\frac{\ell_M^6}{v^2}\,dy^2 + ds_9^2\right). \tag{3.16}$$

This is the metric in string frame. In Einstein frame, and trading ℓ_M for ℓ_s using (3.15), this becomes

$$ds_{IIB,E}^2 = \frac{\sqrt{v}}{L}\left(\frac{L^2\ell_s^4}{v^2}\,dy^2 + ds_9^2\right). \tag{3.17}$$

Let us specialize now to the case $M_9 = \mathbb{R}^{1,2} \times B_6$, with B_6 some Kähler manifold such as the projective space $\mathbb{CP}^3$. Assume moreover that the T^2 depends holomorphically on the coordinates of B_6, that is, that we have an *elliptic fibration* (with a section[7]). If we want this to be a supersymmetric solution, the resulting total space Z_8 must be Calabi–Yau, of complex dimension four. There are many elliptically fibered Calabi–Yau fourfolds known [54]. In elliptic fibrations, τ varies holomorphically over the base B_6 of the fibration, but v remains constant. (This is because the area of a holomorphic 2-cycle (curve) equals the integral of the Kähler form over the curve, and since the Kähler form is closed, this does not change when we slide the curve over the base.) Then we can simply take our conventional scale

$$L \equiv \sqrt{v}, \tag{3.18}$$

and the metric becomes

$$ds_{IIB,E}^2 = -(dx^0)^2 + (dx^1)^2 + (dx^2)^2 + \frac{\ell_s^4}{v}\,dy^2 + ds_{B_6}^2. \tag{3.19}$$

If we send now $v \to 0$ keeping ℓ_s finite, we see that this decompactifies to flat $3+1$ dimensional Minkowski space times B_6, with a nontrivial dilaton profile $\tau(u)$, $u \in B_6$. Since we started with a supersymmetric solution in M-theory, this dual IIB configuration will also be a supersymmetric solution. This gives

---

[7]This means we can globally choose a zero point for the T^2 fiber in a smooth way, as is the case for (3.12).

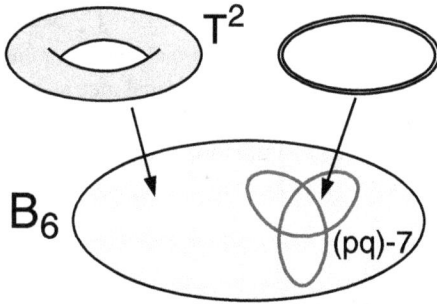

Fig. 3. F-theory realization of (p, q) 7-branes.

an elegant recipe to construct many nontrivial type IIB vacua with varying τ, in which we can use the full power of algebraic geometry applicable to Calabi–Yau manifolds.[8]

Note that remarkably, what was part of the Calabi–Yau fourfold fiber in M-theory, becomes part of noncompact, visible space in type IIB, with full Lorentz invariance in the $v \to 0$ limit! We will later see that this remains true even in the presence of fluxes, when the geometry gets warped and v is no longer constant.

To conclude, "F-theory compactified on an elliptic fibration" should be understood as meaning the type IIB geometry obtained by compactifying M-theory on this elliptic fibration and following the procedure outlined above, in the limit of vanishing elliptic fiber size v. We will discuss a completely explicit example in Section 3.6, but first turn to some further general considerations.

3.2. p-form potentials

Along the same lines, we can deduce F-theory equivalents of other type IIB fields. For example the M-theory 3-form potential C_3 can locally be decomposed as follows in the geometry (3.12):

$$C_3 = C_3' + B_2 \wedge L \, dx + C_2 \wedge L \, dy + B_1 \wedge L \, dx \wedge L \, dy, \qquad (3.20)$$

where the forms C_3', B_2, C_2 and B_1 live on M_9 and L was defined in (3.13) and (3.18). After reduction, T-duality and taking the $L^2 = v \to 0$ limit, B_2

---

[8]We have hidden an issue here. For general v, the actual Calabi–Yau metric will actually depend on the torus coordinates x and y as well (because generic Calabi–Yau manifolds do not have any isometries), so the metric is not quite of the form (3.12). In the IIA theory this will manifest itself as nonzero values for the massive fields whose quanta are D0-branes (thus breaking the $U(1)$ gauge symmetry associated to the $U(1)$ circle isometry). However, in the limit $v \to 0$ of interest here, these fields will become infinitely massive, so one expects them to vanish. Thus, in this limit, the metric ansatz (3.12) is correct.

becomes the NSNS 2-form potential in type IIB, C_2 becomes the RR 2-form potential, C_3' turns into $C_4^{(y)} = C_3' \wedge dy$, i.e. half the components of the self-dual 4-form potential, and B_1 gives rise to off-diagonal metric components mixing the y-direction with the M_9 directions, $g_{iy} = (B_1)_i$.

Note that geometric $SL(2, \mathbb{Z})$ transformations of the T^2 will exactly act as (3.8) on (B_2, C_2), and as (3.7) on τ.

3.3. Branes

Let us see how various branes get mapped between the M-theory and IIB pictures. Of particular interest for our purposes in the case of M-theory on $\mathbb{R}^{1,2} \times Z$ with Z an elliptic fibration over a base B_6 are:

1. The $\mathbb{R}^{1,2}$ space-filling M2 gets mapped to a $\mathbb{R}^{1,3}$ space-filling D3.

2. At special complex codimension 1 loci of the base B_6, the elliptic fiber can degenerate, with generically some 1-cycle of the T^2 collapsing to zero size. If this is the 1-cycle along the x-direction (the A-cycle), this maps to a space-filling D7-brane localized at the degeneration locus in B_6. Note that this is a purely solitonic, geometric object from this point of view. We will see further on how the usual worldvolume degrees of freedom familiar from the perturbative string theory D-brane picture arise in a particular limit identified with the weak coupling limit in type IIB. More generally, if it is the 1-cycle $pA + qB$ which collapses, we get a (p, q) 7-brane. As we will see in a detailed explicit example in Section 3.6, there is an $SL(2, \mathbb{Z})$ monodromy acting on the T^2 fiber—and therefore on the fields τ, B_2 and C_2—when circling around such a degeneration point in the base. In particular around a D7-brane we have $\tau \to \tau + 1$, $B_2 \to B_2$, $C_2 \to C_2 + B_2$.

3. An M5 wrapped on a 4-cycle Σ_4 in Z looks like a $1 + 1$ dimensional domain wall in $\mathbb{R}^{1,2}$, say extended along the (x^0, x^1) directions. The following cases should be distinguished, depending on the nature of Σ_4:

   (a) Σ_4 is an A-cycle fibration over $\Sigma_3 \subset B_6$. Here Σ_3 can be either a closed 3-cycle, or a 3-chain terminating on a locus where the A-cycle vanishes. The latter type of 3-chain in B_6 still produces a closed 4-cycle in Z, since the circle fibers collapse at the boundary of the chain. (This is analogous to the construction of a 2-sphere as a circle fibration over a line segment.) Such an M5 maps in IIB to a D5-brane wrapped on Σ_3, producing a $2 + 1$ dimensional domain wall in $\mathbb{R}^{1,3}$. If Σ_3 is a 3-chain, it maps to a D5 ending on D7 branes.
   A B-cycle fibration over Σ_3 similarly maps to an NS 5-brane on Σ_3. Again, Σ_3 can be a 3-chain, but now with boundary on a vanishing locus of the

B-cycle. This gives an NS5-brane on Σ_3, which may be stretched between $(0, 1)$ 7-branes.

A $(pA + qB)$ circle fibration will map to a (p, q) 5-brane possibly terminating on (p, q) 7-branes.

(b) If Σ_4 is wrapping both cycles of the T^2, i.e. a T^2 fibration over some 2-cycle Σ_2 in the base B_6, we get a D3 wrapping Σ_2 and extending in the (x^0, x^1)-direction. In other words this is a string in four dimensions.

(c) Finally, for Σ_4 completely transversal to the T^2, one gets a KK-monopole extended along Σ_4 and as a string along (x^0, x^1).

4. M5 instantons wrap 6-cycles in Z. The only M5 instantons which retain finite action in the limit $v \to 0$ are those wrapped on the entire elliptic fiber. To see this, note that the M5 instanton action wrapped on n directions in the T^2 fiber $(n = 0, 1, 2)$, has an action

$$S \sim \frac{v^{n/2}}{\ell_M^6} = \frac{v^{n/2}}{L^2 \ell_s^4} = \frac{v^{(n-2)/2}}{\ell_s^4}, \tag{3.21}$$

where we used (3.12), (3.15), and (3.18). So finite action requires $n = 2$. Such M5 instantons map to D3 instantons wrapped on a 4-cycle in B_6.

3.4. Fluxes

We can also turn on magnetic 4-form fluxes $G_4 = dC_3$ on Z. As we will detail in Section 4.4, this will deform the geometry by warping it, but the fourfold metric remains conformal Calabi–Yau. The equations of motion give rise to the selfduality condition $G_4 = *G_4$ where $*$ is the Hodge star in the CY metric without the warp factor. In particular this implies G_4 is harmonic, and is uniquely determined by its (integrally quantized[9]) cohomology class $[G_4]$.

From the discussion below (3.20) we take that the only magnetic fluxes G_4 on Z which will not violate Lorentz invariance of our eventual 4d noncompact space in type IIB are of the form

$$G_4 = H_3 \wedge L\, dx + F_3 \wedge L\, dy, \tag{3.22}$$

where $H_3 = dB_2$ and $F_3 = dC_2$. But one should not make the mistake to conclude from this that all F-theory fluxes suitable for constructing flux vacua are characterized by 3-form cohomology classes $[H_3]$ and $[F_3]$ on the base. In

---

[9]More precisely, $[G_4 - \frac{c_2(Z)}{2}]$ is integrally quantized, where $c_2(Z)$ is the second Chern class of Z [55]. For simplicity of exposition, we will suppress this subtlety for now. We will also always work in de Rham cohomology, to avoid torsion complications.

fact, in many cases the base does not have 3-cohomology at all, while Z has 4-cohomology dimension of the order of ten thousands! The mistake is that as we noted above and will detail in Section 3.6, in the presence of 7-branes, the fields (H_3, F_3) are *not* single valued but undergo $SL(2, \mathbb{Z})$ monodromies around the 7-brane loci. This allows many more topologically nontrivial excitations, matching the large number of 4-form flux cohomology classes we have on Z. Typically only a small fraction of those correspond to "bulk" flux in IIB. As we will see in detail in Section 3.9, the twisting of (H_3, F_3) around the 7-brane loci can produce topologically nontrivial excitations of (H_3, F_3) whose energy and charge densities are localized very close to the 7-brane loci. In the IIB weak coupling limit, these localized excitations can be identified with D7-brane worldvolume fluxes.

Thus, the proper way to think about (H_3, F_3) fluxes topologically is as $SL(2, \mathbb{Z})$-twisted cohomology, but to deal with this the proper way requires a level of formalism which involves more sequences of arrows than these lecture notes can accommodate.

There is however an alternative, more physical and intuitive way to think about these fluxes, topologically at least, and that is to consider the M5 branes (or their IIB duals) which source them, as we will now explain.

As we just noted, fluxes are characterized by their cohomology class $[G_4] \in H^4(Z, \mathbb{Z})$. In general, Poincaré duality canonically relates p-form cohomology classes and $(d - p)$-cycle homology classes, with d the dimension of the space M considered. A concrete way to think about this is as follows. Start with some $(d - p)$-cycle Σ representing the homology class $[\Sigma]$, locally described by equations $f^i = 0$, $i = 1, \ldots, p$. Then the Poincaré dual p-form cohomology class is $\mathrm{PD}_M([\Sigma]) = [\delta_{\Sigma \subset M}]$, where $\delta_{\Sigma \subset M}$ is the p-form current

$$\delta_{\Sigma \subset M} := \delta(f^1) \, df^1 \wedge \cdots \wedge \delta(f^p) \, df^p. \tag{3.23}$$

Conversely, a representative $(d - p)$-cycle of the Poincaré dual to a p-form flux can be thought of as being obtained by squeezing all flux lines maximally together. This interpretation is made clear by the observation that for any p-cycle Σ', the total flux of $G = \mathrm{PD}_M(\Sigma)$ through this cycle is

$$\int_{\Sigma'} G = \int_{\Sigma'} \delta_{\Sigma \subset M} = \#(\Sigma' \cap \Sigma), \tag{3.24}$$

where $\#$ denotes the number of intersection points counted with signs (the signs being determined by the signs of the oriented delta-function at each intersection point). Hence $[\Sigma] = PD_M([G])$ can indeed be thought of as representing the flux lines of G.

Now, magnetic G_4-fluxes on Z are sourced by M5 domain walls wrapping a 4-cycle Σ_4 in Z. Across the domain wall, $[G_4]$ jumps by exactly the Poincaré

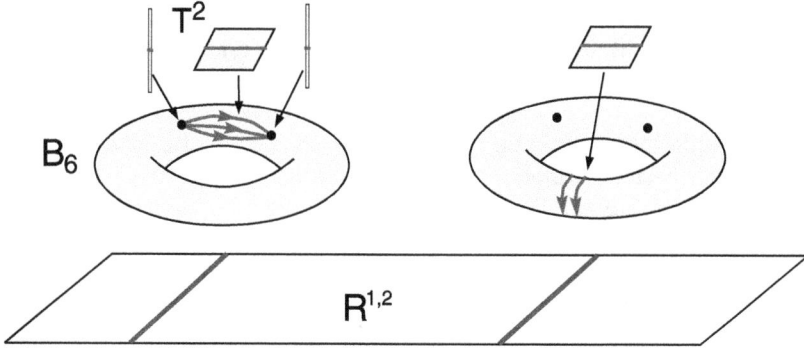

Fig. 4. Domain wall as a source of flux; the domain wall can be thought of as wrapping the fluxlines (or more formally the Poincaré dual) of the flux jump it sources. On the left we have a domain wall producing three units of a brane-type flux; on the right two units of a bulk-type flux.

dual of $[\Sigma_4]$. This can be seen as follows. Assuming the domain wall lies at $x^1 = 0$, the Bianchi identity for G_4 acquires a source term

$$dG_4 = \ell_M^3 \, \delta_{M5} = \ell_M^3 \, \delta_{\Sigma \subset Z} \wedge \delta(x^1) \, dx^1, \tag{3.25}$$

Integrating this over x_1 across the wall gives the jump

$$[G_4]_+ - [G_4]_- = \ell_M^3 \, \mathrm{PD}_Z([\Sigma_4]). \tag{3.26}$$

This is illustrated in Fig. 4.

Thus, to classify magnetic fluxes, we only need to classify possible domain wall wrappings; the wrapped cycle can be thought of as representing the flux lines of the corresponding flux. Now, if we want the flux to preserve Poincaré invariance in four dimensions—which we do if we want to construct vacua—the only allowable sources are M5 domain walls which remain domain walls in IIB.[10] Luckily, we have already analyzed this: These are the M5 domain walls wrapped on 4-cycles which are well-defined $pA + qB$ 1-cycle fibrations over 3-cycles Σ_3, or over 3-chains Γ_3 which terminate on (p, q) 7-branes. So the corresponding fluxes in type IIB are p units of RR 3-form flux and q units of NSNS 3-form flux, with flux lines closing upon themselves in the case of a 3-cycle, and terminating on the 7-branes in the case of a 3-chain.

---

[10]Strictly speaking, domain walls break Poincaré invariance too, of course. But away from a domain wall, all 4d fields attain their vacuum values. Other sources, such as strings, will source nonconstant fluxes with legs (i.e. field gradients) in the noncompact directions; the 4d fields around these objects will be excited away from their vacuum values.

If the 7-branes are D7-branes ($q = 0$) at weak string coupling, the RR 3-form flux emanating from the branes along a 3-chain can be understood in the perturbative string picture as being sourced by worldvolume gauge flux F_2 on the D7-brane (through the coupling $\int_{D7} C_6 \wedge F_2$), where F_2 is Poincaré dual to the boundary of the 3-chain on the 4-cycle wrapped by the D7 in B_6. Keep in mind though that the perturbative D-brane picture is a *different* picture than the F-theory picture we are working in: in F-theory there are no D7-branes added to the geometry, hence no D7 worldvolume fluxes sourcing bulk fields—the 7-branes and all associated flux degrees of freedom emerge purely as solitonic excitations of the fields τ, H_3 and F_3. Nevertheless we should be able to reproduce the perturbative string theory picture from the F-theory picture in a suitable weak coupling limit. This will be explained in detail further on.

3.5. M2/D3 tadpole

The M-theory action (3.11) supplemented with M2-brane sources gives the following equation of motion for G_4:

$$d * G_4 = \frac{1}{2} G_4 \wedge G_4 - \ell_M^6 I_8(R) + \ell_M^6 \sum_i \delta_{M2_i}. \qquad (3.27)$$

Integrating this over the fourfold Z gives

$$\frac{1}{2\ell_M^6} \int_Z G_4 \wedge G_4 + N_{M2} = \frac{\chi(Z)}{24} =: Q_c. \qquad (3.28)$$

Here N_{M2} is the number of $\mathbb{R}^{1,2}$-filling M2-branes and we used the fact that $24\, I_8(R)$ integrates to the Euler characteristic χ on a Calabi–Yau fourfold [53]. What this equation says is that the total M2 charge transversal to Z must vanish—indeed there is nowhere for the flux lines sourced by the M2 charge to go in a compact space. As we will see in chapter 6, this tadpole cancelation condition turns out to be what renders the number of metastable F-theory flux vacua within any compact region of low energy parameter space finite. If it hadn't been for this constraint, string theory would have been infinitely finely tunable.

The type IIB equivalent of this is, using (3.22),

$$\frac{1}{\ell_s^4} \int_{B_6} F_3 \wedge H_3 + N_{D3} = Q_c. \qquad (3.29)$$

Again, as mentioned in the previous subsection, the contributions to the D3 tadpole attributed to D7-brane worldvolume fluxes in the perturbative IIB string picture are in fact already contained in (3.29). We will come back to this in Section 3.9, after determining how to take the weak coupling limit in F-theory.

Before we do this, it is probably useful to consider a simple example.

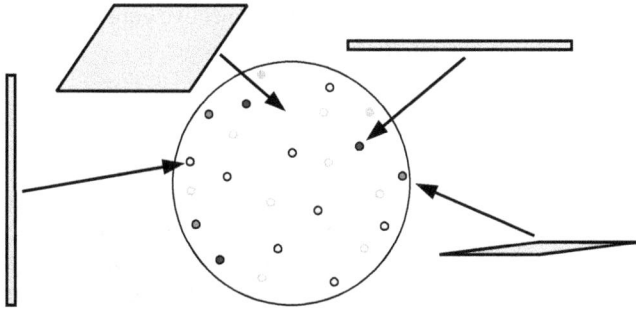

Fig. 5. F-theory on K3 elliptically fibered over a sphere. The dots indicate the 24 degeneration loci of the elliptic fiber.

3.6. Example: K3

We now turn to an explicit example [35], F-theory on an elliptically fibered K3, times (in IIB) some eight dimensional manifold which for definiteness we take to be $\mathbb{R}^{1,7}$, but any other manifold solving the equations of motion would be fine too. (In particular if we wanted to compactify on a fourfold Z, we could take $Z = T^4 \times K3$ or $Z = K3 \times K3$.) From the previous discussion, we know that this means we should consider M-theory on $\mathbb{R}^{1,6} \times K3$. An elliptically fibered K3 can be described by the equation

$$y^2 = x^3 + f(u, v) \, x \, z^4 + g(u, v) \, z^6, \tag{3.30}$$

where $x, y, z, u, v \in \mathbb{C}$, modulo the projective equivalences

$$(u, v, x, y, z) \simeq (\lambda u, \lambda v, \lambda^4 x, \lambda^6 y, z) \tag{3.31}$$
$$\simeq (u, v, \mu^2 x, \mu^3 y, \mu z), \tag{3.32}$$

where $\mu, \lambda \in \mathbb{C}^* = \mathbb{C}\backslash\{0\}$, and $(u, v) \neq (0, 0)$, $(x, y, z) \neq (0, 0, 0)$. The functions f and g are homogeneous polynomials of degree eight and twelve in u, v. Notice that all of these weight assignments are consistent with the embedding equation (3.30), and that it indeed describes a two complex dimensional surface, five coordinates minus one equation minus two equivalence relations. As we will explain systematically in Section 5, the rule to determine whether such a hypersurface is Calabi–Yau is simply that the (weighted) degree of the defining polynomial equals the sum of the weights, for each equivalence. For the first equivalence, the degree is 12, and the sum of the weights is $1 + 1 + 4 + 6 + 0 = 12$. For the second equivalence, the degree is 6, and the sum of the weights is $0 + 0 + 2 + 3 + 1 = 6$. So this is indeed a Calabi–Yau twofold, hence a K3.

In a coordinate patch in which we fix the projective equivalences by putting $z \equiv 1$, $v \equiv 1$, the equation simplifies to

$$y^2 = x^3 + f(u) x + g(u), \tag{3.33}$$

with f and g ordinary polynomials of degree eight and twelve.

To see that we indeed have an elliptic fibration, it is sufficient to note that at fixed (u, v), (3.30) describes a Calabi–Yau onefold, that is, an elliptic curve, that is, a T^2, embedded in (x, y, z) space. Again this follows because the sum of the weights of the remaining equivalence (3.32) equals the degree (six here).

More formally, the projection of the fibration is $\pi : K3 \rightarrow \mathbb{CP}^1 : (x, y, z, u, v) \rightarrow (u, v)$, where $\mathbb{CP}^1 = \{(u, v) \neq (0, 0) | (u, v) \simeq (\lambda u, \lambda v)\}$. You can easily check that this is a well-defined map, in the sense that equivalent points get mapped to equivalent points and no point maps to $(0, 0)$. So, we have an elliptic fibration over base $\mathbb{CP}^1 = S^2$.

How can we relate this algebraic description of a T^2 to the standard representation $T^2 = \mathbb{C}/(\mathbb{Z} \oplus \tau \mathbb{Z})$? This is done by relating holomorphic coordinates. On $T^2 = \mathbb{C}/(\mathbb{Z} \oplus \tau \mathbb{Z})$, the holomorphic coordinate is $z = x + \tau y$, which for any point P can be written as

$$z(P) = \int_0^P \Omega_1, \quad \Omega_1 = dz. \tag{3.34}$$

Up to normalization, Ω_1 is the unique holomorphic 1-form on the torus. In Section 5, we will see how to construct in general the up to normalization unique holomorphic n-form on any algebraic Calabi–Yau n-fold. The result for the T^2 described by (3.33) at fixed u is

$$\Omega_1 = \frac{c \, dx}{y}, \tag{3.35}$$

where c is some normalization constant. Choosing a basis of 1-cycles (A, B) on the algebraic T^2, the modulus τ is then given by

$$\tau = \frac{\oint_B \Omega_1}{\oint_A \Omega_1}. \tag{3.36}$$

The ambiguity in choosing a basis is just the $SL(2, \mathbb{Z})$ S-duality frame ambiguity we expect.

In principle we could now figure out the relation between τ and f and g ourselves by computing period integrals, but it turns out that industrious mathematicians figured this out already more than a century ago, and fortunately left notes. The result can be expressed as

$$j(\tau) = \frac{4 \cdot (24\, f)^3}{\Delta}, \qquad \Delta = 27\, g^2 + 4\, f^3, \tag{3.37}$$

where $j(\tau)$ is the $SL(2, \mathbb{Z})$ modular invariant j-function, $j(\tau) = e^{-2\pi i \tau} + 744 + \mathcal{O}(e^{2\pi i \tau})$.

Although we will not need it, let me mention that once we know $\tau(u)$, we know the exact metric on the base too [57]:

$$ds^2 = \frac{a\, \tau_2(u)\, |\eta(\tau(u))|^4}{|\Delta(u)|^{1/6}}\, du\, d\bar{u}, \tag{3.38}$$

where $\eta(\tau)$ is the Dedekind eta function and a an arbitrary constant setting the size of the $\mathbb{CP}^1$.

The function Δ in (3.37) is called the discriminant of the elliptic curve; when it vanishes, the elliptic curve becomes singular, generically with a 1-cycle collapsing to zero size. It is a homogeneous polynomial of degree 24 on the base $\mathbb{CP}^1$, so Δ has 24 zeros. Generically they will all be distinct, distinct from the zeros of f and g, and distinct from the point $(1, 0)$. Assuming the latter we fix the $\mathbb{CP}^1$ scaling by putting $v \equiv 1$. Near a generic zero $u = u_i$ ($i = 1, \dots, 24$), (3.37) becomes

$$j(\tau(u)) \sim \frac{1}{u - u_i}, \tag{3.39}$$

which is solved as

$$\tau(u) \approx \frac{1}{2\pi i}\, \ln(u - u_i), \tag{3.40}$$

up to $SL(2, \mathbb{Z})$ transformation.

Note that when $u \to u_i$, $\tau \to i\infty$. Geometrically, this means the ratio of A-cycle and B-cycle lengths of the T^2 vanishes. Recalling the physical meaning of τ in type IIB, namely $\tau = C_0 + \frac{i}{g_{\text{IIB}}}$, we see that this corresponds to weak coupling, $g_{\text{IIB}} \to 0$.

Moreover, when circling once around $u = u_i$, following a path $u(\theta) = u_1 + \epsilon\, e^{2\pi i \theta}$, we see that τ undergoes *monodromy* when $\theta : 0 \to 2\pi$:

$$T : \tau \to \tau + 1. \tag{3.41}$$

Equivalently, $C_0 \to C_0 + 1$, or $\oint_{u_i} F_1 = \oint_{u_i} dC_0 = 1$, which means *there is a D7-brane at $u = u_i$*.

This immediately leads to a paradox: now it looks like we have 24 D7-branes in a compact transversal space, $\mathbb{CP}^1$. There can be no net charge in a compact space, since the flux lines have nowhere to go. More directly, the sum of all contour integrals $\sum_{i=1}^{24} \oint_{u_i} F_1$ must vanish, since the total contour is contractible on the sphere. How can this be?

The resolution lies in the innocent looking "up to $SL(2, \mathbb{Z})$ transformation" under (3.40). While it is true that we can always *locally* go to an $SL(2, \mathbb{Z})$ frame where $\tau(u)$ lies in the fundamental domain, we cannot do this *globally*. We can pick one point u_* where we choose $\tau(u_*)$ to lie in the fundamental domain, but once we start walking around on the $\mathbb{CP}^1$, $\tau(u)$ might move off to some other region of the upper half plane. Of course, near any other zero of Δ that we might encounter on our trip, the value of $\tau(u)$ will still be related to (3.40) by an $SL(2, \mathbb{Z})$ transformation M, but then the monodromy (3.41) will be related by conjugation, MTM^{-1} instead of T. As a result, we will in general no longer have a D7-brane there, but a more general (p, q) 7-brane, related to the D7 (i.e. the $(1, 0)$ 7-brane) by the S-duality transformation M. As mentioned earlier already in Section 3.3, a general (p, q) 7-brane can be characterized in F-theory by the vanishing of the $pA + qB$ 1-cycle of the T^2, the image under M of the A-cycle which vanishes for a D7.

In general, it is not an easy task to figure out exactly what kind of (p, q)-branes we have at various points; worse even, this in fact depends on the path we take through the base! More important than the confusion this is bound to instill in anyone who sets out to explore these IIB solutions, is the fact that this makes it entirely impossible to do conventional string perturbation theory on such backgrounds. Even if we make our base $\mathbb{P}^1$ as large as the solar system, and we go to an $SL(2, \mathbb{Z})$ duality frame where the monodromy closest to home is of the D7 form (3.41), and the string coupling is very weak near home, there will always be (p, q) 7-brane monodromies somewhere else with $q \neq 0$, which send g_{IIB} from weak to strong coupling. For example if we have a $(0, 1)$ 7-brane somewhere, this will map $\tau \to -1/\tau$. Equivalently, we can say that if we send off a fundamental string and let it loop around the $(0, 1)$ 7-brane, it will come back to us as a D1 string. So it is not possible to set up conventional perturbation theory for fundamental strings in a globally well defined way.

This is why people say F-theory is intrinsically strongly coupled. Of course, if we are only interested in getting nontrivial solutions of type IIB supergravity, we do not need to care about this; it is only when we need to compute string scattering amplitudes that we get into trouble.

All of this leaves us with a new puzzle: We definitely know there *are* regimes in which type IIB theory in principle has a perturbative string expansion. How can we see this in F-theory?

3.7. The weak coupling limit: orientifolds from F-theory, K3 example

The answer to the puzzle just raised is given by taking a clever limit of the F-theory description, pointed out by Sen [35,56]. For our K3 example, the simplest such limit accomplishing this is as follows. We want to go to a point in the K3 moduli space where $\tau(u)$ is constant and has large imaginary part. We see from (3.37) that constancy requires $f^3/g^2 =$ constant, which is solved by

$$g = p^3, \qquad f = \alpha p^2, \tag{3.42}$$

with α a constant and p a homogeneous polynomial of degree four. Let us go again to a coordinate patch $v \equiv 1$. By a rescaling of y and x we can set the coefficient of u^4 equal to one, so $p(u)$ has the form

$$p = \prod_{i=1}^{4}(u - u_i), \tag{3.43}$$

where the u_i are constants. Plugging this in (3.37), we get

$$\Delta = (4\alpha^3 + 27) \prod_{i=1}^{4}(u - u_i)^6, \qquad j(\tau) = \frac{4 \cdot (24\alpha)^3}{27 + 4\alpha^3}. \tag{3.44}$$

Thus, if we tune

$$\alpha \approx -3/4^{1/3}, \tag{3.45}$$

we get weak IIB string coupling everywhere on the base!

Although τ is now constant everywhere, this does not necessarily mean there is no $SL(2, \mathbb{Z})$ monodromy at all, because there is one nontrivial $SL(2, \mathbb{Z})$ element which acts trivially on τ, namely

$$M = \begin{pmatrix} -1 & 0 \\ 0 & -1 \end{pmatrix}. \tag{3.46}$$

This may seem overly paranoid, but actually it turns out that we do get this monodromy around each of the u_i. To see this, note that after a change of coordinates $x = p\tilde{x}$, $y = p^{3/2}\tilde{y}$ and using (3.42), we can rewrite (3.33) as

$$\tilde{y}^2 = \tilde{x}^3 + \alpha\tilde{x} + 1. \tag{3.47}$$

This makes it completely manifest that the modulus of the torus does not vary with u. However, note that in the new coordinates, Ω_1 defined in (3.35) becomes $\Omega_1 = p^{-1/2}\frac{d\tilde{x}}{\tilde{y}}$. Therefore, when we circle around a zero of $p(u)$ in the u-plane, we map $\Omega_1 \to -\Omega_1$. This implies is particular $\oint_A \Omega_1 \to -\oint_A \Omega_1$, $\oint_B \Omega_1 \to -\oint_B \Omega_1$, and from (3.34), $z \to -z$. In a representation of the torus where we think of Ω_1 as being fixed, such as the standard $T^2 = \mathbb{C}/(\mathbb{Z} \oplus \tau\mathbb{Z})$, this monodromy boils down to

$$(A, B) \rightarrow (-A, -B), \tag{3.48}$$

that is, (3.46). Note that in the type IIB picture, this monodromy implies that the fields (B_2, C_2) are double valued on the $\mathbb{CP}^1$, flipping sign when circling around the zeros of p.

We can conveniently think of this situation in the following way. First we construct a double cover of the base $\mathbb{CP}^1$, which we call X, defined by adding a coordinate $\xi \in \mathbb{C}$ and the equation

$$X : \xi^2 = p(u, v), \qquad (u, v, \xi) \simeq (\lambda u, \lambda v, \lambda^2 \xi), \tag{3.49}$$

with $(u, v, \xi) \neq (0, 0, 0)$. The extension of the projective equivalence of $\mathbb{CP}^1$ is imposed by compatibility with the equation. Notice that this equation again satisfies the Calabi–Yau condition that its degree equals the sum of the weights (four). So this describes again a T^2, but a different one than the fiber T^2 we had before.[11] The original base $\mathbb{CP}^1$ is recovered from X as the quotient $\mathbb{CP}^1 = X/\sigma$, where

$$\sigma : \xi \rightarrow -\xi. \tag{3.50}$$

When circling around the zeros of p on the base, we go from one sheet of the double cover to the other. At the same time, we have the $\mathbb{Z}_2$ transformation (3.46) acting on the T^2 fiber. Thus, on the covering space, everything is single valued; the double valuedness appears in this picture by taking the simultaneous $\mathbb{Z}_2$ quotient of X and the T^2 fiber. Correspondingly, in this limit, we can represent our F-theory K3 as

$$K3 = (T^2 \times T^2)/\mathbb{Z}_2. \tag{3.51}$$

By now, a bell should be ringing: What we have here in the type IIB setting is exactly the same as what one would get from orientifolding X by $\sigma \cdot (-1)^{F_L} \cdot \omega$, where ω denotes worldsheet orientation reversal, i.e. exchange of left- and right-moving modes on the worldsheet, and $(-1)^{F_L}$ changes the sign of the Ramond sector states of the leftmoving sector. You can see this from the way our $\mathbb{Z}_2$ action acts on the different RR and NSNS potentials obtained from F-theory as explained in Section 3.2 and comparing this with the perturbative worldsheet result.

---

[11] Again this T^2 can be mapped to the standard representation; now the holomorphic 1-form in a patch $v \equiv 1$ is $\Omega_1' = du/\xi$. You can check that (3.38) reduces to the flat metric in standard coordinates $z' = \int \Omega_1'$.

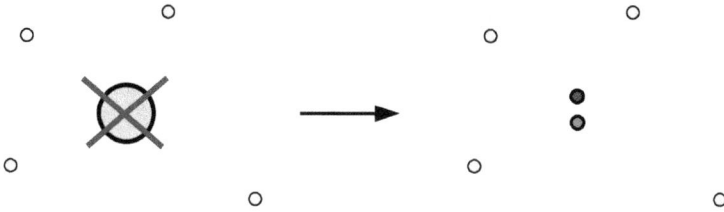

Fig. 6. The naive finite distance orientifold singularity gets resolved in F-theory by splitting the O-plane into two 7-branes.

Thus, the fixed loci of the $\mathbb{Z}_2$ involution σ, that is, the four zeros u_i of p, are identified with O7-planes, which have D7-charge -4 as measured in the base space $\mathbb{CP}^1$. Since there is no monodromy of C_0, there is no net charge at the fixed points, so there must be four D7-branes located on top of the O7-planes.

We expect to be able to move away the D7-branes from the O7-planes. Zooming in on an O7 located say at $u = 0$, and assuming we have moved the four D7-branes which were on top to nearby positions $u = u^{(a)}$, there should now be a D7 T-monodromy (3.41) around each $u^{(a)}$, and a compensating T^{-4} monodromy around the naked O7 which stays behind. So naively, we might expect $\tau(u)$ to be of the form

$$\tau(u) = \tau_0 + \frac{1}{2\pi i}\left(\sum_{a=1}^{4}(u - u^{(a)}) - 4\ln u\right).\tag{3.52}$$

However, upon further reflection, this does not make sense, since too close to $u = 0$, this gives a large *negative* value for Im τ! More precisely for $\mathcal{O}(1)$ values of the $u^{(a)}$ and large Im τ_0 (which we can identify with the large u asymptotic inverse string coupling), this occurs when $|u| < e^{-\frac{\pi}{2}\text{Im}\,\tau_0}$. Note that this is exponentially small at weak coupling, but nevertheless, since nothing fixes the overall size of the base $\mathbb{CP}^1$ at this point, this breakdown could still occur at a distance much larger than the string length.

This is a typical phenomenon occurring for naive supergravity solutions in the presence of orientifold planes: one finds nasty singularities at finite distance from the O-plane. This can be traced back to the fact that these objects have negative tension.

The correct solution obtained from F-theory does not have this pathology, and is completely well-behaved. What actually happens is that at nonzero string coupling $1/\text{Im}\,\tau_0$, the orientifold plane *splits* in two (p, q) 7-branes, with separation of the order of the distance $e^{-\frac{\pi}{2}\text{Im}\,\tau_0}$ where our naive solution breaks down. This follows from (3.44): the limit in which the four D7-branes are coincident with the O7 corresponds to a zero of Δ of multiplicity *six*, that is, it corresponds to a limit in which the 24 generic (p, q) 7-branes coincide in four groups of six. Out of the six, four get identified with D7-branes, and the remaining two must correspond to the orientifold plane. More details can be found [35]. Notice that this splitting cannot be seen in perturbation theory, as it is nonperturbatively small in the string coupling.

3.8. Orientifolds from F-theory: general story

I will now give a refinement and generalization of the orientifold limit we discussed for $K3$, again due to Sen [56]. The refinement consists of allowing the D7-branes to move away from the O7-planes, while retaining weak coupling. The generalization consists of allowing general elliptically fibered Calabi–Yau n-folds.

For definiteness, we will again work with an example, but it will be clear how to generalize it (if not, see [56]). The example is an elliptically fibered Calabi–Yau fourfold, fibered over $\mathbb{CP}^3$, described by the equation, analogous to (3.30),

$$Z : y^2 = x^3 + f(\vec{u}) \, x \, z^4 + g(\vec{u}) \, z^6, \tag{3.53}$$

where $\vec{u} := (u_1, u_2, u_3, u_4)$. We also impose the projective $\mathbb{C}^*$ equivalences

$$(u_1, u_2, u_3, u_4, x, y, z) \simeq (\lambda u_1, \lambda u_2, \lambda u_3, \lambda u_4, \lambda^8 x, \lambda^{12} y, z) \tag{3.54}$$

$$\simeq (u_1, u_2, u_3, u_4, \mu^2 x, \mu^3 y, \mu z), \tag{3.55}$$

where $\vec{u} \neq \vec{0}$ and $(x, y, z) \neq (0, 0, 0)$. In the case at hand, $f(\vec{u})$, $g(\vec{u})$ are homogeneous polynomials of degrees 16 and 24. At fixed u, (3.53) describes an elliptic curve, hence this equation indeed defines an elliptic fibration over $\mathbb{CP}^3$. The sum of the weights equals the degree, so we do have a Calabi–Yau fourfold. The number of complex structure moduli is $h^{3,1}(Z) = 3878$, which can be computed directly by counting the number of coefficients of f and g modulo $GL(4, \mathbb{C})$ coordinate transformations: $\binom{16+3}{3} + \binom{24+3}{3} - 16 = 3878$.

To define the orientifold limit, we first parametrize, without loss of generality, following Sen:

$$\begin{aligned} f &= -3h^2 + \epsilon \eta, \\ g &= -2h^3 + \epsilon h \eta - \epsilon^2 \chi / 12, \end{aligned} \tag{3.56}$$

where h, η and χ are a homogeneous polynomials of degrees 8, 16 and 24 in the u_i, and ϵ is a constant. (Notice that for $\epsilon = 0$, this is essentially the limit discussed in the previous subsection with $\alpha = -3/4^{1/3}$.) When $\epsilon \to 0$ keeping everything else fixed, one finds for the discriminant and $j(\tau)$

$$\Delta \approx -9\,\epsilon^2 h^2 (\eta^2 - h\chi), \qquad j(\tau) \approx \frac{(24)^4}{2} \frac{h^4}{\epsilon^2(\eta^2 - h\chi)}. \tag{3.57}$$

Thus, in this limit,

$$g_{\text{IIB}} \sim -\frac{1}{\log|\epsilon|} \to 0 \tag{3.58}$$

everywhere except near $h = 0$, and the $\epsilon \to 0$ limit can therefore be interpreted as the IIB weak coupling limit. A monodromy analysis similar to what we did in the previous subsection [56] shows that in this limit the two components of $\Delta = 0$ should be identified with an O7-plane and a D7-brane as follows:

$$\text{O7} : h(\vec{u}) = 0, \qquad \text{D7} : \eta(\vec{u})^2 = h(\vec{u})\,\chi(\vec{u}), \tag{3.59}$$

where the orientifolded Calabi–Yau 3-fold is given by the equation

$$X : \xi^2 = h(\vec{u}) \tag{3.60}$$

with $\mathbb{C}^*$ equivalence $(\vec{u}, \xi) \simeq (\lambda\vec{u}, \lambda^4\xi)$, and orientifold involution

$$\sigma : \xi \to -\xi. \tag{3.61}$$

The CY threefold X is a double cover of $\mathbb{CP}^3$ branched over $h(u) = 0$; quotienting by σ gives back $\mathbb{CP}^3$. In the case at hand it has 149 complex structure deformations, given by the coefficients of $h(u)$ modulo $GL(4, \mathbb{C})$ coordinate transformations $\vec{u} \to A\vec{u}$, and one Kähler deformation, its volume. In addition to this, there are D7-brane moduli, counted by the number of inequivalent deformations of the D7 equation in (3.59), i.e. $\binom{16+3}{3} + \binom{24+4}{3} - \binom{8+3}{3} - 1 = 3728$, where the first subtraction comes from the fact that we can shift $\eta \to \eta + h\psi$ with ψ an arbitrary degree 8 polynomial and shift χ accordingly, without changing the form of the D7 equation (3.59), and the last subtraction corresponds to overall rescaling of the coefficients. As a check note that indeed the number of D7 moduli plus the number of 3-fold complex structure moduli plus one for the dilaton-axion modulus ϵ equals 3878, the number of fourfold complex structure moduli.

Observe that the number of D7-brane moduli is vastly larger than the number of bulk moduli.

Finally, for future reference, we relate the holomorphic 4-form living on the Calabi–Yau fourfold Z to the holomorphic 3-form living on the Calabi–Yau threefold X. For an elliptic fibration of the form (3.53), the holomorphic 4-form is, say in a patch $z \equiv 1 \equiv u_4$, $y \neq 0$:

$$\Omega_4 = c \, \frac{dx \wedge du_1 \wedge du_2 \wedge du_3}{y}, \tag{3.62}$$

where c is some normalization constant. (We will see in detail how this expression is obtained in Section 5.) Define a 3-form on the base of the elliptic fibration (here $\mathbb{CP}^3$) by "integrating" out the A-cycle of the T^2:

$$\Omega_3 := \oint_A \Omega_4. \tag{3.63}$$

In general, this would not give a single-valued 3-form on the base, because the A-cycle undergoes various $SL(2, \mathbb{Z})$ monodromies when circling around (p, q) 7-branes. However, in the weak coupling limit, the only monodromy acting on A is $A \to -A$, when circling around the O7 locus $h = 0$, and this disappears altogether when going to the double cover X. This can be seen explicitly by performing the integral in (3.63). To do this, first note that when $\epsilon = 0$, we have

$$y^2 = (x + h)^2 (x - 2h) + \mathcal{O}(\epsilon). \tag{3.64}$$

The A-cycle is the loop in the x-plane collapsing in the limit $\epsilon = 0$, i.e. the loop around the zeros of y which collapse to the double zero $x = -h$ when $\epsilon = 0$. Performing the contour integral, we get

$$\Omega_3 \;=\; c' \, \frac{du_1 \wedge du_2 \wedge du_3}{\sqrt{h}} + \mathcal{O}(\epsilon) \tag{3.65}$$

$$\;=\; c' \, \frac{du_1 \wedge du_2 \wedge du_3}{\xi} + \mathcal{O}(\epsilon), \tag{3.66}$$

where $c' = 2\pi c / \sqrt{3}$. In the last step we used (3.60) and consider Ω_3 to live on X. To leading order in ϵ this is indeed exactly the holomorphic 3-form on the Calabi–Yau 3-fold X. Note that (3.58) implies that the size of the corrections is about $e^{-1/g_{IIB}}$, that is, nonperturbatively small.

If we integrate out the B-cycle instead, we get, by definition of the modular parameter τ:

$$\oint_B \Omega_4 = \tau \, \Omega_3 \approx \left(\tau_0 + \frac{i}{2\pi} \ln \frac{P_{O7}}{P_{D7}} \right) \Omega_{3,CY}, \tag{3.67}$$

where in the last step we used (3.57) and $j(\tau) \approx e^{-2\pi i \tau}$, putting

$$\tau_0 := \frac{i}{2\pi} \ln \frac{288}{\epsilon^2}, \qquad P_{O7} := h^4, \qquad P_{D7} := \eta^2 - h\chi. \tag{3.68}$$

All approximations made here have errors at most of order $\epsilon \sim e^{-\pi/g_s}$, i.e. nonperturbatively small at weak coupling.

A recent explicit study of the weak coupling limit of F-theory can be found in [58], with in particular the example of K3 worked out in detail.

3.9. Localization, fluxes and tadpoles at weak coupling

At weak string coupling $g_{IIB} \to 0$, we expect it to be possible to separate charges, energies and other physical quantities in "bulk background" and "D-brane" contributions. It is instructive to see explicitly how this happens for fluxes.

Let us consider first a local model, F-theory on an elliptic fibration Z over $B_6 = S \times D$, with S an arbitrary Kähler manifold of complex dimension two and D the unit disk, parametrized by a complex coordinate u, with elliptic fiber modulus

$$\tau_1 + i\tau_2 := \tau(u) = \frac{\ln u}{2\pi i}. \tag{3.69}$$

This is a local model for what we have earlier identified as a D7 brane wrapped on a 4-cycle S. The metric on D can be anything conformal to the flat metric. Let g_s be the type IIB string coupling at the boundary of the disk, so we can write in polar coordinates $u =: r\, e^{i\theta}$

$$\tau_1 = \frac{\theta}{2\pi}, \qquad \tau_2 = \frac{1}{g_s} + \frac{\ln(r^{-1})}{2\pi}. \tag{3.70}$$

Using the metric (3.12), you can check that there is a particular normalizable harmonic 2-form on the elliptic fibration over the disk, given by

$$\omega = \frac{1}{g_s} d\left(\frac{dx + \tau_1\, dy}{\tau_2}\right), \tag{3.71}$$

where the normalization is chosen such that $\oint_{\partial D} \int_{y=0}^{1} \omega \equiv 1$. It is anti-self-dual:

$$*\omega = -\omega. \tag{3.72}$$

In fact, this is true even if (3.69) is replaced by any other holomorphic function $\tau(u)$, as you can check by noting that the metric on D is conformal to $d\tau_1^2 + d\tau_2^2$.[12]

---

[12]The most general anti-self-dual form on the elliptic fibration over the disk is of the form $d \operatorname{Re}[f\,(dx + \bar{\tau}dy)/\tau_2]$, with $f(u) = f_0 + f_1 u + f_2 \frac{u^2}{2} + \cdots$ a holomorphic function on the disk. While the constant term (i.e. ω), as we will see, leads to strongly localized energy and charge densities, the $\mathcal{O}(u)$ corrections do not, and should be considered as part of the background in which the D7 is placed.

Define now the following 4-form flux G_4 on Z:

$$G_4 = F_2 \wedge L\,\omega \tag{3.73}$$

where F_2 is some closed 2-form on S. Note that if we take F_2 anti-selfdual too, G_4 will be self-dual, as is required by the classical equations of motion (see Section 4.4). For now we will leave F_2 arbitrary though. Following our usual reduction, in IIB language, this G_4 corresponds to

$$H_3 \;=\; \frac{1}{g_s} F_2 \wedge d\left(\frac{1}{\tau_2}\right), \tag{3.74}$$

$$F_3 \;=\; \frac{1}{g_s} F_2 \wedge d\left(\frac{\tau_1}{\tau_2}\right), \tag{3.75}$$

$$G_3 \equiv F_3 - \tau H_3 \;=\; \frac{1}{g_s} F_2 \wedge \frac{d\tau}{\tau_2}. \tag{3.76}$$

Plugging this in (3.6) and using (3.70), we note that the $|G_3|^2$ part of the Lagrangian density is

$$\frac{2\pi}{\ell_s^8} \frac{1}{2} (F_2 \wedge *F_2) \wedge d\left(\frac{1}{[1 + \frac{g_s}{2\pi}\ln(r^{-1})]^2}\right) \wedge \frac{d\theta}{2\pi}. \tag{3.77}$$

Integrating the last two factors over the disk gives 1, and what remains is exactly the Yang–Mills Lagrangian density for a D7-brane wrapped on S. Note that the radial energy distribution diverges at $r = 0$ as $dr/r\log^3 r$, but in an integrable way. Moreover, in the weak coupling limit $g_s \to 0$, almost all energy is localized exponentially close to $r = 0$, within a radius

$$r_* \sim e^{-2\pi/g_s}. \tag{3.78}$$

This is illustrated in Fig. 7. Similarly, the D3-charge density from the $F_3 \wedge H_3$ term is

$$-\frac{2\pi}{\ell_s^8} \frac{1}{2} (F_2 \wedge F_2) \wedge d\left(\frac{1}{[1 + \frac{g_s}{2\pi}\ln(r^{-1})]^2}\right) \wedge \frac{d\theta}{2\pi}. \tag{3.79}$$

Comparing to the D3 action, we thus see that the total D3-charge is

$$Q_3(D7) = -\frac{1}{\ell_s^4} \int_S \frac{1}{2} F_2 \wedge F_2. \tag{3.80}$$

(This is in conventions in which $\ell_s^{-2} F_2$ is integrally quantized.) This is indeed as expected from the standard D7-brane action.

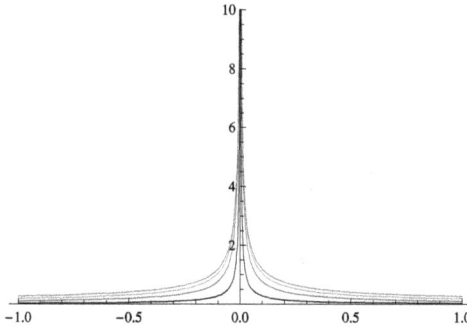

Fig. 7. D7-brane localization of flux energy in the weak coupling limit. The x-axis is the position along a diagonal in the unit disk surrounding the D7, and the y-axis is the charge and energy density normalized to a total of 1. The four curves correspond to four different values of the string coupling constant starting at $g_s = 0.1$ at the bottom and going up in steps of 0.25.

In the language of Section 3.4, a representative of the Poincaré dual to (i.e. the maximally squeezed together flux lines of) the flux G_4 we just constructed is the 4-cycle constructed as a fibration of the A-cycle over the 3-chain consisting of a ray emanating from the origin of the disk times Σ_2, where Σ_2 is the Poincaré dual to F_2 on S. Note that although in Section 3.4 we identified such a cycle topologically with RR flux, H_3 in (3.74) is not zero identically, although it is an exact form on $D\backslash\{0\}$. This is how we can still get a nonzero charge density $F_3 \wedge H_3$. From the point of view of the Poincaré dual cycle to G_4, the charge $\frac{1}{2}\int G_4 \wedge G_4$ is half the self-intersection product of this cycle, which can be seen directly to be equal to half the self-intersection product of Σ_2 on S, in agreement with (3.79).

In contrast, fluxes like (3.74)–(3.75) cannot exist localized on O7-planes. This is because H_3 and F_3 transform with a minus sign under the orientifold involution (equivalently, in the base, they change sign when looping around the orientifold point), which is not satisfied for (3.74)–(3.75). This agrees with the absence of gauge fields on orientifold planes in perturbation theory.

We now extend these local considerations to global constructions. The basic idea is to just patch together these brane localized fluxes. Potential obstructions to this are topological in nature. To think about topological issues, the Poincaré dual picture is particularly useful. Consider an elliptically fibered Calabi–Yau fourfold Z in the weak coupling IIB orientifold limit, and let X be the associated Calabi–Yau threefold (3.60) doubly covering the base B. Let F_2 be a 2-form worldvolume flux class on a D7-brane wrapping a 4-cycle S in X. Then we can associate to this a globally well defined 4-form flux on Z as follows.

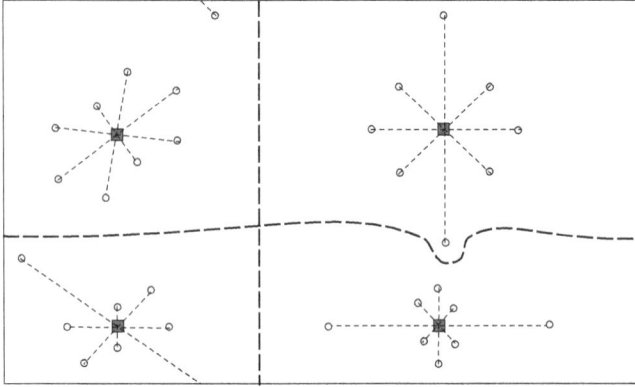

Fig. 8. Example of 3-chains / 3-cycles used to construct a basis of brane /bulk fluxes for the case of $Z = K3 \times S$, with $S = T^4$ or K3. The S part is suppressed in the drawing. What is shown are the corresponding 1-chains and 1-cycles in T^2 which is the CY orientifold double cover of the $\mathbb{CP}^1 = T^2/\mathbb{Z}_2$ base of the elliptically fibered K3. The upper and lower and the left and right boundaries of the rectangle are identified to form the T^2. The 4 red squares are the O7 planes, and the 2×16 yellow circles are the D7-branes, in brane-image-brane pairs. The dotted lines between the D7-image-D7 pairs represent the choice of 1-chains. When combined with a basis of 2-cycles in S, this give the 3-chain basis Γ_α, which in turn determine the 4-cycles Poincaré dual to the brane type fluxes. The wider dashed horizontal and vertical line are the 1-cycles which when combined with 2-cycles in S give 3-cycles determining the bulk type (RR and NSNS) fluxes, by fibering the A- resp. B-cycle of the elliptic fiber. Note that (forgetting about S), this construction gives $16+2+2 = 20$ independent 20-cycles of K3. The "missing" two are the base $\mathbb{CP}^1$ and the elliptic fiber. Since these do not wrap a single 1-cycle of the elliptic fiber, they do not give rise to suitable F-theory fluxes, as discussed in Section 3.4.

First, it is convenient to introduce the auxiliary space $\tilde{Z}$, which we formally construct as the elliptic fibration over X instead of over $B = X/\mathbb{Z}_2$, with fiber at a given point in X given by the fiber at the corresponding point in B.

Let Σ_2 be the Poincaré dual 2-cycle of F_2 in S. The orientifold projects out net D5-brane charge. Therefore Σ_2 although nontrivial in homology on S, must be trivial in homology on X, that is, it must be the boundary of a 3-chain Γ_3. Now let $\tilde{\Sigma}_4$ be the 4-cycle in $\tilde{Z}$ obtained by fibering the A-cycle of the T^2 over it. Because on X, A does not suffer monodromies anywhere, this fibration is guaranteed to be well defined, and produces a closed 4-cycle on $\tilde{Z}$. This projects to a closed 4-cycle Σ_4 in $Z = \tilde{Z}/\mathbb{Z}_2$. The cohomology class of G_4 is defined to be the Poincaré dual to this 4-cycle Σ_4. See Fig. 8.

Locally near the D7, this 4-cycle looks like the local one constructed above. The 4-form flux we constructed will therefore have a part localized on the D7, given by F_2.

Choosing a basis $\{\Sigma_{2,\alpha}\}_\alpha$ of the 2-form flux lattice of S, and corresponding 3-chains Γ_α, and corresponding 4-cycles $\Sigma_{4,\alpha}$, and calling linear combinations of these the (Poincaré duals of) "brane" fluxes, we declare the lattice of "bulk" fluxes to be the fluxes orthogonal to all of the brane fluxes, i.e. $\{G_4| \int_{\Sigma_{4,\alpha}} G_4 = 0\}$. So we can think of the bulk fluxes as those who have Poincaré duals (flux lines) "away" from the D7-branes. These 3-cycles are classified by ordinary 3-homology on X, so the bulk flux space can be thought of as being isomorphic[13] to $H^3(X, \mathbb{Z})$. The construction is illustrated in Fig. 8.

Note however that the precise distinction between bulk and brane fluxes is *not* canonical; it depends on the choice of 3-chains associated to the 2-form fluxes, and there is in general no canonical choice; it may be possible to loop around singularities in the deformation moduli space of S such that a 3-chain does not come back to itself, but to itself plus a closed 3-cycle. For example if we turn on a brane type flux with flux lines stretching between a particular pair of D7-branes in Fig. 8, and we move the pair around a 1-cycle of the torus, then the brane flux lines will transform to the original ones, plus bulk flux lines looping around that 1-cycle.

Bearing this in mind, we can denote the bulk flux cohomology classes as $[H_3]_b$ and $[F_3]_b$, and write the tadpole cancelation condition (3.29) as

$$-\frac{1}{\ell_s^4} \int_S \frac{1}{2} F_2 \wedge F_2 + \frac{1}{\ell_s^4} \int_X [F_3]_b \wedge [H_3]_b + 2 N_{D3} = 2 Q_c, \qquad (3.81)$$

where Q_c is the curvature induced D3-charge measured on B_6, $Q_c = \chi(Z)/24$, and the factor of 2 appears in front of N_{D3} and on the right hand side because we are integrating over the double cover X of B_6 on the left hand side.

When some D7-branes coincide, nonAbelian configurations are possible, and then the first term gets replaced by the second Chern character of the holomorphic vector bundle.

It is possible to write Q_c in terms of curvature induced charges on the orientifold plane and the D7-branes. The naive formula for this is

$$2 Q_c = \frac{\chi(D7) + 4 \chi(O7)}{24}, \qquad (3.82)$$

where $\chi(D7)$ and $\chi(O7)$ are the Euler characteristics of the 4-cycles wrapped by the D7 and the O7. But, recalling from (3.59) and (3.60) that at weak coupling, the 4-cycle wrapped by the D7 is described by the equation $\eta^2 = \xi^2 \chi$ in X, one sees that this complex surface has double point singularities on the complex

---

[13]There is one potential subtlety, and that is that H_3 must vanish on the D7; otherwise F_3 would be ill-defined at the D7 location due to the monodromy $F_3 \rightarrow F_3 + H_3$ around it. The vanishing of H_3 is readily seen to be the case for the localized flux (3.74). In particular therefore the bulk H_3 cohomology class must vanish on the D7. In many cases, the 4-cycles wrapped by the D7 have vanishing 3-cohomology because of the Lefshetz hyperplane theorem, so this is automatically satisfied. We will assume this is the case in what follows.

curve $\xi = \eta = 0$, and additional pinch point singularities on this curve at the points where also $\chi = 0$. This makes the usual notions of Euler characteristic and other topological quantities ambiguous for the D7, and more care has to be taken to define and compute these numbers. This has been analyzed in [59,60].

3.10. Enhanced gauge symmetries and charged matter

When D-branes coincide, one gets enhanced nonAbelian gauge symmetries. For example n D7-branes on X coincident with an $O7^{-14}$ give rise to an $SO(n)$ gauge group, $2n$ D3-branes on an $O7^-$ give rise to $USp(2n)$, and a stack of n coincident branes away from the O7 together with its orientifold image generically gives $SU(n)$. In M/F-theory, coincident D7 or more generally (p, q) 7-branes correspond to a singular elliptic fibration; the massless gauge bosons are M2 branes wrapping collapsed 2-cycles. These 2-cycles can be blown up in M-theory, and the way the resulting blown up 2-cycles intersect each other can be encoded in a Dynkin diagram, which is exactly the Dynkin diagram of the enhanced gauge group. We will briefly revisit this beautiful picture in Section 4.5.3.

The different possible singularities are classified according to the vanishing order of the polynomials f, g and $\Delta = 27\,g^2 + 4\,f^3$ with f and g as in (3.53). The corresponding gauge groups are given in the following table [67]:

ord(f)	ord(g)	ord(Δ)	group
≥ 0	≥ 0	0	none
0	0	n	$SU(n)$
≥ 1	1	2	none
1	≥ 2	3	$SU(2)$
≥ 2	2	4	$SU(3)$
2	≥ 3	$n+6$	$SO(2n+8)$
≥ 2	3	$n+6$	$SO(2n+8)$
≥ 3	4	8	E_6
3	≥ 5	9	E_7
≥ 4	5	10	E_8

More precisely, the above table holds under the assumption that no monodromies act on the collapsing 2-cycles; if such monodromies do occur, the classifications is more complicated and also includes $SO(2n+1)$, USp, F_4 and G_2 gauge groups.

---

[14] The O7-planes we have encountered so far are $O7^-$-planes. $O7^+$-planes also exist, arising from a slightly different representation of the $\mathbb{Z}_2$ on the string worldsheet degrees of freedom. They have positive D7-charge and $2n$ D7-branes coincident with them give rise to a $USp(2n)$ gauge group, while n D3-branes give an $SO(n)$.

Note in particular that more gauge groups are possible in the general F-theory setup than in type IIB at weak coupling. (For some obscenely huge gauge groups, with many exceptional group factors and ranks up to 121328, see [68].)

Massless charged matter on the other hand arises when two stacks of D-branes intersect. More generally, in F-theory, it is associated to singularity enhancement along the singular locus of the elliptic fibration.

The approach we are following here is a little too crude to properly analyze enhanced gauge symmetries and charged matter content, and we will therefore usually assume in what follows that we are at some point in the complex structure moduli space without enhanced gauge symmetry. More information can be found for example in [21, 22, 67, 69, 70].

4. Type IIB/F-theory flux vacua

4.1. Moduli

We are now in the position to determine the four dimensional low energy effective field theory corresponding to F-theory compactified on a Calabi–Yau fourfold Z elliptically fibered over a three complex dimensional base manifold B, or equivalently type IIB on B containing 7-branes, which in the weak coupling orientifold limit can be thought of as the $\mathbb{Z}_2$ orientifold quotient of type IIB on a Calabi–Yau threefold X with O7 and D7 branes. More generally, we could also have O3-planes—these correspond to codimension eight $\mathbb{Z}_2$ singularities in F-theory. We also consider space-filling D3-branes.

The following table shows the massless moduli we have before turning on fluxes, their M/F-theory and weakly coupled type IIB orientifold interpretations and the Hodge numbers counting them:

M/F-theory	# real moduli	IIB orientifold	# real moduli
Kähler	$h^{1,1}(Z) - 1$	Kähler	$h^{1,1}_+(X)$
		Complex structure	$2 h^{2,1}_-(X)$
Complex structure	$2 h^{3,1}(Z)$	D7 deformations	$2 \hat{h}^{2,0}_-(S)$
		Dilaton-axion	1
C_6 axions	$h^{1,1}(Z) - 1$	C_4 axions	$h^{1,1}_+(X)$
C_3 axions	$2 h^{2,1}(Z)$	B_2, C_2 axions	$h^{1,1}_-(X) + h^{1,1}_-(X)$
M2 positions	$6 N_{D3}$	D3 positions	$6 N_{D3}$

The subscripts $\pm$ denote the Hodge numbers counting the even resp. odd parts of the relevant cohomology under the geometrical orientifold involution [71].[15] For simplicity we will assume that $h^{2,1}(Z) = 0$. This implies in particular that there are no C_3 axions in the M-theory picture, and no B_2/C_2 axions in the IIB picture, nor $U(1)$ vectors from the reduction of C_4. Many such $h^{2,1}(Z) = 0$ examples are known [54]. In any case, axions are never really a problem, since they are not control parameters, cannot destabilize compactifications and will generically get lifted as soon as supersymmetry is broken.

4.2. Low energy effective action in F-theory framework

The four dimensional low energy effective theory is $\mathcal{N} = 1$ supergravity, with zero potential classically and to all orders in perturbation theory. The $\mathcal{N} = 1$ supersymmetry constraints imply that the moduli parametrize a Kähler manifold. The complex coordinates in the M-theory representation of F-theory are the complex structure moduli z^a, $a = 1, \ldots, h^{3,1}(Z)$ (which can be thought of as the coefficients of the defining equation, modulo coordinate redefinitions, if the Calabi–Yau is algebraic), the D3 moduli y_i^m, $m = 1, 2, 3$, $i = 1, \ldots, N_{D3}$, and the complexified Kähler moduli

$$T_A = \frac{1}{\ell_M^6} \int_{D_{6,A}} C_6 + i\, dV. \tag{4.1}$$

Here $\{D_{6,A}\}$ is a basis of 6-cycles (divisors) in Z wrapping the T^2 fiber, and dV is the volume element of the 6-cycle. As always, we take the F-theory limit of vanishing fiber area $L^2 = v = \int_{T^2} dV$. According to the general F-theory—type IIB reduction scheme of Section 3.1, using in particular the relation $L/\ell_M^3 = 1/\ell_s^2$, we can also write this in IIB language as

$$T_A = \frac{1}{\ell_s^4} \int_{D_{4,A}} C_4 + i\, dV, \tag{4.2}$$

where now $\{D_{4,A}\}$ is the corresponding basis of 4-cycles (divisors) in the base B (or, if we add a prefactor $\frac{1}{2}$, in X).

In general the D3 moduli and Kähler moduli mix in a rather intricate way in the Kähler potential. To avoid this complication, we will assume there are no D3-branes present for now. See [72–75] for the effective Kähler potential and action for D3-branes, and [76] for concretely applied examples.

---

[15]The hat on $\hat{h}_-^{2,0}(S)$ is there to indicate subtleties in the definition of this number due to the singularities of S [59].

The classical Kähler potential then splits in a Kähler part and a complex structure part:

$$\mathcal{K} = \mathcal{K}_K(T, \bar{T}) + \mathcal{K}_c(z, \bar{z}). \tag{4.3}$$

The Kähler part is determined by the volume:

$$
\begin{aligned}
\mathcal{K}_K &= -2\ln\left(\frac{1}{\ell_M^8}\int_Z dV\right) = -2\ln\left(\frac{1}{\ell_s^6}\int_B dV\right) = -2\ln V(B) \tag{4.4}\\
&= -2\ln\left(\frac{1}{6}\int_B J^3\right) = -2\ln\left(\frac{1}{6}D_{ABC}J^A J^B J^C\right). \tag{4.5}
\end{aligned}
$$

Here $V(B)$ is the volume of B in string units, and the J^A are the components of the Kähler form: $J = J^A D_{4,A}$, where, slightly abusively, we used the same notation for the 4-cycle $D_{4,A}$ (above) and its Poincaré dual (here). The coefficients D_{ABC} are the triple intersection numbers of these divisors:

$$D_{ABC} := \#(D_A \cap D_B \cap D_C) = \int_B D_A \wedge D_B \wedge D_C. \tag{4.6}$$

The J^A are related to the T_A by

$$\mathrm{Im}\, T_A = \partial_{J^A} V(B) = \frac{1}{2}D_{ABC}J^B J^C. \tag{4.7}$$

Inverting $J^A(T, \bar{T})$ may or may not be possible explicitly, depending on the model. Note that $V(B) = V(X)/2$.

The complex structure part is

$$
\begin{aligned}
\mathcal{K}_c &= -\ln\int_Z \Omega_4 \wedge \overline{\Omega_4} \tag{4.8}\\
&= -\ln\left(\Pi_I(z)\, Q^{IJ}\, \overline{\Pi_J(z)}\right). \tag{4.9}
\end{aligned}
$$

Here Ω_4 is the unique holomorphic 4-form on Z, the Π_I are its periods:

$$\Pi_I(z) := \int_{\Sigma_{4,I}} \Omega_4(z), \tag{4.10}$$

with $\{\Sigma_{4,I}\}_I$, $I = 1, \dots, b'_4(Z)$, a basis of 4-cycles wrapping a 1-cycle in the elliptic fiber. Finally, Q^{IJ} is the inverse of Q_{IJ}, and Q_{IJ} is the intersection form of the basis:

$$Q_{IJ} := \Sigma_I \cdot \Sigma_J = \#(\Sigma_{4,I} \cap \Sigma_{4,J}) = \int_Z \Sigma_{4,I} \wedge \Sigma_{4,J}, \tag{4.11}$$

where again we used the same notation for cycle and Poincaré dual form.

Since we have assumed $h^{2,1}(Z) = 0$, there are no further massless fields in four dimensions besides the metric tensor, so our specification of the low energy effective action is complete.

Let us consider a very simple toy model as illustration for the complex structure moduli sector. The model can morally be thought of as $Z = X \times T^2$ with X a rigid Calabi–Yau threefold (i.e. X has no complex structure moduli). Thus, Z has a single complex structure modulus τ, the modular parameter of the T^2. The moduli space of the model is the fundamental domain in the upper half τ-plane. A rigid Calabi–Yau has two independent 3-cycles, so we can make four 4-cycles by combining these with the A and B 1-cycles in the T^2. The nonvanishing periods of Ω_4 are then

$$\Pi_I = (1, \omega, \tau, \omega\tau), \tag{4.12}$$

where ω is some complex number depending on X and our choice of 3-cycles. For simplicity we just put $\omega \equiv i$. If the 3-cycles have intersection product 1, the intersection form for the 4-cycles is

$$Q_{IJ} = \begin{pmatrix} 0 & 0 & 0 & -1 \\ 0 & 0 & 1 & 0 \\ 0 & 1 & 0 & 0 \\ -1 & 0 & 0 & 0 \end{pmatrix}. \tag{4.13}$$

The Kähler potential on the complex structure moduli space is thus

$$\mathcal{K}_c = -\ln \Pi_I(z) \, Q^{IJ} \, \overline{\Pi_J(z)} = -\ln(4 \operatorname{Im} \tau), \tag{4.14}$$

and its complex structure moduli space metric

$$g_{\tau\bar\tau} = \partial_\tau \partial_{\bar\tau} \mathcal{K}_c = \frac{|d\tau|^2}{4(\operatorname{Im} \tau)^2}, \tag{4.15}$$

the standard Poincaré metric on the upper half plane.

4.3. Low energy effective action in IIB weak coupling limit

In the type IIB weak coupling orientifold limit, we can reproduce the structure of the low energy effective action expected from the perturbative string picture as follows. First recall that at the end of Section 3.9, we introduced a (formal) elliptic fibration $\tilde{Z}$ over the CY 3-fold X, the varying field τ on X being the modulus of the elliptic fiber. The space $\tilde{Z}$ can be thought of as a double cover of Z in the orientifold limit. In particular in this limit (ignoring $e^{-\pi/g_s}$ corrections) we have

$$\mathcal{K}_c = -\ln \int_Z \Omega \wedge \bar\Omega = -\ln \frac{1}{2} \int_{\tilde{Z}} \Omega \wedge \bar\Omega. \tag{4.16}$$

We also saw there that we can define bulk and brane 4-cycles on $\tilde{Z}$ (or Z), the brane 4-cycles being A-cycle fibrations over 3-chains ending on the D7 locus S in X, and the bulk cycles being those with zero intersection product with those, which are A- or B-cycle fibrations over 3-cycles in X. Let us denote the chosen basis for the 3-chains by $\{\Gamma_\alpha\}_\alpha$, $\alpha = 1, \ldots, \hat{b}^2_-(S)$, and for the bulk 3-cycles by $\{\Sigma_i\}_i$, $i = 1, \ldots, b^3(X)$. Denote the 4-cycles in $\tilde{Z}$ obtained by fibering the A-cycle over Σ_i by $\Sigma_i \times A$, those obtained by fibering the B-cycle over Σ_i by $\Sigma_i \times B$, and those obtained by fibering the A-cycle over Γ_α by $\Gamma_\alpha \times A$. Then the corresponding periods are, using (3.63) and (3.67), and denoting 3-fold complex structure moduli by ψ and D7 moduli by ϕ, up to $e^{-\pi/g_s}$ corrections:

$$\int_{\Sigma_i \times A} \Omega_4 = \int_{\Sigma_i} \Omega_3(\psi) =: \Pi_i(\psi), \tag{4.17}$$

$$\int_{\Sigma_i \times B} \Omega_4 = \int_{\Sigma_i} \tau \, \Omega_3(\psi) = \int_{\Sigma_i} \left(\tau_0 + \frac{i}{2\pi} \ln \frac{P_{O7}(\psi)}{P_{D7}(\psi, \phi)} \right) \Omega_3(\psi)$$

$$=: \tau_0 \, \Pi_i(\psi) + \chi_i(\psi, \phi), \tag{4.18}$$

$$\int_{\Gamma_\alpha \times A} \Omega_4 = \int_{\Gamma_\alpha(\phi)} \Omega_3(\psi) =: \Pi_\alpha(\psi, \phi), \tag{4.19}$$

where Ω_3 is the holomorphic 3-form on X and $\tau_0 = i/g_s$ is the "bulk" value of τ as defined in (3.68). The dependence of the various terms on the threefold complex structure moduli ψ and the D7 moduli ϕ (up to $e^{-\pi/g_s}$ corrections) is indicated.

Furthermore, the intersection form Q_{IJ} splits in bulk and brane blocks. The nonzero entries are

$$Q_{\alpha\beta} := (\Gamma_\alpha \times A) \cdot (\Gamma_\beta \times A) = -(\partial\Gamma_\alpha) \cdot (\partial\Gamma_\beta)|_S, \tag{4.20}$$

$$Q_{ij} := (\Sigma_i \times A) \cdot (\Sigma_j \times B) = -(\Sigma_i \cdot \Sigma_j)|_X. \tag{4.21}$$

Thus using (4.16) we can write, up to nonperturbative $e^{-\pi/g_s}$ corrections:

$$\mathcal{K}_c = -\ln \frac{1}{2} \left((\tau_0 - \bar{\tau}_0) \, \Pi_i \, Q^{ij} \, \overline{\Pi}_j + \chi_i \, Q^{ij} \, \overline{\Pi}_j + \overline{\chi}_i \, Q^{ij} \, \Pi_j - \Pi_\alpha \, Q^{\alpha\beta} \, \overline{\Pi}_\beta \right). \tag{4.22}$$

In a perturbative $g_s = 1/\mathrm{Im}\, \tau_0$ expansion, this becomes

$$\mathcal{K}_c = \mathcal{K}_{\tau_0} + \mathcal{K}_X(\psi) + g_s \, \mathcal{K}_{D7}(\psi, \phi) + \cdots \tag{4.23}$$

where

$$\mathcal{K}_\tau(\tau_0)(\tau_0) \;=\; -\ln(\mathrm{Im}\,\tau_0), \tag{4.24}$$

$$\mathcal{K}_X(\psi) \;=\; -\ln\!\big(i\,\Pi_i\,Q^{ij}\,\overline{\Pi}_j\big) = -\ln i \int_X \Omega_3 \wedge \overline{\Omega_3}, \tag{4.25}$$

$$\mathcal{K}_{D7}(\psi,\phi) \;=\; \tfrac{1}{2}e^{\mathcal{K}_X}\big(\Pi_\alpha\,Q^{\alpha\beta}\,\overline{\Pi}_\beta + \chi_i\,Q^{ij}\,\overline{\Pi}_j + \overline{\chi}_i\,Q^{ij}\,\Pi_j\big). \tag{4.26}$$

The first two parts of the Kähler potential are the standard dilaton and complex structure Kähler potentials one gets by direct reduction of type IIB on X. The third part governs the D7-brane moduli. Note that it enters at order $g_s = 1/\mathrm{Im}\,\tau_0$ compared to the bulk moduli part; it is nevertheless the first nontrivial order at which the D7 degrees of freedom ϕ appear. This means the backreaction of the D7-branes on the bulk geometry is suppressed by a power of g_s, as it should. At fixed ψ, the χ_i-dependent part of $\mathcal{K}_{D7}$ is merely a Kähler gauge transformation, and therefore does not contribute to the D7 moduli space metric to leading order:

$$g_{r\bar{s}} = \partial_{\phi^r}\bar{\partial}_{\bar{\phi}^s}\mathcal{K}_{D7} = \tfrac{1}{2}e^{\mathcal{K}_X}\partial_r\Pi_\alpha\,Q^{\alpha\beta}\,\bar{\partial}_{\bar{s}}\overline{\Pi}_\beta = \tfrac{1}{2}e^{\mathcal{K}_X}\int_S \omega_r \wedge \overline{\omega}_{\bar{s}}. \tag{4.27}$$

Here $\omega_r := (\Omega_3 \cdot \delta_r n)|_S$, with $\delta_r n$ the holomorphic deformation vector field normal to S corresponding to a variation $\delta\phi^r$ of the moduli of S, and "·" denotes index contraction. The forms ω_r are holomorphic $(2,0)$ forms on S. Notice that the D7 moduli metric does not depend on the choice of 3-chains; only the Kähler potential does.

All of this fits well with what we expect from the perturbative string point of view.

The geometrical structures underlying open-closed string moduli spaces were explored in [77].

4.4. The effect of turning on fluxes

We now consider the effect of turning on F-theory G_4 flux.

4.4.1. Effective potential

We will first work in the Kaluza–Klein approximation, i.e. the M-theory metric remains $ds^2 = -(dx^0)^2 + (dx^1)^2 + (dx^2)^2 + ds_Z^2$, where ds_Z^2 is an unwarped Ricci flat Calabi–Yau metric, and the flux is harmonic. The flux is quantized as

$$\ell_M^{-3}[G_4] = N^I\,\Sigma_{4,I}, \tag{4.28}$$

where $\{\Sigma_I\}$ is a basis of integral 4-form cohomology classes, and N^I is integral modulo a possible half-integral shift equal to $\frac{c_2^I(Z)}{2}$ [55]. The energy density in

$\mathbb{R}^{1,2}$ due to the flux is

$$V_M(G) = \frac{2\pi}{\ell_M^9} \frac{1}{2} \int_Z G_4 \wedge *G_4, \qquad (4.29)$$

where $*$ is the Hodge star on Z. If there were no negative energy contributions to the potential, we would be squarely in the no-go situation described in Section 2.1.2. We already know from (3.11) and more explicitly from (3.28) that the curvature of Z provides negative M2 *charge* $-Q_c = -\chi(Z)/24$. Since consistent Minkowski solutions exist with Q_c space-filling M2-branes canceling this charge, and Q_c space-filling M2-branes have an energy density equal to $\frac{2\pi Q_c}{\ell_M^3}$, this implies there must be additional higher order curvature terms in the action providing an energy density exactly equal to minus this. This is indeed the case [61]. Thus, assuming (3.28) is satisfied, we find for the total potential including contributions from M2-branes, curvature and flux:

$$V_M = \frac{2\pi}{\ell_M^9} \frac{1}{2} \int_Z (G_4 \wedge *G_4 - G_4 \wedge G_4). \qquad (4.30)$$

Splitting G_4 in its self-dual and anti-self-dual part, $G_4 = G_{4,+} + G_{4,-}$, this becomes

$$V_M = \frac{2\pi}{\ell_M^9} \int_Z G_{4,-} \wedge *G_{4,-}. \qquad (4.31)$$

From the general scheme of Section 3.1, it follows that the corresponding energy density in type IIB is

$$V_{\text{IIB}} = \frac{2\pi}{\ell_s^4} \frac{1}{\ell_M^6} \int_Z G_{4,-} \wedge *_Z G_{4,-} = \frac{2\pi}{\ell_s^8} \int_B \frac{1}{\text{Im}\,\tau} G_{3,-} \wedge *_B \overline{G_{3,-}}, \qquad (4.32)$$

where $G_{3,-}$ is the imaginary anti-self-dual part of $G_3 = F_3 - \tau H_3$, i.e. $*G_{3,-} = -iG_{3,-}$. (The i must be there because $*^2 = -1$ on 3-forms in B.)

Whether in IIB or in M-theory, after a Weyl rescaling to bring the 3d/4d Einstein–Hilbert term in canonical form, the potential gets an additional prefactor proportional to an inverse power of the volume, as in (2.3). Therefore to avoid a runaway, we must have $G_{4,-} = 0$ (equivalently $G_{3,-} = 0$) identically, i.e.

$$G_4 = *_Z G_4 \qquad \text{(equivalently } G_3 = i *_B G_3\text{)}. \qquad (4.33)$$

This puts constraints on the complex structure and Kähler moduli of Z. To make this explicit, we need some results about Hodge decompositions, which we develop in the following intermezzo. (This can be skipped by the reader not interested in the general framework.)

4.4.2. Intermezzo: Lefshetz SU(2) and diagonalizing the Hodge star operator
The vector space of harmonic forms on a Kähler manifold of complex dimension
n can be organized according to representations of the *Lefshetz SU(2) algebra*.
Up to normalization, the raising operator $L_+ = L_1 + iL_2$ is given by wedging
with the Kähler form J, the lowering operator L_- by contraction with J, and the
L_3 operator is the form degree up to a constant shift. For a harmonic p-form ω:

$$L_3\,\omega = \frac{p-n}{2}\,\omega, \qquad L_+\omega \sim J \wedge \omega, \qquad L_- = L_+^\dagger, \tag{4.34}$$

where the adjoint is taken with respect to the inner product on forms defined
by the Hodge *-operator. Although in general it is not true that wedging two
harmonic forms together produces a new harmonic form, it is true that wedging
a harmonic form with the Kähler form gives again a harmonic form, so the above
operations are well defined on the space of harmonic forms. One checks that
$L_+ * = *L_-$, $L_- * = *L_+$, and $L_3 * = - * L_3$, so in particular $[\mathbf{L}^2, *] = 0$,
and we can simultaneously diagonalize the Lefshetz spin ℓ and the Hodge $*$.
Explicitly, on a Kähler manifold of complex dimension n, one has for a spin ℓ
harmonic $(n - k, k)$-form ω:

$$*\omega = (-1)^{k+\ell}\,\omega \quad (n\ \text{even}), \qquad *\omega = (-1)^{k+\ell}(-i)\,\omega \quad (n\ \text{odd}). \tag{4.35}$$

For example, for n even, the $(n/2, n/2)$ form $J^{n/2}$ comes in a spin $\ell = n/2$
multiplet $(1, J, J^2, \ldots, J^n)$ and is self-dual.

A p-form has at most spin $\ell = (n - p)/2$. A *primitive p-form* is a form with
spin exactly equal to this. For the middle cohomology this means that it has spin
zero, or $\omega \wedge J = 0$. Thus for a 2-fold for example, primitive (1,1)-forms are
anti-self-dual, while for a 4-fold, primitive (2,2)-forms are self-dual.

For an $SU(4)$ holonomy Calabi–Yau 4-fold, we thus get the decompositions

$$H^4_+ = H^{0,4}_{\ell=0} \oplus H^{2,2}_{\ell=0} \oplus H^{2,2}_{\ell=2} \oplus H^{4,0}_{\ell=0}, \tag{4.36}$$

$$H^4_- = H^{1,3}_{\ell=0} \oplus H^{2,2}_{\ell=1} \oplus H^{3,1}_{\ell=0}. \tag{4.37}$$

There is a unique $\ell = 2$ multiplet given by $(1, J, J^2, J^3, J^4)$, and there are
$h^{1,1} - 1$ independent $\ell = 1$ multiplets $(\omega_k, \omega_k J, \omega_k J^2)$, where $\{\omega_k\}_k$ is a set of
$h^{1,1} - 1$ independent (1,1)-forms such that $\omega_k J^3 = 0$.

It is worth pointing out that for harmonic forms, equations in cohomology are
equivalent to pointwise equations for the differential forms. In particular if say
$J \wedge \omega = 0$ in cohomology, it is zero pointwise. This is because there is always a
unique harmonic representative of a cohomology class.

4.4.3. Superpotential formulation

From the intermezzo we take that the self-duality condition $G_4 = *G_4$ is equivalent to

$$G_4^{1,3} = G_4^{3,1} = 0, \qquad G_4^{2,2}|_{\ell=1} = 0. \tag{4.38}$$

A basis of $H^{3,1}(Z)$ is provided[16] by the covariant derivatives with respect to the complex structure moduli of Z:

$$D_a \Omega_4 := (\partial_a + \partial_a \mathcal{K}_c) \Omega_4. \tag{4.39}$$

Therefore, the first condition is (4.38) is equivalent to G_4 being orthogonal to all $D_a \Omega_4$, i.e.:

$$D_a W(z) = 0, \quad W(z) := \frac{1}{\ell_M^3} \int_Z G_4 \wedge \Omega. \tag{4.40}$$

We recognize this as a superpotential condition. The superpotential $W(z)$ appearing here lives on the complex structure moduli space of Z, and was first derived by Gukov, Vafa and Witten, in [27]. We included the factor ℓ_M^{-3} to make $W(z)$ dimensionless.

This formulation makes it manifest that turning on flux constrains the complex structure moduli. In fact, for sufficiently generic $W(z)$, one expects isolated critical points, suggesting that *all* complex structure moduli can be stabilized in this way. (This is plausible but not completely obvious in the case at hand, because the fluxes are quantized and constrained by the tadpole cancelation condition (3.28).)

Since the $\ell = 1$ part of $H^{2,2}(Z)$, by definition, consists of the (2, 2) forms which are not annihilated by J, minus the (2, 2) forms proportional to $J \wedge J$, the second condition in (4.38) can be written as

$$G_4 \wedge J = c\, J \wedge J \wedge J, \tag{4.41}$$

for some constant c, which can be computed as $c = (\int_Z G_4 \wedge J^2)/(\int_Z J^4)$. This can again be written in a superpotential-like form:

$$D_J \tilde{W}(J) = 0, \quad \tilde{W}(J) := \frac{1}{\ell_M^3} \int_Z G_4 \wedge J \wedge J, \tag{4.42}$$

---

[16] A variation of the complex structure can be thought of as a variation of complex coordinates $\delta y^m = \epsilon f^m(y, \bar{y})$ on the CY fourfold Z, where f is a nonholomorphic function. The resulting variation $\delta \Omega_4$ of the holomorphic (4, 0)-form will thus be of type $(4, 0) + (3, 1)$, with nonzero (3, 1) part. Hence $\partial_a \Omega$ is of type $(4, 0) + (3, 1)$. It is easily checked that going to the Kähler covariant derivative $D_a \Omega$ subtracts off precisely the (4, 0) part. So the $D_a \Omega$ form a set of linearly independent (3, 1) forms. The number of these equals the number of complex structure deformations, which equals $h^{3,1}$ (see e.g. [3] for the analogous case of a CY 3-fold). Thus, the $D_a \Omega$ form a basis of $H^{3,1}$.

where we introduced another covariant derivative

$$D_J \tilde{W} := \partial_J \tilde{W} + (\partial_J \mathcal{K}_J) \tilde{W}, \quad \mathcal{K}_J := -\ln \int_Z \frac{J^4}{4!}. \tag{4.43}$$

However, being a real function, this does not have an actual superpotential interpretation in four dimensions. Instead, (4.41) should be interpreted in four dimensions as a D-term constraint.

Actually, given the form (3.22) of the fluxes G_4 which preserve four dimensional Poincaré invariance in the IIB description, (4.41) is automatically satisfied for harmonic G_4 on smooth, full $SU(4)$ holonomy Calabi–Yau fourfolds Z. To see this, first note that if Z has $SU(4)$ holonomy,[17] $H^2(Z) = H^{1,1}(Z)$, so all 2-cohomology classes have Poincaré dual representatives which are divisors (linear combinations of holomorphic 6-cycles). Let $\{D_M\}_M$ be a basis of $H^{1,1}(Z)$ or equivalently of divisors in Z. Then we claim that $G_4 \wedge D_A$ is zero in cohomology for all D_M, i.e.

$$\int_Z G_4 \wedge D_M \wedge D_N = 0, \quad \forall M, N, \tag{4.44}$$

or equivalently by going to the Poincaré dual representation $\int_{D_M \cap D_N} G_4 = 0$. To prove this, note that for smooth Z at least, all divisors in Z with the exception of the base B itself (more precisely the section of the elliptic fibration) can be thought of as elliptic fibrations over divisors in the base. Hence intersections of divisors $D_M \cap D_N$ are linear combinations of divisors in B and elliptic fibrations over holomorphic curves in B; in particular they wrap the elliptic fiber either completely, or not at all. Fluxes of the form (3.22) integrate to zero on such surfaces, since there are no components with two legs on the elliptic fiber or with no legs on the elliptic fiber at all. This shows that $G_4 \wedge D_M$ is zero in cohomology, so in particular $G_4 \wedge J$ is zero in cohomology. Since we take G_4 to be the harmonic representative in its cohomology class, $G_4 \wedge J$ is harmonic too, and must therefore be zero pointwise. Thus, as claimed,

$$G_4 \wedge J = 0 \tag{4.45}$$

and (4.41) is automatically satisfied.

This should not surprise us, since for a smooth, full $SU(4)$ holonomy CY Z, there are no suitable massless $U(1)$ vectors in the four dimensional effective theory which could generate a D-term.[18] Note that this need not be the case when Z

---

[17]This in contrast to for example $Z = K3 \times K3$ and indeed in this case (4.41) is not automatically satisfied.

[18]Again, this is not true for reduced holonomy CY manifolds such as $K3 \times K3$, where $U(1)$s generating D-terms do survive; in the weak coupling limit, they correspond to the relative $U(1)$s

is singular, so in those cases we might still have a D-term condition to take into account. In particular this will be the case when there is enhanced gauge symmetry or when there are intersecting 7-branes. Our approach has been somewhat too crude to properly deal with singularities however, so we will continue to operate under the assumption of smoothness for now.

This leaves us with (4.40), and when this is satisfied, the effective potential for the remaining moduli, in particular all the Kähler moduli, is flat.

Finally, note that using $G^{2,2}|_{\ell=1} = 0$ and expanding $G^{3,1}$ in the basis $D_a\Omega$, (4.32) can also be written as

$$V_{\text{IIB}} = \frac{m_p^4}{4\pi} e^{\mathcal{K}_c + \mathcal{K}_K} g^{a\bar{b}} D_a W \overline{D_b W}, \tag{4.46}$$

where m_p is the 4d Planck mass defined such that the coefficient of the Einstein–Hilbert term in the 4d action is $\frac{m_p^2}{2}$. This is reminiscent of the standard $\mathcal{N} = 1$ formula for the potential in terms of a superpotential, but seems to be missing a term proportional to $-3|W|^2$. This term is indeed there when *all* moduli are taken into account in the standard formula, but happens to cancel out exactly against the $|DW|^2$ part generated by the moduli different from the complex structure moduli, leaving (4.46) behind.

Because nothing sets the scale of the internal manifold, these compactifications are called "no scale" compactifications. The good thing about them is that this allows us to go to the large radius regime where all of our approximations are justified. The bad thing is that we will need to invoke quantum corrections again to lift this degeneracy.

As an illustration, we return to the toy model introduced at the end of Section 4.2. If we turn on flux with flux quanta $N^I = (A_1, A_2, B_1, B_2) \in \mathbb{Z}^4$, the superpotential becomes

$$W(\tau) = A + B\tau, \quad A := A_1 + iA_2, \quad B := B_1 + iB_2. \tag{4.47}$$

The tadpole cancelation condition (3.28) becomes, using (4.13),

$$\text{Im}\,(\bar{B}A) + N_{\text{D3}} = Q_c. \tag{4.48}$$

Strictly speaking $Q_c = 0$ if $Z = T^2 \times X$, but for the sake of the toy model we will take it to be some arbitrary given number. Using (4.14), we find

$$D_\tau W := (\partial_\tau + \partial_\tau \mathcal{K}_c)W = \frac{A + B\bar{\tau}}{\bar{\tau} - \tau}. \tag{4.49}$$

---

of (disjoint) D7-image-D7 pairs. For smooth, genuine $SU(4)$ holonomy CYs on the other hand, D7-image-D7 pairs get recombined into single branes, and the $U(1)$ is broken.

Hence $V_{\text{IIB}} \sim |A + B\bar{\tau}|^2$. Solving $DW = 0$ for τ gives

$$\tau = -\frac{\bar{A}}{\bar{B}}. \tag{4.50}$$

Note that although this naively looks like an infinite number of vacua, we should not count vacua related by $SL(2, \mathbb{Z})$ transformations separately. To avoid overcounting, we could for example restrict the solutions τ to the fundamental domain. Figure 14 in Section 6.2.5 shows the set of vacua for $Q_c = 150$.

You can find more simple, explicit examples of (bulk) flux vacua in [65, 66]. In particular, in the latter reference some flux vacua for the example (3.53) are constructed. Needless to say though, systematically finding fully explicit examples—let alone enumerating them—for compactifications with many moduli and fluxes becomes effectively intractable. Fortunately it is not necessary to construct things explicitly to find approximate distributions of flux vacua over parameter space. We will get to this in Section 6.

4.4.4. Weak coupling limit
In the IIB weak coupling limit, we can expand the flux in the basis introduced in Section 4.3:

$$\frac{1}{\ell_M^3}[G_4] = \sum_i N^i \, [\Sigma_i \times A] - \sum_i M^i \, [\Sigma_i \times B] + \sum_\alpha N^\alpha \, [\Gamma_\alpha \times A]. \tag{4.51}$$

In terms of the weak coupling type IIB bulk and brane fluxes introduced in Section 3.9, this is

$$\frac{1}{\ell_s^2}[F_3]_b = \sum_i N^i \, [\Sigma_i], \qquad \frac{1}{\ell_s^2}[H_3]_b = \sum_i M^i \, [\Sigma_i],$$

$$\frac{1}{\ell_s^2}[F_2] = \sum_\alpha N^\alpha \, [\partial \Gamma_\alpha]. \tag{4.52}$$

The corresponding Gukov–Vafa–Witten superpotential is then, up to $e^{-\pi/g_s}$ corrections:

$$W(\tau_0, \psi, \phi) \;=\; \sum_i (N^i - \tau_0 M^i) \Pi_i(\psi) \tag{4.53}$$

$$-\sum_i M^i \chi_i(\psi, \phi) + \sum_\alpha N^\alpha \Pi_\alpha(\psi, \phi) \tag{4.54}$$

$$=: \; W_b(\tau_0, \psi) + W_{D7}(\psi, \phi), \tag{4.55}$$

with Π_i, χ_i and Π_α defined in (4.17)–(4.19).

For suitable[19] closed 3-forms F_3 and H_3 representing $[F_3]_b$ resp. $[H_3]_b$, this can also be written as

$$W_b = \int_X (F_3 - \tau_0 H_3) \wedge \Omega_3, \tag{4.56}$$

$$W_{D7} = -\frac{i}{2\pi} \int_X H_3 \wedge \ln \frac{P_{O7}}{P_{D7}} \Omega_3 + \int_{\Gamma(F_2)} \Omega_3, \tag{4.57}$$

where $\partial \Gamma(F_2) = [F_2]$. The bulk superpotential is the same as in the absence of 7-branes. To formally make contact with the general D7 superpotential of [63], put $H_3 = dB_2$. Note that the integrand in the first term has a branch cut 5-chain Γ_5 going between the D7-brane $S : P_{D7} = 0$ and the O7-plane $P_{O7} = 0$, on which it jumps by an amount H_3.[20] Now extend F_2 as a closed form onto this 5-chain Γ_5 (by taking it to be the Poincaré dual of $\Gamma(F_3)$, which we can take to lie in Γ_5), and perform an integration by parts on the first term in W_{D7}. This gives:

$$W_{D7}(\phi) \text{ "} = \text{" } \int_{\Gamma_5} (F_2 - B_2) \wedge \Omega_3, \tag{4.58}$$

reproducing the D7 superpotential of [63]. I have put the equality sign between quotation marks because it is not quite justified to simply put $H_3 = dB_2$, since H_3 is not globally exact. I will not try to make this correspondence more precise; in practice, if one wishes to explicitly compute the superpotential in specific models, the expression involving the periods is computationally superior anyway.

To leading order in g_s, the critical point condition $DW = 0$ splits up as

$$\partial_{\tau_0} W + (\partial_{\tau_0} \mathcal{K}_\tau) W = 0, \tag{4.59}$$

$$\partial_\psi W + (\partial_\psi \mathcal{K}_X) W = 0, \tag{4.60}$$

$$\partial_\phi W_{D7} = 0. \tag{4.61}$$

When the D7-branes coincide with the O7, or more precisely when $\eta^2 - h\chi = h^4$, $W_{D7} = 0$ and the first two equations are equivalent to

$$[G_3]_b^{3,0} = 0, \qquad [G_3]_b^{1,2} = 0, \qquad [G_3]_b := [F_3]_b - \tau_0 [H_3]_b, \tag{4.62}$$

that is, $[G_3]_b$ is of type $(2, 1) + (0, 3)$; a harmonic representative would be imaginary self-dual. This is no longer true when the branes move off the O7. This was to be expected, since the distinction between bulk and brane flux is not canonical:

---

[19] The choice of representative matters for H_3, since the logarithmic branch cut in the integrand of the first term of W_{D7} makes the integral not invariant under $H_3 \to H_3 + d\beta_2$.

[20] The presence of the cut is due to the fact that there is a $SL(2, \mathbb{Z})$ T-monodromy $F_3 \to F_3 + H_3$ around the D7 branes for the original F_3 and H_3 defined in (3.22).

looping around in D7 configuration space can create bulk flux out of brane flux. And certainly, if the bulk flux changes, we expect the bulk flux equations to be changed too.

Taking $[H_3]_b = 0$, the third equation is equivalent to

$$[F_2]^{0,2} = 0 = [F_2]^{2,0}, \tag{4.63}$$

that is, F_2 is of type $(1, 1)$. In addition, note that automatically $[F_2] \wedge [J] = 0$ if the D7 is generic. This is because $[F_2]$ is odd under the orientifold involution, while $[J]$ is even, so $\int F_2 \wedge J = 0$. Genericity of the D7 implies in particular it has a single component, in which case the latter equation implies $[F_2] \wedge [J] = 0$. This fits with our earlier observation that (4.41) is automatically satisfied in our setup. Thus, a harmonic representative F_2 would be anti-self-dual. This is the condition for a D7 configuration with flux to preserve the supersymmetries of the orientifold background, as obtained from the perturbative string description [62].

Explicitly finding flux vacua by computing the periods Π_α, Π_i, χ_i and finding critical points is prohibitively difficult in almost any example. However, the condition (4.63) has a simple geometrical interpretation:

$$[F_2]^{0,2} = 0 \quad \Leftrightarrow \quad [F_2] \text{ is a divisor in S.} \tag{4.64}$$

In other words, in the absence of $[H_3]_b$ and to lowest order in g_s, the D7 embedding must be such that $N^\alpha \partial \Gamma_\alpha$ can be represented in homology as a linear combination of holomorphic curves. It is infinitely much simpler to explicitly construct holomorphic curves as surfaces containing them. Thus, explicitly constructing D7 flux vacua at weak string coupling is in fact more tractable than one might have naively feared.

It is not known if a similar geometrization of flux vacua exists for nonzero bulk fluxes or away from the weak coupling limit.

4.4.5. Supersymmetry

Generically, the flux vacua obtained by solving $D_a W = 0$ break supersymmetry. This is simply because generically $W \neq 0$ at the critical point, and therefore the covariant derivatives with respect to the Kähler moduli, $D_{T_A} W = (\partial_{T_A} \mathcal{K}) W$, do not vanish.

For the flux vacuum to preserve supersymmetry, we need in addition $W = 0$, or equivalently $G_4^{4,0} = G_4^{0,4} = 0$; for the fluxes we are considering, this means G_4 is of type (2,2) and primitive.

Since $W = 0$ is one more constraint than there are variables, we generically do not expect solutions. A notable exception to this is when $D_a W = 0$ alone does not fix all complex structure moduli; in this case we can conceivably move along the flat direction till we hit a zero of W. Of course what we really are after are vacua without any remaining flat directions at all, so this would not be a desirable situation from that point of view.

However it may happen that $W = 0$ "accidentally" even for isolated vacua [64].

4.4.6. Warping

So far we have done our analysis in the effective field theory approximation. It is possible to do better. In the absence of flux and M2-branes, our metric was flat space times the Calabi–Yau metric on Z. In the presence of G_4 flux and/or M2-branes, this metric no longer solves the Einstein equations, but a warped version of it still does [18, 19, 27]. The metric is of the form

$$ds^2 = e^{-w(y)}\left[-(dx^0)^2 + (dx^1)^2 + (dx^2)^2\right] + e^{w(y)/2}ds_Z^2, \tag{4.65}$$

where $w(y)$ is the warp factor, which depends only on the coordinates y^m of the internal space Z. The metric ds_Z^2 is our original Ricci-flat Calabi–Yau metric on Z. The warp factor satisfies the following Poisson equation on Z:

$$d * d(e^{3w/2}) = \frac{1}{2\,\ell_M^6}G_4 \wedge G_4 - I_8(R) + \sum_i \delta_{M2_i}, \tag{4.66}$$

where the Hodge $*$ is with respect to the ds_Z^2 metric. The 4-form G_4 satisfies

$$G = *_Z G, \tag{4.67}$$

again with respect to ds_Z^2. Furthermore $G_{\mu\nu\rho m} = \epsilon_{\mu\nu\rho}\partial_m e^{-3w/2}$.

The metric (4.65) still fits in our original T^2 fibered metric ansatz (3.12), namely

$$ds_M^2 = \frac{v}{\tau_2}\left((dx + \tau_1 dy)^2 + \tau_2^2 dy^2\right) + ds_9^2, \tag{4.68}$$

provided we now allow v to vary over M_9:

$$v = v_0 e^{w/2}, \tag{4.69}$$
$$ds_9^2 = e^{-w}\left[-(dx^0)^2 + (dx^1)^2 + (dx^2)^2\right] + e^{w/2}ds_B^2.$$

Plugging this back into (3.17), taking $L \equiv \sqrt{v_0}$ and defining $x^3 := \frac{\ell_s^2}{\sqrt{v_0}}y$, we find for the Einstein frame IIB metric

$$ds_{IIB}^2 = e^{-3w/4}\left[-(dx^0)^2 + (dx^1)^2 + (dx^2)^2 + (dx^3)^2\right] + e^{3w/4}\,ds_B^2. \tag{4.70}$$

In the F-theory limit $v_0 \to 0$, the periodicity $\ell_s^2/\sqrt{v_0}$ of x^3 goes to infinity, and we retrieve a fully four dimensional Poincaré invariant solution. Note in particular the remarkable fact that the warp factors have combined in just the right way to make full Poincaré invariance possible, despite the different origin of the x^3 direction in M-theory.

The IIB metric obtained here is indeed of the warped form considered in [20] in the weak coupling limit of type IIB.

It was pointed out in [20] that the warping can become very substantial for some flux vacua. A simple local, non-orientifolded, D-brane free model for this is the following. Consider the (noncompact) deformed conifold Calabi–Yau three-fold embedded in $\mathbb{C}^4$:

$$X : u_1^2 + u_2^2 + u_3^2 + u_4^2 = \epsilon^2. \tag{4.71}$$

The holomorphic 3-form is $\Omega_3 = \frac{du_1 du_2 du_3}{u_4}$, and its (α, β) 3-cycle periods are

$$\int_\alpha \Omega_3 = z, \qquad \int_\beta \Omega_3 = z \frac{\log z}{2\pi i} + g(z) =: \mathcal{G}(z), \tag{4.72}$$

where $z \sim \epsilon^2$ and $g(z)$ denotes an order 1 part regular analytic in z which depends on how this local model is embedded into a compact model. Choosing fluxes $F_3 = M\beta$, $H_3 = K\alpha$ (where we use again the same notation for cycles and Poincaré dual fluxes), the superpotential takes the form

$$W(z) = -K\tau z + M\mathcal{G}(z). \tag{4.73}$$

We take $\tau := i/g_s$ fixed (in more complete models it would be fixed by other fluxes). For large K/g_s and small z, we can consistently solve $0 = D_z W \approx \partial_z W$ as

$$z \sim \exp(-2\pi K/g_s M). \tag{4.74}$$

Due to the high concentration of D3-charge $F \wedge H$ near the exponentially small 3-cycle A, such a solution will be strongly warped, the cone of the conifold times $\mathbb{R}^{1,3}$ being deformed into an AdS$_5$ throat capping off at a redshift factor $e^A = e^{-3u/4}$ of order

$$e^{A_{\min}} \sim |z|^{1/3} \sim e^{-2\pi K/3Mg_s}. \tag{4.75}$$

This is the Klebanov–Strassler solution [29].

Computing the four dimensional effective action including warping effects requires more work, see e.g. [75, 76, 78, 79].

4.5. *Quantum corrections*

We have seen that all fourfold complex structure moduli can be stabilized in principle by turning on G_4 flux, but that this leaves the Kähler moduli directions exactly flat at tree level.

To stabilize those, we therefore need to consider quantum corrections to the superpotential and Kähler potential. There can be no Kähler moduli dependent perturbative corrections to the superpotential. This is because the Kähler moduli appear in chiral multiplets whose complex scalar components are given by (4.2):

$$T_A = \frac{1}{\ell_M^6} \int_{D_{6,A}} C_6 + i \, dV = \frac{1}{\ell_s^4} \int_{D_{4,A}} C_4 + i \, dV. \tag{4.76}$$

Shifting the axionic modes by a constant is an exact symmetry of the classical action, so all perturbative corrections will also be invariant under such shifts. Combining this with the holomorphicity of the superpotential then shows that there can be no T-dependent perturbative corrections to W, as such corrections would come as powers of the T_A, which are not invariant under axion shifts.

4.5.1. *Instantons*
However, supersymmetric instantons can give corrections to the superpotential. They are of the form

$$W_{\text{inst}} = \Lambda^3 \, e^{2\pi i n^A T_A}, \tag{4.77}$$

where Λ^3 is some holomorphic function of the other (non-Kähler) scalars in the theory, including the complex structure moduli z. In the type IIB picture, these can be thought of as arising from D3 instantons wrapping divisors $D_4 = n^A D_{4,A}$. In the M-theory picture, these correspond to M5 instantons wrapping the elliptic fiber and D_4, i.e. wrapping $D_6 = n^A D_{6,A}$. As we saw in 3.3, these are indeed the only M5 instantons which have finite action in the F-theory limit of vanishing fiber size.

Some classic references for instantons effects in string theory are [80–84]. The general calculus of instantons was reviewed in [85], and the lecture notes [86] give an introduction to instanton effects in quantum mechanics and field theory. D-brane instanton effects in string theory are an active area of current research and a relatively large literature exists by now. A short recent overview with the relevant references, in particular for IIB applications, can be found in [87].

Every holomorphic divisor D_4 gives rise to an instanton, and there are infinitely many holomorphic divisors. However, the existence of the instanton does not imply there will be an actual nonzero contribution to W: if there are too many fermionic zeromodes, the instanton will not contribute. Guaranteed to contribute are instantons with the absolute minimal amount of fermionic zero-modes, i.e. two of them, corresponding to the two broken supersymmetries of the original four preserved by the vacuum. In particular, such instantons must wrap rigid cycles (cycles without infinitesimal holomorphic deformations), because the superpartners of the deformation moduli provide additional fermionic zeromodes.

A rough sketch of why instantons with exactly two zeromodes contribute to the superpotential goes as follows. Let ψ_A be the fermion in the chiral multiplet with complex scalar T_A. The instanton corrections to the superpotential can in principle be extracted by computing the $(\partial_T^2 W)\psi\psi$ term in the effective action generated by the instanton. If we denote the fermionic zeromodes (or collective coordinates) of the instanton by Θ_i, $i = 1, \ldots, N$, this term is given by the zero momentum correlator

$$\frac{\partial^2 W}{\partial T_A \partial T_B} \sim \langle \psi_A \psi_B \rangle_{\text{inst}} \tag{4.78}$$

$$\sim \int_{\text{inst}} \mathcal{D}\Theta \, \mathcal{D}\psi \, \mathcal{D}(\cdots) e^{-S} \psi_A \psi_B. \tag{4.79}$$

where the path integral is over all fields of the theory, expanded around the instanton background. Expanding the action as

$$S = S_{\text{inst}}^0 + V_A^i \Theta_i \bar{\psi}^A + S'(\Theta, \psi, \ldots), \tag{4.80}$$

where $S_{\text{inst}}^0 = 2\pi i \, n^A T_A$ is the classical instanton action and S' contains all terms not bilinear in $(\Theta, \bar{\psi})$, this becomes

$$\frac{\partial^2 W}{\partial T_A \partial T_B} \sim e^{-S_{\text{inst}}^0} \int \mathcal{D}\Theta \, \mathcal{D}\psi \, \mathcal{D}(\cdots) e^{-V_C^i \Theta_i \bar{\psi}^C - S'} \psi_A \psi_B.$$

If there are exactly two zeromodes Θ_1, Θ_2, integrating over them gives

$$\frac{\partial^2 W}{\partial T_A \partial T_B} \sim e^{-S_{\text{inst}}^0} \int \mathcal{D}\psi \, \mathcal{D}(\cdots) e^{-S'} V_C^1 V_D^2 \bar{\psi}^C \bar{\psi}^D \psi_A \psi_B$$

$$\sim e^{-S_{\text{inst}}^0} \int \mathcal{D}(\cdots) e^{-S'} V_{(A}^1 V_{B)}^2, \tag{4.81}$$

where we contracted $\bar{\psi}$s with ψs. If on the other hand there are more than two zero modes Θ_i, then integrating over them would bring down more than two $\bar{\psi}$s, which is more than there are ψs to contract with, possibly resulting in a zero amplitude.

4.5.2. *Gaugino condensation*

Another nonperturbative effect that can cause a contribution to the effective superpotential in four dimensions is gaugino condensation. For example if N D7-branes wrap a *rigid* divisor $D = n^A D_A$, we get pure $SU(N)$ super Yang–Mills in four dimensions with gauge coupling constant

$$T = \frac{\theta}{2\pi} + \frac{4\pi i}{g_{YM}^2} = n^A T_A. \tag{4.82}$$

Its ground states have a nonzero gaugino condensate, $\langle \lambda_\alpha \lambda^\alpha \rangle \neq 0$, which can be obtained from an effective superpotential as follows. Let $S = \text{Tr}\, W_\alpha W^\alpha$ be the composite chiral superfield with scalar component $s = \text{Tr}\lambda_\alpha \lambda^\alpha$. The effective action for S involves the Veneziano–Yankielowicz superpotential [88, 89]:

$$W(T, S) = 2\pi i T S - N S \left(\ln \frac{S}{\Lambda^3} - 1 \right), \tag{4.83}$$

where Λ is the UV cutoff. Solving $\partial_S W = 0$ gives

$$S = \Lambda^3 e^{2\pi i T/N} \quad \Rightarrow \quad W(T) = N \Lambda^3 e^{2i\pi T/N}. \tag{4.84}$$

This gives the vev of the gluino condensate and the effective superpotential for the Kähler modulus T.

If on the other hand the divisor has deformation moduli, there will be massless adjoint matter coupled to the gauge theory, and no gluino condensation occurs.

4.5.3. *Relation between 4d gaugino condensation and M5 instantons*

The four dimensional gaugino condensate superpotential can be related to a three dimensional M5 instanton superpotential [90]. The idea is as follows. Recall from Section 3.1 that our 4d IIB theory compactified on a circle of length $\ell = \ell_s^2/\sqrt{v} = \ell_M^3/v$ is dual to M-theory on an elliptically fibered Calabi–Yau fourfold with fiber size v. The four dimensional IIB limit is obtained by sending $v \to 0$, but let us not do that now, but instead take v/ℓ_M^2 large, so we are looking at M-theory on a large Calabi–Yau fourfold. Now, imagine we had a pure $\mathcal{N} = 1$ $SU(N)$ Yang–Mills theory in four dimensions, engineered by letting N D7-branes wrapped on a rigid 4-cycle $S = n^A D_{4,A}$ coincide, as just discussed. After T-dualizing on the circle to IIA, these D7-branes become D6-branes. Wilson lines along the circle on the D7-branes become positions of the D6-branes along the dual circle. These moduli correspond to fields in the adjoint of the gauge group. Turning them on breaks the $SU(N)$ gauge symmetry to $U(1)^{N-1}$—i.e. this puts us on the Coulomb branch. In M-theory, D6-branes lift to KK monopoles with

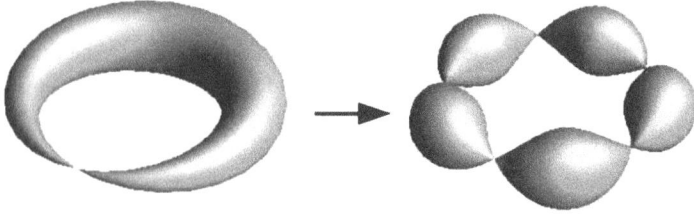

Fig. 9. Moving onto the Coulomb branch: $N = 5$ coincident M-theory KK monopole cores are moved apart along the circle T-dual to the 4d $\rightarrow$ 3d compactification circle. The transversal circle is the M-theory circle.

core at the location of the D6-branes.[21] At the core locus, the M-theory circle shrinks to zero size. Thus we get the situation depicted in Fig. 9: moving apart the N KK centers deforms the degenerate elliptic fiber into N spheres intersecting according to the extended Dynkin diagram of $SU(N)$.

The M5 instantons wrapping S and any of the N spheres S_k^2 generate a superpotential

$$W = \Lambda^3 \sum_{k=1}^{N} e^{2\pi i \rho_k}, \tag{4.85}$$

where Λ can depend on the fourfold complex structure moduli and

$$\rho_k := \frac{1}{\ell_M^6} \int_{S \times S_k^2} C_6 + i \, dV_6. \tag{4.86}$$

Since the sum of the N spheres is homologous to the full elliptic fiber at generic position we have

$$e^{2\pi i \sum_{k=1}^{N} \rho_k} = e^{2\pi i T}, \quad T = n^A T_A, \tag{4.87}$$

$-2\pi i T$ being the instanton action of an M5 wrapping the full elliptic fiber at generic position, or equivalently (as we saw before) the action of a D3 instanton wrapping $S = n^A D_{4,A}$. Extremizing $W(\rho)$ subject to the constraint (4.87) gives N solutions:

$$e^{2\pi i \rho_k} = e^{2\pi i m/N} e^{2\pi i T/N}, \quad m \in \{0, 1, \ldots, N - 1\}, \tag{4.88}$$

---

[21] In the limit of vanishing elliptic fiber size we have been considering before, the localization of these KK monopoles on the elliptic fiber could consistently be neglected, but in the finite size case this is no longer true.

resulting in

$$W = N\Lambda^3 \, e^{2\pi i m/N} \, e^{2\pi i T/N}. \tag{4.89}$$

Note that this is completely independent of v, so we expect this superpotential to survive in the $v \to 0$ limit, despite the fact that the large radius geometric M5-instanton picture is no longer valid in this regime. And indeed, we see that this superpotential exactly coincides with the gluino condensate superpotential (4.84)! The N different solutions we find here correspond to the N vacua we have in 4d, characterized by different complex phases of the gluino condensate (implicit in (4.84)).

This beautiful geometrical unification in M-theory of all nonperturbative effects extends to more complicated gauge groups as well [90].

4.5.4. *Geometric conditions for nonvanishing contributions*
In the absence of flux, a necessary condition for instantons to contribute to the superpotential has been given by Witten [83]. Let $h^{p,0}(D_6)$ be the number of holomorphic $(p, 0)$-forms on the holomorphic divisor D_6 wrapped by an M5 instanton. Then a *necessary* condition for the instanton to contribute is that its holomorphic Euler characteristic (or arithmetic genus) equals 1:

$$1 = \chi_0(D_6) := \sum_{p=0}^{3} (-1)^p \, h^{p,0}(D_6). \tag{4.90}$$

The derivation is based on a $U(1)$ charge selection rule. A sufficient condition to have exactly two fermionic zeromodes and therefore a nonzero contribution to the superpotential is that the divisor is completely rigid; more precisely:

$$h^{0,0} = 1, \qquad h^{0,p} = 0, \quad p \geq 1. \tag{4.91}$$

This satisfies (4.90), of course.

A good thing about the criterion (4.90) is that it involves an index, which is relatively easily computed just from knowledge of the wrapping numbers of the divisor. We will see in detail how this works in Section 5.8. Computing the individual Hodge numbers is also possible, but requires a bit more work.

The necessary condition (4.90) has been derived in the absence of fluxes. In the presence of fluxes, the condition appears to be no longer necessary; essentially this is because fluxes can effectively rigidify previously overly floppy M5-instantons, or, in the IIB picture, rigidify D3 instantons or D7-branes. In the latter case, the physics of this is quite intuitive: As we saw in the previous parts, fluxes can freeze D7-moduli, i.e. give masses to adjoint matter, thus (if all moduli are lifted and no other matter is present) reducing the theory to pure SYM below this mass scale, hence giving rise to strong coupling at low energies and gaugino condensation again.

It has not been completely clarified what replaces (4.90) in the presence of flux. In [91] it was found that the presence of flux effectively modifies the value of $h^{2,0}$ in (4.90) to some lower value $h^{2,0}_{\text{eff}}$ given as the number of solutions to a particular flux-dependent wave equation on D_6. It is unfortunately not known if this number is computed by an efficiently computable index formula. But in any case, since $h^{2,0}_{\text{eff}} \le h^{2,0}$, we have the necessary condition

$$\chi_0 \ge \chi_{0,\text{eff}} = 1. \tag{4.92}$$

More discussion of this issue and more concrete examples of flux modifications of M5 instanton effects can be found in [92–96].

What is not affected by flux is the sufficiency of (4.91) to get a nonzero contribution.

4.5.5. Corrections to the Kähler potential

Corrections to the Kähler potential are much less constrained. In particular, unlike the superpotential, it can receive T_A-dependent perturbative corrections. In the IIB weak coupling limit, the leading correction to (4.5) in an expansion in inverse powers of the volume is [97]

$$\mathcal{K}_K = -2\ln\left(V(B) + \frac{\xi}{g_s^{3/2}}\right), \quad \xi = -\frac{\zeta(3)}{32\,\pi^3}\,\chi(X). \tag{4.93}$$

where $V(B) = \int J_B^3/6 = \frac{1}{2}V(X)$ is the Einstein frame volume of the base B in string units, and X is the Calabi–Yau threefold which is the double cover of B.

More corrections have been considered. A concise recent review can be found in [98], in particular in relation to the large volume scenario we will discuss in Section 4.7.

4.6. The KKLT scenario

The KKLT scenario [4] is a synthesis of all of the elements we introduced so far, providing a way in principle to stabilize all moduli, break supersymmetry and obtain metastable de Sitter vacua in string theory in a reasonably controlled way. The meaning of "reasonable" is somewhat debatable; as always when relying on quantum corrections, the Dine–Seiberg problem (explained in Section 2.1.1) makes strict parametric control impossible and remains an issue that needs to be carefully addressed in specific models.

To outline the basic idea, consider a hypothetical model with one Kähler modulus T, and assume a nonperturbative superpotential is generated depending on T, so the total superpotential is of the form:

$$W(z, T) = W_{\text{flux}}(z) + \Lambda^3 \, e^{2\pi i a T} \tag{4.94}$$

with $a = 1$ if the correction is due to a D3 instanton, and $a = 1/N$ is it is generated by $SU(N)$ gaugino condensation. The flux superpotential was defined in (4.40): $W_{\text{flux}} = \ell_M^{-3} \int_Z G_4 \wedge \Omega_4 = N^I \Pi_I(z)$. Recall we are working in conventions in which W_{flux} is dimensionless; the dimensionful flux effective potential contains a scale-setting factor $\sim e^{\mathcal{K}_K} m_p^4 \sim m_s^4$. The UV scale Λ^3 should therefore be taken to be expressed in string units, and can be expected to be roughly of order 1. In general it may depend on the complex structure moduli of Z.

4.6.1. Complex structure stabilization

We will be interested in vacua for which the second term can self-consistently be considered to be a small perturbation compared to the first one as far as the fourfold complex structure moduli z is concerned. Then we can first solve the classical flux vacua equations of motion for the z^a:

$$D_a W_{\text{flux}}(z) = 0, \tag{4.95}$$

or equivalently (given (4.45)) $G_4 = *G_4$. We assume that for suitable 4-form fluxes G_4, all complex structure moduli z get frozen. The typical mass scale of these moduli will be of order

$$m_z \sim \frac{|G| m_p}{V} \sim \frac{|G| m_s}{V^{1/2}}, \tag{4.96}$$

where $|G|$ is some measure for the size of the flux (given below) and $V = V(B)$ is the volume of the IIB compactification manifold in string units. At weak string coupling there is an additional factor g_s from the $e^{\mathcal{K}_{\tau_0}} \sim g_s$ factor in the potential.

The size of the flux is constrained by the tadpole cancelation condition (3.28):

$$\frac{1}{2} Q_{IJ} N^I N^J + N_{D3} = \frac{\chi(Z)}{24} = Q_c. \tag{4.97}$$

The first term equals $\frac{1}{2\ell_M^6} \int_Z G_4 \wedge G_4$. The intersection product on flux space is not positive definite, so naively it might seem there is an infinite number of possible fluxes, becoming arbitrarily large with positive and negative contributions canceling out in the first term. However, using that $G_4 = *G_4$ on solutions to the equations of motion, one sees that the first term is in fact positive definite for actual flux vacua. The second term is positive as well if we do not introduce anti-D3-branes. Therefore we can estimate

$$|G| \sim \sqrt{Q_c} \tag{4.98}$$

in (4.96).

Moreover, it follows that we can roughly think of the space of possible fluxes as a ball in b'_4-dimensional flux space[22] of radius $\sqrt{2Q_c}$. A rough estimate for the number of flux vacua for sufficiently large Q_c is therefore the volume of this ball:

$$\mathcal{N}_{\text{flux vac}} \sim \frac{(2\pi Q_c)^{b'_4/2}}{(b'_4/2)!}. \tag{4.99}$$

Although this reasoning is very heuristic, it gives essentially the right result (at least for sufficiently large Q_c) [28, 99], as we will see in a much more refined counting analysis in Section 6.

To get an idea of the numbers involved, for the example of Z the elliptic fibration over $\mathbb{CP}^3$ described by (3.53), we have $b'_4 = 23320$, $Q_c = 972$, so according to our estimate

$$\mathcal{N}_{\text{vac}} \sim 10^{1787}. \tag{4.100}$$

The perhaps more famous order of magnitude $\sim 10^{500}$ is obtained by restricting to the much smaller set of *bulk* RR and NSNS fluxes in the IIB weak coupling limit, in which case b'_4 in the above formula gets replaced by $2\,b_3(X) = 600$.

We will also (crucially) assume that $|W_{\text{flux}}|$ can be made extremely small in a (small) fraction of all vacua. For generic vacua this will not be the case, as we expect typical values $|W_{\text{flux}}| \sim |G| \sim \sqrt{Q_c}$. However, the different contributions to $W_{\text{flux}} = N^I \Pi_I(z)$ add up with essentially random complex phases. A small fraction of random walks will happen to end up exponentially close to $W = 0$. The distribution of W_{flux} values will have some broad Gaussian-like structure, but exponentially close to $W = 0$ the density of vacua will be essentially uniform. Hence we expect roughly $\lambda \mathcal{N}_{\text{vac}}$ flux vacua within the region $|W|^2 < \lambda \ll 1$. Given the exponentially large typical values of $\mathcal{N}_{\text{vac}}$, vacua with exponentially small values of $|W_{\text{flux}}|$ should therefore still be abundant in absolute numbers. Again this estimate can be put on a much firmer footing [28].

---

[22]b'_4 is the number of 4-form fluxes with one leg on the elliptic fiber; more formally, it is the dimension of the subspace of $H^4(Z)$ satisfying (4.44); in typical models [54] b'_4 is close to b_4, both being of order 10^4.

4.6.2. *Kähler moduli stabilization*

The effective superpotential for the Kähler moduli after integrating out the complex structure moduli is

$$W(T) = W_0 + \Lambda^3 e^{2\pi i a T}, \tag{4.101}$$

where W_0 is exponentially small and Λ^3 of order 1. Solving $D_T W = 0$ makes the first term balance against the second, resulting in

$$2\pi i a T \sim \ln \frac{W_0^{-1}}{\Lambda^3}. \tag{4.102}$$

Since $\Lambda \sim 1$, W_0 is exponentially small and a is at most 1, this stabilizes the Kähler modulus T at a moderately large value. For example taking $a = 1/5$, $W_0 = 10^{-30}$, $\Lambda = 1$, we get $\operatorname{Im} T \approx 55$, $V \sim T^{3/2} \sim 400$, which is already more than large enough to justify neglecting for example the first correction to the Kähler potential (4.93), and definitely to neglect higher instanton corrections to W. Even larger volumes are possible: Taking $\ln |W_0|^{-1}$ to be of the order of its estimated smallest possible nonzero value in the example given above, i.e. $\ln |W_0|^{-1} \sim 2000$, we get $\operatorname{Im} T \sim 1600$. Note however that the maximal size is bounded by the *logarithm* of the number of vacua, and so the volume will never become exponentially large in this scenario.

The mass scale for the Kähler moduli is (dropping polynomial factors in T):

$$m_T \sim e^{2\pi i a T} m_s \sim |W_0| m_s, \tag{4.103}$$

which is exponentially small compared to the scale of the complex structure moduli masses (4.96), given that we take $|W_0|$ to be exponentially small. This shows it is self-consistent to first integrate out the complex structure moduli; the back-reaction of the Kähler moduli on the complex structure moduli will only give rise to exponentially small corrections to the vacuum values of the z^a.

For realistic applications it should be kept in mind however that m_T should not become too small to be in conflict with observations. (Taking $|W_0| \sim 10^{-30}$, $m_s \sim 10^{18}$ GeV gives $m_T \sim 10^{-3}$ eV, which is the lower bound set by fifth force experiments, and well below the bound set by cosmological considerations.)

Despite the fact that the quantum corrections in this regime give merely exponentially small corrections to the complex structure flux vacua, there is one dramatic qualitative change: instead of a flat Minkowski compactification with exponentially small supersymmetry breaking (due to $D_T W_{\text{flux}} \sim W_{\text{flux}} \neq 0$), we now have an Anti-de Sitter vacuum with exponentially small cosmological constant

$$\Lambda = -3 \frac{m_p^4}{4\pi} e^{\mathcal{K}} |W|^2 \sim -m_s^4 e^{2\pi i a T} \sim -m_s^4 |W_0|^2, \tag{4.104}$$

and supersymmetry restored! In particular, this suggests that thanks to the quantum corrections, such vacua have CFT duals. Some suggestions regarding the nature of these putative CFTs has been made in [100], but they remain clouded in mystery. Understanding them would be a huge step forward in putting these flux vacua on a firm, uncontestable footing as genuine string theory vacua.

4.6.3. Uplifing to dS

The third crucial element in [4] was a way to uplift these vacua with exponentially small negative cosmological constant to vacua with exponentially small positive cosmological constant. As we saw in Section 4.4.6, it is possible to tune fluxes such that regions of exponentially strong warping occur, locally described by a Klebanov–Strassler throat. Using statistical methods, it can moreover be argued that such vacua, which lie exponentially close to a conifold degeneration of the compactification manifold, are relatively common (see Section 6).

Once we have such a region, supersymmetry can be broken by an exponentially small amount by placing an anti-D3 brane at the bottom of the warped throat. This will add an exponentially small positive contribution to the effective potential. There will on the other hand still be an exponentially large discretuum of flux vacua with approximately the same warped throat (we again refer to 6 for justification), so in particular if the number of such vacua is still larger than 10^{120}, there should be at least some of them with a positive cosmological constant of the order of the observed value. The reason we observe such a small cosmological constant is then attributed to environmental selection, giving a concrete realization in string theory of the ideas of [1, 101, 102].

It should be clear now that the existence of the finely spaced discretuum of flux vacua in F-theory/IIB is at the core of being able to circumvent the Dine–Seiberg problem to a certain extent. Although the problem persists for generic vacua, for an exponentially small fraction, but still an exponentially large absolute number, we "accidentally" achieve reasonable control, at least sufficient to argue for existence within the framework of supergravity. In this way, the existence of a landscape is a blessing.

Nevertheless, several of the arguments rely on genericity assumptions. Although explicit models have been constructed realizing AdS complex structure and Kähler moduli stabilization within the supergravity approximation, with satisfactory error estimates for neglected corrections, the same level of confidence has not been achieved for the uplift to de Sitter, in part due to the complications induced by the necessary strong warping. There could in general also always be subtle quantum consistency constraints we have overlooked so far. It would clearly be desirable to establish the existence of these vacua as genuine quantum string vacua more convincingly, perhaps by providing a holographic description.

4.7. The large volume (Swiss cheese) scenario

A drawback of the KKLT scenario is that control over corrections remains relatively marginal, worsening significantly when the number of Kähler moduli goes up. This is because critical points of the superpotential balance off nonperturbative effects $\sim e^{-2\pi V_i}$ and the tree level flux contribution W_0, so $2\pi V_i \sim -\ln|W_0|$. If we want to stabilize the Kähler moduli at masses above 10^{-3} eV to be in agreement with fifth force experiments, we need $W_0 > 10^{-30}$ and therefore $T_i < 10$ or so. If we want the Kähler moduli mass scale to be above the TeV scale, we need $W_0 > 10^{-15}$ and $T_i < 5$ or so. The V_i are 4-cycle volumes, which can be expressed in terms of (positive) areas J^A of a basis of holomorphic 2-cycles C^A as $V_i = \frac{1}{2}n_i^A D_{ABC}J^A J^B$, where the D_{ABC} are triple intersection numbers of the divisors dual to the C^A, which form a basis of the Kähler cone (see Section 5.3). The intersection numbers for a basis of the Kähler cone are nonnegative integers,[23] so if the number of moduli is large, the expression of V_i in terms of the J^A contains a large number of all positive terms. Since V_i is at most of order 10, one thus expects in these cases that at least some of these terms will be small, i.e. that some 2-cycles will be small in string units. On top of that, these are sizes measured in Einstein frame. The string and Einstein frame Kähler moduli are related by $J_s = \sqrt{g_s}\, J_E$. Therefore, if we are at small string coupling, say $g_s \sim 1/10$, we find for the string frame 4-cycle volumes $V_i^s < 1$, and less even for 2-cycle volumes. Clearly, control becomes a serious issue here.

A variant of the KKLT scenario which ameliorates this problem was proposed by Balasubramanian, Berglund, Conlon and Quevedo [5]. The idea here is to consider models with at least two Kähler moduli a large one and a smaller one, and to balance a nonperturbative correction depending exponentially on the smaller one against a perturbative correction depending inversely on the larger one, thus potentially giving rise to exponentially large overall volumes.

More concretely this goes as follows [103]. The dilaton, D7 and complex structure moduli stabilization proceeds as in the KKLT setup, leaving us with an effective superpotential for the large and small Kähler moduli T_L and T_S given by, say,

$$W = W_0 + \Lambda^3 e^{2\pi i a T_S}. \tag{4.105}$$

We assume we have stabilized ourselves in the weak IIB coupling region of the fourfold complex structure moduli space. The Kähler potential for the Kähler

---

[23]This is because such a basis element by definition has only positive intersection numbers with holomorphic curves, and intersections of two such holomorphic basis divisors are holomorphic curves.

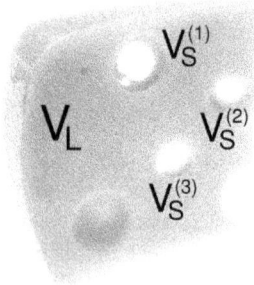

Fig. 10. Swiss cheese volume $V = V_L - \sum_i V_S^{(i)}$.

moduli, including the first perturbative correction, is thus of the form (4.93):

$$\mathcal{K} = -2 \ln\left(V + \frac{\xi}{g_s^{3/2}} \right),$$ (4.106)

where $\xi = -\frac{\zeta(3)}{32\pi^3} \chi(X)$. For the large volume scenario to work, one needs $\xi > 0$, i.e.

$$\chi(X) < 0.$$ (4.107)

For simplicity the expression of the threefold volume in terms of the Kähler moduli is taken to be of "Swiss cheese" form:

$$V \sim (\operatorname{Im} T_L)^{3/2} - (\operatorname{Im} T_S)^{3/2},$$ (4.108)

as illustrated further for multiple small moduli $T_S^{(i)}$ in Fig. 10. Although this seems a rather special Ansatz, several models are known to satisfy it; basically the "hole" contributions are due to blowup modes (see (5.33) for an explicit example).

Focusing on the regime $\operatorname{Im} T_S \ll \operatorname{Im} T_L \sim V^{2/3}$, putting $v_S := 2\pi a \operatorname{Im} T_S$, taking $\Lambda \sim 1$ and setting the axions (consistently) to zero, the effective potential for the Kähler moduli is then of the form

$$U \sim \frac{\sqrt{v_S}\, e^{-2v_S}}{V} - \frac{|W_0|\, v_S\, e^{-v_S}}{V^2} + \frac{\xi\, |W_0|^2}{g_s^{3/2} V^3}.$$ (4.109)

Here and it what follows we are suppressing positive numerical factors.

Minimizing (4.109) with respect to v_s results in

$$e^{-v_S} \sim \frac{|W_0|}{V}.$$ (4.110)

Plugging this back in (4.109):

$$U \sim \frac{W_0^2}{V^3}\left(\left(\ln\frac{V}{|W_0|}\right)^{1/2} - \ln\frac{V}{|W_0|} + \frac{\xi}{g_s^{3/2}}\right), \qquad (4.111)$$

and minimizing with respect to V, we finally get

$$V \sim |W_0|\, e^{\frac{\xi^{2/3}}{g_s}}, \qquad v_S \sim \frac{\xi^{2/3}}{g_s}, \qquad U \sim -\frac{|W_0|^2}{V^3}. \qquad (4.112)$$

Thus, remarkably, even without tuning W_0 exponentially small, we see we can get very large, even exponentially large, volumes V by tuning g_s moderately small, as well as moderately large v_S and very small to exponentially small negative cosmological constant. In string frame, V is still essentially as large as we wish, but now $v_S \sim \xi^{2/3}$—whether this is satisfactorily large depends on the proportionality constant and therefore the model. It should be kept in mind however that this potentially leads us back to the Dine–Seiberg problem in specific models—we can never parametrically escape it. But in any case the situation is significantly better than in the KKLT scenario, where sending $g_s \to 0$ causes all string frame volumes to collapse.

The minimum of the potential is nonsupersymmetric AdS. In principle it can be uplifted to dS by the same mechanism as in the KKLT scenario.

Extensive analysis of various corrections has been done in [103] and [98], and it was found that these large volume compactification models are remarkably robust.

For a recent and more in-depth discussion of realizations of the large volume scenario and several of the ingredients introduced here, see [87].

The large volume scenario has been the starting point for a number of phenomenological explorations, both in particle physics and cosmology. See for instance [104] for a review of some of this work.

4.8. Recap and to do list

So far we have explained in detail the general geometry of IIB/F-theory flux vacua, we have sketched how and under which geometrical conditions quantum corrections can arise and summarized two related scenarios on how these can lead to fully moduli stabilized models with small positive cosmological constant. But, having a scenario is not the same as having an actual model that works. To construct and analyze such models, and in particular to find models that generate the required instanton corrections to the superpotential, we need to develop more sophisticated geometrical techniques. This will be the subject of the next section, which gives a hands-on introduction to various constructions in applied algebraic and toric geometry. We will then apply these constructions to build models meeting all requirements to make one or both of the above scenarios work. In practice however, constructing fully explicit flux vacua in typical F-theory compactifications would require specifying 20,000 or so flux quanta and finding the corresponding critical points in the 3000 or so dimensional complex structure moduli space, hoping to hit the region of parameter space we are interested in (e.g. weak string coupling, tiny cosmological constant, ...). Needless to say, such a task is hopeless. Nevertheless, approximate distributions of vacua over parameter space are relatively easily derived, and from this estimates of how many vacua satisfy properties of interest, without actually having to go through the pain of constructing them explicitly. Developing these statistical techniques will be the subject of section 6.

5. A geometrical toolkit

In this section we will give a hands-on introduction to various constructions in algebraic and toric geometry useful for the construction of explicit examples of moduli stabilized type IIB/F-theory vacua. I have tried to make the exposition as concrete and accessible as possible, with emphasis on computation rather than on abstract formalism and structure. This is at the cost of some rigor and generality, and certainly there are much more sophisticated and powerful treatments, but for the purpose of constructing explicit models to play around with, the elementary approach we will follow here is more than sufficient.

We will only assume a basic familiarity with the differential geometry contained in section 2 of [3].

The outline of this section is as follows.

• In 5.1 we introduce toric varieties as classical ground state manifolds of gauged linear sigma models. Toric varieties and algebraic subspaces of them provide a huge, fully explicit class of possible compactification manifolds for string theory, including moduli stabilized IIB flux compactifications, which is why we introduce them here.

• In 5.2 we define divisors in toric varieties and explain how one can compute their mutual intersection numbers. Once these intersection numbers are known, it is straightforward to compute various quantities of physical interest, such as charges, volumes, Kähler potentials, numbers of moduli, numbers of fermionic zero modes, and so on. They form the basic geometric data of everything that follows.

• In 5.3 we describe the duals of divisors, namely 2-cycles, and we explain how exactly their areas are related to the Fayet–Iliopoulos terms appearing in the definition of the gauged linear sigma model. We also explain how a basis of holomorphic 2-cycles can be constructed and how this allows one to explicitly parametrize the Kähler moduli space.

- In 5.4 we show how volumes of toric varieties and holomorphic subspaces thereof can be explicitly computed as a function of the Kähler moduli. This is needed for example if one wants to compute the Kähler potential for a string compactification, or if one wants to compute instanton actions.
- In 5.5, characteristic classes are introduced. They play an important role in string theory in extracting, from geometrical setups, various physical topological quantities such as RR charges, moduli space and flux lattice dimensions, numbers of fermionic zero modes of instantons, and so on. We show in particular how they can be computed from the divisor intersection products, for any algebraic subspace of a toric variety.
- In 5.6, the concept of Poincaré residue is explained. This is an elegant and useful general construction of holomorphic top forms on algebraic subspaces of toric varieties. As we have seen, periods of holomorphic forms play a crucial role in the computation of super- and Kähler potentials. This section will also make clear where the holomorphic forms on Calabi–Yau manifolds stated in examples came from.
- In 5.7 we focus on Calabi–Yau submanifolds of toric varieties; in particular we consider some examples, one of which is the elliptically fibered CY fourfold over $\mathbb{CP}^3$ introduced before in (3.53). In particular we compute the divisor intersection numbers of this fourfold and its total Chern class. This allows to compute in particular the Euler characteristic of the fourfold, which in turn determines the curvature induced D3 tadpole of F-theory compactified on it, crucial for the existence of flux vacua.
- In 5.8 we list a number of classic index theorems, relating numbers of various bosonic and fermionic zeromodes (e.g. those of M5 instantons) to integrals of characteristic classes, which by now we know how to compute.
- Finally, in 5.9 we show more concretely how these index theorems can be applied to compute Hodge numbers (i.e. numbers of moduli, zeromodes, fluxes, and so on).

By the end of this section, you should be able to construct a gigantic variety of supersymmetric compactifications of F-theory for yourself, tailor them to your liking, and compute all of their physically relevant topological characteristics.

5.1. *Toric varieties as gauged linear sigma model ground states*

Toric varieties can be represented very concretely as supersymmetric moduli spaces of gauged linear sigma models [105]. This is the approach we will follow here. For more advanced introductions to toric varieties, see [3, 106, 107].

Consider n chiral superfields X_i charged under a $U(1)^s$ gauge group with charges Q_i^a, $a = 1, \ldots, s$. In the absence of a superpotential, the potential for the scalar components x_i reads

$$V(x) = \sum_{a=1}^{s} \frac{e_a^2}{2} \left(\sum_{i=1}^{n} Q_i^a |x_i|^2 - \xi^a \right)^2. \tag{5.1}$$

Here the e_a are the $U(1)^s$ coupling constants, and ξ^a are the Fayet–Iliopoulos (FI) parameters. The space $\mathcal{M}$ of classical supersymmetric ground states is given by the zeros of $V(x)$ (i.e. the D-flat configurations), modulo the $U(1)^s$ gauge symmetry:

$$\mathcal{M} = \left\{ x \in \mathbb{C}^n \,\Big|\, \sum_{i=1}^{n} Q_i^a |x_i|^2 = \xi^a \right\} / U(1)^s, \tag{5.2}$$

where $U(1)^s$ acts as

$$x_i \rightarrow e^{i Q_i^a \varphi_a} x_i. \tag{5.3}$$

If the FI parameters ξ_a are such that $d := \dim \mathcal{M} = n - s$, $\mathcal{M}$ is a toric variety.

As a simple example, consider a single $U(1)$ with charges $q_i = 1$ for $i = 1, \ldots, n$ and $\xi > 0$. Then $\mathcal{M} = \mathbb{CP}^{n-1}$. To relate this to the usual description of $\mathbb{CP}^{n-1}$ as $(\mathbb{C}^n - \{0\})/\mathbb{C}^*$ where $\lambda \in \mathbb{C}^*$ acts as $x_i \rightarrow \lambda x_i$, note that the D-flatness condition $\sum_i |x_i|^2 = \xi$ can be thought of as gauge fixing the real rescalings $x_i \rightarrow |\lambda| x_i$ for $x \neq 0$, leaving only the $U(1)$ part to divide out.

In general one can similarly represent a toric variety as $\mathbb{C}^n$ minus a certain set Z quotiented by the complexified gauge group $(\mathbb{C}^*)^s$. The excluded set Z is the set of $x \in \mathbb{C}^n$ which cannot be gauge transformed to a solution of the D-flatness constraints. This can be shown to consist of the union of planes obtained by putting a subset of the coordinates x_i equal to zero, such that the D-flatness constraints cannot be solved. Note that Z thus depends on the choice of FI parameters ξ. The advantage of this description is that it makes holomorphic properties manifest. The advantage of the gauge linear sigma model description on the other hand is that it is very concrete and that specifying a set of FI parameters is in general less complicated than specifying Z. This is the approach we will follow here.

As a less trivial example, consider the space defined by five fields x_i and $U(1) \times U(1)$ gauge group, with charges

$$\begin{pmatrix} Q_i^1 \\ Q_i^2 \end{pmatrix} = \begin{pmatrix} 1 & 1 & 1 & -n & 0 \\ 0 & 0 & 0 & 1 & 1 \end{pmatrix}, \tag{5.4}$$

positive FI parameters (ξ^1, ξ^2), and $n \geq 0$. Thus

$$\mathcal{M}_n = \left\{ x \in \mathbb{C}^5 \mid \begin{array}{l} |x_1|^2 + |x_2|^2 + |x_3|^2 - n|x_4|^2 = \xi_1 \\ |x_4|^2 + |x_5|^2 = \xi_2 \end{array} \right\} / U(1)^2, \tag{5.5}$$

where the $U(1)^2$ act as

$$(x_1, x_2, x_3, x_4, x_5) \rightarrow (e^{i\varphi_1}x_1, e^{i\varphi_1}x_2, e^{i\varphi_1}x_3, e^{i(-n\varphi_1 + \varphi_2)}x_4, e^{i\varphi_2}x_5). \tag{5.6}$$

This is a smooth $\mathbb{CP}^1$ bundle over $\mathbb{CP}^2$, with the "amount of twisting" determined by n.

Any toric variety $\mathcal{M}$ is complex, with local complex coordinates given by $U(1)^s$ invariant combinations of the x_i. For the $\mathbb{CP}^{n-1}$ example such coordinates are e.g. $t_i = x_i/x_n$, $i < n$ in a patch where $x_n \neq 0$. For the second example we can take for example $t_1 = x_2/x_1$, $t_2 = x_3/x_1$, $t_3 = x_4 x_1^n/x_5$ in a patch where $x_1 \neq 0$, $x_5 \neq 0$.

Moreover $\mathcal{M}$ inherits a Kähler form from the standard flat Kähler form on $\mathbb{C}^n$,

$$J = \frac{i}{2\pi} \sum_i dx_i \wedge d\bar{x}_i = \frac{1}{2\pi} \sum_i du_i \wedge d\phi_i, \tag{5.7}$$

where $x_i =: \sqrt{u_i} e^{i\phi_i}$ and the normalization is chosen for later convenience. In the case of $\mathbb{CP}^{n-1}$, this gives the Kähler form of the standard Fubini-Study metric, as can be seen in the coordinate patch parametrized by $t_i = x_i/x_n$ by substituting the D-flatness solution $x_i = \sqrt{\xi} t_i (\sum_j |t_j|^2)^{-1/2}$ in J, where we put $t_n \equiv 1$.

5.2. Divisors, line bundles and intersection numbers

A divisor $D = \sum_I n_I S^I$ is a formal sum of holomorphic hypersurfaces S^I, with (positive or negative) integer coefficients. Physically it can be thought of as a collection of complex codimension one holomorphic branes and anti-branes. The holomorphic hypersurface S^I is described locally on each coordinate patch α by a holomorphic equation $f_\alpha^I(t) = 0$, such that f_α^I/f_β^I has no zeros or poles on the overlap between α and β. To the divisor D we can thus associate in each patch α the meromorphic function $f_{D,\alpha} = \prod_I (f_\alpha^I)^{n_I}$ whose zeros and poles describe the positive resp. negative parts of D. The functions $f_{D,\alpha}/f_{D,\beta}$ can be interpreted as transition functions of a holomorphic line bundle on overlap regions. This construction thus gives a one to one correspondence between the data describing holomorphic line bundles and the data describing divisors. One denotes the line bundle corresponding to the divisor D as $\mathcal{O}(D)$.

Sums and differences of divisors correspond to products and quotients of their defining equations. A divisor given by the zeros and poles of a *globally* defined rational function corresponds to a trivial line bundle (all transition functions are 1) and is trivial in homology. Divisors which differ by such a homologically trivial divisor are called linearly equivalent.

The toric variety $\mathcal{M}$ has a particularly simple set of divisors

$$D_i : x_i = 0. \tag{5.8}$$

More complicated divisors can be constructed as poles and zeros of rational equations in the x_i transforming homogeneously under the gauge transformations. Gauge invariant rational functions of the x_i are globally defined on $\mathcal{M}$ and hence correspond to homologically trivial divisors.

For example on $\mathbb{CP}^{n-1}$, x_i/x_j is gauge invariant for all i, j. Consequently $D_i = D_j$ for all i, j, where the equality should be read here as linear equivalence. So in this case there is only one independent divisor class. Any homogeneous polynomial equation of degree k describes a divisor linearly equivalent to kD_1. For instance $x_1^5 x_2^3 x_3 + x_1^7 x_4^2 = 0$ describes a divisor in class $9D_1$ in $\mathbb{CP}^3$.

In our second example (5.4), we similarly get the relations

$$D_1 = D_2 = D_3, \qquad D_4 = D_5 - n \, D_1. \tag{5.9}$$

More generally, divisor classes are completely characterized by the charges of their defining equation; there will be as many independent divisors D_i as there are $U(1)$ factors, and they generate all divisor classes on $\mathcal{M}$.

The Poincaré dual $\mathrm{PD}_{\mathcal{M}}(D)$ of a divisor class D in $\mathcal{M}$ is an element of $H^2(X, \mathbb{Z})$: If the divisor is locally described by the equation $f(x) = 0$, a representative of the Poincaré dual class is $\delta(f) \, df \wedge \delta(\bar{f}) \, d\bar{f}$.

The *intersection product* of d divisor classes plays a fundamental role in computing just about any topological quantity. It can be defined as

$$D_A \cdots D_B = \int_{\mathcal{M}} \mathrm{PD}(D_A) \wedge \cdots \wedge \mathrm{PD}(D_B) = \#(D_A \cap \cdots \cap D_B), \tag{5.10}$$

where for the last equality we assumed the divisors to be transversally intersecting at regular points. The intersection product is invariant under linear equivalence. Intersection products of less than d divisors are defined similarly, but now represent curves, surfaces and so on rather than numbers or points.

As a first example, say we want to compute $D_1 D_2 D_3$ on $\mathbb{CP}^3$. Setting $x_1 = x_2 = x_3 = 0$ reduces the D-term constraint to $|x_4|^2 = \xi$. The $U(1)$ gauge symmetry can be used to set $x_4 = \sqrt{\xi}$, so the triple intersection is a single (regular) point, i.e. $D_1 D_2 D_3 = 1$. Now, using $D_3 = D_2 = D_1$, this immediately gives $D_1^3 = 1$. Since D_1 generates all divisor classes, this is all we need to know to compute all intersection products.

In general, for $\mathbb{CP}^d$, we have $D_1^d = 1$.

As an example illustrating how to deal with intersections at orbifold singularities, consider the weighted projective space $W\mathbb{CP}^2_{1,2,3}$, defined by 3 fields x_i with charges $Q_i = (1, 2, 3)$, and FI parameter $\xi > 0$. Since x_1^3/x_3 and x_1^2/x_2 are gauge invariant, the toric divisor classes satisfy $D_3 = 3D_1$, $D_2 = 2D_1$. Note that the point $x = (0, \sqrt{\xi}, 0)$ is fixed under a $\mathbb{Z}_2$ subgroup of the gauge group and similarly $x = (0, 0, \sqrt{\xi})$ is fixed by a $\mathbb{Z}_3$ subgroup. Hence these points are $\mathbb{Z}_2$ and $\mathbb{Z}_3$ orbifold singularities, and some special care has to be taken in computing intersection products. To compute D_1D_2, note that $x_1 = x_2 = 0$ is precisely the $\mathbb{Z}_3$ orbifold point. The correct value of D_1D_2 is then $1/3$. One way to see this is to observe that $3D_1D_2 = D_3D_2 = 1$, where the latter equality follows from the fact that $(\sqrt{\xi}, 0, 0)$ is a regular point. Similarly $D_1D_3 = 1/2$, and $D_1^2 = D_1D_3/3 = 1/6$.

In general, for $W\mathbb{CP}^d_Q$, we have $D_j^d = Q_j^d/\prod_i Q_i = Q_j^{d-1}/\prod_{i\neq j} Q_i$.

As a last example, consider (5.4) again, for which we obtained the linear equivalences (5.9). Let us take $\{D_1, D_5\}$ as a basis. All independent triple intersection products can be computed by using the linear equivalences to reduce to intersection products of distinct divisors, which in turn can be directly computed by solving the equations together with the D-term constraints. It is also useful to note that the D-term constraints in (5.4) exclude $(x_1, x_2, x_3) = (0, 0, 0)$ and $(x_4, x_5) = (0, 0)$, so $D_1D_2D_3 = 0$ and $D_4D_5 = 0$. This gives the reduction relation $D_5^2 = D_5(D_4 + nD_1) = nD_1D_5$, hence

$$D_1^3 = D_1D_2D_3 = 0, \tag{5.11}$$

$$D_1^2D_5 = D_1D_2D_5 = 1, \tag{5.12}$$

$$D_1D_5^2 = nD_1^2D_5 = n, \tag{5.13}$$

$$D_5^3 = nD_1^2D_5 = n^2. \tag{5.14}$$

5.3. Curves and Kähler moduli

We will now show that, while the fields x_i correspond to $(n - 2)$-cycles D_i, the charges Q^a correspond to 2-cycles C^a, and that the mutual intersection product between the divisors and these curves is

$$D_i \cdot C^a = Q_i^a, \tag{5.15}$$

and moreover that for the Kähler form J on $\mathcal{M}$ induced by (5.7) we have

$$\int_{C^a} J = \xi^a, \tag{5.16}$$

with ξ^a the FI parameters from (5.2). We will check these claims by constructing the 2-cycle classes C^a explicitly.

A representative 2-cycle C^a can be constructed as the image of a map X : $[0, 1] \times [0, 2\pi] \to \mathcal{M} : (\tau, \sigma) \mapsto x_i = X_i(\tau, \sigma)$ which we build as follows. First split the coordinates x_i in two suitable groups by splitting the index set as the disjoint union $\{1, \ldots, n\} = I_1 \cup I_2$, with the number of elements in I_2 equal to s, the number of $U(1)$'s. Then we put

$$X_i(\tau, \sigma) = \begin{cases} \sqrt{U_i(\tau)} \, \exp(i \, Q_i^a \sigma) & \text{if} \quad i \in I_1, \\ \sqrt{U_i(\tau)} & \text{if} \quad i \in I_2, \end{cases} \tag{5.17}$$

where we choose the path $U_i(\tau)$ such that the following conditions are met:

$$\sum_{i=1}^{n} Q_i^a \, U_i(\tau) = \xi^a \quad \forall \tau \in [0, 1], \, \forall a, \qquad U_i(0) = 0 \quad \forall i \in I_1,$$

$$U_i(1) = 0 \quad \forall i \in I_2. \tag{5.18}$$

The first constraint enforces the X_i to satisfy the D-flatness constraints defining $\mathcal{M}$, while the two last conditions are needed to make the boundary circles $X_i|_{\tau=0,1}$ collapse to a point up to gauge transformation, so the 2-cycle is closed in $\mathcal{M}$. An example for $\mathbb{CP}^2$ is plotted in Fig. 11.

Then we have for $i \in I_1$:

$$D_i \cdot C^a = \frac{1}{2\pi i} \oint_{\sigma=0}^{2\pi} \frac{d X_i}{X_i} \bigg|_{\tau=\epsilon} = Q_i^a. \tag{5.19}$$

Similarly, after doing the proper gauge transformation $X_i \to e^{i Q_i^a \sigma} X_i$, we find $D_i \cdot C^a = Q_i^a$ for $i \in I_2$. This establishes (5.15).

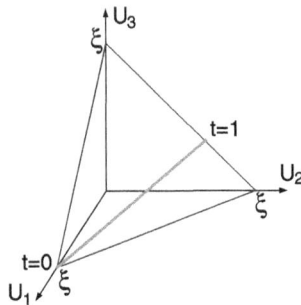

Fig. 11. Example of a path $U_i(\tau)$ for $\mathbb{CP}^2$.

To prove (5.16), note that

$$\int_{C^a} J = \int_{C^a} \frac{1}{2\pi} \sum_{i \in I_1} d(Q_i^a U_i(\tau)) \wedge d\sigma = \xi^a, \tag{5.20}$$

where in the last equality we used (5.18).

This gives a direct connection between the FI parameters and the Kähler moduli of $\mathcal{M}$.

In general the 2-cycles C^a or integer linear combinations thereof will not be holomorphic; the corresponding homology classes might not even have any holomorphic representatives at all. To construct holomorphic curves, one can simply take transversal intersections of $n - 1$ of the divisors D_i. By taking positive linear combinations, these in fact generate the full set of all 2-cycle classes with holomorphic representatives (this set is called the *Mori cone*). The relation with the C^a can be deduced by comparing the intersection products $D_i \cdot C^a$ and $D_i \cdot (D_{i_1} \cdots D_{i_{n-1}})$.

For example for $\mathbb{CP}^{n-1}$, we have $C^1 = D_1^{n-2}$, because indeed $D_1 \cdot (D_1^{n-2}) = 1 = Q_1^1$. For the n-twisted $\mathbb{CP}^1$ bundle over $\mathbb{CP}^2$ defined by (5.4), we have

$$C^1 = D_1 D_4, \qquad C^2 = D_1 D_2, \tag{5.21}$$

so C^1 and C^2 are holomorphic curve classes, and moreover since $D_1 D_5 = C^1 + nC^2$ and $n \geq 0$, they generate (with positive coefficients) the full Mori cone of holomorphic curve classes. (Had we chosen $n \leq 0$, then the Mori cone would be generated by C^2 and $\tilde{C}^1 := C^1 + nC^2 = D_1 D_5$.)

Finally, since for holomorphic curves C the period $\int_C J$ equals the area of C, we must have $\int_C J \geq 0$ for all generators C of the Mori cone. This translates to a set of inequalities on the ξ^a. The space of all $J' \in H^2(\mathcal{M}, \mathbb{R})$ satisfying $\int_C J' \geq 0$ for all C in the Mori cone is called the *Kähler cone*. This corresponds to all possible Kähler classes J which can be obtained by varying the ξ^a without degenerating the space $\mathcal{M}$. The space of these deformations is called the Kähler moduli space.

One can choose the generators of the gauge group to be such that the C^a are generators of the Mori cone. If this is a complete set of generators, then the Kähler cone is given simply by $\xi^a \geq 0$, and if we then choose a basis of divisors K_a dual to the C^a (i.e. $C^a \cdot K_b = \delta_b^a$), we can parametrize the Kähler form as

$$J = \xi^a K_a. \tag{5.22}$$

If there are more than s generators of the Mori cone, there will be additional inequality constraints on the ξ^a. Intersection products of the K_a are always positive.

For $\mathbb{CP}^{n-1}$, we have $K_1 = D_1$ and $J = \xi^1 K_1$ with $\xi^1 \geq 0$. For our n-twisted $\mathbb{CP}^1$ bundle over $\mathbb{CP}^2$, we have $K_1 = D_1$, $K_2 = D_5$, and $J = \xi^1 K_1 + \xi^2 K_2$ with $\xi^a \geq 0$.

At the boundary of the Kähler cone, a 2-cycle collapses to zero area. By keeping on varying the ξ^a (so formally the curve area becomes negative), one often simply continues to a different smooth geometry in this way, still described by (5.2), but with different intersection products, and with its own Kähler cone. This is called a flop transition, with the collapsing curve being flopped. In other cases, there is no transition to a new smooth geometry, although in the full gauged linear sigma model, the physics remains sensible.

5.4. Volumes

Using the parametrization $J = \xi^a K_a$ of (5.22)[24] and the intersection products of the divisors, it is straightforward to compute the volume of $\mathcal{M}$:

$$V_{\mathcal{M}} = \int_{\mathcal{M}} \frac{J^d}{d!} = \frac{1}{d!} K_{a_1} \cdots K_{a_d} \xi^{a_1} \cdots \xi^{a_n}. \tag{5.23}$$

Explicitly, for say $\mathbb{CP}^3$, parametrizing $J = \xi^1 D_1$, this is, using $D_1^3 = 1$:

$$V_{\mathbb{CP}^3} = \frac{(\xi^1)^3}{6}, \tag{5.24}$$

and for the $\mathbb{CP}^1$ fibration over $\mathbb{CP}^2$, parametrizing $J = \xi^1 K_1 + \xi^2 K_2 = \xi^1 D_1 + \xi^2 D_5$, using (5.11)–(5.14):

$$V_{\mathcal{M}_n} = \frac{1}{6} \xi^2 \big(3 (\xi^1)^2 + 3 n \xi^1 \xi^2 + n^2 (\xi^2)^2\big). \tag{5.25}$$

The volume of other holomorphic cycles can be computed similarly. In particular the volume of the divisor K_a is

$$V_{K_a} = \int_{K_a} \frac{J^{d-1}}{(d-1)!} = \int_{\mathcal{M}} \frac{K_a J^{d-1}}{(d-1)!} = \frac{\partial}{\partial \xi^a} V_{\mathcal{M}}(\xi), \tag{5.26}$$

and all other holomorphic divisor volumes can be computed from this by linearity: $V_{\alpha K_a + \beta K_b} = \alpha V_{K_a} + \beta V_{K_b}$. Holomorphic intersections are also easy to compute:

$$V_{K_a \cap \cdots \cap K_b} = \frac{\partial}{\partial \xi^a} \cdots \frac{\partial}{\partial \xi^b} V_{\mathcal{M}}(\xi). \tag{5.27}$$

---

[24]One can of course also consider more general parametrizations $J = J^A E_A$ with $\{E_A\}_A$ some basis of divisor classes; the resulting expressions are completely analogous.

As an example, the volume of D_1 in $\mathbb{CP}^3$ is

$$V_{D_1} = \frac{(\xi^1)^2}{2}. \tag{5.28}$$

In the $\mathbb{CP}^1$ bundle example, we get

$$V_{D_1} = V_{K_1} = \frac{\partial}{\partial \xi_1} V_{\mathcal{M}_n}(\xi) = \frac{\xi^2(2\xi^1 + n\xi^2)}{2}, \tag{5.29}$$

$$V_{D_5} = V_{K_2} = \frac{\partial}{\partial \xi_2} V_{\mathcal{M}_n}(\xi) = \frac{(\xi_1 + n\xi_2)^2}{2}, \tag{5.30}$$

$$V_{D_4} = V_{K_2} - n\, V_{K_1} = \frac{\xi_1^2}{2}. \tag{5.31}$$

More general volumes can be computed easily as well; for example the formal area of the self-intersection of D_4 is

$$\int_{D_4 \cap D_4} J = \left(\frac{\partial}{\partial \xi_2} - n\frac{\partial}{\partial \xi_1}\right)^2 V_{\mathcal{M}}(\xi) = -n\xi_1 = -n\sqrt{2V_{D_4}}. \tag{5.32}$$

Note that this is negative. This simply indicates that the self-intersection of D_4 does not have a holomorphic representative. (In general only transversal inter-sections of holomorphic objects are again holomorphic.) Finally note that in this example, we have

$$V_{\mathcal{M}_n} = \frac{\sqrt{2}}{3n}\left(V_{D_5}^{3/2} - V_{D_4}^{3/2}\right). \tag{5.33}$$

Recalling Section 4.7, this suggests that these manifolds (as base manifolds of elliptically fibered CY fourfolds) might provide examples of the "Swiss cheese" scenario of Section 4.7. We will see in Section 7 that this is indeed the case.

5.5. *Characteristic classes*

Characteristic classes play an important role in string theory in extracting, from geometrical setups, various physical topological quantities such as RR charges, moduli space and flux lattice dimensions, numbers of fermionic zero modes of instantons, and so on. In the following we will first list the general (smooth differential geometric) definitions of the various characteristic classes that appear in this paper, and then specialize to computations of tangent bundle characteristic classes of toric varieties and algebraic submanifolds thereof.

The total *Chern class* $c = c_1 + c_2 + \cdots + c_r$ of a rank r holomorphic vector bundle V with $r \times r$ matrix curvature form F is defined as the cohomology class $\in H^0 + H^2 + \cdots + H^{2r}$ of

$$c(V) = \det\left(1 + \frac{1}{2\pi}F\right) = 1 + \frac{1}{2\pi}\operatorname{Tr}F + \cdots . \tag{5.34}$$

From the properties of the determinant, it immediately follows that $c(V_1 \oplus V_2) = c(V_1)\,c(V_2)$. More generally, if we have three vector bundles U, V, W such that $U = V/W$ (so locally $V = U \oplus W$), then we have the *Whitney sum formula*:

$$c(V) = c(U)\,c(W). \tag{5.35}$$

An important example is given by the tangent and normal bundles of a holomorphic submanifold S of a manifold X. Because the normal bundle NS of S is the quotient of the tangent bundle TX of X restricted to S by the tangent bundle TS of S, we have

$$c(TX)|_S = c(TS)\,c(NS), \tag{5.36}$$

which can be used to compute $c(TS)$ from knowledge of $c(TX)$ and $c(NS)$. (If for example S is given as the intersection of divisors, $S = S_1 \cap S_2 \cap \cdots \cap S_k$, then we simply have $c(NS) = \prod_\alpha(1 + [S_\alpha]|_S)$. When specifically applied to the tangent and normal bundles of a manifold as just shown here, the Whitney sum formula is usually referred to as the *adjunction formula*.

In terms of the eigenvalues λ_m, $m = 1, \ldots, r$, of $\frac{1}{2\pi}F$, we can also write

$$c(V) = \prod_{m=1}^{r}(1 + \lambda_m). \tag{5.37}$$

Thus the λ_m can be thought of as the formal roots of the total Chern class, and for that reason are called the *Chern roots*. They are very useful to define and relate various characteristic classes. The Whitney sum formula given above can be thought of as simply splitting the Chern roots of V into Chern roots of U and Chern roots of W.

The *Euler class* of a holomorphic vector bundle is its top Chern class:

$$e(V) = c_n(V) = \prod_m \lambda_m. \tag{5.38}$$

In particular the Euler characteristic of a complex manifold M equals the integrated Euler class of its holomorphic tangent bundle:

$$\chi(M) = \int_M e(TM). \tag{5.39}$$

Similarly one defines the *Chern character* as

$$\text{ch}(V) := \text{Tr}\, e^F \quad = \quad \sum_m e^{\lambda_m} = r + c_1 + \frac{1}{2}(c_1^2 - 2c_2)$$

$$+ \frac{1}{6}(c_1^3 - 3c_1 c_2 + 3c_3) + \cdots \tag{5.40}$$

and this satisfies the sum and product formulas $\text{ch}(V \oplus W) = \text{ch}(V) + \text{ch}(W)$ and $\text{ch}(V \otimes W) = \text{ch}(V)\,\text{ch}(W)$. The *Todd class* is given by

$$\text{Td}(V) = \prod_m \frac{\lambda_m}{1 - e^{-\lambda_m}} = 1 + \frac{1}{2}c_1 + \frac{1}{12}(c_1^2 + c_2) + \frac{1}{24}c_1 c_2 + \cdots \tag{5.41}$$

and is multiplicative, like the Chern class. Finally, the *Hirzebruch L-genus* is

$$L(V) = \prod_m \frac{\lambda_m}{\tanh \lambda_m} = 1 + \frac{1}{3}\sum_m \lambda_m^2 + \cdots = 1 + \frac{1}{3}(c_1^2 - 2c_2) + \cdots \tag{5.42}$$

and the *A-roof genus* is

$$\widehat{A}(V) \quad = \quad \prod_m \frac{\lambda_m/2}{\sinh(\lambda_m/2)} = 1 - \frac{1}{24}\sum_m \lambda_m^2 + \cdots$$

$$= \quad 1 - \frac{1}{24}(c_1^2 - 2c_2) + \cdots . \tag{5.43}$$

The Chern class of a toric variety is given by the particularly simple expression

$$c(\mathcal{M}) \equiv c(T\mathcal{M}) = c(\oplus_{i=1}^n \mathcal{O}(D_i)) = \prod_{i=1}^n (1 + D_i), \tag{5.44}$$

where D_i should be read in the last expression as the Poincaré dual to the divisor $D_i : x_i = 0$ in $\mathcal{M}$. More generally we get formulas for all characteristic classes defined above by the substitutions $m \to i$, $r \to n$ and

$$\lambda_i \to D_i. \tag{5.45}$$

For example for $\mathcal{M} = \mathbb{CP}^3$ we have, after using the relations between the toric divisors, and putting $H \equiv D_1$:

$$c(\mathcal{M}) = (1 + H)^4 = 1 + 4H + 6H^2 + 4H^3, \tag{5.46}$$

$\chi(\mathcal{M}) = 4$, and $\text{Td}(\mathcal{M}) = 1 + 2H + \frac{11H^2}{6} + H^3$. For weighted projective space $\mathcal{M} = \mathbb{CP}_Q^{n-1}$, defining H by $D_i =: Q_i H$, we have

$$c(\mathcal{M}) = \prod_i (1 + Q_i H). \tag{5.47}$$

For the n-twisted $\mathbb{CP}^1$ bundle over $\mathbb{CP}^2$, we get

$$c = (1 + D_1)^3 (1 + D_5 - n D_1)(1 + D_5), \tag{5.48}$$

so $c_1 = (3 - n)D_1 + 2D_5$, $c_2 = 6C^1 + 3(n + 1)C^2$, and $\chi(\mathcal{M}) = 6$.

It is also straightforward to compute Chern classes of algebraic submanifolds of toric varieties, by making use of the adjunction formula (5.36). For a submanifold S of $\mathcal{M}$ defined by

$$S = S_1 \cap S_2 \cap \cdots \cap S_k, \tag{5.49}$$

where the S_k are hypersurfaces given by polynomial equations in the x_i, this yields

$$\begin{aligned} c(S) &= \left.\frac{c(\mathcal{M})}{\prod_\alpha c(S_\alpha)}\right|_S = \left.\frac{\prod_i (1 + D_i)}{\prod_\alpha (1 + S_\alpha)}\right|_S \\ &= \left.1 + \sum_i D_i - \sum_\alpha S_\alpha + \cdots\right|_S. \end{aligned} \tag{5.50}$$

Similar formulas hold for the other multiplicative characteristic classes. It is important to remember however that these formulas can only be directly applied when the complete intersection S is smooth.

As a classic example, consider the quintic hypersurface in $\mathcal{M} = \mathbb{CP}^4$:

$$S : \sum_{i=1}^5 x_i^5 = 0. \tag{5.51}$$

Then, putting $H \equiv D_1$,

$$c(S) = \left.\frac{(1 + H)^5}{1 + 5H}\right|_S = (1 + 10H^2 - 40H^3)|_S. \tag{5.52}$$

Note that the first Chern class vanishes: the quintic is a Calabi–Yau manifold. Furthermore

$$\chi(S) = \int_S c_3(S) = (5H) \cdot (-40H^3) = -200. \tag{5.53}$$

5.6. Holomorphic forms and Poincaré residues

Periods of holomorphic forms play an important role in the computation of super- and Kähler potentials. An elegant and useful general construction of such forms is as a Poincaré residue, as we now explain.

Any *gauge invariant* meromorphic form on $\mathbb{C}^n$

$$\omega = R(x)\, dx^1 \wedge \cdots \wedge dx^n, \qquad (5.54)$$

where $R(x)$ is a homogeneous rational function of the x^i, descends to a well-defined meromorphic d-form (more precisely a $(d, 0)$ form) on the d-dimensional toric variety $\mathcal{M}$. The reduction goes as follows. Let

$$V^a := \sum_i Q_i^a x_i \frac{\partial}{\partial x_i} \qquad (5.55)$$

be the holomorphic vector fields generating the gauge symmetries (i.e. $\delta_a x_i = i\epsilon V^a x_i = i\epsilon Q_i^a x_i$). Then the contraction of ω with all vector fields,

$$\Omega := \omega \cdot \prod_a V^a \qquad (5.56)$$

is a globally defined meromorphic d-form on $\mathcal{M}$. If $R(x)$ is a polynomial, then Ω is holomorphic.

As a first example, consider $\mathbb{CP}^2$. To make ω gauge invariant, $R(x)$ must have charge -3, so $R(x)$ cannot be polynomial and we do not get any holomorphic 2-forms on $\mathbb{CP}^2$, but e.g. $R(x) = (x_1 x_2 x_3)^{-1}$ will do, leading to the meromorphic 2-form

$$\Omega = \frac{dx_2}{x_2} \wedge \frac{dx_3}{x_3} + \frac{dx_3}{x_3} \wedge \frac{dx_1}{x_1} + \frac{dx_1}{x_1} \wedge \frac{dx_2}{x_2}. \qquad (5.57)$$

For the weighted projective space $\mathcal{M} = W\mathbb{CP}^3_{1,1,1,-n}$, $R(x)$ must have charge $n - 3$. So in particular when $n \geq 3$, there are holomorphic 3-forms on $\mathcal{M}$, and when $n = 3$, there is a unique one up to overall scale, namely R a constant:

$$\begin{aligned}
\Omega = {} & x_1\, dx_2 \wedge dx_3 \wedge dx_4 - x_2\, dx_1 \wedge dx_3 \wedge dx_4 \\
& + x_3\, dx_1 \wedge dx_2 \wedge dx_4 + 3\, x_4\, dx_1 \wedge dx_2 \wedge dx_3.
\end{aligned}$$

Indeed when $n = 3$, the first Chern class is trivial, $c_1 = D_1 + D_2 + D_3 + D_4 = 0$, so $\mathcal{M}$ is a Calabi–Yau 3-fold and has a unique holomorphic 3-form. Note however that it is noncompact. (In fact there are no compact toric Calabi–Yau manifolds.)

It is clear that the above construction will always give a unique holomorphic d-form when $c_1 = \sum_i D_i$ vanishes.

There is also a natural way to construct meromorphic $(p, 0)$-forms on p-dimensional algebraic subspaces of $\mathcal{M}$ defined by a system of homogenous polynomial equations. Consider first the case of a hypersurface $\mathcal{S}$ given by $P(x) = 0$.

Let ω again be as in (5.54), but now we take the charges of $R(x)$ such that $\omega/P(x)$ is gauge invariant instead of ω. Then

$$\Omega := \frac{1}{2\pi i} \oint_{P=0} \frac{\omega \cdot \prod_a V^a}{P}, \tag{5.58}$$

where the contour is taken to be an infinitesimal loop around $P = 0$, defines a globally well defined meromorphic top form on $\mathcal{S}$. The contour integral picks up the so-called *Poincaré residue*. This can be defined more precisely as follows. Let η be a meromorphic d-form in an d-dimensional space with a single pole along a smooth hypersurface $\mathcal{S}$, locally described by the equation $z = 0$. Near $z = 0$ write

$$\eta = \frac{dz}{z} \wedge \rho + \eta_0, \tag{5.59}$$

where ρ and η_0 are locally defined holomorphic d-forms. Then the Poincaré residue of η is the restriction of ρ to $\mathcal{S}$:

$$\text{res}_{\mathcal{S}}\, \eta := \frac{1}{2\pi i} \oint_{z=0} \eta := \rho|_{\mathcal{S}}, \tag{5.60}$$

which is unique and extends globally on $\mathcal{S}$.

For a space $\mathcal{S}$ given by a complete intersection of k divisors S_α given by the polynomial equations $P_\alpha(x) = 0$ in a toric variety, we can generalize (5.58) by picking $R(x)$ to be such that $\omega/\prod_\alpha P_\alpha(x)$ is gauge invariant, and putting

$$\Omega := \frac{1}{2\pi i} \oint_{P_1=0} \cdots \frac{1}{2\pi i} \oint_{P_k=0} \frac{\omega \cdot \prod_a V^a}{\prod_\alpha P_\alpha}. \tag{5.61}$$

Again when $c_1 = \sum_i D_i - \sum_\alpha S_\alpha$ vanishes, so $\mathcal{S}$ is Calabi–Yau, this gives rise to a unique holomorphic top form on $\mathcal{S}$.

As an example consider the quintic hypersurface in $\mathbb{CP}^4$, defined by a degree 5 homogeneous polynomial equation $P(x) = 0$. On it we have the unique holomorphic 3-form

$$\Omega = \frac{1}{2\pi i} \oint_{P=0} \frac{x_1\, dx_2 \wedge dx_3 \wedge dx_4 \wedge dx_5 + \text{cycl.}}{P(x)} \tag{5.62}$$

which in a patch where we gauge fix say $x_1 \equiv 1$ can be evaluated as

$$\Omega = \frac{dx_2 \wedge dx_3 \wedge dx_4}{\partial P/\partial x_5}. \tag{5.63}$$

Although explicit patch-dependent expressions like this one are often easily computed, the gauge invariant integral form of the residue, like (5.62), is often more useful to compute periods and differential equations satisfied by them. For techniques to explicitly compute periods, see e.g. [108].

Another application of the Poincaré residue is the one to one map between holomorphic deformations of a divisor $S : P(x) = 0$ in a Calabi–Yau n-fold and holomorphic $(n-1,0)$-forms on S. For a deformation δP of the polynomial P, the corresponding $(n-1,0)$-form is

$$\omega_{\delta P} = \frac{1}{2\pi i} \oint_{P=0} \Omega \frac{\delta P}{P}, \tag{5.64}$$

where Ω is the holomorphic n-form on the CY. Indeed the number of deformations of a holomorphic divisor in a Calabi–Yau is $h^{n-1,0}(S)$.

5.7. Calabi–Yau submanifolds of toric varieties

Complete intersections in toric varieties with vanishing first Chern class,

$$c_1 = \sum_i D_i - \sum_\alpha S_\alpha = 0, \tag{5.65}$$

provide a large, concrete set of examples of Calabi–Yau manifolds which can be used as target manifolds for string, M or F theory. These manifolds inherit their Kähler moduli spaces from the ambient toric variety, and their complex structure moduli spaces can be identified with the deformations of the defining polynomials modulo coordinate redefinitions.[25]

We consider some examples.

The most general quintic submanifold S of $\mathbb{CP}^4$ is given by an equation of the form $S : P_5(x) = 0$ with P_5 a homogeneous degree 5 polynomial. Such a polynomial has $\binom{5+4}{4} = 126$ coefficients. Polynomials which differ only by a $GL(5,\mathbb{C})$ coordinate transformations of the x_i are isomorphic, so we have $126 - 25 = 101$ independent complex structure moduli.

We can check our moduli counting by computing the Euler characteristic from the Hodge numbers the counting implies and comparing to (5.53). With 1 Kähler modulus and 101 complex structure moduli, the independent Hodge numbers of S are $h^{1,1} = 1$, $h^{2,1} = 101$, so the independent Betti numbers are $b^0 = 1$, $b^1 = 0$, $b^2 = 1$ and $b^3 = 204$, and $\chi(S) = 4 - 204 = -200$, in agreement with (5.53).

The quintic inherits a Kähler class J_S from the Kähler class $J = \xi D_1$ of $\mathbb{CP}^4$, by pulling J back to S. It is usually convenient to express J_S in terms of a basis of $H^{1,1}(S)$. Such a basis is obtained by intersecting the divisor basis of

---

[25] In some cases, there may be additional complex structure deformations which do not correspond to defining polynomial deformations.

the ambient variety with the hypersurface. In this case this consists of the single element $H_S := D_1|_S$. Then $J_S = \xi H_S$. The intersection numbers for this basis follow directly from the intersection numbers of the ambient toric variety:

$$H_S^3 = [S]D_1^3 = (5D_1)D_1^3 = 5. \tag{5.66}$$

A somewhat more complicated example is the Calabi–Yau fourfold elliptically fibered over $\mathbb{CP}^3$ as considered in Section 3.8. To cast this in gauged linear sigma model language, we first define a five complex dimensional toric variety by introducing seven fields x_i, identified with the variables used in Section 3.8 as

$$(x_1, x_2, x_3, x_3, x_5, x_6, x_7) = (u_1, u_2, u_3, u_4, x, y, z). \tag{5.67}$$

We assign these fields the following $U(1) \times U(1)$ charges:

$$\begin{pmatrix} Q_i^1 \\ Q_i^2 \end{pmatrix} = \begin{pmatrix} 1 & 1 & 1 & 1 & 0 & 0 & -4 \\ 0 & 0 & 0 & 0 & 2 & 3 & 1 \end{pmatrix}. \tag{5.68}$$

And take the corresponding FI parameters ξ^1 and ξ^2 positive. The assignment of the charges is uniquely fixed (up to change of basis of $U(1) \times U(1)$ generators) by the Calabi–Yau condition (5.65) and the form of the equation $Z : y^2 = x^3 + fxz^4 + gz^6$. This also fixes the charges of the polynomials f and g. We picked a slightly different basis of $U(1) \times U(1)$ generators compared to (3.54), to make the associated curves C^1 and C^2 to form a basis of the Mori cone. The divisors

$$K_1 = D_1, \qquad K_2 = D_7 + 4D_1 \tag{5.69}$$

dual to C^1, C^2 form a basis for the Kähler cone. Note the relations

$$D_2 = D_3 = D_4 = K_1, \qquad D_5 = 2K_2, \qquad D_6 = 3K_2. \tag{5.70}$$

Using the techniques described in Section 5.2, we find the intersection products

$$K_1^5 = 0, \quad K_1^4 K_2 = 0, \quad K_1^3 K_2^2 = \frac{1}{6}, \quad K_1^2 K_2^3 = \frac{2}{3},$$

$$K_1 K_2^4 = \frac{8}{3}, \quad K_2^5 = \frac{32}{3}. \tag{5.71}$$

There is a $\mathbb{Z}_2$ quotient singularity at $y = 0, z = 0$ and a $\mathbb{Z}_3$ quotient singularity at $x = 0, z = 0$, explaining the fractional intersection numbers. For example the third intersection number is obtained by noting that $6K_1^3 K_2^2 = D_1 D_2 D_3 D_5 D_6$, and that the intersection between those five distinct divisors is given by an up to gauge transformations unique, regular point $(0, 0, 0, \sqrt{\xi_1 + 4\xi_2}, 0, 0, \sqrt{\xi^2})$.

The elliptically fibered Calabi–Yau Z itself is given by (3.53):

$$Z : y^2 = x^3 + f(\vec{u}) x z^4 + g(\vec{u}) z^6. \tag{5.72}$$

This hypersurface avoids the quotient singularities of the ambient toric variety: for example if $x = z = 0$, then (5.72) implies $y = 0$, but the point $(x, y, z) = (0, 0, 0)$ is excluded by the D-term constraints. The homology class of Z is

$$[Z] = 6 K_2, \tag{5.73}$$

and the intersection products of the pullbacks $\tilde{K}_a \equiv K_a|_Z$ are, by the rule $\tilde{K}_a \cdots \tilde{K}_b = [Z] K_a \cdots K_b$:

$$\tilde{K}_1^4 = 0, \quad \tilde{K}_1^3 \tilde{K}_2 = 1, \quad \tilde{K}_1^2 \tilde{K}_2^2 = 4, \quad \tilde{K}_1 \tilde{K}_2^3 = 16, \quad \tilde{K}_2^4 = 64. \tag{5.74}$$

Note in particular that

$$\tilde{K}_2^2 = 4 \tilde{K}_1 \tilde{K}_2. \tag{5.75}$$

The Kähler class on Z is $J_Z = \xi^a \tilde{K}_a$ with $\xi^a > 0$, and hence the volume of Z is

$$V_Z = \frac{1}{24} \int_Z J_Z^4 = \frac{\xi^2}{6} \left((\xi^1)^3 + 6 (\xi^1)^2 \xi^2 + 16 \xi^1 (\xi^2)^2 + 16 (\xi^2)^3 \right). \tag{5.76}$$

A representative elliptic fiber of Z is given by $u_1 = u_2 = u_3 = 0$, so its homology class in Z is $E = \tilde{K}_1^3$. Hence the area of the elliptic fiber is

$$v = \int_E J = \xi^2. \tag{5.77}$$

This is exactly the parameter v introduced in Section 3.1. A section of the base of the fibration is given by $z = 0$, so its homology class is $B = D_7|_Z = \tilde{K}_2 - 4\tilde{K}_1$, and its volume

$$V_B = \int_B \frac{J^3}{6} = \frac{(\xi^1)^3}{6}. \tag{5.78}$$

In the F-theory limit on M-theory, we send $v \to 0$. In this case

$$V_Z \to v V_B, \tag{5.79}$$

as expected from the explicit form of the metric (3.12) in this limit.

The total Chern class of the fourfold is, using (5.50) and (5.70) and the intersection numbers:

$$
\begin{aligned}
c(Z) &= \left. \frac{(1 + K_1)^4 (1 + 2K_2)(1 + 3K_2)(1 + K_2 - 4K_1)}{(1 + 6K_2)} \right|_Z \tag{5.80} \\
&= 1 + (48\tilde{K}_1 \tilde{K}_2 - 10\tilde{K}_1^2) - 20(48\tilde{K}_1^2 \tilde{K}_2 + \tilde{K}_1^3) + 23328 \, \omega_Z. \tag{5.81}
\end{aligned}
$$

We repeatedly used $\tilde{K}_2^2 = 4\tilde{K}_1\tilde{K}_2$, and ω_Z denotes the unit volume element on Z. In particular we thus read off the Euler characteristic of the fourfold:

$$\chi(Z) = 23328. \tag{5.82}$$

Hence for this example, the number Q_c appearing in the tadpole cancelation condition (4.97), i.e. minus the curvature induced D3-charge, equals $Q_c = \chi(Z)/24 = 972$.

5.8. Index formulae

To count various massless string modes (i.e. bosonic and fermionic zero modes), it is very useful to have index formulae relating various indices to expressions involving characteristic classes.

For a k-form ω we define the fermion parity $(-)^F \omega = (-1)^k \omega$. The simplest index is

$$\mathrm{Tr}_{H^*(X)} (-)^F = \sum_k (-1)^k b^k(X) = \chi(X) = \int_X e(X), \tag{5.83}$$

where $b^k(X) = \dim H^k(X)$ and $e(X)$ is the Euler class defined in (5.38).

Twisting this index by the Hodge star operator gives:

$$\mathrm{Tr}_{H^*(X)} (-)^F * = (-1)^d (b_{*+}^d(X) - b_{*-}^d(X)) = \sigma(X) = \int_X L(X). \tag{5.84}$$

Here d is the complex dimension of X, $b_{*\pm}^d(X)$ is the number of (anti-)selfdual harmonic d-forms on X, σ is called the signature of X, and $L(X)$ is the Hirzebruch L-genus of the tangent bundle of X as defined in (5.42). This is the Hirzebruch signature formula.

On Kähler manifolds X there are holomorphic versions of these index theorems, which sum over (bundle valued) $(0, p)$-forms only. These are typically relevant to count brane moduli or fermionic zeromodes. The simplest is the arithmetic genus / holomorphic Euler characteristic formula

$$\mathrm{Tr}_{H^{0,*}(X)}(-)^F = \sum_p (-1)^p h^{p,0} = \int_X \mathrm{Td}(X), \tag{5.85}$$

where the Todd class $\mathrm{Td}(X)$ was defined in (5.41). This formula will allow us to check whether M5 instantons satisfy the necessary condition (4.90) to contribute to the superpotential.

This can be generalized to bundle-valued forms:

$$\text{Tr}_{H^{0,*}(X,V)} (-)^F = \sum_p (-1)^p \, h^{0,p}(V) = \int_X \text{ch}(V)\,\text{Td}(X). \tag{5.86}$$

This is the Hirzebruch–Riemann–Roch theorem.

5.9. Computing Hodge numbers

Hodge numbers $h^{q,p}(X)$, or at least a set of relations between them, can be computed using the above index theorems.

In particular, from (5.86), taking $V = \Omega^q$, i.e. the space of $(q,0)$-forms on X, and using $H^{0,p}(X, \Omega^q) = H^{q,p}(X)$, we get a formula for the arithmetic genera χ_q:

$$\chi_q := \sum_p (-1)^p \, h^{q,p}(X) = \int_X \text{ch}(\Omega^q)\,\text{Td}(X), \tag{5.87}$$

$\text{ch}(\Omega^q)$ can be computed as follows. First note that Ω^1 is just the holomorphic cotangent bundle T^*X, which is dual to the tangent bundle TX, so in terms of Chern roots, if $c(TX) = \prod_i (1 + \lambda_i)$, we have $c(\Omega^1) = \prod_i (1 - \lambda_i)$, and $\text{ch}(\Omega^1) = \sum_i e^{-\lambda_i}$. Now $\Omega^2 = \Omega^1 \wedge \Omega^1$, i.e. the antisymmetrization of $\Omega^1 \otimes \Omega^1$. More physically, one can think of this as the space of 2 particle states built from the 1-particle fermionic states of Ω^1, where the one particle states have curvature eigenvalues $-\lambda_i$. It follows that the 2-particle states have eigenvalues $-\lambda_i - \lambda_j, i < j$, so $\text{ch}(\Omega^2) = \sum_{i<j} e^{-\lambda_i - \lambda_j}$. This reasoning can be continued to higher Ω^q; an efficient way to summarize the result is by the fermionic generating function:

$$\sum_q \text{ch}(\Omega^q(TX))\, y^q = \prod_i (1 + y e^{-\lambda_i}). \tag{5.88}$$

Combining this with (5.87) and (5.41), we obtain the generating function for all arithmetic genera

$$\chi(y) = \sum_q \chi_q \, y^q = \int_X \prod_{i=1}^r (1 + y e^{-\lambda_i}) \frac{\lambda_i}{1 - e^{-\lambda_i}}, \tag{5.89}$$

known as the *Hirzebruch genus*.

This allows us to read off the following results. For $\dim_{\mathbb{C}} X = 2$:

$$\chi_0 = h^{0,0} - h^{0,1} + h^{0,2} = \frac{1}{12} \int_X (c_1^2 + c_2), \tag{5.90}$$

$$\chi_1 = 2 h^{0,1} - h^{1,1} = \frac{1}{6} \int_X (c_1^2 - 5c_2). \tag{5.91}$$

These expressions can be used to determine $h^{0,2}$ and $h^{1,1}$ once $h^{0,1}$ is known. For $\dim_{\mathbb{C}} X = 3$:

$$\chi_0 = h^{0,0} - h^{0,1} + h^{0,2} - h^{0,3} \quad = \quad \frac{1}{24} \int_X c_1 c_2, \tag{5.92}$$

$$\chi_1 = h^{0,1} - h^{1,1} + h^{1,2} - h^{0,2} \quad = \quad \frac{1}{24} \int_X (c_1 c_2 - 12 c_3). \tag{5.93}$$

Note that the formula for χ_0 allows us to rephrase the necessary condition (4.90) for M5 instantons to contribute to the superpotential as $\int_{M5} c_1 c_2 = 24$. Furthermore for subspaces of toric varieties, the formula (5.50) allows to explicitly compute this.

Finally, applied to Calabi–Yau fourfolds (so $c_1(X) = 0$):

$$\chi_0 \quad = \quad h^{0,0} - h^{0,1} + h^{0,2} - h^{0,3} + h^{0,4} = \frac{1}{720} \int_X 3 c_2^2 - c_4, \tag{5.94}$$

$$\chi_1 \quad = \quad h^{0,1} - h^{1,1} + h^{1,2} - h^{1,3} + h^{0,3} = \frac{1}{180} \int_X 3 c_2^2 - 31 c_4, \tag{5.95}$$

$$\chi_2 \quad = \quad 2(h^{0,2} - h^{1,2}) + h^{2,2} = \frac{1}{120} \int_X 3 c_2^2 + 79 c_4. \tag{5.96}$$

For Calabi–Yau fourfolds we also have $h^{0,4} = 1$ (and of course $h^{0,0} = 1$) and if the holonomy is full $SU(4)$, then $h^{0,3} = h^{0,2} = h^{0,1} = 0$. The χ_0 equations then becomes trivial, and the χ_1 and χ_2 equations can be used to determine $h^{1,2}$ and $h^{2,2}$ in terms of $h^{1,1}$ and $h^{1,3}$, i.e. the number of Kähler and complex structure moduli.

For the fourfold example (3.53) discussed above, we have $h^{1,1} = 2$ Kähler moduli and $h^{1,3} = 3878$ complex structure moduli (obtained by direct counting of the number of polynomial deformations modulo $GL(4, \mathbb{C})$ reparametrizations). Furthermore, using the above expressions together with (5.80), we get

$$\chi_0 = 2, \qquad \chi_1 = -3880, \qquad \chi_2 = 15564. \tag{5.97}$$

This determines

$$h^{1,1} = 2, \qquad h^{1,2} = 0, \qquad h^{1,3} = 3878, \qquad h^{2,2} = 15564. \tag{5.98}$$

As we see illustrated in this example, we still need the number of complex structure moduli $h^{1,3}$ (or alternatively some other Hodge number such as $h^{1,2}$) as input. In the case at hand we obtained the correct result by counting polynomial deformations modulo coordinate redefinitions. The problem is that sometimes there are complex structure deformations which are not given by polynomial deformations. In this case, more sophisticated techniques are needed; see e.g. [54] for CY fourfolds in particular.

6. Statistics of flux vacua

We need one last ingredient before we can start our search for explicit realizations of the moduli stabilization scenarios of Sections 4.6 and 4.7: efficient estimates of distributions of tree level flux vacua over parameter space. This is necessary because constructing fully explicit flux vacua in typical F-theory compactifications would require specifying 20,000 or so flux quanta and finding the corresponding critical points in the 3000 or so dimensional complex structure moduli space, hoping to hit the region of parameter space we are interested in (e.g. weak string coupling, tiny cosmological constant, ...)—an effectively intractable task.

The statistical approach to flux vacua was initiated in [109] building on ideas of [102], and further developed in [28, 99, 110] and subsequent works. An extensive review can be found in [26], and a more pedagogical review in [111].

6.1. The Bousso–Polchinski model

It was pointed out in [102] that the freedom one has to turn on various independent flux quanta in string theory compactifications can lead to huge ensembles of vacua with a "discretuum" of low energy effective parameters; like the continuum, the discretuum allows for fine tuning, but *without* the massless moduli necessarily associated to continuously variable parameters.

This is true in particular for the cosmological constant, implying naturally the existence of string vacua with exceedingly small effective four dimensional cosmological constants, such as our own, without the need to invoke any (so far elusive) dynamical mechanism to almost-cancel the vacuum energy.

To see how this comes about, consider the potential induced by some flux G characterized by flux quanta $N^I \in \mathbb{Z}$, $I = 1, \ldots, b$, of the general form we considered in Section 2:

$$V_N(z) =, V_0(z) + \int_Z \|G\|^2 = V_0(z) + g_{IJ}(z) N^I N^J, \tag{6.1}$$

where z denotes the moduli of the compactification manifold Z and $g_{IJ}(z)$ is some positive definite effective metric on the moduli space. Further on we will be interested mainly in F-theory flux vacua, but at this point we just consider the above potential as an abstract starting point for a toy model. In particular we ignore constraints such as tadpole cancelation conditions. The bare potential V_0 is taken to be negative. In the context of string theory, it will be of the order of some typically high fundamental scale, such as the string or KK scale.

Each vacuum of this model is characterized by a choice of flux vector N together with a minimum z_* of $V_N(z)$. As a further drastic simplification however, let us, following [102], simply freeze the moduli by hand at some fixed value

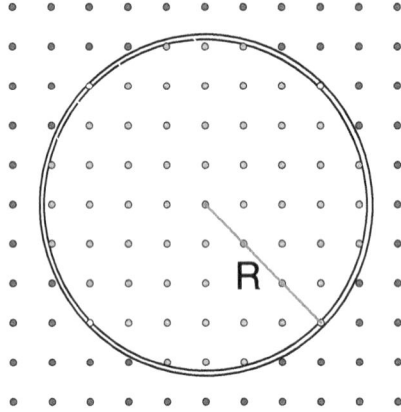

Fig. 12. The number of lattice points within a certain region of flux space can be estimated by the volume of this region. If the dimension is large, even thin shells can contain exponentially many lattice points.

$z = z_0$ and ignore their dynamics altogether. In that case V_N becomes just a quadratic function of N, and it is then easy to compute the distribution of cosmological constant values. The number of vacua with cosmological constant $\Lambda = V_N$ less than Λ_* is now simply given by the number of flux lattice points in a sphere of radius squared $R^2 = |V_0| + \Lambda_*$, measured in the g_{IJ} metric. When R is sufficiently large, this is well-estimated by the volume of this b-dimensional ball, i.e.

$$\mathrm{Vol}_b(R) = \frac{1}{\sqrt{\det g}} \frac{(\pi R^2)^{\frac{b}{2}}}{(\frac{b}{2})!}. \tag{6.2}$$

This leads to the following vacuum number density as a function of Λ:

$$dN_{\mathrm{vac}}(\Lambda) \approx \frac{1}{\sqrt{g}} \frac{\pi^{\frac{b}{2}}(|V_0| + \Lambda)^{\frac{b}{2}-1}}{(\frac{b}{2} - 1)!} d\Lambda \tag{6.3}$$

$$\approx \left(\frac{2\pi e\, (|V_0| + \Lambda)}{\mu^4} \right)^{b/2} \frac{d\Lambda}{|V_0| + \Lambda}, \tag{6.4}$$

where $\mu^4 := (\det g)^{1/b}$ can be interpreted as the typical mass scale of the flux part of the potential. To get the last approximate expression, we assumed large b and used Stirling's formula. Note that in particular at $\Lambda = 0$, for say $|V_0|/\mu^4 \sim \mathcal{O}(10)$, we get a vacuum density $dN_{\mathrm{vac}} \sim 10^b\, d\Lambda/|V_0|$. Hence for b a few hundred, there will be exponentially many vacua with Λ in the observed range

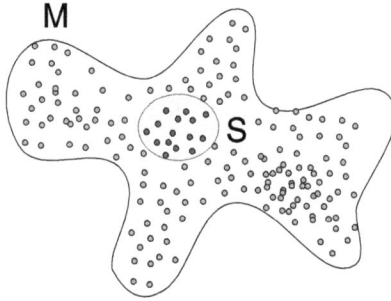

Fig. 13. Counting vacua.

$\Lambda \sim 10^{-120} M_p^4$, even if all fundamental scales setting the parameters of the potential are of order M_p^4!

Thus, in such a model, there is no need to postulate either anomalously large or small numbers, or an unknown dynamical mechanism, to obtain vacua with a small cosmological constant.

However, explicitly finding the flux vectors N^I which give rise to such a small cosmological constant is, even this extremely simplified setting, in general an effectively intractable problem: suitably formalized, this inversion problem can be proven to be NP-hard [112]!

6.2. Distributions of F-theory flux vacua over complex structure moduli space

6.2.1. Setting up the counting problem
We now turn to the problem of counting genuine F-theory flux vacua and computing their distributions over complex structure moduli space. As we have seen in Section 4.4.3, a (tree level) F-theory flux vacuum on an elliptically fibered Calabi–Yau fourfold Z is characterized by

1. A choice of flux quanta N^I, determining the flux G_4 by $[G_4] = N^I \Sigma_I$, satisfying the tadpole cancelation condition (4.97)

$$\frac{1}{2} Q_{IJ} N^I N^J + N_{D3} = \frac{\chi(Z)}{24} = Q_c, \quad Q_{IJ} := \int_Z \Sigma_I \wedge \Sigma_J. \qquad (6.5)$$

If we require $N_{D3} \geq 0$, this imposes the bound

$$\frac{1}{2} Q_{IJ} N^I N^J \leq Q_c. \qquad (6.6)$$

2. A critical point $z^a = z_*^a$ in complex structure moduli space of the superpotential

$$D_a W_N(z_*) = 0, \quad W_N(z) := \int_Z G_4 \wedge \Omega_4 = N^I \Pi_I(z), \tag{6.7}$$

where $D_a W = (\partial_a + \partial_a \mathcal{K})W$, with $\mathcal{K} = -\ln(\Pi_I Q^{IJ} \bar{\Pi}_J) = -\ln \int \Omega \wedge \bar{\Omega}$ being the Kähler potential.

The number of zeros of a function $f(x)$ of one real variable is

$$\#\{x \mid f(x) = 0\} = \int dx \, \delta(f(x)) \, |f'(x)|. \tag{6.8}$$

Similarly, the number of flux vacua in a given region S of complex structure moduli space $\mathcal{M}$ is

$$N_{\text{vac}} = \sum_N \int_S d^{2h} z \, \delta^{2h}(DW) \, |\det D^2 W_N|, \tag{6.9}$$

where $h = h^{3,1}(Z)$ is the complex dimension of the complex structure moduli space and the sum is over all fluxes satisfying (6.6). The determinant factor ensures that each zero of DW contributes $+1$ to the integral, analogous to the $|f'|$ factor in (6.8). In vacua for which $N_{D3} > 0$, there will be residual D3 moduli at tree level. They may however be lifted after inclusion of quantum effects. This will give an additional contribution to the vacuum degeneracy, *not* taken into account in (6.9). Similarly, the Kähler sector, left completely unfixed at tree level, may after inclusion of quantum effects give an additional vacuum degeneracy, or even destabilize the compactification altogether. At this stage however, we wish to focus exclusively on the tree level flux and complex structure sector, so (6.9) is adequate.

Note that the number of vacua without any D3-branes is just

$$N_{\text{vac}}(N_{D3} = 0) = N_{\text{vac}}|_{Q_c} - N_{\text{vac}}|_{Q_c - 1}. \tag{6.10}$$

Of course (6.9) is not terribly useful yet. To make further progress, we will need to make some approximations. First, we will approximate the sum over fluxes by an integral. This is the analog of computing the number of lattice points in the Bousso–Polchinski sphere by computing its volume, and can be expected to be a good approximation in the large Q_c limit. Second, we will drop the absolute value signs around the determinant factor in (6.9). This means we will be counting vacua with signs depending on the number of positive and negative eigenvalues of $D^2 W$; in other words we are computing some sort of index. Strictly speaking this will only give a lower bound on the number of vacua, but in practice the index can be expected to give a good estimate of the order of magnitude of the actual number of vacua in the given region, since there is generically no particular reason for large cancelations.

Before we proceed, we will prove a counting formula in abstract generality which can be applied to many different instances of counting of flux vacua.

6.2.2. A general asymptotic counting formula for zeros of vector field ensembles

Consider a region S in a space with real[26] coordinates x^μ, $\mu = 1, \ldots, m$. Let $P_{I\mu}(x)$, $I = 1, \ldots, b$ be a set of real vector fields, and let A_{IJ} be a nondegenerate symmetric matrix with inverse A^{IJ}. For a choice of integral "flux quanta" N^I satisfying the constraint

$$\frac{1}{2} A_{IJ} N^I N^J \le Q_c, \tag{6.11}$$

we define

$$U_{N,\mu}(x) := N^I P_{I\mu}(x). \tag{6.12}$$

In applications to counting actual flux vacua, this vector field will essentially be the gradient of the superpotential. The set of "flux vacua" we wish to count are labeled by (N, x_*) with

$$U_{N,\mu}(x_*) = 0. \tag{6.13}$$

Hence, similar to (6.9), we wish to estimate

$$N_{\text{vac}} := \sum_N \int_S d^m x \, \delta^m(U_{N,\mu}) \, |\det(\partial_\mu U_{N,\nu})_{\mu\nu}|, \tag{6.14}$$

where the sum is restricted to (6.11). We will approximate this by the continuum index

$$I_{\text{vac}} := \int d^b N \int_S d^m x \, \delta^m(U_{N,\mu}) \, \det(\partial_\mu U_{N,\nu})_{\mu\nu}. \tag{6.15}$$

This can be evaluated as follows. Define a metric on S by

$$g_{\mu\nu} := P_{I\mu} A^{IJ} P_{J\nu}. \tag{6.16}$$

We assume that $g_{\mu\nu}$ is nondegenerate on S. We also define a covariant derivative ∇ such that

$$P_{I\mu} A^{IJ} \nabla_\nu P_{J\rho} \equiv 0, \tag{6.17}$$

---

[26]We switch to real variables here because this makes notation more compact and because it gives a more general formula.

i.e. $\nabla_\nu v_\rho = \partial_\nu v_\rho - \tilde{\Gamma}^\sigma_{\nu\rho} v_\sigma$ with

$$\tilde{\Gamma}^\sigma_{\nu\rho} = g^{\sigma\mu} P_{I\mu} A^{IJ} \partial_\nu P_{J\rho}. \tag{6.18}$$

Then we claim

$$I_{\text{vac}} = \frac{1}{\sqrt{\det A_{IJ}}} \frac{(2\pi Q_c)^{\frac{b}{2}}}{(\frac{b}{2})!} \int_S e(\nabla), \tag{6.19}$$

where $e(\nabla)$ is the Euler density derived from the connection ∇:

$$e(\nabla) = \text{Pf} \left(\frac{\mathcal{R}_{\underline{\mu\nu}}}{2\pi} \right) \tag{6.20}$$

with $\text{Pf}(\cdots)$ is the Pfaffian and $\mathcal{R}_{\underline{\mu\nu}}$ the curvature form in an orthonormal frame with respect to $g_{\mu\nu}$ (underlined indices are frame indices):

$$\mathcal{R}_{\underline{\mu\nu}} = \frac{1}{2} R_{\underline{\mu\nu}\rho\sigma} \, dx^\rho \wedge dx^\sigma, \qquad [\nabla_\rho, \nabla_\sigma] v_\nu =: R^\mu{}_{\nu\rho\sigma} \, v_\mu. \tag{6.21}$$

The proof goes as follows. We define a generating function

$$Z(t) = \int d^b N e^{t \frac{1}{2} N^I A_{IJ} N^J} \int_S d^m x \, \delta^m (U_{N,\mu}) \det(\nabla_\mu U_{N,\nu})_{\mu\nu} \tag{6.22}$$

where now the integral over N is unrestricted. Trading partial derivatives for covariant derivatives in (6.15) or vice versa here does not affect the result, because the difference between $\nabla_\mu U_\nu$ and $\partial_\mu U_\nu$ vanishes when $U_\nu = 0$. The index (6.15) at given Q_c is obtained from the generating function by the contour integral

$$I_{\text{vac}}(Q_c) = \frac{1}{2\pi i} \int \frac{dt}{t} e^{-tQ_c} Z(t) \tag{6.23}$$

where the contour runs over the imaginary axis passing the pole $t = 0$ on the left. (Then if $\frac{1}{2} N^I A_{IJ} N^J - Q_c < 0$, we close the contour on the right and we pick up 1 from the pole, while if $\frac{1}{2} N^I A_{IJ} N^J - Q_c > 0$ we close the contour on the left and the result vanishes. This enforces the constraint (6.11).)

Furthermore we write

$$\delta^m (U_N) = \int d^m \lambda \, e^{2\pi i \lambda^\mu P_{I\mu} N^I}, \tag{6.24}$$

$$\det(\partial U_N) = \int d^m \psi \, d^m \chi \, e^{\psi^\mu \chi^\nu \nabla_\mu P_{I\nu} N^I}, \tag{6.25}$$

where the second integral is over Grassmann variables. Substituting this, the integral over N in (6.22) becomes a simple Gaussian integral,[27] resulting in

$$Z(t) = \frac{1}{t^{b/2}} \frac{(2\pi)^{b/2}}{\sqrt{\det(A_{IJ})}} \int_S d^m x \int d^m \lambda \, d^m \psi \, d^m \chi \, e^{-\frac{1}{2t} f_I A^{IJ} f_J}, \tag{6.26}$$

where

$$f_I = 2\pi i \lambda^\mu P_{I\mu} + \psi^\mu \chi^\nu \nabla_\mu P_{I\nu}. \tag{6.27}$$

Now note that because of (6.17), the λ - $\psi \chi$ cross terms obtained when expanding out $f_I A^{IJ} f_J$ all vanish. The remaining terms are proportional to

$$\lambda^\mu P_{I\mu} A^{IJ} \lambda^\nu P_{J\nu} = g_{\mu\nu} \lambda^\mu \lambda^\nu, \tag{6.28}$$

$$\psi^\mu \chi^\nu \nabla_\mu P_{I\nu} A^{IJ} \psi^\rho \chi^\sigma \nabla_\rho P_{J\sigma} = \psi^\mu \psi^\rho \chi^\nu \chi^\sigma P_{I\nu} A^{IJ} \nabla_{[\mu} \nabla_{\rho]} P_{J\sigma}$$

$$= \tfrac{1}{2} \psi^\mu \psi^\rho \chi^\nu \chi^\sigma P_{I\nu} A^{IJ} R^\tau{}_{\sigma\mu\rho} P_{J\tau}$$

$$= \tfrac{1}{2} \psi^\mu \psi^\rho \chi^\nu \chi^\sigma g_{\nu\tau} R^\tau{}_{\sigma\mu\rho}. \tag{6.29}$$

Performing the Gaussian integrals over λ and ψ, χ then gives, using the Grassmann integral representation of the Pfaffian:

$$Z(t) = \frac{1}{t^{b/2}} \frac{(2\pi)^{b/2}}{\sqrt{\det(A_{IJ})}} \int_S e(\nabla), \tag{6.30}$$

with the Euler density $e(\nabla)$ defined in (6.20). Extracting I_{vac} from the contour integral (6.23) finally gives

$$I_{\text{vac}} = \frac{1}{\sqrt{\det A_{IJ}}} \frac{(2\pi Q_c)^{\frac{b}{2}}}{(\frac{b}{2})!} \int_S e(\nabla), \tag{6.31}$$

as claimed.

The prefactor can morally be thought of as giving the volume of a sphere of radius $\sqrt{2Q_c}$ in flux space. (This is exact when A_{IJ} is positive definite; if not, it is a volume in an analytically continued sense.)

If S is taken to be a compact, closed manifold and $e(\nabla)$ is sufficiently well-behaved, then the integral of the Euler density is a topological quantity, the Euler characteristic of the bundle for which ∇ is a connection. For example when the $P_{I\mu}$ are ordinary sections of T^*S, then $e(\nabla) = e(T^*S) = e(TS)$, the Euler characteristic of S. In this case, our counting formula reproduces the well known

---

[27] In general A_{IJ} need not be positive or negative definite, in which case the Gaussian integral is defined by analytic continuation.

fact that the number of zeros of a vector field on a compact closed manifold, counted with signs, equals the Euler characteristic. However, for our formula, we actually only need the *ensemble* of vector fields to be single valued; there may be monodromies acting on the individual vector fields (as will be the case typically for F-theory flux vacua). Furthermore, S can be any region, and we do not just get the total number of zeros, but their actual distribution as a particular density function $e(\nabla)$.

One interesting general feature following from the expression (6.31) is that flux vacua will tend to cluster anomalously in singular regions where $e(\nabla)$ diverges. We will confirm this below for the example of flux vacua near conifold degenerations, where strong warping occurs.

6.2.3. Application to F-theory flux vacua

Although we derived (6.19 thinking of the x^μ, $P_{I\mu}$ as real variables, we could have thought of them as complex variables z^a, Π_{Ia} as well, by formally setting $x^\mu = z^\mu$ for $\mu = 1, \ldots, h$ and $x^\mu = \bar{z}^{\mu-h}$ for $\mu = h+1, \ldots, 2h$ and similarly $P_{I\mu} = \Pi_{I\mu}$ for $\mu = 1, \ldots, h$ and $P_{I\mu} = \bar{\Pi}_{I,\mu-h}$ for $\mu = h+1, \ldots, 2h$. Everything else would still have gone through, and in particular (6.19) remains true. In case $g_{\mu\nu}$ happens to be a hermitian metric, i.e. $g_{ab} = 0 = g_{\bar{a}\bar{b}}$, the Euler class can also be written in terms of a determinant, as usual for complex varieties with a hermitian metric.

With this in mind, we can immediately apply our result to counting F-theory flux vacua, taking S to be a region in complex structure moduli space and

$$\Pi_{Ia}(z) := e^{\mathcal{K}/2} D_a \Pi_I(z) = e^{\mathcal{K}/2}(\partial_a + \partial_a \mathcal{K})\Pi_I(z). \tag{6.32}$$

where the $\Pi_I(z) = \int \Sigma_I \wedge \Omega$ are the fourfold periods as in Section 6.2.1. Furthermore we take $A_{IJ} = -Q_{IJ}$, with Q_{IJ} the intersection product (6.5).

With these choices, the metric (6.16) has components

$$g_{ab} = e^{\mathcal{K}} \int D_a \Omega \wedge D_b \Omega = 0, \qquad g_{a\bar{b}} = -e^{\mathcal{K}} \int D_a \Omega \wedge D_{\bar{b}} \bar{\Omega} = \partial_a \partial_{\bar{b}} \mathcal{K}. \tag{6.33}$$

The first equation holds because of Griffiths transversality: The derivative $\partial_a \omega$ of a (p, q)-form ω with respect to the complex structure moduli produces a form of type $(p, q) + (p-1, q+1)$. Hence $\partial_a \Omega$ and therefore $D_a \Omega$ is of type $(4, 0) + (3, 1)$. (In fact the covariant derivative D_a is defined in precisely such way that $D_a \Omega$ is exactly of type $(3, 1)$.) In any case the wedge product of $(4, 0) + (3, 1)$ forms is zero, implying $g_{ab} = 0$. The second equation is a consequence of the same Griffiths transversality and the definition of $\mathcal{K}$.

Thus, interestingly, we find that the auxiliary metric (6.16) in this case exactly coincides with the physical metric on complex structure moduli space, which appears in the low energy effective action.

As for the covariant derivative ∇, this is defined in (6.17) by requiring

$$\int (e^{\mathcal{K}/2} D_a \Omega) \wedge \nabla_\mu (e^{\mathcal{K}/2} D_{\bar{b}} \bar{\Omega}) = 0, \tag{6.34}$$

where $\mu = c, \bar{c}$. Again using Griffiths transversality, it can easily be shown that this is satisfied for the standard Levi-Civita and Kähler covariant connection [28] on $TS \otimes \mathcal{L}$ with $\mathcal{L}$ the Kähler line bundle of which the supergravity superpotential is a section:

$$\nabla_a (D_b \Omega) = \partial_a (D_b \Omega) + (\partial_a K)(D_b \Omega) - \Gamma^c_{ab}(D_c \Omega), \tag{6.35}$$

$$\nabla_{\bar{a}} (D_b \Omega) = \partial_{\bar{a}} (D_b \Omega) = g_{b\bar{a}} \Omega. \tag{6.36}$$

Here Γ^c_{ab} is the Levi-Civita connection of $g_{a\bar{b}}$.

Hence we conclude that the continuum index of F-theory flux vacua satisfying (6.6) is

$$I_{\text{vac}} = \frac{1}{\sqrt{\det Q_{IJ}}} \frac{(2\pi Q_c)^{\frac{b}{2}}}{(\frac{b}{2})!} \int_S e(\nabla), \tag{6.37}$$

where the euler density of $TS \otimes \mathcal{L}$ can be written, using the fact that we have a complex structure, as

$$e(\nabla) = \frac{1}{\pi^h} \det(\mathcal{R} + \omega \mathbf{1}). \tag{6.38}$$

Here $\mathcal{R}$ is the curvature form of the holomorphic tangent bundle to S and $\omega = \frac{i}{2} \partial \bar{\partial} \mathcal{K}$ is the Kähler form on S, which is the curvature form of $\mathcal{L}$.

The F-theory flux lattice dimension is given by $b = b_4'$, where b_4' is the number of 4-form fluxes with one leg on the elliptic fiber; more formally, it is the dimension of the subspace of $H^4(Z)$ orthogonal to intersections of divisors, i.e. satisfying (4.44). For the fourfold example (3.53), we read off from (5.98) that $b_4 = 23322$, while (4.44), taking into account the relation (5.75), imposes two independent constraints. Therefore $b_4' = 23320$. Furthermore, as we saw below (5.82), $Q_c = 972$. The intersection form on the full lattice $H^4(X, \mathbb{Z})$, as on any middle cohomology lattice on a compact manifold, is unimodular, i.e. has determinant 1. The sublattice of divisor intersections can be seen to be unimodular too using the results of (5.74), and the orthogonal complement of a unimodular lattice is unimodular. Therefore $\det Q_{IJ} = 1$, and

$$I_{\text{vac}} = 5 \times 10^{1786} \int_S e(\nabla). \tag{6.39}$$

If we let S be the entire complex structure moduli space, then the integral equals the Euler characteristic of $TS \otimes \mathcal{L}$.[28] Since the moduli space is some simple quotient of a projective space (namely the space of coefficients of the defining polynomial modulo coordinate redefinitions), one expects this number to be essentially order 1 compared to the exponential prefactor.

The continuum index of vacua with $N_{D3} = 0$ is, analogous to (6.10):

$$I_{vac}(N_{D3} = 0) = I_{vac}|_{Q_c} - I_{vac}|_{Q_c - 1}. \tag{6.40}$$

In fact, when the number of vacua is exponentially large, almost all flux vacua have $N_{D3} = 0$ according to this estimate; this is related to the fact that for a high dimensional sphere, almost all enclosed volume is located very near its boundary. For our example:

$$\frac{I_{vac}(N_{D3} = 0)}{I_{vac}} = 0.999994. \tag{6.41}$$

This illustrates we have to be particularly careful not to naively apply our low dimensional intuition to high dimensional situations.

We will discuss to what extent the continuum index does (not) give a good estimate for the actual number of vacua, counted with or without signs, in Section 6.3.

A final comment is in order. For a small domain S, I_{vac} can be quite precisely thought of as the volume in flux space of the set of $\vec{N}$ which give rise to a solution of $DW_{\vec{N}} = 0$ located in S. Given the somewhat formal nature of the general computation of I_{ind}, in particular the use of analytic continuation in evaluating the Gaussian integral $\int dN e^{-t\frac{1}{2}NQN}$, one may worry if the result we find does correctly represent this volume in flux space. In particular, given the fact that Q_{IJ} is not positive or negative definite, one might worry that the actual volume is in fact infinite. However, the condition $DW = 0$ effectively renders Q_{IJ} positive definite, since as we saw in Section 4.4.3, $DW = 0$ implies $G_4 = *G_4$, and therefore $N^I Q_{IJ} N^J = \int G \wedge G = \int G \wedge *G \geq 0$. Hence for any finite region S away from singularities, I_{ind} will indeed be finite. Finiteness near singularities and of the actual number of IIB flux vacua has been analyzed in [99, 113, 114]. In [13] the question of finiteness of string vacua was addressed in a much more general setting, and it was argued that, remarkably, in regimes which are in principle under control, the total number is finite as long as one stays bounded away from decompactification limits (characterized by KK modes becoming light).

---

[28] Actually since the moduli space has singularities, the notion of Euler characteristic is ambiguous, and in particular the integral of the above Euler density need not coincide with the topological Euler characteristic. We assume it nevertheless coincides with at least one of the several natural notions of Euler characteristic for singular varieties.

6.2.4. Application to IIB bulk flux vacua

Most of the statistical analysis in the literature has been done purely in simplified models in which one only considers the IIB bulk sector, neglecting D7 degrees of freedom. One can effectively think of these simplified models as F-theory on $Z = T^2 \times X$ with X some Calabi–Yau 3-fold. Note that taken literally, these models have zero Euler characteristic, so $Q_c = 0$ and no flux vacua. However, we can still formally count solutions to $D_a W_N = 0$, by choosing some Q_c by hand. The number of effective F-theory fluxes is now

$$b_4'(Z) = 2\, b_3(X), \qquad (6.42)$$

namely b_3 RR fluxes (G_4 leg on B-cycle T^2) and b_3 NSNS fluxes (G_4 leg on A-cycle T^2). The intersection form is obtained from the symplectic intersection forms on X and T^2, and again unimodular. With these substitutions, all of the above formulae remain valid. These simplified models are presumed to give estimates in some sense for the number of bulk flux vacua in the weak IIB coupling limit, although this has not been made precise. From our considerations in Section 4.4.4, it seems plausible that this makes sense if all D7 branes are coincident with the O7-planes, although turning on bulk NSNS fluxes generically does not appear to keep the D7-branes there.

Taking X to be the Calabi–Yau 3-fold arising in the weak coupling limit of the model (3.53), we get $b_3(X) = 300$, and therefore the continuum index for a region S of the bulk moduli space (complex structures of X and T^2), putting $Q_c = 972$, is

$$I_{\mathrm{vac}} \approx 2 \times 10^{521} \int_S e(\nabla). \qquad (6.43)$$

The restriction to these simplified models is why 10^{500} is such an infamous number, rather than one of the much bigger numbers one gets out of the full F-theory estimates.

6.2.5. Toy model

As a simple illustration, consider again the toy model introduced at the end of Section 4.2 and further analyzed at the end of Section 4.4.3. As we saw there, the critical points can be computed exactly, and all inequivalent flux vacua for a given Q_c can be systematically enumerated [99]. The exact vacua for $Q_c = 150$ are plotted in Fig. 14. The continuum index distribution is also straightforwardly obtained [28, 99]. The Euler density is

$$e(\nabla) = \frac{i}{2\pi} \frac{d\tau \wedge d\bar{\tau}}{(\tau - \bar{\tau})^2}, \qquad (6.44)$$

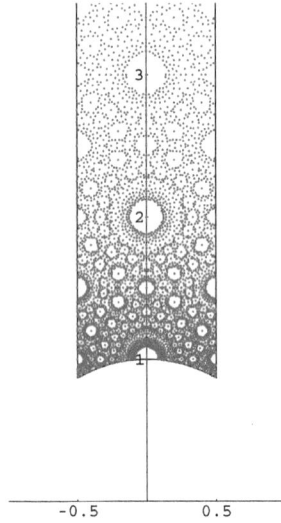

Fig. 14. Values of τ for rigid CY flux vacua with $Q_c = 150$.

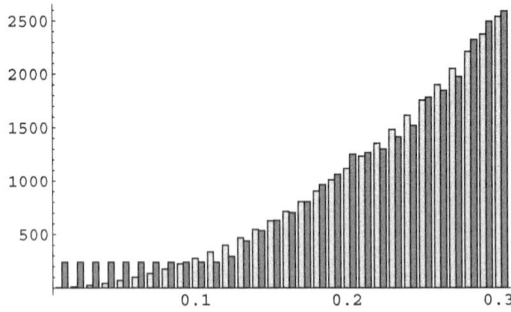

Fig. 15. Number of vacua in a circle of coordinate radius R around $\tau = 2i$, with R increasing in steps $dR = 0.01$. Pink bars give the estimated value, green bars the actual value. The actual number starts at a nonzero value for $R = 0$ because $\tau = 2i$ is multiply degenerate.

hence

$$I_{\text{vac}} = 2\pi Q_c^2 A(S), \tag{6.45}$$

where $A(s)$ is the area of the region under consideration in the Kähler metric (4.15). Letting S be the entire fundamental domain, we get $I_{\text{vac}} = \pi^2 Q_c^2/6$.

Despite the intricate fine structure as evident from Fig. 14 (in particular the striking "voids" around simple complex rational numbers), it is nevertheless true that for large Q_c a disc of sufficiently large area A will contain approximately $2\pi A Q_c^2$ vacua. This is illustrated for $Q_c = 150$ in Fig. 15, where estimated and real numbers of vacua are compared in discs around the center of the largest hole $\tau = 2i$ of stepwise increasing radius.

In more complicated models there are many more fluxes and the periods are highly complex functions. As a result, flux vacua will be much more randomized than in this simple example, and continuum distributions can be expected to become good approximations already at finer scales. Some more comparisons between exact and approximate distributions can be found in [64].

6.3. Regime of validity and improved estimates

We now turn to the question when the continuum index I_{vac} is a good approximation for the actual number of vacua, or at least the actual discrete index of vacua. On general grounds we expect the continuum approximation to be valid in the large Q_c limit, but since Q_c is given to us by the topology of Z, we need to understand better what qualifies as "large".

To get an idea when the approximation certainly fails, we approximate the prefactor of (6.37) using Stirling's formula (and assuming $\det Q = 1$) as

$$\left(\frac{4\pi e Q_c}{b_4'} \right)^{b_4'/2}. \tag{6.46}$$

When $b_4' > 4\pi e Q_c$, this is in fact exponentially small! This is related to the fact that the volume of a sphere of fixed radius goes to zero exponentially when the dimension is sent to infinity. Clearly, in this regime, the approximation breaks down badly. The reason is that in this regime a large fraction of the flux quanta will be zero or some small integer, so the continuum approximation is no longer valid.

In general we have $Q_c = \frac{\chi(Z)}{24}$ and $\chi(Z) = 2 + 2h^{1,1} - 2h^{2,1} + b_4$, so in models with $h^{1,1}, h^{2,1} \ll b_4$ (as is the case for our example and in the models listed in Appendix B.4 of [54]), we have $\chi(Z) \approx b_4 \approx b_4'$. Then

$$\frac{4\pi e Q_c}{b_4'} \approx \frac{\pi e}{6} \approx 1.4, \tag{6.47}$$

so we are barely above the threshold where things go wrong badly. This indicates the continuum index I_{vac} may be a serious underestimate of the actual number of flux vacua in F-theory.

To make this more precise, let us consider as a toy model the problem of counting the number of lattice points $\vec{n} \in \mathbb{Z}^b$ in a sphere of radius $\sqrt{2Q_c}$. In the

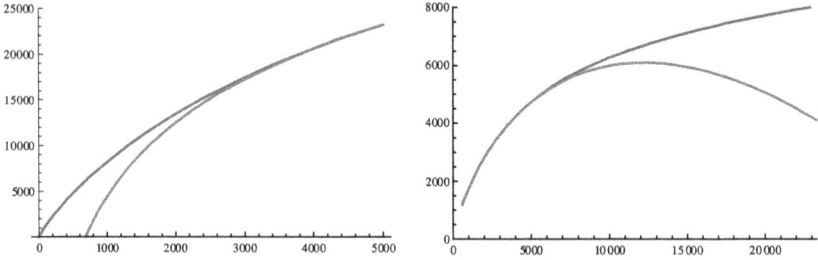

Fig. 16. Left: The blue (upper) line shows $\ln N(b, Q_c)$ as a function of Q_c for $b = 23220$, the red (lower) line shows the continuum estimate. Right: same but as a function of b for $Q_c = 972$.

large Q_c limit, this is the volume of the sphere:

$$N(b, Q_c) \approx \frac{(2\pi Q_c)^{b/2}}{(\frac{b}{2})!} \quad (Q_c \to \infty). \tag{6.48}$$

The exact number can be represented as

$$N(b, Q_c) = \frac{1}{2\pi i} \int \frac{dt}{t} e^{-t Q_c} Z(t), \quad Z(t) := \sum_{\vec{n}} e^{t \vec{n}^2/2} = \left(\vartheta_3(e^t)\right)^b, \tag{6.49}$$

with $\vartheta_3(q) := \sum_{n \in \mathbb{Z}} q^{n^2/2}$, and where we take the contour along the imaginary axis, passing the pole $t = 0$ on the left (compare to (6.23)). For large b, the integral can be computed by saddle point evaluation:

$$\ln N(b, Q_c) \approx S(t_*), \quad \partial_t S(t_*) = 0, \quad S(t) := -\ln t - Q_c t + b \ln \vartheta_3(e^t). \tag{6.50}$$

The results relevant for our usual example are shown in Fig. 16. We see that the continuum approximation becomes very good when Q_c becomes larger than about $b/8$, but that in the regime of interest $Q_c \approx b/24$, the approximation is poor.[29]

In fact when $Q_c \approx b/24$ and b is large, we find $t_* \approx -6.18$ and

$$\ln N(Q_c) \approx 8.27 \times Q_c, \quad N(Q_c) \sim 10^{3.59 \times Q_c}. \tag{6.51}$$

The continuum estimate on the other hand gives $N(Q_c) \sim 10^{1.84 \times Q_c}$. For our example $Q_c = 972$, so $N \sim 10^{3489}$ while the continuum estimate gives a measly $N \sim 10^{1788}$.

---

[29] In the simplified bulk flux models introduced in 6.2.4, which have been the main focus in the literature, this problem typically does not arise, because the number of IIB bulk fluxes is usually much smaller than the total number of F-theory fluxes. In our example, the number of bulk fluxes is $b = 600$, so $Q_c = 972$ is well above $b/8$, and the continuum approximation is excellent.

A natural guess for an improved estimate of the number of F-theory flux vacua would be to replace the volume factor in (6.37) by our toy model $N(b, Q_c)$

$$I'_{\text{vac}} = N(b, Q_c) \int_S e(\nabla). \tag{6.52}$$

However, one should worry that the sparseness of the typical flux vector in the ensemble will have significant effects on the distribution density as well, drastically modifying the $e(\nabla)$ density valid in the continuum approximation. In particular, one could imagine discrete effects such as clustering at enhanced symmetry loci to become more important. This has not been studied yet.

Note that as a rule of thumb, this estimate amounts to about a factor of 10 per moduli space dimension.

6.4. More distributions

6.4.1. Distribution of W

So far we have only discussed estimates for the total number of flux vacua and their distribution over complex structure moduli space. It is not hard to extend this to distributions of other quantities, such as the distribution over the $w := e^{\mathcal{K}/2} W$ plane. To estimate the latter in the continuum index approximation, one can insert an additional $\delta^2(e^{\mathcal{K}/2} W - w)$ in the generating function $Z(t)$ for I_{vac}, rewrite this using Lagrange multipliers, integrate out N and use Griffiths transversality again, to find that the net effect of this additional insertion is (up to some constant factor) $Z(t) \to Z(t) \, t \, e^{t|w|^2}$. So effectively, this amounts to replacing $b'_4 \to b'_4 - 2$ and $Q_c \to Q_c - |w|^2$, and we find for the combined distribution

$$dI_{\text{vac}} \propto N(b'_4 - 2, Q_c - |w|^2) \, d^2w \, e(\nabla), \tag{6.53}$$

where $N(b, Q_c)$ is the usual sphere volume factor in the continuum approximation, or the function $N(b, Q_c)$ introduced in the previous subsection in cases where we believe this to be a better estimate. Note that at large b'_4, due to the exponential dependence of $N(Q_c, b)$ on Q_c, this distribution is approximately Gaussian on the w-plane, peaking at $w = 0$ and cut off at $|w|^2 = Q_c$. This is as one would expect if one thinks of W as being the result of a random addition of a large number of complex numbers. The cutoff can be understood as well, it comes from $Q_c \geq \frac{1}{2} G^2 = |G^{4,0}|^2 + \frac{1}{2}|G^{2,2}|^2 \geq |G^{4,0}|^2 = |w|^2$, where we used that $DW = 0 \Leftrightarrow G^{3,1} = 0$.

The width of the Gaussian is $\sigma \sim (\partial_Q \ln N(b, Q))^{-1/2}$, which in the case of Fig. 16 is somewhat less than one. For applications in constructions of string vacua we are however mainly interested in vacua with $|w|^2 \ll 1$. In this regime, the distribution becomes uniform on the w-plane:

$$dI_{\text{vac}} \sim I_{\text{vac,tot}} \, d^2 w, \tag{6.54}$$

as can be expected on general grounds. In particular this implies we can expect vacua with $|w|^2$ roughly as small as $1/N_{\text{vac}}$.

6.4.2. Distribution of string coupling constants

From the considerations in Section 4.3, we know that in Sen's weak IIB coupling limit, the fourfold complex structure moduli space $\mathcal{M}_Z$ factorizes in a dilaton-axion moduli space, a threefold complex structure moduli space, and (depending on the point in the threefold complex structure moduli space), a D7 moduli space. Hence we get a corresponding factorization of the continuum index density

$$dI_{\text{vac}} \propto e(\nabla) = \omega_\tau \wedge \rho, \tag{6.55}$$

where ρ is some τ-independent density on the threefold complex and D7 moduli spaces, while ω_τ is the Kähler form on the dilaton-axion moduli space (which proportional to the curvature form), i.e.

$$\omega_{T^2} = \frac{i}{2\pi} \frac{d\tau \wedge d\bar{\tau}}{(\tau - \bar{\tau})^2}, \tag{6.56}$$

as in (6.44).

This implies in particular that universally in the weak coupling limit (i.e. $\text{Im}\,\tau$ sufficiently large), τ is uniformly distributed w.r.t. the standard Poincaré metric on the upper half plane. In terms of the string coupling constant $g_s = 1/\text{Im}\,\tau$, this is simply the uniform distribution:

$$dI_{\text{vac}} \propto dg_s. \tag{6.57}$$

The continuum approximation for the distribution is expected to be accurate down to $g_s \sim 1/\sqrt{Q_c}$, where "void" effects like in the toy model might start to get important. (For example in the toy model the actual minimal value of g_s is $1/Q_c$, although the continuum approximation predicts an order $1/Q_c^2$ minimum; the discrepancy can be thought of as being due to the void around $\tau = i\infty$.)

6.4.3. Conifold clustering and distribution of warp factors

Consider a simplified bulk flux model as described in Section 6.2.4, with X a one modulus Calabi–Yau threefold. An example is the mirror quintic, described by

$$X : x_1^5 + x_2^5 + x_3^5 + x_4^5 + x_5^5 - 5\psi x_1 x_2 x_3 x_4 x_5 = 0 \tag{6.58}$$

in $\mathbb{CP}^4$, modulo phase transformations $x_i \rightarrow e^{2\pi i k_i/5}$ leaving this equation invariant. X acquires a conifold singularity when $\psi = 1$. Parametrizing $z \equiv \psi - 1$, the distribution near $z = 0$ in the continuum approximation can be computed from (6.38) to be [28]

$$dN_{\text{vac}}(z) = dI_{\text{vac}}(z) \propto \frac{d^2 z}{|z|^2 \ln^2 |z|^{-1}} \propto d\left(\frac{1}{\ln |z|^{-1}}\right). \tag{6.59}$$

The distribution diverges at $z = 0$, but in an integrable way, and is approximately scale invariant. As a result, there will be a sizable number of flux vacua exponentially close to the conifold point. Now recall from Section 4.4.6 that flux vacua close to conifold points develop a strongly warped KS-type throat. The redshift at the bottom of the throat is given by (4.75): $\mu \sim |z|^{1/3}$, so the above distribution can be viewed as a distribution for warp factors. In the case of the mirror quintic, this gives, taking into account numerical factors and setting $\mu \equiv |z|^{1/3}$, about 3% of all flux vacua has $\mu < 10^{-1}$, 0.7 % has $\mu < 10^{-5}$, and 0.3 % has $\mu < 10^{-12}$.

Similar to the string coupling constant, one expects the continuum distribution to be accurate for $\frac{1}{\ln |z|^{-1}}$ roughly down to $1/\sqrt{Q_c}$.

Having 0.3% of flux vacua with warping $\mu < 10^{-12}$ may not sound like a terribly spectacular enhancement. Admittedly, it isn't. However in generic actual models, there are many more 3-cycles which could potentially shrink to a tiny size, and this may lead to a much higher fraction of vacua with one or more warped throats, as the following simple argument shows. Imagine we have a Calabi–Yau threefold with b 3-cycles which could potentially collapse to zero size. Then the mirror quintic data suggests that a naive rough estimate for the fraction of vacua for which *all* of these 3-cycles remain larger than the size to get $\mu < 10^{-12}$ is equal to something like $(1 - 0.003)^b \approx e^{-0.003 \times b}$. Now when b becomes large, this can become a small fraction. For example if $b \sim 250$, about half of all vacua do have $\mu < 10^{-12}$, and if $b \sim 500$, this goes up to 80%. Of course the actual numbers we used here are just for illustration purposes; but the general idea should be clear. This was studied in more detail in [115].

The above simple argument relies on many poorly justified assumptions though, and is therefore not conclusive. No concrete model has been studied in which these ideas have been tested against actual distributions.

6.4.4. *Distribution of compactification scales*
In the KKLT scenario of moduli stabilization, the compactification radius R is determined by the value of $|w|^2 = e^{\mathcal{K}}|W|^2$:

$$R^4 \sim \ln |w|^{-2}. \tag{6.60}$$

Therefore, from (6.54), in this scenario, the KK scale is distributed as

$$dN_{\text{vac}} \propto de^{-R^4}. \tag{6.61}$$

That is, large volumes are exponentially suppressed, with maximal values of order $R_{\text{max}}^4 \sim \ln N_{\text{vac}} \sim Q_c$.

In the large volume scenario on the other hand, we have according to (4.112)

$$R^6 \sim e^{\xi^{2/3}/g_s}. \tag{6.62}$$

Thence, from (6.57),

$$dN_{\text{vac}} \propto d\left(\frac{1}{\ln R}\right). \tag{6.63}$$

Thus, in this scenario, we have an approximate scale invariant distribution of KK scales.

6.4.5. Distributions of nonsupersymmetric flux vacua

The flux vacua we have been considering so far are generically nonsupersymmetric at tree level, with supersymmetry breaking scale $F^2 \sim |w|^2 m_s^4$, due to the fact that $D_T W \neq 0$. However in for example the KKLT scenario, supersymmetry gets restored by T-dependent quantum corrections. There could be other minima of the full effective potential where supersymmetry is still broken, by some generic $F_a = D_a W \neq 0$. Note that at tree level $D_a W \neq 0$ is forbidden by the equations of motion, so in order to get such minima, the full quantum corrected effective potential must be considered. This is in general a complicated problem.

A slightly simplified model is to consider again the flux superpotential W and the supergravity potential $V = e^{\mathcal{K}}(|DW|^2 - 3|W|^2)$, but now without including any contributions from the Kähler moduli—in fact pretending there are no Kähler moduli whatsoever in the game. In particular solutions to $D_a W = 0$ will now have negative V, because the covariant derivatives with respect to the Kähler moduli are no longer there and so no longer kill off the $-3|W|^2$ term.

By "nonsupersymmetric flux vacua" we mean in this model the minima of V which have $F_a \sim D_a W \neq 0$. We are interested in the regime $|F| \ll 1$, that is supersymmetry breaking well below the fundamental scale. Then it was shown in [110], and more intuitively explained in [116], that for generic flux vacua the distribution of supersymmetry breaking scales F and cosmological constants Λ goes as

$$dN_{\text{vac}}(F, \Lambda) \propto F^5 dF \, d\Lambda. \tag{6.64}$$

This "favors" high scale supersymmetry breaking. In [116] the possibility was considered that other branches of the landscape could exist where low scale breaking was favored. The issue whether string theory favors high or low susy breaking remains inconclusive, and will remain so as long as we have no clue what "favored" means.

6.5. *Metastability, landscape population and probabilities*

So far we have only considered *number* distributions of flux vacua over parameter space. It is of course tempting to wonder if there is any sense in which one could, as in statistical mechanics, compute *probability* distributions on parameter space. This, of course, immediately runs into a heap of conceptual problems. To begin with, it already unclear of *what* exactly these would be probabilities. At a vague level, one would imagine it to be the probability of finding ourselves in a particular vacuum, but making this precise, given the fact that we already found ourselves here and that there is no way to repeat the experiment, is challenging to say the least.

But regardless of the precise definition of these probabilities, it is clear that to determine them, cosmological considerations will come into play in an important way. In particular, it is necessary to consider the mechanism by which vacua actually come into existence. As we saw in Section 3.4, fluxes are sourced by 5-brane domain walls wrapping internal cycles; the flux jumps across such domain walls. Now, if we are in a flux vacuum with positive cosmological constant, quantum fluctuations can cause nucleation of flux-changing domain wall bubbles [102, 117–119], by a tunneling mechanism similar to Coleman-de Luccia bubble nucleation in scalar potential landscapes. If the cosmological constant inside the bubble is positive again, it will itself inflate and eventually nucleate new bubbles, and so on (see Fig. 17 for a psychedelic impression of this). This is a version of eternal inflation (see e.g. [121] for a brief review, and Steve Shenker's lectures at this school).

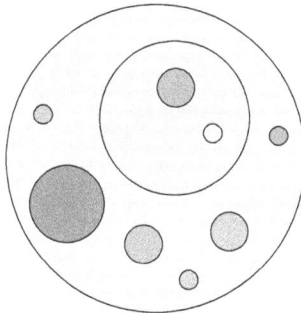

Fig. 17. Bubbles in bubbles.

Even if we are not interested in computing probabilities, such bubble nucleation processes are still of crucial importance to determine to what extent particular vacua are metastable.

According to [117, 118],[30] the nucleation rate per unit 4d spacetime volume for a bubble with tension T, cosmological constant Λ_o outside and cosmological constant Λ_i inside is given by

$$\Gamma \sim e^{-12\pi^2 B}, \tag{6.65}$$

where (in units with $m_p \equiv 1$, which we will use in the remainder of this section):

$$B = \frac{T\rho^3}{6} - \frac{1 - \sigma_i(1 - \frac{\Lambda_i \rho^2}{3})^{3/2}}{\Lambda_i} + \frac{1 - \sigma_o(1 - \frac{\Lambda_o \rho^2}{3})^{3/2}}{\Lambda_o}. \tag{6.66}$$

Here $\sigma_{i,o} = \text{sign} [\pm 3\, T^2 + 4(\Lambda_o - \Lambda_i)]$ and ρ is the bubble radius, which must be evaluated at the stationary[31] point of $B(\rho)$:

$$\rho = \frac{12\, T}{[9\, T^4 + 24\, T^2(\Lambda_i + \Lambda_o) + 16\,(\Lambda_i - \Lambda_o)^2]^{1/2}}. \tag{6.67}$$

If $\Lambda_o > 0$, there is always a nonzero nucleation rate. If $\Lambda_o < 0$ and the initial space is AdS, one needs in addition $\Lambda_i < \Lambda_o - 3T^2/4$ and the argument of the square root of (6.67) to be positive (which is automatic if $\Lambda_o > 0$).

Consistency of the semiclassical approximation requires $\rho \gg 1$ and therefore exponentially small decay rates, as usual with instantons. In particular any well-controlled domain wall bubble will almost tautologically give rise to an extremely small decay rate. Note that ρ is infinite and the decay rate zero when $T = \frac{2}{\sqrt{3}}|\sqrt{-\Lambda_o} \pm \sqrt{-\Lambda_i}|$; this is the case when the domain wall is BPS saturated, interpolating between two supersymmetric vacua.

When $T^2 \ll (\Delta\Lambda)^2/\bar{\Lambda}$ with $\bar{\Lambda} := \Lambda_i + \Lambda_o$ and $\Delta\Lambda := \Lambda_o - \Lambda_i > 0$, this becomes

$$\Gamma \sim \exp\left(-\frac{27\pi^2}{2} \frac{T^4}{(\Delta\Lambda)^3}\right) \tag{6.68}$$

and when $T^2 \gg \Lambda_o, \Lambda_i$:

$$\Gamma \sim \exp\left(-\frac{24\pi^2}{\Lambda_o} + \frac{64\pi^2}{T^2}\right). \tag{6.69}$$

---

[30]Their analysis does not take into account moduli dynamics and is intrinsically done in a thin-wall approximation as the membranes are taken to be infinitely thin. The resulting formulas should therefore be taken to be estimates rather than exact results for F-theory flux vacua.

[31]This is a minimum iff $3T^2 - 4|\Lambda_i - \Lambda_o| > 0$.

Notice that after restoring powers of m_p, (6.68) does not involve Newton's constant—the result is indeed identical to the rate of bubble nucleation in the absence of gravity, in the thin wall approximation. The second expression does depend on Newton's constant. This rate is extremely suppressed for vacua with $\Lambda_o \ll m_p^4$ such as our own (although it is always larger than the Poincaré recurrence rate $e^{-24\pi^2/\Lambda_0}$). Thus for the stability of a vacua, the most dangerous domain wall bubbles are those with small tension but sizable change of cosmological constant.

Decay rates such as those give here can be taken as starting point to try to find sensible probability measures on the landscape, as explained by Steve Shenker at this school.

7. Explicit realizations of moduli stabilization scenarios

In this section we will finally get to building explicit models of moduli stabilized F-theory flux vacua, drawing on all of the techniques developed in earlier chapters. From the section on statistics, we already take that typically, there will be a fine discretuum of vacua which we can use to tune various physical parameters to our liking, and in particular generate large scale hierarchies through warping. This allows us to consider controlled regimes.

The main remaining challenge is to make sure all Kähler moduli are stabilized. In both the KKLT and the large volume scenarios, this hinges on the existence of suitable nonperturbative corrections to the superpotential. The first concrete models satisfying the necessary geometrical requirements for this (in the KKLT scenario) were proposed in [123], and a simpler and more explicit model was given in [124] and subsequently generalized in [125, 126]. Various powerful mathematical criteria for M5 instantons to have the right zeromode structure to contribute to the superpotential were systematically developed in [122]. We will however stick to the more elementary methods we have developed in these lectures. We will in these lectures also not show explicitly the existence of a Kähler stabilized minimum of the effective potential, but only show that models exist where we do get the necessary structure of the Kähler potential and the right kind of contributions to the superpotential to make in particular the large volume scenario of Section 4.7 work. But once these conditions are met, the existence of large volume minima of the effective potential in this scenario is guaranteed by the general analysis of [5, 87, 103, 104]. (However as noted in Section 4.7 one should still check if the "small" Kähler moduli T_{S_i} in string units can be made sufficiently large to trust the geometrical picture; we will not do this here.)

7.1. The elliptic fibration over $\mathbb{CP}^3$

Let us first see if we can turn our basic example (3.53) in an explicit realization of one of the moduli stabilization scenarios outlined in Sections 4.6 and 4.7.
The large volume scenario needs at least two Kähler moduli, so this is excluded,
leaving only the KKLT scenario. To make this scenario work we need some non-
perturbative contributions to the superpotential, and the existence of classical flux
vacua with exponentially small $e^{\mathcal{K}}|W|^2$. As explained in Section 6, according to
the distribution estimates, there is certainly no shortage of the latter. Generating
nonperturbative corrections to W is more subtle. As noted in Section 4.5.4, in
the M-theory picture, all nonperturbative effects can be thought of as being generated by holomorphic M5 instantons wrapping the elliptic fiber and a divisor in
the base. A necessary condition for this instanton to contribute in the absence of
fluxes is the arithmetic genus $\chi_0 = 1$ condition (4.90). In the presence of fluxes
this gets replaced by the weaker condition (4.92): $\chi_0 \geq 1 (= \chi_{0,\text{eff}})$.

In the notation of Section 5.7 where we studied this case as an example, the
most general holomorphic divisor D in Z wrapping the elliptic fiber is given
by some degree k polynomial equation $P_k(\vec{u}) = 0$ on the base, so $D = k\tilde{K}_1$,
$k \in \mathbb{Z}^+$. To compute $\chi_0(D)$, we use the index formula (5.92): $\chi_0 = \frac{1}{24}\int_D c_1 c_2$.
Here the Chern classes c_1 and c_2 are those of the tangent bundle TD of D, which
can be computed from the Chern classes of the ambient CY fourfold using the
adjunction formula (5.36): $c(TD) = c(TZ)/c(ND) = c(TZ)/(1 + D)$. But we
know $c(TZ)$ already; it is given by (5.80). Expanding out the adjunction formula
quotient, we find

$$c_1(TD) = -k\tilde{K}_1, \qquad c_2(TD) = \left((k^2 - 10)\tilde{K}_1 + 48\,\tilde{K}_2\right)\tilde{K}_1, \qquad (7.1)$$

and from this, using the formula for χ_0 just quoted and the intersection numbers (5.74):

$$\chi_0(D) = \frac{1}{24}k\tilde{D}_1\,c_1(TD)\,c_2(TD) = -2k^2. \qquad (7.2)$$

In particular this is always *negative*, so even the weak condition $\chi_0 \geq 1$ is not
satisfied.

We conclude that neither the large volume, nor the KKLT scenario for this
model works.[32]

---

[32] Actually, if we tune the complex structure moduli to a locus of enhanced gauge symmetry as
discussed in section 3.10, so Z becomes singular, there could still be nonperturbative contributions
associated to M5 instantons wrapping the divisors obtained by blowing up the singularity (i.e. going
to the Coulomb branch), as explained in Section 4.5.3. The blown up fourfold will have different
Hodge numbers than the original Z, and as a result different flux lattice dimensions and D3 tadpole.
Whether we still consider this to be the same model is a matter of semantics. We will consider it to
be a different model here.

7.2. The elliptic fibration over $\mathcal{M}_n$

We consider now the CY elliptic fibration with as base manifold B our example (5.4), which we denoted by $\mathcal{M}_n$, the $\mathbb{CP}^1$ bundle over $\mathbb{CP}^2$ with twist $n \geq 0$.[33] This was defined by five fields, which we will now call u_i, and $U(1) \times U(1)$ gauge group, with charges

u_1	u_2	u_3	u_4	u_5
1	1	1	$-n$	0
0	0	0	1	1

and positive FI parameters (ξ^1, ξ^2). Recall from (5.33) that the volume of $\mathcal{M}_n$, $n > 0$ is indeed of Swiss cheese type.

We consider again a CY elliptic fibration of the form

$$Z : y^2 = x^3 + f(\vec{u}) \, x z^4 + g(\vec{u}) \, z^6 = 0 \tag{7.3}$$

over $\mathcal{M}_n$, and the Calabi–Yau condition $\sum_i D_i = [Z]$ fixes the charges of the fields and polynomials to be

u_1	u_2	u_3	u_4	u_5	x	y	z	f	g
1	1	1	$-n$	0	0	0	$n-3$	$4(3-n)$	$6(3-n)$
0	0	0	1	1	0	0	-2	8	12
0	0	0	0	0	2	3	1	0	0

The corresponding D-term constraints are, explicitly:

$$|u_1|^2 + |u_2|^2 + |u_3|^2 - n\,|u_4|^2 + (n-3)\,|z|^2 = \xi^1, \tag{7.4}$$
$$|u_4|^2 + |u_5|^2 - 2\,|z|^2 = \xi^2, \tag{7.5}$$
$$2\,|x|^2 + 3\,|y|^2 + |z|^2 = \xi^3. \tag{7.6}$$

In accord with the F-theory limit of vanishing elliptic fiber, we take the third FI parameter ξ^3 much smaller than ξ^1, ξ^2.

It may seem like we have constructed an infinite number of Calabi–Yau four-folds, labeled by n. This is not true. We should keep in mind that we have made the implicit assumption (by using the formula $c_1 = \sum_i D_i - [Z]$) that Z is smooth. If this is not the case, we should in principle first resolve the singularities before applying this formula, or use a modification of the formula appropriate for

---

[33]The cases $n < 0$ are isomorphic to $n > 0$ by exchanging u_4 and u_5.

singular spaces. Now, from the $U(1)^3$ charges of the polynomials f and g given above, we see that if $n > 3$, f and g become negatively charged under the first $U(1)$ and so must necessarily contain an overall factor equal to a power of u_4. More precisely $f(u) = u_4^k \tilde{f}(u)$, $g(u) = u_4^l \tilde{g}(u)$ where k is the smallest integer $\geq 4(1 - \frac{3}{n})$ and l the smallest integer $\geq 6(1 - \frac{3}{n})$. So in this case f, g and the discriminant $\Delta = 27 g^2 + 4 f^3$ vanish as some power of u_1 on the divisor $D_4 : u_1 = 0$, and hence the fourfold is singular along the locus $\Delta = 0$. For n not too large, the singularities are harmless in the sense that they can be resolved while preserving the $c_1 = 0$ condition, and moreover they have a clean physical interpretation as loci of enhanced gauge symmetry, as mentioned in Section 3.10. For example for $n = 4$, we generically have $f \sim u_4$, $g \sim u_4^2$ and $\Delta \sim u_4^3$, so from the table in Section 3.10 we read off that we get an $SU(2)$ gauge group enhancement. For $n = 18$, we have $f \sim u_4^4$, $g \sum u_4^5$, $\Delta \sim u_4^{10}$ and we get an E_8 gauge group enhancement. For $n > 18$, we fall off the table; at this point the singularity becomes so bad that it cannot be resolved preserving the CY condition. This puts a cutoff on n.

At any rate, we will focus on the cases without gauge symmetry enhancement, i.e. $n \leq 3$, for which the analysis is most straightforward.

From the charge assignments above, we read off the following relations between the divisors:

$$D_1 = D_2 = D_3, \qquad D_5 - D_4 = n\, D_1, \tag{7.7}$$

$$[Z] = 3D_x = 2D_y = 6D_z + (3 + n)D_1 + 2D_4, \tag{7.8}$$

where in the last line $[Z]$ is the homology class of our Calabi–Yau Z.

An independent set of divisors is given for instance by D_4, D_5, D_z. Their pullbacks to Z are denoted by $\tilde{D}_4$, $\tilde{D}_5$, $\tilde{D}_z$. The first two are divisors wrapped on the elliptic fiber and a divisor in the base. The third one is a section of the elliptic fibration, i.e. the base itself. Using the techniques of Section 5, we find the following nonzero intersection numbers between these divisors:

$$\tilde{D}_4^3 \tilde{D}_z = n^2, \quad \tilde{D}_4^2 \tilde{D}_z^2 = (3 - n)n, \quad \tilde{D}_4 \tilde{D}_z^3 = (3 - n)^2, \tag{7.9}$$

$$\tilde{D}_5^3 \tilde{D}_z = n^2, \quad \tilde{D}_5^2 \tilde{D}_z^2 = -(3 + n)n, \quad \tilde{D}_5 \tilde{D}_z^3 = (3 + n)^2, \tag{7.10}$$

$$\tilde{D}_z^4 = -2(n^2 + 24). \tag{7.11}$$

This data allows us to compute volumes, characteristic classes, indices and so on. (A basis for the Kähler cone is given by $\tilde{K}_1 = \tilde{D}_1$, $\tilde{K}_2 = \tilde{D}_5$, $\tilde{K}_3 = [Z]_Z$, but we will continue to work in the above divisor basis in what follows.)

Again we need some nonperturbative contributions to W, associated to holomorphic M5 instantons wrapping the elliptic fiber and a divisor in $\mathcal{M}_n$. The most general such divisor D is given by some polynomial equation $P(\vec{u}) = 0$, so

$$D = a D_4 + b D_5 = (a + b) D_4 + bn D_1, \tag{7.12}$$

where $a + b \in \mathbb{Z}^+$, $bn \in \mathbb{Z}^+$.

As in the previous example, we can compute the holomorphic Euler characteristic χ_0 and find (assisted by Mathematica to do the series expansions of characteristic classes and to substitute the intersection numbers):

$$\chi_0(D) = -\frac{1}{2} n \big((n - 3) a^2 + (n + 3) b^2 \big). \tag{7.13}$$

When $n = 0$ or $n \geq 3$, this is nonpositive, and therefore even the weak necessary condition $\chi_0 \geq 1$ is not satisfied.[34] On the other hand the Diophantine equation $\chi_0(D) = 1$ has infinitely many solutions for $n = 1, 2$. For definiteness let us specialize to

$$n \equiv 1 \tag{7.14}$$

from now on. Then to find divisors of arithmetic genus one, we have to solve $a^2 - 2b^2 = 1$ for $a + b$, b nonnegative integers. This is explicitly solved as

$$a = \frac{(3 + 2\sqrt{2})^k + (3 - 2\sqrt{2})^k}{2}, \qquad b = \frac{(3 + 2\sqrt{2})^k - (3 - 2\sqrt{2})^k}{2\sqrt{2}}, \tag{7.15}$$

$k \geq 0$. The first few solutions are $(a, b) = \{(1, 0), (3, 2), (17, 12), (99, 70), \cdots\}$.

In particular for $(a, b) = (1, 0)$, i.e. $D = D_4 : u_4 = 0$, the instanton is completely rigid and has exactly two zeromodes, i.e. $h^{1,0} = h^{2,0} = h^{3,0} = 0$. This can be seen as follows. First, it is clear from the charge assignments of the fields that $u_4 = 0$ is the unique holomorphic representative in its homology class—there are no other polynomials with the same charges as u_1. From the one to one correspondence between holomorphic deformations and elements of $H^{3,0}(D)$, this implies $h^{3,0} = 0$. Furthermore, from the Lefschetz hyperplane theorem, it follows that $b_1(D) = b_1(Z) = 0$, and therefore $h^{1,0} = 0$. Hence $1 = \chi_0 = 1 + h^{2,0}$, and therefore also $h^{2,0} = 0$.

Thus, flux or no flux, D will always contribute to the superpotential, and given (5.33) this is moreover exactly the kind of contribution we need for the large volume scenario to work! Note also that unlike in the KKLT scenario, we only need one instanton correction to stabilize all Kähler moduli.[35]

---

[34] However as we just saw when $n > 3$ we need to consider more divisors, namely those obtained from resolving the enhanced gauge singularities, but we will stick to the smooth cases here.

[35] Given the infinite set of solutions (7.15) it is possible that we have more instanton contributions, but these will be exponentially suppressed compared to $(a, b) = (1, 0)$.

For completeness we give some further topological data for this model, obtained in a way similar to what we did for the example of the elliptic fibration over $\mathbb{CP}^3$. The Hodge data of $Z|_{n=1}$ is

$$h^{1,1} = 3, h^{2,1} = 0, h^{3,1} = 3397, h^{2,2} = 13644. \tag{7.16}$$

This implies in particular $b_4 = 20440$ and a curvature induced D3 tadpole

$$Q_c = \frac{\chi(Z)}{24} = 852. \tag{7.17}$$

According to the estimate (6.51), this yields a discretuum of about 10^{3000} flux vacua.

Following Section 3.8, we find that the IIB weak coupling limit is a $v \to -v$ orientifold of the Calabi–Yau hypersurface:

$$X : v^2 = h(\vec{u}), \tag{7.18}$$

in a toric variety with fields $(u_1, \ldots, u_5, v)$ and the following charge assignments:

u_1	u_2	u_3	u_4	u_5	v	h
1	1	1	-1	0	2	4
0	0	0	1	1	2	4

Computing the third Chern class in the usual way, we find $\chi(X) = -260$, so (using $h^{1,1} = 2$) $h^{2,1} = 132$. This also determines the number ξ defined in (4.93): $\xi \approx 0.315$.

Thus we conclude that in this model, the large volume scenario can indeed be realized.

Acknowledgements

I would like to thank the organizers of this excellent school, Costas Bachas, Laurent Baulieu, Michael Douglas, Elias Kiritsis, Eliezer Rabinovici, Pierre Vanhove and Paul Windey, for providing me the opportunity to teach on this topic. I am very much indebted to my collaborators, Andres Collinucci, Michael Douglas, Mboyo Esole, Bogdan Florea, Antonella Grassi, Shamit Kachru and Greg Moore, whose insights and work are reflected throughout these lectures. Thanks also to the schools' students and fellow lecturers for questions and discussions and providing such a pleasant environment, and to the company responsible for brewing Desperados. Finally special thanks to Alessandro Tomasiello for discussions related to the contents of these lectures, and to Andres Collinucci, Jonathan Ruel and Erik Plauschinn for helpful comments on the manuscript.

References

[1] L. Susskind, The anthropic landscape of string theory, arXiv:hep-th/0302219.

[2] T. Banks, M. Dine and E. Gorbatov, Is there a string theory landscape? *JHEP* **0408** (2004) 058 [arXiv:hep-th/0309170].

[3] B.R. Greene, String theory on Calabi–Yau manifolds, arXiv:hep-th/9702155.

[4] S. Kachru, R. Kallosh, A. Linde and S.P. Trivedi, De Sitter vacua in string theory, *Phys. Rev. D* **68** (2003) 046005 [arXiv:hep-th/0301240].

[5] V. Balasubramanian, P. Berglund, J.P. Conlon and F. Quevedo, Systematics of moduli stabilisation in Calabi–Yau flux compactifications, *JHEP* **0503** (2005) 007 [arXiv:hep-th/0502058].

[6] L. Susskind, *The cosmic landscape: String theory and the illusion of intelligent design*, New York, USA: Little, Brown (2005) 403 p.

[7] http://en.wikipedia.org/wiki/Rube_Goldberg_machine

[8] E. Witten, *Nucl. Phys. B* **443** (1995) 85 [arXiv:hep-th/9503124].

[9] C. Vafa, Evidence for F-Theory, *Nucl. Phys. B* **469** (1996) 403 [arXiv:hep-th/9602022].

[10] M. Dine and N. Seiberg, Is the superstring weakly coupled? *Phys. Lett. B* **162** (1985) 299.

[11] J. Polchinski and A. Strominger, New vacua for type II string theory, *Phys. Lett. B* **388** (1996) 736 [arXiv:hep-th/9510227].

[12] P.G.O. Freund and M.A. Rubin, Dynamics of dimensional reduction, *Phys. Lett. B* **97** (1980) 233.

[13] B.S. Acharya and M.R. Douglas, A finite landscape? arXiv:hep-th/0606212.

[14] J.M. Maldacena and C. Nunez, Supergravity description of field theories on curved manifolds and a no go theorem, *Int. J. Mod. Phys. A* **16** (2001) 822 [arXiv:hep-th/0007018].

[15] B. de Wit, D.J. Smit and N.D. Hari Dass, Residual supersymmetry of compactified $D = 10$ supergravity, *Nucl. Phys. B* **283** (1987) 165.

[16] D.H. Wesley, New no-go theorems for cosmic acceleration with extra dimensions, arXiv:0802.2106 [hep-th].

[17] A. Tomasiello, New string vacua from twistor spaces, arXiv:0712.1396 [hep-th].

[18] K. Becker and M. Becker, M-theory on eight-manifolds, *Nucl. Phys. B* **477** (1996) 155 [arXiv:hep-th/9605053].

[19] K. Dasgupta, G. Rajesh and S. Sethi, M theory, orientifolds and G-flux, *JHEP* **9908** (1999) 023 [arXiv:hep-th/9908088].

[20] S.B. Giddings, S. Kachru and J. Polchinski, Hierarchies from fluxes in string compactifications, *Phys. Rev. D* **66** (2002) 106006 [arXiv:hep-th/0105097].

[21] R. Donagi and M. Wijnholt, Model building with F-theory, arXiv:0802.2969 [hep-th].

[22] C. Beasley, J.J. Heckman and C. Vafa, GUTs and exceptional branes in F-theory—I, arXiv:0802.3391 [hep-th].

[23] O. DeWolfe, A. Giryavets, S. Kachru and W. Taylor, Type IIA moduli stabilization, *JHEP* **0507** (2005) 066 [arXiv:hep-th/0505160].

[24] B.S. Acharya, F. Benini and R. Valandro, Fixing moduli in exact type IIA flux vacua, *JHEP* **0702** (2007) 018 [arXiv:hep-th/0607223].

[25] M. Grana, Flux compactifications in string theory: A comprehensive review, *Phys. Rept.* **423** (2006) 91 [arXiv:hep-th/0509003].

[26] M.R. Douglas and S. Kachru, Flux compactification, *Rev. Mod. Phys.* **79** (2007) 733 [arXiv:hep-th/0610102].

[27] S. Gukov, C. Vafa and E. Witten, CFT's from Calabi–Yau four-folds, *Nucl. Phys. B* **584** (2000) 69 [Erratum-ibid. *B* **608** (2001) 477] [arXiv:hep-th/9906070].

[28] F. Denef and M.R. Douglas, Distributions of flux vacua, *JHEP* **0405** (2004) 072 [arXiv:hep-th/0404116].

[29] I.R. Klebanov and M.J. Strassler, Supergravity and a confining gauge theory: Duality cascades and chiSB-resolution of naked singularities, *JHEP* **0008** (2000) 052 [arXiv:hep-th/0007191].

[30] L. Randall and R. Sundrum, A large mass hierarchy from a small extra dimension, *Phys. Rev. Lett.* **83** (1999) 3370 [arXiv:hep-ph/9905221].

[31] S. Kachru, R. Kallosh, A. Linde, J.M. Maldacena, L.P. McAllister and S.P. Trivedi, Towards inflation in string theory, *JCAP* **0310** (2003) 013 [arXiv:hep-th/0308055].

[32] D. Baumann, A. Dymarsky, I.R. Klebanov, L. McAllister and P.J. Steinhardt, A delicate universe, *Phys. Rev. Lett.* **99** (2007) 141601 [arXiv:0705.3837 [hep-th]].

[33] D. Baumann, A. Dymarsky, I.R. Klebanov and L. McAllister, Towards an explicit model of D-brane inflation, arXiv:0706.0360 [hep-th].

[34] R. Blumenhagen, B. Kors, D. Lust and S. Stieberger, Four-dimensional string compactifications with D-branes, orientifolds and fluxes, *Phys. Rept.* **445** (2007) 1 [arXiv:hep-th/0610327].

[35] A. Sen, F-theory and orientifolds, *Nucl. Phys. B* **475** (1996) 562 [arXiv:hep-th/9605150].

[36] N. Seiberg and E. Witten, Gauge dynamics and compactification to three dimensions, arXiv:hep-th/9607163.

[37] T. Banks and K. van den Broek, Massive IIA flux compactifications and U-dualities, *JHEP* **0703** (2007) 068 [arXiv:hep-th/0611185].

[38] M.P. Hertzberg, S. Kachru, W. Taylor and M. Tegmark, Inflationary constraints on type IIA string theory, *JHEP* **0712** (2007) 095 [arXiv:0711.2512 [hep-th]].

[39] E. Silverstein, Simple de sitter solutions, arXiv:0712.1196 [hep-th].

[40] C.M. Hull, Massive string theories from M-theory and F-theory, *JHEP* **9811** (1998) 027 [arXiv:hep-th/9811021].

[41] B.S. Acharya, A moduli fixing mechanism in M theory, arXiv:hep-th/0212294.

[42] G.W. Gibbons, S.A. Hartnoll and C.N. Pope, Bohm and Einstein–Sasaki metrics, black holes and cosmological event horizons, *Phys. Rev. D* **67** (2003) 084024 [arXiv:hep-th/0208031].

[43] C.P. Boyer and K. Galicki, Sasakian geometry, hypersurface singularities, and Einstein metrics, arXiv:math/0405256.

[44] D. Martelli, J. Sparks and S.T. Yau, Sasaki–Einstein manifolds and volume minimisation, arXiv:hep-th/0603021.

[45] S. Kachru and A.K. Kashani-Poor, Moduli potentials in type IIA compactifications with RR and NS flux, *JHEP* **0503** (2005) 066 [arXiv:hep-th/0411279].

[46] P. Horava and E. Witten, Eleven-dimensional supergravity on a manifold with boundary, *Nucl. Phys. B* **475** (1996) 94 [arXiv:hep-th/9603142].

[47] V. Braun and B.A. Ovrut, Stabilizing moduli with a positive cosmological constant in heterotic M-theory, *JHEP* **0607** (2006) 035 [arXiv:hep-th/0603088].

[48] J. Shelton, W. Taylor and B. Wecht, Nongeometric flux compactifications, *JHEP* **0510** (2005) 085 [arXiv:hep-th/0508133].

[49] K. Becker, M. Becker, C. Vafa and J. Walcher, Moduli stabilization in non-geometric backgrounds, *Nucl. Phys. B* **770** (2007) 1 [arXiv:hep-th/0611001].

[50] S. Hellerman and I. Swanson, Charting the landscape of supercritical string theory, *Phys. Rev. Lett.* **99** (2007) 171601 [arXiv:0705.0980 [hep-th]].

[51] M. Henneaux and C. Teitelboim, Dynamics of chiral (selfdual) P forms, *Phys. Lett. B* **206** (1988) 650.

[52] D. Belov and G.W. Moore, Holographic action for the self-dual field, arXiv:hep-th/0605038.

[53] M.J. Duff, J.T. Liu and R. Minasian, Eleven-dimensional origin of string/string duality: A one-loop test, *Nucl. Phys. B* **452** (1995) 261 [arXiv:hep-th/9506126].

[54] A. Klemm, B. Lian, S.S. Roan and S.T. Yau, Calabi–Yau fourfolds for M- and F-theory compactifications, *Nucl. Phys. B* **518** (1998) 515 [arXiv:hep-th/9701023].

[55] E. Witten, On flux quantization in M-theory and the effective action, *J. Geom. Phys.* **22** (1997) 1 [arXiv:hep-th/9609122].

[56] A. Sen, Orientifold limit of F-theory vacua, *Phys. Rev. D* **55** (1997) 7345 [arXiv:hep-th/9702165].

[57] B.R. Greene, A.D. Shapere, C. Vafa and S.T. Yau, Stringy cosmic strings and noncompact Calabi–Yau manifolds, *Nucl. Phys. B* **337** (1990) 1.

[58] A.P. Braun, A. Hebecker and H. Triendl, D7-brane motion from M-theory cycles and obstructions in the weak coupling limit, arXiv:0801.2163 [hep-th].

[59] A. Collinucci, F. Denef and M. Esole, D-brane deconstructions in type IIB orientifolds, to appear.

[60] P. Aluffi and M. Esole, Chern class identities from tadpole matching in type IIB and F-theory, arXiv:0710.2544 [hep-th].

[61] M. Haack and J. Louis, M-theory compactified on Calabi–Yau fourfolds with background flux, *Phys. Lett. B* **507** (2001) 296 [arXiv:hep-th/0103068].

[62] M. Marino, R. Minasian, G.W. Moore and A. Strominger, Nonlinear instantons from supersymmetric p-branes, *JHEP* **0001** (2000) 005 [arXiv:hep-th/9911206].

[63] L. Martucci, D-branes on general $N = 1$ backgrounds: Superpotentials and D-terms, *JHEP* **0606** (2006) 033 [arXiv:hep-th/0602129].

[64] O. DeWolfe, A. Giryavets, S. Kachru and W. Taylor, Enumerating flux vacua with enhanced symmetries, *JHEP* **0502** (2005) 037 [arXiv:hep-th/0411061].

[65] S. Kachru, M.B. Schulz and S. Trivedi, Moduli stabilization from fluxes in a simple IIB orientifold, *JHEP* **0310** (2003) 007 [arXiv:hep-th/0201028].

[66] A. Giryavets, S. Kachru, P.K. Tripathy and S.P. Trivedi, Flux compactifications on Calabi–Yau threefolds, *JHEP* **0404** (2004) 003 [arXiv:hep-th/0312104].

[67] D.R. Morrison and C. Vafa, Compactifications of F-Theory on Calabi–Yau Threefolds—II, *Nucl. Phys. B* **476** (1996) 437 [arXiv:hep-th/9603161].

[68] P. Candelas, E. Perevalov and G. Rajesh, Toric geometry and enhanced gauge symmetry of F-theory/heterotic vacua, *Nucl. Phys. B* **507** (1997) 445 [arXiv:hep-th/9704097].

[69] D.R. Morrison and C. Vafa, Compactifications of F-theory on Calabi–Yau threefolds—I, *Nucl. Phys. B* **473** (1996) 74 [arXiv:hep-th/9602114].

[70] M. Bershadsky, K.A. Intriligator, S. Kachru, D.R. Morrison, V. Sadov and C. Vafa, Geometric singularities and enhanced gauge symmetries, *Nucl. Phys. B* **481** (1996) 215 [arXiv:hep-th/9605200].

[71] I. Brunner and K. Hori, Orientifolds and mirror symmetry, *JHEP* **0411** (2004) 005 [arXiv:hep-th/0303135].

[72] O. DeWolfe and S.B. Giddings, Scales and hierarchies in warped compactifications and brane worlds, *Phys. Rev. D* **67** (2003) 066008 [arXiv:hep-th/0208123].

[73] M. Grana, T.W. Grimm, H. Jockers and J. Louis, Soft supersymmetry breaking in Calabi–Yau orientifolds with D-branes and fluxes, *Nucl. Phys. B* **690** (2004) 21 [arXiv:hep-th/0312232].

[74] T.W. Grimm and J. Louis, The effective action of $N = 1$ Calabi–Yau orientifolds, *Nucl. Phys. B* **699** (2004) 387 [arXiv:hep-th/0403067].

[75] S.B. Giddings and A. Maharana, Dynamics of warped compactifications and the shape of the warped landscape, *Phys. Rev. D* **73** (2006) 126003 [arXiv:hep-th/0507158].

[76] D. Baumann, A. Dymarsky, I.R. Klebanov, J.M. Maldacena, L.P. McAllister and A. Murugan, On D3-brane potentials in compactifications with fluxes and wrapped D-branes, *JHEP* **0611** (2006) 031 [arXiv:hep-th/0607050].

[77] W. Lerche, P. Mayr and N. Warner, $N = 1$ special geometry, mixed Hodge variations and toric geometry, arXiv:hep-th/0208039.

[78] M.R. Douglas, J. Shelton and G. Torroba, Warping and supersymmetry breaking, arXiv:0704.4001 [hep-th].

[79] G. Torroba et al., to appear.

[80] M. Dine, N. Seiberg, X.G. Wen and E. Witten, Nonperturbative effects on the string world sheet, *Nucl. Phys. B* **278** (1986) 769.

[81] M. Dine, N. Seiberg, X.G. Wen and E. Witten, Nonperturbative effects on the string world sheet. 2, *Nucl. Phys. B* **289** (1987) 319.

[82] K. Becker, M. Becker and A. Strominger, Five-branes, membranes and nonperturbative string theory, *Nucl. Phys. B* **456** (1995) 130 [arXiv:hep-th/9507158].

[83] E. Witten, Non-perturbative superpotentials in string theory, *Nucl. Phys. B* **474** (1996) 343 [arXiv:hep-th/9604030].

[84] J.A. Harvey and G.W. Moore, Superpotentials and membrane instantons, arXiv:hep-th/9907026.

[85] N. Dorey, T.J. Hollowood, V.V. Khoze and M.P. Mattis, The calculus of many instantons, *Phys. Rept.* **371** (2002) 231 [arXiv:hep-th/0206063].

[86] S. Vandoren and P. van Nieuwenhuizen, Lectures on instantons, arXiv:0802.1862 [hep-th].

[87] R. Blumenhagen, S. Moster and E. Plauschinn, Moduli stabilisation versus chirality for MSSM like type IIB orientifolds, *JHEP* **0801** (2008) 058 [arXiv:0711.3389 [hep-th]].

[88] G. Veneziano and S. Yankielowicz, An effective Lagrangian for the pure $N = 1$ supersymmetric Yang–Mills theory, *Phys. Lett. B* **113** (1982) 231.

[89] T.R. Taylor, G. Veneziano and S. Yankielowicz, Supersymmetric QCD and its massless limit: An effective Lagrangian analysis, *Nucl. Phys. B* **218** (1983) 493.

[90] S.H. Katz and C. Vafa, Geometric engineering of $N = 1$ quantum field theories, *Nucl. Phys. B* **497** (1997) 196 [arXiv:hep-th/9611090].

[91] R. Kallosh, A.K. Kashani-Poor and A. Tomasiello, Counting fermionic zero modes on M5 with fluxes, *JHEP* **0506** (2005) 069 [arXiv:hep-th/0503138].

[92] L. Gorlich, S. Kachru, P.K. Tripathy and S.P. Trivedi, Gaugino condensation and nonperturbative superpotentials in flux compactifications, *JHEP* **0412** (2004) 074 [arXiv:hep-th/0407130].

[93] N. Saulina, Topological constraints on stabilized flux vacua, *Nucl. Phys. B* **720** (2005) 203 [arXiv:hep-th/0503125].

[94] D. Tsimpis, Fivebrane instantons and Calabi–Yau fourfolds with flux, *JHEP* **0703** (2007) 099 [arXiv:hep-th/0701287].

[95] D. Lust, S. Reffert, W. Schulgin and P.K. Tripathy, Fermion zero modes in the presence of fluxes and a non-perturbative superpotential, *JHEP* **0608** (2006) 071 [arXiv:hep-th/0509082].

[96] R. Blumenhagen, M. Cvetic, R. Richter and T. Weigand, Lifting D-instanton zero modes by recombination and background fluxes, *JHEP* **0710** (2007) 098 [arXiv:0708.0403 [hep-th]].

[97] K. Becker, M. Becker, M. Haack and J. Louis, Supersymmetry breaking and alpha'-corrections to flux induced potentials, *JHEP* **0206** (2002) 060 [arXiv:hep-th/0204254].

[98] M. Berg, M. Haack and E. Pajer, Jumping through loops: On soft terms from large volume compactifications, *JHEP* **0709** (2007) 031 [arXiv:0704.0737 [hep-th]].

[99] S. Ashok and M.R. Douglas, Counting flux vacua, *JHEP* **0401** (2004) 060 [arXiv:hep-th/0307049].

[100] E. Silverstein, AdS and dS entropy from string junctions or the function of junction conjunctions, arXiv:hep-th/0308175.

[101] S. Weinberg, Anthropic bound on the cosmological constant, *Phys. Rev. Lett.* **59** (1987) 2607.

[102] R. Bousso and J. Polchinski, Quantization of four-form fluxes and dynamical neutralization of the cosmological constant, *JHEP* **0006** (2000) 006 [arXiv:hep-th/0004134].

[103] J.P. Conlon, F. Quevedo and K. Suruliz, Large-volume flux compactifications: Moduli spectrum and D3/D7 soft supersymmetry breaking, *JHEP* **0508** (2005) 007 [arXiv:hep-th/0505076].

[104] J.P. Conlon, Moduli stabilisation and applications in IIB string theory, *Fortsch. Phys.* **55** (2007) 287 [arXiv:hep-th/0611039].

[105] E. Witten, Phases of $N = 2$ theories in two dimensions, *Nucl. Phys.* B **403** (1993) 159 [arXiv:hep-th/9301042].

[106] M. Kreuzer, Toric geometry and Calabi–Yau compactifications, [arXiv:hep-th/0612307].

[107] K. Hori et al., *Mirror symmetry*, Providence, USA: AMS (2003) 929 p.

[108] P. Berglund, P. Candelas, X. De La Ossa, A. Font, T. Hubsch, D. Jancic and F. Quevedo, Periods for Calabi–Yau and Landau–Ginzburg vacua, *Nucl. Phys.* B **419** (1994) 352 [arXiv:hep-th/9308005].

[109] M.R. Douglas, The statistics of string/M theory vacua, *JHEP* **0305** (2003) 046 [arXiv:hep-th/0303194].

[110] F. Denef and M.R. Douglas, Distributions of nonsupersymmetric flux vacua, *JHEP* **0503** (2005) 061 [arXiv:hep-th/0411183].

[111] F. Denef, M.R. Douglas and S. Kachru, Physics of string flux compactifications, *Ann. Rev. Nucl. Part. Sci.* **57** (2007) 119 [arXiv:hep-th/0701050].

[112] F. Denef and M.R. Douglas, Computational complexity of the landscape. I, arXiv:hep-th/0602072.

[113] T. Eguchi and Y. Tachikawa, Distribution of flux vacua around singular points in Calabi–Yau moduli space, *JHEP* **0601** (2006) 100 [arXiv:hep-th/0510061].

[114] G. Torroba, Finiteness of flux vacua from geometric transitions, *JHEP* **0702** (2007) 061 [arXiv:hep-th/0611002].

[115] A. Hebecker and J. March-Russell, The ubiquitous throat, *Nucl. Phys.* B **781** (2007) 99 [arXiv:hep-th/0607120].

[116] M. Dine, D. O'Neil and Z. Sun, Branches of the landscape, *JHEP* **0507** (2005) 014 [arXiv:hep-th/0501214].

[117] J.D. Brown and C. Teitelboim, Dynamical neutralization of the cosmological constant, *Phys. Lett.* B **195** (1987) 177.

[118] J.D. Brown and C. Teitelboim, Neutralization of the cosmological constant by membrane creation, *Nucl. Phys.* B **297** (1988) 787.

[119] J.L. Feng, J. March-Russell, S. Sethi and F. Wilczek, Saltatory relaxation of the cosmological constant, *Nucl. Phys.* B **602** (2001) 307 [arXiv:hep-th/0005276].

F. Denef

[120] S.R. Coleman and F. De Luccia, Gravitational effects on and of vacuum decay, *Phys. Rev. D* **21** (1980) 3305.

[121] A.H. Guth, Inflation and eternal inflation, *Phys. Rept.* **333** (2000) 555 [arXiv:astro-ph/0002156].

[122] A. Grassi, Divisors on elliptic Calabi–Yau 4-folds and the superpotential in F-theory—I, *Journal of Geometry and Physics* **28** (1998) 289 [arXiv:alg-geom/9704008].

[123] F. Denef, M.R. Douglas and B. Florea, Building a better racetrack, *JHEP* **0406** (2004) 034 [arXiv:hep-th/0404257].

[124] F. Denef, M.R. Douglas, B. Florea, A. Grassi and S. Kachru, Fixing all moduli in a simple F-theory compactification, *Adv. Theor. Math. Phys.* **9** (2005) 861 [arXiv:hep-th/0503124].

[125] D. Lust, S. Reffert, W. Schulgin and S. Stieberger, Moduli stabilization in type IIB orientifolds. I: Orbifold limits, *Nucl. Phys. B* **766** (2007) 68 [arXiv:hep-th/0506090].

[126] D. Lust, S. Reffert, E. Scheidegger, W. Schulgin and S. Stieberger, Moduli stabilization in type IIB orientifolds. II, *Nucl. Phys. B* **766** (2007) 178 [arXiv:hep-th/0609013].

Course 13

GAUGE–STRING DUALITIES AND SOME APPLICATIONS

Marcus K. Benna and Igor R. Klebanov

Joseph Henry Laboratories and Princeton Center for Theoretical Physics,
Princeton University, Princeton, NJ 08544, USA

C. Bachas, L. Baulieu, M. Douglas, E. Kiritsis, E. Rabinovici, P. Vanhove, P. Windey
and L.F. Cugliandolo, eds.
Les Houches, Session LXXXVII, 2007
String Theory and the Real World: From Particle Physics to Astrophysics
© *2008 Published by Elsevier B.V.*

Contents

1. Introduction

String theory is well known to be the leading prospect for quantizing gravity and unifying it with other interactions. One may also take a broader view of string theory as a description of string-like excitations that arise in many different physical systems, such as the superconducting flux tubes or the chromo-electric flux tubes in non-Abelian gauge theories. From the point of view of quantum field theories describing the physical systems where these string-like objects arise, they are "emergent" rather than fundamental. However, thanks to the AdS/CFT correspondence [1–3] and its extensions, we now know that at least some field theories have dual formulations in terms of string theories in curved backgrounds. In these examples, the strings that are "emergent" from the field theory point of view are dual to fundamental or D-strings in the string theoretic approach. Besides being of great theoretical interest, such dualities are becoming a useful tool for studying strongly coupled gauge theories. These ideas also have far-reaching implications for building connections between string theory and the real world.

These notes, based on five lectures delivered by I.R.K. at Les Houches in July 2007, are not meant to be a comprehensive review of what has become a vast field. Instead, they aim to present a particular path through it, which begins with old and well-known concepts, and eventually leads to some recent developments. The first part of these lectures is based on an earlier brief review [4]. It begins with a bit of history and basic facts about string theory and its connections with strong interactions. Comparisons of stacks of Dirichlet branes with curved backgrounds produced by them are used to motivate the AdS/CFT correspondence between superconformal gauge theory and string theory on a product of Anti-de Sitter space and a compact manifold. The ensuing duality between semi-classical spinning strings and gauge theory operators carrying large charges is briefly reviewed. We go on to describe recent tests of the AdS/CFT correspondence using the Wilson loop cusp anomaly as a function of the coupling, which also enters dimensions of high-spin operators. Strongly coupled thermal SYM theory is explored via a black hole in 5-dimensional AdS space, which leads to explicit results for its entropy and shear viscosity.

The second part of these lectures (Sections 7–11) focuses on the gauge-string dualities that arise from studying D-branes on the conifold geometry. The $AdS_5 \times$

$T^{1,1}$ background appears as the type IIB dual of an $SU(N) \times SU(N)$ supercon-formal gauge theory with bi-fundamental fields. A warped resolved conifold is then presented as a description of holographic RG flow from this theory to the $\mathcal{N} = 4$ supersymmetric gauge theory produced by giving a classical value to one of the bi-fundamentals. The warped deformed conifold is instead the dual of the cascading $SU(kM) \times SU((k+1)M)$ gauge theory which confines in the infrared. Some features of the bound state spectrum are discussed, and the su-pergravity dual of the baryonic branch, the resolved warped deformed conifold, is reviewed. We end with a brief update on possible cosmological applications of these backgrounds.

2. Strings and QCD

String theory was born out of attempts to understand the strong interactions. Empirical evidence for a string-like structure of hadrons comes from arranging mesons and baryons into approximately linear Regge trajectories. Studies of πN scattering prompted Dolen, Horn and Schmid [5] to make a duality conjec-ture stating that the sum over s-channel exchanges equals the sum over t-channel ones. This posed the problem of finding the analytic form of such dual ampli-tudes. Veneziano [6] found the first, and very simple, expression for a manifestly dual 4-point amplitude:

$$A(s, t) \sim \frac{\Gamma(-\alpha(s))\Gamma(-\alpha(t))}{\Gamma(-\alpha(s) - \alpha(t))} \tag{2.1}$$

with an exactly linear Regge trajectory $\alpha(s) = \alpha(0) + \alpha' s$. Soon after, its open string interpretation was proposed [7–9] (for detailed reviews of these early de-velopments, see [10]). In the early 70's this led to an explosion of interest in string theory as a description of strongly interacting particles. The basic idea is to think of a meson as an open string with a quark at one end-point and an anti-quark at the other. Then various meson states arise as rotational and vibrational excitations of such an open string. The splitting of such a string describes the decay of a meson into two mesons.

The string world sheet dynamics is governed by the Nambu–Goto area action

$$S_{\mathrm{NG}} = -T \int d\sigma d\tau \sqrt{-\det \partial_a X^\mu \partial_b X_\mu}, \tag{2.2}$$

where the indices a, b take two values ranging over the σ and τ directions on the world sheet. The string tension is related to the Regge slope through $T^{-1} = 2\pi\alpha'$. The quantum consistency of the Veneziano model requires that the Regge intercept is $\alpha(0) = 1$, so that the spin 1 state is massless but the spin 0 is a tachyon. But the ρ meson is certainly not massless, and the presence of a tachyon in the spectrum indicates an instability. This is how the string theory of strong interactions started to run into problems.

Calculation of the string zero-point energy gives $\alpha(0) = (d - 2)/24$. Hence the model has to be defined in 26 space-time dimensions. Consistent supersymmetric string theories were discovered in 10 dimensions, but their relation to strong interactions in 4 dimensions was initially completely unclear. Most importantly, the Asymptotic Freedom of strong interactions was discovered [11], singling out Quantum Chromodynamics (QCD) as the exact field theory of strong interactions. At this point most physicists gave up on strings as a description of strong interactions. Instead, since the graviton appears naturally in the closed string spectrum, string theory emerged as the leading hope for unifying quantum gravity with other forces [12, 13].

Now that we know that a non-Abelian gauge theory is an exact description of strong interactions, is there any room left for string theory in this field? Luckily, the answer is positive. At short distances, much smaller than 1 fermi, the quark anti-quark potential is approximately Coulombic due to the Asymptotic Freedom. At large distances the potential should be linear due to formation of a confining flux tube [14]. When these tubes are much longer than their thickness, one can hope to describe them, at least approximately, by semi-classical Nambu strings [15]. This appears to explain the existence of approximately linear Regge trajectories. For the leading trajectory, a linear relation between angular momentum and mass-squared

$$J = \alpha' m^2 + \alpha(0), \tag{2.3}$$

is provided by a semi-classical spinning relativistic string with massless quark and anti-quark at its endpoints. In case of baryons, one finds a di-quark instead of an anti-quark at one of the endpoints. A semi-classical string approach to the QCD flux tubes is widely used, for example, in jet hadronization algorithms based on the Lund String Model [16].

Semi-classical quantization around a long straight Nambu string predicts the quark anti-quark potential [17]

$$V(r) = Tr + \mu + \frac{\gamma}{r} + O(1/r^2). \tag{2.4}$$

The coefficient γ of the universal Lüscher term depends only on the space-time dimension d and is proportional to the Regge intercept: $\gamma = -\pi(d - 2)/24$. Lattice calculations of the force vs. distance for probe quarks and anti-quarks [18] produce good agreement with this value in $d = 3$ and $d = 4$ for $r > 0.7$ fm. Thus, long QCD strings appear to be well described by the Nambu–Goto area

action. But quantization of short, highly quantum QCD strings, that could lead
to a calculation of light meson and glueball spectra, is a much harder problem.

The connection of gauge theory with string theory is strengthened by 't Hooft's
generalization of QCD from 3 colors ($SU(3)$ gauge group) to N colors ($SU(N)$
gauge group) [19]. The idea is to make N large, while keeping the 't Hooft
coupling $\lambda = g_{YM}^2 N$ fixed. In this limit each Feynman graph carries a topological factor N^{χ}, where χ is the Euler characteristic of the graph. Thus, the sum
over graphs of a given topology can perhaps be thought of as a sum over world
sheets of a hypothetical "QCD string." Since the spheres (string tree diagrams)
are weighted by N^2, the tori (string one-loop diagrams) by N^0 etc., we find that
the closed string coupling constant is of order N^{-1}. Thus, the advantage of taking N to be large is that we find a weakly coupled string theory. In the large N
limit the gauge theory simplifies in that only the planar diagrams contribute. But
directly summing even this subclass of diagrams seems to be an impossible task.
From the dual QCD string point of view, it is not clear how to describe this string
theory in elementary terms.

Because of the difficulty of these problems, between the late 70's and the mid-
90's many theorists gave up hope of finding an exact gauge-string duality. One
notable exception is Polyakov who in 1981 proposed that the string theory dual to
a 4-d gauge theory should have a 5-th hidden dimension [20]. In later work [21]
he refined this proposal, suggesting that the 5-d metric must be "warped."

3. The geometry of Dirichlet branes

In the mid-90's Dirichlet branes, or D-branes for short, brought string theory
back to gauge theory. D-branes are soliton-like "membranes" of various internal
dimensionalities contained in theories of closed superstrings [22]. A Dirichlet
p-brane (or Dp-brane) is a $p + 1$ dimensional hyperplane in $9 + 1$ dimensional
space-time where strings are allowed to end. A D-brane is much like a topological defect: upon touching a D-brane, a closed string can open up and turn
into an open string whose ends are free to move along the D-brane. For the endpoints of such a string the $p + 1$ longitudinal coordinates satisfy the conventional
free (Neumann) boundary conditions, while the $9 - p$ coordinates transverse to
the Dp-brane have the fixed (Dirichlet) boundary conditions; hence the origin
of the term "Dirichlet brane." In a seminal paper [22] Polchinski showed that a
Dp-brane preserves 1/2 of the bulk supersymmetries and carries an elementary
unit of charge with respect to the $p + 1$ form gauge potential from the Ramond-
Ramond sector of type II superstring.

For our purposes, the most important property of D-branes is that they realize
gauge theories on their world volume. The massless spectrum of open strings liv-

ing on a Dp-brane is that of a maximally supersymmetric $U(1)$ gauge theory in $p + 1$ dimensions. The $9 - p$ massless scalar fields present in this supermultiplet are the expected Goldstone modes associated with the transverse oscillations of the Dp-brane, while the photons and fermions provide the unique supersymmetric completion. If we consider N parallel D-branes, then there are N^2 different species of open strings because they can begin and end on any of the D-branes. N^2 is the dimension of the adjoint representation of $U(N)$, and indeed we find the maximally supersymmetric $U(N)$ gauge theory in this setting.

The relative separations of the Dp-branes in the $9 - p$ transverse dimensions are determined by the expectation values of the scalar fields. We will be interested in the case where all scalar expectation values vanish, so that the N Dp-branes are stacked on top of each other. If N is large, then this stack is a heavy object embedded into a theory of closed strings which contains gravity. Naturally, this macroscopic object will curve space: it may be described by some classical metric and other background fields. Thus, we have two very different descriptions of the stack of Dp-branes: one in terms of the $U(N)$ supersymmetric gauge theory on its world volume, and the other in terms of the classical charged p-brane background of the type II closed superstring theory. The relation between these two descriptions is at the heart of the connections between gauge fields and strings that are the subject of these lectures.

Parallel D3-branes realize a $3 + 1$ dimensional $U(N)$ gauge theory, which is a maximally supersymmetric "cousin" of QCD. Let us compare a stack of coincident D3-branes with the Ramond-Ramond charged black 3-brane classical solution whose metric assumes the form [23]:

$$ds^2 = h^{-1/2}(r) \left[-f(r)(dx^0)^2 + \sum_{i=1}^{3} (dx^i)^2 \right]$$
$$+ h^{1/2}(r) \left[f^{-1}(r)dr^2 + r^2 d\Omega_5^2 \right], \tag{3.1}$$

where

$$h(r) = 1 + \frac{L^4}{r^4}, \qquad f(r) = 1 - \frac{r_0^4}{r^4}.$$

Here $d\Omega_5^2$ is the metric of a unit 5 dimensional sphere, $\mathbf{S}^5$.

In general, a d-dimensional sphere of radius L may be defined by a constraint

$$\sum_{i=1}^{d+1} (X^i)^2 = L^2 \tag{3.2}$$

on $d + 1$ real coordinates X^i. It is a positively curved maximally symmetric space with symmetry group $SO(d + 1)$. Similarly, the d-dimensional Anti-de Sitter space, AdS_d, is defined by a constraint

$$(X^0)^2 + (X^d)^2 - \sum_{i=1}^{d-1}(X^i)^2 = L^2, \tag{3.3}$$

where L is its curvature radius. AdS_d is a negatively curved maximally symmetric space with symmetry group $SO(2, d-1)$. There exists a subspace of AdS_d called the Poincaré wedge, with the metric

$$ds^2 = \frac{L^2}{z^2}\left(dz^2 - (dx^0)^2 + \sum_{i=1}^{d-2}(dx^i)^2\right), \tag{3.4}$$

where $z \in [0, \infty)$. In these coordinates the boundary of AdS_d is at $z = 0$.

The event horizon of the black 3-brane metric (3.1) is located at $r = r_0$. In the extremal limit $r_0 \to 0$ the 3-brane metric becomes

$$ds^2 = h^{-1/2}(r)\eta_{\mu\nu}dx^\mu dx^\nu + h^{1/2}(r)\left(dr^2 + r^2 d\Omega_5^2\right), \tag{3.5}$$

where $\eta_{\mu\nu}$ is the $3 + 1$ dimensional Minkowski metric. Just like the stack of parallel, ground state D3-branes, the extremal solution preserves 16 of the 32 supersymmetries present in the type IIB theory. Introducing $z = L^2/r$, one notes that the limiting form of (3.5) as $r \to 0$ factorizes into the direct product of two smooth spaces, the Poincaré wedge (3.4) of AdS_5, and $\mathbf{S}^5$, with equal radii of curvature L:

$$ds^2 = \frac{L^2}{z^2}\left(dz^2 + \eta_{\mu\nu}dx^\mu dx^\nu\right) + L^2 d\Omega_5^2. \tag{3.6}$$

The 3-brane geometry may thus be viewed as a semi-infinite throat of radius L which for $r \gg L$ opens up into flat $9 + 1$ dimensional space. Thus, for L much larger than the string length scale, $\sqrt{\alpha'}$, the entire 3-brane geometry has small curvatures everywhere and is appropriately described by the supergravity approximation to type IIB string theory.

The relation between L and $\sqrt{\alpha'}$ may be found by equating the gravitational tension of the extremal 3-brane classical solution to N times the tension of a single D3-brane, and one finds

$$L^4 = g_{\text{YM}}^2 N\alpha'^2. \tag{3.7}$$

Thus, the size of the throat in string units is $\lambda^{1/4}$. This remarkable emergence of the 't Hooft coupling from gravitational considerations is at the heart of the success of the AdS/CFT correspondence. Moreover, the requirement $L \gg \sqrt{\alpha'}$ translates into $\lambda \gg 1$: the gravitational approach is valid when the 't Hooft coupling is very strong and the perturbative field theoretic methods are not applicable.

4. The AdS/CFT correspondence

Studies of massless particle absorption by the 3-branes [24] indicate that, in the low-energy limit, the $AdS_5 \times \mathbf{S}^5$ throat region ($r \ll L$) decouples from the asymptotically flat large r region. Similarly, the $\mathcal{N} = 4$ supersymmetric $SU(N)$ gauge theory on the stack of N D3-branes decouples in the low-energy limit from the bulk closed string theory. Such considerations prompted Maldacena [1] to make the seminal conjecture that type IIB string theory on $AdS_5 \times \mathbf{S}^5$, of radius L given in (3.7), is dual to the $\mathcal{N} = 4$ SYM theory. The number of colors in the gauge theory, N, is dual to the number of flux units of the 5-form Ramond-Ramond field strength.

It was further conjectured in [2,3] that there exists a one-to-one map between gauge invariant operators in the CFT and fields (or extended objects) in AdS_5. The dimension Δ of an operator is determined by the mass of the dual field φ in AdS_5. For example, for scalar operators one finds that $\Delta(\Delta - 4) = m^2 L^2$. For the fields in AdS_5 that come from the type IIB supergravity modes, including the Kaluza–Klein excitations on the 5-sphere, the masses are of order $1/L$. Hence, it is consistent to assume that their operator dimensions are independent of L, and therefore independent of λ. This is due to the fact that such operators commute with some of the supercharges and are therefore protected by supersymmetry. Perhaps the simplest such operators are the chiral primaries which are traceless symmetric polynomials made out of the six scalar fields Φ^i: $\operatorname{tr} \Phi^{(i_1} \dots \Phi^{i_k)}$. These operators are dual to spherical harmonics on $\mathbf{S}^5$ which mix the graviton and RR 4-form fluctuations. Their masses are $m_k^2 = k(k - 4)/L^2$, where $k = 2, 3, \dots$. These masses reproduce the operator dimensions $\Delta = k$ which are the same as in the free theory. The situation is completely different for operators dual to the massive string modes: $m_n^2 = 4n/\alpha'$. In this case the AdS/CFT correspondence predicts that the operator dimension grows at strong coupling as $2n^{1/2}\lambda^{1/4}$.

Precise methods for calculating correlation functions of various operators in a CFT using its dual formulation were formulated in [2,3] where a gauge theory quantity, W, was identified with a string theory quantity, Z_{string}:

$$W[\varphi_0(\vec{x})] = Z_{\text{string}}[\varphi_0(\vec{x})]. \tag{4.1}$$

W generates the connected Euclidean Green's functions of a gauge theory operator $\mathcal{O}$,

$$W[\varphi_0(\vec{x})] = \left\langle \exp \int d^4 x \varphi_0 \mathcal{O} \right\rangle. \tag{4.2}$$

Z_{string} is the string theory path integral calculated as a functional of φ_0, the boundary condition on the field φ related to $\mathcal{O}$ by the AdS/CFT duality. In the large N limit the string theory becomes classical, which implies

$$Z_{\text{string}} \sim e^{-I[\phi_0(\vec{x})]}, \tag{4.3}$$

where $I[\phi_0(\vec{x})]$ is the extremum of the classical string action calculated as a functional of ϕ_0. If we are further interested in correlation functions at very large 't Hooft coupling, then the problem of extremizing the classical string action reduces to solving the equations of motion in type IIB supergravity whose form is known explicitly.

If the number of colors N is sent to infinity while $g_{\text{YM}}^2 N$ is held fixed and large, then there are small string scale corrections to the supergravity limit [1–3] which proceed in powers of $\alpha'/L^2 = \lambda^{-1/2}$. If we wish to study finite N, then there are also string loop corrections in powers of $\kappa^2/L^8 \sim N^{-2}$. As expected, taking N to infinity enables us to take the classical limit of the string theory on $AdS_5 \times \mathbf{S}^5$.

Immediate support for the AdS/CFT correspondence comes from symmetry considerations [1]. The isometry group of AdS_5 is $SO(2, 4)$, and this is also the conformal group in $3 + 1$ dimensions. In addition we have the isometries of $\mathbf{S}^5$ which form $SU(4) \sim SO(6)$. This group is identical to the R-symmetry of the $\mathcal{N} = 4$ SYM theory. After including the fermionic generators required by supersymmetry, the full isometry supergroup of the $AdS_5 \times \mathbf{S}^5$ background is $PSU(2, 2|4)$, which is identical to the $\mathcal{N} = 4$ superconformal symmetry.

The fact that after the compactification on $\mathbf{S}^5$ the string theory is 5-dimensional supports earlier ideas on the necessity of the 5-th dimension to describe 4-d gauge theories [20]. The z-direction is dual to the energy scale of the gauge theory: small z corresponds to the UV domain of the gauge theory, while large z to the IR.

In the AdS/CFT duality, type IIB strings are dual to the chromo-electric flux lines in the gauge theory, providing a string theoretic set-up for calculating the quark anti-quark potential [25]. The quark and anti-quark are placed near the boundary of Anti-de Sitter space ($z = 0$), and the fundamental string connecting them is required to obey the equations of motion following from the Nambu action. The string bends into the interior ($z > 0$), and the maximum value of the z-coordinate increases with the separation r between quarks. An explicit calculation of the string action gives an attractive $q\bar{q}$ potential [25]:

$$V(r) = -\frac{4\pi^2\sqrt{\lambda}}{\Gamma\left(\frac{1}{4}\right)^4 r}. \tag{4.4}$$

Its Coulombic $1/r$ dependence is required by the conformal invariance of the theory. Historically, a dual string description was hoped for mainly in the cases of confining gauge theories, where long confining flux tubes have string-like properties. In a pleasant surprise, we now see that a string description applies to non-confining theories too, due to the presence of extra dimensions with a warped metric.

5. Semiclassical spinning strings vs. highly charged operators

A few years ago it was noted that the AdS/CFT duality becomes particularly powerful when applied to operators with large quantum numbers. One class of such single-trace "long operators" are the BMN operators [26] that carry a large R-charge in the SYM theory and contain a finite number of impurity insertions. The R-charge is dual to a string angular momentum on the compact space $\mathbf{S}^5$. So, in the BMN limit the relevant states are short closed strings with a large angular momentum, and a small amount of vibrational excitation [27]. Furthermore, by increasing the number of impurities the string can be turned into a large semiclassical object moving in $AdS_5 \times \mathbf{S}^5$. Comparing such objects with their dual long operators has become a very fruitful area of research [28]. Work in this direction has also produced a great deal of evidence that the $\mathcal{N} = 4$ SYM theory is exactly integrable (see [29] for reviews).

Another familiar example of an operator with a large quantum number is the twist-2 operator carrying a large spin S,

$$\text{Tr } F_{+\mu} D_+^{S-2} F_+^{\mu}. \tag{5.1}$$

In QCD, such operators play an important role in studies of deep inelastic scattering [30]. In general, the anomalous dimension of a twist-2 operator grows logarithmically [31] for large spin S:

$$\Delta - S = f(g)\ln S + O(S^0), \tag{5.2}$$

where $g = \sqrt{g_{YM}^2 N/4\pi}$. This was demonstrated early on at 1-loop order [30] and at two loops [32], where a cancellation of $\ln^3 S$ terms occurs. There are solid arguments that (5.2) holds to all orders in perturbation theory [31, 33]. The function of coupling $f(g)$ also measures the anomalous dimension of a cusp in a light-like Wilson loop [31], and is of definite physical interest in QCD.

There has been significant interest in determining $f(g)$ in the $\mathcal{N} = 4$ SYM theory, in which case we can consider operators in the $SL(2)$ sector, of the form

$$\text{Tr } D^S Z^J + \cdots, \tag{5.3}$$

where Z is one of the complex scalar fields, the R-charge J is the twist, and the dots serve as a reminder that the operator is a linear combination of such terms with the covariant derivatives acting on the scalars in all possible ways. The object dual to such a high-spin twist-2 operator is a folded string [27] spinning around the center of AdS_5; its generalization to large J was found in [34]. The result (5.2) is generally applicable when J is held fixed while S is sent to infinity [35]. While the function $f(g)$ for $\mathcal{N} = 4$ SYM is not the same as the cusp anomalous dimension for QCD, its perturbative expansion is in fact related to that in QCD by the conjectured "transcendentality principle" [36].

The perturbative expansion of $f(g)$ at small g can be obtained from gauge theory, but calculations in the planar $\mathcal{N} = 4$ SYM theory are quite formidable, and until recently were available only up to 3-loop order [36, 37]:

$$f(g) = 8g^2 - \frac{8}{3}\pi^2 g^4 + \frac{88}{45}\pi^4 g^6 + O(g^8). \tag{5.4}$$

This $\mathcal{N} = 4$ answer can be extracted [36] from the corresponding QCD calculation [38] using the transcendentality principle, which states that each expansion coefficient has terms of definite degree of transcendentality (namely the exponent of g minus two), and that the QCD result contains the same terms (in addition to others which have lower degree of transcendentality).

The AdS/CFT correspondence relates the large g behavior of $f(g)$ to the energy of a folded string spinning around the center of a weakly curved AdS_5 space [27]. This gives the prediction that $f(g) \to 4g$ at strong coupling. The same result was obtained from studying the cusp anomaly using string theory methods [39]. Furthermore, the semi-classical expansion for the spinning string energy predicts the following correction [34]:

$$f(g) = 4g - \frac{3\ln 2}{\pi} + O(1/g). \tag{5.5}$$

An interesting problem is to smoothly match these explicit predictions of string theory for large g to those of gauge theory at small g. During the past few years methods of integrability in AdS/CFT [40] (for reviews and more complete references, see [29]) have led to major progress in addressing this question. In an impressive series of papers [41–43] an integral equation that determines $f(g)$

was proposed. This equation was obtained from the asymptotic Bethe ansatz for the $SL(2)$ sector by considering the limit $S \to \infty$ with J finite, and extracting the piece proportional to $\ln S$, which is manifestly independent of J. Taking the spin to infinity, the discrete Bethe equations can be rewritten as an integral equation for the density of Bethe roots in rapidity space (though the resulting equation is most concisely expressed in terms of the variable t that arises from performing a Fourier-transform).

The cusp anomalous dimension $f(g)$ is related to value of the fluctuation density $\hat{\sigma}(t)$ at $t = 0$ [41, 43, 44]:

$$f(g) = 16g^2 \hat{\sigma}(0), \tag{5.6}$$

where $\hat{\sigma}(t)$ is determined by the integral equation

$$\hat{\sigma}(t) = \frac{t}{e^t - 1}\left(K(2gt, 0) - 4g^2 \int_0^\infty dt' K(2gt, 2gt')\hat{\sigma}(t') \right). \tag{5.7}$$

The kernel $K(t, t') = K^{(m)}(t, t') + K^{(d)}(t, t')$ is the sum of the so-called main scattering kernel $K^{(m)}(t, t') = K_0(t, t') + K_1(t, t')$ and the "dressing kernel"

$$K^{(d)}(t, t') = 8g^2 \int_0^\infty dt'' K_1(t, 2gt'') \frac{t''}{e^{t''} - 1} K_0(2gt'', t'). \tag{5.8}$$

Here K_0 and K_1 can be expressed as the following sums of Bessel functions, which are even and odd, respectively, under change of sign of both t and t':

$$K_0(t, t') = \frac{2}{t\, t'} \sum_{n=1}^\infty (2n - 1) J_{2n-1}(t) J_{2n-1}(t'), \tag{5.9}$$

$$K_1(t, t') = \frac{2}{t\, t'} \sum_{n=1}^\infty (2n) J_{2n}(t) J_{2n}(t'). \tag{5.10}$$

Including the dressing kernel [43] is of crucial importance since it takes into account the dressing phase in the asymptotic two-particle world-sheet S-matrix, which is the only function of rapidities appearing in the S-matrix that is not fixed a priori by the $PSU(2, 2|4)$ symmetry of $\mathcal{N} = 4$ SYM theory, and whose general form was deduced in [45, 46]. The perturbative expansion of the phase starts at the 4-loop order, and at strong coupling coincides with the earlier results from string theory [42, 45, 47–49]. The important requirement of crossing symmetry [50] is satisfied by this phase, and it also obeys the transcendentality principle of [36]. Thus, there is strong evidence that this phase describes the exact magnon S-matrix at any coupling, which constitutes important progress in the understanding of the $\mathcal{N} = 4$ SYM theory, and the AdS/CFT correspondence.

An immediate check of the validity of the integral equation, which gives the expansion

$$f(g) = 8g^2 - \frac{8}{3}\pi^2 g^4 + \frac{88}{45}\pi^4 g^6 - 16\left(\frac{73}{630}\pi^6 + 4\zeta(3)^2\right)g^8 + \mathcal{O}(g^{10}), \quad (5.11)$$

was provided by the gauge theory calculation of the 4-loop, $\mathcal{O}(g^8)$ term. In a remarkable paper [51], which independently arrived at the same conjecture for the all-order expansion of $f(g)$ as [43], the 4-loop coefficient was calculated numerically by on-shell unitarity methods in SYM theory, and was used to conjecture the analytic result of (5.11). Subsequently, the numerical precision was improved to produce agreement with the analytic value within 0.001 percent accuracy [52].

In fact the perturbative expansion of $f(g)$ can be obtained from the integral equation (5.7) to arbitrary order. Although the expansion has a finite radius of convergence, as is customary in planar theories, it is expected to determine the function completely. But solving (5.7) for $\hat{\sigma}(t)$ at values of the coupling constant g beyond the perturbative regime is not an easy task. By expanding the fluctuation density as a Neumann series of Bessel functions,

$$\hat{\sigma}(t) = \frac{t}{e^t - 1}\sum_{n\geq 1} s_n \frac{J_n(2gt)}{2gt}, \quad (5.12)$$

one can reduce the problem to an (infinite-dimensional) matrix problem, which can be consistently truncated and is thus amenable to numerical solution [53]. This was shown to give $f(g)$ with high accuracy up to rather large values of g, and the function was found to be monotonically increasing, smooth, and in excellent agreement with the linear asymptotics predicted by string theory (5.5). The cross-over from perturbative to the linear behavior takes place at around $g \sim 1/2$, which is comparable to the radius of convergence $|g| = 1/4$.

Subsequently, the leading term in the strong coupling asymptotic expansion of the fluctuation density $\hat{\sigma}(t)$ was derived analytically [54, 55] from the integral equation. More recently, the complete asymptotic expansion of $f(g)$ was determined in an impressive paper [56] (for further work, see [57]). This expansion obeys its own transcendentality principle; in particular, the coefficient of the $1/g$ term in (5.5) is

$$-\frac{K}{4\pi^2} \approx -0.0232, \quad (5.13)$$

in agreement with the numerical work of [53] (K is the Catalan constant). As a further check, this coefficient was reproduced analytically from a two-loop calculation in string sigma-model perturbation theory [58].

Thus, the cusp anomaly $f(g)$ is an example of a non-trivial interpolation function for an observable not protected by supersymmetry that smoothly connects weak and strong coupling regimes, and tests the AdS/CFT correspondence at a very deep level. The final form of $f(g)$ was arrived at using inputs from string theory, perturbative gauge theory, and the conjectured exact integrability of planar $\mathcal{N} = 4$ SYM. Further tests of this proposal may be carried out perturbatively: in fact, the planar expansion (5.11) makes an infinite number of explicit analytic predictions for higher loop coefficients. This shows how implications of string theory may sometimes be tested via perturbative gauge theory.

6. Thermal gauge theory from near-extremal D3-branes

Entropy

An important black hole observable is the Bekenstein-Hawking (BH) entropy, which is proportional to the area of the event horizon, $S_{BH} = A_h/(4G)$. For the 3-brane solution (3.1), the horizon is located at $r = r_0$. For $r_0 > 0$ the 3-brane carries some excess energy E above its extremal value, and the BH entropy is also non-vanishing. The Hawking temperature is then defined by $T^{-1} = \partial S_{BH}/\partial E$.

Setting $r_0 \ll L$ in (3.1), we obtain a near-extremal 3-brane geometry, whose Hawking temperature is found to be $T = r_0/(\pi L^2)$. The small r limit of this geometry is $\mathbf{S}^5$ times a certain black hole in AdS_5. The 8-dimensional "area" of the event horizon is $A_h = \pi^6 L^8 T^3 V_3$, where V_3 is the spatial volume of the D3-brane (i.e. the volume of the x^1, x^2, x^3 coordinates). Therefore, the BH entropy is [59]

$$S_{BH} = \frac{\pi^2}{2} N^2 V_3 T^3.$$ (6.1)

This gravitational entropy of a near-extremal 3-brane of Hawking temperature T is to be identified with the entropy of $\mathcal{N} = 4$ supersymmetric $U(N)$ gauge theory (which lives on N coincident D3-branes) heated up to the same temperature.

The entropy of a free $U(N)$ $\mathcal{N} = 4$ supermultiplet, which consists of the gauge field, $6N^2$ massless scalars and $4N^2$ Weyl fermions, can be calculated using the standard statistical mechanics of a massless gas (the black body problem), and the answer is

$$S_0 = \frac{2\pi^2}{3} N^2 V_3 T^3.$$ (6.2)

It is remarkable that the 3-brane geometry captures the T^3 scaling characteristic of a conformal field theory (in a CFT this scaling is guaranteed by the extensivity of the entropy and the absence of dimensionful parameters). Also, the N^2 scaling indicates the presence of $O(N^2)$ unconfined degrees of freedom, which is exactly what we expect in the $\mathcal{N} = 4$ supersymmetric $U(N)$ gauge theory. But what is the explanation of the relative factor of $3/4$ between S_{BH} and S_0? In fact, this factor is not a contradiction but rather a *prediction* about the strongly coupled $\mathcal{N} = 4$ SYM theory at finite temperature. As we argued above, the supergravity calculation of the BH entropy, (6.1), is relevant to the $\lambda \to \infty$ limit of the $\mathcal{N} = 4$ $SU(N)$ gauge theory, while the free field calculation, (6.2), applies to the $\lambda \to 0$ limit. Thus, the relative factor of $3/4$ is not a discrepancy: it relates two different limits of the theory. Indeed, on general field theoretic grounds, in the 't Hooft large N limit the entropy is given by [60]

$$S = \frac{2\pi^2}{3} N^2 f_e(\lambda) V_3 T^3. \tag{6.3}$$

The function f_e is certainly not constant: Feynman graph calculations valid for small $\lambda = g_{YM}^2 N$ give [61]

$$f_e(\lambda) = 1 - \frac{3}{2\pi^2}\lambda + \frac{3 + \sqrt{2}}{\pi^3}\lambda^{3/2} + \cdots. \tag{6.4}$$

The BH entropy in supergravity, (6.1), is translated into the prediction that

$$\lim_{\lambda \to \infty} f_e(\lambda) = \frac{3}{4}. \tag{6.5}$$

A string theoretic calculation of the leading correction at large λ gives [60]

$$f_e(\lambda) = \frac{3}{4} + \frac{45}{32}\zeta(3)\lambda^{-3/2} + \cdots. \tag{6.6}$$

These results are consistent with a monotonic function $f_e(\lambda)$ which decreases from 1 to $3/4$ as λ is increased from 0 to ∞. The $1/4$ deficit compared to the free field value is a strong coupling effect predicted by the AdS/CFT correspondence.

It is interesting that similar deficits have been observed in lattice simulations of deconfined non-supersymmetric gauge theories [62–64]. The ratio of entropy to its free field value, calculated as a function of the temperature, is found to level off at values around 0.8 for T beyond 3 times the deconfinement temperature T_c. This is often interpreted as the effect of a sizable coupling. Indeed, for $T = 3T_c$, the lattice estimates [63] indicate that $g_{YM}^2 N \approx 7$. This challenges an old prejudice that the quark-gluon plasma is inherently weakly coupled. We now turn to calculations of the shear viscosity where strong coupling effects are more pronounced.

Shear viscosity

The shear viscosity η may be read off from the form of the stress-energy tensor in the local rest frame of the fluid where $T_{0i} = 0$:

$$T_{ij} = p\delta_{ij} - \eta\left(\partial_i u_j + \partial_j u_i - \frac{2}{3}\delta_{ij}\partial_k u_k\right),\tag{6.7}$$

where u_i is the 3-velocity field. The viscosity can be also determined [65] through the Kubo formula

$$\eta = \lim_{\omega\to 0}\frac{1}{2\omega}\int dt d^3x e^{i\omega t}\langle[T_{xy}(t,\vec{x}), T_{xy}(0,0)]\rangle.\tag{6.8}$$

For the $\mathcal{N} = 4$ supersymmetric YM theory this 2-point function may be computed from absorption of a low-energy graviton h_{xy} by the 3-brane metric [24]. Using this method, it was found [65] that at very strong coupling

$$\eta = \frac{\pi}{8}N^2 T^3,\tag{6.9}$$

which implies

$$\frac{\eta}{s} = \frac{\hbar}{4\pi}\tag{6.10}$$

after $\hbar$ is restored in the calculation (here $s = S/V_3$ is the entropy density). This is much smaller than what perturbative estimates imply. Indeed, at weak coupling η/s is very large, $\sim \hbar\lambda^{-2}/\ln(1/\lambda)$ [66], and there is evidence that it decreases monotonically as the coupling is increased [67].

The saturation of η/s at some value of order $\hbar$ is reasonable on general physical grounds [68]. The shear viscosity η is of order the energy density times quasi-particle mean free time τ. So η/s is of order of the energy of a quasi-particle times its mean free time, which is bounded from below by the uncertainty principle to be some constant times $\hbar$. These considerations prompted a suggestion [68] that $\hbar/(4\pi)$ is the lower bound on η/s. However, the generality of this bound was called into question in [69]. Recently it was shown [70] that for special large N gauge theories whose AdS duals contain D7-branes, the leading $1/N$ correction to (6.10) is negative; thus, the bound can be violated even for theories that have a holographic description.

Nevertheless, (6.10) applies to the strong coupling limit of a large class of gauge theories that have gravity duals. Its important physical implication is that a low value of η/s is a consequence of strong coupling. For known fluids (e.g. helium, nitrogen, water) η/s is considerably higher than (6.10). On the other hand, the quark-gluon plasma produced at RHIC is believed to have a very low η/s [71, 72], with some recent estimates [73] suggesting that it is below (6.10). Lattice studies of pure glue gauge theory [74]) also lead to low values of η/s. This suggests that, at least for T near T_c, the theory is sufficiently strongly coupled. Indeed, a new term, sQGP, which stands for "strongly coupled quark-gluon plasma," has been coined to describe the deconfined state observed at RHIC [75, 76] (a somewhat different term, "non-perturbative quark-gluon plasma", was advocated in [77]). As we have reviewed, the gauge-string duality is a theoretical laboratory which allows one to study some extreme examples of such a new state of matter. These include the thermal $\mathcal{N} = 4$ SYM theory at very strong 't Hooft coupling, as well as other gauge theories which have less supersymmetry and exhibit confinement at low temperature. An example of such a gauge theory, whose dual is the warped deformed conifold, will be discussed in Section 8.

Lattice calculations indicate that the deconfinement temperature T_c is around 175 MeV, and the energy density is ≈ 0.7 GeV/fm^3, around 6 times the nuclear energy density. RHIC has reached energy densities around 14 GeV/fm^3, corresponding to $T \approx 2T_c$. Furthermore, heavy ion collisions at the LHC are expected to reach temperatures up to $5T_c$. Thus, RHIC and LHC should provide a great deal of new information about the sQGP, which can be compared with calculations based on gauge-string duality. For a more detailed discussion of the recent theoretical work in this direction, see U. Wiedemann's lectures in this volume.

7. Warped conifold and its resolution

To formulate the AdS/CFT duality at zero temperature, but with a reduced amount of supersymmetry, we may place the stack of D3-branes at the tip of a 6-dimensional Ricci flat cone, [78–80] whose base is a 5-dimensional compact Einstein space Y_5. The metric of such a cone is $dr^2 + r^2 ds_5^2$; therefore, the 10-d metric produced by the D3-branes is obtained from (3.5) by replacing $d\Omega_5^2$, the metric on $\mathbf{S}^5$, by ds_Y^2, the metric on Y_5. In the $r \to 0$ limit we then find the space $AdS_5 \times Y_5$ as the candidate dual of the CFT on the D3-branes placed at the tip of the cone. The isometry group of Y_5 is smaller than $SO(6)$, but AdS_5 is the "universal" factor present in the dual description of any large N CFT, making the $SO(2, 4)$ conformal symmetry geometric.

To obtain gauge theories with $\mathcal{N} = 1$ superconformal symmetry the Ricci flat cone must be a Calabi–Yau 3-fold [80, 81] whose base Y_5 is called a Sasaki-Einstein space. Among the simplest examples of these is $Y_5 = T^{1,1}$. The corresponding Calabi–Yau cone is called the conifold. Much of the work on gauge-string dualities originating from D-branes on the conifold is reviewed in the 2001 Les Houches lectures [82]. Here we will emphasize the progress that has taken place since then, in particular on issues related to breaking of the $U(1)$ baryon number symmetry.

The conifold is a singular non-compact Calabi–Yau three-fold [83]. Its importance arises from the fact that the generic singularity in a Calabi–Yau three-fold locally looks like the conifold, described by the quadratic equation in $\mathbf{C}^4$:

$$z_1^2 + z_2^2 + z_3^2 + z_4^2 = 0. \tag{7.1}$$

This homogeneous equation defines a real cone over a 5-dimensional manifold. For the cone to be Ricci-flat the 5d base must be an Einstein manifold ($R_{ab} = 4g_{ab}$). For the conifold [83], the topology of the base can be shown to be $\mathbf{S}^2 \times \mathbf{S}^3$ and it is called $T^{1,1}$, with the following Einstein metric [84]:

$$
\begin{aligned}
d\Omega_{T^{1,1}}^2 &= \frac{1}{9}(d\psi + \cos\theta_1 d\phi_1 + \cos\theta_2 d\phi_2)^2 \\
&\quad + \frac{1}{6}(d\theta_1^2 + \sin^2\theta_1 d\phi_1^2) + \frac{1}{6}(d\theta_2^2 + \sin^2\theta_2 d\phi_2^2).
\end{aligned} \tag{7.2}
$$

$T^{1,1}$ is a homogeneous space, being the coset $SU(2) \times SU(2)/U(1)$. The metric on the cone is then $ds^2 = dr^2 + r^2 d\Omega_{T^{1,1}}^2$.

We may introduce two other types of complex coordinates on the conifold, w_a and a_i, b_j, as follows:

$$
\begin{aligned}
Z &= \begin{pmatrix} z_3 + iz_4 & z_1 - iz_2 \\ z_1 + iz_2 & -z_3 + iz_4 \end{pmatrix} = \begin{pmatrix} w_1 & w_3 \\ w_4 & w_2 \end{pmatrix} = \begin{pmatrix} a_1 b_1 & a_1 b_2 \\ a_2 b_1 & a_2 b_2 \end{pmatrix} \\
&= r^{\frac{3}{2}} \begin{pmatrix} -c_1 s_2\, e^{\frac{i}{2}(\psi+\phi_1-\phi_2)} & c_1 c_2\, e^{\frac{i}{2}(\psi+\phi_1+\phi_2)} \\ -s_1 s_2\, e^{\frac{i}{2}(\psi-\phi_1-\phi_2)} & s_1 c_2\, e^{\frac{i}{2}(\psi-\phi_1+\phi_2)} \end{pmatrix},
\end{aligned} \tag{7.3}
$$

where $c_i = \cos\frac{\theta_i}{2}$, $s_i = \sin\frac{\theta_i}{2}$ (see [83] for other details on the w, z and angular coordinates). The equation defining the conifold is now $\det Z = 0$.

The a, b coordinates above will be of particular interest to us because the symmetries of the conifold are most apparent in this basis. The conifold equation has $SU(2) \times SU(2) \times U(1)$ symmetry since under these symmetry transformations,

$$\det LZR^\dagger = \det e^{i\alpha} Z = 0. \tag{7.4}$$

This is also a symmetry of the metric presented above where each $SU(2)$ acts on θ_i, ϕ_i, ψ (thought of as Euler angles on $\mathbf{S}^3$) while the $U(1)$ acts by shifting ψ. This symmetry can be identified with $U(1)_R$, the R-symmetry of the dual gauge theory, in the conformal case. The action of the $SU(2) \times SU(2) \times U(1)_R$

symmetry on a_i, b_j defined in (7.3) is given by:

$$SU(2) \times SU(2) \text{ symmetry:} \quad \begin{pmatrix} a_1 \\ a_2 \end{pmatrix} \to L \begin{pmatrix} a_1 \\ a_2 \end{pmatrix}, \tag{7.5}$$

$$\begin{pmatrix} b_1 \\ b_2 \end{pmatrix} \to R \begin{pmatrix} b_1 \\ b_2 \end{pmatrix}, \tag{7.6}$$

$$\text{R-symmetry:} \quad (a_i, b_j) \to e^{i\frac{\alpha}{2}}(a_i, b_j), \tag{7.7}$$

i.e. a and b transform as $(1/2, 0)$ and $(0, 1/2)$ under $SU(2) \times SU(2)$ with R-charge $1/2$ each. We can thus describe the singular conifold as the manifold parametrized by a, b, but from (7.3), we see that there is some redundancy in the a, b coordinates. Namely, the transformation

$$a_i \to \lambda \, a_i, \quad b_j \to \frac{1}{\lambda} b_j, \quad (\lambda \in \mathbb{C}), \tag{7.8}$$

give the same z, w in (7.3). We impose the constraint $|a_1|^2 + |a_2|^2 - |b_1|^2 - |b_2|^2 = 0$ to fix the magnitude in the above transformation. To account for the remaining phase, we describe the singular conifold as the quotient of the a, b space with the above constraint by the relation $a \sim e^{i\alpha}a, b \sim e^{-i\alpha}b$.

The importance of the coordinates a_i, b_j is that in the gauge theory on D3-branes at the tip of the conifold they are promoted to chiral superfields. The low-energy gauge theory on N D3-branes is a $\mathcal{N} = 1$ supersymmetric $SU(N) \times SU(N)$ gauge theory with bi-fundamental chiral superfields A_i, B_j $(i, j = 1, 2)$ in the $(\mathbf{N}, \overline{\mathbf{N}})$ and $(\overline{\mathbf{N}}, \mathbf{N})$ representations of the gauge groups, respectively [80, 81]. The superpotential for this gauge theory is

$$W \sim \text{Tr} \det A_i B_j = \text{Tr} \, (A_1 B_1 A_2 B_2 - A_1 B_2 A_2 B_1). \tag{7.9}$$

The continuous global symmetries of this theory are $SU(2) \times SU(2) \times U(1)_B \times U(1)_R$, where the $SU(2)$ factors act on A_i and B_j respectively, $U(1)_B$ is a baryonic symmetry under which the A_i and B_j have opposite charges, and $U(1)_R$ is the R-symmetry with charges of the same sign $R_A = R_B = \frac{1}{2}$. This assignment ensures that W is marginal, and one can also show that the gauge couplings do not run. Hence this theory is superconformal for all values of gauge couplings and superpotential coupling [80, 81].

A simple way to understand the resolution of the conifold is to deform the modulus constraint above into

$$|b_1|^2 + |b_2|^2 - |a_1|^2 - |a_2|^2 = u^2, \tag{7.10}$$

where u is a real parameter which controls the resolution. The resolution corresponds to a blow up of the $\mathbf{S}^2$ at the bottom of the conifold. In the dual gauge theory turning on u corresponds to a particular choice of vacuum [85]. After promoting the a, b fields into the bi-fundamental chiral superfields of the dual gauge theory, we can define the operator $\mathcal{U}$ as

$$\mathcal{U} = \frac{1}{N} \text{Tr} \, (B_1^\dagger B_1 + B_2^\dagger B_2 - A_1^\dagger A_1 - A_2^\dagger A_2). \tag{7.11}$$

Thus, the warped singular conifolds correspond to gauge theory vacua where $\langle \mathcal{U} \rangle = 0$, while the warped resolved conifolds correspond to vacua where $\langle \mathcal{U} \rangle \neq 0$. In the latter case, some VEVs for the bi-fundamental fields A_i, B_j must be present. Since these fields are charged under the $U(1)_B$ symmetry, the warped resolved conifolds correspond to vacua where this symmetry is broken [85].

A particularly simple choice is to give a diagonal VEV to only one of the scalar fields, say, B_2. As seen in [86], this choice breaks the $SU(2) \times SU(2) \times U(1)_B \times U(1)_R$ symmetry of the CFT down to $SU(2) \times U(1) \times U(1)$. The string dual is given by a warped resolved conifold

$$ds^2 = h^{-1/2} \eta_{\mu\nu} dx^\mu dx^\nu + h^{1/2} ds_6^2. \tag{7.12}$$

The explicit form of the Calabi–Yau metric of the resolved conifold is given by [87]

$$ds_6^2 = K^{-1} dr^2 + \frac{1}{9} K r^2 (d\psi + \cos\theta_1 d\phi_1 + \cos\theta_2 d\phi_2)^2$$
$$+ \frac{1}{6} r^2 (d\theta_1^2 + \sin^2\theta_1 d\phi_1^2) + \frac{1}{6} (r^2 + 6u^2)(d\theta_1^2 + \sin^2\theta_2 d\phi_2^2), \tag{7.13}$$

where

$$K = \frac{r^2 + 9u^2}{r^2 + 6u^2}. \tag{7.14}$$

The N $D3$-branes sourcing the warp factor are located at the north pole of the finite S^2, i.e. at $r = 0, \theta_2 = 0$. The warp factor is the Green function on the resolved conifold with this source [86]:

$$h(r, \theta_2) = L^4 \sum_{l=0}^{\infty} (2l + 1) H_l(r) P_l(\cos\theta_2). \tag{7.15}$$

Here $L^4 = \frac{27\pi}{4} g_s N(\alpha')^2$, $P_l(\cos\theta)$ is the l-th Legendre polynomial, and

$$H_l = \frac{2C_\beta}{9u^2 r^{2+2\beta}} \, {}_2F_1\left(\beta, 1 + \beta, 1 + 2\beta; -\frac{9u^2}{r^2}\right), \tag{7.16}$$

with the coefficients C_β and β given by

$$C_\beta = \frac{(3u)^{2\beta}\Gamma(1+\beta)^2}{\Gamma(1+2\beta)}, \qquad \beta = \sqrt{1 + \frac{3}{2}l(l+1)}. \tag{7.17}$$

Far in the IR the gauge theory flows to the $\mathcal{N} = 4\ SU(N)$ SYM theory, as evidenced by the appearance of an $AdS_5 \times \mathbf{S}^5$ throat near the location of the stack of the $D3$-branes. This may be verified in the gauge theory as follows. The condensate $B_2 = u\,1_{N\times N}$ breaks the $SU(N) \times SU(N)$ gauge group down to $SU(N)$, all the chiral fields now transforming in the adjoint of this diagonal group. After substituting this classical value for B_2, the quartic superpotential (7.9) reduces to the cubic $\mathcal{N} = 4$ form,

$$W \sim \mathrm{Tr}(A_1[B_1, A_2]). \tag{7.18}$$

This confirms that the gauge theory flows to the $\mathcal{N} = 4\ SU(N)$ SYM theory. The gauge theory also contains an interesting additional sector coupled to this infrared CFT; in particular, it contains global strings due to the breaking of the $U(1)_B$ symmetry [88].

Baryonic condensates and Euclidean D3-branes

The gauge invariant order parameter for the breaking of $U(1)_B$ is $\det B_2$. Let us review the calculation of this baryonic VEV using the dual string theory on the warped resolved conifold background [86].

The objects in $AdS_5 \times T^{1,1}$ that are dual to baryonic operators are D3-branes wrapping 3-cycles in $T^{1,1}$ [89]. Classically, the 3-cycles dual to the baryons made out of the B's are located at fixed θ_2 and ϕ_2, while quantum mechanically one has to carry out collective coordinate quantization and finds wave functions of spin $N/2$ on the 2-sphere.

To calculate VEVs of such baryonic operators we need to consider the action of a Euclidean D3-brane whose world volume ends at large r on the 3-sphere at fixed θ_2 and ϕ_2. The D3-brane action should be integrated up to a radial cut-off r, and we identify $e^{-S(r)}$ with the field $\varphi(r)$ dual to the baryonic operator. Close to the boundary, a field φ dual to an operator of dimension Δ in the AdS/CFT correspondence behaves as

$$\varphi(r) \to \varphi_0\, r^{\Delta-4} + A_\varphi\, r^{-\Delta}. \tag{7.19}$$

Here A_φ is the operator expectation value [90], and φ_0 is the source for it. There are no sources added for baryonic operators, hence we will find that $\varphi_0 = 0$, but the term scaling as $r^{-\Delta}$ is indeed present in $e^{-S(r)}$.

The Born–Infeld action of the D3-brane is given by

$$S_{BI} = T_3 \int d^4\xi \sqrt{g}, \tag{7.20}$$

where $g_{\mu\nu}$ is the metric induced on the D3 world-volume. The smooth 4-chain which solves the BI equations of motion subject to our boundary conditions is located at fixed θ_2 and ϕ_2, and spans the r, θ_1, ϕ_1 and ψ directions. Using the D3-brane tension

$$T_3 = \frac{1}{g_s (2\pi)^3 (\alpha')^2}, \tag{7.21}$$

we find

$$S_{BI} = \frac{3N}{4} \int_0^r d\tilde{r}\,\tilde{r}^3 h(\tilde{r}, \theta_2). \tag{7.22}$$

Using the expansion (7.15), we note that the $l = 0$ term needs to be evaluated separately since it contains a logarithmic divergence:

$$\int_0^r d\tilde{r}\,\tilde{r}^3 H_0(r) = \frac{1}{4} + \frac{1}{2} \ln\left(1 + \frac{r^2}{9u^2}\right). \tag{7.23}$$

For the $l > 0$ terms the integral converges and we find the simple result [86]

$$\int_0^\infty d\tilde{r}\,\tilde{r}^3 \sum_{l=1}^\infty H_l(\tilde{r}) P_l(\cos\theta_2) = \frac{2}{3}(-1 - 2\ln[\sin(\theta_2/2)]). \tag{7.24}$$

This expression is recognized as the Green's function on a sphere. Combining the results, and taking $r \gg u$, we find

$$e^{-S_{BI}} = \left(\frac{3e^{5/12}u}{r}\right)^{3N/4} \sin^N(\theta_2/2). \tag{7.25}$$

In [89] it was argued that the wave functions of θ_2, ϕ_2, which arise though the collective coordinate quantization of the D3-branes wrapped over the 3-cycle (ψ, θ_1, ϕ_1), correspond to eigenstates of a charged particle on S^2 in the presence of a charge N magnetic monopole. Taking the gauge potential $A_\phi = N(1 + \cos\theta)/2$, $A_\theta = 0$ we find that the ground state wave function $\sim \sin^N(\theta_2/2)$ carries the $J = N/2$ and $m = -N/2$ quantum numbers.[1] These are the $SU(2)$ quantum numbers of $\det B_2$. Therefore, the angular dependence of e^{-S} identifies

---

[1] In a different gauge this wave function would acquire a phase. In the string calculation the phase comes from the purely imaginary Chern–Simons term in the Euclidean D3-brane action.

det B_2 as the only operator that acquires a VEV, in agreement with the gauge theory.

The power of r indicates that the operator dimension is $\Delta = 3N/4$, which is indeed the exact dimension of the baryonic operators [89]. The VEV depends on the parameter u as $\sim u^{3N/4}$. This is not the same as the classical scaling that would give det $B_2 \sim u^N$. The classical scaling is not obeyed because we are dealing with a strongly interacting gauge theory where the baryonic operator acquires an anomalous dimension.

Goldstone bosons and global strings

Since the $U(1)_B$ symmetry is broken spontaneously, the spectrum of the theory must include a Goldstone boson. Its gravity dual is a normalizable RR 4-form fluctuation around the warped resolved conifold [88]:

$$\delta F^{(5)} = (1 + \star)d(a_2(x) \wedge W). \tag{7.26}$$

Here W is a closed 2-form inside the warped resolved conifold,

$$W = \sin\theta_2 d\theta_2 \wedge d\phi_2 + d(f_1 g^5 + f_2 \sin\theta_2 d\varphi_2), \tag{7.27}$$

where f_1, f_2 are functions of r, θ_2. The equations of motion reduce to

$$d \star_4 da_2 = 0, \tag{7.28}$$

provided W satisfies

$$d(h \star_6 W) = 0, \tag{7.29}$$

where $\star_4$, $\star_6$ are the Hodge duals with respect to the unwarped Minkowski and resolved conifold metrics, respectively. This gives coupled PDE's for f_1 and f_2. Their solution can be obtained through minimizing a positive definite functional subject to certain boundary conditions [88]. Introducing the Goldstone boson field $p(x)$ through $\star_4 da_2 = dp$, we note that the fluctuation in the 5-form field strength reads

$$\delta F^{(5)} = da_2 \wedge W + dp \wedge h \star_6 W. \tag{7.30}$$

The corresponding fluctuation of the 4-form potential is

$$\delta C^{(4)} = a_2(x) \wedge W + p \wedge h \star_6 W. \tag{7.31}$$

In addition to the existence of a Goldstone boson, a hallmark of a broken $U(1)$ symmetry is the appearance of "global" strings. The Goldstone boson has a non-trivial monodromy around such a string, thus giving it a logarithmically divergent energy density. On the string side of the duality these global strings are $D3$-branes wrapping the 2-sphere at the bottom of the warped resolved conifold [88]. The Goldstone bosons and the global strings interact with the $\mathcal{N} = 4$ SYM theory that appears far in the infrared. The coupling of such an extra sector and an infrared CFT is an interesting fact reminiscent of the unparticle physics scenarios [91].

8. Deformation of the conifold

We have seen above that the singularity of the cone over $T^{1,1}$ can be replaced by an $\mathbf{S}^2$ through resolving the conifold (7.1) as in (7.10). An alternative supersymmetric blow-up, which replaces the singularity by an $\mathbf{S}^3$, is the deformed conifold [83]

$$z_1^2 + z_2^2 + z_3^2 + z_4^2 = \varepsilon^2. \tag{8.1}$$

To achieve the deformation, one needs to turn on M units of RR 3-form flux. This modifies the dual gauge theory to $\mathcal{N} = 1$ supersymmetric $SU(N) \times SU(N+M)$ theory coupled to chiral superfields A_1, A_2 in the $(\mathbf{N}, \overline{\mathbf{N}+\mathbf{M}})$ representation, and B_1, B_2 in the $(\overline{\mathbf{N}}, \mathbf{N}+\mathbf{M})$ representation. Indeed, in type IIB string theory D5-branes source the 7-form field strength from the Ramond-Ramond sector, which is Hodge dual to the 3-form field strength. Therefore, the M wrapped D5-branes create M flux units of this field strength through the 3-cycle in the conifold; this number is dual to the difference between the numbers of colors in the two gauge groups. Thus, unlike the resolution, the deformation cannot be achieved in the context of the $SU(N) \times SU(N)$ gauge theory.

The 10-d metric takes the following form [92]:

$$ds_{10}^2 = h^{-1/2}(\tau) \, \eta_{\mu\nu} dx^\mu dx^\nu + h^{1/2}(\tau) \, ds_6^2, \tag{8.2}$$

where ds_6^2 is the Calabi–Yau metric of the deformed conifold:

$$ds_6^2 = \frac{\varepsilon^{4/3}}{2} K(\tau) \left[\sinh^2\left(\frac{\tau}{2}\right) \left[(g^1)^2 + (g^2)^2 \right] + \cosh^2\left(\frac{\tau}{2}\right) \left[(g^3)^2 + (g^4)^2 \right] \right.$$
$$\left. + \frac{1}{3K^3(\tau)} \left[d\tau^2 + (g^5)^2 \right] \right], \tag{8.3}$$

where

$$K(\tau) = \frac{(\sinh \tau \cosh \tau - \tau)^{1/3}}{\sinh \tau}. \tag{8.4}$$

For $\tau \gg 1$ we may introduce another radial coordinate r defined by

$$r^2 = \frac{3}{2^{5/3}}\varepsilon^{4/3}e^{2\tau/3}, \tag{8.5}$$

and in terms of this coordinate we find $ds_6^2 \to dr^2 + r^2 ds_{T^{1,1}}^2$.

The basis one-forms g^i in terms of which this metric is diagonal are defined by

$$g^1 \equiv \frac{e_2 - \epsilon_2}{\sqrt{2}}, \quad g^2 \equiv \frac{e_1 - \epsilon_1}{\sqrt{2}}, \tag{8.6}$$

$$g^3 \equiv \frac{e_2 + \epsilon_2}{\sqrt{2}}, \quad g^4 \equiv \frac{e_1 + \epsilon_1}{\sqrt{2}}, \tag{8.7}$$

$$g^5 \equiv \epsilon_3 + \cos\theta_1 d\phi_1, \tag{8.8}$$

where the e_i are one-forms on $\mathbf{S}^2$

$$e_1 \equiv d\theta_1, \qquad e_2 \equiv -\sin\theta_1 d\phi_1, \tag{8.9}$$

and the ϵ_i a set of one-forms on $\mathbf{S}^3$

$$\epsilon_1 \equiv \sin\psi \sin\theta_2 d\phi_2 + \cos\psi d\theta_2, \tag{8.10}$$

$$\epsilon_2 \equiv \cos\psi \sin\theta_2 d\phi_2 - \sin\psi d\theta_2, \tag{8.11}$$

$$\epsilon_3 \equiv d\psi + \cos\theta_2 d\phi_2. \tag{8.12}$$

The NSNS two-form is given by

$$B^{(2)} = \frac{g_s M\alpha'}{2}\frac{\tau\coth\tau - 1}{\sinh\tau}\left[\sinh^2\left(\frac{\tau}{2}\right)g^1 \wedge g^2 + \cosh^2\left(\frac{\tau}{2}\right)g^3 \wedge g^4\right] \tag{8.13}$$

and the RR fluxes are most compactly written as

$$F^{(3)} = \frac{M\alpha'}{2}\left\{g^3 \wedge g^4 \wedge g^5 + d\left[\frac{\sinh\tau - \tau}{2\sinh\tau}(g^1 \wedge g^3 + g^2 \wedge g^4)\right]\right\}, \tag{8.14}$$

$$\tilde{F}^{(5)} = dC^{(4)} + B^{(2)} \wedge F^{(3)} = (1 + *)(B^{(2)} \wedge F^{(3)}). \tag{8.15}$$

Note that the complex three-form field of this BPS supergravity solution is imaginary self dual:

$$\star_6 G_3 = iG_3, \quad G_3 = F_3 - \frac{i}{g_s}H_3, \tag{8.16}$$

where $\star_6$ again denotes the Hodge dual with respect to the unwarped metric ds_6^2. This guarantees that the dilaton is constant, and we set $\phi = 0$.

The above expressions for the NSNS- and RR-forms follow by making a simple ansatz consistent with the symmetries of the problem, and solving a system of differential equations, which owing to the supersymmetry of the problem are only first order [92]. The warp factor is then found to be completely determined up to an additive constant, which is fixed by demanding that it go to zero at large τ:

$$h(\tau) = (g_s M\alpha')^2 2^{2/3} \varepsilon^{-8/3} I(\tau), \tag{8.17}$$

$$I(\tau) \equiv 2^{1/3} \int_\tau^\infty dx \frac{x \coth x - 1}{\sinh^2 x} (\sinh x \cosh x - x)^{1/3}. \tag{8.18}$$

For small τ the warp factor approaches a finite constant since $I(0) \approx 0.71805$. This implies confinement because the chromo-electric flux tube, described by a fundamental string at $\tau = 0$, has tension

$$T_s = \frac{1}{2\pi\alpha' \sqrt{h(0)}}. \tag{8.19}$$

The KS solution [92] is $SU(2) \times SU(2)$ symmetric and the expressions above can be written in an explicitly $SO(4)$ invariant way. It also possesses a $\mathbb{Z}_2$ symmetry $\mathcal{I}$, which exchanges (θ_1, ϕ_1) with (θ_2, ϕ_2) accompanied by the action of $-I$ of SL(2, $\mathbb{Z}$), changing the signs of the three-form fields.

Examining the metric ds_6^2 for $\tau = 0$ we see that it degenerates into

$$d\Omega_3^2 = \frac{1}{2} \varepsilon^{4/3} (2/3)^{1/3} \left[\frac{1}{2} (g^5)^2 + (g^3)^2 + (g^4)^2 \right], \tag{8.20}$$

which is the metric of a round $\mathbf{S}^3$, while the $\mathbf{S}^2$ spanned by the other two angular coordinates, and fibered over the $\mathbf{S}^3$, shrinks to zero size. In the ten-dimensional metric (8.2) this appears multiplied by a factor of $h^{1/2}(\tau)$, and thus the radius squared of the three-sphere at the tip of the conifold is of order $g_s M\alpha'$. Hence for $g_s M$ large, the curvature of the $\mathbf{S}^3$, and in fact everywhere in this manifold, is small and the supergravity approximation reliable.

The field theoretic interpretation of the KS solution is unconventional. After a finite distance along the RG flow, the $SU(N + M)$ group undergoes a Seiberg duality transformation [93]. After this transformation, and an interchange of the two gauge groups, the new gauge theory is $SU(\tilde{N}) \times SU(\tilde{N} + M)$ with the same bi-fundamental field content and superpotential, and with $\tilde{N} = N - M$. The self-similar structure of the gauge theory under the Seiberg duality is the crucial fact that allows this pattern to repeat many times. For a careful field theoretic discussion of this quasi-periodic RG flow, see [94]. If $N = (k + 1)M$, where k is an integer, then the duality cascade stops after k steps, and we find a

$SU(M) \times SU(2M)$ gauge theory. This IR gauge theory exhibits a multitude of interesting effects visible in the dual supergravity background, which include the confinement and chiral symmetry breaking.

9. Normal modes of the warped throat

As we shall see below, the $U(1)$ baryonic symmetry of the warped deformed conifold is in fact spontaneously broken, since baryonic operators acquire expectation values. The corresponding Goldstone boson is a massless pseudoscalar supergravity fluctuation which has non-trivial monodromy around D-strings at the bottom of the warped deformed conifold [95]. Like fundamental strings they fall to the bottom of the conifold (corresponding to the IR of the field theory), where they have non-vanishing tension. But while F-strings are dual to confining strings, D-strings are interpreted as global solitonic strings in the dual cascading $SU(M(k+1)) \times SU(Mk)$ gauge theory.

Thus the warped deformed conifold naturally incorporates a supergravity description of the supersymmetric Goldstone mechanism. Below we review the supergravity dual of a pseudoscalar Goldstone boson, as well as its superpartner, a massless scalar glueball [95]. In the gauge theory they correspond to fluctuations in the phase and magnitude of the baryonic condensates, respectively.

The Goldstone mode

A D1-brane couples to the three-form field strength F_3, and therefore we expect a four-dimensional pseudo-scalar $p(x)$, defined so that $\star_4 dp = \delta F_3$, to experience monodromy around the D-string.

The following ansatz for a linear perturbation of the KS solution

$$\delta F^{(3)} = \star_4 dp + f_2(\tau)\, dp \wedge dg^5 + f_2'(\tau)\, dp \wedge d\tau \wedge g^5, \tag{9.1}$$

$$\delta F^{(5)} = (1 + \star)\delta F_3 \wedge B_2 = \left(\star_4 dp - \frac{\varepsilon^{4/3}}{6K^2(\tau)} h(\tau)\, dp \wedge d\tau \wedge g^5 \right) \wedge B_2,$$

where $f_2' = df_2/d\tau$, falls within the general class of supergravity backgrounds discussed by Papadopoulos and Tseytlin [96]. The metric, dilaton and $B^{(2)}$-field remain unchanged. This can be shown to satisfy the linearized supergravity equations [95], provided that $d \star_4 dp = 0$, i.e. $p(x)$ is massless, and $f_2(\tau)$ satisfies

$$-\frac{d}{d\tau}[K^4 \sinh^2 \tau f_2'] + \frac{8}{9K^2} f_2$$

$$= \frac{(g_s M\alpha')^2}{3\varepsilon^{4/3}} (\tau \coth \tau - 1) \left(\coth \tau - \frac{\tau}{\sinh^2 \tau} \right). \tag{9.2}$$

The normalizable solution of this equation is given by [95]

$$f_2(\tau) = -\frac{2c}{K^2 \sinh^2 \tau} \int_0^\tau dx \, h(x) \sinh^2 x, \tag{9.3}$$

where $c \sim \varepsilon^{4/3}$. We find that $f_2 \sim \tau$ for small τ, and $f_2 \sim \tau e^{-2\tau/3}$ for large τ.

As we have remarked above, the $U(1)$ baryon number symmetry acts as $A_k \to e^{i\alpha} A_k$, $B_j \to e^{-i\alpha} B_j$. The massless gauge field in AdS_5 dual to the baryon number current originates from the RR 4-form potential [80, 97]:

$$\delta C^{(4)} \sim \omega_3 \wedge \tilde{A}. \tag{9.4}$$

The zero-mass pseudoscalar glueball arises from the spontaneous breaking of the global $U(1)_B$ symmetry [98], as seen from the form of δF_5 in (9.1), which contains a term $\sim \omega_3 \wedge dp \wedge d\tau$ that leads us to identify $\tilde{A} \sim dp$.

If N is an integer multiple of M, the last step of the cascade leads to a $SU(2M) \times SU(M)$ gauge theory coupled to bifundamental fields A_i, B_j (with $i, j = 1, 2$). If the $SU(M)$ gauge coupling were turned off, then we would find an $SU(2M)$ gauge theory coupled to $2M$ flavors. In this $N_f = N_c$ case, in addition to the usual mesonic branch there exists a baryonic branch of the quantum moduli space [99]. This is important for the gauge theory interpretation of the KS background [92, 98]. Indeed, in addition to mesonic operators $(N_{ij})^\alpha_\beta \sim (A_i B_j)^\alpha_\beta$, the IR gauge theory has baryonic operators invariant under the $SU(2M) \times SU(M)$ gauge symmetry, as well as the $SU(2) \times SU(2)$ global symmetry rotating A_i, B_j:

$$
\begin{aligned}
\mathcal{A} &\sim \epsilon_{\alpha_1 \alpha_2 \ldots \alpha_{2M}} (A_1)_1^{\alpha_1} (A_1)_2^{\alpha_2} \ldots (A_1)_M^{\alpha_M} \\
&\qquad (A_2)_1^{\alpha_{M+1}} (A_2)_2^{\alpha_{M+2}} \ldots (A_1)_M^{\alpha_{2M}}, \\
\mathcal{B} &\sim \epsilon_{\alpha_1 \alpha_2 \ldots \alpha_{2M}} (B_1)_1^{\alpha_1} (B_1)_2^{\alpha_2} \ldots (B_1)_M^{\alpha_M} \\
&\qquad (B_2)_1^{\alpha_{M+1}} (B_2)_2^{\alpha_{M+2}} \ldots (B_1)_M^{\alpha_{2M}}.
\end{aligned}
\tag{9.5}
$$

These operators contribute an additional term to the usual mesonic superpotential:

$$W = \lambda (N_{ij})^\alpha_\beta (N_{k\ell})^\beta_\alpha \epsilon^{ik} \epsilon^{j\ell} + X \left(\det[(N_{ij})^\alpha_\beta] - \mathcal{AB} - \Lambda_{2M}^{4M} \right), \tag{9.6}$$

where X can be understood as a Lagrange multiplier. The supersymmetry-preserving vacua include the baryonic branch:

$$X = 0; \qquad N_{ij} = 0; \qquad \mathcal{AB} = -\Lambda_{2M}^{4M}, \tag{9.7}$$

where the $SO(4)$ global symmetry rotating A_i, B_j is unbroken. In contrast, this global symmetry is broken along the mesonic branch $N_{ij} \neq 0$. Since the supergravity background of [92] is $SO(4)$ symmetric, it is natural to assume that the dual of this background lies on the baryonic branch of the cascading theory. The expectation values of the baryonic operators spontaneously break the $U(1)$ baryon number symmetry $A_k \rightarrow e^{i\alpha} A_k$, $B_j \rightarrow e^{-i\alpha} B_j$. The KS background corresponds to a vacuum where $|\mathcal{A}| = |\mathcal{B}| = \Lambda_{2M}^{2M}$, which is invariant under the exchange of the A's with the B's accompanied by charge conjugation in both gauge groups. This gives a field theory interpretation to the $\mathcal{I}$-symmetry of the warped deformed conifold background. As noted in [98], the baryonic branch has complex dimension one, and it can be parametrized by ξ as follows

$$\mathcal{A} = i\xi \Lambda_{2M}^{2M}, \qquad \mathcal{B} = \frac{i}{\xi} \Lambda_{2M}^{2M}. \tag{9.8}$$

The pseudo-scalar Goldstone mode must correspond to changing ξ by a phase, since this is precisely what a $U(1)_B$ symmetry transformation does.

Thus the non-compact warped deformed conifold exhibits a supergravity dual of the Goldstone mechanism due to breaking of the global $U(1)_B$ symmetry [95,98]. On the other hand, if one considered a warped deformed conifold throat embedded in a flux compactification, $U(1)_B$ would be gauged, the Goldstone boson $p(x)$ would combine with the $U(1)$ gauge field to form a massive vector, and therefore in this situation we would find a manifestation of the supersymmetric Higgs mechanism [95].

The scalar zero-mode

By supersymmetry the massless pseudoscalar is part of a massless $\mathcal{N} = 1$ chiral multiplet. and therefore there must also be a massless scalar mode and corresponding Weyl fermion, with the scalar corresponding to changing ξ by a positive real factor. This scalar zero-mode comes from a metric perturbation that mixes with the NSNS 2-form potential.

The warped deformed conifold preserves the $\mathbf{Z}_2$ interchange symmetry $\mathcal{I}$. However, the pseudo-scalar mode we found breaks this symmetry: from the form of the perturbations (9.1) we see that $\delta F^{(3)}$ is even under the interchange of (θ_1, ϕ_1) with (θ_2, ϕ_2), while $F^{(3)}$ is odd; similarly $\delta F^{(5)}$ is odd while $F^{(5)}$ is even. Therefore, the scalar mode must also break the $\mathcal{I}$ symmetry because in the field theory it breaks the symmetry between expectation values of $|\mathcal{A}|$ and of $|\mathcal{B}|$. The necessary translationally invariant perturbation that preserves the $SO(4)$ but breaks the $\mathcal{I}$ symmetry is given by the following variation of the NSNS 2-form and the metric:

$$\delta B_2 = \chi(\tau) \, dg^5, \qquad \delta G_{13} = \delta G_{24} = \lambda(\tau), \tag{9.9}$$

where, for example $\delta G_{13} = \lambda(\tau)$ means adding $2\lambda(\tau) g^{(1} g^{3)}$ to ds_{10}^2. To see that these components of the metric break the $\mathcal{I}$ symmetry, we note that

$$(e_1)^2 + (e_2)^2 - (\epsilon_1)^2 - (\epsilon_2)^2 = g^1 g^3 + g^3 g^1 + g^2 g^4 + g^4 g^2. \tag{9.10}$$

Defining $\lambda(\tau) = h^{1/2} K \sinh(\tau) z(\tau)$ one finds [95] that all the linearized supergravity equations are satisfied provided that

$$\frac{((K \sinh(\tau))^2 z')'}{(K \sinh(\tau))^2} = \left(2 + \frac{8}{9} \frac{1}{K^6} - \frac{4}{3} \frac{\cosh(\tau)}{K^3}\right) \frac{z}{\sinh(\tau)^2}, \tag{9.11}$$

and

$$\chi' = \frac{1}{2} g_s M z(\tau) \frac{\sinh(2\tau) - 2\tau}{\sinh^2 \tau}. \tag{9.12}$$

The solution of (9.11) for the zero-mode profile is remarkably simple:

$$z(\tau) = s \frac{(\tau \coth(\tau) - 1)}{[\sinh(2\tau) - 2\tau]^{1/3}}, \tag{9.13}$$

with s a constant. Like the pseudo-scalar perturbation, the large τ asymptotic is again $z \sim \tau e^{-2\tau/3}$. We note that the metric perturbation has the simple form $\delta G_{13} \sim h^{1/2}[\tau \coth(\tau) - 1]$. The perturbed metric $d\tilde{s}_2^6$ differs from the metric of the deformed conifold (8.3) by

$$\sim (\tau \coth \tau - 1)(g^1 g^3 + g^3 g^1 + g^2 g^4 + g^4 g^2), \tag{9.14}$$

which grows as $\ln r$ in the asymptotic radial variable r.

The scalar zero-mode is actually an exact modulus: there is a one-parameter family of supersymmetric solutions which break the $\mathcal{I}$ symmetry but preserve the $SO(4)$ (an ansatz with these properties was found in [96], and its linearization agrees with (9.9)). These backgrounds, the resolved warped deformed conifolds, will be reviewed in the next section. We add the word resolved because both the resolution of the conifold, which is a Kähler deformation, and these resolved warped deformed conifolds break the $\mathcal{I}$ symmetry. In the dual gauge theory turning on the $\mathcal{I}$ breaking corresponds to the transformation $\mathcal{A} \to (1 + s)\mathcal{A}$, $\mathcal{B} \to (1 + s)^{-1}\mathcal{B}$ on the baryonic branch. Therefore, s is dual to the $\mathcal{I}$ breaking parameter of the resolved warped deformed conifold.

The presence of these massless modes is a further indication that the infrared dynamics of the cascading $SU(M(k + 1)) \times SU(Mk)$ gauge theory, whose supergravity dual is the warped deformed conifold, is richer than that of the pure glue $\mathcal{N} = 1$ supersymmetric $SU(M)$ theory. The former incorporates a Goldstone supermultiplet, which appears due to the $U(1)_B$ symmetry breaking, as well as solitonic strings dual to the D-strings placed at $\tau = 0$ in the supergravity background.

Massive glueballs

Let us comment on massive normal modes of the warped deformed conifold. These can by found by studying linearized perturbations with four-dimensional momentum k_μ, and looking for the eigenvalues of $-k_\mu^2 = m_4^2$ at which the resulting equations of motion admit normalizable solutions. Some early results on the massive glueball spectra were obtained in [100, 101]. More recently, several families of such massive radial excitations which arise from a subset of the deformations contained in the PT ansatz were discussed in [102]. Interestingly, in all cases the mass-squared grows quadratically with the mode number n:

$$m_4^2 = An^2 + \mathcal{O}(n). \tag{9.15}$$

The quadratic dependence on n, which is characteristic of Kaluza–Klein theory, is a general feature of strongly coupled gauge theories that have weakly curved 10-d gravity duals (see [103] for a discussion). It was also observed that the coefficient A of the n^2 term is approximately universal in the sense that the values it takes for different towers of excitations are numerically very close to each other [102].

The towers of massive glueball states based on the pseudoscalar Goldstone boson and its scalar superpartner were recently studied in [104]. The necessary generalization of the scalar ansatz (9.9) to non-zero 4-d momentum is

$$\delta B^{(2)} = \chi(x, \tau) \, dg^5 + \partial_\mu \sigma(x, \tau) \, dx^\mu \wedge g^5, \tag{9.16}$$
$$\delta G_{13} = \delta G_{24} = \lambda(x, \tau). \tag{9.17}$$

A gauge equivalent way of writing (9.16) is

$$\delta B^{(2)} = (\chi - \sigma) \, dg^5 - \sigma' \, d\tau \wedge g^5. \tag{9.18}$$

Such an ansatz is more general than the generalized PT ansatz used in [102] in that it contains an extra function, σ. After some transformations we find that the linearized supergravity equations of motion reduce to two coupled equations that determine the glueball masses:

$$\tilde{z}'' - \frac{2}{\sinh^2 \tau} \tilde{z} + \tilde{m}^2 \frac{I(\tau)}{K^2(\tau)} \tilde{z} = \tilde{m}^2 \frac{9}{4 \cdot 2^{2/3}} K(\tau) \, \tilde{w}, \tag{9.19}$$

$$\tilde{w}'' - \frac{\cosh^2 \tau + 1}{\sinh^2 \tau} \tilde{w} + \tilde{m}^2 \frac{I(\tau)}{K^2(\tau)} \tilde{w} = \frac{16}{9} K(\tau) \, \tilde{z}, \tag{9.20}$$

where

$$\tilde{z} = h^{-1/2} \lambda, \qquad \tilde{w} = \frac{\epsilon^{4/3}}{g_s M \alpha'} K^5 \sinh(\tau)^2 \sigma', \tag{9.21}$$

and χ is determined by the solution for σ. Here the dimensionless eigenvalue $\tilde{m}^2$ is related to the mass-squared through

$$\tilde{m}^2 = m_4^2 \frac{2^{2/3}(g_s M \alpha')^2}{6\,\epsilon^{4/3}},\tag{9.22}$$

These coupled equations were solved numerically in [104] yielding radial excitation spectra of the asymptotic form (9.15) with the coefficients of quadratic terms close to those found in [102].

Note that the scale of these glueball mass-squared m_4^2, calculated in the limit $g_s M \gg 1$, is parametrically lower than the confining string tension (8.19). Using (9.22) and (8.17), we find that the coefficient A of the n^2 term is

$$A \sim \frac{T_s}{g_s M}.\tag{9.23}$$

Thus, for radial excitation numbers $n \ll \sqrt{g_s M}$, these modes are much lighter than the string tension scale, and therefore much lighter than all glueballs with spin > 2. Such anomalously light bound states appear to be special to gauge theories that stay very strongly coupled in the UV, such as the cascading gauge theory; they do not appear in asymptotically free gauge theories. Therefore, the anomalously light low-spin glueballs could perhaps be used as a special "signature" of gauge theories with weakly curved gravity duals, if they are realized in nature.

10. The baryonic branch

Since the global baryon number symmetry $U(1)_B$ is broken by expectation values of baryonic operators, the spectrum contains the Goldstone boson found above. The zero-momentum mode of the scalar superpartner of the Goldstone mode leads to a Lorentz-invariant deformation of the background which describes a small motion along the baryonic branch. In this section we shall extend the discussion from linearized perturbations around the warped deformed conifold solution to finite deformations, and describe the supergravity backgrounds dual to the complete baryonic branch. These are the resolved warped deformed conifolds, which preserve the $SO(4)$ global symmetry but break the discrete $\mathcal{I}$ symmetry of the warped deformed conifold.

The full set of first-order equations necessary to describe the entire moduli space of supergravity backgrounds dual to the baryonic branch, also called the resolved warped deformed conifolds, was derived and solved numerically in [105] (for a further discussion, see [106]). This continuous family of supergravity solutions is parameterized by the modulus of ξ (the phase of ξ is not manifest in these backgrounds). The corresponding metric can be written in the form of the Papadopoulos-Tseytlin ansatz [96] in the string frame:

$$ds^2 = h^{-1/2}\eta_{\mu\nu}dx^\mu dx^\nu + e^x ds_{\mathcal{M}}^2 = h^{-1/2}dx_{1,3}^2 + \sum_{i=1}^{6} G_i^2, \tag{10.1}$$

where

$$G_1 \equiv e^{(x+g)/2}e_1, \qquad G_3 \equiv e^{(x-g)/2}(\epsilon_1 - ae_1), \tag{10.2}$$

$$G_2 \equiv \frac{\cosh\tau + a}{\sinh\tau}e^{(x+g)/2}e_2 + \frac{e^g}{\sinh\tau}e^{(x-g)/2}(\epsilon_2 - ae_2), \tag{10.3}$$

$$G_4 \equiv \frac{e^g}{\sinh\tau}e^{(x+g)/2}e_2 - \frac{\cosh\tau + a}{\sinh\tau}e^{(x-g)/2}(\epsilon_2 - ae_2), \tag{10.4}$$

$$G_5 \equiv e^{x/2}v^{-1/2}d\tau, \qquad G_6 \equiv e^{x/2}v^{-1/2}g^5. \tag{10.5}$$

These one-forms describe a basis that rotates as we move along the radial direction, and are particularly convenient since they allow us to write down very simple expressions for the holomorphic $(3, 0)$ form

$$\Omega = (G_1 + iG_2) \wedge (G_3 + iG_4) \wedge (G_5 + iG_6), \tag{10.6}$$

and the fundamental $(1, 1)$ form

$$J = \frac{i}{2}\big[(G_1 + iG_2) \wedge (G_1 - iG_2) + (G_3 + iG_4) \wedge (G_3 - iG_4)$$
$$+ (G_5 + iG_6) \wedge (G_5 - iG_6)\big]. \tag{10.7}$$

While in the warped deformed conifold case there was a single warp factor $h(\tau)$, now we find several additional functions $x(\tau), g(\tau), a(\tau), v(\tau)$. The warp factor $h(\tau)$ is deformed away from (8.17) when $|\xi| \neq 1$.

The background also contains the fluxes

$$B^{(2)} = h_1(\epsilon_1 \wedge \epsilon_2 + e_1 \wedge e_2) + \chi(e_1 \wedge \epsilon_2 - \epsilon_1 \wedge e_2)$$
$$+ h_2(\epsilon_1 \wedge e_2 - \epsilon_2 \wedge e_1), \tag{10.8}$$

$$F^{(3)} = -\frac{1}{2}g_5 \wedge \big[\epsilon_1 \wedge \epsilon_2 + e_1 \wedge e_2 - b(\epsilon_1 \wedge e_2 - \epsilon_2 \wedge e_1)\big]$$
$$-\frac{1}{2}d\tau \wedge \big[b'(\epsilon_1 \wedge e_1 + \epsilon_2 \wedge e_2)\big], \tag{10.9}$$

$$\tilde{F}^{(5)} = \mathcal{F}^{(5)} + *_{10}\mathcal{F}^{(5)}, \tag{10.10}$$

$$\mathcal{F}^{(5)} = -(h_1 + bh_2)e_1 \wedge e_2 \wedge \epsilon_1 \wedge \epsilon_2 \wedge \epsilon_3, \tag{10.11}$$

parameterized by functions $h_1(\tau), h_2(\tau), b(\tau)$ and $\chi(\tau)$. In addition, since the 3-form flux is not imaginary self-dual for $|\xi| \neq 1$ (i.e. $\star_6 G_3 \neq i G_3$), the dilaton ϕ now also depends on the radial coordinate τ.

The functions a and v satisfy a system of coupled first order differential equations [105] whose solutions are known in closed form only in the warped deformed conifold and the Chamseddine–Volkov–Maldacena–Nunez (CVMN) [108] limits. All other functions $h, x, g, h_1, h_2, b, \chi, \phi$ are unambiguously determined by $a(\tau)$ and $v(\tau)$ through the relations

$$h = \gamma U^{-2}(e^{-2\phi} - 1), \qquad \gamma = 2^{10/3}(g_s M\alpha')^2 \varepsilon^{-8/3}, \tag{10.12}$$

$$e^{2x} = \frac{(bC-1)^2}{4(aC-1)^2} e^{2g+2\phi}(1 - e^{2\phi}), \tag{10.13}$$

$$e^{2g} = -1 - a^2 + 2a\,C, \qquad b = \frac{\tau}{S}, \tag{10.14}$$

$$h_2 = \frac{e^{2\phi}(bC-1)}{2S}, \qquad h_1 = -h_2\,C, \tag{10.15}$$

$$\chi' = a(b-C)(aC-1)\,e^{2(\phi-g)}, \tag{10.16}$$

$$\phi' = \frac{(C-b)\,(a\,C-1)^2}{(b\,C-1)\,S}\,e^{-2g}, \tag{10.17}$$

where $C \equiv -\cosh\tau$, $S \equiv -\sinh\tau$, and we require $\phi(\infty) = 0$. In writing these equations we have specialized to the baryonic branch of the cascading gauge theory by imposing appropriate boundary conditions at infinity [106], which guarantee that the background asymptotes to the warped conifold solution [107]. The full two parameter family of SU(3) structure backgrounds discussed in [105] also includes the CVMN solution [108], which however is characterized by linear dilaton asymptotics that are qualitatively different from the backgrounds discussed here. The baryonic branch family of supergravity solutions is labelled by one real "resolution parameter" U [106]. While the leading asymptotics of all supergravity backgrounds dual to the baryonic branch are identical to those of the warped deformed conifold, terms subleading at large τ depend on U. As required, this family of supergravity solutions preserves the $SU(2) \times SU(2)$ symmetry, but for $U \neq 0$ breaks the $\mathbb{Z}_2$ symmetry $\mathcal{I}$.

On the baryonic branch we can consider a transformation that takes ξ into ξ^{-1}, or equivalently U into $-U$. This transformation leaves $v(\tau)$ invariant and changes $a(\tau)$ as follows

$$a \to -\frac{a}{1 + 2a\cosh\tau}. \tag{10.18}$$

It is straightforward to check that ae^{-g} is invariant while $(1 + a\cosh\tau)e^{-g}$ changes sign. This transformation also exchanges $e^{g} + a^2e^{-g}$ with e^{-g} and therefore it is equivalent to the exchange of (θ_1, ϕ_1) and (θ_2, ϕ_2) involved in the $\mathcal{I}$-symmetry.

The baryonic condensates have been calculated on the string theory side of the duality [109] by identifying the Euclidean D5-branes wrapped over the deformed conifold, with appropriate gauge fields turned on, as the appropriate object dual to the baryonic operators in the sense of gauge/string duality [98]. Similarly to the case of baryonic operators on the warped resolved conifold discussed in Section 7, the field corresponding to a baryon is recognized as the (semi-classical) equivalent of $e^{-S_{D5}(r)}$, where $S_{D5}(r)$ is the action of a Euclidean D5-brane wrapping the Calabi–Yau coordinates up to the radial coordinate cut-off r. The different baryon operators $\mathcal{A}$, $\mathcal{B}$, and their conjugates $\overline{\mathcal{A}}$, $\overline{\mathcal{B}}$, are distinguished by the two possible D5-brane orientations, and the two possible κ-symmetric choices for the world volume gauge field that has to be turned on inside the D5-brane.

This identification is legitimate since in the cascading theory, which is near-AdS in the UV, equation (7.19) holds modulo powers of $\ln r$ [110, 111]. Due to the absence of sources for baryonic operators we again have $\varphi_0 = 0$, and $e^{-S_{D5}(r)}$ gives the dimensions of the baryon operators, and the values of their condensates.

In contrast to the simpler case of the warped resolved conifold, a world-volume gauge bundle $F^{(2)} = dA^{(1)}$ is required by κ-symmetry in this case, which leads to the conditions [112] that $\mathcal{F} \equiv F^{(2)} + B^{(2)}$ be a $(1, 1)$-form, and that

$$\frac{1}{2!}J \wedge J \wedge \mathcal{F} - \frac{1}{3!}\mathcal{F} \wedge \mathcal{F} \wedge \mathcal{F} = \mathfrak{g}\left(\frac{1}{3!}J \wedge J \wedge J - \frac{1}{2!}J \wedge \mathcal{F} \wedge \mathcal{F}\right). \tag{10.19}$$

Here $\mathfrak{g}$ would simply be a constant if the internal manifold were Calabi–Yau, but since we are dealing with a generalized Calabi–Yau with fluxes, $\mathfrak{g}$ becomes coordinate dependent, a function of τ in our case.

The $SU(2) \times SU(2)$ invariant ansatz for the gauge potential is given by

$$A^{(1)} = \zeta(\tau)g^5, \tag{10.20}$$

which together with (10.19) implies that ζ has to satisfy the differential equation

$$\zeta' = \frac{e^x(\mathfrak{g}\mathfrak{a} + \mathfrak{b})}{v(\mathfrak{a} - \mathfrak{g}\mathfrak{b})}, \tag{10.21}$$

where we have defined

$$\mathfrak{a}(\zeta, \tau) \equiv e^{-2x}[e^{2x} + h_2^2\sinh^2(\tau) - (\zeta + \chi)^2],$$
$$\mathfrak{b}(\zeta, \tau) \equiv 2e^{-x-g}\sinh(\tau)[a(\zeta + \chi) - h_2(1 + a\cosh(\tau))]. \tag{10.22}$$

Using the explicitly known Killing spinors of the baryonic branch backgrounds (or from an equivalent argument starting from the Dirac–Born–Infeld equations of motion) one can show [109] that

$$\mathfrak{g} = -\frac{e^{-x+g}h_2 \sinh(\tau)}{(1+a\cosh(\tau))} = \frac{e^{\phi}}{\sqrt{1-e^{2\phi}}}, \tag{10.23}$$

which determines ζ and thus the action of the Euclidean D5-brane:

$$S_{D5} \sim \int d\tau\, e^{-\phi}\sqrt{\det G + \mathcal{F}} = \int d\tau\, \frac{e^{-\phi}e^{3x}\sqrt{1+\mathfrak{g}^2}\,(\mathfrak{a}^2+\mathfrak{b}^2)}{v|\mathfrak{a}-\mathfrak{g}\mathfrak{b}|}. \tag{10.24}$$

The dimension of the baryon operators in the KS background can be extracted from the divergent terms of the D5-brane action as a function of the radial cut-off r, which leads to

$$\Delta(r) = r\frac{dS_{D5}(r)}{dr} = \frac{27g_s^2 M^3}{16\pi^2}(\ln(r))^2 + \mathcal{O}(\ln(r)). \tag{10.25}$$

To compare this with the cascading gauge theory, we use the fact that the baryon operators of the $SU(M(k+1)) \times SU(Mk)$ theory have the schematic form $\mathcal{A} \sim (A_1 A_2)^{k(k+1)M/2}$ and $\mathcal{B} \sim (B_1 B_2)^{k(k+1)M/2}$, with appropriate contractions described in [98]. For large k, their dimensions are $\Delta(k) \approx 3Mk(k+1)/4$. If we remember that the radius at which the k-th Seiberg duality is performed is given by

$$r(k) \sim \varepsilon^{2/3}\exp\left(\frac{2\pi k}{3g_s M}\right), \tag{10.26}$$

we find that the leading term in the operator dimension agrees with (10.25).

The baryon expectation values as a function of U can be computed by evaluating the finite terms in the action (10.24). The baryonic condensates calculated in this fashion [109] satisfy the important condition that $\langle\mathcal{A}\rangle\langle\mathcal{B}\rangle = $ const. along the whole baryonic branch. This leads to a precise relation between the baryonic branch modulus $|\xi|$ in the gauge theory (9.8) and the modulus U in the dual supergravity description.

Furthermore, pseudoscalar perturbations around the warped deformed conifold background are seen explicitly to shift the phase of the baryon expectation value, through the Chern–Simons term in the D5-brane action, as required for consistency.

11. Cosmology in the throat

A promising framework for realizing cosmological inflation [113] in string theory is D-brane inflation (for reviews and more complete references, see [114]). The original proposal [115] was to consider a D3-brane and a $\overline{\text{D3}}$-brane separated by some distance along the compactified dimensions, and with their world volumes spanning the 4 observable coordinates x^μ. From the 4-d point of view, the distance r between the branes is a scalar field that is identified with the inflaton. However, in flat space the Coulomb potential $\sim 1/r^4$ is typically too steep to support slow-roll inflation. An ingenious proposal to circumvent this problem [116] is to place the brane-antibrane pair in a warped throat region of a flux compactification, of which the warped deformed conifold is an explicitly known and ubiquitous [117] example.

The $\overline{\text{D3}}$-brane breaks supersymmetry and experiences potential a $2T_3/h(\tau)$ attracting it to the bottom of the conifold, $\tau = 0$. Its energy density there, $2T_3/h(0)$, plays the important role of "uplifting" the negative 4-d cosmological constant to a positive value in the KKLT model for moduli stabilization [118]. When a mobile D3-brane is added, it perturbs the background warp factor. The energy density of the $\overline{\text{D3}}$-brane becomes

$$V(r) = \frac{2T_3}{h(0) + \delta h(0, r)} \approx \frac{2T_3}{h(0)} - 2T_3 \frac{\delta h(0, r)}{h(0)^2}, \tag{11.1}$$

where $\delta h(0, r)$ is the perturbation of the warped factor at the position of the $\overline{\text{D3}}$-brane $r = 0$, caused by the D3-brane at radial coordinate r. For a D3-brane far from the tip of the throat, with radial coordinate $r \gg \varepsilon^{2/3}$, $\delta h(0, r) \approx 27/(32\pi^2 T_3 r^4)$ [119]. Thus, the potential assumes the form [116, 120] (note that the definition of r^2 here differs by a factor of $3/2$ from that in [119])

$$V(r) = \frac{2T_3}{h(0)} - \frac{27}{16\pi^2 h(0)^2 r^4}. \tag{11.2}$$

Thus, the force on the D3-brane is suppressed by a small factor $h(0)^{-2}$ compared to the force in unwarped space used in the original model [115]. We recall that

$$h(0) = a_0 (g_s M\alpha')^2 2^{2/3} \varepsilon^{-8/3}, \quad a_0 \approx 0.71805. \tag{11.3}$$

In flux compactifications containing a long KS throat, $h(0)$ is of order $e^{8\pi K/3g_s M}$ [121]. The flattening of the brane-antibrane potential by exponential warping is an important factor in constructing realistic brane inflation scenarios.

There are possibilities other than a $\overline{\text{D3}}$-brane at the bottom of the throat for creating a slow-varying potential for the mobile D3-brane. As pointed out in

[106], if the throat is taken to be a *resolved* warped deformed conifold, which was reviewed in Section 10, then the potential experienced by the D3-brane is

$$V(\tau) = T_3 h^{-1}(\tau)(e^{-\phi(\tau)} - 1). \tag{11.4}$$

The first term comes from the Born–Infeld term and has a factor of $e^{-\phi(\tau)}$; the second term, originating from the interaction with the background 4-form C_{0123}, does not have this factor. For the KS solution ($U = 0$), $\phi(\tau) = 0$; therefore, the potential vanishes and the D3-brane may be located at any point on the deformed conifold. For $U \neq 0$ we may use (10.12) to write

$$V(\tau) = \frac{T_3}{\gamma} \frac{U^2}{e^{-\phi(\tau)} + 1}. \tag{11.5}$$

Since $\phi(\tau)$ is a monotonically increasing function, the D3-brane is attracted to $\tau = 0$.

For large enough τ, the potential becomes

$$V(\tau) = \frac{T_3}{\gamma} \left[\frac{U^2}{2} - \frac{3U^4}{256}(4\tau - 1)e^{-4\tau/3} + \cdots \right]. \tag{11.6}$$

Since $e^{-4\tau/3} \sim \varepsilon^{8/3} r^{-4}$, this is rather similar to the brane-antibrane potential (11.2), and has the additional feature that the resolution parameter U may be varied. In a complete treatment, U should be determined by the details of the compactification.

Cosmic strings

In addition to suppressing the D3-brane potential, the large value of $h(0)$ is responsible for the viability of cosmic strings in flux compactifications containing long warped throats. Cosmic strings with Planckian tensions are ruled out by the CMB spectrum and other astrophysical observations (for a review, see [122]). The current constraints on cosmic string tension μ suggest $G\mu \ll 10^{-7}$, and they continue to improve. In traditional models where the string scale is not far from the 4-d Planck scale, the tension of a fundamental string far from the warped throat, $1/(2\pi\alpha')$, badly violates this constraint. However, at the bottom of the throat the tension is reduced to $1/(2\pi\alpha'\sqrt{h(0)})$ which can be consistent with the constraint [123]. It is remarkable that such a cosmic string has a dual description as a confining string in the cascading gauge theory dual to the throat. This shows that the phenomenon of color confinement may have implications reaching far beyond the physics of hadrons.

Type IIB flux compactifications with long warped throats may contain a variety of species of cosmic strings. q fundamental strings at the bottom of the throat may form a bound state, which is described by a D3-brane wrapping a 2-sphere within the 3-sphere [124]. A D-string at the bottom of the throat is dual to a certain solitonic string in the gauge theory [95]. Furthermore, p D-strings and q F-strings can bind into a (p, q) string [125]. Networks of such (p, q) cosmic strings have various characteristic features in their evolution and interaction probabilities which could distinguish them from other cosmic string models. Importantly, such strings are copiously produced during the brane-antibrane annihilation that follows the brane inflation [126]. However, in models where some D3-branes remain at the bottom of the throat after inflation (for example, [106]), long cosmic strings cannot exist because they break on the D-branes [122]. Thus, non-observation of cosmic strings would not rule out D-brane inflation.

Compactification effects

The potentials (11.2), (11.6) are very flat for large r, approaching a constant that arises due to D-term breaking of supersymmetry. However, additional contributions to the potential that arise due to moduli stabilization effects tend to destroy this flatness and generally render slow-roll inflation impossible. Indeed, in the compactified setting, the contribution of the brane-antibrane interaction to the 4-d "Einstein frame" potential is

$$V_D(\rho, r) = \frac{V(r)}{U^2(\rho, r)}. \tag{11.7}$$

The extra factor comes from the DeWolfe-Giddings Kähler potential [127] which depends both on the volume modulus, ρ, and the D3-brane position z_α, $\alpha = 1, 2, 3$:

$$\kappa^2 \mathcal{K}(\rho, \bar{\rho}, z_\alpha, \overline{z_\alpha}) = -3\log[\rho + \bar{\rho} - \beta k(z_\alpha, \overline{z_\alpha})] \equiv -3\log U. \tag{11.8}$$

Here $k(z_\alpha, \overline{z_\alpha})$ denotes the Kähler potential of the Calabi–Yau manifold, which in the throat reduces to

$$k = \frac{3}{2}\left(\sum_{i=1}^{4}|z_i|^2\right)^{2/3} = r^2, \tag{11.9}$$

where we ignored the deformation ε. The normalization constant β in (11.8) may be expressed as

$$\beta \equiv \frac{\sigma_0}{3}\frac{T_3}{M_P^2}, \tag{11.10}$$

where $2\sigma_0 \equiv 2\sigma_\star(0) = \rho_\star(0) + \bar\rho_\star(0)$ is the stabilized value of the Kähler modulus when the D3-brane is near the tip of the throat. For a D3-brane far from the tip, we may ignore the deformation of the conifold and use

$$U(\rho, r) = \rho + \bar\rho - \beta r^2. \tag{11.11}$$

The r-dependence from the factor $U^{-2}(\rho, r)$ spoils the flatness of the potential even in the region where $V(r)$ is very flat.

The complete inflaton potential

$$V_{tot} = V_F(\rho, z_\alpha) + V_D(\rho, r) \tag{11.12}$$

also includes the F-term contribution whose standard expression in $\mathcal{N} = 1$ supergravity is

$$V_F = e^{\kappa^2 \mathcal{K}} \big[D_\Sigma W \mathcal{K}^{\Sigma\bar\Omega} \overline{D_\Omega W} - 3\kappa^2 W \overline{W} \big], \quad \kappa^2 = M_P^{-2} \equiv 8\pi G, \tag{11.13}$$

where $\{Z^\Sigma\} \equiv \{\rho, z_\alpha; \alpha = 1, 2, 3\}$ and $D_\Sigma W = \partial_\Sigma W + \kappa^2(\partial_\Sigma \mathcal{K})W$. The superpotential W has the structure

$$W(\rho, z_\alpha) = W_0 + A(z_\alpha)e^{-a\rho}, \quad a \equiv \frac{2\pi}{n}, \tag{11.14}$$

where the second, nonperturbative term arises either from strong gauge dynamics on a stack of $n > 1$ D7-branes or from Euclidean D3-branes (with $n = 1$). We assume that either sort of brane supersymmetrically wraps a four-cycle in the warped throat that is specified by a holomorphic embedding equation $f(z_\alpha) = 0$. The warped volume of the four-cycle governs the magnitude of the nonperturbative effect, by affecting the gauge coupling on the D7-branes (equivalently, the action of Euclidean D3-branes) wrapping this four-cycle. The presence of a D3-brane gives rise to a perturbation to the warp factor, and this leads to a correction to the warped four-cycle volume. This correction depends on the D3-brane position and is responsible for the prefactor $A(z_\alpha)$ [128]. In [119], D3-brane backreaction on the warped four-cycle volume was calculated leading to a simple formula

$$A(z_\alpha) = A_0 \left(\frac{f(z_\alpha)}{f(0)} \right)^{1/n}. \tag{11.15}$$

The canonical inflaton φ is proportional to r, the radial location of the D3-brane. Using (11.12) to compute the slow-roll parameter

$$\eta \equiv M_P^2 \frac{V_{,\varphi\varphi}}{V}, \tag{11.16}$$

one finds

$$\eta = \frac{2}{3} + \Delta\eta(\varphi), \tag{11.17}$$

where the 2/3 arises from the Kähler potential (11.8), (11.11), and $\Delta\eta$ from the variation of $A(z_\alpha)$. A simple model studied in [120] was based on the Kuperstein embedding of a stack of D7-branes

$$f(z_1) = \mu - z_1 = 0. \tag{11.18}$$

Explicit calculation of the full inflaton potential in this model [120, 129] shows that it is possible to fine-tune the parameters to achieve an inflection point in the vicinity of which slow-roll inflation is possible.

Other supergravity and string theory constructions where such Inflection Point Inflation may be achieved were proposed in [130–133]. While such models are fine-tuned, and the initial conditions have to be chosen carefully to prevent the field from running through the inflection point with high speed (see, however, [132, 134] for possible ways to circumvent this problem) the model is fairly predictive. In particular, the spectral index n_s is around 0.93 in the limit that the total number of e-folds is large during inflation. This fact may render this model distinguishable from others by more precise observations of the CMB.

12. Summary

Throughout its history, string theory has been intertwined with the theory of strong interactions. The AdS/CFT correspondence [1–3] has succeeded in making precise connections between conformal 4-dimensional gauge theories and superstring theories in 10 dimensions. This duality leads to a multitude of dynamical predictions about strongly coupled gauge theories. While many of these predictions are difficult to check, recent applications of methods of exact integrability to planar $\mathcal{N} = 4$ SYM theory have produced some impressive tests of the correspondence for operators with high spin. When extended to theories at finite temperature, the correspondence serves as a theoretical laboratory for studying a novel state of matter: a gluonic plasma at very strong coupling. This appears to have surprising connections to the new state of matter, sQGP, which was observed at RHIC and will be further studied at the LHC.

Breaking symmetries in the AdS/CFT correspondence is important for bringing it closer to the real world. Some of the supersymmetry may be broken by considering D3-branes at conical singularities; the case of the conifold is discussed in detail in these lectures. In this set-up, breaking of gauge symmetry typically leads to a resolution of the singularity. The associated breaking of global symmetry leads to the appearance of Goldstone bosons and global strings. Extensions of the gauge-string duality to confining gauge theories provide new geometrical viewpoints on such important phenomena as chiral symmetry breaking and dimensional transmutation, which are encoded in the dual smooth warped throat background. Embedding of the throat into flux compactifications of string theory allows for an interesting interplay between gauge-string duality and models of particle physics and cosmology. For example, D3-branes rolling in the throat might model inflation while various strings attracted to the bottom of the throat may describe cosmic strings. All of this raises hopes that the new window into strongly coupled gauge theory opened by the discovery of gauge-string dualities will one day lead to new striking connections between string theory and the real world.

Acknowledgements

We thank the organizers of the Les Houches 2007 summer school "String Theory and the Real World" for hospitality and for organizing a very stimulating school. We are very grateful to D. Baumann, S. Benvenuti, A. Dymarsky, S. Gubser, C. Herzog, J. Maldacena, L. McAllister, A. Murugan, A. Peet, A. Polyakov, D. Rodriguez-Gomez, A. Scardicchio, A. Solovyov, M. Strassler, A. Tseytlin, J. Ward and E. Witten for collaboration on some of the papers reviewed here. We also thank D. Baumann, C. Herzog and A. Murugan for helpful comments on the manuscript. This research is supported in part by the National Science Foundation Grant No. PHY-0243680.

References

[1] J.M. Maldacena, The large N limit of superconformal field theories and supergravity, *Adv. Theor. Math. Phys.* **2** (1998) 231 [arXiv:hep-th/9711200].

[2] S.S. Gubser, I.R. Klebanov and A.M. Polyakov, Gauge theory correlators from non-critical string theory, *Phys. Lett. B* **428** (1998) 105 [arXiv:hep-th/9802109].

[3] E. Witten, Anti-de Sitter space and holography, *Adv. Theor. Math. Phys.* **2** (1998) 253 [arXiv:hep-th/9802150].

[4] I.R. Klebanov, QCD and string theory, *Int. J. Mod. Phys. A* **21** (2006) 1831 [arXiv:hep-ph/0509087].

[5] R. Dolen, D. Horn and C. Schmid, Finite energy sum rules and their application to Pi N charge exchange, *Phys. Rev.* **166** (1968) 1768.

[6] G. Veneziano, Construction of a crossing—symmetric, Regge behaved amplitude for linearly rising trajectories, *Nuovo Cim. A* **57** (1968) 190.

[7] Y. Nambu, Quark model and the factorization of the Veneziano amplitude, in: *Symmetries and Quark Models*, R. Chand (ed.), Gordon and Breach (1970).

[8] H.B. Nielsen, An almost physical interpretation of the integrand of the n-point Veneziano amplitude, submitted to the 15th International Conference on High Energy Physics, Kiev (1970).

[9] L. Susskind, Dual-symmetric theory of hadrons, *Nuovo Cim.* **69A** (1970) 457.

[10] P. Di Vecchia and A. Schwimmer, The beginning of string theory: a historical sketch, arXiv:0708.3940 [physics.hist-ph]; P. Di Vecchia, The birth of string theory, arXiv:0704.0101 [hep-th].

[11] D.J. Gross and F. Wilczek, Ultraviolet behavior of non-Abelian gauge theories, *Phys. Rev. Lett.* **30** (1973) 1343; H.D. Politzer, Reliable perturbative results for strong interactions? *Phys. Rev. Lett.* **30** (1973) 1346.

[12] J. Scherk and J. Schwarz, Dual models for non-hadrons, *Nucl. Phys.* **B81** (1974) 118.

[13] T. Yoneya, Connection of dual models to electrodynamics and gravidynamics, *Prog. Theor. Phys.* **51** (1974) 1907.

[14] K.G. Wilson, Confinement of quarks, *Phys. Rev. D* **10** (1974) 2445.

[15] Y. Nambu, QCD and the string model, *Phys. Lett. B* **80** (1979) 372.

[16] B. Andersson, G. Gustafson, G. Ingelman and T. Sjostrand, Parton fragmentation and string dynamics, *Phys. Rept.* **97** (1983) 31.

[17] M. Luscher, K. Symanzik and P. Weisz, Anomalies of the free loop wave equation in the Wkb approximation, *Nucl. Phys. B* **173** (1980) 365.

[18] M. Luscher and P. Weisz, Quark confinement and the bosonic string, *JHEP* **0207** (2002) 049 [arXiv:hep-lat/0207003].

[19] G. 't Hooft, A planar diagram theory for strong interactions, *Nucl. Phys. B* **72** (1974) 461.

[20] A.M. Polyakov, Quantum geometry of bosonic strings, *Phys. Lett. B* **103** (1981) 207;

[21] A.M. Polyakov, String theory and quark confinement, *Nucl. Phys. Proc. Suppl.* **68** (1998) 1 [arXiv:hep-th/9711002].

[22] J. Polchinski, Dirichlet-branes and Ramond-Ramond charges, *Phys. Rev. Lett.* **75** (1995) 4724 [arXiv:hep-th/9510017].

[23] G.T. Horowitz and A. Strominger, Black strings and P-branes, *Nucl. Phys. B* **360** (1991) 197.

[24] I.R. Klebanov, World-volume approach to absorption by non-dilatonic branes, *Nucl. Phys. B* **496** (1997) 231 [arXiv:hep-th/9702076]; S.S. Gubser, I.R. Klebanov and A.A. Tseytlin, String theory and classical absorption by three-branes, *Nucl. Phys. B* **499** (1997) 217 [arXiv:hep-th/9703040]; S.S. Gubser and I.R. Klebanov, Absorption by branes and Schwinger terms in the world volume theory, *Phys. Lett. B* **413** (1997) 41 [arXiv:hep-th/9708005].

[25] J.M. Maldacena, Wilson loops in large N field theories, *Phys. Rev. Lett.* **80** (1998) 4859 [arXiv:hep-th/9803002];
S.J. Rey and J.T. Yee, Macroscopic strings as heavy quarks in large N gauge theory and anti-de Sitter supergravity, *Eur. Phys. J. C* **22** (2001) 379 [arXiv:hep-th/9803001].

[26] D. Berenstein, J.M. Maldacena and H. Nastase, Strings in flat space and pp waves from $N = 4$ super Yang–Mills, *JHEP* **0204** (2002) 013 [arXiv:hep-th/0202021].

[27] S.S. Gubser, I.R. Klebanov and A.M. Polyakov, A semi-classical limit of the gauge/string correspondence, *Nucl. Phys. B* **636** (2002) 99 [arXiv:hep-th/0204051].

[28] A.A. Tseytlin, Semiclassical strings and AdS/CFT, arXiv:hep-th/0409296.

[29] A.V. Belitsky, V.M. Braun, A.S. Gorsky and G.P. Korchemsky, Integrability in QCD and beyond, *Int. J. Mod. Phys. A* **19** (2004) 4715 [arXiv:hep-th/0407232]; N. Beisert, The dilatation operator of $N = 4$ super Yang–Mills theory and integrability, *Phys. Rept.* **405** (2005) 1 [arXiv:hep-th/0407277].

[30] D.J. Gross and F. Wilczek, Asymptotically free gauge theories. 2, *Phys. Rev. D* **9** (1974) 980; H. Georgi and H.D. Politzer, Electroproduction scaling in an asymptotically free theory of strong interactions, *Phys. Rev. D* **9** (1974) 416.

[31] G.P. Korchemsky, Asymptotics of the Altarelli–Parisi–Lipatov evolution kernels of parton distributions, *Mod. Phys. Lett. A* **4** (1989) 1257; G.P. Korchemsky and G. Marchesini, Structure function for large x and renormalization of Wilson loop, *Nucl. Phys. B* **406** (1993) 225 [arXiv:hep-ph/9210281].

[32] E.G. Floratos, D.A. Ross, Christopher T. Sachrajda, Higher order effects in asymptotically free gauge theories. 2. Flavor singlet Wilson operators and coefficient functions, *Nucl. Phys.* **B152** (1979) 493.

[33] G. Sterman and M.E. Tejeda-Yeomans, Multi-loop amplitudes and resummation, *Phys. Lett. B* **552** (2003) 48 [arXiv:hep-ph/0210130].

[34] S. Frolov and A.A. Tseytlin, Semiclassical quantization of rotating superstring in $AdS(5) \times S(5)$, *JHEP* **0206** (2002) 007 [arXiv:hep-th/0204226].

[35] A.V. Belitsky, A.S. Gorsky and G.P. Korchemsky, Logarithmic scaling in gauge/string correspondence, *Nucl. Phys. B* **748** (2006) 24 [arXiv:hep-th/0601112].

[36] A.V. Kotikov, L.N. Lipatov, A.I. Onishchenko and V.N. Velizhanin, Three-loop universal anomalous dimension of the Wilson operators in $N = 4$ SUSY Yang–Mills model, *Phys. Lett. B* **595** (2004) 521 [arXiv:hep-th/0404092].

[37] Z. Bern, L.J. Dixon and V.A. Smirnov, Iteration of planar amplitudes in maximally supersymmetric Yang–Mills theory at three loops and beyond, arXiv:hep-th/0505205.

[38] S. Moch, J.A. M. Vermaseren and A. Vogt, The three-loop splitting functions in QCD: The non-singlet case, *Nucl. Phys. B* **688** (2004) 101 [arXiv:hep-ph/0403192].

[39] M. Kruczenski, A note on twist two operators in $N = 4$ SYM and Wilson loops in Minkowski signature, *JHEP* **0212** (2002) 024 [arXiv:hep-th/0210115].

[40] J.A. Minahan and K. Zarembo, The Bethe-ansatz for $N = 4$ super Yang–Mills, *JHEP* **0303** (2003) 013 [arXiv:hep-th/0212208]; N. Beisert, C. Kristjansen and M. Staudacher, The dilatation operator of $N = 4$ super Yang–Mills theory, *Nucl. Phys. B* **664** (2003) 131 [arXiv:hep-th/0303060]; N. Beisert and M. Staudacher, The $N = 4$ SYM integrable super spin chain, *Nucl. Phys. B* **670** (2003) 439 [arXiv:hep-th/0307042].

[41] B. Eden and M. Staudacher, Integrability and transcendentality, *J. Stat. Mech.* **0611**, P014 (2006) [arXiv:hep-th/0603157].

[42] N. Beisert, R. Hernandez and E. Lopez, A crossing-symmetric phase for $AdS(5) \times S * *5$ strings, *JHEP* **0611** (2006) 070 [arXiv:hep-th/0609044].

[43] N. Beisert, B. Eden and M. Staudacher, Transcendentality and crossing, *J. Stat. Mech.* **0701**, P021 (2007) [arXiv:hep-th/0610251].

[44] A.V. Belitsky, Long-range $SL(2)$ Baxter equation in $N = 4$ super-Yang–Mills theory, *Phys. Lett. B* **643** (2006) 354 [arXiv:hep-th/0609068].

[45] G. Arutyunov, S. Frolov and M. Staudacher, Bethe ansatz for quantum strings, *JHEP* **0410** (2004) 016 [arXiv:hep-th/0406256].

[46] N. Beisert and T. Klose, Long-range $gl(n)$ integrable spin chains and plane-wave matrix theory, *J. Stat. Mech.* **0607**, P006 (2006) [arXiv:hep-th/0510124].

[47] R. Hernandez and E. Lopez, Quantum corrections to the string Bethe ansatz, *JHEP* **0607** (2006) 004 [arXiv:hep-th/0603204].

[48] L. Freyhult and C. Kristjansen, A universality test of the quantum string Bethe ansatz, *Phys. Lett. B* **638** (2006) 258 [arXiv:hep-th/0604069].

[49] D.M. Hofman and J.M. Maldacena, Giant magnons, *J. Phys. A* **39** (2006) 13095 [arXiv:hep-th/0604135].

[50] R.A. Janik, The $AdS(5) \times S**5$ superstring worldsheet S-matrix and crossing symmetry, *Phys. Rev. D* **73** (2006) 086006 [arXiv:hep-th/0603038].

[51] Z. Bern, M. Czakon, L.J. Dixon, D.A. Kosower and V.A. Smirnov, The four-loop planar amplitude and cusp anomalous dimension in maximally supersymmetric Yang–Mills theory, *Phys. Rev. D* **75** (2007) 085010 [arXiv:hep-th/0610248].

[52] F. Cachazo, M. Spradlin and A. Volovich, Four-loop cusp anomalous dimension from obstructions, *Phys. Rev. D* **75** (2007) 105011 [arXiv:hep-th/0612309].

[53] M.K. Benna, S. Benvenuti, I.R. Klebanov and A. Scardicchio, A test of the AdS/CFT correspondence using high-spin operators, *Phys. Rev. Lett.* **98** (2007) 131603 [arXiv:hep-th/0611135].

[54] L.F. Alday, G. Arutyunov, M.K. Benna, B. Eden and I.R. Klebanov, On the strong coupling scaling dimension of high spin operators, *JHEP* **0704** (2007) 082 [arXiv:hep-th/0702028].

[55] I. Kostov, D. Serban and D. Volin, Strong coupling limit of Bethe ansatz equations, *Nucl. Phys. B* **789** (2008) 413 [arXiv:hep-th/0703031].

[56] B. Basso, G.P. Korchemsky and J. Kotanski, Cusp anomalous dimension in maximally supersymmetric Yang–Mills theory at strong coupling, arXiv:0708.3933 [hep-th].

[57] I. Kostov, D. Serban and D. Volin, Functional BES equation, arXiv:0801.2542 [hep-th].

[58] R. Roiban, A. Tirziu and A.A. Tseytlin, Two-loop world-sheet corrections in $AdS_5 \times S^5$ superstring, *JHEP* **0707** (2007) 056 [arXiv:0704.3638 [hep-th]]; R. Roiban and A.A. Tseytlin, Strong-coupling expansion of cusp anomaly from quantum superstring, *JHEP* **0711** (2007) 016 [arXiv:0709.0681 [hep-th]].

[59] S.S. Gubser, I.R. Klebanov and A.W. Peet, Entropy and temperature of black 3-branes, *Phys. Rev. D* **54** (1996) 3915 [arXiv:hep-th/9602135]; I.R. Klebanov and A.A. Tseytlin, Entropy of near-extremal black p-branes, *Nucl. Phys. B* **475** (1996) 164 [arXiv:hep-th/9604089].

[60] S.S. Gubser, I.R. Klebanov and A.A. Tseytlin, Coupling constant dependence in the thermodynamics of $N = 4$ supersymmetric Yang–Mills theory, *Nucl. Phys. B* **534** (1998) 202 [arXiv:hep-th/9805156].

[61] A. Fotopoulos and T.R. Taylor, Comment on two-loop free energy in $N = 4$ supersymmetric Yang–Mills theory at finite temperature, *Phys. Rev. D* **59** (1999) 061701 [arXiv:hep-th/9811224]; M.A. Vazquez-Mozo, A note on supersymmetric Yang–Mills thermodynamics, *Phys. Rev. D* **60** (1999) 106010 [arXiv:hep-th/9905030]; C.J. Kim and S.J. Rey, Thermodynamics of large-N super Yang–Mills theory and AdS/CFT correspondence, *Nucl. Phys. B* **564** (2000) 430 [arXiv:hep-th/9905205].

[62] F. Karsch, Lattice QCD at high temperature and density, *Lect. Notes Phys.* **583** (2002) 209 [arXiv:hep-lat/0106019].

[63] R.V. Gavai, S. Gupta and S. Mukherjee, A new method to determine the equation of state, specific heat, and speed of sound above and below the transition temperature in QCD, arXiv:hep-lat/0506015.

[64] B. Bringoltz and M. Teper, The pressure of the $SU(N)$ lattice gauge theory at large-N, *Phys. Lett. B* **628** (2005) 113 [arXiv:hep-lat/0506034].

[65] G. Policastro, D.T. Son and A.O. Starinets, The shear viscosity of strongly coupled $N = 4$ supersymmetric Yang–Mills plasma, *Phys. Rev. Lett.* **87** (2001) 081601 [arXiv:hep-th/0104066].

[66] S.C. Huot, S. Jeon and G.D. Moore, Shear viscosity in weakly coupled $N = 4$ super Yang–Mills theory compared to QCD, *Phys. Rev. Lett.* **98** (2007) 172303 [arXiv:hep-ph/0608062].

[67] A. Buchel, J.T. Liu and A.O. Starinets, Coupling constant dependence of the shear viscosity in $N = 4$ supersymmetric Yang–Mills theory, *Nucl. Phys. B* **707** (2005) 56 [arXiv:hep-th/0406264].

[68] P. Kovtun, D.T. Son and A.O. Starinets, Holography and hydrodynamics: Diffusion on stretched horizons, *JHEP* **0310** (2003) 064 [arXiv:hep-th/0309213]; Viscosity in strongly interacting quantum field theories from black hole physics, *Phys. Rev. Lett.* **94** (2005) 111601 [arXiv:hep-th/0405231].

[69] T.D. Cohen, Is there a 'most perfect fluid' consistent with quantum field theory? *Phys. Rev. Lett.* **99** (2007) 021602 [arXiv:hep-th/0702136].

[70] Y. Kats and P. Petrov, Effect of curvature squared corrections in AdS on the viscosity of the dual gauge theory, arXiv:0712.0743 [hep-th]; M. Brigante, H. Liu, R.C. Myers, S. Shenker and S. Yaida, Viscosity bound violation in higher derivative gravity, arXiv:0712.0805 [hep-th].

[71] D. Teaney, Effect of shear viscosity on spectra, elliptic flow, and Hanbury Brown-Twiss radii, *Phys. Rev. C* **68** (2003) 034913.

[72] T. Hirano and M. Gyulassy, Perfect fluidity of the quark gluon plasma core as seen through its dissipative hadronic corona, arXiv:nucl-th/0506049.

[73] P. Romatschke and U. Romatschke, Viscosity information from relativistic nuclear collisions: How perfect is the fluid observed at RHIC? *Phys. Rev. Lett.* **99** (2007) 172301 [arXiv:0706.1522 [nucl-th]].

[74] H.B. Meyer, A calculation of the shear viscosity in $SU(3)$ gluodynamics, *Phys. Rev. D* **76** (2007) 101701 [arXiv:0704.1801 [hep-lat]].

[75] M. Gyulassy and L. McLerran, New forms of QCD matter discovered at RHIC, *Nucl. Phys. A* **750** (2005) 30 [arXiv:nucl-th/0405013].

[76] E.V. Shuryak, What RHIC experiments and theory tell us about properties of quark-gluon plasma? *Nucl. Phys. A* **750** (2005) 64 [arXiv:hep-ph/0405066].

[77] R. Pisarsky, talk available at http://quark.phy.bnl.gov/ pisarski/talks/unicorn.pdf.

[78] S. Kachru and E. Silverstein, 4d conformal theories and strings on orbifolds, *Phys. Rev. Lett.* **80** (1998) 4855 [arXiv:hep-th/9802183];
A.E. Lawrence, N. Nekrasov and C. Vafa, On conformal field theories in four dimensions, *Nucl. Phys. B* **533** (1998) 199 [arXiv:hep-th/9803015].

[79] A. Kehagias, New type IIB vacua and their F-theory interpretation, *Phys. Lett. B* **435** (1998) 337 [arXiv:hep-th/9805131].

[80] I.R. Klebanov and E. Witten, Superconformal field theory on threebranes at a Calabi–Yau singularity, *Nucl. Phys. B* **536** (1998) 199 [arXiv:hep-th/9807080].

[81] D. Morrison and R. Plesser, Non-spherical horizons, I, *Adv. Theor. Math. Phys.* **3** (1999) 1, [arXiv:hep-th/9810201].

[82] C.P. Herzog, I.R. Klebanov and P. Ouyang, D-branes on the conifold and $N = 1$ gauge/gravity dualities, arXiv:hep-th/0205100.

[83] P. Candelas and X.C. de la Ossa, Comments on conifolds, *Nucl. Phys. B* **342** (1990) 246.

[84] L. Romans, New compactifications of chiral $N = 2$, $d = 10$ supergravity, *Phys. Lett.* **B153** (1985) 392.

[85] I.R. Klebanov and E. Witten, AdS/CFT correspondence and symmetry breaking, *Nucl. Phys. B* **556** (1999) 89 [arXiv:hep-th/9905104].

[86] I.R. Klebanov and A. Murugan, Gauge/gravity duality and warped resolved conifold, *JHEP* **0703** (2007) 042 [arXiv:hep-th/0701064].

[87] L.A. Pando Zayas and A.A. Tseytlin, 3-branes on resolved conifold, *JHEP* **0011** (2000) 028 [arXiv:hep-th/0010088].

[88] I.R. Klebanov, A. Murugan, D. Rodriguez-Gomez and J. Ward, Goldstone bosons and global strings in a warped resolved conifold, arXiv:0712.2224 [hep-th].

[89] S.S. Gubser and I.R. Klebanov, Baryons and domain walls in an $N = 1$ superconformal gauge theory, *Phys. Rev. D* **58** (1998) 125025 [arXiv:hep-th/9808075].

[90] I.R. Klebanov and E. Witten, AdS/CFT correspondence and symmetry breaking, *Nucl. Phys. B* **556** (1999) 89 [arXiv:hep-th/9905104].

[91] H. Georgi, Unparticle physics, *Phys. Rev. Lett.* **98** (2007) 221601 [arXiv:hep-ph/0703260].

[92] I.R. Klebanov and M.J. Strassler, Supergravity and a confining gauge theory: Duality cascades and chiSB-resolution of naked singularities, *JHEP* **0008** (2000) 052 [arXiv:hep-th/0007191].

[93] N. Seiberg, Electric–magnetic duality in supersymmetric non-Abelian gauge theories, *Nucl. Phys. B* **435** (1995) 129 [arXiv:hep-th/9411149].

[94] M.J. Strassler, The duality cascade, arXiv:hep-th/0505153.

[95] S.S. Gubser, C.P. Herzog and I.R. Klebanov, Symmetry breaking and axionic strings in the warped deformed conifold, *JHEP* **0409** (2004) 036 [arXiv:hep-th/0405282]; Variations on the warped deformed conifold, *Comptes Rendus Physique* **5** (2004) 1031 [arXiv:hep-th/0409186].

[96] G. Papadopoulos and A.A. Tseytlin, Complex geometry of conifolds and 5-brane wrapped on 2-sphere, *Class. Quant. Grav.* **18** (2001) 1333 [arXiv:hep-th/0012034].

[97] A. Ceresole, G. Dall'Agata, R. D'Auria and S. Ferrara, Spectrum of type IIB supergravity on $AdS(5) \times T(11)$: Predictions on $N = 1$ SCFT's, *Phys. Rev. D* **61** (2000) 066001 [arXiv:hep-th/9905226].

[98] O. Aharony, A note on the holographic interpretation of string theory backgrounds with varying flux, *JHEP* **0103** (2001) 012 [arXiv:hep-th/0101013].

[99] N. Seiberg, Exact results on the space of vacua of four-dimensional susy gauge theories, *Phys. Rev. D* **49** (1994) 6857 [arXiv:hep-th/9402044].

[100] M. Krasnitz, A two point function in a cascading $N = 1$ gauge theory from supergravity, arXiv:hep-th/0011179; Correlation functions in a cascading $N = 1$ gauge theory from supergravity, *JHEP* **0212** (2002) 048 [arXiv:hep-th/0209163].

[101] E. Caceres, A brief review of glueball masses from gauge/gravity duality, *J. Phys. Conf. Ser.* **24** (2005) 111.

[102] M. Berg, M. Haack and W. Mueck, Bulk dynamics in confining gauge theories, *Nucl. Phys. B* **736** (2006) 82 [arXiv:hep-th/0507285]; Glueballs vs. gluinoballs: Fluctuation spectra in non-AdS/non-CFT, *Nucl. Phys. B* **789** (2008) 1 [arXiv:hep-th/0612224].

[103] A. Karch, E. Katz, D.T. Son and M.A. Stephanov, Linear confinement and AdS/QCD, *Phys. Rev. D* **74** (2006) 015005 [arXiv:hep-ph/0602229].

[104] M.K. Benna, A. Dymarsky, I.R. Klebanov and A. Solovyov, On normal modes of a warped throat, arXiv:0712.4404 [hep-th].

[105] A. Butti, M. Grana, R. Minasian, M. Petrini and A. Zaffaroni, The baryonic branch of Klebanov–Strassler solution: A supersymmetric family of $SU(3)$ structure backgrounds, *JHEP* **0503** (2005) 069 [arXiv:hep-th/0412187].

[106] A. Dymarsky, I.R. Klebanov and N. Seiberg, On the moduli space of the cascading $SU(M+p) \times SU(p)$ gauge theory, *JHEP* **0601** (2006) 155 [arXiv:hep-th/0511254].

[107] I.R. Klebanov and A.A. Tseytlin, Gravity duals of supersymmetric $SU(N) \times SU(N + M)$ gauge theories, *Nucl. Phys. B* **578** (2000) 123 [arXiv:hep-th/0002159].

[108] A.H. Chamseddine and M.S. Volkov, Non-Abelian BPS monopoles in $N = 4$ gauged supergravity, *Phys. Rev. Lett.* **79** (1997) 3343 [arXiv:hep-th/9707176]; Non-Abelian solitons in $N = 4$ gauged supergravity and leading order string theory, *Phys. Rev. D* **57** (1998) 6242 [arXiv:hep-th/9711181]; J.M. Maldacena and C. Nunez, Towards the large N limit of pure $N = 1$ super Yang–Mills, *Phys. Rev. Lett.* **86** (2001) 588 [arXiv:hep-th/0008001].

[109] M.K. Benna, A. Dymarsky and I.R. Klebanov, Baryonic condensates on the conifold, *JHEP* **0708** (2007) 034 [arXiv:hep-th/0612136].

[110] O. Aharony, A. Buchel and A. Yarom, Holographic renormalization of cascading gauge theories, *Phys. Rev. D* **72** (2005) 066003 [arXiv:hep-th/0506002].

[111] O. Aharony, A. Buchel and A. Yarom, Short distance properties of cascading gauge theories, *JHEP* **0611** (2006) 069 [arXiv:hep-th/0608209].

[112] M. Marino, R. Minasian, G.W. Moore and A. Strominger, Nonlinear instantons from super-symmetric p-branes, *JHEP* **0001** (2000) 005 [arXiv:hep-th/9911206].

[113] A.H. Guth, The inflationary universe: A possible solution to the horizon and flatness problems, *Phys. Rev. D* **23** (1981) 347; A.D. Linde, A new inflationary universe scenario: A possible solution of the horizon, flatness, homogeneity, isotropy and primordial monopole problems, *Phys. Lett. B* **108** (1982) 389; A. Albrecht and P.J. Steinhardt, Cosmology for grand unified theories with radiatively induced symmetry breaking, *Phys. Rev. Lett.* **48** (1982) 1220.

[114] A. Linde, Inflation and string cosmology, *eConf* **C040802**, L024 (2004) [*J. Phys. Conf. Ser.* **24** (2005 PTPSA,163,295-322.2006) 151] [arXiv:hep-th/0503195]; S.H. Henry Tye, Brane infla-tion: String theory viewed from the cosmos, arXiv:hep-th/0610221; J.M. Cline, String cosmol-ogy, arXiv:hep-th/0612129; R. Kallosh, On inflation in string theory, arXiv:hep-th/0702059. L. McAllister and E. Silverstein, String cosmology: a review, arXiv:0710.2951 [hep-th].

[115] G.R. Dvali and S.H. H. Tye, Brane inflation, *Phys. Lett. B* **450** (1999) 72 [arXiv:hep-ph/9812483].

[116] S. Kachru, R. Kallosh, A. Linde, J.M. Maldacena, L.P. McAllister and S.P. Trivedi, Towards inflation in string theory, *JCAP* **0310** (2003) 013 [arXiv:hep-th/0308055].

[117] A. Hebecker and J. March-Russell, The ubiquitous throat, *Nucl. Phys. B* **781** (2007) 99 [arXiv:hep-th/0607120].

[118] S. Kachru, R. Kallosh, A. Linde and S.P. Trivedi, De Sitter vacua in string theory, *Phys. Rev. D* **68** (2003) 046005 [arXiv:hep-th/0301240].

[119] D. Baumann, A. Dymarsky, I.R. Klebanov, J.M. Maldacena, L.P. McAllister and A. Murugan, On D3-brane potentials in compactifications with fluxes and wrapped D-branes, *JHEP* **0611** (2006) 031 [arXiv:hep-th/0607050].

[120] D. Baumann, A. Dymarsky, I.R. Klebanov and L. McAllister, Towards an explicit model of D-brane inflation, *JCAP* **0801** (2008) 024 [arXiv:0706.0360 [hep-th]]; D. Baumann, A. Dy-marsky, I.R. Klebanov, L. McAllister and P.J. Steinhardt, A delicate universe: Compactifi-cation obstacles to D-brane inflation, *Phys. Rev. Lett.* **99** (2007) 141601 [arXiv:0705.3837 [hep-th]].

[121] S.B. Giddings, S. Kachru and J. Polchinski, Hierarchies from fluxes in string compactifica-tions, *Phys. Rev. D* **66** (2002) 106006 [arXiv:hep-th/0105097].

[122] J. Polchinski, Introduction to cosmic F- and D-strings, arXiv:hep-th/0412244.

[123] E.J. Copeland, R.C. Myers and J. Polchinski, Cosmic F- and D-strings, *JHEP* **0406** (2004) 013 [arXiv:hep-th/0312067].

[124] C.P. Herzog and I.R. Klebanov, On string tensions in supersymmetric $SU(M)$ gauge theory, *Phys. Lett. B* **526** (2002) 388 [arXiv:hep-th/0111078].

[125] H. Firouzjahi, L. Leblond and S.H. Henry Tye, The (p, q) string tension in a warped deformed conifold, *JHEP* **0605** (2006) 047 [arXiv:hep-th/0603161].

[126] S. Sarangi and S.H. H. Tye, Cosmic string production towards the end of brane inflation, *Phys. Lett. B* **536** (2002) 185 [arXiv:hep-th/0204074].

[127] O. DeWolfe and S.B. Giddings, Scales and hierarchies in warped compactifications and brane worlds, *Phys. Rev. D* **67** (2003) 066008 [arXiv:hep-th/0208123].

[128] S.B. Giddings and A. Maharana, Dynamics of warped compactifications and the shape of the warped landscape, *Phys. Rev. D* **73** (2006) 126003 [arXiv:hep-th/0507158].

[129] A. Krause and E. Pajer, Chasing brane inflation in string-theory, arXiv:0705.4682 [hep-th].

[130] R. Holman, P. Ramond and G.G. Ross, Supersymmetric inflationary cosmology, *Phys. Lett. B* **137** (1984) 343.

[131] R. Allahverdi, K. Enqvist, J. Garcia-Bellido and A. Mazumdar, Gauge invariant MSSM inflaton, *Phys. Rev. Lett.* **97** (2006) 191304 [arXiv:hep-ph/0605035]; R. Allahverdi, K. Enqvist, J. Garcia-Bellido, A. Jokinen and A. Mazumdar, MSSM flat direction inflation: Slow roll, stability, fine tunning and reheating, *JCAP* **0706** (2007) 019 [arXiv:hep-ph/0610134]; J.C. Bueno Sanchez, K. Dimopoulos and D.H. Lyth, A-term inflation and the MSSM, *JCAP* **0701** (2007) 015 [arXiv:hep-ph/0608299].

[132] N. Itzhaki and E.D. Kovetz, Inflection point inflation and time dependent potentials in string theory, arXiv:0708.2798 [hep-th]; see also N. Itzhaki's talk at the Conference String Theory: Achievements and Perspectives, Jerusalem and Tel-Aviv, April 2007, http://stringfest.tau.ac.il

[133] A. Linde and A. Westphal, Accidental inflation in string theory, arXiv:0712.1610 [hep-th].

[134] B. Underwood, Brane inflation is attractive, arXiv:0802.2117 [hep-th].